Quantum Scattering and Spectral Theory

Techniques of Physics

Editor

N.H. MARCH

Department of Theoretical Chemistry; University of Oxford, Oxford, England

Techniques of physics find wide application in biology, medicine, engineering and technology generally. This series is devoted to techniques which have found and are finding application. The aim is to clarify the principles of each technique, to emphasize and illustrate the applications, and to draw attention to new fields of possible employment.

1. D.C. Champeney: *Fourier Transforms and their Physical Applications*
2. J.B. Pendry: *Low Energy Electron Diffraction*
3. K.G. Beauchamp: *Walsh Functions and their Applications*
4. V. Cappellini, A.G. Constantinides and P. Emiliani: *Digital Filters and their Applications*
5. G. Rickayzen: *Green's Functions and Condensed Matter*
6. M.C. Hutley: *Diffraction Gratings*
7. J.F. Cornwell: *Group Theory in Physics*
8. N.H. March and B.M. Deb: *The Single-Particle Density in Physics and Chemistry*
9. D.B. Pearson: *Quantum Scattering and Spectral Theory*

Quantum Scattering and Spectral Theory

D. B. PEARSON

Department of Applied Mathematics
University of Hull, UK

1988

ACADEMIC PRESS
Harcourt Brace Jovanovich, Publishers
London San Diego New York
Boston Sydney Tokyo Toronto

Academic Press Limited
24–28 Oval Road
London NW1

US edition published by
Academic Press Inc. San Diego, CA 92101

British Library Cataloguing in Publication Data

Pearson, D.B.
Quantum scattering and spectral theory.
1. Scattering (Physics) 2. Quantum
theory
I. Title
539.7′54 QC794.6.S3

ISBN 0-12-548260-4

Photoset by Interprint Ltd. Malta
Printed in Northern Ireland by The Universities Press (Belfast) Ltd.

Preface

This book deals with the foundations of the quantum theory of scattering. Scattering theory may be regarded either as a branch of mathematical physics or as a branch of mathematics worthy of independent study in its own right. The imporance of spectral analysis to the theory is central; every modern text on scattering theory makes reference to the methods and ideas of spectral analysis, and conversely any comprehensive treatment of spectral theory will refer to methods and ideas drawn from applications to quantum theory, and to quantum scattering in particular. I have therefore found it necessary to deal extensively with those areas of spectral theory that are of especial relevance to the subject.

In writing the book, I have had in mind three principal types of reader, and I have tried to make the book self-contained for all three. The first is the physicist or mathematical/theoretical physicist with some knowledge of quantum mechanics and who wishes to explore the subject of scattering theory at a deeper level. Such a reader will almost certainly have made some acquaintance with the subject already. Practically all elementary texts and courses in quantum mechanics have a section on elementary scattering theory. Usually this involves simple one-dimensional problems with square-well potentials or potential barriers, and may go on to treat more advanced parts of the subject, such as the Lippmann–Schwinger equation or the Coulomb problem. This traditional approach is often strongly physically motivated, and makes little use of more advanced mathematical techniques.

The physicist wishing to pursue the subject further will not normally have the appropriate mathematical background to embark on this task. Scattering theory makes use of techniques of Hilbert space, ordinary and partial differential equations, measure theory and so on. The physicist can hardly be expected to have the commitment or time to extract the required material on these and many other branches of modern mathematical analysis, each of which may command an advanced textbook on its own. I have therefore attempted in Chapter 2 to draw together much of this material and provide a self-contained and readable introduction to the mathematics of scattering theory. The only prerequisite for understanding this chapter is a basic knowledge of analysis, as taught in most university undergraduate physics courses. In this and in the remainder of the book, most of the material should be accessible to the physicist who is prepared to make an effort to master the techniques. To make things more familiar to the physicist, I have made the notation as consistent as possible with that to be found in standard physics texts. Space vectors will usually be in 3, and not n, dimensions. The Lippmann–Schwinger equation will look like the Lippmann–Schwinger equation. (However, I shall show how to derive this equation from a formula for the wave operator as a spectral integral, a result *not* to be found in most books on quantum mechanics.) I hope, too, that the physicist who has read one of the standard works on scattering theory from the physicist's standpoint will also learn much from these pages. I have tried to deal with the subject from a more fundamental point of view than is usual, and to show how mathematical and physical developments go hand in hand.

The second kind of reader for whom this book was written is the mathematician who wishes to enter a rich and deeply rewarding branch of mathematics. Such a reader will no doubt be able to omit some of the mathematical preliminaries, but even here there may be areas with which he or she is unfamiliar. The pure mathematician who is interested in the use of functional analysis or C^*-algebras in quantum theory is frequently at a disadvantage in entering this field through a lack of knowledge of the physical background. I have therefore attempted to give what may be regarded as a pure mathematician's account of the foundations of quantum theory. This should allow the reader to acquire at least a minimal understanding of the role of modern analysis in this area, and of the close relationship between mathematical and physical ideas. Much of this book is, indeed, devoted to closer examination of this relationship and what it implies for scattering theory. The mathematician will often be content to start from the standard definition of the unmodified wave operator as a strong limit, to prove that such limits exist, in certain circumstances, and to obtain consequences of the existence of such limits. I am not myself satisfied by this approach. I remember too well my own introduction to the subject. Having been assured that there were sound reasons, both mathematical and physical, why (unmodified) wave operators had to exist, it was only later that I learnt these operators did not exist in the profoundly important case of the Coulomb potential. So I have developed the theory of short- and long-range wave operators together, as far as possible, and deduced their existence in each case from more fundamental basic principles. The mathematician may or may not be happy with my use of the traditional Lemma,

Proof, Theorm format. I have, however, tried always to explain why I do what I do, rather than just doing it. Whether I am successful must be left to the reader to decide. I have sought for a systematic presentation of the theory, proceeding frequently in the direction from the more abstract to the more concrete. (The book ends with a collection of examples.)

The third reader for whom I have catered is the specialist in the subject, or the intending specialist. Much of the material, while relating in my view to important aspects of the theory, is new or is presented for the first time in book form. I have striven to look afresh at the subject, so that even well-known results may perhaps be seen again in a new light. I have adopted a sceptical attitude to what is supposed to be obvious. Cross-sections need not be finite unless they are proved to be so; the scattering operator need not always be unitary. An operator will be suspected guilty of having a singular continuous spectrum until proved innocent. I have taken the theory further than is customary in this kind of work, perhaps to some readers stretching its domain of application to an intolerable degree. It is thus that we learn the boundaries between what is possible and what is impossible. In one important area, however, I have not taken the theory far enough. I have not provided, in this volume, a proof of asymptotic completeness for systems of three or more particles, although I have provided references where these proofs may be found. The question of asymptotic completeness for multiparticle systems is an area in which much current progress is being made, but the problem is not yet, in my view, resolved to the degree that a definitive and comprehensive proof should be developed in these pages. Rather let those whose interests lie in that direction grapple themselves with the formidable mathematical challenges of this area. The study of the foundations of relativistic scattering theory, within the context of quantum field theory, is also left to other authors more ambitious than myself.

The results described in this book represent the work of many workers over a number of years. I have found it impossible to decide, in most cases, who is the inventor of a particular method or discovery. I have chosen to collect references towards the end of the book, rather than to distribute them throughout the text. These are not at all intended to be complete, nor to cover even in outline all branches of the subject. Sometimes I have drawn attention to a whole range of new developments and further reading through just a single reference. The importance of an individual's contribution to scattering theory will *not* therefore be in direct proportion to the number of references to that author. To think otherwise would in any case pander to the attitude that has overtaken many universities in the UK that the academic worth of a contribution is to be measured in numerical (even financial) terms.

Equation (*a.b.c*) refers to the *c*th equation of Section *a.b*. Section *a.b* refers to the *b*th section of Chapter *a*; Chapter *a* refers to the *a*th chapter of the book. Theorem *a.b* is the *b*th theorem of Chapter *a*, independently of the section in which the theorem occurs; Lemmas and Definitions are numbered similarly. The symbol ■ denotes the end of a proof.

So many have contributed, indirectly, to the writing of this volume, that it would be difficult to name them all. I owe a profound debt of gratitude to Josef M. Jauch, who for two short years before his death in 1974 was such an inspiration to myself and to others who were fortunate enough to know him.

It is a pleasure also to express my thanks to Werner O. Amrein, for his continuing support and kindness over many years.

Without Josef Jauch and Werner Amrein, this book would never have been written, but as Ann and I know on a more personal level, so much would have been lost besides.

Finally, I must express my appreciation of the magnificent work of Gill Chilton, who typed the manuscript, and my thanks to Academic Press, whose patience rivalled even that of my parents.

D.B. PEARSON

Contents

Chapter 1

Introduction

In the Preface I have referred to three kinds of reader whom I had in mind when this book was written. Therefore not every reader may wish to approach scattering theory in the same way. Those wishing to gain a quick initial overview of the subject matter might do well to consult Chapter 14 first of all.

Chapter 2 is the longest chapter, and deals with the necessary mathematical preliminaries. Those who are already familiar with, say, the elements of measure theory, can skip the appropriate section, after making sure that they have the required background for any more specialized applications mentioned in the section. The whole of the contents of Chapter 2 is important, but those who are unused to Hilbert-space theory should make an effort to master the theory of linear operators, domains, closures, self-adjoint extensions, sesquilinear forms and so on. On the other hand, the pure mathematician who is unaware of the connection between pure states and vectors, self-adjoint operators and observables, in quantum mechanics, should read the relevant sections. Even those who have a good knowledge of differential equations may profit from reading the section on the Schrödinger equation, which, in addition to establishing the basic groundwork, includes an analysis of asymptotic behaviour of solutions both at real and complex energies.

Chapter 3 examines the consequences of the asymptotic condition of scattering theory, in a general context. The statement of the asymptotic condition is little more than a refinement of the idea that, at large positive and negative times, evolution should look like some kind of free evolution in which particle momenta are conserved. I have taken care to formulate the condition in such terms as allow for possible modifications to the free evolution. In this way, both long- and short-range interactions can be dealt with within a common framework. I have also allowed for the possibility of several scattering channels, so that many-particle scattering of bound states is included.

Chapter 4 *starts* from the existence of wave operators, and may be a good point to begin for those who are prepared to accept the existence of these operators, and who wish to go on to deduce some of their principal properties. The progression from Chapters 3 and 4 to Chapters 5 and 6 is in the direction of decreasing abstraction. Whereas Chapters 3 and 4 are concerned with pairs of self-adjoint operators acting in an abstract Hilbert space, in the later chapters we have to deal specifically with the kinetic-energy operator and its perturbations in non-relativistic quantum mechanics. Chapters 6 and 7 present a systematic account of the one-dimensional Schrödinger operator in an interval, leading to a spectral analysis based primarily on the notion of subordinacy. In Chapter 8 I have shown how to determine short- and long-range wave and scattering operators in the three-dimensional scattering problem, with spherical symmetry. The emphasis here is on scattering theory in a single partial-wave subspace. Of course scattering experiments are carried out neither in a single space dimension, nor do scattering states propagate at fixed angular momentum. Chapter 9 is a treatment of potential scattering in full three-dimensional physical space. It is here that I have forged the link between time-dependent scattering theory and the traditional physicists' treatment in three dimensions. The physicisit will see how to define cross-sections and to relate cross-sections to the mathematical structures of the theory. A further link is established—that between cross-sections and other observable quantities with their partial-wave counterparts.

Chapter 10, on localization theory, describes a number of results relating spectral properties to an analysis of the degree to which states can be localized simultaneously in position and energy. The whole chapter might be thought of as a challenge to the Heisenberg uncertainty principle!

Chapter 11 provides a suitable starting point for those seeking an entry into the more mathematical aspects of the theory. The trace theory has a long and honourable history, and is perhaps the most direct route by which the question of asymptotic completeness may be approached. Though unable to achieve the strongest possible results, the trace method may be taken considerably further than is commonly supposed.

Chapter 12 introduces the geometric method of Enss, which has been so succesful in treating a wide variety of problems in potential scattering. Not only does the method give just about the strongest results for short-range potentials, but extensions can be carried out to long-range and many particle problems as well. Probably the key to success of Enss' method lies in the felicitous blend of mathematical and physical ideas that are so much a feature of this approach. The proof of asymptotic completeness for long-range potentials that is presented here combines the Enss/Mourre/Perry theory with estimates of boundary values of the resolvent operator. A study of the details of these arguments demands more than the usual commitment on the part of the reader, although I hope that I have made the broad outlines of the proof clear enough. The specialist in scattering theory will find a great deal in common with the ideas and methods of other authors, although the precise shape of the arguments is new.

The dedicated reader of this book may be surprised, on reaching Chapter 13, to encounter precise definitions of bound states and scattering states only at this late stage. My aim here is the ambitious one of presenting a complete asymptotic analysis of the large-time behaviour of states, applicable to scattering by short-range potentials that may have arbitrary local singularities. The characterization of bound states, scattering states, and a further mode of asymptotic evolution giving rise to absorbed states, relies on the ideas and method of the previous chapter. The examples provided in the final chapter (14) should convince the reader of the richness and diversity of this subject.

Chapter 2

Mathematical Foundations

The quantum theory of scattering has seen in recent years (i.e. since the mid 1950s) an increasingly fruitful interaction with a number of branches of mathematics; for example with aspects of the theory of ordinary and partial differential equations, functional analysis and von Neumann algebras. Progress in the understanding of the physics has frequently proceeded hand in hand with the development of associated mathematical ideas, each discipline in turn stimulating the other. As a recent example of this cross-fertilization of ideas, and one that transcends the boundaries of its original field, may be cited the inverse method of reconstructing a differential operator from its spectral function. Motivated in part by attempts to determine a potential from given scattering data, this method has in turn elucidated a wide variety of nonlinear phenomena in classical continuum mechanics.

The case for a proper mathematical treatment of this particular branch of physics should not, however, be misunderstood. It would not, I think, be claimed that purely mathematical developments have led to the formulation of new physical laws and principles. Rather, one is enabled to view existing physical theory in a new context, to establish a coherent and precise framework in which the essentials of the theory may be brought out more clearly, and to understand from a fundamental and unified standpoint that which had been obscured by incomplete and heuristic arguments lacking in mathematical precision. An approach to the foundations of scattering theory and its associated mathematics should be justified in this way, and mathematical physicists may safely leave it to others to discover the number of different types of quark in the universe, or to tell us where to dig on the seabed to recover the magnetic monopole.

What then, is this complex of mathematical ideas on which scattering theory is to be based? Primarily, of course, we are concerned with the mathematics of quantum mechanics, of Hilbert space, operator and spectral theory and the rest. We begin this survey with measure, at once the basis of probability and a vital ingredient of spectral analysis.

2.1 MEASURE

Measure is a generalization of the idea of length. A collection of subsets of a given set X is said to form a *σ-algebra* whenever

(i) both X itself and the empty set $\emptyset$ belong to the collection,
(ii) the complement $X \setminus A$ of any set A of the collection again belongs to the collection.
(iii) if $\{A_n\}$ is any sequence of sets, each belonging to the collection, then the union $\bigcup_n A_n$ is a member of the collection.

Usually in applications X will be $\mathbb{R}$ or $\mathbb{R}_+$ or $\mathbb{R}^n$. ($\mathbb{R}_+ = \{x \in \mathbb{R}; x > 0\}$. It is not, in general, possible to assign a measure to every subset of X, and the σ-algebra will be the collection of sets to which a measure may be assigned, i.e. the collection of *measurable* sets.

In the case $X = \mathbb{R}$ we should like the σ-algebra at least to contain all intervals. There is a smallest σ-algebra that contains all the intervals. This σ-algebra is called the *Borel algebra*, and may be defined as the intersection of all σ-algebras that contain the intervals. The Borel algebra is a very large collection of sets indeed, containing much more than simply unions and intersections of countably many intervals. It is difficult to give a precise characterization of the Borel algebra, apart from merely restating its defining property, and indeed one has to work quite hard to construct subsets of $\mathbb{R}$ that do *not* belong to the Borel algebra. (Though in a sense "most" sets are of this category.) Any set that belongs to the Borel algebra is referred to as a *Borel set*. It is often useful to observe that any open set is a Borel set, since any open set is the union of countably many (disjoint) open intervals.

This term "measurable" may also be applied to functions. The function f from (the whole of) X to $\mathbb{R}$ is said to be measurable if for every real number α the set

$$\{x \in X;\ f(x) > \alpha\}$$

is measurable. In other words, the inverse image of every open inerval. (α, ∞) should be measurable. There are several equivalent ways of restating this definition. For example, the open intervals may be replaced by closed intervals $[\alpha, \infty)$, or by $(-\infty, \alpha)$ or $(-\infty, \alpha]$, or one may simply say that the inverse image of every open set is to be measurable. Sums and products of measurable functions are measurable, as is the absolute value $|f|$ of any measurable function f, and any constant multiple of f. If A is any measurable set then the *characteristic function* $\chi_A(x)$ of A, defined by

$$\chi_A(x) = \begin{cases} 1 & \text{if } x \in A, \\ 0 & \text{otherwise,} \end{cases}$$

is measurable. Probabilists often refer to χ_A as the *indicator function* of the set A. (They reserve the term "characteristic function" for the Fourier transform of a probability measure.)

It is sometimes convenient to define an *extended* real-valued function on X. Such a function takes values on the extended real line $\mathbb{R}\cup\{\infty\}$. obtained by appending to $\mathbb{R}$ a single point dentoed by $+\infty$. An extended real-valued function is said to be measurable provided that

(i) the inverse image of (α, ∞) is measurable for any $\alpha\in\mathbb{R}$, and
(ii) the inverse image of the single point $+\infty$ is measurable.

A function f from X to $\mathbb{C}$ is said to be measurable provided that both real and imaginary parts of f are measurable.

In the case $X=\mathbb{R}$ with the Borel algebra, a measurable function is said to be *Borel-measurable*. Examples of Borel-measurable functions are continuous functions, monotonic functions, and indeed it is quite difficult to define functions that are *not* Borel measurable (although, again, they are in the majority!).

A *measure* assigns to each set A of the σ-algebra an extended non-negative real number $\mu(A)$, the *measure* of the set A. To say that a real number is *extended* means that $\mu(A)=\infty$ is allowed. The measure μ must be countably additive.

More formally, a measure on a set X is an extended real-valued function on a σ-algebra of subsets of X, satisfying

(i) $\mu(\emptyset)=0$ and $\mu(A)\geq 0$ for all A in the σ-algebra (of course the point ∞ appended to $\mathbb{R}$ is regarded as greater than any real number),
(ii) if $\{A_n\}$ is any disjoint finite or countable sequence of sets in the σ-algebra (i.e. $A_i\cap A_j=\emptyset$ for $i\neq j$) then

$$\mu\left(\bigcup_n A_n\right)=\sum_n \mu(A_n)$$

(on the right-hand side we write $\Sigma_n\mu(A_n)=\infty$ if and only if either one of the A_n has measure ∞ or the $\{A_n\}$ define an infinite sequence for which the series $\Sigma_n\mu(A_n)$ diverges).

The measure μ is called *finite* if $\mu(X)<\infty$, and σ-finite if X is a union of countably many sets, each having finite measure.

The measure generalizing in the most direct way the idea of length of a subset of $\mathbb{R}$ is called the Borel measure, and may be defined as follows. The Borel measure of an interval $(a,b]$ is defined to be the length of the interval, viz

$$l\{(a,b]\}=b-a. \tag{2.1.1}$$

Since single points have zero Borel measure, it is immaterial whether in this formula we close the interval at a or open it at b. We write $(a, b]$ for convenience, since disjoint intervals of this type fit together in a nice way. Identifying the σ-algebra with the "Borel algebra", the Borel measure of an arbitrary Borel set A may be defined to be

$$l(A) = \inf \sum_{n=1}^{\infty} l(A_n), \tag{2.1.2}$$

where the infimum is extended over all sequences $\{A_n\}$ of intervals $A_n = (a_n, b_n]$ such that $A_n \subseteq \bigcup_n A_n$. Again, in this definition one may equivalently replace the A_n by open intervals (a_n, b_n). For an arbitrary open set B, one may write

$$l(B) = \sum_n l(E_n), \tag{2.1.3}$$

where $A = \bigcup_n E_n$ has been expressed as a union of finitely many, or countably many *disjoint* open intervals. The measure of an arbitrary Borel set A is then given by

$$l(A) = \inf l(B), \tag{2.1.4}$$

where the infimum is extended over all open sets B such that $A \subseteq B$.

The Borel measure is the unique measure on the Borel algebra that is given by (2.1.1) for intervals. Of course, this is not the only measure that may be defined on the Borel algebra. Given any measure μ on the Borel algebra, we set

$$\mu\{(a, b]\} = \rho(b) - \rho(a), \tag{2.1.5}$$

so that the function ρ is uniquely determined up to an additive constant. Let us suppose that no finite interval has infinite μ-measure, in which case ρ is a real-valued function rather than an extended real-valued function. Then ρ has the two properties

(i) $\rho(\lambda)$ is a non-decreasing function of λ,
(ii) $\rho(\lambda)$ is continuous from the right, i.e.

$$\lim_{\varepsilon \to 0+} \rho(\lambda + \varepsilon) = \rho(\lambda). \tag{2.1.6}$$

(The first statement follows from $\mu \geq 0$; the second follows from the fact that $\mu\{(a, b]\}$ converges to zero as b approaches a.) Conversely, given any real-valued function ρ satisfying (i) and (ii) above, one may prove the existence of a unique measure μ on the Borel algebra that satisfies (2.1.5) for intervals $(a, b]$. Thus there is therefore a correspondence between measures on the

Borel subsets of $\mathbb{R}$ and functions satisfying conditions (i) and (ii). We shall refer to μ as the Borel–Stieltjes measure generated by the function ρ. The Borel measure corresponds to the special case $\rho(\lambda)=\lambda$. Again, for Borel–Stieltjes measure μ, (2.1.2) holds with l replaced by μ (infimum over sequences of intervals $(a_n, b_n]$, or equivalently over sequences of open intervals (a_n, b_n), with $A \subseteq \bigcup_n A_n$), as do (2.1.3) and (2.1.4) (infimum over open sets B with $A \subseteq B$).

For a Borel–Stieltjes measure, a single point λ_0 will have strictly positive measure if and only if the function $\rho(\lambda)$ is discontinuous at $\lambda=\lambda_0$. The measure of the single point λ_0 is given by

$$\mu\{\lambda_0\}=\rho(\lambda_0)-\lim_{\varepsilon\to 0+}\rho(\lambda-\varepsilon). \tag{2.1.7}$$

Points having strictly positive measure will be referred to as the discrete points of the measure. For a Borel–Stieltjes measure there are at most countably many discrete points; if there are no discrete points, so that the function $\rho(\lambda)$ is continuous, we refer to a continuous measure.

Any countable set of points has Borel measure zero. More generally, for any continuous Borel–Stieltjes measure the measure of any countable set of points will be zero. There will, however, always be many uncountable sets of points having measure zero. A property P is said to hold μ-almost everywhere (μ-a.e.) in $\mathbb{R}$ if the property holds for all λ belonging to some set Σ, where the measure μ of the complement of Σ is zero. Thus "μ-almost everywhere" depends very much on the measure μ under consideration. One feature of Borel–Stieltjes measure that is sometimes inconvenient is that we may frequently encounter subsets of sets of measure zero that are not themselves measurable. We should like to be able to say that any subset of a set having measure zero is itself measurable, and hence has measure zero. A measure having this property is said to be complete, and a Borel–Stieltjes measure is incomplete. The way to get round this difficulty is to enlarge the collection of measurable sets, i.e. to enlarge the σ-algebra. This is done by appending to the original σ-algebra all sets of the form $\Sigma\cup\Sigma'$, where Σ belongs to the original σ-algebra and Σ' is a subset of some set having μ-measure zero. This enlarged collection of sets may be verified to constitute a σ-algebra, and with the above notation we extend the measure to the new σ-algebra by defining

$$\mu(\Sigma\cup\Sigma')=\mu(\Sigma).$$

Any Borel–Stieltjes measure extended in this way is called a Lebesgue–Stieltjes measure, and is complete. The completion of Borel measure is referred to simply as Lebesgue measure. The corresponding σ-algebras are referred to respectively as the collections of Lebesgue–Stieltjes-measurable sets and of Lebesgue-measurable sets. Where the measure is understood, we refer simply to measurable sets. We shall also talk of the Lebesgue–Stieltjes measure generated by the function $\rho(\lambda)$. Note that, for these completed measures, the collection of measurable sets depends very much on the choice of the function $\rho(\lambda)$. For example, if $\rho(\lambda)$ is constant on some open interval then any subset of that interval will be measurable, and indeed will have measure zero. We may say that there is no contribution to the measure from points of constancy of $\rho(\lambda)$. A property P is said to hold almost everywhere (a.e.) if the property holds μ-almost everywhere where μ is Lebesgue measure (or equivalently where μ is Borel measure). If, on the other hand, μ is a Lebesgue–Stieltjes measure, we shall sometimes say that property P holds almost everywhere with respect to the measure μ.

In view of the important concept of sets of measure zero, a basic notion is that of absolute continuity of one measure with respect to another. We shall say that μ_1 is absolutely continuous with respect to μ_2 if any set having μ_2-measure zero also has μ_1-measure zero, and this will be written $\mu_1 \ll \mu_2$. If $\mu_1 \ll \mu_2$ and $\mu_2 \ll \mu_1$ we shall say that the two measures μ_1 and μ_2 are equivalent. Any two equivalent measures μ_1 and μ_2 will have the same collection of sets having measure zero, and property P will hold μ_1-almost everywhere if and only if it holds μ_2-almost everywhere. A Lebesgue–Stieltjes or Borel–Stieltjes measure that is absolutely continuous with respect to Lebesgue measure (respectively Borel measure) is said simply to be absolutely continuous.

A related notion to that of absolute continuity is the relation between two measures that obtains when one measure is singular with respect to another. For two measures μ_1 and μ_2 defined on subsets of a set X we shall say that μ_1 is singular with respect to μ_2 if there exist two disjoint sets Σ_1 and Σ_2, with $X = \Sigma_1 \cup \Sigma_2$, where

$$\mu_1(\Sigma_2) = 0, \qquad \mu_2(\Sigma_1) = 0. \tag{2.1.8}$$

Since this relation is symmetric, we can equivalently say that μ_2 is singular with respect to μ_1, or that μ_1 and μ_2 are mutually singular. We then write $\mu_1 \perp \mu_2$. A Lebesgue–Stieltjes (or Borel–Stieltjes) measure that is singular with respect to Lebesgue measure (respectively Borel measure) is described simply as a singular measure.

We shall say that a measure μ is concentrated on a set Σ if the μ-measure of the complement $X \setminus \Sigma$ of Σ is zero. From (2.1.8) we see that μ_1 will be singular with respect to μ_2 if μ_1 is concentrated on Σ_1 and μ_2 on Σ_2, where $\Sigma_1 \cap \Sigma_2 = \emptyset$. In other words, mutually singular measures are concentrated on mutually disjoint sets.

A measure μ is said to be discrete if μ is concentrated on a finite or countable set of points (the discrete points of the measure). Such a measure will certainly be singular, since this set of points will necessarily have Lebesgue measure zero. Discrete measures are not, however, the only examples of singular measures, and even a continuous measure may be singular, in which case it is described as singular continuous. Since singular continuity is of some interest in scattering theory, if only as an extreme of behaviour for a measure which one is often at pains to avoid, it may be helpful at this stage to see an example.

The Cantor measure

The Cantor measure is concentrated on the so-called Cantor set C, consisting of all points in $[0, 1]$ having a "decimal" expansion, to the base 3

$$x = 0 \cdot x_1 x_2 x_3 \cdots \qquad \left(or \ x = \sum_{i=1}^{\infty} x_i/3^i \right),$$

in which each x_i is either 0 or 2. There are uncountably many such points. An alternative characterization of the Cantor set is to start from the interval $[0, 1]$ and remove the middle one-third interval, i.e. $(\frac{1}{3}, \frac{2}{3})$. From what remains (two subintervals), we remove the middle one third of each subinterval, i.e. $(\frac{1}{9}, \frac{2}{9})$ and $(\frac{7}{9}, \frac{8}{9})$. Then we remove the middle third of the four subintervals that remain, and so on. What remains after an infinite sequence of these excisions is the Cantor set, which, being the complement in $[0, 1]$ of an open set, is closed.

We define a function f, periodic with period 1, by

$$f(\lambda) = \begin{cases} 1 & (0 \leq \lambda \leq \frac{1}{3}), \\ 0 & (\frac{1}{3} < \lambda < \frac{2}{3}), \\ 1 & (\frac{2}{3} \leq \lambda \leq 1). \end{cases} \tag{2.1.9}$$

Then the characteristic function of the Cantor set is

$$\chi_C(\lambda) = \prod_{k=1}^{\infty} f(3^{k-1}\lambda),$$

and the Cantor measure (or its restriction to the interval $[0, 1]$) is the Lebesgue–Stieltjes measure generated by the function

$$\rho_C(\lambda) = \lim_{n \to \infty} \int_0^{\lambda} dx \prod_{k=1}^{n} f_k(x), \tag{2.1.10}$$

where

$$f_k(x) = \tfrac{3}{2} f(3^{k-1}x).$$

Thus for any subinterval Σ of $[0,1]$ we have

$$\mu(\Sigma)=\lim_{n\to\infty}\int_{\Sigma}\mathrm{d}x\prod_{k=1}^{n}f_k(x). \tag{2.1.10$'$}$$

We can write this as $\mu(\Sigma)=\lim_{n\to\infty}\mu_n(\Sigma)$, where the measure μ_n is concentrated on the set C_n obtained from $[0,1]$ by n consecutive processes of excision. The measure μ_n takes the Lebesgue measure 1 for the entire interval $[0,1]$, and spreads it uniformly over the set C_n. Thus the limiting measure $\mathrm{d}\rho_C(\lambda)$ may be regarded as derived from "Lebesgue measure spread uniformly over the Cantor set". The singular nature of Cantor measure resides in the fact that the Cantor set, on which the measure is concentrated, has Lebesgue measure zero ($l(C)=\lim_{n\to\infty}l(C_n)=\lim_{n\to\infty}(\frac{2}{3})^n=0$). The measure is continuous since, for intervals Σ, $l(\Sigma)\leq(\frac{1}{3})^n\Rightarrow\mu_C(\Sigma)<(\frac{1}{2})^n$, so that for any point λ_0, which is certainly contained in an interval of length $\leq(\frac{1}{3})^n$, we have $\mu_C\{\lambda_0\}\leq\lim_{n\to\infty}(\frac{1}{2})^n=0$.

Once measure has been defined, it is relatively straightforward to define the notion of integral. Take the Lebesgue–Stieltjes measure μ generated by $\rho(\lambda)$, and consider first a non-negative (measurable) simple function $h(\lambda)$. A simple function is a function that takes on only a finite number of distinct values. The integral of h with respect to the measure μ, taken over the entire real line, is defined to be

$$\int h\,\mathrm{d}\mu=\int h(\lambda)\,\mathrm{d}\rho(\lambda)=\sum_k c_k\mu(E_k), \tag{2.1.11}$$

where $h=\Sigma_k c_k\chi_{E_k}$ has been expressed as a linear combination of finitely many characteristic functions, and the E_k are measurable. (The right-hand side of (2.1.11) is independent of the particular representation of h.) Note that in (2.1.11) the c_k are finite. For an arbitrary non-negative measurable function $f_+(\lambda)$, the integral $\int f_+(\lambda)\,\mathrm{d}\rho(\lambda)$ may be defined to be the supremum, over non-negative simple functions h such that $h(\lambda)\leq f_+(\lambda)$ for all λ, of $\int h(\lambda)\,\mathrm{d}\rho(\lambda)$. The same definition applies if f_+ is allowed to be an extended real-valued function. Note that the integral may be $+\infty$, but if it is finite then we shall say that f_+ is integrable. If f is an arbitrary measurable function, we may define f_+ and f_- by

$$f_+(\lambda)=\begin{cases}f(\lambda) & \text{if } f(\lambda)\geq 0,\\ 0 & \text{if } f(\lambda)<0,\end{cases}$$

and

$$f_-(\lambda)=\begin{cases}f(\lambda) & \text{if } f(\lambda)\leq 0,\\ 0 & \text{if } f(\lambda)>0.\end{cases}$$

Then $f=f_+ + f_-$. If f_+ and $-f_-$ are both integrable then we shall say that f is integrable, and define

$$\int f(\lambda)\,\mathrm{d}\rho(\lambda)=\int f_+(\lambda)\,\mathrm{d}\rho(\lambda)-\int(-f_-(\lambda))\,\mathrm{d}\rho(\lambda).$$

We refer to this as the Lebesgue–Stieltjes integral of f with respect to the measure $\mu=\mathrm{d}\rho(\lambda)$. If μ is a Lebesgue measure then we refer simply to the Lebesgue integral. If f is an extended real-valued function such that the set of points λ at which $f(\lambda)=\infty$ has ρ-measure zero then $\int f(\lambda)\,\mathrm{d}\rho(\lambda)$ is the same as $\int f_0(\lambda)\,\mathrm{d}\rho(\lambda)$, where

$$f_0(\lambda)=\begin{cases} f(\lambda) & \text{if } f(\lambda)<\infty,\\ 0 & \text{if } f(\lambda)=\infty.\end{cases}$$

More generally, two functions that differ only on a set of ρ-measure zero (i.e. that are equal ρ-almost everywhere) will have the same integral.

A function f will be integrable if and only if $\int|f|\,\mathrm{d}\mu<\infty$. Thus, in the Lebesgue theory, integrability means *absolute* integrability.

Let E be a μ-measurable set and f_+ a non-negative real-valued measurable function. We define $\int_E f_+(\lambda)\,\mathrm{d}\rho(\lambda)$, the integral of f_+ over E, with respect to μ, to be $\int\chi_E(\lambda)f_+(\lambda)\,\mathrm{d}\rho(\lambda)$, where χ_E is the characteristic function of E. Again, this integral may be extended to arbitrary integrable functions f, where f is said to be integrable over E provided that $\int_E|f(\lambda)|\,\mathrm{d}\rho(\lambda)<\infty$. The function f may be an extended real-valued function if the set of points λ in E at which $f(\lambda)=\infty$ has ρ-measure zero. Note that $\mu(E)=\int_E 1\,\mathrm{d}\rho(\lambda)$.

In the special cases $E=(a,b]$ and $E=\mathbb{R}$ we shall often write

$$\left.\begin{aligned}\int_E f(\lambda)\,\mathrm{d}\rho(\lambda)&\equiv\int_a^b f(\lambda)\,\mathrm{d}\rho(\lambda),\\ \int_R f(\lambda)\,\mathrm{d}\rho(\lambda)&\equiv\int_{-\infty}^{\infty} f(\lambda)\,\mathrm{d}\rho(\lambda).\end{aligned}\right\}\tag{2.1.12}$$

Given a Lebesgue–Stieltjes or Borel–Stieltjes measure μ, generated by $\rho(\lambda)$, together with a non-negative function f_+, integrable with respect to μ over any finite interval, one may definite a measure μ_{f_+}, for μ-measurable sets E, by

$$\mu_{f_+}(E)=\int_E f_+(\lambda)\,\mathrm{d}\rho(\lambda).\tag{2.1.13}$$

The measure μ_{f_+} will be absolutely continuous with respect to μ, since the integral on the right-hand side is zero whenever E has μ-measure zero. The converse of this result is known as the Radon–Nikodym Theorem, and may be extended to pairs of σ-finite measures. It says that if v is an arbitrary measure that is absolutely continuous with respect to μ and that is finite for finite intervals then there exists a non-negative measurable function f_+ such that (2.1.13) holds with $\mu_{f_+} = v$. Given v and μ, the function f_+ is determined μ-almost everywhere. (That is, any two possible f_+'s will agree except at a set of points λ having μ-measure zero.) The function f_+ in (2.1.13) is often described as the Radon–Nikodym derivative of v with respect to μ. Note that f_+ is not necessarily integrable, but will be integrable with respect to μ if and only if v is a finite measure (i.e. $v(\mathbb{R}) < \infty$). It will, however, be true under the stated conditions that f_+ is integrable over finite intervals.

A special case of this result is that any absolutely continuous measure (i.e. absolutely continuous with respect to Lebesgue measure) is of the form

$$v(E) = \int_E f_+(\lambda)\,\mathrm{d}\lambda. \tag{2.1.13$'$}$$

We shall refer to f_+ in this case as a *density function* for the measure v. Thus absolutely continuous measures are precisely those that are derivable from a density function, which will be integrable whenever the measure is finite. The density function for Lebesgue measure is $f_+(\lambda) = 1$. If, in each case, the measure is finite, then μ_{f_+} in (2.1.13) is the measure generated by $\int_{-\infty}^{\lambda} f_+(\lambda)\,\mathrm{d}\rho(\lambda)$, and in (13)$'$ by $\int_{-\infty}^{\lambda} f_+(\lambda)\,\mathrm{d}\lambda$. If it is not necessarily finite, then, the measure is generated by $\int_c^{\lambda} f_+(\lambda)\,\mathrm{d}\rho(\lambda)$ or by $\int_c^{\lambda} f_+(\lambda)\,\mathrm{d}\lambda$ respectively, where we adopt the conventions $\int_a^b = -\int_b^a$ and $\int_0^0 = 0$.

For the Cantor measure, which is singular, it is not difficult to understand why a density function fails to exist. The most obvious candidate for $f_+(x)$ in (2.1.10)$'$ would be $\Pi_{k=1}^{\infty} f_k(x)$. However, this infinite product is divergent for all x in the Cantor set, and elsewhere is identically zero. The measure is, in fact, too unevenly distributed to be absolutely continuous.

Many of the results of the theory of measure and integral extend in a simple way from measures to *charges*. (Here we shall discuss finite charges only.) A charge ω is a real-valued function on a σ-algebra of subsets of a set X, satisfying

(i) $\omega(\emptyset) = 0$,

(ii) if $\{A_n\}$ is any disjoint finite or countable sequence of sets in the σ-algebra then

$$\omega\left(\bigcup_n A_n\right) = \sum_n \omega(A_n).$$

(the sum on the right-hand side must be absolutely convergent).

Unlike a measure, $\omega(A)$ is allowed to be negative. Any (finite) charge may be represented as the difference of two finite measures, and in fact one can write

$$\omega = \mu_1 - \mu_2, \tag{2.1.14}$$

where μ_1 and μ_2 are concentrated respectively on disjoint sets, and so are mutually singular. Integration with respect to a charge may then be defined by

$$\int f \, d\omega = \int f \, d\mu_1 - \int f \, d\mu_2, \tag{2.1.15}$$

for any function f that is integrable with respect to μ_1 and μ_2. In (2.1.14), if μ_1 is generated by ρ_1, and μ_2 by ρ_2, we can say that the charge ω is generated by the function $\rho_1 - \rho_2$, and define

$$\int f \, d\rho(\lambda) = \int f \, d\rho_1(\lambda) - \int f \, d\rho_2(\lambda). \tag{2.1.15$'$}$$

The function $\rho = \rho_1 - \rho_2$ has *bounded variation.* (A function ρ is said to be of bounded variation if there exists a constant M such that $\Sigma_{i=1}^{n-1} |\rho(x_{i+1}) - \rho(x_i)| \leq M$ for all n and for all increasing sequences $x_1, x_2, \ldots, x_n$. A function of bounded variation may alternatively be characterized as the *difference* between two bounded non-decreasing functions.) Conversely, the integral on the left-hand side of (2.1.15)′ makes sense for any function ρ of bounded variation, since for such a function we have $\rho = \rho_1 - \rho_2$, where ρ_1 and ρ_2 generate finite measures μ_1 and μ_2. The integral is independent of the choice of the functions ρ_1 and ρ_2. (There is an "optimal" choice if μ_1 and μ_2 are taken, as above, to be mutually singular.) Thus the theory of integration may be extended to define integrals with respect to an arbitrary function of bounded variation. One may even allow $\rho(\lambda)$ to be complex, as is often convenient in spectral analysis, since real and imaginary parts may be taken.

Closely related to absolutely continuous measures is the idea of an *absolutely continuous function.* A real-valued function h on an open interval I is said to be *locally absolutely continuous* in I if there exists a real-valued function f, integrable (with respect to Lebesgue measure) over every closed subinterval of I, such that, for $x, y \in I$,

$$h(y) - h(x) = \int_x^y f(\lambda) \, d\lambda. \tag{2.1.16}$$

One then has, for almost all λ in I, $f(\lambda)=(\mathrm{d}/\mathrm{d}\lambda)h(\lambda)$, so that an absolutely continuous function is almost-everywhere differentiable. The converse is false; for example, a monotonic increasing function will be differentiable almost everywhere, but even if continuous it need not be absolutely continuous. However, a function differentiable at *all* points of I is always locally absolutely continuous.

One can also see, from (2.1.16), that h is of bounded variation in each closed subinterval of I; we shall say that h is *locally* of bounded variation. On each closed subinterval, h will generate a charge, with density function f. This charge ω will be absolutely continuous, in the sense that $\omega=\mu_1-\mu_2$, where the measures μ_1 and μ_2 are absolutely continuous, and if h is non-decreasing it generates an absolutely continuous *measure.*

There is a corresponding theory extending Lebesgue measure and integral to $\mathbb{R}^n$. It is unnecessary here to pursue the theory any further in this direction. Some of the relevant aspects for function spaces will be considered in Section 2.2. Let us conclude this section with two results on limits, of measures and of integrals respectively, and one that allows exchange of orders of integration in repeated integrals.

1. Let $\{E_n\}$ denote an increasing sequence of measurable sets (i.e. $E_m\subseteq E_n$ for $m<n$). Then

$$\mu\left\{\bigcup_n E_n\right\}=\lim_{n\to\infty}\mu(E_n). \tag{2.1.17}$$

Let $\{F_n\}$ denote a decreasing sequence of measurable sets. Then

$$\mu\left\{\bigcap_n F_n\right\}=\lim_{n\to\infty}\mu(F_n). \tag{2.1.17$'$}$$

The proof of these results is a straightforward exercise, using countable additivity.

2. (*Lebesgue Dominated-Convergence Theorem*) Let μ be a σ-finite measure, and $\{f_n\}$ a sequence of measurable functions such that $|f_n|\leq f$, where f is integrable with respect to μ. Suppose $\lim_{n\to\infty}f_n(x)=g(x)$, for μ-almost all x. Then

$$\lim_{n\to\infty}\int f_n\,\mathrm{d}\mu=\int g\,\mathrm{d}\mu. \tag{2.1.18}$$

3. Let μ_1 and μ_2 be Stieltjes measures, generated respectively by ρ_1 and ρ_2, and let the function $F(x,y)$ satisfy

$$\int\left\{\int|F(x,y)|\,\mathrm{d}\rho_2(y)\right\}\mathrm{d}\rho_1(x)<\infty$$

or

$$\int\left\{\int |F(x,y)|\,\mathrm{d}\rho_1(x)\right\}\mathrm{d}\rho_2(y)<\infty.$$

(Both integrals then exist and are equal.) Then

$$\int\left\{\int F(x,y)\,\mathrm{d}\rho_2(y)\right\}\mathrm{d}\rho_1(x)=\int\left\{\int F(x,y)\,\mathrm{d}\rho_1(x)\right\}\mathrm{d}\rho_2(y),$$

provided that each function integrated is measurable with respect to ρ_1 or ρ_2, as appropriate. (Actually what is required is measurability with respect to the product measure $\mathrm{d}\rho_1(x)\,\mathrm{d}\rho_2(y)$.)

2.2 SPACES OF FUNCTIONS

When we deal with functions collectively, we speak of spaces of functions, or function spaces. All of the spaces with which we shall be concerned will be *linear* spaces. That is, if ϕ_1 and ϕ_2 belong to the space then so does $c_1\phi_1+c_2\phi_2$. Here c_1 and c_2 are any two real numbers, if we are referring to a space over the field of real numbers (the elements ϕ_1 and ϕ_2 of the space are then real-valued functions), or c_1 and c_2 are any two complex numbers, for a space over the complex field (ϕ_1 and ϕ_2 are then complex-valued functions). With the usual rules for addition of functions, and multiplication of functions by numbers, our function spaces will be *vector spaces*. In almost all applications these vector spaces will be infinite-dimensional.

To deal effectively with infinite-dimensional spaces, we need some notion of continuity, in order to define more precisely the idea of one element of the space being close to another element. This can often be done by introducing a *norm*. To do this, we assign a non-negative real number $\|\phi\|$ to each element ϕ of the space, such that

(i) $\|\phi\|\geq 0$, with $\|\phi\|=0$ if and only if $\phi=0$ (the zero element of a function space is the zero function, defined by $\phi(x)=0$ for all x),
(ii) $\|c\phi\|=|c|\cdot\|\phi\|$,
(iii) $\|\phi_1+\phi_2\|\leq\|\phi_1\|+\phi_2\|$ for any two elements of the space. This is the so-called triangle inequality.

A vector space equipped with a norm $\|\cdot\|$ in this way is called a normed space. The notion of continuity is easily defined in a normed space—we can say that ϕ_1 is close to ϕ_2 if $\|\phi_1-\phi_2\|$ is small. A one-parameter family $\{\phi_\lambda\}$ ($\lambda\in\mathbb{R}$) of elements of the space is said to be *strongly continuous* in λ whenever, for each $\lambda_0\in$ R, $\lim_{\lambda\to\lambda_0}\|\phi_\lambda-\phi_{\lambda_0}\|=0$. In the same way, we say that a sequence $\{\phi_n\}$ ($n=1,2,3,\ldots$) *converges strongly* to an element ϕ of the space whenever $\lim_{n\to\infty} \|\phi_n - \phi\| = 0$. We then say that ϕ is the *strong limit* of the sequence $\{\phi_n\}$, and write this as

$$\operatorname*{s-lim}_{n\to\infty} \phi_n = \phi. \tag{2.2.1}$$

(It is readily verified, using properties (i)–(iii) of the norm, that a given sequence cannot have two different limits.)

A sequence $\{\psi_n\}$ $(n = 1, 2, 3, \ldots)$ of elements of a normed space is said to be a *Cauchy sequence* if $\lim_{m,n\to\infty} \|\psi_m - \psi_n\| = 0$. Any sequence $\{\phi_n\}$ satisfying (1) for some ϕ is a Cauchy sequence. However, in general a Cauchy sequence need have no strong limit. A normed space such that *every* Cauchy sequence has a strong limit (the limit belonging to the normed space) is said to be *complete*, and a complete normed space is called a *Banach space*. Any normed space that is not complete may be augmented to form a Banach space, in much the same way as the rational numbers are augmented to form the reals. The elements of the enlarged space are defined to be equivalence classes of Cauchy sequences of elements of the original space, where two Cauchy sequences $\{\phi_n\}$ and $\{\psi_n\}$ belong to the same equivalence class whenever $\lim_{n\to\infty} \|\phi_n - \psi_n\| = 0$. Addition and multiplication by numbers are defined in a natural way. Thus $c\{\phi_n\} = \{c\phi_n\}$ and $\{\phi_n\} + \{\psi_n\} = \{\phi_n + \psi_n\}$. The norm is defined by $\|\{\phi_n\}\| = \lim_{n\to\infty} \|\phi_n\|$, where the limit may be shown to exist for each Cauchy sequence, and to be the same for Cauchy sequences belonging to the same equivalence class.

Any element ϕ of the original space corresponds, in the augmented space, to the equivalence class consisting of all Cauchy sequences that converge strongly to ϕ. Indeed, usually no harm is done if we simply *identify* ϕ with this equivalence class. However, if the original space is incomplete then the augmented space (which may be shown to be complete) will contain additional elements that do not correspond to any element of the original space. These additional elements correspond, roughly, to limits outside the original space of sequences of elements within that space. The Banach space constructed in this way is referred to as the *completion* of the original space.

We see, then, that, given a norm on a function space, there is a natural way to make the space complete, if it is not so already. Thus one may assume a normed space to be complete without loss of generality.

Let us now consider some examples of function spaces, some of which come equipped with a natural norm and some of which do not. At the same time we shall introduce the notation by which we intend to refer to these spaces.

Broadly speaking, functions may be categorized by their behaviour with respect to continuity and differentiability, by bounds on their integrals, and sometimes by their support properties. We shall not, at first, take the trouble to distinguish in the notation between spaces of real functions and spaces of complex functions; all our definitions apply equally to either case.

We denote by $C[a, b]$ the space of continuous functions on an interval $[a, b]$ (i.e. continuous functions from $[a, b]$ to $\mathbb{R}$ or from $[a, b]$ to $\mathbb{C}$, depending on whether we are dealing with real-valued or complex-valued functions.) We equip this space with a norm (the so-called sup norm) defined by

$$\|\phi\| = \sup_{x \in [a,b]} |\phi(x)|.$$

This norm is always finite, because a continuous function on a (finite) closed interval is necessarily bounded. Moreover, $C[a, b]$ is complete with this norm, and is therefore a Banach space.

More generally, if E is any closed bounded subset of $\mathbb{R}^n$, we can define a space of continuous functions on E. We shall denote this space by $C(E)$, again with the sup norm defined by

$$\|\phi\| = \sup_{\boldsymbol{r} \in E} |\phi(\boldsymbol{r})|,$$

and $C(E)$ is complete with this norm.

Given a Banach space $\mathscr{B}$, a subset $\mathscr{A}$ of $\mathscr{B}$ is said to be dense in $\mathscr{B}$ if, for each f in $\mathscr{B}$, and for any $\varepsilon > 0$, there exists an element ϕ of $\mathscr{A}$ satisfying $\|\phi - f\| < \varepsilon$. In other words, for $\mathscr{A}$ to be dense, each element of the space $\mathscr{B}$ should have elements of $\mathscr{A}$ arbitrarily close to it. It is often extremely useful to identify dense subsets of a space, in particular because by arguments based on continuity it may be possible to extend results that hold initially only for the dense set $\mathscr{A}$ to the entire space $\mathscr{B}$. We shall consider examples of the use of this idea shortly. The space $\mathscr{B}$ is said to be *separable* if there exists a dense subset $\mathscr{A}$ containing only countably many elements.

In the case of space $C(E)$, where E is a closed bounded subset of $\mathbb{R}^n$, the *Stone–Weierstrass Theorem* gives sufficient conditions for a subset $\mathscr{A}$ of $C(E)$ to be *dense*. Namely, $\mathscr{A}$ is dense in $C(E)$ provided that

(i) $\phi_1, \phi_2 \in \mathscr{A} \Rightarrow c_1\phi_1 + c_2\phi_2 \in \mathscr{A}$ and $\phi_1\phi_2 \in \mathscr{A}$ (c_1 and c_2 are arbitrary numbers),
(ii) $\boldsymbol{r}_1, \boldsymbol{r}_2 \in E \Rightarrow \exists \phi \in \mathscr{A}$ such that $\phi(\boldsymbol{r}_1) \neq \phi(\boldsymbol{r}_2)$, unless $\boldsymbol{r}_1 = \boldsymbol{r}_2$,
(iii) $\boldsymbol{r} \in E \Rightarrow \exists \phi \in \mathscr{A}$ such that $\phi(\boldsymbol{r}) \neq 0$, and
(iv) $\phi \in \mathscr{A} \Rightarrow \bar{\phi} \in \mathscr{A}$ (of course this condition is automatically satisfied if we are dealing with a space of real functions). Hence $\bar{\phi}$ is the complex conjugate of ϕ.

A particularly straightforward, but useful, application of the Stone–Weierstrass Theorem shows that polynomials are dense in $C[a, b]$. Since every polynomial in $C[a, b]$ may be approximated arbitrarily closely, in norm, by a polynomial having rational coefficients, and since the set of polynomials having rational coefficients is countable, it also follows that $C[a, b]$ has a countable dense subset, i.e. the space is separable.

A subset of a Banach space $\mathscr{B}$ is said to be *closed* if, for any sequence of elements of the subset converging strongly to a limit, say ϕ, in $\mathscr{B}$, the element ϕ is also in the subset. Thus a closed subset of $\mathscr{B}$ is closed under the operation of taking limits. A closed subset of $\mathscr{B}$ that is also closed under the operations of addition and of multiplication by numbers (i.e. such that $c_1\phi_1 + c_2\phi_2$ belongs to the subset whenever ϕ_1 and ϕ_2 do) is said to be a *subspace* of $\mathscr{B}$. Any subspace of $\mathscr{B}$ may be regarded as a Banach space in its own right. A simple example of a subspace of $C[a, b]$ is the set of functions in $C[a, b]$ that vanish at the midpoint of the interval $[a, b]$.

The *dual space* of a Banach space is defined to be the space of all continuous linear functionals on the Banach space. A *linear functional* is a linear mapping from the Banach space to $\mathbb{C}$ (for a complex Banach space) or to $\mathbb{R}$ (for a real Banach space). An example of a linear functional on $C[a, b]$ is the mapping that sends ϕ into $\phi(\xi)$, where ξ is some (fixed) number in the interval $[a, b]$. A linear functional $\mathscr{L}$ on a Banach space $\mathscr{B}$ is said to be *continuous* if there exists some number M such that $|\mathscr{L}\phi| \le M\|\phi\|$ for all $\phi \in \mathscr{B}$. In this case we also speak of a *bounded* linear functional, and define the *norm* of $\mathscr{L}$ by

$$\|\mathscr{L}\| = \sup_{\phi \in \mathscr{B}, \phi \neq 0} \frac{|\mathscr{L}\phi|}{\|\phi\|}.$$

The dual space of a Banach space, with this norm and with the obvious definition of addition, and multiplication by numbers, is itself a Banach space. So duality allows us to construct new Banach spaces from old.

What is the dual space of $C[a, b]$? If μ is any Borel–Stieltjes measure on $[a, b]$ (i.e. μ is the restriction to Borel subsets of $[a, b]$ of a Borel–Stieltjes measure on $\mathbb{R}$) then μ is a finite measure, and a corresponding continuous linear functional $\mathscr{L}_\mu$ on $C[a,b]$ may be defined by the mapping

$$\phi \to \int_{[a,b]} \phi \, d\mu. \tag{2.2.2}$$

Thus $\mathscr{L}_\mu$ is an element of the dual space of $C[a,b]$. Moreover, $\mathscr{L}_\mu$ is a *positive* linear functional, in the sense that $\mathscr{L}_\mu\phi \ge 0$ for any ϕ satisfying $\phi(x) \ge 0$ $(x \in [a,b])$. There is an important representation theorem for measures, which states that *every* positive linear functional on $[a, b]$ is of the form (2.2.2) for some Borel–Stieltjes measure μ on $[a, b]$. (Notice that every *positive* linear functional is necessarily bounded. For example, if ϕ is real then we have, for $\mathscr{L}$ positive, $\mathscr{L}(|\phi| \pm \phi) \ge 0$, so that

$|\mathscr{L}\phi| = \pm\mathscr{L}\phi \leq \mathscr{L}|\phi|$. Hence $|\mathscr{L}\phi| \leq \mathscr{L}|\phi| \leq \mathscr{L}||\phi|| = ||\phi||\mathscr{L}(1)$, and it follows that $||\mathscr{L}|| \leq \mathscr{L}(1)$. Since, clearly, $\mathscr{L}(1) \leq ||\mathscr{L}||$, we always have $||\mathscr{L}|| = \mathscr{L}(1)$ for positive linear functionals.)

The following lemma states the representation theorem rather more generally (the result can be generalized still further). We have left undefined the notion of Borel–Stieltjes measure on spaces more general than $\mathbb{R}^n$, but the interested reader is referred to the literature quoted in the notes for this chapter (see Chapter 14).

Lemma 2.1

Let E be a closed bounded subset of $\mathbb{R}^n$. Then there is a one-to-one correspondence between positive linear functionals $\mathscr{L}$ on $C(E)$ and Borel–Stieltjes measures μ on E, through the correspondence

$$\mathscr{L}\phi = \int_E \phi \, \mathrm{d}\mu. \tag{2.2.2$'$}$$

The representation theorem still does not answer the original question regarding the dual space of $C[a,b]$, because not every bounded linear functional is positive. However, it may be shown that every bounded linear functional $\mathscr{L}$ on the real space $C[a,b]$, and more generally on $C(E)$, is the *difference* between two positive linear functionals, which we may denote by $\mathscr{L}_{\pm}$. Thus $\mathscr{L} = \mathscr{L}_{+} - \mathscr{L}_{-}$, where $\mathscr{L}_{\pm}$ may be chosen such that $||\mathscr{L}|| = ||\mathscr{L}_{+}|| + ||\mathscr{L}_{-}||$, and are then unique.

Corollary (*C*(*E*) the *real* function space)

Let E be a closed bounded subset of $\mathbb{R}^n$. Then every bounded linear functional $\mathscr{L}$ on $C(E)$ may be represented as

$$\mathscr{L}\phi = \int_E \phi \, \mathrm{d}\mu_1 - \int_E \phi \, \mathrm{d}\mu_2, \tag{2.2.2$''$}$$

where μ_1 and μ_2 are (unique) Borel–Stieltjes measures on E satisfying

$$||\mathscr{L}|| = \mu_1(E) + \mu_2(E). \tag{2.2.3}$$

Note that for $||\phi|| = 1$ the maximum of $|\mathscr{L}\phi|$ occurs when $\phi = \phi_1$, with $\phi_1 = 1$ μ_1-almost everywhere and $\phi_1 = -1$ μ_2-almost everywhere, corresponding to the value of $||\mathscr{L}||$ in (2.2.3). Since the sets of points at which respectively $\phi_1 = 1$ and $\phi_1 = -1$ clearly do not overlap, it follows that the measures μ_1 and μ_2 are mutually singular. Conversely, any pair of mutually singular Borel–Stieltjes measures on E define, through (2.2.2)′, a bounded linear functional on $C(E)$. Thus the corollary furnishes us with a characterization of the dual space of $C(E)$, and of $C[a, b]$ in particular.

Let us turn now to spaces characterized by the behaviour of functions under the operation of *differentiation*. It will be convenient here to start with an *open* subset Ω of $\mathbb{R}^n$, so that, at least for suitably smooth functions ϕ, it makes sense to talk about the derivatives of ϕ.

Let $\boldsymbol{r} = (x_1, x_2, \ldots x_n)$ be the position vector of a general point in $\mathbb{R}^n$, and let $\alpha = \{\alpha_1, \alpha_2, \ldots \alpha_n\}$ be a sequence of non-negative integers with $|\alpha| \equiv \Sigma_i \alpha_i$. We shall denote by D^α the partial derivative of order $|\alpha|$.

$$D^\alpha = \frac{\partial^{|\alpha|}}{\partial x_1^{\alpha_1} \partial x_2^{\alpha_2} \ldots \partial x_n^{\alpha_n}}. \tag{2.2.4}$$

For example, if $n = 3$ and $\alpha = \{1, 0, 3\}$ then

$$D^\alpha = \frac{\partial^4}{\partial x_1 \partial x_3^3},$$

which is a fourth-order partial derivative.

The space $C^m(\bar{\Omega})$, where $\bar{\Omega}$ is the *closure* of the open set Ω, is defined as containing all (real- or complex-valued) functions ϕ on Ω such that, for each α with $|\alpha| \leq m$, $D^\alpha\phi$ exists and is bounded and uniformly continuous on Ω. A function ψ is said to be uniformly continuous on Ω if, given any $\varepsilon > 0$, there exists $\delta > 0$ such that, for all $\boldsymbol{r}_1$ and $\boldsymbol{r}_2$ in Ω satisfying $|\boldsymbol{r}_1 - \boldsymbol{r}_2| < \delta$, one has $|\psi(\boldsymbol{r}_1) - \psi(\boldsymbol{r}_2)| < \varepsilon$. Uniformly continuous functions on Ω are exactly those functions that may be extended to continuous function on the *closed* set $\bar{\Omega}$. Thus every function ϕ in $C^m(\Omega)$ defines (uniquely) a corresponding bounded continuous function on $C^m(\bar{\Omega})$, and similarly, for $|\alpha| \leq m$, there is a natural way of defining a bounded continuous function $D^\alpha\phi$ on $\bar{\Omega}$.

$C^m(\bar{\Omega})$ becomes a Banach space if we define a norm by

$$\|\phi\| = \max_{0 \leq |\alpha| \leq m} \sup |D^\alpha \phi(\boldsymbol{r})|. \tag{2.2.5}$$

Here the supremum is taken over all $\boldsymbol{r} \in \Omega$ (equivalently $D^\alpha\phi(\boldsymbol{r})$ may be extended to $\bar{\Omega}$ and the supremum taken; if $\bar{\Omega}$ is bounded then the supremum will be the maximum modulus attained in $\bar{\Omega}$) and the maximum is taken as α runs over all possible sequences $\{\alpha_1, \alpha_2, \ldots \alpha_n\}$ of non-negative integers such that $|\alpha| \leq m$.

We shall usually write the space $C^0(\bar{\Omega})$ simply as $C(\bar{\Omega})$. If $\bar{\Omega} \equiv E$ is bounded then this corresponds to the space $C(E)$ defined previously. (As $\bar{\Omega} \neq \Omega$ for bounded sets, the two spaces are technically different, since $C(\bar{\Omega})$ is a space of functions on Ω whereas $C(E)$ is a space of functions on $\bar{\Omega}$. However, functions in $C(E)$ are in one-to-one correspondence with functions in $C(\bar{\Omega})$ through continuous extension onto the boundary of Ω.)

The space $C^\infty(\bar{\Omega})$ is defined by

$$C^\infty(\bar{\Omega}) = \bigcap_m C^m(\bar{\Omega}). \tag{2.2.6}$$

Thus functions in $C^\infty(\bar{\Omega})$ are infinitely differentiable, and all derivatives are bounded. We shall not, here, deal with the notion of a *topological space* (see Folland, 1984) which is called for in treating the topological properties of spaces such as $C^\infty(\bar{\Omega})$. What is really necessary for our purposes is to define continuity in such spaces, and we shall do this shortly for some function spaces that are contained within $C^\infty(\bar{\Omega})$. Suffice it to note here that the space $C^\infty(\bar{\Omega})$ is both too large to admit a norm by lifting the restriction $|\alpha| \leq m$ in (2.2.5), and at the same time too small to become a complete normed space compatible with the topology that one would wish to impose on such a function space.

Functions can also be characterized by their support properties. The *support* of a (real- or complex-valued) function ϕ is defined to be the closure of the set of points $\boldsymbol{r}$ at which $\phi(\boldsymbol{r}) \neq 0$. The support of a function is then a closed subset of $\mathbb{R}^n$, and ϕ is said to be of *compact support* if $\operatorname{supp} \phi$ is bounded. (In the present context, "compact" can be taken to mean "closed and bounded".)

Given an open subset Ω of $\mathbb{R}^n$, the set $C_0^m(\Omega)$ is defined to consist of those functions ϕ on Ω having compact support, and such that $D^\alpha \phi$ is continuous (and therefore uniformly continuous, since the support is bounded) on Ω for all α satisfying $|\alpha| \leq m$. For $m=0$, $C_0^0(\Omega)$ will usually be written $C_0(\Omega)$. As in (2.2.6), $C_0^\infty(\Omega)$ is defined by

$$C_0^\infty(\Omega) = \bigcap_m C_0^m(\Omega), \tag{2.2.6$'$}$$

and consists of infinitely differentiable functions having compact support in Ω. $C_0^m(\Omega)$ and $C_0^\infty(\Omega)$ may be considered either as subsets of $C^m(\bar{\Omega})$ and $C^\infty(\bar{\Omega})$ respectively, or as function spaces in their own right, endowed with an appropriate topology. First of all, let us consider $C_0^m(\Omega)$ as a subset of $C^m(\bar{\Omega})$. $C_0^m(\Omega)$ *is* a subset because, in the first place, for $\phi \in C_0^m(\Omega)$, $D^\alpha \phi$ is uniformly continuous on Ω ($|\alpha| \leq m$), and secondly any continuous function on a compact set (i.e. on the support of $D^\alpha \phi$ in this case) is necessarily bounded. Regarded as a subset of $C^m(\bar{\Omega})$, $C_0^m(\Omega)$ may be endowed with a norm that is inherited from the larger space. However, $C_0^m(\Omega)$ is not a *subspace*, since the property of completeness is lacking. Nevertheless, the closure of $C_0^m(\Omega)$ *is* a subspace of $C^m(\bar{\Omega})$, and it is of some interest to consider what this subspace is. (The *closure* of a subset $\mathscr{S}$ of a Banach space is the set of vectors of the form $\phi = \text{s-lim}_{n\to\infty} \phi_n$, where $\phi_n \in \mathscr{S}$. If $\mathscr{S}$ is closed under addition, and multiplication by numbers, then the closure of $\mathscr{S}$ is the smallest subspace $\bar{\mathscr{S}}$ such that $\bar{\mathscr{S}} \supseteq \mathscr{S}$.)

In order to answer this question, we consider the case $m=0$ and rely on the following corollary to the Stone–Weierstrass Theorem.

Let E be a compact (i.e. closed and bounded) subset of $\mathbb{R}^n$, and let E' be an arbitrary subset of E. Define the subspace $\mathscr{B}'$ of the Banach space $\mathscr{B}=C(E)$ to consist of all $\phi \in C(E)$ such that $\phi(\boldsymbol{r})=0$ for $\boldsymbol{r} \in E'$ (i.e. $\mathscr{B}'$ contains all functions in $C(E)$ that vanish on E'). Let $\mathscr{A}'$ be a subset of $\mathscr{B}'$. Then $\mathscr{A}'$ is dense in $\mathscr{B}'$ provided that

(i) $\phi_1, \phi_2 \in \mathscr{A}' \Rightarrow c_1\phi_1 + c_2\phi_2 \in \mathscr{A}'$ and $\phi_1\phi_2 \in \mathscr{A}'$,

(ii) $\boldsymbol{r}_1, \boldsymbol{r}_2 \in E \setminus E' \Rightarrow \exists \phi \in \mathscr{A}'$ such that $\phi(\boldsymbol{r}_1) \neq \phi(\boldsymbol{r}_2)$, $(\boldsymbol{r}_1 \neq \boldsymbol{r}_2)$,

(iii) $\boldsymbol{r} \in E \setminus E' \Rightarrow \exists \phi \in \mathscr{A}'$ such that $\phi(\boldsymbol{r}) \neq 0$,

(iv) $\phi \in \mathscr{A}' \Rightarrow \bar{\phi} \in \mathscr{A}'$.

If Ω is a bounded open subset of $\mathbb{R}^n$ then take $E=\bar{\Omega}$ (closure of Ω) and $E'=\bar{\Omega}\setminus\Omega$ (boundary of Ω), with $\mathscr{A}'=C_0(\Omega)$, in this version of the Stone–Weierstrass Theorem. It follows that the closure of $C_0(\Omega)$, considered as a subspace of $C(\bar{\Omega})$, is the space consisting of all $\phi \in C(\bar{\Omega})$ that vanish on the boundary of Ω (more accurately, all $\phi \in C(\bar{\Omega})$ such that $\phi(\boldsymbol{r})$ has zero boundary value as $\boldsymbol{r}$ approaches the boundary of Ω). If Ω is an unbounded open subset of $\mathbb{R}^n$ then the closure of $C_0(\Omega)$ will be the space of all $\phi \in C(\bar{\Omega})$ that vanish both on the boundary of Ω and in the limit as $|\boldsymbol{r}| \to \infty$. To see how this comes about, we consider first the special case in which $\Omega=(a,b)$ is a finite subinterval of $\mathbb{R}$. Then $C_0(a,b)$ is the set of continuous functions that vanish identically near each endpoint a, b. The closure of $C_0(a,b)$ in the sup-norm topology is the subspace of $C[a,b]$, consisting of continuous functions that vanish at each endpoint. For an infinite interval, say $(0,\infty)$, $C_0(0,\infty)$ is the set of continuous functions that vanish identically as functions of x both for x close to zero and for x sufficiently large (i.e. close to ∞). Through the change of variable $x=(1-t)/t$, with functions of x becoming functions of t, $C_0(0,\infty)$ is transformed into $C_0(0,1)$. Since the closure of $C_0(0,1)$ is the subspace of $C[0,1]$, consisting of continuous functions that vanish at $t=0$ and $t=1$, the closure of $C_0(0,\infty)$ will be the subspace of $C[0,\infty)$ consisting of continuous functions that vanish at $x=0$ and in the limit as $x\to\infty$.

What is the dual space of $C_0(\Omega)$? Since every continuous linear functional on $C_0(\Omega)$ may be extended by *continuity* onto the closure of $C_0(\Omega)$ (i.e. if $\phi_n \in C_0(\Omega)$ and $\phi_n \to \phi$ strongly then one defines $\mathscr{L}\phi = \lim_{n\to\infty} \mathscr{L}\phi_n$) one may as well ask what is the dual space of the *closure* of $C_0(\Omega)$. The *Hahn–Banach Theorem* asserts that any bounded linear functional $\mathscr{L}$ on a subspace of a

Banach space $\mathscr{B}$ may be extended, without changing the value of $||\mathscr{L}||$, to a bounded linear functional on the whole of $\mathscr{B}$. If Ω is bounded then we may use the Corollary to Lemma 2.1 to deduce that every bounded linear functional $\mathscr{L}$ on the real space $C_0(\Omega)$ or its closure may be represented as

$$\mathscr{L}\phi = \int_\Omega \phi \, d\mu_1 - \int_\Omega \phi \, d\mu_2, \tag{2.2.7}$$

where μ_1 and μ_2 are Borel measures on Ω satisfying

$$||\mathscr{L}|| = \mu_1(\Omega) + \mu_2(\Omega).$$

(*A priori* μ_1 and μ_2 should be measures on $\bar{\Omega}$ rather than Ω, but there will be no contribution to the integrals over the boundary of Ω, where we know that ϕ vanishes.) This result also extends to the case where Ω is unbounded.

One may also consider spaces of continuous functions of compact support as linear spaces in their own right, equipped with an appropriate topology. Since our primary interest will be with the dual spaces, and since the smaller the space the larger will be its dual (in the sense that every continuous linear functional on the larger space will automatically define a continuous linear functional on the smaller space, but not conversely if the spaces we are dealing with are not normed) we shall consider immediately $C_0^\infty(\Omega)$ equipped with its new topology, and its dual. The new topology differs so fundamentally from the old that we shall use a new notation $\mathscr{D}(\Omega)$ (instead of $C_0^\infty(\Omega)$) to denote the space, and $\mathscr{D}'(\Omega)$ to denote its dual.

Thus $\mathscr{D}(\Omega)$ consists of infinitely differentiable functions having compact support, where Ω is some open subset of $\mathbb{R}^n$. We shall say that a sequence of functions $\phi_n \in \mathscr{D}(\Omega)$ converges (in the new topology) to $\phi \in \mathscr{D}(\Omega)$ if there is a compact subset K of Ω such that both ϕ_n and ϕ have support inside K and ϕ_n converges (strongly) to ϕ in the sup norm. A linear functional $\mathscr{L}$ on $\mathscr{D}(\Omega)$ is said to be *continuous* if $\mathscr{L}\phi_n$ converges to $\mathscr{L}\phi$ whenever ϕ_n converges (in the new topology) to ϕ. The dual space $\mathscr{D}'(\Omega)$ is the space of all continuous linear functionals on $\mathscr{D}(\Omega)$. Elements of the dual space are called *generalized functions* or distributions.

This criterion for a linear functional to be continuous is difficult to verify directly. However, an equivalent statement is that a linear functional $\mathscr{L}$ on $\mathscr{D}(\Omega)$ is continuous if and only if for each compact subset K of Ω there is a constant C and an integer m such that

$$|\mathscr{L}\phi| \leq C \max_{0 \leq |\alpha| \leq m} \sup |D^\alpha \phi(\boldsymbol{r})| \tag{2.2.8}$$

for all $\phi \in \mathscr{D}(\Omega)$ having support contained in K. Comparing with (2.2.5), this means that the restriction of $\mathscr{L}$ to functions having support in K should be bounded as a linear functional in $C^m(\bar{\Omega})$.

Examples of distributions

(i) Let f be any continuous function on Ω (more generally, any function locally L^1 in Ω; see later in this section) and define a distribution $\mathscr{L}$ by

$$\mathscr{L}\phi = \int_{\Omega} f\phi \, \mathrm{d}^n \boldsymbol{r}. \tag{2.2.9}$$

(This integral will be finite since the support of the integral is bounded, and hence has finite volume.) A distribution generated in this way is called a *regular* distribution (the regular distribution defined by f).

(ii) In $\mathscr{D}'(\mathbb{R})$ define a distribution δ_a by

$$\delta_a \phi = \phi(a). \tag{2.2.10}$$

δ_a is *not* a regular distribution (i.e. (2.2.9) does not hold for any f with $\mathscr{L} = \delta_a$). Despite this, one often introduces a fictitious delta function $\delta(x-a)$, with the convention that

$$\delta_a \phi = \phi(a) = \int_{-\infty}^{\infty} \delta(x-a)\phi(x)\,\mathrm{d}x. \tag{2.2.11}$$

(iii) Given any generalized function $\mathscr{L}$, one can define a derivative of $\mathscr{L}$, of order $|\alpha|$, to be the generalized function given by

$$(\mathrm{D}^\alpha \mathscr{L})(\phi) = (-1)^{|\alpha|} \mathscr{L}(\mathrm{D}^\alpha \phi). \tag{2.2.12}$$

This definition is motivated by the requirement that, for any $f \in C^\alpha(\bar{\Omega})$, if $\mathscr{L}$ is the regular distribution defined by f then $\mathrm{D}^\alpha \mathscr{L}$ is the regular distribution defined by $\mathrm{D}^\alpha f$. (Use integration by parts to show this.) On the other hand, the derivative of a regular distribution need not itself be regular. In $\mathscr{D}'(\mathbb{R})$, if $\mathscr{L}$ is the regular distribution defined by the step function

$$f(x) = \begin{cases} 1 & (x \geq a), \\ 0 & (x < a) \end{cases}$$

then

$$\frac{\mathrm{d}}{\mathrm{d}x} \mathscr{L} = \delta_a \qquad \text{(verify this!)}. \tag{2.2.13}$$

We also have

$$\frac{\mathrm{d}^2}{\mathrm{d}x^2} |x| = 2\delta_0.$$

A framework (distribution theory) within which practically any "function" can be differentiated certainly has its attractions, and distribution theory plays an important role in the theory of partial differential equations, where it is often useful to look for "solutions" of the maximum degree of generality.

(iv) Given a Borel measure μ on $\mathbb{R}$, a corresponding distribution $\mathscr{L}_\mu$ may be defined by

$$\mathscr{L}_\mu \phi = \int_{-\infty}^{\infty} \phi \, \mathrm{d}\mu. \tag{2.2.14}$$

It follows from previous results that every *positive* distribution can be generated in this way from a measure. There is, however, in distribution theory no representation comparable to (2.2.7) for non-positive distributions in general. For example, the derivative of the δ-distribution, namely

$$(\delta'_a \phi) = -\phi'(a),$$

cannot be expressed as the difference between two positive distributions, and hence in this case there is no representation as the difference between two distributions, each of which is generated by a measure.

A sequence of functions $\{f_n\}$ is said to converge to $\mathscr{L}$ *in the sense of distributions* if

$$\lim_{n\to\infty} \mathscr{L}_n \phi = \mathscr{L}\phi,$$

for each $\phi \in \mathscr{D}(\Omega)$, where $\mathscr{L}_n$ is the regular distribution defined by f_n. In other words, one has

$$\lim_{n\to\infty} \int_\Omega f_n \phi \, \mathrm{d}^n \boldsymbol{r} = \mathscr{L}\phi. \tag{2.2.15}$$

If $\mathscr{L} \in \mathscr{D}'(\mathbb{R})$ is the regular distribution defined by a function f then one may easily verify that the distributional derivative $\mathrm{d}\mathscr{L}/\mathrm{d}x$ of $\mathscr{L}$ is the limit, in the sense of distributions, as $h \to 0$, of $(f(x+h)-f(x))/h$, where this limiting process is extended in an obvious way to limits of a family depending continuously on a parameter. In this situation, one sometimes refers to $\mathrm{d}\mathscr{L}/\mathrm{d}x$ as the derivative of f in the sense of distributions.

Let us turn now to the use of criteria based on *integrability* in order to define function spaces. If E is any (Lebesgue-measurable) subset of $\mathbb{R}^n$ then the space $L^p(E)$ $(p \geq 1)$ consists of all functions f on E satisfying

$$\int_E |f|^p \, \mathrm{d}^n \boldsymbol{r} < \infty.$$

Here two functions f_1 and f_2 in the space are identified if $f_1(\boldsymbol{r}) = f_2(\boldsymbol{r})$ except when $\boldsymbol{r}$ belongs to a set of Lebesgue measure zero. (It is perhaps better to regard the elements of $L^p(E)$ as *equivalence classes* of functions, such that two functions that agree except on a set of measure zero belong to the same equivalence class.)

$L^p(E)$ becomes a Banach space, with the norm

$$\|f\|=\left(\int_E |f|^p \, \mathrm{d}^n \boldsymbol{r}\right)^{1/p}. \tag{2.2.16}$$

We shall denote this norm by $\|f\|_p$, the set E being understood. It is a non-trivial result that $L^p(E)$ is complete and separable. That $\|f\|_p$ defines a norm follows from the Minkowski inequality

$$\|f+g\|_p \leq \|f\|_p + \|g\|_p, \tag{2.2.17}$$

and we also have the Hölder inequality

$$\|fg\|_r \leq \|f\|_p \, \|g\|_q, \tag{2.2.18}$$

with $p^{-1}+q^{-1}=r^{-1}$ (p, q, r all ≥ 1).

All of these results and definitions extend to the space $L^p(E, \mathrm{d}\mu)$, where μ is a general measure on E and the norm is defined by

$$\|f\|_p=\left(\int_E |f|^p \, \mathrm{d}\mu\right)^{1/p} \qquad (E \textit{ and } \mu \text{ understood}).$$

For $p=q=2$, $r=1$, (2.2.18) is called the Schwarz inequality.

Many of the results regarding dense sets, closures, etc., that we have met extend in a straightforward manner to L^p spaces. We cannot list all such results here, but as an example it is often useful to note that, for any open subset Ω of $\mathbb{R}^n$, the set of functions in $C_0^\infty(\Omega)$ (i.e. infinitely differentiable with compact support) is dense in $L^p(\Omega)$.

There is a particularly simple description of *dual spaces* of the L^p-spaces. A simple application of the Hölder inequality (2.2.18), with $r=1$, shows that if g is an arbitrary function in $L^q(\mathbb{R}^n)$ then a corresponding element $\mathscr{L}$ of the dual space of $L^p(\mathbb{R}^n)$ may be defined by

$$\mathscr{L}_g \phi = \int \phi g \, \mathrm{d}^n \boldsymbol{r} \qquad (p^{-1}+q^{-1}=1). \tag{2.2.19}$$

($\mathscr{L}$ is bounded because $|\mathscr{L}\phi| \leq \|\phi g\|_1 \leq \|\phi\|_p \, \|g\|_q$.) Conversely, *every* linear functional $\mathscr{L}$ in the dual space of $L^p(\mathbb{R}^n)$ has the form (2.2.19) for some $g \in L^q(\mathbb{R}^n)$. There is therefore a one-to-one correspondence between elements of the dual space of $L^p(\mathbb{R}^n)$ and elements of the space $L^q(\mathbb{R}^n)$ ($p^{-1}+q^{-1}=1$; $p, q \geq 1$), and the mapping $g \to \mathscr{L}_g$ is linear. Actually, a mapping that arises more naturally in this context is the mapping $g \to \mathscr{L}_{\bar{g}}$, which is *antilinear*.) Without too much abuse of notation, the dual space of L^p may be *identified* with L^q, where $p^{-1}+q^{-1}=1$. An important special case is $p=2$; the dual space of $L^2(\mathbb{R}^n)$ is again $L^2(\mathbb{R}^n)$. This comes from the fact that, in L^2 alone, an *inner product* can be defined. In $L^2(\mathbb{R}^n)$ the inner product of any two functions will be defined by

$$\langle f, g\rangle = \int \bar{f} g \,\mathrm{d}^n \boldsymbol{r}. \tag{2.2.20}$$

Note that we adopt the convention of complex-conjugating the function on the *left* rather than the right of the inner product. Which convention is used makes, of course, no difference if we are dealing with a real function space. More generally, in $L^2(E, \mathrm{d}\mu)$ we define

$$\langle f, g\rangle = \int \bar{f} g \,\mathrm{d}\mu. \tag{2.2.20$'$}$$

On a general linear space, an *inner product* assigns to any two vectors f and g a number $\langle f, g\rangle$ (real for a vector space over $\mathbb{R}$ and complex for a vector space over $\mathbb{C}$) with properties

(i) $\langle f+g, h\rangle = \langle f, h\rangle + \langle g, h\rangle$,
(ii) $\langle f, g\rangle = \overline{\langle g, f\rangle}$,
(iii) $\langle f, cg\rangle = c\langle f, g\rangle$,
(iv) $\langle f, f\rangle \geq 0$, with $\langle f, f\rangle = 0 \Leftrightarrow f=0$.

A linear space equipped with an inner product is called an *inner-product space*. The space may also be given a norm, by taking $\|f\| = \langle f, f\rangle^{1/2}$, so that an inner-product space is a normed space with extra structure. (Not every normed space admits an inner product; for example $C[0, 1]$, with sup norm, has no corresponding inner product, although of course this set of functions can be assigned an inner product that is associated with a different norm (the L^2 norm)).

A complete, separable, complex inner-product space is called a Hilbert space, about which we shall have more to say shortly in a more general context. Suffice it to observe here that all of the spaces $L^2(E, \mathrm{d}\mu)$ are examples of Hilbert spaces. For Hilbert space $\mathscr{H}$, any member of the dual space is a functional of the form $f \to \langle \phi, f\rangle$ for some $\phi \in \mathscr{H}$, a result known as the Riesz Representation Theorem.

Another L^p space that should be mentioned here is the space $L^\infty(\mathbb{R}^n)$, consisting of (equivalence classes of) measurable functions f on $\mathbb{R}^n$ such that $|f|$ is bounded. We define a norm $\|f\|_\infty$ to be the smallest value of M such that $|f(\boldsymbol{r})| \leq M$ almost everywhere for $\boldsymbol{r} \in \mathbb{R}^n$. $L^\infty(\mathbb{R}^n)$ is a Banach space with this norm, and may be regarded (loosely) as the limit of the space $L^p(\mathbb{R}^n)$ as $p \to \infty$. The dual space of L^1 is L^∞, as might be expected; however, the dual space of L^∞ is *not* L^1, but a much larger space. (This is because many *generalized functions* may be extended to continuous linear functionals on L^∞). L^∞ also differs from L^p in not being separable.

The space $L^\infty(E)$ may be defined in a similar way for any measurable subset E of $\mathbb{R}^n$. Some of the Banach spaces referred to above may be regarded as subspaces of $L^\infty(E)$. For example, $C[0,1]$ is a subspace of $L^\infty(0,1)$, $\|\cdot\|_\infty$ coinciding, for continuous functions, with the sup norm that we have assigned to $C[0,1]$.

Before leaving these examples of function spaces, we should briefly mention the so-called *Sobolev spaces*. These spaces are Banach spaces of functions satisfying L^p conditions on their derivatives. Consider, for example, a function u belonging to $L^p(\Omega)$ for some open subset Ω of $\mathbb{R}^n$. There is a regular distribution corresponding to u, and $D^\alpha u$ may be defined in the sense of distributions. Suppose further that $D^\alpha u$ is itself a regular distribution for each multi-index α satisfying $|\alpha|\leq m$, and that the corresponding function that defines this distribution is again in the space $L^p(\Omega)$. We shall write, simply,

$$D^\alpha u\in L^p(\Omega) \quad \text{for } 0\leq|\alpha|\leq m,$$

and the space of such functions is the Sobolev space $W^{m,p}(\Omega)$. Thus

$$W^{m,p}(\Omega)=\{u\in L^p(\Omega);\ D^\alpha u\in L^p(\Omega) \quad \text{for } 0\leq|\alpha|\leq m\}. \tag{2.2.21}$$

$W^{m,p}(\Omega)$ becomes a Banach space with the norm

$$\|u\|_{m,p}=\left\{\sum_{|\alpha|\leq m}(\|D^\alpha u\|_p)^p\right\}^{1/p}. \tag{2.2.21$'$}$$

An important subspace of $W^{m,p}(\Omega)$ is $W_0^{m,p}(\Omega)$, defined to be the closure, in the above norm, of the infinitely differentiable functions of compact support. There is another way of introducing the function space $W^{m,p}(\Omega)$. We start with the set of functions $u(\boldsymbol{r})$ on Ω, such that $u(\boldsymbol{r})$ and $D^\alpha u(\boldsymbol{r})$ exist for all $\boldsymbol{r}\in\Omega$ (this is a more restrictive condition than demanding the existence of these derivatives *in the sense of distributions*), for all α with $|\alpha|\leq m$, and such that u and $D^\alpha u$ belong to the space $L^p(\Omega)$. Then $W^{m,p}(\Omega)$ is the closure of this set of functions, in the norm given by (2.2.21)′.

Our brief survey of some of the function spaces that are encountered in mathematical physics, and in particular in scattering theory, will, it is hoped, give the reader some preliminary idea of ways in which functions may be categorized and described. The function spaces of quantum mechanics are, of course, generally Hilbert spaces, and it will be important for us (Section 2.5) to study Hilbert space in more depth and at a more abstract level. It should not be forgotten, however, that in most specific calculations in quantum mechanics it will be necessary to descend from the pure heights of abstraction and consider Hilbert space in one of its various concrete realizations as a function space. It is when it is necessary to consider local (continuity, differentiability, etc.) and global (L^∞, L^p, etc.) properties of the functions that emerge that we are led to some of the spaces considered above.

Moreover, a knowledge of the dense sets, bases, etc. of a particular space will often allow us to extend our results to a greater degree of generality. As an example of this, we consider the following *proof by extension*.

It is required to prove the identity

$$\lim_{y\to 0+} \int_{-\infty}^{\infty} \frac{y f(x)\,\mathrm{d}x}{(\lambda - x)^2 + y^2} = \pi f(\lambda)\,. \tag{2.2.22}$$

It is not clear *a priori* for what class of functions we should expect the identity (2.2.22) to hold. It *is* clear, however, that if (2.2.22) holds for f_1 and f_2 then the identity holds also for $c_1 f_1 + c_2 f_2$. In other words, the set of functions for which (2.2.22) holds forms a *linear* space of functions.

In order to start with the simplest possible case, we assume $f \in C_0^1(R)$. Thus f has compact support, which allows us to escape, in the most definitive way possible, from the problem that the left-hand side of (2.2.22) might not even define a convergent integral; the integration becomes effectively over a finite interval. The requirement that f have continuous first derivative makes the function behave sufficiently well locally to make the result at least plausible. (Certainly if, say, f could be the characteristic function of the interval (0, 1) or of [0, 1] then we should have some difficulty in interpreting the $f(\lambda)$ of the right-hand side).

It turns out (for details see the proof of Lemma 2.3 (i)) that (2.2.22) may be proved fairly directly in the case $f \in C_0^1(\mathbb{R})$, and that in that case convergence in the limit as $\varepsilon \to 0+$ is uniform in λ. Now uniform convergence is precisely convergence in the sup norm. This suggests that, having proved (2.2.22) for $f \in C_0^1(\mathbb{R})$, the identity can be extended to limits, in the sup norm, of such f. To test this idea, we let h be the strong limit, in the sup norm, of a sequence $\{f_n\}$, where $f_n \in C_0^1(\mathbb{R})$.

Given $\varepsilon > 0$, arbitrary, we choose n such that

$$\|h - f_n\| \equiv \sup |h(x) - f_n(x)| < \varepsilon/3\pi.$$

Then for $y > 0$ sufficiently small we have

$$\left| \int_{-\infty}^{\infty} \frac{y f_n(x)\,\mathrm{d}x}{(\lambda - x)^2 + y^2} - \pi f_n(\lambda) \right| < \frac{\varepsilon}{3}.$$

This implies that, for $y>0$ sufficiently small,

$$\left|\int_{-\infty}^{\infty}\frac{yh(x)\,dx}{(\lambda-x)^2+y^2}-\pi h(\lambda)\right|$$

$$=\left|\int_{-\infty}^{\infty}\frac{y(h(x)-f_n(x))\,dx}{(\lambda-x)^2+y^2}+\left\{\int_{-\infty}^{\infty}\frac{yf_n(x)\,dx}{(\lambda-x)^2+y^2}-\pi f_n(\lambda)\right\}\right.$$

$$\left.+\{\pi(f_n(\lambda)-h(\lambda))\}\right|$$

$$\leq\|h-f_n\|\int_{-\infty}^{\infty}\frac{y\,dx}{(\lambda-x)^2+y^2}+\frac{\varepsilon}{3}+\pi\|h-f_n\|$$

$$=2\pi\|h-f_n\|+\varepsilon/3\leq\varepsilon.$$

Since $\varepsilon>0$ is arbitrary, we have extended (2.2.22) to functions h that are limits of C_0^1 functions, in the sup norm. Hence in (2.2.22) f may be any function in the closure of $C_0^1(\mathbb{R})$, considered as a subset of the space of continuous functions on $\mathbb{R}$. We have already seen that the closure of $C_0(\mathbb{R})$ is the space of continuous functions vanishing in the limits as $x\to\pm\infty$, this being a special case of the closure of $C_0(\Omega)$ for an unbounded open set Ω. Exactly the same argument, using the Stone–Weierstrass Theorem, tells us that the closure of $C_0^1(\mathbb{R})$ is just the same space of functions, i.e. continuous functions vanishing at infinity. So proof by extension allows us to assert the identity (2.2.22) for a much wider class of functions than before, and, moreover, convergence is again uniform in λ. We shall state these results formally as Lemma 2.3 (i).

It is also possible to extend (2.2.22) by taking closures in the L^1 norm. Indeed, if $f\in C_0^1(\mathbb{R})$ then (2.2.22) may be shown to hold (for details see the proof of Lemma 2.3 (ii)) if the limit on the left-hand side is interpreted as the limit of a function of λ in L^1 norm. Note also that the set of functions $C_0^1(\mathbb{R})$ (even $C_0^\infty(\mathbb{R})$, which is a smaller set of functions) is dense in $L^1(\mathbb{R})$.

Proceeding as before, we now let h be the strong limit, in the L^1 norm, of a sequence $\{f_n\}$, where $f_n\in L^1(\mathbb{R})$.

Given $\varepsilon>0$, arbitrary, we choose n such that

$$\|h-f_n\|_1<\varepsilon/3\pi.$$

Then for $y>0$ sufficiently small, we have, taking norms of functions of λ,

$$\left\| \int_{-\infty}^{\infty} \frac{y f_n(x)\,dx}{(\lambda-x)^2+y^2} - \pi f_n(\lambda) \right\|_1 < \frac{\varepsilon}{3}$$

The proof that (2.2.22) holds for h, with the limit in the L^1 norm, now proceeds as before if we note, on exchanging orders of integration, that

$$\left\| \int_{-\infty}^{\infty} \frac{y(h(x)-f_n(x))\,dx}{(\lambda-x)^2+y^2} \right\|_1$$

$$= \int_{-\infty}^{\infty} d\lambda \left| \left\{ \int_{-\infty}^{\infty} \frac{y(h(x)-f_n(x))\,dx}{(\lambda-x)^2+y^2} \right\} \right|$$

$$\leqq \int_{-\infty}^{\infty} dx \int_{-\infty}^{\infty} d\lambda \, \frac{y\,|h(x)-f_n(x)|}{(\lambda-x)^2+y^2}$$

$$= \pi \int_{-\infty}^{\infty} |h(x)-f_n(x)|\,dx = \pi \|h-f_n\|_1 .$$

Hence (2.2.22) holds (with L^1 limit) for all $f \in L^1(\mathbb{R})$; this result will be stated formally as Lemma 2.3 (ii).

Proof by extension then enables us to ensure for our results the maximum degree of validity. Give us a norm, and we shall extend the result! This assumes, of course, that the norm is appropriate to the problem under consideration. The example of (2.2.22) serves to show us that, in this case, there is more than one norm that will do, and it should come as no surprise that, in Lemma 2.3, the identity will be proved for $f \in L^2(\mathbb{R})$ as well.

2.3 ANALYSIS OF MEASURES

Let μ_1 and μ_2 be Borel–Stieltjes measures. Although it is not true in general that $\mu_1 \ll \mu_2$, it does follow immediately from the definition of absolute continuity that

$$\mu_1 \ll \mu_1 + \mu_2 .$$

Hence the existence follows, by the Radon–Nikodym Theorem, of a non-negative $(\mu_1+\mu_2)$-measurable function f such that, for any Borel set E,

$$\mu_1(E)=\int_E f(\lambda)\,\mathrm{d}(\rho_1(\lambda)+\rho_2(\lambda)), \tag{2.3.1}$$

where ρ_1 and ρ_2 generate the measures μ_1 and μ_2 respectively. We denote by A the Borel set consisting of all λ such that $f(\lambda)<1$, and we let $\mu_1|_A$ denote the restriction of μ_1 to A; i.e. $(\mu_1|_A)(E)=\mu_1(A\cap E)$.

We let E_ε be the set of $\lambda\in E$ at which $f(\lambda)\leq 1-\varepsilon$. From (2.3.1) we have, with $0<\varepsilon<1$.

$$\begin{aligned}\mu_1(E_\varepsilon)&=\int_{E_\varepsilon} f(\lambda)\,\mathrm{d}(\rho_1+\rho_2)\leq\int_{E_\varepsilon}(1-\varepsilon)\,\mathrm{d}(\rho_1+\rho_2)\\&=(1-\varepsilon)(\mu_1(E_\varepsilon)+\mu_2(E_\varepsilon)).\end{aligned} \tag{2.3.2}$$

Hence

$$\mu_1(E_\varepsilon)\leq\frac{1-\varepsilon}{\varepsilon}\mu_2(E_\varepsilon).$$

In particular, if $\mu_2(E)=0$ then $\mu_1(E_\varepsilon)=0$, so that

$$(\mu_1|_A)(E)=\mu_1\left(\bigcup_\varepsilon E_\varepsilon\right)=\lim_{\varepsilon\to 0+}\mu_1(E_\varepsilon)=0.$$

This means that $\mu_1|_A$ is absolutely continuous with respect to μ_2.

Now we let F_ε denote the set of points λ at which $f(\lambda)\geq 1+\varepsilon$. Then, as in (2.3.2), we have

$$\mu_1(F_\varepsilon)\geq(1+\varepsilon)(\mu_1(F_\varepsilon)+\mu_2(F_\varepsilon)),$$

from which we have $\mu_1(F_\varepsilon)=\mu_2(F_\varepsilon)=0$. Hence again, almost everywhere with respect to μ_1 and μ_2, we have $f(\lambda)\leq 1$. Since points λ at which $f(\lambda)>1$ will not then contribute to the right-hand side of (2.3.1), we may assume without loss of generality that the complement A^c of A consists of those λ at which $f(\lambda)=1$. We then have

$$\begin{aligned}\mu_1(A^c)&=\int_{A^c} f(\lambda)\,\mathrm{d}(\rho_1+\rho_2)=\int_{A^c}1\,\mathrm{d}(\rho_1+\rho_2)\\&=\mu_1(A^c)+\mu_2(A^c).\end{aligned}$$

so that

$$\mu_2(A^c)=0.$$

Hence $\mu_1|_{A^c}$ is concentrated on a set A^c having zero μ_2-measure.

We can write, then,

$$\mu_1 = \mu_1|_A + \mu_1|_{A^c}, \tag{2.3.3}$$

where the measures on the right-hand side are respectively absolutely continuous and singular with respect to μ_2. This decomposition of μ_1 into its absolutely continuous and singular components with respect to μ_2 is unique.

The most important application of this idea to scattering theory is the decomposition of a given Borel–Stieltjes measure μ into its absolutely continuous and singular parts with respect to Lebesgue measure.

We shall write, in this case,

$$\mu = \mu_{ac} + \mu_s. \tag{2.3.4}$$

The singular component μ_s may be decomposed further into its singular continuous and discrete components respectively. Thus μ_{sc} is a singular measure that is also continuous in the sense that single points have zero measure, or equivalently $\mu_{sc}(a, x]$ is a continuous function of x. On the other hand, the discrete component μ_d of the measure μ is concentrated on those points (finite or countable in number) that have strictly positive measure. These are the so-called discrete points of the measure.

Writing

$$\mu_s = \mu_{sc} + \mu_d,$$

we then have the complete decomposition

$$\mu = \mu_{ac} + \mu_{sc} + \mu_d. \tag{2.3.5}$$

Note that (2.3.5) allows us to write $\mathbb{R} = \Sigma_{ac} \cup \Sigma_{sc} \cup \Sigma_d$ where the Σs are three disjoint Borel sets on which the corresponding components of the measure are concentrated. Though these Borel sets are non-unique (for example, an arbitrary set of Lebesgue measure zero might be removed from Σ_{ac} and added to Σ_{sc}), they do help to show that the decomposition (2.3.5) involves a splitting of the real line into sets of points that may be characterized respectively as absolutely continuous, singular continuous and discrete. The following lemma leads to a way of describing this classification via a local comparison of the measure μ with Lebesgue measure ℓ. Although we deal with the most widely applicable case in which the second measure is Lebesgue measure, the proof may immediately be generalized to the case where the second measure is an arbitrary Borel–Stieltjes measure.

Lemma 2.2

Let μ be a Borel–Stieltjes measure on $\mathbb{R}$, and let $\mathcal{S}_m$ be a Borel set such that, for each $\lambda \in \mathscr{S}_m$ and for $\delta > 0$ sufficiently small (how small depends on λ),

$$\mu(\mathscr{I}) \le m\ell(\mathscr{I}) \tag{2.3.6}$$

for every subinterval $\mathscr{I}$, containing the point λ, of the interval $[\lambda - \delta, \lambda + \delta]$. Then

$$\mu(\mathscr{S}_m) \le m\ell(\mathscr{S}_m). \tag{2.3.6$'$}$$

Proof

(i) Consider first of all the case in which the set $\mathscr{S}_m$ is closed and bounded. Then (cf. Section 2.1), given $\varepsilon > 0$, there is an open set $\mathscr{S}_0$ such that $\mathscr{S}_m \subseteq \mathscr{S}_0$ and

$$\ell(\mathscr{S}_0) \le \ell(\mathscr{S}_m) + \varepsilon. \tag{2.3.7}$$

For each $\lambda \in \mathscr{S}_m$, we find an open interval $\mathscr{I}_\lambda$ containing λ such that (2.3.6) is satisfied for subintervals $\mathscr{I}$, containing λ, of $\mathscr{I}_\lambda$. Without loss of generality, it may be assumed that $\mathscr{I}_\lambda$ is taken sufficiently small that $\mathscr{I}_\lambda \subseteq \mathscr{S}_0$ in each case. The collection of intervals $\{\mathscr{I}_\lambda\}$ constitute a covering of $\mathscr{S}_m$ by open intervals, and by the Heine–Borel theorem the set $\mathscr{S}_m$ will be covered by a *finite* subset of the $\mathscr{I}_\lambda$. By removing intervals if necessary, this finite covering may be assumed to be *minimal*, in the sense that no further intervals may be removed without uncovering some point of $\mathscr{S}_m$. It is then possible to shrink each interval (if necessary), while retaining the property $\lambda \in \mathscr{I}_\lambda$, in such a way that the intervals do not overlap, and such that $\mathscr{S}_m$ is still completely covered. (The shrunk intervals need no longer be open, however.)

For example, if $\mathscr{I}_{\lambda_1} = (\lambda_1 - \varepsilon_1, \lambda_1 + \varepsilon_1)$ and $\mathscr{I}_{\lambda_2} = (\lambda_2 - \varepsilon_2, \lambda_2 + \varepsilon_2)$ are two overlapping intervals with $\lambda_1 < \lambda_2$ then they could be replaced by $\mathscr{I}_{\lambda_1} = (\lambda_1 - \varepsilon_1, \frac{1}{2}(\lambda_1 + \lambda_2)]$ and $\mathscr{I}_{\lambda_2} = (\frac{1}{2}(\lambda_1 + \lambda_2), \lambda_2 + \varepsilon_2)$, which do not overlap, but which together cover the same set of points.

Now, for each of the shrunken intervals, we have

$$\mu(\mathscr{I}_\lambda) \le m\ell(\mathscr{I}_\lambda).$$

Adding this inequality over the $\mathscr{I}_\lambda$, we find that the total μ-measure of the finite covering of $\mathscr{S}_m$ cannot exceed m times its Lebesgue measure. Since the finite covering is also contained in $\mathscr{S}_0$, we have

$$\mu(\mathscr{S}_m) \le m\ell(\mathscr{S}_0),$$

so that, from (2.3.7),

$$\mu(\mathscr{S}_m) \le m(\ell(\mathscr{S}_m) + \varepsilon).$$

But this inequality holds for all $\varepsilon > 0$, and we may deduce (2.3.6)$'$.

(ii) We now let $\mathscr{S}_m$ be an arbitrary Borel set satisfying the hypothesis of the lemma. Since

$$\mu(\mathcal{S}_m) = \lim_{N\to\infty} \mu(\mathcal{S}_m \cap [-N, N])$$

and

$$\ell(\mathcal{S}_m) = \lim_{N\to\infty} \ell(\mathcal{S}_m \cap [-N, N]),$$

in order to prove (2.3.6)′ it will be sufficient to consider sets that are contained in a finite closed interval $[-N, N]$.

We suppose then that $\mathcal{S}_m \subseteq [-N, N]$, and let $\mathcal{S}_m^c$ denote the complement of $\mathcal{S}_m$ with respect to $[-N, N]$ (i.e. $\mathcal{S}_m^c = [-N, N] \setminus \mathcal{S}_m$). Given $\varepsilon > 0$, we cover $\mathcal{S}_m^c$ by an open set $\mathcal{S}'_0$ such that

$$\mu(\mathcal{S}'_0) \leq \mu(\mathcal{S}_m^c) + \varepsilon. \tag{2.3.8}$$

If $(\mathcal{S}'_0)^c$ denotes the complement of $\mathcal{S}'_0$ with respect to $[-N, N]$ then $(\mathcal{S}'_0)^c \subseteq \mathcal{S}_m$, and is a closed bounded set satisfying the hypothesis of the lemma. Hence by (i) above we have

$$\mu((\mathcal{S}'_0)^c) \leq m\ell((\mathcal{S}'_0)^c). \tag{2.3.9}$$

Also,

$$\mu[-N, N] = \mu(\mathcal{S}_m) + \mu(\mathcal{S}_m^c) \leq \mu(\mathcal{S}'_0) + \mu((\mathcal{S}'_0)^c).$$

From (2.3.8) and (2.3.9) we now have

$$\mu(\mathcal{S}_m) \leq \mu((\mathcal{S}'_0)^c) + \varepsilon \leq m\ell((\mathcal{S}'_0)^c) + \varepsilon.$$

But $(\mathcal{S}'_0)^c \subseteq \mathcal{S}_m$, so that

$$\mu(\mathcal{S}_m) \leq m\ell(\mathcal{S}_m) + \varepsilon.$$

Since this inequality holds for all $\varepsilon > 0$, (2.3.6)′ follows, and the lemma is proved. ■

In the following lemma, we shall say that $\mu(\mathcal{I}_\lambda)/\ell(\mathcal{I}_\lambda)$ is bounded, for some fixed λ, if there exists $m > 0$, which may be different for different λ, such that $\mu(\mathcal{I}_\lambda) \leq m\ell(\mathcal{I}_\lambda)$ for all intervals $\mathcal{I}_\lambda$ containing λ such that $\ell(\mathcal{I}_\lambda) < \text{const}$. We shall say that $\lim_{\ell(\mathcal{I}_\lambda)\to 0} \mu(\mathcal{I}_\lambda)/\ell(\mathcal{I}_\lambda) = c_\lambda$, say, if this limit exists and is the same for every sequence of intervals $\mathcal{I}_\lambda^{(n)}$ such that $\lim_{n\to\infty} \ell(\mathcal{I}_\lambda^{(n)}) = 0$. It is easy to see that this will be so if and only if

$$\lim_{\varepsilon\to 0+} \frac{\mu\{(\lambda, \lambda+\varepsilon]\}}{\varepsilon} = \lim_{\varepsilon\to 0+} \frac{\mu\{(\lambda-\varepsilon, \lambda]\}}{\varepsilon}$$

$$= \lim_{\varepsilon\to 0+} \frac{\mu\{(\lambda-\varepsilon, \lambda+\varepsilon]\}}{2\varepsilon} = c_\lambda. \tag{2.3.10}$$

In other words, for any constant c, the function $\mu\{(c, \lambda]\}$ should be differentiable at the point λ. We have already observed that this function is locally of bounded variation and hence differentiable for almost all values of λ. It follows that $\mu(\mathcal{I}_\lambda)/\ell(\mathcal{I}_\lambda)$ converges to a limit, and hence in particular it is bounded, for almost all λ.

We are now ready to give a local characterization of the respective supports of the measures μ_{ac} and μ_s. The basic result of this section is given by the following theorem.

Theorem 2.1

For a Borel–Stieltjes measure μ let $\mathcal{S}$ be the set of values of λ at which $\mu(\mathcal{I}_\lambda)/\ell(\mathcal{I}_\lambda)$ is bounded. Then

(i) the absolutely continuous part of μ is given by

$$\mu_{ac} = \mu|_{\mathcal{S}} \qquad \text{(restriction of } \mu \text{ to } \mathcal{S}\text{);}$$

μ_{ac} has density function $f(\lambda)$ given (for almost all λ) by

$$f(\lambda) = \lim_{\ell(\mathcal{I}_\lambda)\to 0} \mu(\mathcal{I}_\lambda)/\ell(\mathcal{I}_\lambda), \tag{2.3.11}$$

and any Borel set of points λ for which this limit is zero has μ-measure zero;

(ii) the singular part of μ is given by

$$\mu = \mu|_{\mathcal{S}^c} \qquad \text{(restriction of } \mu \text{ to the complement of } \mathcal{S}\text{).}$$

Moreover, the measure μ is singular (i.e. $\mu = \mu_s$) if and only if

$$\lim_{\ell(\mathcal{I}_\lambda)\to 0} \mu(\mathcal{I}_\lambda)/\ell(\mathcal{I}_\lambda) = 0 \tag{2.3.11$'$}$$

for almost all values of λ.

Proof

That $\mathcal{S}$ is a Borel set will follow later when we shall describe an alternative characterization of $\mathcal{S}$. We next prove that the restriction of μ to $\mathcal{S}$ is absolutely continuous. The restriction to $\mathcal{S}$ is defined by

$$(\mu|_{\mathcal{S}})(E) = \mu(E \cap \mathcal{S}).$$

and to demonstrate absolutely continuity it will be sufficient to show that, for any subset A of $\mathcal{S}$.

$$\ell(A) = 0 \Rightarrow \mu(A) = 0. \tag{2.3.12}$$

Suppose then that $\ell(A) = 0$. Consider $\mathcal{S}_m$ as in Lemma 2.2. Then we can write $\mathcal{S} = \bigcup_m \mathcal{S}_m$ as m runs over the set of positive integers. Moreover,

$\ell(A \cap \mathscr{S}_m)=0$, where the set $A \cap \mathscr{S}_m$ satisfies the hypothesis of Lemma 2.2, so that

$$\mu(A \cap \mathscr{S}_m) \leq m\ell(A \cap \mathscr{S}_m).$$

It follows that $\mu(A \cap \mathscr{S}_m)=0$. However, from Section 2.1, we have

$$\mu(A)=\mu(A \cap \mathscr{S})=\lim_{m \to \infty} \mu(A \cap \mathscr{S}_m)=0,$$

so that (2.3.12) is satisfied and $\mu|_{\mathscr{S}}$ is absolutely continuous.

Since $\mu(\mathscr{I}_\lambda)/\ell(\mathscr{I}_\lambda)$ is bounded for almost all λ, it follows that the complement of $\mathscr{S}$ has Lebesgue measure zero. Hence $\mu|_{\mathscr{S}^c}$ is a measure concentrated on a set of Lebesgue measure zero, and is therefore singular with respect to Lebesgue measure. It follows immediately that

$$\mu=\mu_{\text{ac}}+\mu_{\text{s}},$$

$$\text{with } \mu_{\text{ac}}=\mu|_{\mathscr{S}} \text{ and } \mu_{\text{s}}=\mu|_{\mathscr{S}^c}.$$

We now consider a Borel set E of points λ for which the limit on the right-hand side of (2.3.11) is zero. Then Lemma 2.2 may be applied to the set E, with the value of m arbitrarily small in (2.3.6). From (2.3.6)′ in the limit $m \to 0$, with $\mathscr{S}_m=E$, we have $\mu(E)=0$.

The next step in the proof is to show that, for almost all λ,

$$\lim_{\ell(\mathscr{I}_\lambda) \to 0} \mu_{\text{s}}(\mathscr{I}_\lambda)/\ell(\mathscr{I}_\lambda)=0. \tag{2.3.13}$$

Let $\mathscr{S}_{\text{s}}$ be the set of values of λ such that $\mu_{\text{s}}(\mathscr{I}_\lambda)/\ell(\mathscr{I}_\lambda)$ is bounded. From the first part of the theorem, applied to μ_{s}, we know that the absolutely continuous part of μ_{s} is $\mu_{\text{s}}|_{\mathscr{S}_{\text{s}}}$. However, μ_{s} is purely singular, so that $\mu_{\text{s}}|_{\mathscr{S}_{\text{s}}}=0$, or, in other words, $\mu_{\text{s}}(\mathscr{S}_{\text{s}})=0$.

Consider the set E_m of points λ at which the limit on the left-hand side of (2.3.13) exceeds m^{-1}. Then for each $\lambda \in E_m$, and for any $\theta>0$,

$$\mu_{\text{s}}(\mathscr{I}_\lambda)/\ell(\mathscr{I}_\lambda)>(m+\theta)^{-1},$$

for $\ell(\mathscr{I}_\lambda)$ sufficiently small. Inverting this inequality then gives

$$\ell(\mathscr{I}_\lambda)<(m+\theta)\mu_{\text{s}}(\mathscr{I}_\lambda).$$

Lemma 2.2 may now be applied, with the roles of ℓ and μ_{s} exchanged, and we have

$$\ell(E_m) \leq (m+\theta)\mu_{\text{s}}(E_m).$$

However, $E_m \subseteq \mathscr{S}_{\text{s}}$ and $\mu_{\text{s}}(\mathscr{S}_{\text{s}})=0$, so that $\ell(E_m)=0$. This means that the limit on the left-hand side of (2.3.13), which exists for almost all λ, is less than or equal to m^{-1} for almost all λ, independently of the (positive) value of m. The Lebesgue measure of the set of points at which this limit differs from zero is $\ell(\bigcup_m E_m)$, where m runs over the set of positive integers. Since $\ell(\bigcup_m E_m)=\lim_{m\to\infty}\ell(E_m)=0$, we have proved (2.3.13) for almost all λ.

The density function for μ_{ac} is almost everywhere given by

$$f(\lambda)=\frac{\mathrm{d}}{\mathrm{d}\lambda}\rho_{ac}(\lambda),$$

where ρ_{ac} is the function that generates the measure μ_{ac}. Thus $f(\lambda)=\lim_{\ell(\mathscr{I}_\lambda)\to 0}\mu_{ac}(\mathscr{I}_\lambda)/\ell(\mathscr{I}_\lambda)$, which gives (2.3.11) on setting $\mu_{ac}=\mu-\mu_s$ and using (2.3.13). If (2.3.11)′ holds then $f(\lambda)=0$ for almost all λ, in which case $\mu_{ac}=0$, and the measure μ is purely singular. On the other hand, if μ is singular, so that $\mu=\mu_s$, then (2.3.13) $\Rightarrow$ (2.3.11)′. So (2.3.11)′ is a necessary and sufficient condition for a singular measure, and the proof is complete. ■

Besides furnishing a necessary and sufficient condition for a measure to be singular, the theorem also lays down a sufficient condition for the measure to be absolutely continuous. Namely, μ will be purely absolutely continuous provided $\mu(\mathscr{I}_\lambda)/\ell(\mathscr{I}_\lambda)$ is bounded for all values of λ. If, on the other hand, $\mu(\mathscr{I}_\lambda)/\ell(\mathscr{I}_\lambda)$ is bounded for all except a *countable* set of values of λ, one may deduce $\mu_{sc}=0$, since a singular *continuous* measure, restricted to a discrete set of points, must vanish. In that case μ decomposes into an absolutely continuous and discrete component.

Very often, and especially for measures relating to second-order differential operators, information concerning a measure is to be obtained indirectly, often via a transform of the measure. We consider here two particularly important transforms, and in each case we describe how an analysis of the measure may be carried out through the transform. The first of these we shall describe as the *resolvent transform*, through its connection with the *resolvent* of a self-adjoint operator (cf. Section 2.6). The second is the familiar Fourier transform, in perhaps a slightly unfamiliar guise.

The resolvent transform

Consider a Borel–Stieltjes measure μ on $\mathbb{R}$, generated by $\rho(\lambda)$, and suppose that

$$\int_{-\infty}^{\infty}(\lambda^2+1)^{-1}\,\mathrm{d}\rho(\lambda)<\infty. \tag{2.3.14}$$

The resolvent transform of μ is the function v of x and y, defined for $y>0$ by

$$v=\int_{-\infty}^{\infty}\frac{y\,\mathrm{d}\rho(\lambda)}{(\lambda-x)^2+y^2}. \tag{2.3.15}$$

(This actually corresponds to the imaginary part of the transform. We shall later write down the real part.)

We are interested especially in the limit of v as y approaches zero. Now the expression $y/(\lambda-x)^2+y^2$, regarded as a function of x for fixed λ, converges pointwise to zero as $y\to 0$, except at $x=\lambda$, at which point the function diverges in the limit. For small values of y, we have a function that is small except in the neighbourhood of $x=\lambda$, and for all positive values of y we have

$$\int_{-\infty}^{\infty} \frac{y}{(\lambda-x)^2+y^2}\,\mathrm{d}x = \left[\tan^{-1}\left(\frac{x-\lambda}{y}\right)\right]_{-\infty}^{\infty} = \pi.$$

In other words, as $y\to 0$, $y/(\lambda-x)^2+y^2$ looks like the δ-function $\pi\delta(\lambda-x)$. In that case, we expect to find, for a large class of functions f,

$$\lim_{y\to 0+} \int_{-\infty}^{\infty} \frac{yf(x)\,\mathrm{d}x}{(\lambda-x)^2+y^2} = \pi f(\lambda) \tag{2.3.16}$$

The following lemma shows that there are at least three senses in which (2.3.16) is justified.

Lemma 2.3

(i) Equation (2.3.16) holds for any continuous function f such that $\lim_{x\to\pm\infty} f(x)=0$. Moreover, for given f, convergence is uniform in λ in this case.

(ii) Equation (2.3.16) holds for any $f\in L^1(\mathbb{R})$, where the limit is defined in the sense of L^1-norm convergence.

(iii) Equation (2.3.16) holds for any $f\in L^2(\mathbb{R})$, where the limit is defined in the sense of L^2-norm convergence.

Proof

(i) This is our canonical example of proof by extension. It will be sufficient (see (2.2.22) *et seq.*) to take $f\in C_0^1(\mathbb{R})$. Such a function satisfies an inequality of the form

$$|f(x_1)-f(x_2)| \le \text{const}\,|x_1-x_2|^{1/2}. \tag{2.3.17}$$

We suppose that $f(x)$ vanishes outside the interval $[-N, N]$. We know that $(f(x_1)-f(x_2))/(x_1-x_2)=f'(\alpha)$ for some $\alpha\in[x_1, x_2]$. Since f' is bounded, this means $|f(x_1)-f(x_2)|\le \text{const}|x_1-x_2| \le \text{const}|x_1-x_2|^{1/2}$ if we assume

$|x_1 - x_2| \le 2N$, in which case $|x_1 - x_2| \le (2N)^{1/2}|x_1 - x_2|^{1/2}$. On the other hand, if $|x_1 - x_2| > 2N$ then (2.3.17) is a consequence of the fact that f is bounded.

Making the change of variable $x = \lambda + yt$ on the left-hand side of (2.3.16) gives

$$\int_{-\infty}^{\infty} \frac{f(\lambda + yt)}{t^2 + 1} \, dt.$$

The difference between this and the right-hand side of (2.3.16) is just

$$\int_{-\infty}^{\infty} \frac{f(\lambda + yt) - f(\lambda)}{t^2 + 1} \, dt.$$

Using (2.3.17) with $x_1 = \lambda + yt$ and $x_2 = \lambda$, this integral is bounded in absolute value by

$$\text{const}\, y^{1/2} \int_{-\infty}^{\infty} \frac{|t|^{1/2}\, dt}{t^2 + 1},$$

and hence converges uniformly to zero as $y \to 0+$.

(ii) Again it is sufficient to take $f \in C_0^1(\mathbb{R})$. In this case, we take some γ in the interval $\frac{1}{2} < \gamma < 1$, and use the inequality

$$|f(x_1) - f(x_2)| \le \text{const} \, \frac{|x_1 - x_2|^\gamma}{(1 + |x_1|)^\gamma (1 + |x_2|)^\gamma}. \tag{2.3.17$'$}$$

(As before, this is easily shown if $|x_1 - x_2| \le 2N$. If $|x_1 - x_2| > 2N$, we assume, for example, that $x_1 \in [-N, N]$ but $x_2 \notin [-N, N]$. (If neither x_1 nor x_2 belong to this interval then $f(x_1) = f(x_2) = 0$ and there is nothing to prove.) Then the right-hand side of (2.3.17)′ is bounded below, for this range of values of x_1 and x_2, by a positive constant, so that once more (2.3.17)′ is consequence of the boundedness of f.)

In this case, for the difference between the right- and left-hand sides of (2.3.16), we have the estimate

$$|\text{difference}| \le \text{const}\; y^\gamma \int_{-\infty}^{\infty} \frac{t^\gamma \, dt}{(t^2 + 1)(1 + |\lambda + yt|)^\gamma (1 + |\lambda|)^\gamma}.$$

We want an estimate of the L^1-norm of this difference, so, on integrating with respect to λ (the orders of integration with respect to t and λ may be exchanged, since the double integral is absolutely convergent) and using the inequality

$$\int_{-\infty}^{\infty} \frac{d\lambda}{(1+|\lambda+yt|)^{\gamma}(1+|\lambda|)^{\gamma}}$$

$$\leq \left(\int_{-\infty}^{\infty} \frac{d\lambda}{(1+|\lambda+yt|)^{2\gamma}} \right)^{1/2} \left(\int_{-\infty}^{\infty} \frac{d\lambda}{(1+|\lambda|)^{2\gamma}} \right)^{1/2} \qquad \text{(Schwarz inequality)}$$

$$= \int_{-\infty}^{\infty} \frac{d\lambda}{(1+|\lambda|)^{2\gamma}} < \infty,$$

we have

$$\|\cdot\|_1 \leq \text{const } y^{\gamma} \int_{-\infty}^{\infty} \frac{t^{\gamma}\, dt}{t^2+1} = \text{const } y^{\gamma}.$$

Hence (2.3.16) certainly holds with convergence in L^1-norm.

(iii) Suppose that $f \in L^2(\mathbb{R})$. Regarding the integral on the left-hand side of (2.3.16) as defining an integral operator on $L^2(\mathbb{R})$, with kernel $k(\lambda, x) = y/[(\lambda - x)^2 + y^2]$, we know that this integral operator corresponds to the multiplication of $\hat{f}(k)$, the Fourier transform of f, by the Fourier transform (with respect to x) of $(2\pi)^{1/2}\, y/(x^2+y^2)$. This Fourier transform may be evaluated (for example by contour integration) to give

$$\pi e^{-ky} \quad (k>0),$$

$$\pi e^{ky} \quad (k<0).$$

(See Section 2.5 for the Fourier transform as a linear operator, and p. 93 for integral operators.)

Hence the L^2-norm of the difference between the left- and right-hand sides of (2.3.16), which is invariant under Fourier transform, is given by

$$(\|\cdot\|_2)^2 = \int_{-\infty}^{0} |\pi\hat{f}(k)(1-e^{ky})|^2\, dk + \int_{0}^{\infty} |\pi\hat{f}(k)(1-e^{-ky})|^2\, dk,$$

which, by the Lebesgue Dominated-Convergence Theorem, tends to zero in the limit $y \to 0+$. Hence also (2.3.16) holds with convergence in L^3-norm.

■

The reader should, by now, be convinced of the validity of (2.3.16) in a wide variety of different contexts—perhaps even convinced enough to try out the formula obtained by replacing the Lebesgue integral by the more general Lebesgue–Stieltjes integral. What can we say about (2.3.15) in the limit as $y \to 0+$? ((2.3.16) corresponds to the special case in which the measure μ generated by ρ is both finite and absolutely continuous; we can then write $\mathrm{d}\rho(\lambda) = f(\lambda)\,\mathrm{d}\lambda$ for some $f \in L^1$. The notation, for reasons that will become clear later, is a little different on comparing (2.3.15) and (2.3.16); the roles of λ and x have been exchanged.) This time we shall look particularly at *pointwise* convergence, rather than convergence in some function space. An educated guess would suggest that, as $y \to 0+$, the right-hand side of (2.3.15) converges to $\pi\,\mathrm{d}\rho(x)/\mathrm{d}x$ whenever this limit exists. Taking from (2.3.10) the final expression for $\mathrm{d}\rho(\lambda)/\mathrm{d}\lambda$, this would imply that, for any x such that $\mathrm{d}\rho(x)/\mathrm{d}x$ exists,

$$\lim_{y \to 0+} \int_{-\infty}^{\infty} \frac{y\,\mathrm{d}\rho(\lambda)}{(\lambda - x)^2 + y^2} = \pi \lim_{\varepsilon \to 0+} \mu \frac{\{(x-\varepsilon, x+\varepsilon]\}}{2\varepsilon}. \tag{2.3.18}$$

Actually, the right-hand side of (2.3.18) may exist even if $\mathrm{d}\rho(x)/\mathrm{d}x$ does not (let μ be the Stieltjes measure generated by

$$\rho(x) = \begin{cases} 0 & (x < 0), \\ x & (x \geq 0); \end{cases}$$

then $\mathrm{d}\rho(x)/\mathrm{d}x$ does not exist at $x = 0$, although the limit on the right-hand side of (2.3.18) *does* exist and is equal to $\frac{1}{2}$.) So (2.3.18) is, perhaps, a stronger result than we have a right to expect. Nevertheless the following lemma provides a justification of (2.3.18).

Lemma 2.4

Let μ be a Borel–Stieltjes measure on $\mathbb{R}$, generated by $\rho(\lambda)$, and suppose that (2.3.14) holds. Then, for given x,

(i)
$$\int_{-\infty}^{\infty} \frac{y\,\mathrm{d}\rho(\lambda)}{(\lambda - x)^2 + y^2}$$

is *bounded* (say for $0 < y < 1$) if and only if $\mu\{(x-\varepsilon, x+\varepsilon]\}/2\varepsilon$ is bounded (say for $0 < \varepsilon < 1$);

(ii) (2.3.18) holds, in the sense that the existence of either limit implies that both limits exist and are equal.

Proof

(i) Suppose first that $\int_{-\infty}^{\infty} (y\,\mathrm{d}\rho(\lambda)/[(\lambda - x)^2 + y^2])$ is bounded as y approaches zero. Then

$$\int_{\lambda=x-\varepsilon}^{\lambda=x+\varepsilon} \frac{\varepsilon\,\mathrm{d}\rho(\lambda)}{(\lambda-x)^2+\varepsilon^2} \leq \int_{-\infty}^{\infty} \frac{\varepsilon\,\mathrm{d}\rho(\lambda)}{(\lambda-x)^2+\varepsilon^2},$$

which is bounded. Now, in the interval $x-\varepsilon<\lambda\leq x+\varepsilon$ we have $(\lambda-x)^2+\varepsilon^2\leq 2\varepsilon^2$, so that

$$\int_{\lambda=x-\varepsilon}^{\lambda=x+\varepsilon} \frac{\varepsilon\,\mathrm{d}\rho(\lambda)}{2\varepsilon^2} = \frac{\mu\{(x-\varepsilon, x+\varepsilon]\}}{2\varepsilon}$$

is bounded as ε approaches zero.

Conversely, suppose that $\mu\{(x-\varepsilon, x+\varepsilon]\}/2\varepsilon$ is bounded for some x. It will be convenient to choose $\rho(\lambda)$, which is determined only up to an additive constant, to satisfy $\rho(x)=0$, in which case it follows that $\rho(\lambda)/(\lambda-x)$ is both non-negative and bounded (for $|\lambda-x|\leq 1$, say).

Let us write

$$0\leq \rho(\lambda)/(\lambda-x)\leq K. \tag{2.3.19}$$

Now

$$\int_{R\setminus(x-1,x+1]} \frac{y\,\mathrm{d}\rho(\lambda)}{(\lambda-x)^2+y^2} \quad \text{is bounded by} \quad y\int_{R\setminus(x-1,x+1]} \frac{\mathrm{d}\rho(\lambda)}{(\lambda-x)^2},$$

and hence is bounded (even converges to zero) in the limit $y\to 0$. It therefore remains to consider only $\int_{(x-1,x+1]} y\,\mathrm{d}\rho(\lambda)/[(\lambda-x)^2+y^2]$, which on integrating by parts reduces to

$$\left[\frac{y\rho(\lambda)}{(\lambda-x)^2+y^2}\right]_{\lambda=x-1}^{\lambda=x+1} + \int_{(x-1,x+1]} \rho(\lambda)\frac{\partial}{\partial\lambda}\left\{\frac{-y}{(\lambda-x)^2+y^2}\right\}\mathrm{d}\lambda.$$

The first contribution is just $y(\rho(x+1)-\rho(x-1))/(1+y^2)$, which is bounded (even converges to zero). The second contribution, on using (2.3.19) and the fact that the partial derivative is positive for $\lambda>x$ and negative for $\lambda<x$, is bounded by

$$\int_{(x-1,x+1]} K(\lambda-x)\,\frac{\partial}{\partial\lambda}\left\{\frac{-y}{(\lambda-x)^2+y^2}\right\}\mathrm{d}\lambda.$$

This may be further integrated by parts, giving a bounded contribution (indeed one that converges to zero), together with $\int_{x-1}^{x+1} Ky\,\mathrm{d}\lambda/[(\lambda-x)^2+y^2]$, which is itself less that πK. So part (i) of the lemma is proved.

(ii) Suppose that $\lim_{\varepsilon\to 0+}\mu\{(x-\varepsilon,\ x+\varepsilon]\}/2\varepsilon=\beta$ exists. Given $\delta>0$, and setting $\lambda=x+\varepsilon$, $2x-\lambda=x-\varepsilon$, we have, for λ sufficiently close to x (say for $x\leq\lambda\leq x+h$),

$$2(\beta-\delta)(\lambda-x)\leq\rho(\lambda)-\rho(2x-\lambda)\leq 2(\beta+\delta)(\lambda-x). \tag{2.3.20}$$

By the same argument as in (i) above, $\int_{-\infty}^{\infty} y\,\mathrm{d}\rho(\lambda)/[\lambda-x)^2+y^2]$ may be replaced, in the limit of small positive y, by

$$\int_{(x-h,x+h]}\rho(\lambda)\frac{\partial}{\partial\lambda}\left\{\frac{-y}{(\lambda-x)^2+y^2}\right\}\mathrm{d}\lambda,$$

which, on making the change of variable $\lambda\to 2x-\lambda$ in the integral from $x-h$ to x, becomes

$$\int_x^{x+h}(\rho(\lambda)-\rho(2x-\lambda))\frac{\partial}{\partial\lambda}\left\{\frac{-y}{(\lambda-x)^2+y^2}\right\}\mathrm{d}\lambda.$$

According to (2.3.20), this integral lies between

$$2(\beta-\delta)\int_x^{x+h}(\lambda-x)\frac{\partial}{\partial\lambda}\left\{\frac{-y}{(\lambda-x)^2+y^2}\right\}\mathrm{d}\lambda$$

and

$$2(\beta+\delta)\int_x^{x+h}(\lambda-x)\frac{\partial}{\partial\lambda}\left\{\frac{-y}{(\lambda-x)^2+y^2}\right\}\mathrm{d}\lambda.$$

Integrating by parts in each case, and omitting boundary terms, which vanish in the limit as $y\to 0+$, these bounds are respectively

$$2(\beta-\delta)\int_x^{x+h}\frac{y\,\mathrm{d}\lambda}{(\lambda-x)^2+y^2}\qquad\text{and}\qquad 2(\beta+\delta)\int_x^{x+h}\frac{y\,\mathrm{d}\lambda}{(\lambda-x)^2+y^2},$$

which converge respectively, as $y\to 0+$, to

$$\pi(\beta-\delta)\qquad\text{and}\qquad\pi(\beta+\delta).$$

(Note that $\lim_{y\to 0+}\int_{x+h}^{\infty} y\,\mathrm{d}\lambda/[(\lambda-x)^2+y^2]=0$, so that the limits of integration $(x, x+h]$ may be replaced by (x,∞) in the limit.)

In other words, as $y\to 0+$, the integral on the left-hand side of (2.3.18) consists of a term within $\pi\delta$ of $\pi\beta$, together with a term that is vanishingly small. Since δ is arbitrary positive, it follows that the integral tends to $\pi\beta$ in the limit, and (2.3.18) is proved (assuming the existence of the limit on the right-hand side).

Let us now prove (2.3.18) under the hypothesis that the limit exists on the left-hand side of (2.3.18). On making the substitution $\lambda = x + yt$, the limit becomes

$$\lim_{y\to 0+} \int_{-\infty}^{\infty} \frac{1}{y} \frac{d\rho(x+yt)}{t^2+1},$$

in which $\rho(x+yt)$ is to be regarded as a function of t. Thus for functions $F(t)$ we may consider the linear functional $\mathscr{L}$ defined by

$$\mathscr{L}(F) = \lim_{y\to 0+} \int_{-\infty}^{\infty} \frac{F(t)}{y} \frac{d\rho(x+yt)}{t^2+1} \tag{2.3.21}$$

(compare the proof of Lemma 2.3).

Here we know that $\mathscr{L}(F)$ is defined for $F(t) \equiv 1$. On the other hand, $\mathscr{L}(F)$ is unchanged if, on the right-hand side of (2.3.21), we replace the integration variable t by $t/\beta^{1/2}$ for some $\beta > 0$, and if we replace y by $\beta^{1/2} y$. Taking $F(t) \equiv 1$, this implies

$$\mathscr{L}(1) = \lim_{y\to 0+} \int_{-\infty}^{\infty} \frac{\beta^{1/2}\, d\rho(x+yt)}{y(t^2+\beta)}.$$

Comparing with (2.3.21), we have proved that $\mathscr{L}(F)$ exists for

$$F(t) = \frac{t^2+1}{t^2+\beta}, \qquad \text{with} \qquad \mathscr{L}\left(\frac{t^2+1}{t^2+\beta}\right) = \beta^{-1/2}\mathscr{L}(1).$$

A simple calculation shows that, in both this case and with $F(t) \equiv 1$,

$$\mathscr{L}(F) = \frac{\mathscr{L}(1)}{\pi} \int_{-\infty}^{\infty} \frac{F(t)\, dt}{t^2+1}. \tag{2.3.21$'$}$$

Equations (2.3.21) and (2.3.21)$'$ may be extended, as for Lemma 2.3, to a larger class of functions F, by taking uniform limits of the F's. For example,

$$\frac{\partial}{\partial\beta}(t^2+\beta)^{-1} = -(t^2+\beta)^{-2} = \lim_{h\to 0} \frac{1}{h}\left[\frac{1}{(t+h)^2+\beta} - \frac{1}{t^2+\beta}\right].$$

and for fixed $\beta>0$ this limit is uniform in t. Hence (2.3.21)′ may be extended to the function $F(t)=(t^2+\beta)^{-2}$. (Note that $\mathscr{L}(1/(t^2+\beta))=(1-\beta)^{-1}\ \{\mathscr{L}((t^2+1)/(t^2+\beta))-\mathscr{L}(1)\}$ if $\beta\neq 1$, and $\mathscr{L}(1/(t^2+1))=\lim_{\beta\to 1}\mathscr{L}(1/(t^2+\beta))$.)

In the same way, by further differentiation, the case $F(t)=(t^2+\beta)^{-k}$ may be treated, where k is any positive integer, and (2.3.21)′ may be verified in each case, using differentiation under the integral sign. One can then deal with $F(t)=t^{2N}(t^2+\beta)^{-k}$ for $N=0, 1, 2, \dots, k$. (Use induction, with $t^{2(N+1)}/(t^2+\beta)^k=t^{2N}/(t^2+\beta)^{k-1}-\beta t^{2N}/(t^2+\beta)^k$.)

Thus (2.3.21) and (2.3.21)′ are established for *polynomials* $F(t)$ in $(t^2+1)/(t^2+\beta)$. If we take, for example, $\beta=2$ then the transformation

$$z=\frac{t^2+1}{t^2+2}, \qquad t^2=\frac{2z-1}{1-z}$$

defines a one-to-one correspondence between these polynomials and polynomials in z, for $\frac{1}{2}\leq z\leq 1$. Now polynomials are dense in $C[\frac{1}{2},1]$. Hence our corresponding polynomials in $(t^2+1)/(t^2+2)$ are dense in the space that corresponds to $C[\frac{1}{2}, 1]$, namely the space of *even* continuous functions on $-\infty<t<\infty$ that have a limit as $t\to\pm\infty$ (this limit corresponds to $z=1$). In particular, $F(t)$ in (2.3.21) and (2.3.21)′ can be an arbitrary even continuous function having compact support.

The most convenient choice of $F(t)$ is to take $F(t)=(t^2+1)G(t)$, where

$$G(t)\equiv 1 \quad (-1\leq t\leq 1),$$

and to let $G(t)$ decrease smoothly to zero between $t=1$ and $t=1+\delta$, say. (See Fig. 2.1). From (2.3.21)′ we see that, in this case

$$\mathscr{L}(F)\leq\frac{2(1+\delta)}{\pi}\mathscr{L}(1).$$

In the integral on the right-hand side of (2.3.21), we set $F(t)=(t^2+1)\chi(t)$, where $\chi(t)$ is the characteristic function of the interval $-1<t\leq 1$. Certainly $\chi(t)\leq G(t)$, so that for y sufficiently small and positive we have

$$\int_{-\infty}^{\infty}\frac{\chi(t)}{y}\,\mathrm{d}\rho(x+yt)\leq\frac{2(1+\delta)}{\pi}\mathscr{L}(1).$$

A similar argument gives the lower bound

$$\int_{-\infty}^{\infty}\frac{\chi(t)}{y}\,\mathrm{d}\rho(x+yt)\geq\frac{2}{\pi}\mathscr{L}(1).$$

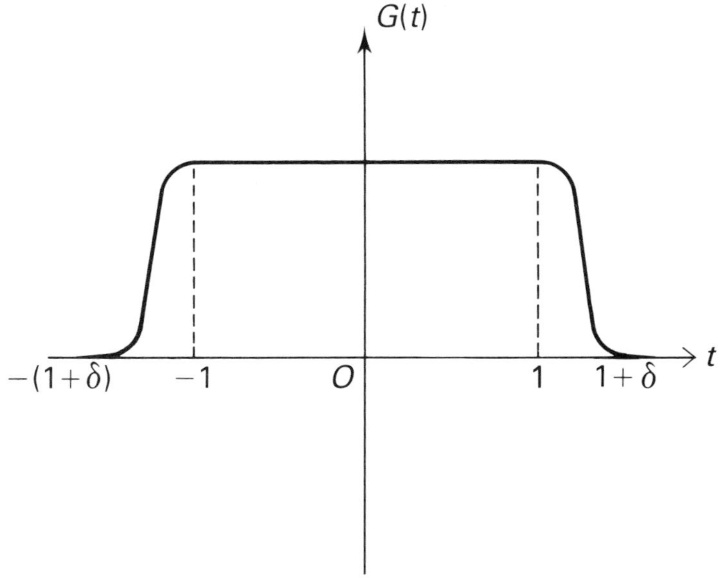

Figure 2.1

Since $\delta > 0$ is arbitrary, we may deduce

$$\lim_{y\to 0+} \int_{-\infty}^{\infty} \frac{\chi(t)\,\mathrm{d}\rho(x+yt)}{y} = \frac{2}{\pi}\mathscr{L}(1).$$

The integral on the left-hand side is

$$\frac{\rho(x+y)-\rho(x-y)}{y} = \frac{\mu\{(x-y,x+y]\}}{y}$$

and $\mathscr{L}(1)$ is exactly the left-hand side of (2.3.18). Replacing y by ε, we have now verified (2.3.18) under the hypothesis, in (2.3.18), of the existence of the limit on the left-hand side. ■

In Section 2.4 we shall follow up the close relationship that exists between Borel–Stieltjes measures and the boundary values of analytic functions. Let μ be a Borel–Stieltjes measure, generated by $\rho(\lambda)$, and suppose that

$$\int_{-\infty}^{\infty} \frac{\mathrm{d}\rho(\lambda)}{|\lambda|+1} < \infty.$$

Then a function analytic in the entire complex plane apart from the axis Im $(z)=0$ may be defined by the formula

$$m(z)=\int_{-\infty}^{\infty}\frac{\mathrm{d}\rho(\lambda)}{\lambda-z}. \tag{2.3.22}$$

Setting $z=x+\mathrm{i}y$, we have

$$\mathrm{Im}\,(m(z))=\int_{-\infty}^{\infty}\frac{y\,\mathrm{d}\rho(\lambda)}{(\lambda-x)^2+y^2}, \tag{2.3.23}$$

$$\mathrm{Re}\,(m(z))=\int_{-\infty}^{\infty}\frac{(\lambda-x)\,\mathrm{d}\rho(\lambda)}{(\lambda-x)^2+y^2}. \qquad (2.3.23)'$$

Thus (2.3.18) is a formula for the boundary value of the imaginary part of $m(z)$, as $z=x+\mathrm{i}y$ approaches the real z-axis from above. We note that, for a given Stieltjes measure, the condition (2.3.14) for the integral in (2.3.23) to be finite does not imply the finiteness of the integral in (2.3.23)′, which requires the integrability of $(|\lambda|+1)^{-1}$.

It is of some interest to evaluate the boundary value of the real part of $m(z)$, using the same methods as for the proof of Lemma 2.4. The following lemma expresses this boundary value as a principal-value integral. The principal-value integral of a function $\Phi(t)$ is defined to be

$$\mathrm{P}\int_{-\infty}^{\infty}\Phi(t)\,\mathrm{d}t=\lim_{h\to 0+}\int_{|t|>h}\Phi(t)\,\mathrm{d}t.$$

In other words,

$$\mathrm{P}\int_{-\infty}^{\infty}\Phi(t)\,\mathrm{d}t=\lim_{h\to 0+}\left\{\int_{-\infty}^{-h}\Phi(t)\,\mathrm{d}t+\int_{h}^{\infty}\Phi(t)\,\mathrm{d}t\right\}.$$

Unlike the Lebesgue integral, which requires *absolute* integrability, the principal-value integral allows cancellations between contributions to the integral near $t=0$, from $t<0$ and $t>0$ respectively. Thus $\mathrm{P}\int_{-1}^{1}\mathrm{d}t/t=\lim_{h\to 0+}\int_{|t|>h}\mathrm{d}t/t=0$, since the integrand is an odd function. The principal-value integral is defined in a similar way for integrands that are singular at points other than $t=0$. For example, consider the function

$(t-1)^{-1}\ (t^2+1)^{-1}$; it is singular at $t=1$, and we define

$$\mathrm{P}\int_{-\infty}^{\infty}\frac{\mathrm{d}t}{(t-1)(t^2+1)}=\lim_{h\to0+}\int_{|t-1|>h}\frac{\mathrm{d}t}{(t-1)(t^2+1)}.$$

For an integrand singular at $t=a$, a convenient way of expressing the principal-value integral is to write

$$\mathrm{P}\int_{-\infty}^{\infty}\Phi(t)\,\mathrm{d}t=\int_{a}^{\infty}\{\Phi(t)-\Phi(2a-t)\}\,\mathrm{d}t,$$

whenever the right-hand side exists as a Lebesgue integral.

We are now ready to evaluate $\mathrm{Re}\,(m(z))$ in the limit $\mathrm{Im}\,(z)\to0$.

Lemma 2.5

With $m(z)$ given by (2.3.22), suppose that, for given x, the boundary value $\lim_{y\to0+} m(x+\mathrm{i}y)\equiv m_+(x)$ exists. Then

$$\mathrm{Re}\,(m_+(x))=\lim_{y\to0+}\int_{-\infty}^{\infty}\frac{(\lambda-x)\,\mathrm{d}\rho(\lambda)}{(\lambda-x)^2+y^2}=\mathrm{P}\int_{-\infty}^{\infty}\frac{\mathrm{d}\rho(\lambda)}{\lambda-x}. \tag{2.3.24}$$

Proof

Substitute $\lambda=x+yt$, so that

$$\lim_{y\to0+}\int_{-\infty}^{\infty}\frac{t\,\mathrm{d}\rho(x+yt)}{y(t^2+1)}=\gamma,\quad \text{say.}$$

As in the proof of Lemma 2.4 (ii), we replace t by $\beta^{-1/2}t$ and y by $\beta^{1/2}y$, giving, for $\beta>0$,

$$\lim_{y\to0+}\int_{-\infty}^{\infty}\frac{t\,\mathrm{d}\rho(x+yt)}{y(t^2+\beta)}=\gamma.$$

Subtracting these two limits, we now have

$$\lim_{y\to0+}\int_{-\infty}^{\infty}\frac{t\,\mathrm{d}\rho(x+yt)}{y(t^2+1)(t^2+\beta)}=0, \tag{2.3.25}$$

which for $F(t)\equiv 1$ may be written

$$\lim_{y\to 0+}\int_{-\infty}^{\infty}\frac{tF(t)\,\mathrm{d}\rho(x+yt)}{y(t^2+1)(t^2+\beta)}=0. \tag{2.3.26}$$

Equations (2.3.21) and (2.3.21)′ may be used to estimate the integral in (2.3.26) over $(-\infty, 0]$ and $(0, \infty)$ respectively. (In these equations, we replace $F(t)$ by $|t|F(t)(t^2+\beta)^{-1}$.) In particular,

$$\int_{0}^{\pm\infty}\frac{t\,\mathrm{d}\rho(x+yt)}{y(t^2+1)(t^2+\beta)}\leq\int_{-\infty}^{\infty}\frac{|t|\,\mathrm{d}\rho(x+yt)}{y(t^2+1)(t^2+\beta)},$$

which is bounded in the limit $y\to 0+$. This is enough to ensure that (2.3.26) may be extended, as in Lemma 2.4. by taking uniform limits of the $F(t)$ and that (2.3.26) holds even with $\beta=1$. As before, we may take, for $F(t)$, any *even* continuous function that tends to a limit as $t\to\pm\infty$, and here the most suitable choice is $F(t)=t^{-1}G(t)(t^2+1)(t^2+\beta)-(t^2+1)$, where $G(t)$ is sketched in Fig. 2.2. Note that $G(t)$ is odd, and that

$$G(t)=\begin{cases}0 & (|t|\leq a),\\ 1/t & (|t|\geq 1).\end{cases}$$

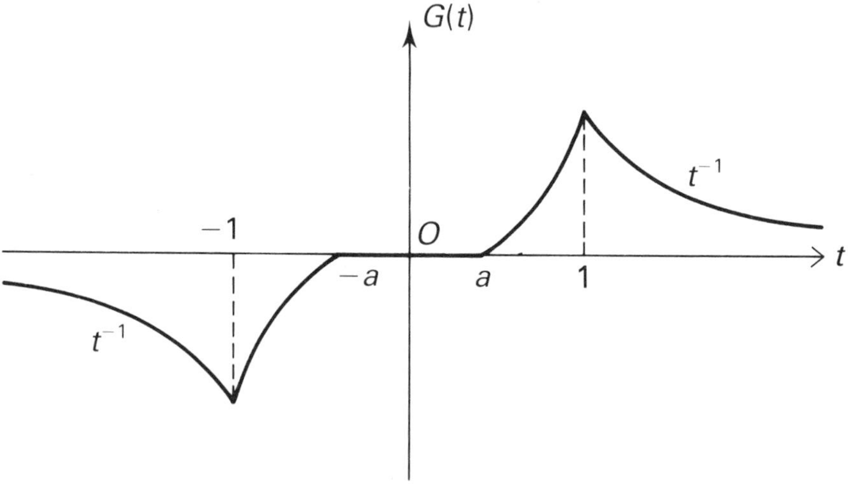

Figure 2.2

The above estimates may be used to show that the contribution to the integral from the interval $a \le t \le 1$ becomes arbitrarily small, as a approaches 1, in the limit as $y \to 0+$. Hence we may even take

$$G(t)=\begin{cases}0 & (|t|<1),\\ 1/t & (|t|>1).\end{cases}$$

On substituting this into (2.3.26), with the given formula for $F(t)$, we have

$$\lim_{y\to 0+}\int_{|t|>1}\frac{d\rho(x+yt)}{ty}-\lim_{y\to 0+}\int_{-\infty}^{\infty}\frac{t\,d\rho(x+yt)}{y(t^2+\beta)}=0,$$

so that

$$\lim_{y\to 0+}\int_{|t|>1}\frac{d\rho(x+yt)}{ty}=\gamma=\operatorname{Re}(m_+(x)).$$

Making the original substitution $\lambda = x + yt$ in reverse now gives

$$\operatorname{Re}(m_+(x))=\lim_{y\to 0+}\int_{|\lambda-x|>y}\frac{d\rho(\lambda)}{\lambda-x},$$

which is the principal-value integral in (2.3.24). ■

We shall defer until Section 2.4 further consideration of the relation between the resolvent transform and the boundary values of analytic functions. Let us note here, however, that the resolvent transform can be a useful method of proving measurability of certain sets that we have already encountered. For example, we consider the set $\mathscr{S}$ defined in Theorem 2.1, where, as in the proof of this theorem, we can define $\mathscr{S}_m$ to be the set of points such that $\mu(\mathscr{I})/\ell(\mathscr{I})$ is bounded by m, for intervals $\mathscr{I}$ containing the point in question. According to Lemma 2.4(i), $\mathscr{S}$ will consist precisely of these points x such that

$$v(x,y)\equiv\int_{-\infty}^{\infty}\frac{y\,d\rho(\lambda)}{(\lambda-x)^2+y^2}$$

is bounded (say for $0<y<1$). If we define, corresponding to the $\mathscr{S}_m$, a family of sets $\mathscr{U}_m$, where $\mathscr{U}_m$ consists of those x satisfying

$$v(x,y)\le m \quad (0<y<1),$$

then we have $\mathscr{S}=\bigcup_m \mathscr{U}_m$, where m runs through the set of positive integers. Hence $\mathscr{S}$ will be measurable provided $\mathscr{U}_m$ is measurable. Moreover $\mathscr{U}_m$ itself may be defined by the inequalities

$$v(x, y_i) \leq m \quad (i=1, 2, 3, \ldots) \tag{2.3.27}$$

as $\{y_i\}$ is the sequence of *rational* numbers between 0 and 1, arranged in some order. (It is sufficient to take rational numbers since $v(x, y)$ depends continuously on y.) Thus $\mathscr{U}_m$ is a countable intersection of the sets defined, respectively for each y_i, by the inequalities (2.3.27). Since $v(x, y_i)$ depends continuously on x (for example by the Lebesgue Dominated-Convergence Theorem), it is straightforward to verify that we have a sequence of closed, and hence measurable, sets. It follows that each $\mathscr{U}_m$, and therefore finally $\mathscr{S}$, is measurable. Measurability for the sets $\mathscr{S}_m$ may be proved in a similar way.

The Fourier transform of a finite measure

The second transform of particular importance in the analysis of measures is the Fourier transform. The Fourier transform of a finite Borel–Stieltjes measure μ, generated by $\rho(\lambda)$, is the function $F_\mu(t)$ defined by

$$F_\mu(t) = \int_{-\infty}^{\infty} \mathrm{e}^{-\mathrm{i}\lambda t}\,\mathrm{d}\rho(\lambda). \tag{2.3.28}$$

We omit the standard numerical factor $(2\pi)^{-1/2}$, which occurs in the definition of the Fourier transform as a unitary linear operator from $L^2(\mathbb{R})$ to $L^2(\mathbb{R})$ (for which see Section 2.6), since here F_μ does not, in general, belong naturally to any particular L^2 space.

F_μ is a continuous function from $\mathbb{R}$ to $\mathbb{C}$, satisfying the bound

$$|F_\mu(t)| \leq \int_{-\infty}^{\infty} \mathrm{d}\rho(\lambda) = \mu(\mathbb{R}) < \infty.$$

The following result shows how the Fourier transform may be used to estimate the discrete part μ_{d} of the measure μ.

Lemma 2.6

Let $F_\mu(t)$ be the Fourier transform of a finite measure μ, and suppose that $\lambda_1, \lambda_2, \ldots$ are the discrete points of μ. Then

$$\lim_{T\to\infty} \frac{1}{T}\int_0^T \mathrm{d}t\,|F_\mu(t)|^2 = \sum_n \mu(\{\lambda_n\})^2. \tag{2.3.29}$$

In particular, if the limit on the left-hand side of (2.3.29) is zero then $\mu_d = 0$, so that μ is purely continuous.

Proof

We have

$$\frac{1}{T}\int_0^T dt\,|F_\mu(t)|^2$$

$$=\frac{1}{T}\int_0^T dt\left\{\int_{-\infty}^{\infty} e^{-i\lambda t}\, d\rho(\lambda)\right\}\left\{\int_{-\infty}^{\infty} e^{i\lambda' t}\, d\rho(\lambda')\right\}$$

$$=\frac{1}{T}\int_0^T dt \int_{-\infty}^{\infty} d\rho(\lambda) \int_{-\infty}^{\infty} d\rho(\lambda')\, e^{-i(\lambda-\lambda')t}.$$

Since, for a finite measure μ, this triple integral is absolutely convergent, it does not matter in which order the integrations are carried out. Integrating first with respect to t, we obtain

$$\int_{-\infty}^{\infty} d\rho(\lambda) \int_{-\infty}^{\infty} d\rho(\lambda') F_T(\lambda, \lambda'),$$

where

$$F_T(\lambda, \lambda') = \begin{cases} \dfrac{i(e^{-i(\lambda-\lambda')T}-1).}{(\lambda-\lambda')T} & (\lambda \neq \lambda') \\ 1 & (\lambda = \lambda'). \end{cases}$$

Hence

$$\lim_{T\to\infty} F_T(\lambda, \lambda') = \chi(\lambda, \lambda'),$$

where

$$\chi(\lambda, \lambda') = \begin{cases} 1 & (\lambda = \lambda'), \\ 0 & \text{otherwise} \end{cases}.$$

Applying the Lebesgue Dominated-Convergence Theorem to the left-hand side of (2.3.29), we obtain, in the limit $T\to\infty$,

$$\int_{-\infty}^{\infty} \mathrm{d}\rho(\lambda) \int_{-\infty}^{\infty} \mathrm{d}\rho(\lambda')\chi(\lambda,\lambda').$$

Holding λ' fixed and carrying out the integration with respect to λ, there will be no contribution unless λ' is a discrete point of the measure. For example, if $\lambda'=\lambda_n$, the λ integration gives

$$\int_{-\infty}^{\infty} \mathrm{d}\rho(\lambda)\chi(\lambda,\lambda_n)=\mu(\{\lambda_n\}).$$

Integrating now with respect to λ', and summing the contributions of each discrete point to the integral, we obtain (2.3.29). If the limit in (2.3.29) is zero then there can be no discrete points of the measure μ, which in that case is purely continuous. ■

The left-hand-side of (2.3.29) may be regarded as a *time average* of $|F_\mu(t)|^2$, over a time interval $[0, T]$, and in the limit $T\to\infty$. The lemma tells us in particular that, for a continuous measure, $|F_\mu(t)|^2$ will tend to zero *in time average* for large times. A frequently used consequence of this result is that, for a continuous measure μ, a sequence $t_1, t_2, t_3, \ldots$ can always be found such that $t_n\to\infty$ and $|F_\mu(t_n)|\to 0$. Since, by the Schwarz inequality, one has

$$\int_0^T \mathrm{d}t\,|F_\mu(t)| \leq \left(\int_0^T \mathrm{d}t\,|F_\mu(t)|^2\right)^{1/2} \left(\int_0^T 1\,\mathrm{d}t\right)^{1/2},$$

it will also be true, for a continuous measure, that

$$\lim_{T\to\infty} \frac{1}{T}\int_0^T \mathrm{d}t|F_\mu(t)|=0. \tag{2.3.30}$$

In that case, $|F_\mu(t)|$ as well as $|F_\mu(t)|^2$ will approach zero in time average. Similar results hold in the limit as $T\to-\infty$.

The question arises as to whether one can go further and prove that $F_\mu(t)$ converges to zero in the limit as $t\to\infty$ (i.e. without time average) for any continuous measure. Such a result is not possible, since it turns out that there are singular continuous measures, for example the Cantor measure defined in Section 2.1, for which the Fourier transform does not converge to zero. The following lemma gives us the result we want if we restrict attention to measures that are absolutely continuous. We assume here the standard result that the Fourier transform of an L^1 function tends to zero at $\pm\infty$ (Riemann–Lebesgue Lemma).

Lemma 2.7

Let μ be a finite absolutely continuous measure (i.e. $\mu_{sc}=\mu_d=0$). Then the Fourier transform of μ, defined by (2.3.28), converges to zero in the limits $t\to\pm\infty$.

Proof

Since μ is absolutely continuous, we can define a density function $f(\lambda)$, where $f(\lambda)=(\mathrm{d}/\mathrm{d}\lambda)\rho(\lambda)$, and

$$F_\mu(t)=\int_{-\infty}^{\infty} \mathrm{e}^{-\mathrm{i}\lambda t} f(\lambda)\,\mathrm{d}\lambda.$$

Since μ is a finite measure, we have $f\in L^1(\mathbb{R})$, and the lemma follows immediately on applying the Riemann–Lebesgue Lemma. ∎

Since there are singular continuous measures for which $\lim_{t\to\pm\infty} F_\mu(t)=0$, this result gives us a necessary but not sufficient condition for a finite measure to be absolutely continuous. To obtain a sufficient condition, we need to assume rather more on the *rate* at which the Fourier transform converges to zero. The following result shows that a power decay better that $t^{-1/2}$ will do.

Corollary

Let $F_\mu(t)$ be the Fourier transform of a finite measure μ, and suppose that

$$\int_0^{\infty} \mathrm{d}t\,|F_\mu(t)|^2<\infty. \tag{2.3.31}$$

Then μ is absolutely continuous.

Proof

Suppose that (2.3.31) is satisfied. Since $|F_\mu(-t)|=|\bar{F}_\mu(t)|=|F_\mu(t)|$, it follows that

$$\int_{-\infty}^{\infty} \mathrm{d}t\,|F_\mu(t)|^2<\infty.$$

For arbitrary $\phi \in C_0^\infty(\mathbb{R})$, define $\tilde{\phi}(t)$, the inverse Fourier transform of ϕ, apart from a multiplicative constant, by

$$\tilde{\phi}(t) = \frac{1}{2\pi} \int_{-\infty}^{\infty} e^{i\lambda t} \phi(\lambda)\, d\lambda.$$

Then

$$\int_{-\infty}^{\infty} e^{-i\lambda t} \tilde{\phi}(t)\, dt = \phi(\lambda).$$

Multiplying (2.3.28) by $\tilde{\phi}(t)$ and integrating with respect to t, the double integral on the right-hand side is absolutely convergent, and we obtain

$$\int_{-\infty}^{\infty} \tilde{\phi}(t) F_\mu(t)\, dt = \int_{-\infty}^{\infty} \phi(\lambda)\, d\rho(\lambda). \tag{2.3.32}$$

On the left-hand side of (2.3.32) we have $\|\tilde{\phi}\| = \|\phi\|/(2\pi)^{1/2}$. So an application of the Schwarz inequality gives, using (2.3.31), $|\int_{-\infty}^{\infty} \tilde{\phi}(t) F_\mu(t)\, dt| \leq \text{const. } \|\phi\|$ (see p. 28). Hence the right-hand side of (2.3.32) defines a bounded linear functional from $C_0^\infty(\mathbb{R})$ to $\mathbb{C}$, which may be extended by continuity to a bounded linear functional on $L^2(\mathbb{R})$. By the Riesz Representation Theorem, we have, for some $f \in L^2(\mathbb{R})$, and for every $\phi \in C_0^\infty(\mathbb{R})$.

$$\int_{-\infty}^{\infty} \phi(\lambda)\, d\rho(\lambda) = \int_{-\infty}^{\infty} \phi(\lambda) f(\lambda)\, d\lambda. \tag{2.3.33}$$

Now (2.3.29) and (2.3.31) together imply that the measure μ can have no discrete points. We can take a limit of functions ϕ converging pointwise to the characteristic function of an interval $(a, b]$ and apply the Lebesgue Dominated-Convergence Theorem to the limit of (2.3.33), obtaining

$$\mu\{(a, b]\} = \int_a^b d\rho(\lambda) = \int_a^b f(\lambda)\, d\lambda. \tag{2.3.34}$$

Thus f is a density function for the measure μ. *A priori* we know only $f \in L^2(\mathbb{R})$. However, this implies $f \in L^2(a, b)$ for finite intervals. Since necessarily $f(\lambda) \geq 0$ a.e., we also have, from (2.3.34),

$$\mu(\mathbb{R}) = \int_{-\infty}^{\infty} f(\lambda)\,\mathrm{d}\lambda = \int_{-\infty}^{\infty} |f(\lambda)|\,\mathrm{d}\lambda < \infty,$$

so that $f \in L^1(\mathbb{R})$.

We then have $\mathrm{d}\rho(\lambda)/\mathrm{d}\lambda = f(\lambda)$, and μ is absolutely continuous. ■

We conclude this survey of the application of the resolvent and Fourier transforms to the analysis of measures by considering the relation between these two transforms.

For arbitrary $x \in \mathbb{R}$ and $y > 0$, we multiply (2.3.28) by $\mathrm{e}^{\mathrm{i}(x+\mathrm{i}y)t}$ and integrate with respect to t, to obtain

$$\mathrm{i} \int_0^{\infty} \mathrm{e}^{\mathrm{i}(x+\mathrm{i}y)t}\, F_\mu(t)\,\mathrm{d}t = \int_{-\infty}^{\infty} \frac{\mathrm{d}\rho(\lambda)}{\lambda - x - \mathrm{i}y}. \tag{2.3.35}$$

In particular, if $F_\mu \in L^1(0.\ \infty)$, we can take the imaginary part of (2.3.35) to deduce that

$$\int_0^{\infty} |F_\mu(t)|\,\mathrm{d}t \geq \int_{-\infty}^{\infty} \frac{y\,\mathrm{d}\rho(\lambda)}{(\lambda - x)^2 + y^2}. \tag{2.3.36}$$

Following our previous analysis of the resolvent transform, an L^1 estimate in (2.3.36) for F_μ may be used to obtain a pointwise estimate for the density function of an absolutely continuous measure. In scattering theory the absolutely continuous part of a spectral measure is of great significance in the description of scattering states, and the Fourier transform allows a connection between the time development of states and spectral analysis.

2.4 ANALYTICITY

Let us begin by looking more closely at the relationship (Section 2.3) between the resolvent transform of a measure and the set of functions analytic in (say) the upper half plane $\mathrm{Im}\,(z) > 0$. We note from (2.3.23) that in the resolvent transform we have to deal with a function analytic in the upper half-plane, and having positive imaginary part. The imaginary part (or equivalently the real part) of an analytic function is called a *harmonic* function. Harmonic functions may also be defined as solutions of the two-dimensional Laplace equation.

There is a well-known characterization (Herglotz's Theorem) of positive harmonic functions on the open unit disc $|\omega|<1$ $(\omega=x+iy)$. Given any such function $\mathscr{V}(\omega)$, there exists a non-decreasing function $\alpha(\phi)$, $0<\phi\leq 2\pi$, continuous from the right such that

$$\mathscr{V}(\omega)=\int_0^{2\pi} K(e^{i\phi},\omega)\,d\alpha(\phi). \tag{2.4.1}$$

where

$$K(e^{i\phi},\omega)=\mathrm{Re}\left(\frac{e^{i\phi}+\omega}{e^{i\phi}-\omega}\right). \tag{2.4.1$'$}$$

Conversely, given any such $\alpha(\phi)$, (2.4.1) and (2.4.1)$'$ define a positive function $\mathscr{V}(\omega)$ harmonic on the open unit disc. Given $\mathscr{V}(\omega)$, $\alpha(\phi)$ is determined up to an additive constant. (In the special case that $\mathscr{V}(\omega)\equiv\mathscr{V}(re^{i\theta})$ is continuous on the closed unit disc $|\omega|\leq 1$, one can write $d\alpha(\phi)=(2\pi)^{-1}\mathscr{V}(e^{i\phi})\,d\phi$. Equation (2.4.1) then becomes

$$\mathscr{V}(re^{i\theta})=\frac{1}{2\pi}\int_0^{2\pi}\frac{(1-r^2)\mathscr{V}(e^{i\phi})\,d\phi}{1-2r\cos(\theta-\phi)+r^2},$$

which is the so-called Poisson integral formula expressing $\mathscr{V}$ at any point in the unit disc in terms of the boundary values of $\mathscr{V}$ as ω approaches the unit circle.)

Through the transformation

$$\omega=\frac{iz+1}{iz-1},$$

the open unit disc $|\omega|<1$ is mapped conformally onto the upper half-plane $\mathrm{Im}(z)>0$. We then have a one-to-one correspondence between positive harmonic functions $\mathscr{V}(\omega)$ in the open unit disc and positive harmonic functions $v(z)$ in the upper half-plane, where we set

$$\mathscr{V}(\omega)=\mathscr{V}\left(\frac{iz+1}{iz-1}\right)=v(z). \tag{2.4.2}$$

What becomes of the formula (2.4.1)?

If we make the change of integration variable

$$e^{i\phi}=\frac{i\lambda+1}{i\lambda-1}\qquad(0<\phi<2\pi),$$

and write $\alpha(\phi)=\beta(\lambda)$, then $\beta(\lambda)$ is monotonic non-decreasing over the entire real line $-\infty<\lambda<\infty$. Substituting in (2.4.1) and (2.4.1)′ for $e^{i\phi}$ in terms of λ and for ω in terms of z, after a little straightforward algebra, we find

$$K(e^{i\phi},\omega)=\mathrm{Re}\left(\frac{-i(1+\lambda z)}{\lambda-z}\right)$$

$$=\mathrm{Re}\left(\frac{-i(1+\lambda^2)}{\lambda-z}\right)$$

$$=\mathrm{Im}\left(\frac{1+\lambda^2}{\lambda-z}\right);$$

so that (2.4.1) becomes

$$v(z)=A\,\mathrm{Im}\,(z)+\int_{-\infty}^{\infty}\mathrm{Im}\left(\frac{1+\lambda^2}{\lambda-z}\right)\mathrm{d}\beta(\lambda). \tag{2.4.3}$$

Here the additional term $A\ \mathrm{Im}\,z$ comes from the contribution of the single point $\phi=2\pi$ to the integral in (2.4.1), observing that

$$K(1,\omega)=\mathrm{Re}\left(\frac{1+\omega}{1-\omega}\right)=\mathrm{Im}\ (z).$$

There is no corresponding value of λ to $\phi=2\pi$, so that this additional term cannot be incorporated into the λ integration (unless this is done by introducing a point at infinity $\lambda=\infty$, which would then have positive β-measure). Equation (2.4.3) gives a complete characterization of positive harmonic functions on the upper half-plane; i.e. any such function $v(z)$ may be represented in the form (2.4.3), where $A>0$ and the non-decreasing function $\beta(\lambda)$ generates a *finite* Borel–Stieltjes measure on the real line. If we now write $\mathrm{d}\rho(\lambda)=(1+\lambda^2)\,\mathrm{d}\beta(\lambda)$ (i.e. we define a measure μ by $\mu(\Sigma)=\int_\Sigma(1+\lambda^2)\,\mathrm{d}\beta(\lambda)$), we have

$$v(z)=A\,\mathrm{Im}\,(z)+\int_{-\infty}^{\infty}\mathrm{Im}\left(\frac{1}{\lambda-z}\right)\mathrm{d}\rho(\lambda)$$

$$=Ay+\int_{-\infty}^{\infty}\frac{y\,\mathrm{d}\rho(\lambda)}{(\lambda-x)^2+y^2}, \tag{2.4.4}$$

where $z=x+iy$. The measure μ generated by $\rho(\lambda)$ is then no longer a *finite* measure, but does satisfy

$$\int_{-\infty}^{\infty} \frac{d\rho(\lambda)}{1+\lambda^2} < \infty. \tag{2.4.5}$$

By the Lebesgue Dominated-Convergence Theorem,

$$\lim_{y\to\infty} \int_{-\infty}^{\infty} \frac{d\rho(\lambda)}{(\lambda-x)^2+y^2} = 0,$$

so that the constant A in (2.4.4) is given by $A=\lim_{y\to\infty} v(x+iy)/y$. In particular, A will vanish provided that $v(x+iy)=o(y)$ in the limit as $y\to\infty$. In that case, we have

$$v(z) = \int_{-\infty}^{\infty} \frac{y\, d\rho(\lambda)}{(\lambda-x)^2+y^2}, \tag{2.4.6}$$

so that (cf. (2.3.15)) v is then the resolvent transform of the measure μ generated by $\rho(\lambda)$. We shall shortly show that (2.4.6) may be inverted; thus, given $v(z)$, the measure μ may be constructed. We then see that the resolvent transform, through (2.4.6), establishes a one-to-one correspondence between the family of measures μ, generated by functions $\rho(\lambda)$ such that (2.4.5) holds, and the family of positive harmonic functions v in the upper half-plane, for which $v(x+iy)=o(y)$ as $y\to\infty$.

There is a close connection between the asymptotic behaviour of $v(x+iy)$ for large y and the corresponding large-λ behaviour of the measure μ. Suppose, for example, that $yv(x+iy)$ is bounded as $y\to\infty$, say $yv(x+iy)\le c$. Then, from (2.4.6) by an application of the Lebesgue Dominated-Convergence Theorem, we have, for any $N>0$,

$$\int_{-N}^{N} d\rho(\lambda) = \lim_{y\to\infty} y \int_{-N}^{N} \frac{y\, d\rho(\lambda)}{(\lambda-x)^2+y^2} \le c.$$

Hence, in that case, $\mu(\mathbb{R})=\int_{-\infty}^{\infty} d\rho(\lambda)\le c$, so that μ is a finite measure. Conversely, if μ is a finite measure then it is easy to see from (2.4.6) that $yv(x+iy)$ is bounded (even bounded uniformly in x). So *finite* measures are exactly those measures having resolvent transform satisfying the condition that $yv(x+iy)$ is bounded. For a finite measure μ, one also has, from (2.4.6), the identity

$$\mu(\mathbb{R}) = \pi^{-1} \int_{-\infty}^{\infty} v(x+iy)\, dx. \tag{2.4.7}$$

Conversely, one may wish to verify that if v is positive harmonic in the upper half-plane and satisfies $\int_{-\infty}^{\infty} v(x+\mathrm{i}y)\,\mathrm{d}x<\infty$ for some $y>0$ then v is the resolvent transform of a finite measure. Equation (2.4.7) then implies that, in fact, $\int_{-\infty}^{\infty} v(x+\mathrm{i}y)\,\mathrm{d}x<\infty$ for *all* $y>0$, and that the integral is independent of the value of x.

The following lemma relates to three possible ways of defining boundary values of $v(z)$ as z approaches the real axis; namely, by pointwise limits, limits in the sense of distributions, and L^1 limits respectively.

Lemma 2.8

Let μ be a finite Borel–Stieltjes measure, generated by $\rho(\lambda)$, and define, through (2.4.6), the resolvent transform $v(z)$ of μ. Then

(i) for almost all x,

$$\lim_{y\to 0+} v(x+\mathrm{i}y)=\pi\frac{\mathrm{d}\rho(x)}{\mathrm{d}x}, \tag{2.4.8}$$

(ii) if ϕ is any real continuous function such that $\lim_{x\to\pm\infty}\phi(x)=0$ then

$$\lim_{y\to 0+}\int_{-\infty}^{\infty} v(x+\mathrm{i}y)\phi(x)\,\mathrm{d}x=\pi\int_{-\infty}^{\infty}\phi(\lambda)\,\mathrm{d}\rho(\lambda), \tag{2.4.9}$$

(iii) if μ is absolutely continuous, with density function $f(x)$, then, as $y\to 0+$, $v(x+\mathrm{i}y)$ converges to $\pi f(x)$ in L^1, norm, and also pointwise for almost all x.

Proof

(i) The function ρ is monotonic non-decreasing, and hence differentiable for almost all x. For each x at which ρ is differentiable, the right-hand side of (2.3.18) exists and is $\pi\,\mathrm{d}\rho(x)/\mathrm{d}x$ (cf. also Theorem 2.1). The result (2.4.8) now follows from Lemma 2.4 (ii).

(ii) From (2.4.6), on interchanging orders of integration,

$$\int_{-\infty}^{\infty} v(x+\mathrm{i}y)\phi(x)\,\mathrm{d}x=\int_{-\infty}^{\infty}\left(\int_{-\infty}^{\infty}\frac{y\phi(x)\,\mathrm{d}x}{(\lambda-x)^2+y^2}\right)\mathrm{d}\rho(\lambda).$$

By Lemma 2.3(i), the integral within parentheses converges uniformly to $\pi\phi(\lambda)$ in the limit as $y\to 0+$. Equation (2.4.9) is then a simple consequence.

(iii) Suppose that μ is absolutely continuous, with density function

$f(x)$. Then the right-hand side of (2.4.6) may be written as $v(z)=\int_{-\infty}^{\infty} y f(\lambda)\,d\lambda/[(\lambda-x)^2+y^2]$, with $f\in L^1(\mathbb{R})$. Then the convergence of $v(x+iy)$ to $\pi f(x)$, in L^1 norm, follows, with minor change of notation, from Lemma 2.3(ii). The pointwise convergence follows from (i) above, using the fact that, for almost all x, $f(x)$ is the derivative of $\rho(x)$. ■

Corollary

Under the hypothesis of the lemma, if $(a,b]$ is any finite interval such that neither a nor b have positive μ-measure then

$$\mu\{(a,b]\}=\lim_{y\to 0+}\frac{1}{\pi}\int_a^b v(x+iy)\,dx. \tag{2.4.10}$$

Proof

Equation (2.4.10) follows from (2.4.9) if we are allowed to take, for ϕ, the characteristic function of the interval $(a,b]$. However, such a choice for ϕ would not be continuous, and the best we can do is to take, for ϕ, a smooth function that satisfies $0\le\phi(x)\le 1$, and (for small positive ε)

$$\phi(x)=\begin{cases}1 & (a\le x\le b),\\ 0 & (x\le a-\varepsilon),\\ 0 & (x\ge b+\varepsilon).\end{cases}$$

The contributions to the right-hand side of (2.4.9) from integration over the intervals $(a-\varepsilon,a]$ and $(b,b+\varepsilon]$ respectively vanish in the limit as $\varepsilon\to 0$, and to prove (2.4.10) it remains only to show that, as ε and y *both* tend to zero,

$$\lim\int_{a-\varepsilon}^{a} v(x+iy)\,dx=\lim\int_{b}^{b+\varepsilon} v(x+iy)\,dx=0.$$

But this is again a consequence of applying (2.4.9) respectively to functions $\phi=\phi_a$ and ϕ_b satisfying $0\le\phi(x)\le 1$ and

$$\phi_a(x)=\begin{cases}1 & (a-\varepsilon\le x\le a),\\ 0 & (x\le a-2\varepsilon)\\ 0 & (x\ge a+2\varepsilon).\end{cases}$$

$$\phi_b(x)=\begin{cases}1 & (b\le x\le b+\varepsilon),\\ 0 & (x\le b-2\varepsilon),\\ 0 & (x\ge b+2\varepsilon).\end{cases}$$

We have, for example,

$$\lim \int_{a-\varepsilon}^{a} v(x+\mathrm{i}y)\,\mathrm{d}x \le \lim \int_{-\infty}^{\infty} v(x+\mathrm{i}y)\phi_a(x)\,\mathrm{d}x$$

$$= \pi \lim \int_{-\infty}^{\infty} \phi_a(\lambda)\,\mathrm{d}\rho(\lambda)$$

$$\le \pi \lim_{\varepsilon \to 0+} \int_{a-2\varepsilon}^{a+2\varepsilon} \mathrm{d}\rho(\lambda) = 0,$$

since a does not belong to the discrete set of points of the measure μ. Similarly $\lim \int_b^{b+\varepsilon} v(x+\mathrm{i}y)\,\mathrm{d}x = 0$. These two results allow us to replace ϕ in (2.4.9) by the characteristic function of $(a, b]$, and (2.4.10) follows. ■

2.5 QUANTUM MECHANICS AND HILBERT SPACE

The mathematics of quantum mechanics is, to a great extent, the mathematics of Hilbert space. How does this come about? One answer to this question, not the only answer nor even an answer that would command universal support among mathematical physicists, is to cite the close correspondence between the *propositions* of a quantum system and the *subspaces* of Hilbert space.

The *propositions* of a quantum system are defined by the questions that may be asked of the system, and in particular by those questions that give rise to an answer "yes" or "no". For example, for a single particle, one may ask "is the kinetic energy of the particle greater than 750 MeV?" or "is the z-component of orbital angular momentum one of the values $\hbar$, $2\hbar$?" Such questions define propositions. The questions are asked of the system by means of *observations*, the result of each observation being interpreted as an answer "yes" or "no" for some proposition or set of propositions. Some observations (measurement of the z-component of orbital angular momentum) provide the answers to a large number of questions (is this value $\hbar$?, or $2\hbar$?, or is the value negative?); the result of the observation will depend, of course, on the current *state* of the system when the observation is made.

Two propositions may be identified (take equivalence classes of propositions) if, in any conceivable experiment on the system in any possible state, "yes" ("no") is obtained for one of the propositions if and only if the same answer is obtained for the other. For example "is the z-component of orbital angular momentum $\hbar$?" would correspond to the same proposition as "does the z-component of orbital angular momentum lie between $\frac{1}{2}\hbar$ and $\frac{3}{2}\hbar$?"

Given a proposition a, we may define a proposition $a^\perp$, the *complement* of a, such that an observation gives yes/no for $a^\perp$ if and only if it gives no/yes for a.

Two propositions a and b may sometimes be related to the extent that any observation giving "yes" for a necessarily also gives "yes" for b. We write this $a \subseteq b$, and assume also that "no" for b gives "no" for a. This means that $a \subseteq b$ if and only if $b^\perp \subseteq a^\perp$. Also, $a \subseteq b$, $b \subseteq a$ together imply $a = b$. As an example, let a be "is the kinetic energy greater than 750 MeV?", and let b be "is the kinetic energy greater than 700 MeV?" Then $a \subseteq b$ in this case, and $a^\perp$ and $b^\perp$ are respectively "is the kinetic energy less than or equal to 750 MeV?", and "is the kinetic energy less than or equal to 700 MeV?".

$a \subseteq b$ defines an *order relation*; in particular, $a \subseteq b$, $b \subseteq c$ together imply $a \subseteq c$.

We can introduce two ideal propositions, $\emptyset$ (always "no") and I (always "yes"). For example, if a is the proposition "is the kinetic energy less than zero?" then one may legitimately write $a = \emptyset$, $a^\perp = I$.

Keeping an analogy with traditional logic, one may suppose, given any two propositions a and b, that there is a proposition $a \cap b$, pronounced "a and b". How may such a proposition $a \cap b$ be defined? Certainly "yes" for $a \cap b$ must imply both "yes" for a and "yes" for b, that is

(i) $a \cap b \subseteq a, \qquad a \cap b \subseteq b.$

What would be observational evidence for $a \cap b$? Let y be any proposition such that "yes" for this proposition allows us to conclude both "yes" for a and "yes" for b. Then "yes" for y must certainly be interpreted as experimentally implying "yes" for our new proposition $a \cap b$. In other words,

(ii) $y \subseteq a$ and $y \subseteq b$ together imply $y \subseteq a \cap b$.

Properties (i) and (ii) determine $a \cap b$ uniquely, assuming such a proposition to exist. Note that $a \cap b$ will be defined even when a and b are observationally incompatible propositions, such that the measurement of a interferes with the measurement of b. The point is that one does *not* measure $a \cap b$ by measuring both a and b, which may in general be impossible without interference. If a and b are incompatible so that one never has "yes" for a and b simultaneously then one can have only $y = \emptyset$ in (ii) above, in which case (i) and (ii) together imply $a \cap b = \emptyset$. As is common in all branches of applied mathematics and mathematical physics, the mathematics is taken much further than direct analogy with the physics would support; here we assume that corresponding to *any* indexed family

$\{a_i\}$ of propositions, whether finite or infinite, there is a proposition $\bigcap_i a_i$ satisfying properties corresponding to (i) and (ii) above in the case of just two propositions. In view of (i) and (ii) we may regard $\bigcap_i a_i$ as the "greatest lower bound" of the family of propositions. Given any two propositions a and b we may now define $a \cup b$ (*a or b*) by

$$a \cup b = \bigcap_{\substack{a \subseteq y \\ b \subseteq y}} y.$$

Thus we have "yes" for $a \cup b$ if and only if we have "yes" for each proposition y that is implied both by a and by b (interpreting $a \subseteq y$ and $b \subseteq y$ as some observationally sanctioned form of implication).

Clearly the set of propositions of a quantum system, with their associated relations and operations, presents us with a very rich mathematical structure, and the traditional interpretation of quantum mechanics implies that this structure is intimately related to that of the family of subspaces of a Hilbert space $\mathscr{H}$.

Thus to every proposition a may be associated, uniquely, a corresponding subspace $\mathscr{M}_a$ of $\mathscr{H}$, in such a way that

(i) $\emptyset$ corresponds to the subspace containing only the zero vector, and I to the space $\mathscr{H}$ itself,
(ii) $a \cap b$ corresponds to the subspace $\mathscr{M}_a \cap \mathscr{M}_b$ (intersection of $\mathscr{M}_a$ with $\mathscr{M}_b$),
(iii) $a \cup b$ corresponds to $\mathscr{M}_a + \mathscr{M}_b$, which is defined to be the *closure* of the linear set obtained by adding vectors in $\mathscr{M}_a$ to vectors in $\mathscr{M}_b$,
(iv) $a^\perp$ corresponds to $\mathscr{M}_a^\perp$, the subspace consisting of vectors that are orthogonal to every vector in $\mathscr{M}_a$ (two vectors f and g are said to be *mutually orthogonal* if $\langle f, g \rangle = 0$), and
(v) $a \subseteq b$ if and only if $\mathscr{M}_a \subseteq \mathscr{M}_b$, i.e. if and only if subspace $\mathscr{M}_a$ is contained in subspace $\mathscr{M}_b$.

That such a correspondence exists between propositions of a quantum system and subspaces of a Hilbert space is a fundamental assumption that is implicit in virtually all of the accepted treatments of the foundations of quantum mechanics. Whether one treats this assumption as the starting point of the theory, or derives it from more basic postulates, or sets up the theory from some quite different standpoint, is a question of individual preference. One of the advantages of the approach through propositions is that the mathematical interpretation of states and observables may thereby be clarified. Before seeing how this is so, we need to develop some of the mathematics of Hilbert space a little further.

As defined in Section 2.2, a Hilbert space $\mathscr{H}$ will be, for us, a complete, separable, complex inner-product space. A sequence $\{f_n\}$ of vectors in $\mathscr{H}$ is said to converge strongly to the vector f if $\lim_{n\to\infty}\|f_n-f\|=0$, where $\|h\|=\langle h,h\rangle^{1/2}$. A subspace $\mathcal{M}$ of $\mathcal{H}$ is a collection of elements (or vectors) of $\mathcal{H}$ that is closed under the operations of addition and of multiplication by complex numbers, and that is also closed in the sense that strong limits of vectors in $\mathcal{M}$ are themselves in $\mathcal{M}$. A sequence $\{f_n\}$ of vectors in $\mathcal{H}$ is said to converge *weakly* to a limit f if, for all $h\in\mathcal{H}$,

$$\lim_{n\to\infty}\langle h,f_n\rangle=\langle h,f\rangle.$$

If we use the *Schwarz inequality* (c.f. p. 28)

$$|\langle h,g\rangle|\le\|h\|\,\|g\|, \tag{2.5.1}$$

which holds for any pair of vectors h, g in $\mathcal{H}$, we see that if $f_n\to f$ strongly then

$$\begin{aligned}|\langle h,f_n\rangle-\langle h,f\rangle|&=|\langle h,f_n-f\rangle|\\&\le\|h\|\,\|f_n-f\|\to 0,\end{aligned}$$

so that $f_n\to f$ weakly. Hence strongly convergent sequences converge weakly as well. The converse is false. (In $L^2(\mathbb{R})$ let $f_n(x)$ be the characteristic function of the interval $(n, n+1]$. Then

$$\begin{aligned}|\langle h,f_n\rangle|=\left|\int_n^{n+1}\bar{h}(x)\,\mathrm{d}x\right|&\le\int_n^{n+1}|h(x)|1\,\mathrm{d}x\\&\le\left(\int_n^{n+1}|h|^2\,\mathrm{d}x\right)^{1/2},\end{aligned}$$

by the Schwarz inequality applied to the functions $|h|$ and 1 in the space $L^2(n, n+1)$. Since

$$\sum_{n=-\infty}^{\infty}\int_n^{n+1}|h|^2\,\mathrm{d}x=\int_{-\infty}^{\infty}|h|^2\,\mathrm{d}x=\|h\|^2<\infty,$$

we know that $\lim_{n\to\infty}\int_n^{n+1}|h|^2\,\mathrm{d}x=0$. Hence $\lim_{n\to\infty}\langle h,f_n\rangle=0$. However, $\|f_n\|=1$, so that certainly f_n cannot converge strongly to zero.)

A sequence $\{f_n\}$ convergent weakly to a limit f will, however, converge strongly to f provided that

$$\lim_{n\to\infty}\|f_n\|=\|f\|, \tag{2.5.2}$$

since in that case

$$\begin{aligned}||f_n-f||^2&=\langle f_n-f,f_n-f\rangle\\&=\langle f_n,f_n\rangle-\langle f_n,f\rangle-\langle f,f_n\rangle+\langle f,f\rangle\\&=||f_n||^2-2\,\mathrm{Re}\,\langle f,f_n\rangle+||f||^2\to||f||^2-2||f||^2+||f||^2=0.\end{aligned}$$

Note in particular that (2.5.2) holds for any sequence $\{f_n\}$ converging strongly to f.

A linear operator T in $\mathscr{H}$ is a transformation from elements of $\mathscr{H}$ to elements of $\mathscr{H}$ satisfying

$$\left.\begin{aligned}T(f_1+f_2)&=Tf_1+Tf_2,\\T(cf)&=cTf\end{aligned}\right\}\tag{2.5.3}$$

The domain of T (set of elements on which the transformation is defined) will be denoted by $D(T)$. Thus $D(T)$ is a linear set of vectors (closed under addition, and under multiplication by complex numbers), and (2.5.3) is assumed to hold for all f_1, f_2 and f in $D(T)$, and for all (complex) numbers c.

Although we shall not suppose, in general, that $D(T)$ is the whole of $\mathscr{H}$, we *shall* insist, unless otherwise stated, that $D(T)$ is dense in $\mathscr{H}$. This means that the closure of $D(T)$ (in norm) must be the whole of $\mathscr{H}$. Another way of stating this is to say that every element h of $\mathscr{H}$ must have elements of $D(T)$ arbitrarily close (in norm).

A linear operator T in $\mathscr{H}$ is said to be *bounded* provided that there exists a positive number M such that

$$||Tf||\leq M||f||\tag{2.5.4}$$

for all $f\in D(T)$. The smallest such M is called the *norm* of T and written $||T||$. In other words,

$$||T||=\sup||Tf||/||f||,\tag{2.5.5}$$

the supremum being taken over all $f\in D(T)$ such that $f\neq 0$. A bounded linear operator (defined on a domain dense in $\mathscr{H}$) can always be *extended by continuity* to an operator defined on the entire space. To do this, we define Th (or more accurately $T'h$, where T' is the *extension* of T; i.e. T' is obtained from T by enlarging the domain) by

$$Th=\operatorname*{s-lim}_{n\to\infty}Tf_n,\tag{2.5.6}$$

where $\{f_n\}$ is a sequence of elements in $D(T)$ that converge to h. That such a sequence exists follows from the fact that $D(T)$ is dense in $\mathscr{H}$. Note that, using (2.5.5),

$$\begin{aligned}||Tf_m - Tf_n|| &= ||T(f_m - f_n)|| \leq ||T||\ ||f_m - f_n|| \\ &= ||T||\ ||(f_m - f) - (f_n - f)|| \\ &\leq ||T||\ ||f_m - f|| + ||T||\ ||f_n - f|| \to 0 \quad \text{as } m, n \to \infty.\end{aligned}$$

Hence $\{Tf_n\}$ is a Cauchy sequence, which, by completeness of the space, converges strongly to a limit. The right-hand side of (2.5.6) is readily seen to be independent, for given h, of which particular sequence $\{f_n\}$ converging to h is taken.

Bounded linear operators may therefore be assumed, without loss of generality, to act on the whole of $\mathscr{H}$. (We often then speak of a linear operator *on* $\mathscr{H}$, rather than a linear operator *in* $\mathscr{H}$, to emphasize that $D(T) = \mathscr{H}$ in this case.)

For linear operators, addition and multiplication are defined in an obvious way; thus $(cT)f = c(Tf)$, $(T_1 + T_2)f = T_1 f + T_2 f$, $(T_1 T_2)f = T_1(T_2 f)$. For operators not defined on the whole of $\mathscr{H}$, it will often be necessary to be very careful about domains. For example, $T_1 + T_2$ is only defined *a priori* on $D(T_1) \cap D(T_2)$, and it is possible for both T_1 and T_2 to have dense domains, and yet their respective domains to intersect in the zero vector only! We shall defer these domain questions to later chapters, when we shall be dealing with various special classes of operators.

For bounded linear operators, we have

$$||cT|| = |c|\ ||T||, \qquad ||T_1 + T_2|| \leq ||T_1|| + ||T_2||, \qquad ||T_1 T_2|| \leq ||T_1||\ ||T_2||.$$

The first two results justify the use of the word "norm" for linear operators. Indeed, the bounded linear operators on $\mathscr{H}$, with their *operator norm*, are the elements of a Banach space. (Actually, of course, the structure is richer still, since the algebraic operation of multiplication is also defined, and we may talk of a *Banach algebra*.)

We shall use the notation $S \supseteq T$ to indicate that a linear operator S is an extension of a linear operator T. If $S \supseteq T$ then it follows that $Sf = Tf$ for all $f \in D(T)$, and that either $S = T$, or there are elements in the domain of S that are not in the domain of T, in which case one says that S is a *proper* extension of T and writes $S \supset T$.

A linear operator T is said to be *closed* if it has the property that, for every sequence $\{f_n\}$ of elements in $D(T)$ such that both f_n and Tf_n converge strongly, say $f_n \to h$ and $Tf_n \to g$, one has $h \in D(T)$ and $Th = g$. A linear operator that is not closed may or may not have a closed extension; an operator that does have a closed extension is said to be closable. (The problem that prevents an operator from being closable is that there may be *two* sequences, say $\{f_n\}$ and $\{g_n\}$, both converging strongly to an element f, but such that s-lim Tf_n and s-lim Tg_n exist and are different.)

If a linear operator T *is* closable then one may obtain a closed extension by taking the *closure* of T. The domain of cl (T) consists of all $f \in \mathscr{H}$ such that there is a sequence $\{f_n\}$, converging strongly to f, for which s-lim Tf_n exists. One then defines, for $f \in D(\mathrm{cl}\,(T))$,

$$(\mathrm{cl}\,(T))f = \operatorname*{s-lim}_{n\to\infty} Tf_n,$$

the limit on the right-hand side being independent of the particular choice of the sequence $\{f_n\}$, since T is closable. The closure of T is the closed extension of T having the smallest domain, in the sense that if T' is *any* closed extension of T then $T' \supseteq \mathrm{cl}\,(T)$.

We note that any closed linear operator T will have proper closed extensions, unless of course $D(T) = \mathscr{H}$. For let T be closed, and let ϕ be any element of $\mathscr{H}$ that is not in $D(T)$. We define a linear operator T' having domain $D(T')$ consisting of all vectors of the form $f + \alpha\phi$, where $f \in D(T)$, $\alpha \in \mathbb{C}$, and set

$$T'(f+\alpha\phi) = Tf + \alpha\psi, \tag{2.5.7}$$

for any fixed $\psi \in \mathscr{H}$.

To show that T' is closed, we consider a sequence $\{f_n + \alpha_n\phi\}$ of vectors in $D(T')$ such that both $\{f_n + \alpha_n\phi\}$ and $T'(f_n + \alpha_n\phi)$ converge strongly. Explicitly, we suppose that $f_n + \alpha_n\phi \to h$ and $Tf_n + \alpha_n\psi \to h'$.

We observe that $|\alpha_n|$ is bounded. For if it were not, we could assume, by choosing a subsequence if necessary, that $|\alpha_n| \to \infty$, in which case $(f_n + \alpha_n\phi)/\alpha_n$ and $(Tf_n + \alpha_n\psi)/\alpha_n$ would both converge to zero. This would imply $\text{s-lim}_{n\to\infty} f_n/\alpha_n = -\phi$, $\text{s-lim}_{n\to\infty}\, T(f_n/\alpha_n) = -\psi$, giving $\phi \in D(T)$ since T is closed, and this contradicts our original assumption that $\phi \notin D(T)$.

Then, since $|\alpha_n|$ is bounded, we may assume, by choosing a subsequence if necessary, that $\lim_{n\to\infty} \alpha_n \equiv \alpha$ exists. (Any bounded sequence of complex numbers has a convergent subsequence). This gives the strong limits $f_n + \alpha\phi \to h$, $Tf_n + \alpha\psi \to h'$. Hence $f_n \to h - \alpha\phi$, $Tf_n \to h' - \alpha\psi$, and since T is closed we have $h' - \alpha\psi = T(h - \alpha\phi)$. Writing $f = h - \alpha\phi$ and comparing with (2.5.7), we see that $h' = T'h$, and we have shown that T' is indeed closed.

Operators that are closed but unbounded share some useful properties with bounded operators regarding convergence. For example, if $\{f_n\}$ is some strongly convergent sequence and T is bounded then the right-hand side of (2.5.6) will always exist as a strong limit, since, as we have seen, the sequence $\{Tf_n\}$ is Cauchy. On the other hand, if T is only known to be closed then (2.5.6) will hold *provided* that the right-hand side exists as a strong limit, which it may be possible to verify in particular instances.

The following result, known as the Closed-Graph Theorem, is further evidence of the importance of the closure property for operators.

Closed-Graph Theorem

A closed linear operator T defined on the whole space $(D(T)=\mathscr{H})$ is necessarily bounded.

A linear operator T is said to be *symmetric* provided that

$$\langle Tf, g\rangle = \langle f, Tg\rangle, \tag{2.5.8}$$

for all $f, g \in D(T)$. (Some authors describe T in this case as *Hermitian*). Given an arbitrary linear operator T, we shall define the *adjoint operator* T^* as follows. A vector $h \in \mathscr{H}$ will belong to the domain of T^* if and only if there exists a vector $h^* \in \mathscr{H}$ such that

$$\langle Tf, h\rangle = \langle f, h^*\rangle \tag{2.5.9}$$

for all $f \in D(T)$. We then write $h^* = T^*h$, so that the adjoint operator T^* is defined by the requirement

$$\langle Tf, h\rangle = \langle f, T^*h\rangle \tag{2.5.10}$$

for all $f \in D(T)$. Equation (2.5.10) defines T^*h uniquely. For if $\langle f, T^*h\rangle = \langle f, h_1^*\rangle = \langle f, h_2^*\rangle$ for all $f \in D(T)$ then we have, on taking strong limits of such f and using the fact that $D(T)$ is dense, $\langle f, h_1^*\rangle = \langle f, h_2^*\rangle$ for *all* $f \in \mathscr{H}$. Taking $f = h_1^* - h_2^*$, we may then deduce $h_1^* - h_2^* = 0$, so that $h_1^* = h_2^*$.

The domain of T^* is thus very precisely specified; $h \in D(T^*)$ if and only if (2.5.9) holds for some h^*, and for all $f \in D(T)$. In the special case that T is bounded then so is T^*. In fact $\|T\| = \|T^*\|$ in that case.

For arbitrary T (with $D(T)$ dense in $\mathscr{H}$) it need not necessarily follow that $D(T^*)$ is dense in $\mathscr{H}$. However, in the important special case that T is symmetric, $D(T^*)$ *will* be dense. This is because (2.5.10) follows from (2.5.8) in that case, with $T^*h = Th$, for any $h \in D(T)$. In other words, if T is symmetric then $T^* \supseteq T$.

We note also that the adjoint of any linear operator is always closed. We let $\{h_n\}$ be any sequence of vectors in $D(T^*)$ such that both h_n and T^*h_n converge strongly, say $h_n \to h'$ and $T^*h_n \to h''$. From (2.5.10) with $h = h_n$, we have $\langle Tf, h_n\rangle = \langle f, T^*h_n\rangle$, which on taking the limit $n \to \infty$ (strong convergence $\Rightarrow$ weak convergence!) becomes $\langle Tf, h'\rangle = \langle f, h''\rangle$, for all $f \in D(T)$. Hence $h'' = T^*h'$, and we have shown that T^* is closed.

In particular, for any symmetric operator $T^* \supseteq T$, so that any symmetric operator has a closed extension. In other words, every symmetric operator is closable. If T is symmetric, then $(f, Tf >$ is real, and any eigenvalues of T must be read. If $Tf_1 = \lambda_1 f_1$, $Tf_2 = \lambda_2 f_2$, then $\langle f_1, f_2 > = 0 (\lambda_1 \mp \lambda_2)$.

Lemma 2.9

Let T by any symmetric linear operator, or more generally any linear operator such that $D(T^*)$ is dense in $\mathscr{H}$. Then the closure of T is T^{**} $(\equiv (T^*)^*)$.

Proof

It is a straightforward exercise to verify that, for two linear operators T_1 and T_2,

$$T_2 \supseteq T_1 \Rightarrow T_1^* \supseteq T_2^*. \tag{2.5.11}$$

To show that cl (T) exists, we have to show that if $\{f_n\}$ and $\{g_n\}$ both converge strongly to an element f, and if the strong limits $\lim_{n\to\infty} Tf_n$ and $\lim_{n\to\infty} Tg_n$ both exist, then these two limits are equal. By considering $h_n = f_n - g_n$, it is sufficient to show that s-$\lim_{n\to\infty} h_n = 0$ and s-$\lim_{n\to\infty} Th_n = k$ together imply $k=0$. We then have, for any $h \in D(T^*)$,

$$\langle k, h\rangle = \lim_{n\to\infty} \langle Th_n, h\rangle = \lim_{n\to\infty} \langle h_n, T^*h\rangle = 0.$$

Since $D(T^*)$ is dense, we can take strong limits of such h to show that $\langle k, h\rangle = 0$ for *all* $h \in \mathscr{H}$, which implies $k=0$ on taking $h=k$.

Therefore cl (T) exists and is a closed extension of T, from which (2.5.11) implies $T^* \supseteq [\text{cl }(T)]^*$. Moreover, if $h \in D(T^*)$, so that (2.5.10) holds for $f \in D(T)$, then for $\tilde{f} \in D(\text{cl}(T))$ we can write $\tilde{f} = \text{s-}\lim_{n\to\infty} f_n$, cl $(T)\tilde{f} =$ s-$\lim_{n\to\infty} Tf_n$. Taking the limit as $n \to \infty$ of the equation

$$\langle Tf_n, h\rangle = \langle f_n, h^*\rangle,$$

we then have

$$\langle \text{cl }(T)\tilde{f}, h\rangle = \langle \tilde{f}, h^*\rangle.$$

This holds for all $\tilde{f}$ in the domain of cl (T), so that we are entitled to write $h^* = [\text{cl }(T)]^*h$. This means that $[\text{cl }(T)]^* \supseteq T^*$. We have already shown that $T^* \supseteq [\text{cl }(T)]^*$. Hence T and cl (T) have the same adjoint. We then have, $T^{**} = [\text{cl }(T)]^{**}$, so that in proving the lemma we may assume without loss of generality that we are dealing with a closed operator.

Assuming, in view of this, that T is closed, we let $\mathscr{H} \times \mathscr{H}$ be the Hilbert space of ordered pairs $\{f_1, f_2\}$, where $f_1 \in \mathscr{H}$, $f_2 \in \mathscr{H}$ and the inner product is

$$\langle \{g_1, g_2\}, \{f_1, f_2\}\rangle = \langle g_1, f_1\rangle + \langle g_2, f_2\rangle.$$

We let $\mathscr{M}$ denote the subspace of $\mathscr{H} \times \mathscr{H}$ consisting of all elements of the form $\{f, Tf\}$, for $f \in D(T)$. (That $\mathscr{M}$ is closed follows from the closure property of T.) Referring to (2.5.10), we see that the subspace $\mathscr{M}^\perp$ orthogonal to $\mathscr{M}$ consists of all elements of the form $\{T^*h, -h\}$, for $h \in D(T^*)$. By the same argument, $(\mathscr{M}^\perp)^\perp$ consists of all elements of the form $\{g, T^{**}g\}$, for $g \in D(T^{**})$. The equality of T^{**} with T in this case now follows from the identity $(\mathscr{M}^\perp)^\perp = \mathscr{M}$. ■

A linear operator T is said to be *self-adjoint* if $T^* = T$. It is trivial that every self-adjoint operator is symmetric. Moreover, since T^* is necessarily closed, every self-adjoint operator is closed. But not every closed symmetric operator is self-adjoint (we shall meet examples in Chapter 6). A symmetric operator may or may not have self-adjoint extensions. A symmetric operator of which the closure is self-adjoint is said to be *essentially self-adjoint*. Since $T^* = [\mathrm{cl}(T)]^* = \mathrm{cl}\,(T)$ in this case, for an operator that is essentially self-adjoint we have $\mathrm{cl}\ (T) = T^* = T^{**}$. Any *bounded* symmetric operator is self-adjoint.

If T is self-adjoint and $S \supseteq T$, where S is symmetric, we have $S^* \supseteq S \supseteq T$. Moreover, from (2.5.11), applied to S and T, we have $T = T^* \supseteq S^*$. Hence $T = S = S^*$ in this case, so that we may deduce that a self-adjoint operator *has no proper symmetric extension.* A self-adjoint operator, in other words, has the largest possible domain compatible with remaining symmetric.

Given a symmetric operator T, how do we attempt to construct self-adjoint extensions? First of all, we assume that T is closed (if not, we take the closure of T), but not self-adjoint. For $\phi \in D(T^*) \setminus D(T)$, either $\mathrm{Im}\,\langle \phi, T^*\phi \rangle > 0$ always, or $\mathrm{Im}\,\langle \phi, T^*\phi \rangle < 0$ always, or there exist ϕ for which $\mathrm{Im}\,\langle \phi, T^*\phi \rangle = 0$. (For, if $\mathrm{Im}\,\langle \phi_0, T^*\phi_0 \rangle > 0$ and $\mathrm{Im}\,\langle \phi_1, T^*\phi_1 \rangle < 0$, then, with $\phi_t = t\phi_1 + (1-t)\phi_0$, $\mathrm{Im}\,\langle \phi_t, T^*\phi_t \rangle = 0$ for some t in the interval $0 < t < 1$. One may verify $\phi_t \notin D(T)$.)

If $\mathrm{Im}\,\langle \phi, T^*\phi \rangle > 0$ for all $\phi \in D(T^*) \setminus D(T)$ then T will have no proper symmetric extension, and hence certainly no self-adjoint extension, since such an extension T' would require $T' \subseteq T^*$, in which case $\langle \phi, T'\phi \rangle = \langle \phi, T^*\phi \rangle$ is real for some ϕ. Similarly if $\mathrm{Im}\,\langle \phi, T^*\phi \rangle < 0$

We suppose then that $\langle \phi, T^*\phi \rangle$ is real for some $\phi \in D(T^*) \setminus D(T)$. Then we use (2.5.7), with $\psi = T^*\phi$, to define a closed symmetric extension T' of T.

This process of extending T may lead eventually to a self-adjoint extension. As $D(T)$ enlarges, so $D(T^*)$ contracts, until we reach (we hope) $T = T^*$. On the other hand, it may be that no self-adjoint extension exists, or that, extending the domain by one linearly independent vector at a time, an infinite number of such extensions is required. Section 2.7 will present a more systematic treatment of this question.

How is the relation between the observables of a quantum-mechanical system and the family of self-adjoint operators in a Hilbert space $\mathscr{H}$ to be understood? We consider first the very special case of an observable whose measurement gives rise, in any state of the quantum system, to one of a *finite* number of possible values $\lambda_1, \lambda_2, \ldots \lambda_n$.

Corresponding to each of these values, λ_i say, will be a proposition a_i, associated with the question "does the given observable have value λ_i?" Again, corresponding to each proposition a_i, will be a subspace $\mathscr{M}_i$ of the underlying Hilbert space. We let E_i denote the (orthogonal) projection onto

$\mathcal{M}_i$. Any vector $f \in \mathcal{H}$ may be expressed uniquely in the form

$$f = f_i + f_i^{\perp},$$

where $f_i \in \mathcal{M}_i$ and $f_i^{\perp} \in \mathcal{M}_i^{\perp}$, and E_i is defined by

$$E_i f = f_i.$$

It may readily be verified that the projection operator E onto any subspace $\mathcal{M}$ satisfies $E^* = E$ (E is self-adjoint) and $E^2 = E$. Indeed, any linear operator satisfying these two properties is a projection operator onto some subspace $\mathcal{M}$ ($\mathcal{M}$ is the range of E).

Hence $E_i^* = E_i$ and $E_i^2 = E_i$ in this case. Since, with $i \neq j$, "yes" for a_i implies "no" for a_j, in terms of the corresponding subspaces, we have

$$\mathcal{M}_i \subseteq \mathcal{M}_j^{\perp}.$$

In other words, the family $\{\mathcal{M}_i\}$ of subspaces are mutually orthogonal. For the associated projection operators this means

$$E_i E_j = E_j E_i = 0 \quad (i \neq j). \tag{2.5.12}$$

Corresponding to the given observable, we define a self-adjoint operator by

$$T = \sum_{i=1}^{n} \lambda_i E_i. \tag{2.5.13}$$

Conversely, given T, the values λ_i and the projection operators E_i (and hence the subspaces onto which the E_i project) can be reconstructed. The λ_i are the eigenvalues of T, and the E_i are projection operators onto the respective subspaces $\mathcal{M}_i$, consisting of eigenvectors of T with eigenvalue λ_i. The self-adjoint operator corresponding to any function F of the given observable is easily seen to be

$$F(T) = \sum_{i=1}^{n} F(\lambda_i) E_i. \tag{2.5.14}$$

Of course we have considered here a very special kind of observable (finite number of possible values) and obtain correspondingly a very special kind of self-adjoint operator—a self-adjoint operator having a finite discrete spectrum; this means that (i) T has a finite number of distinct eigenvalues, and (ii) the eigenvectors of T span the space, in the sense that any vector in $\mathcal{H}$ is a (finite) linear combination of eigenvectors. It will be part of the aim of the following section to establish the correspondence between observables and self-adjoint operators quite generally. This will require a treatment of unbounded self-adjoint operators, and self-adjoint operators having continuous spectrum.

How is the state of a quantum–mechanical system to be defined? A *pure state* may be defined as a maximal collection of propositions $\{b_i\}$ that are known to be "true" (for the system in that particular state). A proposition is said to be true if a measurement of the proposition will give "yes" with certainty. The collection of propositions is maximal in the sense that $\{b' \subset b_i$ for all $i\} \Rightarrow b' = \emptyset$. For if a non-trivial proposition b' existed, with $b' \subset b_i$ for all i, then a measurement of b' giving the answer "yes" would lead to more information about the system, contradicting the assumption that the state should embody the maximum available degree of knowledge of the system. (One could, of course, measure an arbitrary proposition c, but in general this measurement will interfere with propositions b_i previously known to be true.)

A pure state of the system could just as well be represented by the single proposition $b = \bigcap_i b_i$, which is "true" if and only if each b_i is individually true. This representative proposition b will satisfy $b' \subset b \Rightarrow b' = \emptyset$. Translating this property into a statement about the subspace $\mathcal{M}_b$ to which b corresponds, we see that $\mathcal{M}_b$ must contain no proper subspace except the trivial subspace consisting of the zero vector. Such a subspace $\mathcal{M}_b$ can only be one-dimensional, since, for example, a two-dimensional subspace would contain within itself an infinite number of one-dimensional subspaces. We find, then, that pure states correspond to one-dimensional subspaces of the Hilbert space $\mathcal{H}$. It is usual to select a single normalized vector f that spans this subspace, and to represent the pure state by a single vector f. ($e^{i\alpha} f, \alpha \in \mathbb{R}$, would do just as well.) We then talk, loosely, of the system being "in the (pure) state f", meaning that the state of the system is represented by the proposition b that is in correspondence with the one-dimensional subspace spanned by f. The projection operator E_f onto this subspace is given by

$$E_f h = \langle f, h \rangle f.$$

Two common notations for this projection operator are

$$E_f = \langle f, \cdot \rangle f \qquad \text{and} \qquad E_f = |f\rangle\langle f|.$$

For a system in state f, the fact that the state represents a complete description of the physical system implies that f determines a function $p(a)$ representing the probability, in state f, that a measurement of proposition a will give the answer "yes". Since propositions a are in one-to-one correspondence with subspaces $\mathcal{M}_a$, and hence with the projection operators E onto these subspaces, it is notationally more convenient to write $p(E)$ for this probability, where E projects onto the subspace $\mathcal{M}_a$ corresponding to a.

Clearly $0 \le p(E) \le 1$, $p(0) = 0$, $p(I) = 1$, where 0 stands for the zero operator and I for the identity operator. For a pair of mutually exclusive propositions (i.e. $\mathcal{M}_1 \subseteq \mathcal{M}_2^{\perp}$ for the corresponding subspaces, and $E_1 E_2 = E_2 E_1 = 0$ for the corresponding projection operators), we also require the additive law for probabilities:

$$p(E_1+E_2)=p(E_1)+p(E_2). \tag{2.5.15}$$

(It should be noted that, for projections onto mutually orthogonal subspaces $\mathcal{M}_1$ and $\mathcal{M}_2$, the projection onto $\mathcal{M}_1+\mathcal{M}_2$ is just E_1+E_2.) It turns out, and is quite a deep result, that the only assignment of the probability function $p(E)$ compatible with these properties, and satisfying the further requirement $p(E_f)=1$ ("yes" for b with certainty) is

$$p(E)=\langle f, Ef\rangle. \tag{2.5.16}$$

Equation (2.5.16) gives, then, a formula, in the state f, for the probability of "yes" for an arbitrary proposition of the system. The expectation value $\langle T\rangle$ of the observable corresponding to the self-adjoint operator $T=\Sigma_{i=1}^{n}\lambda_i E_i$ then becomes

$$\langle T\rangle=\sum_{i=1}^{n}\lambda_i p(E_i)=\sum_{i=1}^{n}\lambda_i\langle f, E_i f\rangle=\langle f, Tf\rangle, \tag{2.5.17}$$

which is the familiar expression. The identity $\langle T\rangle=\langle f, Tf\rangle$ extends to arbitrary self-adjoint operators, which may have continuous spectrum—though in that case sums must be replaced by integrals (see Section 2.6).

The correspondence between states and one-dimensional subspaces, and between observables and self-adjoint operators, are part of the justification for the study of Hilbert space in the context of quantum mechanics. Many of the most important theoretical ideas within Hilbert space find their physical interpretation in quantum mechanics, and scattering theory is especially fruitful in this respect.

The self-adjoint operator H corresponding to the total energy of a quantum system is called the (total) Hamiltonian. The operator H may also be defined as the generator of time evolution. Thus the family $\{U_t\}$ of unitary operators given by $U_t=e^{-iHt/h}$ relates the state $f_t=e^{-iHt/h}f$ at time t to the state f at time $t=0$. In Section 2.6 we shall see how to define such functions of a self-adjoint operator; normally we shall adopt units such that $\hbar=1$. The family $\{U_t\}$ satisfies $U_0=1$ and $U_s U_t=U_{s+t}$. It is a theorem of Stone that any such unitary family, depending continuously on a parameter t, defines a unique self-adjoint operator H, through the formula $U_t=e^{-iHt}$.

For a general self-adjoint operator T, the expectation value at time t is given by

$$\langle T\rangle_t=\langle e^{-iHt}f, Te^{-iHt}f\rangle=\langle f, T_1 f\rangle \tag{2.5.18}$$

where $T_t=e^{iHt}Te^{-iHt}$ gives the variation of an observable with time in the *Heisenberg representation*; in the Heisenberg representation, through (2.5.18), one may treat the state vector f as fixed, whilst in general observables will depend on time.

For quantum mechanics and scattering theory, the closest connections between physical theory and mathematical interpretation are to be found in operator analysis, to which we now turn.

2.6 OPERATOR ANALYSIS AND SPACES OF OPERATORS

Given a pair of linear operators T_1, T_2 on a Hilbert space $\mathscr{H}$, we shall write $T_1 \leq T_2$ whenever

$$\langle f, T_1 f\rangle \leq \langle f, T_2 f\rangle$$

for all $f \in \mathscr{H}$. For a pair of projection operators E_1, E_2, it is not difficult to verify that if E_1 projects onto $\mathscr{M}_1$ and E_2 onto $\mathscr{M}_2$ then $E_1 \leq E_2$ if and only if $\mathscr{M}_1 \subseteq \mathscr{M}_2$. A necessary and sufficient condition for $E_1 \leq E_2$ is that $E_1 E_2 = E_2 E_1 = E_1$.

A one-parameter family $\{E_\lambda\}$ of projection operators is called a *resolution of the identity* if

$$\left.\begin{array}{ll}\text{(i)} & \left.\begin{array}{l}\underset{\lambda\to-\infty}{\text{s-lim}}\, E_\lambda f = 0, \quad \underset{\lambda\to\infty}{\text{s-lim}}\, E_\lambda f = f, \\ \underset{\varepsilon\to 0+}{\text{s-lim}}\, E_{\lambda+\varepsilon} f = E_\lambda f\end{array}\right\}\text{for all } f \in \mathscr{H}, \\ \text{(ii)} & E_{\lambda_1} \leq E_{\lambda_2} \quad \text{for } \lambda_1 \leq \lambda_2 \quad (\lambda_1, \lambda_2 \in \mathbb{R}).\end{array}\right\} \tag{2.6.1}$$

Condition (ii) may be rewritten as

$$E_{\lambda_1} E_{\lambda_2} = E_{\lambda_2} E_{\lambda_1} = E_{\lambda_1} \quad (\lambda_1 \leq \lambda_2). \tag{2.6.1$'$}$$

Given any f, $g \in \mathscr{H}$, it is a matter of straightforward algebra to express $\langle g, E_\lambda f\rangle$ as a linear combination of the four inner products $\langle f+g,\ E_\lambda(f+g)\rangle$, $\langle f-g,\ E_\lambda(f-g)\rangle$, $\langle f+ig,\ E_\lambda(f+ig)\rangle$ and $\langle f-ig,\ E_\lambda(f-ig)\rangle$. Since each of these four inner products is a monotonic non-decreasing function of λ, continuous on the right, we may use these functions to generate associated Lebesgue–Stieltjes measures. This allows us to give a meaning, for any (Lebesgue) measurable set A, to $\int_A d\langle g, E_\lambda f\rangle$. Since, for example

$$\begin{aligned}\int_A d\langle f+g, E_\lambda(f+g)\rangle &\leq \int_{-\infty}^{\infty} d\langle f+g, E_\lambda(f+g)\rangle \\ &= \langle f+g, (E_\infty - E_{-\infty})(f+g)\rangle \\ &= \|f+g\|^2 \leq (\|f\| + \|g\|)^2,\end{aligned}$$

there will be a constant C (depending on f) such that if $\|g\|=1$ then $|\int_A d\langle g, E_\lambda f\rangle| \leq C$. Hence, quite generally,

$$\left|\int_A d\langle g, E_\lambda f\rangle\right| \leq C\|g\|,$$

so that the complex conjugate of $\int_A \mathrm{d}\langle g, E_\lambda f\rangle$ defines a bounded linear functional on elements g of $\mathscr{H}$. By the Riesz Representation Theorem, there then exists a vector, which we shall denote by $E_A f$, such that, for all $g \in \mathscr{H}$,

$$\langle g, E_A f\rangle = \int_A \mathrm{d}\langle g, E_\lambda f\rangle. \tag{2.6.2}$$

We shall often write

$$E_A f = \int_A \mathrm{d}E_\lambda f. \tag{2.6.3}$$

From (2.6.2), since E_λ is symmetric, so too is E_A. E_A commutes with each projection E_λ, since, for example,

$$\langle g, E_A E_\mu f\rangle = \int_A \mathrm{d}_\lambda \langle g, E_\lambda E_\mu f\rangle = \int_A \mathrm{d}_\lambda \langle g, E_\mu E_\lambda f\rangle$$

$$= \int_A \mathrm{d}_\lambda \langle E_\mu g, E_\lambda f\rangle = \langle E_\mu g, E_A f\rangle = \langle g, E_\mu E_A f\rangle.$$

We can also write, with $\chi_A \equiv$ characteristic function of the set A,

$$\langle g, E_A E_\mu f\rangle = \int_{\mathbb{R}} \chi_A(\lambda)\, \mathrm{d}_\lambda \langle g, E_\mu E_\lambda f\rangle = \int_{-\infty}^{\mu} \chi_A(\lambda)\, \mathrm{d}\langle g, E_\lambda f\rangle,$$

on using (2.6.1)′. Hence

$$\langle g, E_A^2 f\rangle = \int_{\mathbb{R}} \chi_A(\mu)\, \mathrm{d}\langle g, E_A E_\mu f\rangle$$

$$= \int_{\mathbb{R}} \chi_A(\mu)\, \mathrm{d}_\mu \int_{-\infty}^{\mu} \chi_A(\lambda)\, \mathrm{d}\langle g, E_\lambda f\rangle$$

$$= \int_{\mathbb{R}} (\chi_A(\mu))^2\, \mathrm{d}\langle g, E_\mu f\rangle$$

$$= \int_{\mathbb{R}} \chi_A(\mu)\, \mathrm{d}\langle g, E_\mu f\rangle = \langle g, E_A f\rangle,$$

so that

$$E_A^2 = E_A.$$

Since E_A is also symmetric, it follows that each E_A is a projection operator. Let E_A project onto the subspace $\mathcal{M}_A$.

Any resolution of the identity, then, defines a mapping from measurable subsets of the real line to projection operators, or, equivalently, to the associated subspaces. This corresponds exactly to the mathematical representation of an arbitrary observable in quantum mechanics. For any given observable, we may associate to each measurable set A a corresponding proposition "does the value of the observable lie within the set A?" And to this proposition there will be a corresponding subspace of the underlying Hilbert space $\mathcal{H}$. So both observables and resolutions of the identity define mappings $A \to E_A$. Moreover, the properties of these mappings, derivable from (2.6.1) (i) and (ii), are just those properties that we should demand for observables on the basis of the postulates of quantum mechanics. Note that the original family $\{E_\lambda\}$ of projection operators may be identified with $\{E_{(-\infty,\lambda]}\}$, so that the subspace $\mathcal{M}_\lambda$ onto which E_λ projects corresponds to the question "does the value of the observable lie within the interval $(-\infty, \lambda]$?"

Given a resolution of the identity $\{E_\lambda\}$, we may define a self-adjoint operator $T = \int_{-\infty}^{\infty} \lambda \, dE_\lambda$. This requires a little more care than the construction of the projections E_A, since T will generally be unbounded and cannot in that case be defined on the whole space $\mathcal{H}$. The self-adjoint operator T is defined by

$$\langle g, Tf \rangle = \int_{-\infty}^{\infty} \lambda \, d\langle g, E_\lambda f \rangle. \tag{2.6.4}$$

with domain $D(T)$ given by $f \in D(T)$ if and only if

$$\|Tf\|^2 \equiv \int_{-\infty}^{\infty} \lambda^2 \, d\langle f, E_\lambda f \rangle < \infty. \tag{2.6.5}$$

That is, f will belong to $D(T)$ whenever the integral $\int_{-\infty}^{\infty} \lambda^2 \, d\langle f, E_\lambda f \rangle$ is convergent. In that case, the right-hand side of (2.6.4), complex-conjugated, defines a bounded linear functional on elements g of $\mathcal{H}$, and Tf is defined by (2.6.4), through the Riesz Representation Theorem.

The *spectral theorem* for self-adjoint operators, conversely, asserts that to every self-adjoint operator T may be associated, uniquely, a resolution of the identity $\{E_\lambda\}$ such that $T = \int_{-\infty}^{\infty} \lambda \, dE_\lambda$. The correspondence between observables and resolutions of the identity implies that there is a one-to-one correspondence, in quantum mechanics, between observables and self-adjoint operators.

Functions of self-adjoint operators may be defined by an obvious extension of these ideas. If F is any measurable function, we define $F(T)=\int F(\lambda)\,\mathrm{d}E_\lambda$. That is,

$$\langle g, F(T)f\rangle = \int_{-\infty}^{\infty} F(\lambda)\,\mathrm{d}\langle g, E_\lambda f\rangle, \tag{2.6.6}$$

where $f\in D(F(T))$ if and only if

$$\|F(T)f\|^2 \equiv \int_{-\infty}^{\infty} |F(\lambda)|^2\,\mathrm{d}\langle f, E_\lambda f\rangle < \infty. \tag{2.6.7}$$

In particular, if $|F|$ is bounded then the integral on the right-hand side will always converge, so that *bounded* measurable functions of T are defined as bounded linear operators on the entire space $\mathscr{H}$. Note that $\|F(T)\|\leq\sup_\lambda|F(\lambda)|$. Particular cases of functions of T are

(i) the zero operator and the identity operator, corresponding respectively to $F(\lambda)\equiv 0$ and $F(\lambda)\equiv 1$,
(ii) the self-adjoint operator T itself, $F(\lambda)=\lambda$,
(iii) the projection operator E_A, for which $F(\lambda)=\chi_A(\lambda)$, and
(iv) the projection operator E_μ, for which $F(\lambda)=\chi_{(-\infty,\mu]}(\lambda)$.

Whenever we wish to indicate explicitly the dependence of E_λ and E_A on the self-adjoint operator T, we shall write $E_\lambda(T)$ and $E_A(T)$ respectively.

Note the identity (subject to qualifications regarding domains, if $F_1(\lambda)$ or $F_2(\lambda)$ are unbounded)

$$(F_1F_2)(T)=F_1(T)F_2(T)=F_2(T)F_1(T). \tag{2.6.8}$$

If $F(\lambda)\neq 0$ then we have

$$\left(\frac{1}{F}\right)(T)=[F(T)]^{-1}. \tag{2.6.9}$$

More generally, if there is a set B of zeros of the function $F(\lambda)$, but if $E_B(T)=0$, then (2.6.9) remains valid. A particular case of this result is that $T^{-1}=\int\lambda^{-1}\,\mathrm{d}E_\lambda$ whenever T does not have zero as an eigenvalue.

The eigenvalues of T are values of λ for which $E_{\{\lambda\}}(T)\neq 0$, where $\{\lambda\}$

denotes the set consisting of the single number λ. Since

$$E_{\{\lambda\}}(T) = \text{s-}\lim_{\varepsilon \to 0+} (E_\lambda(T) - E_{\lambda-\varepsilon}(T)),$$

the eigenvalues are the values of λ at which E_λ is (left-)discontinuous. The eigenvalues are countable in number, and $E_{\{\lambda\}}$ is the projection onto the eigenspace having eigenvalue λ. For a self-adjoint operator having finite discrete spectrum, E_λ is a constant projection operator (i.e. independent of λ) between consecutive eigenvalues, and $E_{\{\lambda_i\}}$ is the discontinuity in E_λ at the ith eigenvalue λ_i. The only contributions to the integral $T = \int \lambda \, \mathrm{d}E_\lambda$ then come from the eigenvalues, and (2.5.13) may be recovered in this case if we set (slightly at variance with the present notation) $E_i = E_{\{\lambda_i\}}$.

How is the spectrum of an arbitrary self-adjoint operator to be defined? We define, for complex z, the *resolvent* operator $R(z) = (T - z)^{-1}$. For $\operatorname{Im}(z) \neq 0$, $R(z)$ is a bounded linear operator defined on the whole space. One way to see this is to use the spectral theorem. Thus

$$\|R(z)\| = \left\| \int (\lambda - z)^{-1} \, \mathrm{d}E_\lambda \right\| \leq \sup |(\lambda - z)^{-1}| = |\operatorname{Im}(z)|^{-1}.$$

Alternatively, we observe first of all that $(T-z)^{-1}$ exists as a bona fide operator, since $(T-z)g = 0 \Rightarrow g = 0$ (see any elementary quantum-mechanics text; $\operatorname{Im}\langle g, (T-z)g\rangle = -\operatorname{Im}(z)\|g\|^2 = 0$). Now, for $f \in D(T)$, and assuming $\operatorname{Re}(z) \geq 0$ (if $\operatorname{Re}(z) < 0$, we replace T by $-T$), we have

$$\begin{aligned}
\|(T-z)f\|^2 &= \langle (T-z)f, (T-z)f\rangle \\
&= \|Tf\|^2 - (z + \bar{z}) \langle f, Tf\rangle + |z|^2 \|f\|^2 \\
&\geq \|Tf\|^2 - 2\operatorname{Re}(z)\|Tf\| \, \|f\| + |z|^2 \|f\|^2 \qquad \text{(Schwarz inequality)} \\
&= (\|Tf\| - \operatorname{Re}(z)\|f\|)^2 + (\operatorname{Im}(z))^2 \|f\|^2 \\
&\geq (\operatorname{Im}(z))^2 \|f\|^2.
\end{aligned}$$

Hence $\|(T-z)f\| \geq |\operatorname{Im}(z)| \, \|f\|$, so that with $h = (T-z)f$ we have

$$\|(T-z)^{-1}\| = \sup_{h \neq 0} \frac{\|(T-z)^{-1}h\|}{\|h\|} = \sup_{f \neq 0} \frac{\|f\|}{\|(T-z)f\|} \leq \frac{1}{|\operatorname{Im}(z)|}.$$

Therefore $(T-z)^{-1}$ is bounded. Moreover, since $T-z$ is closed, so is $(T-z)^{-1}$; it then follows as an elementary exercise that the domain of $(T-z)^{-1}$, or equivalently the range of $T-z$, is a subspace of $\mathscr{H}$. If $\langle h, (T-z)g\rangle = 0$ for all $g \in D(T)$ then, by the definition of adjoint, we have $T^*h = Th = \bar{z}h$, which implies $h = 0$. Since the orthogonal complement of the range of $T-z$ consists of the zero vector only, we may conclude that this range is the whole of $\mathscr{H}$. Hence, indeed, for $\operatorname{Im}(z) \neq 0$, $R(z) = (T-z)^{-1}$ is bounded and defined on the whole space.

The *resolvent set* of T is defined to be the set consisting of all $z \in \mathbb{C}$ for which $R(z)$ exists as a bounded linear operator defined on the whole space, and the *spectrum* of T, denoted by $\sigma(T)$, is the complement (in $\mathbb{C}$) of the resolvent set. Thus, for a self-adjoint operator T, $\sigma(T)$ is a subset of $\mathbb{R}$. Note also that

(i) $\sigma(T)$ is a closed set,
(ii) the eigenvalues of T belong to $\sigma(T)$,
(iii) if T is bounded then $\|T\| = \sup_{\lambda \in \sigma(T)} |\lambda|$.

In order to characterize a given real number λ as belonging to $\sigma(T)$, it is sufficient to find a sequence $\{f_n\}$, $f_n \in D(T)$, $\|f_n\| = 1$, such that $\lim_{n\to\infty} \|(T-\lambda)f_n\| = 0$. The sequence $\{f_n\}$ may be regarded as a sequence of approximate eigenvectors of T. It is not difficult to verify that, in the limit as $n \to \infty$, the *expectation value* $\langle T \rangle = \langle f_n, Tf_n \rangle$ of T converges to λ, and the *uncertainty* $\delta T = [\langle (T - \langle T \rangle)^2 \rangle]^{1/2}$ of T converges to zero.

(An alternative, and perhaps more transparent, way of defining $\sigma(T)$ is as the smallest closed set A for which $E_A = I$. This means that $\sigma(T)$ is the smallest closed set for which the measurement of the corresponding observable will give, with certainty, a value within the given set.)

Note the connection between the resolvent operator and the the resolvent transform considered in Sections 2.3 and 2.4. Thus if $\rho(\lambda) = \langle f, E_\lambda f \rangle$ then, with $z = x + iy$, the resolvent transform $v(z)$ of the measure generated by ρ is just

$$v(z) = \mathrm{Im}\left\{ \int (\lambda - z)^{-1} \, \mathrm{d} \langle f, E_\lambda f \rangle \right\}$$
$$= \mathrm{Im}\, \{ \langle f, (T-z)^{-1} f \rangle \}.$$

We may also write this as

$$v(z) = (2\mathrm{i})^{-1} \langle f, \{(T-z)^{-1} - (T-\bar{z})^{-1}\} f \rangle.$$

Using (2.4.10) to reconstruct a measure from its transform, we can write, if a and b are not eigenvalues,

$$\langle f, E_{(a,b]} f \rangle = \lim_{\varepsilon \to 0+} \frac{1}{2\pi \mathrm{i}} \int_a^b \langle f, (T-(x+\mathrm{i}y))^{-1} - (T-(x-\mathrm{i}y))^{-1} f \rangle \, \mathrm{d}x,$$

More generally, we have

$$\langle g, E_{(a,b]} f \rangle = \lim_{\varepsilon \to 0+} \frac{1}{2\pi \mathrm{i}} \int_a^b \langle g, (T-(x+\mathrm{i}y))^{-1} - (T-(x-\mathrm{i}y))^{-1} f \rangle \, \mathrm{d}x.$$

Thus

$$E_{(a,b]} = \lim_{\varepsilon \to 0+} \frac{1}{2\pi i} \int_a^b \{(T-(x+iy))^{-1} - (T-(x-iy))^{-1}\}\, dx \qquad (2.6.10)$$

where the integral and limit on the right-hand side are to be understood in the weak sense. This equation allows, in principle, the reconstruction of the resolution of the identity from the corresponding self-adjoint operator, just as the formula $T = \int \lambda\, dE_\lambda$ expresses T in terms of the spectral family $\{E_\lambda\}$. Through the connection between resolvent and resolvent transform, we shall be able to relate properties of the *spectral measures* $d\langle f, E_\lambda f\rangle$ to the behaviour of the resolvent for values of z near to the real axis.

An important point to appreciate in dealing with the analysis of self-adjoint operators is that the decomposition of measures leads to a decomposition of the underlying Hilbert space $\mathscr{H}$. Consider, for example, the decomposition of a measure into its discrete and singular parts.

Given a vector f in $\mathscr{H}$, we shall say that f lies in the (spectrally) *continuous subspace* $\mathscr{M}_c(T)$ of the self-adjoint operator T if $\langle f, E_\lambda(T) f\rangle$ is a continuous function of λ. Since, for any λ, the discontinuity at λ of $\langle f, E_\lambda f\rangle$ is $\lim_{\varepsilon \to 0+} \langle f, (E_\lambda - E_{\lambda-\varepsilon}) f\rangle = \langle f, E_{[\lambda]} f\rangle = \|E_{[\lambda]} f\|^2$, we see that $f \in \mathscr{M}_c(T) \Leftrightarrow E_{[\lambda]} f = 0$ for all λ. (This also guarantees that $\mathscr{M}_c(T)$ is a subspace, since $\mathscr{M}_c(T)$ is clearly closed under addition and under multiplication by numbers, and it is easy to verify that $\mathscr{M}_c(T)$ is closed in the Hilbert-space norm.) Now $E_{\{\lambda\}} = 0$ except if λ is an eigenvalue of T, in which case $E_{\{\lambda\}}$ is the projection onto the corresponding eigenspace $\mathscr{M}_{\{\lambda\}}$. If λ *is* an eigenvalue then $E_{\{\lambda\}} f = 0$ is just the condition that f be orthogonal to $\mathscr{M}_{\{\lambda\}}$. And if $E_{\{\lambda\}} f = 0$ for *all* λ then f will be orthogonal to every eigenvector of T. Let us define $\mathscr{M}_d(T)$, the *discrete subspace* of T, consisting of eigenvectors of T, their linear combinations, and strong limits of their linear combinations. Then $\mathscr{M}_d(T)$ will be the smallest subspace of $\mathscr{H}$ containing all the eigenvectors of T. (Note that $h \in \mathscr{M}_d(T)$ does *not* imply that h is an eigenvector, since, in general, linear combinations of eigenvectors are not eigenvectors.) If T has no eigenvectors then $\mathscr{M}_d(T)$ contains only the zero vector.

We have shown, then, that

$$\mathscr{M}_c(T) = \mathscr{M}_d^\perp(T), \qquad (2.6.11)$$

and the entire space decomposes as the direct sum

$$\mathscr{H} = \mathscr{M}_c(T) \oplus \mathscr{M}_d(T).$$

Let $E_c(T)$ and $E_d(T)$ denote the projection operators onto the subspaces

$\mathscr{M}_c(T)$ and $\mathscr{M}_d(T)$ respectively. Then, given any $f \in \mathscr{H}$, we can write

$$\langle f, E_\lambda f\rangle = \langle f, E_\lambda E_c f\rangle + \langle f, E_\lambda E_d f\rangle,$$

where, on the right-hand side the measure generated by the first term is purely a continuous measure, and the measure generated by the second term is purely discrete. Thus the two projection operators accomplish, for any vector f, the decomposition of the measure $d\langle f, E_\lambda f\rangle$ into its continuous and discrete parts.

The subspaces $\mathscr{M}_c(T)$ and $\mathscr{M}_d(T)$ each *reduce* T. A subspace $\mathscr{M}$ is said to *reduce* a linear operator T if $f \in D(T)$ implies $Ef \in D(T)$, $E^\perp f \in D(T)$, $TEf \in \mathscr{M}$ and $TE^\perp f \in \mathscr{M}^\perp$, where E is the projection operator onto $\mathscr{M}$ and $E^\perp = 1 - E$ is the projection operator onto $\mathscr{M}^\perp$. In this case, we can write

$$T = TE + TE^\perp,$$

where each of the operators TE and $TE^\perp$ are self-adjoint, having ranges contained respectively in $\mathscr{M}$ and $\mathscr{M}^\perp$. In other words, T may be expressed as the sum of a self-adjoint operator acting in $\mathscr{M}$ and a self-adjoint operator in $\mathscr{M}^\perp$. Denoting these operators in $\mathscr{M}$ and $\mathscr{M}^\perp$ by T_1 and T_2 respectively, we may write $T = T_1 \oplus T_2$, and describe T as the *direct sum* of T_1 and T_2. Note that the range of any projection $E_A(T)$, where A is a measurable subset of $\mathbb{R}$, is a reducing subspace for T.

The subspace of continuity $\mathscr{M}_c(T)$ for a self-adjoint operator T may be decomposed still further, in a way that we shall see to be of great importance for scattering theory. Given a vector $f \in \mathscr{H}$, we shall say that f lies in the *absolutely continuous subspace* $\mathscr{M}_{ac}(T)$ if $E_A f = 0$ for each set A having Lebesgue measure zero. Since any single point $\{\lambda\}$ has Lebesgue measure zero, it follows that, for $f \in \mathscr{M}_{ac}(T)$, $E_{\{\lambda\}} f = 0$. Thus

$$\mathscr{M}_{ac}(T) \subseteq \mathscr{M}_c(T).$$

We shall say that a vector $h \in \mathscr{H}$ belongs to the *singular* subspace $\mathscr{M}_s(T)$ if there exists a set B (*a priori* depending on h), having Lebesgue measure zero, such that $(1 - E_B)h = 0$. It is easy to see that $\mathscr{M}_{ac}(T)$ is a subspace. To verify that $\mathscr{M}_s(T)$ is a subspace, the only non-trivial step is to prove that $\mathscr{M}_s(T)$ is closed. But this follows from the fact that, if $\{B_n\}$ is a sequence of measurable sets such that $\ell(B_n) = 0$, and $E_{B_n} h_n = h_n$, then

$$\operatorname*{s-lim}_{n\to\infty} h_n = h \quad \text{implies} \quad E_B h = \lim_{n\to\infty} E_{\cup_n B_n} h_n = \lim_{n\to\infty} h_n = h,$$

where $B = \cup_n B_n$ and $\ell(B) = 0$.

Moreover, the subspaces $\mathscr{M}_{ac}$ and $\mathscr{M}_s$ are mutually orthogonal, since if $f \in \mathscr{M}_{ac}$, $h \in \mathscr{M}_s$, and B is a set such that $\ell(B) = 0$ and $E_B h = h$, we have

$$\langle f, h\rangle = \langle f, E_B h\rangle = \langle E_B f, h\rangle = 0.$$

Now we let h be an arbitrary vector in $\mathscr{H}$. Then the measure μ generated by $\langle h, E_\lambda(T)h\rangle h$ may be split into its absolutely continuous and singular parts μ_{ac} and μ_s, and a set B may be found having zero Lebesgue measure such that $\mu_s(B^c)\equiv\mu_s(\mathbb{R}\setminus B)=0$. Clearly, $E_B h\in\mathscr{M}_s(T)$.

The restriction of the measure μ to B^c is just μ_{ac}, and is absolutely continuous, so that if A is an arbitrary set having Lebesgue measure zero then we have

$$\|E_A E_{B^c}h\|^2=\|E_{A\cap B^c}h\|^2=\mu(A\cap B^c)=0.$$

From the definition of $\mathscr{M}_{ac}$, it follows that

$$E_{B^c}h=(1-E_B)h\in\mathscr{M}_{ac}(T).$$

We have, then, decomposed a general element h of $\mathscr{H}$ into the sum of a vector $E_B h$ in the singular subspace and a vector $(1-E_B)h$ in the absolutely continuous subspace.

We are therefore, entitled to write

$$\mathscr{M}_{ac}(T)=\mathscr{M}_s^{\perp}(T) \tag{2.6.12}$$

and

$$\mathscr{H}=\mathscr{M}_{ac}(T)\oplus\mathscr{M}_s(T).$$

It also follows that a vector h will belong to $\mathscr{M}_{ac}(T)$ if and only if the measure generated by $\langle h, E_\lambda(T)h\rangle$ is absolutely continuous (with respect to Lebesgue measure), and to $\mathscr{M}_s(T)$ if and only if this measure is singular. The splitting of h into its absolutely continuous and singular components corresponds exactly to the splitting of the measure $d\langle h, E_\lambda(T)h\rangle$ into its absolutely continuous and singular components.

The set B such that $\ell(B)=0$, introduced in the definition of the singular subspace, may be made independent of the vector h. We let $\{h_i\}$ be an orthonormal basis of $\mathscr{M}_s(T)$, and let $\{B_i\}$ be a corresponding sequence of sets such that $\ell(B_i)=0$ and $E_{B_i}(T)h_i=h_i$. Then, with $B=\bigcup_i B_i$, we have $\ell(B)=0$ and $E_B(T)h_i=h_i$, so that $E_B(T)h=h$ for *every* vector in $\mathscr{M}_s(T)$. Since $E_B(T)f=0$ for all $f\in\mathscr{M}_{ac}(T)$, it follows that $E_B(T)$ is precisely the projection onto $\mathscr{M}_s(T)$. Thus the projection onto $\mathscr{M}_s(T)$ may be identified with the spectral projection of T associated with some set B having Lebesgue measure zero, and the projection onto $\mathscr{M}_{ac}(T)$ is the same as the spectral projection of T for the complement of B. It follows at once (and may be proved independently) that both $\mathscr{M}_{ac}(T)$ and $\mathscr{M}_s(T)$ are reducing subspaces for T.

The orthogonal subspace to $\mathscr{M}_{ac}(T)$ within $\mathscr{M}_c(T)$ is called the *subspace of singular continuity*, and is denoted by $\mathscr{M}_{sc}(T)$. Then $g\in\mathscr{M}_{sc}(T)$ if and only if the measure $d<g, E_\lambda(T)g\rangle$ is singular continuous. $\mathscr{M}_{sc}(T)$ is contained in

$\mathcal{M}_s(T)$, and $\mathcal{M}_s = \mathcal{M}_{sc} \oplus \mathcal{M}_d$. The complete decomposition of the Hilbert space, for a given self-adjoint operator T, is

$$\mathcal{H} = \mathcal{M}_d(T) \oplus \mathcal{M}_{ac}(T) \oplus \mathcal{M}_{sc}(T). \tag{2.6.13}$$

In the context of potential scattering, where T is the total Hamiltonian H, we shall find a physical interpretation to each of these three subspaces. One characterization of $\mathcal{M}_{ac}$ in that context, which relates to the evolution operator e^{-iHt}, is as follows:

$$f \in \mathcal{M}_{ac}(H) \Leftrightarrow \text{there is a dense set of vectors}$$

$$\phi \in \mathcal{H} \text{ such that } \int_{-\infty}^{\infty} |\langle \phi, e^{-iHt} f \rangle|^2 \, dt < \infty.$$

First we note that, with $H = \int \lambda \, dE_\lambda$, $\langle \phi, e^{-iHt} f \rangle = \int_{-\infty}^{\infty} e^{-i\lambda t} \, d\langle \phi, E_\lambda f \rangle$. By the corollary to Lemma 2.7, if $\langle \phi, e^{-iHt} f \rangle \in L^2(\mathbb{R})$ then it follows that the charge $d\langle \phi, E_\lambda f \rangle$ is absolutely continuous. (The original result applied to a measure rather than a charge, but may easily be extended to cover charges as well.) Hence if a set A has Lebesgue measure zero then we have $\langle \phi, E_A f \rangle = 0$. Since the ϕ are dense in $\mathcal{H}$, this implies $E_A f = 0$, so that $f \in \mathcal{M}_{ac}(H)$.

Conversely, let us suppose that $f \in \mathcal{M}_{ac}(H)$, and ψ is an arbitrary vector in $\mathcal{M}_{ac}(H)$. We denote by B_n the subset of $\mathbb{R}$ consisting of those λ for which $(d/d\lambda) \langle \psi, E_\lambda \psi \rangle \leq n$. We let $\psi_n = E_{B_n} \psi$, where E_{B_n} is the spectral projection of H associated with the set B_n. Then, using the Lebesgue Dominated-Convergence theorem, we may show that $\psi_n \to \psi$ strongly as $n \to \infty$. Moreover, $(d/d\lambda) \langle \psi_n, E_\lambda \psi_n \rangle \leq n$, so that, writing derivatives as limits,

$$\left| \frac{d}{d\lambda} \langle \psi_n, E_\lambda f \rangle \right|^2 \leq \frac{d}{d\lambda} \langle \psi_n, E_\lambda \psi_n \rangle \frac{d}{d\lambda} \langle f, E_\lambda f \rangle$$

$$\leq n \frac{d}{d\lambda} \langle f, E_\lambda f \rangle,$$

and

$$\int_{-\infty}^{\infty} \left| \frac{d}{d\lambda} \langle \psi_n, E_\lambda f \rangle \right|^2 d\lambda \leq n \int_{-\infty}^{\infty} d \langle f, E_\lambda f \rangle = n \|f\|^2.$$

By standard properties of the Fourier transform, this gives

$$\|\langle \psi_n, e^{-iHt} f \rangle\|_2^2 = \left\| \int_{-\infty}^{\infty} e^{-i\lambda t} \frac{d}{d\lambda} \langle \psi_n, E_\lambda f \rangle \, d\lambda \right\|_2^2 \leq 2\pi n \|f\|^2,$$

where the ψ_n are dense in $\mathcal{M}_{ac}(H)$. On the other hand, for any vector $\phi \in \mathcal{M}_{ac}^{\perp}(H)$, we have $\langle \phi, e^{-iHt} f \rangle = 0$, so that trivially $\int_{-\infty}^{\infty} |\langle \phi, e^{-iHt} f \rangle|^2 \, dt < \infty$ in this case also. This proves the converse, namely that $f \in \mathcal{M}_{ac}(H) \Rightarrow \langle \phi, e^{-iHt} f \rangle \in L^2(\mathbb{R})$ for ϕ dense in $\mathcal{H}$. An alternative characterization of $\mathcal{M}_{ac}(H)$, which may be verified by the same methods, is that $\mathcal{M}_{ac}(H)$ is the *closure* of the set of vectors f for which $\langle f, e^{-iHt} f \rangle \in L^2(\mathbb{R})$ as a function of t. In Chapter 13 we shall see how these results may be used to describe the time evolution of states.

We have referred a number of times to the Fourier transform of a measure. The standard context, of course, in which the student of quantum mechanics meets the Fourier transform is that of the *linear operator* that effects the transform from position space to momentum space. For $f \in L^2(\mathbb{R})$, we define the Fourier transform $\hat{f}(k)$ of f by

$$\hat{f}(k) = \frac{1}{(2\pi)^{1/2}} \int_{-\infty}^{\infty} e^{-ikx} f(x) \, dx, \tag{2.6.14}$$

and the inverse transform is given by

$$f(x) = \frac{1}{(2\pi)^{1/2}} \int_{-\infty}^{\infty} e^{ikx} \hat{f}(k) \, dk. \tag{2.6.14$'$}$$

In (2.6.14) the integral on the right-hand side should be interpreted as a strong limit, in the Hilbert space $L^2(\mathbb{R})$, of $\int_{-n}^{n} e^{-ikx} f(x) \, dx$ as $n \to \infty$. This limit exists for all $f \in L^2(\mathbb{R})$, and defines $\hat{f}(k)$ as an element of $L^2(\mathbb{R})$. In $L^2(\mathbb{R}^3)$ the formulae corresponding to (2.6.14) and (2.6.14)′ are

$$\hat{f}(\boldsymbol{k}) = \left(\frac{1}{2\pi}\right)^{3/2} \int e^{-i\boldsymbol{k}\cdot\boldsymbol{r}} f(\boldsymbol{r}) \, d^3\boldsymbol{r}, \tag{2.6.15}$$

and

$$f(\boldsymbol{r}) = \left(\frac{1}{2\pi}\right)^{3/2} \int e^{i\boldsymbol{k}\cdot\boldsymbol{r}} \hat{f}(\boldsymbol{k}) \, d^3\boldsymbol{k}, \tag{2.6.15$'$}$$

where, in (2.6.15), the integral is evaluated as a strong limit of the integral over the region $|\boldsymbol{r}| < n$, with a similar interpretation in (2.6.15)′. We write down these formulae primarily to establish the notation, and assume the reader to be already familiar with elementary properties of the Fourier transform.

The Fourier transform is an important example of an *integral operator*. We shall encounter numerous examples of integral operators in the sequel, among which those that describe *compact operators* will be of especial interest.

A linear operator T, acting on a Hilbert space $\mathscr{H}$, is said to be *compact* if T sends any *weakly* convergent sequence of vectors into a *strongly* convergent sequence. In other words, T is compact provided that

$$f_n \to f \text{ weakly } \Rightarrow Tf_n \to Tf \text{ strongly.}$$

Clearly the zero operator is compact, but the identity operator is not (except in a finite-dimensional space), since $f_n \to f$ weakly need not imply $f_n \to f$ strongly. A less trivial example of a compact operator is any linear operator of finite rank, where T being of finite rank means that T has finite-dimensional range. For example, any linear operator of rank 1 is of the form $T = \langle \psi, \cdot \rangle \phi$, with $\phi, \psi \in \mathscr{H}$, meaning that $Tg = \langle \psi, g \rangle \phi$, for some fixed pair of vectors ϕ, ψ. An alternative notation is $T = |\phi\rangle\langle\psi|$. For such an operator, if $f_n \to f$ weakly then we have $Tf_n = \langle \psi, f_n \rangle \phi \to \langle \psi, f \rangle \phi$ strongly, since $\|Tf_n - Tf\| = \|\langle \psi, f_n - f \rangle \phi\| = |\langle \psi, f_n - f \rangle| \, \|\phi\| \to 0$. More generally, any linear operator of finite rank may be expressed in the form $T = \Sigma_{j=1}^{n} \langle \psi_j, \cdot \rangle \phi_j$, and such an operator is compact.

It is a result of considerable importance in the theory of compact operators that the space of compact operators is closed in operator norm, and that the set of linear operators of finite rank is dense in that space. Thus any compact operator on a Hilbert space may be approximated arbitrarily closely, in operator norm, by (compact) operators having finite rank. If T is compact then one can find a sequence $\{T_n\}$ of operators of finite rank such that $\lim_{n\to\infty} \|T - T_n\| = 0$. Moreover, if T is any linear operator such that $\|T - T_n\| \to 0$ for some sequence $\{T_n\}$ of compact operators then T is compact.

In an infinite-dimensional Hilbert space $\mathscr{H}$ a compact linear operator is close, then, to an operator that maps $\mathscr{H}$ into a subspace of finite dimension. For this reason, we may expect a compact operator to behave in some respects like an operator acting in a finite-dimensional space.

There are several alternative and equivalent definitions of compact operators. For example, T is compact if and only if T sends any *bounded* sequence $\{f_n\}$ of vectors into a sequence having a strongly convergent subsequence. In other words, T is compact provided that

$$\|f_n\| \leq \text{const for all } n \Rightarrow Tf_n \text{ has a subsequence converging strongly to a limit.}$$

The equivalence of this to our previous definition comes from the fact that any bounded sequence of vectors has a *weakly* convergent subsequence.

It is often useful to consider the special case of a sequence $\{f_n\}$ converging weakly to zero. In that case, for T compact we know that $Tf_n \to 0$ strongly. In particular, any orthonormal sequence $\{f_n\}$ converges weakly to zero ($\langle g, f_n \rangle \to 0$ since $\Sigma_{n=1}^{\infty} |\langle g, f_n \rangle|^2 \leq \|g\|^2 < \infty$).

Knowing that an operator is compact, then, gives us a way of constructing strongly convergent sequences. One can even prove convergence in operator norm under suitable conditions, as the following result shows.

Lemma 2.10

Let A be a compact operator, and $\{B_n\}$ a uniformly bounded sequence of operators such that $B_n^* \to B^*$ strongly. (This means $B_n^* f \to B^* f$ for all $f \in \mathscr{H}$.) Then $AB_n \to AB$ in operator norm.

Proof

We have to prove that $\|A(B_n - B)\|$ converges to zero. Since $\|B_n - B\| \leq$ const., and since A may be arbitrarily closely approximated, in operator norm, by an operator of finite rank, it will suffice to prove the result in the case for which A is of finite rank.

Suppose then that $A = \Sigma_j \langle \psi_j, \cdot \rangle \phi_j$ (finite sum). Then

$$A(B_n - B)f = \sum_j \langle \psi_j, (B_n - B)f \rangle \phi_j$$
$$= \sum_j \langle (B_n^* - B^*)\psi_j, f \rangle \phi_j.$$

Hence $\|A(B_n - B)\| \leq \Sigma_j \|(B_n^* - B^*)\psi_j\| \ \|\phi_j\| \to 0$ as $n \to \infty$, since $B_n^* \to B^*$ strongly. ■

It is important in this result to have $B_n^* \to B^*$ rather than $B_n \to B$, which does not suffice, in general, to obtain the conclusion.

Compact operators are also a useful tool in spectral analysis. Given a self-adjoint operator T, we define the *essential spectrum* of T, $\sigma_{\text{ess}}(T)$, by $\lambda \in \sigma_{\text{ess}}(T) \Leftrightarrow$ there exists an orthonormal sequence $f_1, f_2, f_3, \ldots$ such that $(T - \lambda)f_n \to 0$ strongly. The reader should verify that the essential spectrum is a subset of the spectrum of T. Indeed, $\sigma_{\text{ess}}(T)$ consists of all points of the spectrum, apart from isolated eigenvalues of finite multiplicity. An important consequence of our definition of compactness is that the essential spectrum is unaltered by the addition of a compact operator, since if A is compact and $\{f_n\}$ is orthonormal, we have s-lim $(T + A - \lambda)f_n =$ s-lim$(T - \lambda)f_n$. A special case of this result is that a self-adjoint compact operator A will have the same essential spectrum as the zero operator. Thus the spectrum of A will consist of isolated eigenvalues of finite multiplicity, except for $\lambda = 0$, which will be either an eigenvalue of infinite multiplicity, or a limit point of eigenvalues, or possibly both. The result extends easily to the resolvents of self-adjoint operators T_1 and T_2. Thus if $(T_1 - z)^{-1} - (T_2 - z)^{-1}$ is compact for some z then T_1 and T_2 will have the same essential spectrum.

We note, in our survey of the theory of compact operators, the following elementary properties:

(i) a compact operator is bounded;
(ii) if A is compact then so are cA and A^*, where c is an arbitrary complex number;
(iii) if A is compact and B is bounded then both AB and BA are compact;
(iv) if A^*A or AA^* is compact then so are A, $(A^*A)^{1/2}$ and $(AA^*)^{1/2}$;
(v) the sum of two compact operators is compact.

The spectrum of a compact operator A, whether self-adjoint or not, consists of isolated eigenvalues of finite multiplicity, except for $\lambda=0$, which may be an eigenvalue of infinite multiplicity or a limit point of eigenvalues. If, in addition, A is self-adjoint, the eigenvectors of A will span the entire Hilbert space, and all eigenvalues will be real.

The latter result allows us to write down a canonical representation of an arbitrary operator compact A as a norm limit of operators of finite rank. We let A be compact, and define the self-adjoint operator A^*A. We let β_j^2 $(j=1,2,3\ldots)$ be the non-zero eigenvalues of A^*A, corresponding to the orthonormal sequence $\{\phi_j\}$ of eigenvectors. We know that $\beta_j^2>0$, since $A^*A\geq 0$, and we take β_j to be positive, with eigenvalues in the sequence repeated according to multiplicity. For the operator A^*A, we have the representation

$$A^*A=\sum_{j=1}^{\infty}\beta_j^2\langle\phi_j,\cdot\rangle\phi_j, \tag{2.6.16}$$

as may be verified by allowing the left- and right-hand sides to operate on any eigenvector. It may happen that there are only finitely many non-zero eigenvalues, in which case we have a finite number of terms on the right-hand side. In general, the right-hand side of (2.6.16) may be interpreted as a norm limit of the sum from $j=1$ to n, as $n\to\infty$. In fact,

$$\left\|\sum_{j=n+1}^{\infty}\beta_j^2\langle\phi_j,\cdot\rangle\phi_j\right\|=\sup_{j=n+1}^{\infty}\beta_j^2\to 0$$

as $n\to\infty$, since zero is the only limit point of eigenvalues. (Note that

$$\left\|\sum_{j=n+1}^{\infty}\beta_j^2\langle\phi_j,f\rangle\phi_j\right\|^2\leq\sup_{j=n+1}^{\infty}\beta_j^4\left\|\sum_{j=n+1}^{\infty}\langle\phi_j,f\rangle\phi_j\right\|^2$$

$$=\sup_{j=n+1}^{\infty}\beta_j^4\sum_{j=n+1}^{\infty}|\langle\phi_j,f\rangle|^2\to 0\quad\text{as } n\to\infty,$$

since

$$\sum_{j=1}^{\infty}|\langle\phi_j,f\rangle|^2\leq\|f\|^2<\infty.$$

Now we define a sequence $\{\psi_j\}$, $j=1,2,\ldots$, by $\psi_j = A\phi_j/\beta_j$. Then $\langle \psi_i, \psi_j \rangle = (\beta_i\beta_j)^{-1} \langle \phi_i, A^*A\phi_j \rangle = (\beta_j/\beta_i) \langle \phi_i, \phi_j \rangle$, so that the sequence $\{\psi_j\}$ is orthonormal. It is now a simple matter to verify the identity

$$A = \sum_{j=1}^{\infty} \beta_j \langle \phi_j, \cdot \rangle \psi_j. \tag{2.6.17}$$

Equation (2.6.17) is a canonical representation of an arbitrary compact operator as a norm limit of operators of finite rank. (The sum of the right-hand side includes only a finite number of terms if A^*A has only finitely many non-zero eigenvalues.) The β's are the non-zero eigenvalues of $(A^*A)^{1/2}$, or, equivalently, of $(AA^*)^{1/2}$.

For $p \geq 1$, we define the so-called C_p class of compact operators by

$$A \in C_p \Leftrightarrow \sum_{j=1}^{\infty} (\beta_j)^p < \infty. \tag{2.6.18}$$

Each C_p class is a linear space of operators on which we define a norm by $\|A\|_p = (\Sigma_{j=1}^{\infty} (\beta_j)^p)^{1/p}$ (with sum ranging from 1 to N if there are N non-vanishing β's, taking due account of multiplicity). Note that the operator norm cannot exceed the C_p norm; thus $\|A\| \leq \|A\|_p$. Each C_p class is closed with respect to the C_p norm.

These spaces of operators have the following elementary properties;

(i) if $A \in C_p$ then $cA \in C_p$ and $A^* \in C_p$, where $\|cA\|_p = |c| \, \|A\|_p$, and $\|A^*\|_p = \|A\|_p$;
(ii) if $A \in C_p$ and B is bounded then $AB \in C_p$, $BA \in C_p$;
(iii) if $A \in C_p$ and $B \in C_q$ then $AB \in C_r$, where $p^{-1} + q^{-1} = r^{-1}$;
(iv) for $p \leq q$ we have $C_p \subset C_q$.

The cases $p=1$ and $p=2$ are of particular importance. Operators belonging to the class C_1 are said to be of trace class. If $A \in C_1$ and $\{f_n\}$ is an orthonormal basis of $\mathscr{H}$ then we define the *trace* of A by the formula

$$\operatorname{tr}(A) = \sum_{n=1}^{\infty} \langle f_n, Af_n \rangle.$$

Then $\operatorname{tr}(A)$ is independent of the choice of orthonormal basis, as may be verified from (2.6.17). If A is self-adjoint then $\operatorname{tr}(A)$ is just the sum of the eigenvalues of A, with due account taken of multiplicity.

Operators belonging to the class C_2 are said to be of Hilbert–Schmidt class. The Hilbert–Schmidt norm $\|A\|_2$ is often denoted by $\|A\|_{\mathrm{HS}}$. A linear operator A will belong to the Hilbert–Schmidt class if and only if

$$(\|A\|_{\mathrm{HS}})^2 \equiv \sum_{n=1}^{\infty} \|Af_n\|^2 < \infty,$$

where $\{f_n\}$ is an orthonormal basis. Note that if A and B are Hilbert–Schmidt then AB is of trace class, with $\|AB\|_1 \leq \|A\|_2 \, \|B\|_2$. Conversely, any linear operator of trace class is the product of two Hilbert–Schmidt operators. Chapter 11 will be devoted to the application to scattering theory of trace-class operators.

In L^2 spaces compact operators may often be represented as *integral operators*. Consider, for example, the integral operator T on $L^2(\mathbb{R})$ with kernel $K(x, y)$; thus

$$(Tf)(x) = \int_{-\infty}^{\infty} K(x, y) f(y) \, \mathrm{d}y \tag{2.6.19}$$

Let $\{f_n(x)\}$ be an orthonormal basis of $L^2(\mathbb{R})$. Then $(Tf_n)(x)$ may be thought of as an inner product, for fixed x, of $\tilde{K}(x, y)$ with $f_n(y)$. We write this as

$$(Tf_n)(x) = \langle \bar{K}(x, \cdot), f_n \rangle.$$

Hence

$$\sum_{n=1}^{\infty} \|Tf_n\|^2 = \sum_{n=1}^{\infty} \int_{-\infty}^{\infty} \mathrm{d}x |\langle \bar{K}(x, \cdot), f_n \rangle|^2.$$

Assuming that the summation/integration is absolutely convergent, we can write this as

$$\sum_{n=1}^{\infty} \|Tf_n\|^2 = \int_{-\infty}^{\infty} \mathrm{d}x \sum_{n=1}^{\infty} |\langle \bar{K}(x, \cdot), f_n \rangle|^2. \tag{2.6.20}$$

Since, for arbitrary $g \in L^2(\mathbb{R})$, $\|g\|^2 = \Sigma_{n=1}^{\infty} |\langle g, f_n \rangle|^2$, the right-hand side of (2.6.20) becomes $\int_{-\infty}^{\infty} \mathrm{d}x \int_{-\infty}^{\infty} \mathrm{d}y \; |K(x, y)|^2$, and we have shown, for an integral operator T defined by (2.6.19), that

$$(\|T\|_{\mathrm{HS}})^2 = \int_{-\infty}^{\infty} \int_{-\infty}^{\infty} \mathrm{d}x \, \mathrm{d}y \, |K(x, y)|^2. \tag{2.6.21}$$

Hence the integral operator is Hilbert–Schmidt, if and only if the double integral on the right-hand side of (2.6.21) is convergent. A similar result applies to an integral operator in $L^2(\mathbb{R}^3)$ with kernel $K(\mathbf{r}, \mathbf{r}')$, for which the result corresponding to (2.6.21) is

$$(\|T\|_{\mathrm{HS}})^2 = \iint \mathrm{d}^3\mathbf{r} \, \mathrm{d}^3\mathbf{r}' \, |K(\mathbf{r}, \mathbf{r}')|^2. \tag{2.6.22}$$

A useful application of these results, and one that forms the basis of the standard Sturm–Liouville theory of differential operators, is the proof, for a class of regular differential operators, that the resolvent is an integral operator. One may then use (2.6.21) to show, under suitable conditions, that the resolvent operator is compact, and use this to carry out a spectral analysis of the original differential operator. We shall make some use of these ideas in Chapter 7, while recognizing that, for scattering theory, it is the absolutely continuous rather than the discrete spectrum that plays the central role.

2.7 SELF-ADJOINT EXTENSIONS—OPERATORS AND FORMS

We return to the question: given a symmetric operator T, how can we construct self-adjoint extensions of T? Indeed, when can we be sure that self-adjoint extensions exist, and how many extensions can we expect there to be? We shall consider here two methods appropriate to this problem: the Caley transform and the method of sesquilinear forms.

Since every symmetric operator has a closure, we need consider only the case where T is closed. Given a (fixed) complex number z with $\mathrm{Im}\,(z)\neq 0$, we define a linear operator U by the equations

$$\left.\begin{aligned}(T-\bar{z})h&=f,\\(T-z)h&=Uf.\end{aligned}\right\}\tag{2.7.1}$$

Thus the domain of U is the range of $T-\bar{z}$, and we may write $U=(T-z)(T-\bar{z})^{-1}$. U is called the *Caley transform* of the symmetric linear operator T. Since T is closed, the range of $T-\bar{z}$ is a (closed) subspace of $\mathscr{H}$, which may or may not equal the whole of $\mathscr{H}$. So U need not be densely defined. It is straightforward to verify that

$$\langle Uf, Ug\rangle=\langle f, g\rangle\tag{2.7.2}$$

for all $f, g\in D(U)$. A linear operator satisfying (2.7.2) is said to be *isometric*, or we say that U is an *isometry*. The range of U is the same as the range of $T-z$, which again is a subspace. Let h be a vector orthogonal to the range of $T-z$. Then

$$\langle (T-z)f, h\rangle=0$$

for all $f\in D(T)$, so that by the definition of adjoint we see that h must satisfy the equation

$$(T^*-\bar{z})h=0.$$

In particular, in the special case when T is self-adjoint, we have $Th=\bar{z}h$, implying that $h=0$, since a self-adjoint operator can have only real eigenvalues. We have thus shown that if T is self-adjoint then the orthogonal subspace to the range of $T-z$ will consist of the zero vector. In other words, in that case the range of $T-z$, and similarly of $T-\bar{z}$, is the entire Hilbert space $\mathscr{H}$.

Let us define, therefore, the *deficiency indices* (m, n) of an arbitrary symmetric linear operator T as follows. The first deficiency index m is the dimension of the subspace orthogonal to the range of $T-\bar{z}$ (that is, the *codimension* of the range of $T-\bar{z}$), and the second deficiency index n is the dimension of the subspace orthogonal to the range of $T-z$ (codimension of the range of $T-z$). Another way to characterize these deficiency indices is that m and n are respectively the numbers of linearly independent solutions h of the equations $(T^*-z)h=0$ and $(T^*-\bar{z})h=0$. We allow m or $n=\infty$. It follows from our remarks concerning the special case of self-adjoint operators that a self-adjoint operator will have deficiency indices $(0, 0)$. Conversely, if we suppose that T is any closed symmetric operator having deficiency indices $(0, 0)$ then for $g \in D(T^*)$, with $g^*=T^*g$, we have, for all $h \in D(T)$,

$$\langle Th, g \rangle = \langle h, g^* \rangle.$$

so that

$$\langle (T-z)h, g \rangle = \langle h, g^* - \bar{z}g \rangle. \tag{2.7.3}$$

Since the range of $T-\bar{z}$ is the entire space, we can write

$$g^* - \bar{z}g = (T-\bar{z})k \quad \text{for some } k \in \mathscr{H}.$$

so that from (2.7.3) we may deduce, for all $h \in D(T)$, that

$$\langle (T-z)h, g \rangle = \langle (T-z)h, k \rangle. \tag{2.7.4}$$

Moreover the range of $T-z$ is also the entire space, so that (2.7.4) implies $g=k$. Hence $g^*=\bar{z}g+(T-\bar{z})k=Tg$. This shows, for T having deficiency indices $(0, 0)$, that $T^* \subseteq T$, and since (for symmetric T) $T^* \supseteq T$ always, it follows that $T^*=T$ in this case. We have thus proved that T has deficiency indices $(0, 0)$ $\Leftrightarrow$ T is self-adjoint.

Whenever T is self-adjoint, the Caley transform U of T will not only be isometric but also *unitary*. (A unitary operator is an isometric operator such that both the domain and the range are equal to the entire space $\mathscr{H}$. If U is unitary, it is a simple exercise to prove that $UU^*=U^*U=1$. On the other hand, an isometric operator, even if defined on the whole space, need satisfy only $U^*U=1$.) This result indicates a way of obtaining self-adjoint extensions more generally. Let T_0 be any closed, symmetric operator having *equal* deficiency indices (n, n), and let U_0 denote the Caley transform of T_0.

Then U_0 is isometric, with domain $\mathcal{M}$ (range of $T-\bar{z}$) and range $\bar{\mathcal{M}}$ (range of $T-z$), where $\mathcal{M}$ and $\bar{\mathcal{M}}$ have the same codimension. If T_0 is not self-adjoint then $\mathcal{M}^\perp$ and $\bar{\mathcal{M}}^\perp$ have non-zero dimension n, and we may define (in many different ways) a unitary mapping $U^\perp$ from $\mathcal{M}^\perp$ to $\bar{\mathcal{M}}^\perp$. (This means an isometric mapping from $\mathcal{M}^\perp$ onto the whole of $\bar{\mathcal{M}}^\perp$. Such a mapping may be defined by taking orthonormal bases $\{e_k\}$ and $\{f_k\}$ of $\mathcal{M}^\perp$ and $\mathcal{M}^\perp$ respectively, and setting $U^\perp e_k = f_k$ for all k. Of course, this only works if $\mathcal{M}^\perp$ and $\bar{\mathcal{M}}^\perp$ have the same dimension, so that the two orthonormal bases have the same number of elements. $U^\perp$ may be extended by linearity and continuity onto the whole of $\mathcal{M}^\perp$, and it is not difficult to show that $U^\perp$ is then unitary.) We may then extend U_0 to a unitary operator U defined on the entire space by

$$U(f+f^\perp)=U_0 f+U^\perp f^\perp \tag{2.7.5}$$

for $f\in\mathcal{M}$, $f^\perp\in\mathcal{M}^\perp$. Moreover, U is the Caley transform of a unique self-adjoint operator T, which is an extension of T_0. Indeed, the relation between U and T is given by the equations

$$\left.\begin{aligned} U&=(T-z)(T-\bar{z})^{-1}, \\ T&=(\bar{z}U-z)(U-1)^{-1}. \end{aligned}\right\} \tag{2.7.6}$$

It follows from (2.7.1), with $U=U_0$, $T=T_0$, that the range of U_0-1 (domain of T_0) is dense in $\mathcal{H}$. Hence the range of $U-1$ is also dense. Thus only the zero vector is orthogonal to the range of $U-1$. Another way of stating this is to say that $(U^*-1)h=0 \Rightarrow h=0$. In particular, if $Ug=g$ for some vector g then we have $(U^*-1)Ug=g-Ug=0$, so that $Ug=0$ and $g=0$. Hence $Ug=g$ has only the trivial solution $g=0$. In this case, $(U-1)^{-1}$ exists as a densely defined operator in $\mathcal{H}$, and the right-hand side of the formula for T in (2.7.6) is well defined.

There is thus a one-to-one correspondence between the self-adjoint extensions T of a closed symmetric operator T_0 having equal deficiency indices (n, n) and the unitary operators $U^\perp$ from one space of dimension n ($\mathcal{M}^\perp$) onto another ($\bar{\mathcal{M}}^\perp$). We consider, for example, the case in which T_0 has deficiency indices (1.1). Then a unitary operator between two spaces each of dimension 1 may be represented as $e_1 \to \exp(i\alpha) f_1$, where e_1 is a normalized vector in the first space and f_1 is one in the second, and α is real. Thus the self-adjoint extensions of T_0 in this case may be labelled, modulo 2π, by a real parameter α. Another way of stating this is to say that the self-adjoint extensions may be parametrized by a point on the unit circle. We shall see in Chapter 6 how for Schrödinger operators this parametrization corresponds to the fixing of boundary conditions.

Self-adjoint operators may often be defined via the use of sesquilinear forms. This is really an extension of the idea, to be found in elementary treatments of quantum mechanics, of defining a linear operator by means of its "matrix elements". For example, the matrix element of the identity operator between vectors f and g is simply $\langle f,g\rangle$, and for suitable wave functions ϕ,ψ in $L^2(\mathbb{R})$ the matrix elements of the position and the kinetic-energy operator may be written respectively as

$$\int_{-\infty}^{\infty} x\bar{\phi}(x)\psi(x)\,dx \quad \text{and} \quad \int_{-\infty}^{\infty} \frac{\hbar^2}{2m}\bar{\phi}'(x)\psi'(x)\,dx.$$

The problem then is that of constructing the operator from its matrix elements, or, as we shall say, from its (sesquilinear) form.

Given a linear subset of $\mathscr{H}$, called the *form domain* and denoted by $Q(\tau)$, a form τ is a mapping $f,g\to\tau(f,g)$ from $Q(\tau)\times Q(\tau)$ to $\mathbb{C}$ that is *Hermitian* and is linear in g and conjugate-linear in f. In other words, given any pair of vectors f,g in $Q(\tau)$, a corresponding complex number $\tau(f,g)$ is defined, satisfying

(i) $\bar{\tau}(f,g)=\tau(g,f)$,

(ii) $\tau(f,\alpha g_1+\beta g_2)=\alpha\tau(f,g_1)+\beta\tau(f,g_2)$,

(iii) $\tau(\alpha f_1+\beta f_2,g)=\bar{\alpha}\tau(f_1,g)+\bar{\beta}\tau(f_2,g)$,

where α and β are arbitrary complex numbers. Since most (but not all) of the applications of forms are to semibounded forms, we shall restrict our attention to this case. A form τ is said to be *semi-bounded*, or *bounded below*, if there is a real number C such that

$$\tau(f,f)\geq C\|f\|^2 \tag{2.7.7}$$

for all $f\in Q(\tau)$. We shall suppose also that $Q(\tau)$ is dense in $\mathscr{H}$. (Strictly, of course, *semibounded* should also allow any form that is bounded *above*; however, in that case we can consider $-\tau$ instead of τ.)

A form τ is said to be *closed* if, for any sequence $\{f_n\}$ of vectors in $Q(\tau)$ such that f_n converges strongly to a limit f, and such that $\lim_{m,n\to\infty}\tau(f_m-f_n, f_m-f_n)=0$, one has $f\in Q(\tau)$ and $\lim_{n\to\infty}\tau(f_n-f,f_n-f)=0$.

A (semibounded) form τ is said to be *closeable* if τ has a closed semibounded extension. That is, τ is closeable if there is a closed semi-bounded form τ' such that $Q(\tau')\supseteq Q(\tau)$ and such that $\tau'(f,g)=\tau(f,g)$ for all $f,g\in Q(\tau)$. These definitions are a natural extension to forms of the idea of closure that we have already seen applied to operators.

We now let τ be any semibounded form. Then we may assume, without loss of generality, by adding to $\tau(f,g)$ a constant multiple of $\langle f,g\rangle$ if necessary, that (2.7.7) holds with $C=1$. Hence we assume that

$$\tau(f,f)\geq\|f\|^2 \tag{2.7.7$'$}$$

for all $f \in Q(\tau)$. Now we define, for any pair of vectors in $Q(\tau)$, an inner product $\langle \cdot,\cdot \rangle_\tau$ by

$$\langle f, g \rangle_\tau = \tau(f, g). \tag{2.7.8}$$

(It is a simple exercise to verify that this does indeed define an inner product.)

The corresponding norm is

$$\|f\|_\tau = (\langle f, f \rangle_\tau)^{1/2} = (\tau(f, f))^{1/2},$$

so that (2.7.7)′ implies that

$$\|f\| \leq \|f\|_\tau. \tag{2.7.9}$$

This construction of an inner product $\langle \cdot,\cdot \rangle_\tau$ makes $Q(\tau)$ into an inner-product space, which is, however, not necessarily complete. To investigate completeness, we consider a sequence $\{f_n\}$ of vectors in $Q(\tau)$ that is Cauchy in $\|\cdot\|_\tau$, in the sense that $\lim_{m,n\to\infty} \|f_m - f_n\|_\tau = 0$. According to (2.7.9), $\{f_n\}$ is also a Cauchy sequence in the usual sense, that is $\lim_{m,n\to\infty} \|f_m - f_n\| = 0$. Hence, by completeness of $\mathscr{H}$, there exists a vector f such that $f_n \to f$ strongly. This vector f is the only candidate for a vector satisfying $\lim_{n\to\infty} \|f_n - f\|_\tau = 0$, or equivalently $\lim_{n\to\infty} \tau(f_n - f, f_n - f) = 0$. We see that if τ is a closed form then indeed $f \in \bar{Q}(\tau)$ and $\lim_{n\to\infty} \|f_n - f\|_\tau = 0$. More succinctly, for a closed form τ, any sequence $\{f_n\}$ of vectors in $Q(\tau)$ that is Cauchy in $\|\cdot\|_\tau$ converges in $\|\cdot\|_\tau$ to a limit. Moreover, we see that this condition is necessary and sufficient for a form to be closed. That is, τ satisfying (2.7.7)′ is closed if and only if $Q(\tau)$, with inner product given by (2.7.8), is a (complete) Hilbert space.

A similar argument gives us a necessary and sufficient condition for a semibounded form to be closable. Namely, τ is closable if and only if, for any sequence $\{g_n\}$ of vectors in $Q(\tau)$ satisfying $\text{s-lim}_{n\to\infty} g_n = 0$ and $\lim_{m,n\to\infty} \tau(g_m - g_n, g_m - g_n) = 0$, one has $\lim_{n\to\infty} \tau(g_n, g_n) = 0$. Adding to the form a multiple of $\langle f, g \rangle$ so that (2.7.7)′ is satisfied, this result may be stated as follows: τ is closable if and only if, with $g_n \in Q(\tau)$, $\|g_n\| \to 0$ and $\|g_m - g_n\|_\tau \to 0$ together imply $\|g_n\|_\tau \to 0$.

Proof

We suppose that τ is closable and we let τ' be a closed semibounded extension, which, as for τ, may be assumed to satisfy $\|f\| \leq \|f\|_{\tau'}$. We let $g_n \in Q(\tau)$ satisfy $\|g_n\| \to 0$ and $\|g_m - g_n\|_\tau \to 0$. Since τ' is an extension of τ, we then have $\|g_m - g_n\|_{\tau'} \to 0$. Since the inner-product space $Q(\tau')$ with $\langle \cdot,\cdot \rangle_{\tau'} = \tau'(\cdot,\cdot)$ is complete, it follows that, for some $g \in Q(\tau')$, $\lim_{n\to\infty} \|g_n - g\|_{\tau'} = 0$. This implies that $\lim_{n\to\infty} \|g_n - g\| = 0$, so that $g = 0$, and $\lim_{n\to\infty} \|g_n\|_{\tau'} = 0$. But $g_n \in Q(\tau) \subseteq Q(\tau')$, so that $\lim_{n\to\infty} \|g_n\|_\tau = 0$. Conversely, we suppose, for

$g_n \in Q(\tau)$, that $\|g_n\| \to 0$ and $\|g_m - g_n\|_\tau \to 0$ together imply $\|g_n\|_\tau \to 0$. (We again take (2.7.7)′ to be satisfied.) We denote by $\mathscr{H}_\tau$ the Hilbert space that is the completion of $Q(\tau)$ with inner product $\langle \cdot, \cdot \rangle_\tau$. Elements of $\mathscr{H}_\tau$ are (equivalence classes of) sequences $\{f_n\}$ of vectors in $Q(\tau)$ that are Cauchy in $\|\cdot\|_\tau$. Such sequences converge in $\|\cdot\|$; $f_n \to f$ strongly, say. Moreover, two such sequences both converging to the same element f of $\mathscr{H}$ belong to the same equivalence class, since $\|f_n^{(1)} - f\| \to 0$ and $\|f_n^{(2)} - f\| \to 0$, where $\{f_n^{(1)}\}$ and $\{f_n^{(2)}\}$ are Cauchy in $\|\cdot\|_\tau$, imply that $\|f_n^{(2)} - f_n^{(2)}\| \to 0$, where $\{f_n^{(1)} - f_n^{(2)}\}$ is Cauchy in $\|\cdot\|_\tau$, so that $\|f_n^{(1)} - f_n^{(2)}\|_\tau \to 0$. It follows that elements of $\mathscr{H}_\tau$ are in one-to-one correspondence with a linear subset (but not in general a *subspace*) of elements in $\mathscr{H}$. It does no harm to *identify* elements of $\mathscr{H}_\tau$ with their corresponding elements of $\mathscr{H}$, provided that we remember that in the Hilbert space $\mathscr{H}_\tau$, we have an inner product derived from $\langle \cdot, \cdot \rangle_\tau$ rather than from $\langle \cdot, \cdot \rangle$. Indeed, we can denote this inner product by $\langle \cdot, \cdot \rangle_{\tau'}$, set $Q(\tau') = \mathscr{H}_\tau$, and define a closed semibounded form τ' by $\tau'(f, g) = \langle f, g \rangle_{\tau'}$ for $f, g \in \mathscr{H}_\tau$. The form τ' is closed because $\mathscr{H}_{\tau'}$ is a Hilbert space, and τ' is an extension of τ. ■

Given a closable semibounded form, the form τ' obtained by the above construction is called the *closure* of τ. The form domain $Q(\tau')$ of the closure of τ consists of all vectors $f \in \mathscr{H}$ that are strong limits of sequences $\{f_n\}$ of vectors in $Q(\tau)$ satisfying $\lim_{m,n\to\infty} \tau(f_m - f_n, f_m - f_n) = 0$. That not all semibounded forms are closable follows from a simple example. In the Hilbert space $L^2(\mathbb{R})$ let $Q(\tau)$ consist of all *continuous* square integrable functions, and define the form τ by $\tau(f, g) = \bar{f}(0)g(0)$. Then $\tau(f, f) \geq 0$, so that τ is bounded below. Let $g_n(x) = \exp(-nx^2)$. Then certainly $\|g_n\| \to 0$, and we have $\lim_{m,n\to\infty} \tau(g_m - g_n, g_m - g_n) = \lim_{m,n\to\infty} 0 = 0$. However, in this case $\lim_{n\to\infty} \tau(g_n, g_n) = \lim_{n\to\infty} 1 = 1$, so that the condition for closability is violated, and we may infer that τ here is not closable.

One might expect, from this example, that the difficulty lies in the fact that $\bar{f}(0)g(0)$ is not the matrix element of any linear operator. Formally, we have in this case $\tau(f, g) = \langle f, \delta g \rangle$, where δ denotes multiplication by the Dirac delta function $\delta(x)$. However, multiplication by $\delta(x)$ does not define a bona fide operator in $L^2(\mathbb{R})$. The following result shows that any semibounded form that *is* the matrix element of a linear operator is closable.

Lemma 2.11

Let τ be a semibounded form, and suppose that there is a symmetric linear operator T_0 having domain $D(T_0) = Q(\tau)$ such that

$$\tau(f, g) = \langle f, T_0 g \rangle = \langle T_0 f, g \rangle \tag{2.7.10}$$

for all $f, g \in Q(\tau)$. Then τ is closable.

Proof

We assume as usual that (2.7.7)′ is satisfied (if necessary, we add a multiple of the identity operator to T_0). Let us verify our condition for closability under these assumptions. We let $\{g_n\}$ be any sequence in $Q(\tau)$ such that $||g_n||\to 0$ and $||g_m-g_n||_\tau\to 0$. We have to show that $||g_n||_\tau\to 0$.

Given any $\varepsilon>0$, we can find N_0 sufficiently large that $||g_m-g_n||_\tau<\varepsilon$ provided that $m,n>N_0$. Hence, by the Schwarz inequality applied to the inner product $\langle\cdot,\cdot\rangle_\tau$, we have, for arbitrary $h\in Q(\tau)$, and for $m,n>N_0$,

$$|\langle h, g_m-g_n\rangle_\tau|\leq\varepsilon||h||_\tau. \tag{2.7.11}$$

However, $\lim_{m\to\infty}\langle h,g_m\rangle_\tau=\lim_{m\to\infty}\langle T_0 h,g_m\rangle=0$ since $g_m\to 0$ strongly. Taking the limit $m\to\infty$ in (2.7.11) now gives

$$|\langle h,g_n\rangle_\tau|\leq\varepsilon||h||_\tau \tag*{(2.7.11)′}$$

for $n>N_0$ and $h\in Q(\tau)$. In particular, we can set $h=g_n$ and deduce that $||g_n||_\tau\leq\varepsilon$ for $n>N_0$. We have thus shown that $\lim_{n\to\infty}||g_n||_\tau=0$ as required. ■

In view of the connection between symmetric operators and closable forms, it is natural to explore the consequences of generating forms by self-adjoint, rather than just symmetric operators. If we are looking for closed forms generated by self-adjoint operators T then the correct form domain turns out to be not $D(T)$ but rather $D(T^{1/2})$ (if $T\geq 0$; or $D(T+c)^{1/2}$ more generally).

Lemma 2.12

Let T be a self-adjoint operator such that $T\geq 0$ (meaning $\langle f, Tf\rangle\geq 0$ for all $f\in D(T)$). Define a semibounded form τ by

$$\tau(f,g)=\langle T^{1/2}f, T^{1/2}g\rangle \tag{2.7.12}$$

for all $f,g\in Q(\tau)=D(T^{1/2})$. Then τ is closed.

Proof

We have to show that, for any sequence $\{f_n\}$ of vectors in $D(T^{1/2})$ such that f_n converges strongly to a limit f, and such that $\lim_{m,n\to\infty}\langle T^{1/2}(f_m-f_n), T^{1/2}(f_m-f_n)\rangle=0$, we have $f\in D(T^{1/2})$ and $\lim_{n\to\infty}\langle T^{1/2}(f_n-f), T^{1/2}(f_n-f)\rangle=0$. Paraphrasing, we have to show that if $f_n\to f$ strongly and $T^{1/2}f_n$ is a Cauchy sequence, and hence also converges to a limit, then $\text{s-lim}_{n\to\infty} T^{1/2}f_n=T^{1/2}f$. This is precisely the requirement that $T^{1/2}$ be closed. But $T^{1/2}$ is self-adjoint, and hence closed. ■

The following result shows that *all* closed semibounded forms can be generated in this way from self-adjoint operators. We assume that τ is positive; otherwise we add a multiple of $\langle f,g\rangle$.

Lemma 2.13

Let τ be a closed form such that $\tau(f,f)\geq 0$ for all $f\in Q(\tau)$. Then there exists a self-adjoint operator T with $T\geq 0$, such that $D(T^{1/2})=Q(\tau)$ and (2.7.12) holds for all $f\in Q(\tau)=D(T^{1/2})$.

Proof

We start, as usual, by assuming (2.7.7)′. Afterwards we shall see how to reduce the general problem to this case.

We let $\mathscr{H}_\tau=Q(\tau)$ with inner product given by (2.7.8). For arbitrary fixed $h\in\mathscr{H}$, we define on $\mathscr{H}_\tau$ a linear functional by $f\to\langle h,f\rangle$. This is a bounded linear functional (in $\mathscr{H}_\tau$ as a *Hilbert space*) since

$$|\langle h,f\rangle|\leq\|h\|\;\|f\|\leq\|h\|\;\|f\|_\tau,$$

by (2.7.9). By the Riesz Representation Theorem, there is a vector in $\mathscr{H}_\tau$, which we shall write as Ah, such that

$$\langle Ah,f\rangle_\tau=\langle h,f\rangle \tag{2.7.13}$$

for all $f\in\mathscr{H}_\tau$. As the notation implies, Ah is linearly related to h, so that A is a linear operator from $\mathscr{H}$ to $\mathscr{H}_\tau$. From (2.7.13), with $f=Ah$, we have

$$(\|Ah\|_\tau)^2=\langle h,Ah\rangle\leq\|h\|\;\|Ah\|\leq\|h\|\;\|Ah\|_\tau,$$

where we have used (2.7.9) together with the Schwarz inequality. Hence $\|Ah\|_\tau\leq\|h\|$, so that $\|Ah\|\leq\|Ah\|_\tau\leq\|h\|$. That is, regarding A now as a linear operator from $\mathscr{H}$ into $\mathscr{H}$, we have $\|A\|\leq 1$. In particular, A is a bounded operator from $\mathscr{H}$ into $\mathscr{H}$.

Replacing f by Af in (2.7.13), we have

$$\langle Ah,Af\rangle_\tau=\langle h,Af\rangle \tag{2.7.14}$$

for all $f,h\in Q(\tau)=\mathscr{H}_\tau$. Interchanging f and h in (2.7.14) gives also

$$\langle Af,Ah\rangle_\tau=\langle f,Ah\rangle. \tag{2.7.14$'$}$$

Hence for $f,h\in Q(\tau)$ we have

$$\overline{\langle Ah,f\rangle}=\langle Af,Ah\rangle_\tau=\overline{\langle h,Af\rangle},$$

implying that

$$\langle Ah,f\rangle=\langle h,Af\rangle. \tag{2.7.15}$$

This result, initially proved for $f, h \in Q(\tau)$, may be extended by continuity to all, $f, h \in \mathscr{H}$, using the fact that $Q(\tau)$ is dense in $\mathscr{H}$ and A is bounded. Hence A is a bounded *symmetric* operator from $\mathscr{H}$ into $\mathscr{H}$, and is therefore self-adjoint.

In (2.7.13), if $Ah=0$ then we should have $\langle h,f\rangle=0$ for all $f\in Q(\tau)$. Since $Q(\tau)$ is dense in $\mathscr{H}$, this would imply $h=0$. So $Ah=0 \Rightarrow h=0$. It follows that A^{-1} exists as a self-adjoint operator in $\mathscr{H}$. We set $T=A^{-1}$. From (2.7.14) with $h=f$, we have $\langle f, Af\rangle \geq 0$, initially for all $f\in Q(\tau)$, but by completion for all $f\in \mathscr{H}$. Since $\|A\|\leq 1$, the spectrum of A is contained in $[0, 1]$, so that the spectrum of $T=A^{-1}$ is contained in $[1, \infty)$. Hence $T\geq 1$.

Paraphrasing (2.7.13), the defining equation for T is

$$\langle y,f\rangle_\tau=\tau(y,f)=\langle Ty,f\rangle \tag{2.7.16}$$

for all $f\in Q(\tau)$. Thus if the vectors y and z satisfy

$$\langle y,f\rangle_\tau=\tau(y,f)=\langle z,f\rangle$$

for all $f\in Q(\tau)$ then we have $y\in D(T)$ and $Ty=z$. Note that the domain of T is the range of A, and

$$D(T)\subseteq Q(\tau), \qquad r(T)=\mathscr{H}.$$

We now let f be an arbitrary vector in $D(T^{1/2})$. Then we can construct a sequence $\{f_n\}$ of vectors in $D(T)$ such that $\text{s-lim}_{n\to\infty} f_n=f$ and $\text{s-lim}_{n\to\infty} T^{1/2} f_n=T^{1/2} f$. (For example, we let $f_n=E_n f$, where E is the spectral projection of T associated with the interval $[1, n]$. The fact that this sequence has the required properties may be verified from the spectral theorem.) Hence, as $m, n\to\infty$,

$$(\|f_m-f_n\|_\tau)^2=\langle T(f_m-f_n), (f_m-f_n)\rangle=\|T^{1/2}(f_m-f_n)\|^2\to 0,$$

so that $\{f_n\}$ is Cauchy in $\|\cdot\|_\tau$. Since $Q(\tau)=\mathscr{H}_\tau$ is complete, it follows that f_n converges in $\|\cdot\|_\tau$, and this limit must be f. Taking such limits for a pair of vectors $f, g\in D(T^{1/2})$, we have

$$\begin{aligned}\tau(f,g)=\langle f,g\rangle_\tau&=\lim_{n\to\infty}\langle f_n,g_n\rangle_\tau=\lim_{n\to\infty}\langle T^{1/2}f_n, T^{1/2}g_n\rangle\\&=\langle T^{1/2}f, T^{1/2}g\rangle.\end{aligned}$$

Therefore we have verified (2.7.12) for all $f, g\in D(T^{1/2})$. Indeed, $D(T^{1/2})$ may be regarded as a subspace of the Hilbert space $\mathscr{H}_\tau$. (If $\{f_n\}$ is a sequence of vectors in $D(T^{1/2})$ that is Cauchy in $\|\cdot\|_\tau$ then $T^{1/2}f_n$ is Cauchy in $\|\cdot\|$, so that, by the closure property of $T^{1/2}$, $\|f_n-f\|_\tau\to 0$, where $f\in D(T^{1/2})$.) If $f\in\mathscr{H}_\tau$ is orthogonal in $\langle\cdot,\cdot\rangle_\tau$ to this subspace then

$$\langle y,f\rangle_\tau=0 \quad \text{for all } y\in D(T^{1/2}).$$

In particular, since $D(T)\subseteq D(T^{1/2})$, $\langle y,f\rangle_\tau=0$ for all $y\in D(T)$. From (2.7.16) we then have $\langle Ty, f\rangle = 0$ for all $y \in D(T)$. Since the range of T is the entire space, it follows that $f=0$. Therefore the orthogonal subspace, within $\mathscr{H}_\tau$, of $D(T^{1/2})$, is the zero subspace, implying that $D(T^{1/2}) = \mathscr{H}_\tau = Q(\tau)$.

This completes the proof under the assumption that (2.7.7)′ holds. More generally, any positive form may be expressed as $\tau(f,g) - \langle f,g\rangle$ for some τ satisfying (2.7.7)′, and hence has a representation $\langle T^{1/2} f, T^{1/2} g\rangle - \langle f,g\rangle$ for some self-adjoint T with $T \geq 1$. But this is just $\langle (T-1)^{1/2} f, (T-1)^{1/2} g\rangle$. (Note, from the spectral theorem, that $D((T-1)^{1/2}) = D(T^{1/2})$, since for $\lambda \geq 1$, $(\lambda-1)^{1/2} \leq \lambda^{1/2}$ and $\lambda^{1/2} \leq (\lambda-1)^{1/2} + 1$. Therefore form domains are unaltered by going from $T^{1/2}$ to $(T-1)^{1/2}$.)

Hence the theorem holds quite generally, with $T-1$ replacing T. ■

We can now combine this result with Lemma 2.11 to construct a self-adjoint extension of an arbitrary semibounded symmetric operator.

Lemma 2.14

Let τ be a semibounded form and suppose that there is a symmetric linear operator T_0 having domain $D(T_0) = Q(\tau)$ such that

$$\tau(f,g) = \langle f, T_0 g\rangle = \langle T_0 f, g\rangle$$

for all $f, g \in Q(\tau)$. Let τ' be the closure of τ. Then T_0 has a self-adjoint extension T such that

$$D(T) = Q(\tau') \cap D(T_0^*). \tag{2.7.17}$$

Moreover, if τ is a positive form then $T \geq 0$ and $D(T^{1/2}) = Q(\tau')$.

Proof

From Lemma 2.11 we know that τ is closable. We let T be the self-adjoint operator constructed from τ' as in Lemma 2.13. Then certainly $\tau \geq 0 \Rightarrow T \geq 0$ and $D(T^{1/2}) = Q(\tau')$. For the rest of the proof, we may assume without loss of generality (2.7.7)′ is satisfied.

We let f be an arbitrary element of $D(T)$. Then certainly $f \in D(T^{1/2}) = \mathscr{H}_\tau$. From the construction of $\mathscr{H}_\tau$ as the completion of $Q(\tau)$ in $\|\cdot\|_\tau$, there exists a sequence $\{f_n\}$ of vectors in $D(T_0)$ such that $\text{s-lim}_{n\to\infty} f_n = f$ and $\lim_{m,n\to\infty} \|f_m - f_n\|_\tau = 0$, For $g \in D(T_0)$ we then have

$$\tau'(f,g) = \lim_{n\to\infty} \tau(f_n, g) = \lim_{n\to\infty} \langle f_n, T_0 g\rangle = \langle f, T_0 g\rangle.$$

But, from (2.7.16) applied to the closed form $\tau'(f,g)$, we also have

$$\tau'(f,g) = \langle Tf, g\rangle.$$

Comparing these equations for $\tau'(f,g)$, we find

$$\langle Tf, g\rangle = \langle f, T_0 g\rangle \tag{2.7.18}$$

for $f \in D(T)$, $g \in D(T_0)$, and it follows that $Tf = T_0^* f$ for $f \in D(T)$. Thus $T \subseteq T_0^*$, and, since $D(T) \subseteq D(T^{1/2})$, we also have

$$D(T) \subseteq D(T^{1/2}) \cap D(T_0^*) = Q(\tau') \cap D(T_0^*). \tag{2.7.19}$$

We now let f be an arbitrary vector in $D(T^{1/2}) \cap D(T_0^*)$. Then, as before, we have

$$\tau'(f,g) = \langle f, T_0 g\rangle \tag{2.7.20}$$

for all $g \in D(T_0) \subseteq D(T^{1/2}) = Q(\tau')$.

Again, for any $h \in D(T^{1/2})$, we can find a sequence $\{g_n\}$ of vectors in $D(T_0)$ such that $\text{s-lim}_{n\to\infty} g_n = h$, $\text{s-lim}_{n\to\infty} T^{1/2} g_n = T^{1/2} h$, and taking such limits in (2.7.20), we find

$$\tau'(f,h) = \langle T^{1/2} f, T^{1/2} h\rangle = \langle T_0^* f, h\rangle \tag{2.7.21}$$

for all $h \in D(T^{1/2})$. Hence $T^{1/2} f \in D((T^{1/2})^*) = D(T^{1/2})$, and

$$T^{1/2} T^{1/2} f = Tf = T_0^* f.$$

One consequence of this is that $f \in D(T_0) \Rightarrow Tf = T_0^* f = T_0 f$, so that T is a self-adjoint extension of T_0. Moreover, we have shown that $D(T^{1/2}) \cap D(T_0^*) \subseteq D(T)$ which, together with (2.7.19), implies (2.7.17). ■

The extension T of the semibounded symmetric operator T_0 is called the *Friedrichs extension* of T_0. The domain $D(T)$ of the Friedrichs extension consists of all vectors f belonging to $D(T_0^*)$ such that a sequence $\{f_n\}$ of vectors in $D(T_0)$ exists satisfying

$$\text{s-}\lim_{n\to\infty} f_n = f \qquad \text{and} \qquad \lim_{m,n\to\infty} \langle (f_m - f_n), T_0(f_m - f_n)\rangle = 0.$$

This characterization of $D(T)$ may often be used to give useful estimates of elements of the domain of this self-adjoint extension of T_0.

Let us note also that a semibounded self-adjoint operator is uniquely determined by its form. We suppose, for example, that T_1, T_2 are two self-adjoint operators such that $T_1 \geq 0$, $T_2 \geq 0$, and that

$$\langle T_1^{1/2} f, T_1^{1/2} f\rangle = \langle T_2^{1/2} f, T_2^{1/2} f\rangle \tag{2.7.22}$$

for all $f \in D(T_1^{1/2}) = D(T_2^{1/2})$. From (2.7.22) it follows, by replacing f by $f+g$, $f-g$, $f+ig$, $f-ig$ respectively, and taking linear combinations, that

$$\langle T_1^{1/2} g, T_1^{1/2} f\rangle = \langle T_2^{1/2} g, T_2^{1/2} f\rangle \tag{2.7.23}$$

for all $f, g \in D(T_1^{1/2}) = D(T_2^{1/2})$.

We may categorize $f \in D(T_1)$ by the condition

$$|\langle T_1^{1/2} g, T_1^{1/2} f \rangle| \leq \text{const} \, ||g|| \quad \text{for all } g \in D(T_1^{1/2}),$$

since in that case $T_1^{1/2} f$ is in the domain of $T_1^{1/2}$ (actually the adjoint of this operator, which is the same thing.) Moreover, for $f \in D(T_1)$, $T_1 f$ is defined by the equation

$$\langle T_1^{1/2} g, T_1^{1/2} f \rangle = \langle g, T_1 f \rangle$$

for all $g \in D(T_1^{1/2})$. Because of (2.7.23), none of this is changed if we replace T_1 by T_2 throughout. Hence $T_1 f = T_2 f$ for all $f \in D(T_1) = D(T_2)$, so that the self-adjoint operators T_1 and T_2 may be identified. A similar argument applies to an arbitrary pair of semibounded operators having the same forms. The fact that such operators are uniquely defined by their forms is a justification of the common practice, in elementary quantum-mechanics texts, of regarding a self-adjoint operator as defined by its "matrix elements".

2.8 THE SCHRÖDINGER EQUATION

For a particle moving in a potention $V(|\boldsymbol{r}|)$, the time-independent Schrödinger equation, at energy λ, for the radial wave function $f(r)$, is the differential equation

$$-\frac{\mathrm{d}^2 f}{\mathrm{d}r^2} + V(r) f + \frac{l(l+1)}{r^2} f - \lambda f = 0, \tag{2.8.1}$$

where, in suitable units, l is the quantum number of total angular momentum. Equation (2.8.1) has to be solved, for example, in the case $V(r) = -e/r$, to determine the energy eigenvalues λ for the bound states of the hydrogen atom. In that case, we are interested in solutions for which $f \in L^2(0, \infty)$, and such that the appropriate boundary condition $f(0) = 0$ is satisfied at the origin. We take $\lambda \in \mathbb{R}$.

In scattering theory, on the other hand, we are interested in the solutions of (2.8.1) usually for a range of values of λ lying in the continuous spectrum of the Hamiltonian. For these values of λ there may be no square-integrable solutions for f. In very few cases indeed can an exact non-trivial solution be found. However, most of the information that can be gleaned from (2.8.1) and that is of interest in scattering theory comes not from an exact solution but rather from an asymptotic analysis of the behaviour of solutions for large and small values of r.

A sufficient condition, which we shall assume throughout this section, for (2.8.1) to make sense as a differential equation, is that the potential $V(r)$

belongs to $L^1_{\text{loc}}(0, \infty)$. This means that V is integrable over any closed bounded subinterval of $(0, \infty)$. However, we shall allow $V(r)$ to be arbitrarily singular at $r=0$. In most applications, $V(r)$ converges to zero in the limit as $r \to \infty$, and it may be important to know at what rate this convergence takes place. It is in this context, for example, that we may draw the distinction between short-range and long-range potentials.

We have mentioned that a potential may be singular at $r=0$. Roughly, $V(r)$ is said to be singular at $r=0$ if $|V(r)|$ increases as $r \to 0$ at least as fast as const$/r^2$, and non-singular otherwise. If we want a rather more technical definition, and one more in line with an approach that imposes L^1 conditions on the potential, we can say that a non-singular potential is one for which $\int_0^1 r|V(r)|\,dr < \infty$. Note that the Coulomb potential is to be regarded as non-singular.

The principal aim of this section will be to provide ways of estimating the asymptotic behaviour of solutions of the radial Schrödinger equation. Such estimates will play a crucial role in the spectral analysis of Chapter 7, in our treatment of localization (Chapter 10) and much of the subsequent development of scattering theory. In the case of non-singular potentials, the situation near $r=0$ is very simple. For non-singular potentials with $l=0$, $f(r)$ looks like a solution of the $V=0$ equation, as the following result shows.

Lemma 2.15

Suppose that $V(r)$ is non-singular in the sense that

$$\int_0^1 r|V(r)|\,dr < \infty. \tag{2.8.2}$$

Then, for $l=0$, (2.8.1) has two linearly independent solutions f_1 and f_2 such that $f_1 \sim r$ and $f_2 \sim 1$ as $r \to 0$.

Proof

By absorbing the constant λ into $V(r)$, we may consider without loss of generality the equation

$$\frac{d^2 f}{dr^2} = V(r) f. \tag{2.8.3}$$

To obtain one solution of (2.8.3), we replace the differential equation by an integral equation

$$f(r)=r+r\int_0^r V(t)f(t)\,dt-\int_0^r tV(t)f(t)\,dt. \tag{2.8.4}$$

Equation (2.8.4) may now be iterated. We define a sequence of functions $f^{(0)}, f^{(1)}, f^{(2)},\ldots$, where $f^{(0)}(r)=0$, $f^{(1)}(r)=r$, and

$$f^{(n+1)}(r)=r+r\int_0^r V(t)f^{(n)}(t)\,dt-\int_0^r tV(t)f^{(n)}(t)\,dt. \tag{2.8.5}$$

Thus

$$f^{(2)}(r)=r+r\int_0^r tV(t)\,dt-\int_0^r t^2V(t)\,dt,$$

so that

$$|f^{(2)}(r)-f^{(1)}(r)| \le 2r\int_0^r t|V(t)|\,dt=o(r) \quad \text{as } r\to 0, \qquad \text{by (2.8.2).}$$

If $|f^{(n)}-f^{(n-1)}|\le A_n r$ then we find

$$|f^{(n+1)}-f^{(n)}|\le 2A_n r\int_0^r t|V(t)|\,dt\le A_{n+1}r,$$

where $A_{n+1}=\delta A_n$, with δ arbitrarily small as r approaches zero. For r sufficiently small, this gives $\Sigma_n A_n<\infty$, and the iteration converges to a solution $f_1(r)$ of (2.8.4), which is readily verified to satisfy the differential equation (2.8.3). Moreover, the iterative solution satisfies the inequality

$$|f_1(r)-r|\le\sum_{n=2}^{\infty}A_n r=o(r) \quad \text{as } r\to 0.$$

Hence $\lim_{r\to 0} f_1(r)/r=1$ as required.

Given that $f_1(r)$ satisfies (2.8.3), it is easily verified that a second solution of (2.8.3) is

$$f_2(r)=f_1(r)\int_r^1\left(\frac{1}{f_1(t)}\right)^2 dt. \tag{2.8.6}$$

Since $f_1(r)\sim r$, we find $\int_r^1 (1/f_1(t))^2\,dt\sim r^{-1}$ as $r\to 0$, which, from (2.8.6), implies $f_2(r)\sim 1$ as required. (Suppose, for $r<c(\varepsilon)$, say, that f_1 satisfies

$$\frac{1-\varepsilon}{t^2}\le\left(\frac{1}{f_1(t)}\right)^2\le\frac{1+\varepsilon}{t^2}.$$

Then

$$\int_r^1\left(\frac{1}{f_1(t)}\right)^2 dt=\int_r^c\left(\frac{1}{f_1(t)}\right)^2 dt+\int_c^1\left(\frac{1}{f_1(t)}\right)^2 dt,$$

where, for fixed c, the second integral on the right-hand side is a constant, and the first integral lies between

$$(1-\varepsilon)\left(\frac{1}{r}-\frac{1}{c}\right)\quad\text{and}\quad(1+\varepsilon)\left(\frac{1}{r}-\frac{1}{c}\right).$$

Hence $\lim_{r\to 0} r\int_r^1(1/f_1(t))^2\,dt$ lies between $1-\varepsilon$ and $1+\varepsilon$. Since $\varepsilon>0$ is arbitrary, it follows that $\lim_{r\to 0} r\int_r^1(1/f_1(t))^2\,dt=1$, from which, with (2.8.6), we may deduce $\lim_{r\to 0} f_2(r)=1$.) ■

The proof of the above estimates is a typical application, in a simple case, of iterative techniques for solving differential equations through their associated Volterra integral equations. The basic idea, in writing down an appropriate integral equation, is to start from a pair of linearly independent solutions of a differential equation with potential close to the one in question. In this instance, (2.8.4) may be derived from the method of variation of constants applied to the inhomogeneous differential equation $d^2f/dr^2=g$, starting from the solutions r, 1 of the homogeneous equation $d^2f/dr^2=0$, which corresponds to $V(r)\equiv 0$. Near $r=0$, solutions of (2.8.1) behave like solutions of the Schrödinger equation in the absence of a potential. A similar result holds for $l\neq 0$. In this case, again absorbing the constant λ into $V(r)$, the Schrödinger equation becomes

$$\frac{d^2f}{dr^2}=V(r)f+\frac{l(l+1)}{r^2}f. \tag{2.8.7}$$

Solutions for $V(r)\equiv 0$ are then r^{l+1} and r^{-l}, and for one solution of (2.8.7) we may write down the integral equation

$$f(r)=r^{l+1}+(2l+1)^{-1}r^{l+1}\int_0^r t^{-l}V(t)f(t)\,dt$$

$$-(2l+1)^{-1}r^{-l}\int_0^r t^{l+1}V(t)f(t)\,dt. \tag{2.8.8}$$

Equation (2.8.8) may be iterated as before to obtain a solution $f_1(r)$, which on applying (2.8.6) yields a second solution, to give the following corollary.

Corollary

Suppose that $V(r)$ is non-singular in the sense of (2.8.2). Then (2.8.1) has two linearly independent solutions f_1 and f_2, such that $f_1 \sim r^{l+1}$ and $f_2 \sim r^{-l}$ as $r \to 0$.

Let us now turn to a consideration of the behaviour near $r=0$ of solutions of the Schrödinger equation in the case of a singular potential. It is quite impossible, here, to devise an asymptotic formula that will hold for arbitrary singular potentials $V(r)$. There is an infinite variety of different possibilities, depending on the form of $V(r)$; indeed in Chapter 14 we shall meet some examples of quite unexpected asymptotic behaviour, for which the resulting spectral analysis and scattering theory are also unusual. Here we shall content ourselves with a limited treatment, which will nevertheless extend to most of the singular potentials that are encountered in practice.

The transition between singular and non-singular potentials is realized by the examples $V(r)=g/r^2$, where g is a constant; $g<0$ for an attractive potential and $g>0$ for a repulsive potential.

For arbitrary l in such cases, if the so-called centrifugal term $l(l+1)/r^2$ in (2.8.1) is taken into account then we have an effective potential $V_{\text{eff}}(r)=[g+l(l+1)]/r^2$. This case has essentially been dealt with already, except that we now find $l(l+1)$ replaced by $g+l(l+1)$. By the same method as before, we obtain two linearly independent solutions of (2.8.1), behaving near $r=0$ like $r^{l'+1}$ and $r^{-l'}$ respectively, where

$$l'(l'+1)=g+l(l+1). \tag{2.8.9}$$

For $g>-\frac{1}{4}(2l+1)^2$ the value of l' is real, and gives rise to two solutions f_1 and f_2, where $f_1 \sim r^{\alpha}$, $f_2 \sim r^{\beta}$, and $\alpha>\frac{1}{2}, \beta<\frac{1}{2}$, with $\alpha+\beta=1$. For $g=-\frac{1}{4}(2l+1)^2$, (2.8.9) for l' has double roots $l'=-\frac{1}{2}$. We then obtain a single solution $f_1 \sim r^{1/2}$, but through (2.8.6) a second solution $f_2 \sim r^{1/2}\log r$ is obtained.

For $g<-\frac{1}{4}(2l+1)^2$ the value of l' must be complex. On taking real and imaginary parts of the resulting solutions, we have, in this case, $f_1 \sim r^{1/2} \cos(\gamma \log r)$ and $f_2 \sim r^{1/2} \sin(\gamma \log r)$. It will be of interest, in making the limit-point/limit-circle classification of boundary points to be discussed in Chapter 6, to consider the number of linearly independent solutions of (2.8.1) that belong to $L^2(0,1)$. For $V(r)=g/r^2$ there is one such L^2 solution provided that $g+l(l+1)\geq\frac{3}{4}$, and there are two such solutions provided that $g+l(l+1)<\frac{3}{4}$. Another critical value occurs when $g+l(l+1)=-\frac{1}{4}$. For this value of g, the differential operator $-\mathrm{d}^2/\mathrm{d}r^2+V(r)+l(l+1)/r^2$, with $V=g/r^2$, is bounded below, whereas for any smaller value of g any self-adjoint extension of this operator will have eigenvalues of arbitrarily large, negative energies. The question of whether or not a Schrödinger operator is bounded below will be considered more fully in Chapter 5, both in the one-dimensional and three-dimensional cases.

What of operators that are more singular near $r=0$ than g/r^2? Typically, one can consider $V(r)=g/r^n$, where $n>2$, and $g<0$ for an attractive potential, $g>0$ for a repulsive potential. We consider first the attractive case. A set of conditions holding in the case $V(r)=g/r^n$ for $g<0$, $n>2$, and allowing a complete asymptotic analysis, is

$$V(r)<0 \quad \text{for } 0<r\leq c, \qquad \lim_{r\to 0} V(r)=-\infty, \tag{2.8.10}$$

$$\int_0^c \left(\frac{\mathrm{d}}{\mathrm{d}r}|V(r)|^{-1/4}\right)^2 \mathrm{d}r<\infty, \tag{2.8.11}$$

$$\frac{\mathrm{d}}{\mathrm{d}r}|V(r)|^{-1/2} \quad \text{is of bounded variation} \quad (0<r\leq c). \tag{2.8.12}$$

where c is some positive constant.

Assuming (2.8.10)–(2.8.12), let us write

$$J_\varepsilon=\int_\varepsilon^c r^{-2}|V|^{-1/2}\,\mathrm{d}r$$

$$=[-r^{-1}|V|^{-1/2}]_\varepsilon^c+\int_\varepsilon^c r^{-1}\frac{\mathrm{d}}{\mathrm{d}r}(|V|^{-1/2})\,\mathrm{d}r. \tag{2.8.13}$$

Now $(\mathrm{d}/\mathrm{d}r)|V|^{-1/2}$, being of bounded variation, has a finite limit as $r\to 0$, so that, by l'Hôpital's rule, $r^{-1}|V|^{-1/2}$ also has a finite limit. Thus $[r^{-1}|V|^{-1/2}]_\varepsilon^c$ remains bounded in the limit as $\varepsilon\to 0$. Moreover, on the right-hand side of (2.8.13) the integrand is $2r^{-1}|V|^{-1/4}(\mathrm{d}/\mathrm{d}r)|V|^{-1/4}$, so that the Cauchy–

Schwarz inequality gives a bound

$$2\left[\left(\int_{\varepsilon}^{c} r^{-2}|V|^{-1/2}\,\mathrm{d}r\right)\left(\int_{\varepsilon}^{c}\left(\frac{\mathrm{d}}{\mathrm{d}r}|V|^{-1/4}\right)^2\mathrm{d}r\right)\right]^{1/2}$$

for the integral. Using (2.8.11) we then have from (2.8.13)

$$J_{\varepsilon} \leq A + BJ_{\varepsilon}^{1/2},$$

from which it follows that J_{ε} is bounded as $\varepsilon \to 0$.

Hence $\int_0^c r^{-2}|V|^{-1/2}\,\mathrm{d}r < \infty$, and it follows that $\lim_{r\to 0} r^{-1}|V|^{-1/2} = 0$, since a positive value for this limit would make the integral diverge logarithmically. This confirms that the assumptions (2.8.10)–(2.8.12) together imply $\lim_{r\to 0} r^2|V(r)| = \infty$, which means that we are indeed dealing with potentials more singular that g/r^2 at the origin. Assumptions (2.8.10)–(2.8.12) allow potentials even more singular than any inverse power of r, such as $V(r) = -\exp(1/r^2)$.

The following result shows that, in these circumstances, a solution $f(r)$ must converge rapidly to zero in the limit $r \to 0$.

Lemma 2.16

Let $V(r)$ be a negative singular potential satisfying (2.8.10)–(2.8.12). Then any solution $f(r)$ of (2.8.1) satisfies the bound

$$|f(r)| \leq \text{const}\,|V(r)|^{-1/4} \quad (0 < r \leq c). \tag{2.8.14}$$

Proof

For $0 < r \leq c$ we make the change of variable

$$x = \int_r^c |V(t)|^{1/2}\,\mathrm{d}t, \qquad \frac{\mathrm{d}x}{\mathrm{d}r} = -|V|^{1/2}. \tag{2.8.15}$$

We write $f(r) = |V(r)|^{-1/4}u(r)$, and by substituting for r in terms of x we suppose that $u(r) = v(x)$. It is then not difficult to show that (2.8.1) becomes, with this change of variable,

$$\frac{\mathrm{d}^2 v(x)}{\mathrm{d}x^2} + (1 + Q(x))v(x) = 0, \tag{2.8.16}$$

where

$$Q(x) = |V(r)|^{-3/4}\frac{\mathrm{d}^2}{\mathrm{d}r^2}(|V(r)|^{-1/4}) + \left[\lambda - \frac{l(l+1)}{r^2}\right]|V(r)|^{-1}, \tag{2.8.17}$$

with r expressed in terms of x on the right-hand side through the change of variables. We have, moreover,

$$\int_0^\infty |Q(x)|\,\mathrm{d}x = -\int_0^c |Q(x)|\frac{\mathrm{d}x}{\mathrm{d}r}\,\mathrm{d}r$$

$$\leq \int_0^c (|V(r)|)^{-1/4}\,\frac{\mathrm{d}^2}{\mathrm{d}r^2}(|V(r)|^{-1/4})\,\mathrm{d}r$$

$$+|\lambda|\int_0^c |V|^{-1/2}\,\mathrm{d}r + l(l+1)\int_0^c r^{-2}|V|^{-1/2}\,\mathrm{d}r. \qquad (2.8.18)$$

On the right-hand side of (2.8.18) the first integral converges because

$$|V|^{-1/4}\,\frac{\mathrm{d}^2}{\mathrm{d}r^2}|V|^{-1/4} = \frac{1}{2}\,\frac{\mathrm{d}}{\mathrm{d}r}\left(\frac{\mathrm{d}}{\mathrm{d}r}|V|^{-1/2}\right) - \left(\frac{\mathrm{d}}{\mathrm{d}r}|V|^{-1/4}\right)^2,$$

where each term is integrable, by (2.8.11) and (2.8.12). The second integral on the right-hand side converges because $|V|^{-1/2}$ is bounded, and we have already shown, through estimates of J_ε, that the final integral on the right-hand side converges. We have thus shown that

$$\int_0^\infty |Q(x)|\,\mathrm{d}x < \infty. \qquad (2.8.19)$$

We are now ready to obtain an iterative solution of (2.8.16) starting from the integral equation

$$v(x) = a\cos x + b\sin x$$

$$+\int_x^\infty \mathrm{d}t\,\sin(x-t)\,Q(t)v(t). \qquad (2.8.20)$$

Using (2.8.19) and proceeding as in our previous solutions of integral equations in this section, there is no difficulty in proving convergence of the iteration, for x sufficiently large. This gives us, for arbitrary a and b, a solution $v(x)$ of (2.8.16) that has the asymptotic behaviour, as $x \to \infty$,

$$v(x) = (a + o(1))\cos x + (b + o(1))\sin x.$$

Substituting for x in terms of r, from (2.8.15), and evaluating f, this gives in the limit as $r \to 0$,

$$f(r) = (a + o(1))|V(r)|^{-1/4} \cos \int_r^c |V(t)|^{1/2}\, dt$$

$$+ (b + o(1))|V(r)|^{-1/4} \sin \int_r^c |V(t)|^{1/2}\, dt, \tag{2.8.21}$$

from which the conclusion of the lemma follows immediately. Note that every solution of (2.8.1), in this case, vanishes at $r=0$. Indeed, the more singular the potential, the more rapidly $f(r)$ converges to zero in the limit as $r \to 0$. ■

It is instructive to reconsider the example $V(r) = g/r^n$ for $g<0$ and $n>2$, for which, after redefinition of a and b,

$$f(r) = r^{n/4} \left\{ (a + o(1)) \cos\left(\frac{|g|^{1/2}}{(\frac{1}{2}n - 1) r^{n/2 - 1}} \right) \right.$$

$$\left. + (b + o(1)) \sin\left(\frac{|g|^{1/2}}{(\frac{1}{2}n - 1) r^{n/2 - 1}} \right) \right\}. \tag{2.8.22}$$

Note how, although converging to zero, $f(r)$ is extremely rapidly oscillating near $r=0$.

With minor modifications, the same analysis may be carried out for positive (i.e. repulsive), singular potentials.

Lemma 2.17

Let $V(r)$ be a repulsive potential satisfying $V(r)>0$ for $0<r\le c$, $\lim_{r\to 0} V(r) = +\infty$, and such that (2.8.11) and (2.8.12) hold. Then there is just one solution $f_1(r)$ of (2.8.1) having the asymptotic behaviour, as $r \to 0$,

$$f_1(r) \sim [V(r)]^{-1/4} \exp\left\{ -\int_r^c [V(t)]^{1/2}\, dt \right\}. \tag{2.8.23}$$

Moreover, any solution of (2.8.1) that is not a constant multiple of $f_1(r)$ satisfies

$$f_2(r) \sim \text{const}\, [V(r)]^{-1/4} \exp \int_r^c [V(t)]^{1/2}\, dt. \tag{2.8.23$'$}$$

Proof

We proceed as above, with the change of variable (2.8.15). We write $f(r) = [V(r)]^{-1/4} u(r)$, again with $u(r) = v(x)$. Then $v(x)$ satisfies the equation

$$\frac{d^2 v(x)}{dx^2} - (1 + Q(x))v(x) = 0, \tag{2.8.24}$$

for some $Q(x)$ satisfying $\int_0^\infty |Q(x)|\, dx < \infty$. The differential equation may be written as an integral equation for some solution $v_1(x)$ that satisfies $v_1(x) \sim e^{-x}$ as $x \to \infty$. The corresponding solution $f_1(r)$ of (2.8.1) then satisfies the estimate (2.8.23). For a second solution we can take $v_2 = v_1(x) \int_{\text{const}}^x [v_1(t)]^{-2}\, dt$, with $v_2(x) \sim e^x/2$ as $x \to \infty$. Any solution of (2.8.24) that is not a multiple of $v_1(x)$ will then give rise to a solution $f_2(r)$ of (2.8.1), satisfying (2.8.23)′. ■

It is instructive to compare the two solutions f_1 and f_2 in their behaviour near $r = 0$. On the one hand, $f_1(r)$ vanishes extremely rapidly as r approaches zero. For example, $V(r) = 1/r^4$ gives $f_1(r) \sim r \exp(-1/r)$, which vanishes more rapidly at $r = 0$ than any positive power of r. That this will be so in general, for the class of potentials under discussion, follows from the fact that $|V| > \text{const}/r^2$ as $r \to 0$, where the positive multiplicative constant may be arbitrarily chosen. If $|V| > \beta^2/r^2$ then we have $\exp\{-\int_r^c [V(t)]^{1/2}\, dt\} < \text{const } r^\beta$. Correspondingly, $f_2(r)$ is rapidly divergent in the limit as $r \to 0$. In the example cited, we have $f_2(r) \sim r \exp(1/r)$, which diverges more rapidly than any inverse power of r. More generally, under the assumption of the lemma, we cannot have $f_2 \in L^2(0, c)$. To see this, we note first that $V^{1/4} \exp\{-\frac{1}{2} \int_r^c [V(t)]^{1/2}\, dt\} \in L^2(0, c)$, by direct evaluation of the L^2 norm. Therefore if we had $V^{-1/4} \exp \int_r^c (V(t))^{1/2}\, dt \in L^2(0, c)$ then we should obtain on multiplication of these two functions that $\exp\{\frac{1}{2} \int_r^c [V(t)]^{1/2}\, dt\} \in L^1(0, c)$, which cannot be so since this function increases more rapidly than any inverse power of r. It follows from (2.8.23)′ that $f_2 \notin L^2(0, c)$, and indeed the same argument tells us that $r^\beta f_2 \notin L^2(0, c)$ for all $\beta > 0$. $f_2(r)$ is therefore highly singular at $r = 0$, and there is just one linearly independent solution of (2.8.1) in this case that is square-integrable near $r = 0$.

Let us now turn to an asymptotic analysis of solutions of (2.8.1) in the limit as r tends to infinity. The following result applies to a wide class of potentials, and illustrates the difference in behaviour of positive- and negative-energy solutions, for potentials that decay at infinity.

Lemma 2.18

Suppose for some $c > 0$ that $V(r) = V_1(r) + V_2(r)$, where V_1 is continuous and of bounded variation in $[c, \infty)$ and $V_2 \in L^1(c, \infty)$.

(i) Suppose that $\lambda=k^2>0$ and $\sup_{r\geq a}V_1(r)<k^2$. Then every solution of eq (2.8.1) has the asymptotic behaviour, as $r\to\infty$,

$$f(r)=(a+o(1))\cos\int_c^r (k^2-V_1(t))^{1/2}\,\mathrm{d}t$$

$$+(b+o(1))\sin\int_c^r (k^2-V_1(t))^{1/2}\,\mathrm{d}t. \tag{2.8.25}$$

(ii) Suppose that $\lambda=-k^2<0$ and $\inf_{r\geq c}V_1(r)>-k^2$. Then there is just one solution $f_1(r)$ of (2.8.1) having the asymptotic behaviour, as $r\to\infty$,

$$f_1(r)\sim\exp\left\{-\int_c^r (k^2+V_1(t))^{1/2}\,\mathrm{d}t\right\}. \tag{2.8.26}$$

Moreover, any solution of (2.8.1) that is not a constant multiple of $f_1(r)$ satisfies

$$f_2(r)\sim\text{const}\exp\left\{\int_c^r (k^2+V_1(t))^{1/2}\,\mathrm{d}t\right\}. \tag{2.8.26$'$}$$

Proof

(i) Absorbing the $l(l+1)/r^2$ contribution to the potential into $V_2(r)$, the Schrödinger equation may be written as

$$-\frac{\mathrm{d}^2f}{\mathrm{d}r^2}+V_2f=(k^2-V_1)f. \tag{2.8.27}$$

Let us make the change of variables

$$x=\int_c^r (k^2-V_1(t))^{1/2}\,\mathrm{d}t \tag{2.8.28}$$

Then

$$\frac{\mathrm{d}f}{\mathrm{d}r}=(k^2-V_1(r))^{1/2}\frac{\mathrm{d}f}{\mathrm{d}x},$$

which leads to

$$\frac{d^2f}{dr^2}=(k^2-V_1(r))\frac{d^2f}{dx^2}-\tfrac{1}{2}(k^2-V_1(r))^{-1/2}\frac{dV_1}{dr}\frac{df}{dx}$$

$$=(k^2-V_1(r))\frac{d^2f}{dx^2}-\frac{1}{2}\frac{dV_1}{dx}\frac{df}{dx}.$$

The differential equation (2.8.27) now becomes

$$-\frac{d^2f}{dx^2}+\tfrac{1}{2}(k^2-V_1(r))^{-1}\frac{dV_1}{dx}\frac{df}{dx}+\frac{V_2(r)f}{k^2-V_1(r)}=f, \tag{2.8.29}$$

where, in principle, the change of variable may be used to express r in terms of x.

Since $(k^2-V_1)^{-1}$ is bounded and V_1 is of bounded variation, we have

$$\int_0^\infty \left|(k^2-V_1)^{-1}\frac{dV_1}{dx}\right| dx<\infty.$$

Moreover,

$$\int_c^\infty \left|\frac{V_2(r)}{k^2-V_1(r)}\right| dx=\int_c^\infty \left|\frac{V_2(r)}{(k^2-V_1(r))^{1/2}}\right| dr<\infty,$$

since $V_2\in L^1(c,\infty)$. Hence the differential equation (2.8.29) is of the form

$$-\frac{d^2f}{dx^2}+q_1(x)\frac{df}{dx}+q_2(x)f=f, \tag{2.8.30}$$

where $\int_0^\infty |q_1(x)|\,dx<\infty$ and $\int_0^\infty |q_2(x)|\,dx<\infty$. If we now write $df/dx=g$, we may replace (2.8.30) by a pair of integral equations, to be solved simultaneously for $f(x)$ and $g(x)$: thus

$$f(x)=a\cos x+b\sin x+\cos x\int_x^\infty [q_1(t)g(t)+q_2(t)f(t)]\sin t\,dt$$

$$-\sin x\int_x^\infty [q_1(t)g(t)+q_2(t)f(t)]\cos t\,dt,$$

$$g(x)=-a\sin x+b\cos x-\sin x\int_x^\infty [q_1(t)g(t)+q_2(t)f(t)]\sin t\,dt$$

$$-\cos x\int_x^\infty [q_1(t)g(t)+q_2(t)f(t)]\cos t\,dt.$$

These equations may be iterated, starting from the zeroth approximation

$$\{f_0(x), g_0(x)\} = \{a\cos x + b\sin x, -a\sin x + b\cos x\},$$

and expressing $\{f_{n+1}, g_{n+1}\}$ in the obvious way in terms of $\{f_n, g_n\}$. For sufficiently large values of x, the L^1 estimates for q_1 and q_2 are enough to guarantee the convergence of the iteration, and we have, in the limit as $x\to\infty$,

$$f(x) = (a + o(1))\cos x + (b + o(1))\sin x,$$
$$g(x) = f'(x) = -(a + o(1))\sin x + (b + o(1))\cos x.$$

The limit $x\to\infty$ is equivalent to the limit $r\to\infty$, and, using (2.8.28) to return to the original variable, (2.8.25) is established.

(ii) In the case $\lambda = -k^2 < 0$, the appropriate change of variables is $x = \int_c^r (k^2 + V_1(t))^{1/2}\,dt$. The resulting integral equations in this case may again be iterated, starting with $\{f_0, g_0\} = \{e^{-x}, -e^{-x}\}$, to obtain a solution $f_1(x)$ satisfying $f_1 \sim e^{-x}$ as $x\to\infty$. A second solution may be defined as

$$f_2(x) = f_1(x)\int_\beta^x \frac{dt}{P(t)[f_1(t)]^2},$$

where

$$P(t) = \exp\left\{\int_t^\infty q_1(t)\,dt\right\}.$$

Since $P(t)\to 1$ as $t\to\infty$, and using the known asymptotic formula for $f_1(x)$, we find $f_2 \sim \frac{1}{2}e^x$ in the limit as $x\to\infty$. Any solution f_2 that is not a multiple of f_1 will satisfy $f_2 \sim \text{const}\, e^x$ as $x\to\infty$, and (2.8.26)′ follows on returning to the original variable r. ∎

In the above proof we supplemented a condition of bounded variation for V_1 by supposing V_1 to be differentiable. Strictly, this proof does not go through as stated unless V_1 is absolutely continuous. However, in the more general case where V_1 is assumed to be continuous and of bounded variation, but not necessarily absolutely continuous, integrals involving dV_1/dx may be interpreted as Stieltjes integrals with respect to the function V_1. The integral equations following (2.8.30) may then be iterated as before. In general, $g(x)$ need not therefore be absolutely continuous, although $f(x)$ is. However, $f(r)$ and $(d/dr)f(r)$ may be verified to be absolutely continuous, and we obtain, as in the more restricted case, a solution of the differential equation having the required asymptotic behaviour. A similar modification may be carried out to the proof of the previous lemma.

We distinguish between potentials of *short range* and *long range*. A potential $V(r)$ is said to be of short range if $\int_c^\infty |V(r)|\,\mathrm{d}r<\infty$, and of long range otherwise. Thus $V(r)=g/r^n$ is short range for $n>1$ and long range for $n\leq 1$. Note that the transitional potential $V(r)=g/r$, the Coulomb potential, is of long range. The asymptotic behaviour, for solutions of (2.8.1) in the limit as $r\to\infty$, is quite different in the two cases. For a short-range potential we can set $V_1(r)\equiv 0$ in (2.8.25), (2.8.26) and (2.8.26)′, and take $c=0$. The resulting asymptotic behaviour gives solutions behaving like $a\cos kr+b\sin kr$ for $\lambda=k^2>0$, and like e^{-kr} or e^{kr} for $\lambda=-k^2<0$. This asymptotic behaviour has to be modified for long-range potentials, owing to the slow rate of decrease of the potential at infinity.

We consider, as an example, the case of the Coulomb potential $V(r)=g/r$. We take $V_1(r)=g/r$ and $V_2(r)\equiv 0$. Then

$$(k^2-V_1(t))^{1/2}=\left(k^2-\frac{g}{t}\right)^{1/2}=k-\frac{g}{2kt}+O(t^{-2})\quad\text{as } t\to\infty.$$

Hence

$$\int_c^r (k^2-V_1(t))^{1/2}\,\mathrm{d}t=kr-\frac{g}{2k}\log r+O(r^{-1}),$$

apart from an additive constant. So for $\lambda=k^2>0$ we have the asymptotic behaviour in this case, for solutions of (2.8.1),

$$f(r)=(a+o(1))\cos\left(kr-\frac{g}{2k}\log r\right)$$
$$+(b+o(1))\sin\left(kr-\frac{g}{2k}\log r\right). \tag{2.8.31}$$

Thus the tail of a Coulomb potential results in additive logarithmic contributions to the arguments of the sine and cosine functions. Similarly, we may use (2.8.26) to show that, for $\lambda=-k^2<0$, the exponential decay must be modified, for large r, by a multiplicative power of r. It is for this reason that the bound-state wave functions for the hydrogen atom are not exponential functions, but rather exponentials multiplied by polynomials.

Let us mention, in passing, the case $\lambda=0$, which has so far been omitted from the analysis. For $\lambda=0$ the effect of the centrifugal contribution $l(l+1)/r^2$ to the effective potential can no longer be ignored in general. Indeed, if $V(r)$ decreases at infinity more rapidly than const$/r^2$, in the sense that $\int_c^\infty r|V(r)|\,\mathrm{d}r<\infty$, the centrifugal term becomes dominant. In that case, we arrive at two solutions for $\lambda=0$, behaving for large r like r^{l+1} and r^{-l}

respectively. If on the other hand, the potential $V(r)$ dominates the $l(l+1)/r^2$ contribution for $\lambda=0$, an analysis similar to that of the preceeding lemma may be carried out, the asymptotic behaviour of solutions depending on the particular form of the potential.

For the class of potentials analysed above, it will be noticed that, for $\lambda>0$ there are no solutions of (2.8.1) belonging to $L^2(c,\infty)$, whereas for $\lambda<0$ there is just one such (linearly independent) solution. In no case do we find that all solutions are square-integrable. The following result shows that, to find potentials giving rise to two linearly undependent solutions in $L^2(c,\infty)$, we have to consider potentials that are extremely large and negative as $r\to\infty$. Such potentials are of little interest in scattering theory.

Lemma 2.19

For some positive constant C, suppose that $V(r)\geq -Cr^2$ for sufficiently large r. Then, for arbitrary (real) λ, there is at most one linearly independent solution of (2.8.1) belonging to $L^2(c,\infty)$ for $c>0$.

Proof

We suppose that $f(r)$ is such a solution belonging to $L^2(c,\infty)$. Then as $r\to\infty$, and assuming without loss of generality $f(r)$ to be real,

$$\text{const}\geq C\int_c^r f^2(t)\,\mathrm{d}t\geq\int_c^r -\frac{V(t)f^2(t)}{t^2}\,\mathrm{d}t$$

$$=\int_c^r \frac{f(t)}{t^2}\left(-f''(t)+\frac{l(l+1)}{t^2}f(t)-\lambda f(t)\right)\mathrm{d}t.$$

Since $f\in L^2(c,\infty)$, this gives $-\int_c^r(f(t)f''(t)/t^2)\,\mathrm{d}t\leq$const., where the integral may be rewritten as

$$-\left[\frac{ff'}{t^2}+\frac{f^2}{t^3}\right]_c^r+\int_c^r\left(\frac{f'^2}{t^2}-\frac{3f^2}{t^4}\right)\mathrm{d}t.$$

Since $\int_c^\infty(f^2/t^4)\,\mathrm{d}t<\infty$, we now have

$$-\left[\frac{ff'}{t^2}+\frac{f^2}{t^3}\right]_c^r+\int_c^r\frac{f'^2}{t^2}\,\mathrm{d}t\leq\text{const}. \tag{2.8.32}$$

This inequality implies that $\int_c^\infty (f'^2/t^2)\,\mathrm{d}t=\infty$ would entail that $\lim_{r\to\infty}\ (f(r)f'(r)/r^2+f^2(r)/r^3)=+\infty$. Since $r^{-1}f\in L^2(c,\infty)$, this last equation would lead to $\int_c^\infty f(r)f'(r)\,\mathrm{d}r=[\frac{1}{2}f^2]_c^\infty=\infty$. This is only possible if $f^2(r)\to\infty$ as $r\to\infty$, which contradicts the original assumption that $f\in L^2(c,\infty)$. So from (2.8.32), we must have $\int_c^\infty (f'^2/t^2)\,\mathrm{d}t<\infty$.

Now we let f and g be two linearly independent solutions of (2.8.1), both belonging to $L^2(c,\infty)$. We then have $f\in L^2(c,\infty)$, $g\in L^2(c,\infty)$, and our argument shows that $f'(r)/r\in L^2(c,\infty)$, $g'(r)/r\in L^2(c,\infty)$. By the Schwarz inequality, we then have

$$\frac{fg'}{r}\in L^1(c,\infty), \qquad \frac{gf'}{r}\in L^1(c,\infty).$$

We now define the *Wronskian* of the two solutions as $W(f,g)=fg'-gf'$, so that

$$\begin{aligned}\frac{\mathrm{d}}{\mathrm{d}r}W(f,g)&=fg''-gf''\\&=f\left(Vg+\frac{l(l+1)}{r^2}g-\lambda g\right)-g\left(Vf+\frac{l(l+1)}{r^2}f-\lambda f\right)\\&=0,\end{aligned}$$

with $W(f,g)=0$ only if f and g are linearly dependent. Hence $W(f,g)=\text{const}\neq 0$ in this case, and we have $(fg'-gf')/r=\text{const}/r\in L^1(c,\infty)$, which is a contradiction. It follows that (2.8.1) can have at most a single linearly independent solution that is square-integrable over (c,∞). ■

We conclude this section by referring to the possibility of allowing λ in (2.8.1) to be complex. The resulting Schrödinger equation, at complex energy z, becomes

$$-\frac{\mathrm{d}^2 f}{\mathrm{d}r^2}+V(r)f+\frac{l(l+1)}{r^2}f-zf=0 \quad (0<r<\infty) \tag{2.8.33}$$

The motivation for allowing z to be complex will become apparent in our treatment of the Schrödinger operator, and the subsequent development of spectral analysis and scattering theory.

Near $r=0$ the change from real to complex energy makes very little difference to the asymptotic behaviour of solutions, at least in the cases that we have analysed so far. (There are, however, examples of highly singular and oscillating potentials for which the term λf or zf has a profound impact on spectral analysis. Such examples will be considered, in a different context, in Chapter 14.)

In the limit as $r \to \infty$ the situation is different, since for potentials that decay at infinity we can expect the term zf in (2.8.33) to dominate. The following lemma extends the results of Lemma 2.18 to allow complex energies, and at the same time leads to solutions that depend analytically on z.

Lemma 2.20

Suppose, for some $c > 0$, that $V(r) = V_1(r) + V_2(r)$, where V_1 is continuous and of bounded variation in $[c, \infty)$ and $V_2 \in L^1(c, \infty)$. Let $V_{\min}$ and $V_{\max}$ be constants such $V_{\min} \leq 0$. $V_{\max} \geq 0$, and $V_{\min} \leq V_1(r) \leq V_{\max}$ for $r \in [c, \infty)$. Then solutions $f(r, z)$ of (2.8.33) may be found, such that the following hold.

(i) $f(r, z)$ is defined and satisfies (2.8.33) for each $z \in \mathscr{R}$, where $\mathscr{R}$ is the complex plane $\mathbb{C}$ with the interval $[V_{\min}, \infty)$ of the real axis removed (i.e. $\mathscr{R} = \mathbb{C} \setminus \{[V_{\min}, \infty)\}$.) Moreover, for fixed $r > 0$, $f(r, z)$ is an analytic function of z, for $z \in \mathscr{R}$.

(ii) As $r \to \infty$, $f(r, z)$ has the asymptotic behaviour

$$f(r, z) \sim \exp\left\{ i \int_c^r (z - V_1(t))^{1/2} \, dt \right\}, \tag{2.8.34}$$

where the square root $w^{1/2}$, for complex w, is defined according to the convention $(re^{i\theta})^{1/2} = r^{1/2} e^{i\theta/2}$, $0 \leq \theta < 2\pi$ (thus for $\mathrm{Im}(w) \neq 0$, $w^{1/2}$ is defined to have positive imaginery part).

(iii) As z approaches the real axis *from above*, such that $z \to k^2 > V_{\max}$, the boundary value of $f(r, z)$ is given by the formula

$$\lim_{z \to k^2} f(r, z) = f(r, k^2), \tag{2.8.35}$$

where $f(r, k^2)$ is the solution of the real-energy Schrödinger equation (2.8.1), with the asymptotic behaviour, as $r \to \infty$,

$$f(r, k^2) \sim \exp\left\{ i \int_c^r (k^2 - V_1(t))^{1/2} \, dt \right\}. \tag{2.8.36}$$

Proof

The idea of the proof follows that of Lemma 2.18. However, with k^2 replaced by complex z, we are no longer permitted to make the change of variable (2.8.28), since the new independent variable x would not be real.

Instead, we formally write down an integral equation, subject to appropriate asymptotic behaviour, for the solution of (2.8.30), and transform back to the proper variable r. We shall again deal with the case where V_1 is absolutely continuous. Suitable modifications to the argument may be carried out in the general case. We start therefore from the integral equations, to be solved simultaneously for $f(r)$ and $g(r)=(\mathrm{d}/\mathrm{d}r)f(r)$:

$$\begin{aligned} f(r) = {} & \exp\left[\mathrm{i}\int_c^r (z-V_1(t))^{1/2}\,\mathrm{d}t\right] \\ & -(2\mathrm{i})^{-1}\exp\left[\mathrm{i}\int_c^r (z-V_1(t))^{1/2}\,\mathrm{d}t\right] \\ & \times\int_r^\infty \mathrm{d}t\left\{G(t,z)\exp\left[-i\int_c^t (z-V_1(s))^{1/2}\,\mathrm{d}s\right]\right\} \\ & +(2\mathrm{i})^{-1}\exp\left[-\mathrm{i}\int_c^r (z-V_1(t))^{1/2}\,\mathrm{d}t\right] \\ & \times\int_r^\infty \mathrm{d}t\left\{G(t,z)\exp\left[\mathrm{i}\int_c^t (z-V_1(s))^{1/2}\,\mathrm{d}s\right]\right\}, \end{aligned}$$

$$\begin{aligned} g(r) = {} & \mathrm{i}(z-V_1(r))^{1/2}\exp\left[\mathrm{i}\int_c^r (z-V_1(t))^{1/2}\,\mathrm{d}t\right] \\ & -\tfrac{1}{2}(z-V_1(r))^{1/2}\exp\left[\mathrm{i}\int_c^r (z-V_1(t))^{1/2}\,\mathrm{d}t\right] \\ & \times\int_r^\infty \mathrm{d}t\left\{G(t,z)\exp\left[-\mathrm{i}\int_c^t (z-V_1(s))^{1/2}\,\mathrm{d}s\right]\right\} \\ & -\tfrac{1}{2}(z-V_1(r))^{1/2}\left\{\exp\left[-\mathrm{i}\int_c^r (z-V_1(t))^{1/2}\,\mathrm{d}t\right]\right\} \\ & \times\int_r^\infty \mathrm{d}t\left\{G(t,z)\exp\left[\mathrm{i}\int_c^t (z-V_1(s))^{1/2}\,\mathrm{d}s\right]\right\}. \end{aligned}$$

We have defined $G(t, z)$ in these equations by

$$G(t,z)=\tfrac{1}{2}(z-V_1(t))^{-3/2}\frac{\mathrm{d}V_1(t)}{\mathrm{d}t}g(t)+(z-V_1(t))^{-1/2}V_2(t)f(t),$$

where we use the convention for $w^{-3/2}$ and $w^{-1/2}$ which is consistent with that for $w^{1/2}$ defined above. The reader should verify by explicit differentiation that $\mathrm{d}f/\mathrm{d}r=g$ and that the two integral equations together imply (2.8.33). (The centrifugal term $l(l+1)/r^2$ is absorbed into the function $V_2(r)$.) In the more general case where V_1 is merely assumed to be continuous, and of bounded variation, it can be verified that $g=-\int(z-V_1-V_2)f\,\mathrm{d}r$ $+\text{const.}$, which again implies (2.8.33).

We now iterate the integral equations for f and g, starting with the zeroth approximation

$$\{f_0, g_0\}=\left\{\exp\left[\mathrm{i}\int_c^r (z-V_1(t))^{1/2}\,\mathrm{d}t\right], \mathrm{i}(z-V_1(r))^{1/2}\exp\left[\mathrm{i}\int_c^r (z-V_1(t))^{1/2}\,\mathrm{d}t\right]\right\}.$$

The kind of estimates that we made in the case of real energy also apply in this case, and it is straightforward to prove the convergence of the iteration not only for $z\in\mathscr{R}$ but also for values of z in the interval $(V_{\max}, \infty)$ of the real axis. The important points to bear in mind when proving convergence are as follows.

(a) $\operatorname{Im}[(z-V_1(t))^{1/2}]\geq 0$, so that $\operatorname{Im}[\int_c^r(z-V_1(t))^{1/2}\,\mathrm{d}t]$ is either increasing or zero (the latter in the case $z\in(V_{\max}, \infty)$); this implies that $|\exp[\mathrm{i}\int_c^r(z-V_1(t))^{1/2}\,\mathrm{d}t]|$ is either a decreasing function of r or is unity, and in either case cannot exceed 1.

(b) Using (a), an integral such as

$$\int_r^\infty \mathrm{d}t\,(z-V_1(t))^{-1/2}\exp\left[2\mathrm{i}\int_c^t (z-V_1(s))^{1/2}\,\mathrm{d}s\right]V_2(t)$$

may be bounded in absolute value by

$$\left|\exp\left[2\mathrm{i}\int_c^r (z-V_1(s))^{1/2}\,\mathrm{d}s\right]\right|\int_r^\infty |z-V_1(t)|^{-1/2}|V_2(t)|\,\mathrm{d}t.$$

Since, for a fixed value of z belonging to $\mathscr{R}\cup(V_{\max},\ \infty)$, $|z-V_1(t)|^{-1/2}$ is bounded, we also have

$$\int_r^\infty |z-V_1(t)|^{-1/2}|V_2(t)|\mathrm{d}t \le \text{const} \int_r^\infty |V_2(t)|\,\mathrm{d}t,$$

so that the integral converges to zero in the limit as $r\to\infty$.

(c) Each step in the iteration defines a pair of functions $\{f_n(r,z),\ g_n(r,z)\}$ that is differentiable with respect to z, and hence analytic, for $z\in\mathscr{R}$ with r fixed. Moreover, each function depends continuously on z, for $z\in\mathscr{R}\cup(V_{\max},\ \infty)$. (We interpret continuity at $z=k^2>V_{\max}$ to mean continuity from above.)

(d) Convergence of the iteration is *uniform*, for z belonging to any closed bounded subset of $\mathscr{R}\cup(V_{\max},\ \infty)$. Since a uniform limit of analytic functions is analytic, and a uniform limit of continuous functions is continuous, the iterative solution $\{f(r,z),\ g(r,z)\}$ defines a pair of functions that is analytic on $\mathscr{R}$ and continuous on $\mathscr{R}\cup(V_{\max},\ \infty)$. This proves (2.8.35) of the lemma, showing that $f(r,k^2)$ is the *upper* boundary value of $f(r,z)$ as $z\to k^2$. The asymptotic formulae (2.8.34) and (2.8.36) follow by standard estimates of integrals in the limit as $r\to\infty$.

It should be noted finally that we claim that (2.8.35) holds, and that $f(r,z)$ is analytic, for *all* $r>0$, whereas in proving these results we rely on an iterative technique that converges only for values of r sufficiently large, for $r\ge r_0$, say. However, there is no difficulty in proving (for example again by iterative techniques), that for fixed r in the interval $0<r<r_0$, $f(r,z)$ and $g(r,z)$ depend linearly on $f(r_0,z)$ and $g(r_0,z)$, through coefficients that are analytic in z even in the entire complex plane. Hence the analyticity of f and g, for $z\in\mathscr{R}$, and the continuity of f and g as functions of z, in $\mathscr{R}\cup(V_{\max},\ \infty)$ may be extended to *all* values of r belonging to $(0,\ \infty)$. The proof of the lemma is now complete. ■

The importance of these results is that they establish a link between a family of solutions $f(r,z)$ of (2.8.33) that are square-integrable for large r and the solution $f(r,k^2)$ having the asymptotic behaviour (2.8.36). Note that, for a given z, $f(r,z)$ is the only linearly independent solution of (2.8.33) that is square-integrable at infinity. We shall see in Chapter 6 how to define analytic families of solutions that belong to $L^2(c,\ \infty)$ under conditions of greater generality than those discussed here. And it is precisely the boundary values $f(r,k^2)$ of these L^2 solutions that in Chapter 7 will play a central role in spectral analysis. In Chapter 8 we shall use the solutions $f(r,k^2)$ to construct the scattering phase shifts. Analytically continuing from $z=k^2>0$ to $z=-k^2<0$, we obtain the real solutions $f(r,-k^2)$, which decay exponentially for large r, and which, subject to a boundary condition at $r=0$, define the eigenfunctions of the Hamiltonian $H=-\mathrm{d}^2/\mathrm{d}r^2+V+l(l+1)/r^2$. Solutions $f(r,z)$ for $\mathrm{Im}\,(z)\neq 0$ also decay exponentially in absolute value. For such z, any solution that is not a constant multiple of $f(r,z)$ will *increase* exponentially as $r\to\infty$, and we have, in that case, the asymptotic behaviour $f\sim\text{const}\,\exp\,[-\mathrm{i}\int_c^r(z-V_1(t))^{1/2}\,\mathrm{d}t]$. For $z=k^2>0$ we can obtain a second solution by complex conjugation. Thus the general solution of (2.8.1) in that case is a linear combination of $f(r,k^2)$ and $\bar{f}(r,k^2)$. The asymptotic behaviour (2.8.36), applied to f and $\bar{f}$, corresponds to (2.8.25), with complex exponentials replacing the sine and cosine functions.

The case of a short-range potential is particularly simple. If $V(r)$ is of short range then we may take $V_1(r)\equiv 0$, and $c=0$ in (2.8.34). Solutions $f(r,z)$ then have the asymptotic behaviour

$$f(r,z)\sim\exp\,(\mathrm{i}z^{1/2}), \tag{2.8.37}$$

and (2.8.36) becomes

$$f(r,k^2)\sim\mathrm{e}^{\mathrm{i}kr}. \tag{2.8.38}$$

The application of our results to long-range potentials encounters one difficulty that has to be overcome. This is that the asymptotic formulae (2.8.34) and (2.8.36), the region $\mathscr{R}$ of analyticity in z, and the range of real values of k^2 for which we can take boundary values according to (2.8.35), all depend on the value of the constant c. We should like to remove this dependence on c, and at the same time ensure the largest possible domain of analyticity and range of k^2 values. This can be carried out under the conditions of the following lemma, which applies to most long-range potentials of interest in scattering theory.

Lemma 2.21

Suppose for all sufficiently large c that $V(r)=V_1(r)+V_2(r)$, where V_1 is continuous and of bounded variation in $[c,\infty)$, with $\lim_{r\to\infty}V_1(r)=0$ and $V_2\in L^1(c,\infty)$. Let $\mathscr{R}$ denote the complex plane $\mathbb{C}$ with the positive real axis removed (i.e. $\mathscr{R}=\mathbb{C}\setminus\{[0,\infty)\}$), and let $\mathscr{R}_c$ denote the complex plane with the interval $[V_{\min},\infty)$ removed, where $V_{\min}=\inf_{r\geq c}V_1(r)$. Suppose that a function $F(r,z)$ exists, independent of c, and analytic in $\mathscr{R}$, such that

$$G_c(z)\equiv\lim_{r\to\infty}\left\{F(r,z)-\int_c^r(z-V_1(t))^{1/2}\,\mathrm{d}t\right\} \tag{2.8.39}$$

exists as a uniform limit on closed, bounded subsets of $\mathscr{R}_c \cup (V_{\max}, \infty)$. (For $z = k^2 > V_{\max}$ we interpret $F(r, k^2)$ to mean the *upper* boundary value of $F(r, z)$ as $z \to k^2$.) Then solutions $f(r, z)$ of (2.8.33) may be found, such that

(i) $f(r, z)$ is defined and satisfies (2.8.33) for each $z \in \mathscr{R}$; moreover, for fixed $r > 0$, $f(r, z)$ is an analytic function of z for $z \in \mathscr{R}$;

(ii) As $r \to \infty$, $f(r, z)$ has the asymptotic behaviour

$$f(r, z) \sim \exp[iF(r, z)]; \tag{2.8.40}$$

(iii) as z approaches the positive real axis from above, the boundary value of $f(r, z)$ is given by

$$\lim_{z \to k^2} f(r, z) = f(r, k^2),$$

where $f(r, k^2)$ is the solution of the real-k^2 Schrödinger equation with the asymptotic behaviour, as $r \to \infty$,

$$f(r, k^2) \sim \exp[iF(r, k^2)]. \tag{2.8.41}$$

Proof

We let $f_c(r, z)$ denote the solution defined by Lemma 2.20, for given c. Then $f_c(r, z)$ is an analytic function of z, for $z \in \mathscr{R}_c$. The function $G_c(z)$ defined by (2.8.39), being a uniform limit of analytic functions, is also analytic in $\mathscr{R}_c$. We define, for $z \in \mathscr{R}_c$, $f(r, z)$ by

$$f(r, z) = f_c(r, z) \exp[iG_c(z)]. \tag{2.8.42}$$

From (2.8.39), and given the asymptotic behaviour of $f_c(r, z)$ in the limit as $r \to \infty$, we may deduce (2.8.40). Since there is just one linearly independent solution of (2.8.33), for given z, that belongs to $L^2(c, \infty)$, we know that (2.8.40) defines $f(r, z)$ uniquely. For this reason, since F is independent of c, we have dropped the suffix c on the left-hand side of (2.8.42).

Moreover, by construction, $f(r, z)$ is analytic for $z \in \mathscr{R}_c$. Since $\mathscr{R}$ is the intersection of the $\mathscr{R}_c$ as c tends to infinity, it follows that $f(r, z)$ is analytic for $z \in \mathscr{R}$.

In the same way, we define $f(r, k^2)$ by

$$f(r, k^2) = f_c(r, k^2) \exp[iG_c(k^2)],$$

where again $G_c(k^2)$ denotes an upper boundary value. Then $f(r, k^2)$ is an upper boundary value of $f(r, z)$ for $k^2 > V_{\max}$. Since $V_{\max} \to 0$ as $c \to \infty$, $f(r, k^2)$ is an upper boundary value of $f(r, z)$ for all $k^2 > 0$. It remains only to note the asymptotic formula (2.8.41), and the proof is complete. ■

The case of the Coulomb potential is worthy of special attention. We therefore take $V_2(r) \equiv 0$ and $V_1(r) = g/r$. By the binomial theorem, we have the existence of the limit

$$\lim_{r \to \infty} \int_c^r \left\{ z^{1/2} \left(1 - \frac{g}{2tz} \right) - (z - V_1(t))^{1/2} \right\} dt.$$

The limit is uniform over closed bounded subsets of $\mathscr{R}_c \cup (V_{\max}, \infty)$. Since we want $F(r, z)$ in (2.8.39) to be independent of c, we take

$$F(r, z) = rz^{1/2} - \frac{g \log r}{2z^{1/2}},$$

leading to the asymptotic behaviour

$$f(r, z) \sim \exp\left[i \left(rz^{1/2} - \frac{g \log r}{2z^{1/2}} \right) \right]. \tag{2.8.43}$$

For $z = k^2 > 0$ we can see that this formula is consistent with our previous result (2.8.31), where now we are permitting the constants a and b of that formula to be complex.

Chapter 3

Time-Dependent Scattering Theory—First Steps

3.1 TYPES OF ASYMPTOTIC EVOLUTION

Time-dependent scattering theory, to which a major part of this book is devoted, is concerned with the development of quantum systems at large positive and negative times. In view of the very small time intervals involved (typically 10^{-10} s for interactions on the atomic scale, and 10^{-24} s for the so-called "strong" interactions) it will be legitimate to study this large-time behaviour by means of the limits $t \to +\infty$ and $t \to -\infty$ respectively.

The theory deals then with the evolution of states in these limits, and in particular seeks to relate the asymptotic states at $t = +\infty$ to those at $t = -\infty$. The evolution of states in quantum mechanics is, of course, governed by the Hamiltonian H of the system, via the solution of the time-dependent Schrödinger equation. So we know, at least in principle, how to determine the variation of a state with time. What will happen in the limits as $t \to \pm\infty$ will depend on H.

One possibility is that nothing at all may happen! If the state vector f at time $t = 0$ is an eigenstate of H then the state at a general time t is given by $f_t = e^{-iHt} f = e^{-i\lambda t} f$, where λ is the eigenvalue. In that case only the phase multiplying the state vector changes, and the state itself is independent of time. From the point of view of scattering theory, this possibility might appear to be uninteresting, but nevertheless it cannot be ignored.

Eigenstates of the Hamiltonian are frequently referred to as "bound states". This terminology is largely motivated by the theory of potential scattering, where for a large class of potentials it is possible to identify eigenstates of the total Hamiltonian, and linear combinations of them, with those states that remain localized (in some suitable sense) close to the scattering centre. This identification is not always valid, even in potential scattering. We shall defer until Chapter 13 further study of this important question. For the moment, and in the present general context, it is more accurate and appropriate to refer to eigenstates of H as the *stationary states* of the theory. A further question arises in this connection. In addition to the eigenstates of H, can there be *asymptotically stationary states*?

Definition 3.1

Given a Hamiltonian H, the evolution of a normalized state vector f is said to be asymptotically stationary in the limit as $t \to +\infty$ if, for any projection operator E, $\langle e^{-iHt}f, E\, e^{-iHt}f\rangle$ converges to a limit as $t \to +\infty$.

In other words, whatever property of the system we care to measure, described by a proposition or by its corresponding projection operator E, the probability of any particular result of the observation will tend to some limiting value for large times. The state approaches a limiting state (described probabilistically) as $t \to +\infty$. (Such stationary states are familiar from statistical mechanics. The difference is that here we are starting with a *pure* state.) The following theorem tells us that, in fact, asymptotically stationary states give us nothing new.

Theorem 3.1

The only asymptotically stationary states are stationary states, i.e. eigenstates of H.

Proof

We suppose that f is asymptotically stationary, and take $\phi \in \mathscr{H}$, $\|\phi\| = 1$. Then, setting E equal to the one-dimensional projection $|\phi\rangle\langle\phi|$ shows that

$$C(\phi) \equiv \lim_{t+\infty} |\langle\phi, e^{-iHt}f\rangle| \quad \text{exists.}$$

We use this as the definition of $C(\phi)$ even if ϕ is not normalized. Since $\langle\phi, e^{-iHt}f\rangle$ is continuous, we may define a phase $e^{i\beta_\phi(t)}$ such that

$$\lim_{t\to\infty} e^{i\beta_\phi(t)}\langle\phi, e^{-iHt}f\rangle = C(\phi), \tag{3.1.1}$$

where β_ϕ depends continuously on t. We also have, for any vector $\psi \in \mathscr{H}$.

$$\lim_{t\to\infty} e^{i\beta_\psi(t)}\langle\psi, e^{-iHt}f\rangle = C(\psi). \tag{3.1.2}$$

Now, choosing $a, b \in \mathbb{R}$, we know also that $|\langle a\phi + b\psi, e^{-iHt}f\rangle|^2$ converges to a limit. Applying (3.1.1) and (3.1.2) to the inner products $\langle\phi, e^{-iHt}f\rangle$ and $\langle\psi, e^{-iHt}f\rangle$ implies that $|ae^{-i\beta_\phi(t)}C(\phi)+be^{-i\beta_\psi(t)}C(\psi)|^2$ converges to a limit, and, evaluating this expression, we may deduce that, whenever $C(\phi)\neq 0$ and $C(\psi)\neq 0$, $\cos(\beta_\phi(t)-\beta_\psi(t))$ tends to a limit. Using the continuity of the β-functions, we now have

$$\beta_\phi(t)-\beta_\psi(t)\to\text{limit}.$$

This means that in (3.1.2) the limit still exists if on the left-hand side we replace β_ψ by β_ϕ. In other words, the phases in (3.1.1) and (3.1.2) may be chosen to be the same, though this will involve a redefinition of the value of $C(\psi)$. This still applies if either $C(\phi)$ or $C(\psi)$ vanishes, since in that case one at least of the phases may be assigned arbitrarily. In this way we may identify a common phase $e^{i\beta(t)}$ for *all* $\phi\in\mathscr{H}$, so that $\lim_{t\to\infty}\langle\phi, e^{-iHt}e^{i\beta(t)}f\rangle$ exists for all $\phi\in\mathscr{H}$. By the Riesz Representation Theorem, we can write, for some $g\in\mathscr{H}$, and for all $\phi\in\mathscr{H}$.

$$\lim_{t\to\infty}\langle\phi, e^{-iHt}e^{i\beta(t)}f\rangle=\langle\phi, g\rangle. \tag{3.1.3}$$

We can first of all dispose of the case $g=0$. In this case, $e^{-iHt}f$ converges weakly to zero, and a straightforward development of the arguments of Theorem 3.2 will show that a projection E may then be constructed such that $\langle e^{-iHt}f, Ee^{-iHt}f\rangle$ does *not* converge to a limit. This contradicts the hypothesis that the state is asymptotically stationary. We therefore have $g\neq 0$. We now replace t by $t+\tau$ in (3.1.3). Then

$$\begin{aligned}\langle\phi, g\rangle &=\lim_{t\to\infty}\langle\phi, e^{-iH(t+\tau)}e^{i\beta(t+\tau)}f\rangle\\ &=\lim_{t\to\infty} e^{i\beta(t+\tau)}e^{-i\beta(t)}\langle e^{iHt}\phi, e^{-iHt}e^{i\beta(t)}f\rangle\\ &=\lim_{t\to\infty} e^{i\beta(t+\tau)}e^{-i\beta(t)}\langle\phi, e^{-iH\tau}g\rangle,\end{aligned}$$

on using (3.1.3) for a second time with ϕ replaced by $e^{iH\tau}\phi$.

Setting

$$\lim_{t\to\infty} e^{i\beta(t+\tau)}e^{-i\beta(t)}=e^{i\gamma(\tau)}, \tag{3.1.4}$$

we now have

$$\langle\phi, g\rangle=\langle\phi, e^{-iH\tau}e^{i\gamma(\tau)}g\rangle, \tag{3.1.5}$$

from which it follows that

$$e^{-iH\tau}g=e^{-i\gamma(\tau)}g. \tag{3.1.6}$$

Equation (3.1.6) already implies that g is a stationary state (or multiple of one). In fact for $h\in D(H)$, $\langle h, e^{-iH\tau}g\rangle=\langle e^{iH\tau}h, g\rangle$ is differentiable with respect to τ. Hence $e^{-i\gamma(\tau)}\langle h, g\rangle$ is also differentiable, so that $e^{-i\gamma(\tau)}$ is itself a differentiable function.

Differentiating (3.1.6) w.r.t. τ and setting $\tau = 0$ now gives an equation of the form

$$Hg = \lambda g. \tag{3.1.7}$$

Therefore g is an eigenvector. Setting $\phi = g$ in (3.1.3) gives

$$\lim_{t\to\infty} e^{i\beta(t)} e^{-i\lambda t}\langle g, f\rangle = \|g\|^2.$$

Hence we can write $f = (\lim_{t\to\infty} e^{-i(\beta(t)-\lambda t)})g + g^{\perp}$, where $g^{\perp}$ is orthogonal to g. Substituting this expression for f into (3.1.3), we find that $\lim_{t\to\infty}\langle \phi, e^{-iHt}e^{i\beta(t)}g^{\perp}\rangle = 0$, so that $e^{-iHt}g^{\perp}$ converges weakly to zero. By exactly the same argument as used above, but applied to the evolution of $g^{\perp}$ in the subspace orthogonal to g, we may construct a projection E' acting in this subspace such that $\lim_{t\to\infty}\langle e^{-iHt}g^{\perp}, E' e^{-iHt}g^{\perp}\rangle$ does not exist. This argument fails, of course, only if $g^{\perp} = 0$. Hence indeed $g^{\perp} = 0$, and we have proved that f is a constant multiple of g. Since g is stationary, so must f be. ■

The case of stationary states, or eigenstates, represents one of a wide variety of possible asymptotic modes of development of a quantum system at large times. We shall be exploring a number of these possibilities in the pages that follow. The most important case, and certainly that which has always attracted by far the greatest attention in the literature of scattering theory, is that in which we describe the asymptotic behaviour by saying that, for large times, the system behaves like one or more freely moving particles.

Consider for example a single particle moving in a field of force that decreases at large distances. If we look at the system for large times, one possibility is to find the particle receding to infinity, having escaped from the influence of the field of force. This is the canonical example of so-called asymptotically free evolution. In momentum space, we expect there to be some limiting probability distribution over momentum, reflecting the fact that, far from the centre of the force field, momentum should be approximately conserved.

We again consider a three-particle system, interacting via short-range forces between each pair of particles. Asymptotically, the system might look very much like a two-particle system, one particle being a bound state of particles 1 and 2, and the other, particle 3, moving far from the (1,2) state. Both the (1,2) bound state and particle 3 could be freely moving in such a way that the system is asymptotically indistinguishable from a non-interacting two-particle system. Again, there would be a limiting probability distribution for the (1,2) and 3 pair of momenta. For a system of three particles there may be several different ways in which particles could combine to form bound states, and there will be correspondingly several possible modes of asymptotic behaviour, each of which may be qualified by the description "asymptotically free evolution".

In view of the wide applicability of this idea, we shall formulate the notion of asymptotically free evolution in a somewhat general and abstract way. To start with, what do we mean by "free evolution", to which the behaviour of our system is supposedly asymptotic? And how can one evolution ($e^{-iHt}f$) "look like" another (free evolution) for large times?

Consider a Hilbert space $\mathscr{H}^{(0)}$, the so-called free Hilbert space, in which the free momentum operator $\boldsymbol{P}^{(0)}$ acts. We leave open, for the moment, the number of components that $\boldsymbol{P}^{(0)}$ is supposed to have. If we want to describe free evolution of a single particle then $\boldsymbol{P}^{(0)}$ will have three components, whereas for a freely evolving pair of particles there will be six components, three for each particle of the pair. Suppose in any case that the components of $\boldsymbol{P}^{(0)}$ form a complete set of commuting operators in the space $\mathscr{H}^{(0)}$, so that $\mathscr{H}^{(0)}$ can be represented by some L^2 space (momentum space) over the appropriate number of dimensions, where $\boldsymbol{P}^{(0)}$ is just multiplication by $\boldsymbol{p}^{(0)}$. Free-particle states will be represented by normalized wave functions $\psi(\boldsymbol{p}^{(0)})$, and free evolution of these states will be described by 1-parameter families $\{\psi_t(\boldsymbol{p}^{(0)})\}$. For freely evolving states (as opposed to states that are *asymptotically free*) we want conservation of momentum to hold *exactly*. In other words for any region Λ of momentum space $\int_\Lambda |\psi_t(\boldsymbol{p}^{(0)}|^2\, d^3\boldsymbol{p}^{(0)}$ should be independent of time t. The fundamental postulates of quantum mechanics also imply that free evolution should be a *linear* property. Given freely evolving states ϕ_t and ψ_t, $\alpha\phi_t+\beta\psi_t$ should also evolve freely, so that $\int_\Lambda |\alpha\phi_t+\beta\psi_t|^2\, d^3\boldsymbol{p}^{(0)}$ should be time-independent, for any pair of numbers α, β. Taking various different values of α and β, this implies that $\int_\Lambda \bar{\phi}_t\psi_t\, d^3\boldsymbol{p}^{(0)}$ is time-independent. It is reasonable to suppose that the families $\{\phi_t\}$ and $\{\psi_t\}$ can be chosen to be continuous functions of momentum, for each value of t, in which case, by shrinking the region Λ to a single point, it follows that $\bar{\phi}_t(\boldsymbol{p}^{(0)})\psi_t(\boldsymbol{p}^{(0)})$, regarded as a function of $\boldsymbol{p}^{(0)}$, is independent of t. A special case of this is that $\bar{\phi}_t(\boldsymbol{p}^{(0)})\phi_t(\boldsymbol{p}^{(0)})$ is t-independent, so that, combining these results, we have the t-independence of $\psi_t(\boldsymbol{p}^{(0)})/\phi_t(\boldsymbol{p}^{(0)})$. It follows that $\psi_t(\boldsymbol{p}^{(0)})/\psi_0(\boldsymbol{p}^{(0)})$ must be a function of $\boldsymbol{p}^{(0)}$ and t alone, denoted by $F(\boldsymbol{p}^{(0)}, t)$. Since each ψ_t is normalized, we must have $|F(\boldsymbol{p}^{(0)}, t)| \equiv 1$. Free evolution may be expressed in operator-theoretic language by the equation

$$\psi_t = F(\boldsymbol{P}^{(0)}, t)\psi, \tag{3.1.8}$$

where $\psi=\psi_0$. Before giving a formal definition of free evolution, we need to derive some further properties of the family of operators $F(\boldsymbol{P}^{(0)}, t)$. As a preliminary to this, we consider the second question posed above, namely how does one evolution look like another at large times? Before we can answer this, we have to establish a connection between our Hilbert space, describing the states of a quantum system, and the free Hilbert space $\mathscr{H}^{(0)}$.

We consider a projection operator acting in $\mathscr{H}^{(0)}$, corresponding to a proposition or yes–no experiment for the free system. If we are to experimentally compare evolution in $\mathscr{H}$ with free evolution, there must be a corresponding projection operator acting in $\mathscr{H}$, and more generally there must be a correspondence between the projections in $\mathscr{H}^{(0)}$ and projections (but not *all* projections) in $\mathscr{H}$. These projections in $\mathscr{H}$ that correspond to projections in $\mathscr{H}^{(0)}$ will be described as the *free projections* of $\mathscr{H}$. Not all projections in $\mathscr{H}$ will be free projections, since the set of observations that may be carried out on a system in a given mode of asymptotic evolution form, in general, a small subclass of the set of all possible observations on the system. Free projections may be regarded as acting in some subspace $\mathscr{D}_0$ of $\mathscr{H}$. $\mathscr{D}_0$ may be defined as the linear span of the range, in $\mathscr{H}$, of all those free projections that correspond to one-dimensional projections of $\mathscr{H}^{(0)}$. Within $\mathscr{D}_0$, the lattice structure of the free projections must correspond to the structure of projections, or equivalently propositions, in $\mathscr{H}^{(0)}$. Mathematically, this means that the correspondence is implemented by a unitary transformation (an anti-unitary transformation is also theoretically possible). We denote this transformation by U_0, so that

$$E \to U_0 E U_0^{-1} \tag{3.1.9}$$

for projections E in $\mathscr{H}^{(0)}$. The operator U_0 maps from $\mathscr{H}^{(0)}$ to $\mathscr{D}_0$, so that in $\mathscr{D}_0$ we have, within $\mathscr{H}$, an exact replica of the free Hilbert space $\mathscr{H}^{(0)}$. For this reason, it is sometimes convenient, without too great a risk of confusion, to identify the spaces $\mathscr{D}_0$ and $\mathscr{H}^{(0)}$.

We define $\boldsymbol{P}_0$, acting in $\mathscr{D}_0$, by

$$\boldsymbol{P}_0 = U_0 \boldsymbol{P}^{(0)} U_0^{-1}, \tag{3.1.10}$$

so that

$$F(\boldsymbol{P}_0, t) = U_0 F(\boldsymbol{P}^{(0)}, t) U_0^{-1}. \tag{3.1.11}$$

We now let $g \in \mathscr{H}$ be normalized. We want to make precise the notion that, for large t, the state $e^{-iHt}g$ "looks like" the freely evolving state $F(\boldsymbol{P}^{(0)}, t)\psi$, where ψ is a normalized vector belonging to $\mathscr{H}^{(0)}$.

Let us turn the question around. Experimentally, how might the evolution $e^{-iHt}g$ be *distinguished* from the evolution $F(\boldsymbol{P}^{(0)}, t)$? Obviously, we could not distinguish between these two evolutions by a single observation, made at a single instant of time. It would be necessary to make a sequence of observations, at times $t_1, t_2, \ldots$, say, and to find some property of the free evolution that was not, asymptotically, shared by the evolution $e^{-iHt}g$. We can formalize this process as follows.

Definition 3.2

Let $\{g_t\}$ and $\{h_t\}$ be two normalized one-parameter families of vectors in a Hilbert space. We shall say that $\{g_t\}$ is *asymptotically distinguishable* from $\{h_t)$ as $t\to\infty$ if there exists an increasing sequence $\{t_n\}$ and a projection operator E in the space such that $t_n\to\infty$ and

$$\langle h_{t_n}, Eh_{t_n}\rangle = 1, \tag{3.1.12}$$

whereas

$$\lim_{n\to\infty} \langle g_{t_n}, Eg_{t_n}\rangle \neq 1. \tag{3.1.13}$$

Here the projection E represents some property that obtains for every element of the sequence $\{h_{t_n}\}$, but that does *not* hold, even asymptotically, for the sequence $\{g_{t_n}\}$. The family $\{g_t\}$ is distinguishable from the family $\{h_t\}$, as far as the physical interpretation is concerned, because by a sequence of observations at times $t_1, t_2, \ldots$, we find that property E is always satisfied for the evolution h_t, but not satisfied for the evolution g_t as $t\to\infty$. Note that, by taking subsequences, we can always replace (3.1.13) by

$$\lim_{n\to\infty} \langle g_{t_n}, Eg_{t_n}\rangle = \beta < 1. \tag{3.1.13$'$}$$

There is also no loss of generality in supposing that E is the projection onto the subspace spanned by the sequence $\{h_{t_n}\}$, in which case (3.1.12) holds automatically.

In the context of scattering theory, property E will hold, for the free evolution at time t, provided that

$$\langle F(\boldsymbol{P}^{(0)}, t)\psi, EF(\boldsymbol{P}^{(0)}, t)\psi\rangle = 1 \tag{3.1.14}$$

for $t = t_n$. Corresponding to E is the free projection $U_0 E U_0^{-1}$, acting in $\mathscr{H}$, and the property will hold asymptotically as $t = t_n \to \infty$ for the evolution $e^{-iHt}g$, provided that

$$\lim_{n\to\infty} \langle U_0^{-1} e^{-Ht_n} g, EU_0^{-1} e^{-Ht_n} g\rangle = 1 \tag{3.1.15}$$

This motivates the following definition, which formalizes the notion that, asymptotically as $t\to\infty$, the evolution $e^{-iHt}g$ looks like a free evolution.

Definition 3.3

The evolution $e^{-iHt}g$ is said to satisfy the *asymptotic condition* (for $t\to+\infty$), defined by the family of operators $F(\boldsymbol{P}^{(0)}, t)$, acting in the free Hilbert space $\mathscr{H}^{(0)}$, if there exists $\psi \in \mathscr{H}^{(0)}$ such that, as $t\to+\infty$, $U_0^{-1}e^{-iHt}g$ is asymptotically indistinguishable from the free evolution $F(\boldsymbol{P}^{(0)}, t)\psi$.

Even within the confines of the potential scattering of a single particle in non-relativistic quantum mechanics, we should not wish to commit ourselves from the start to a particular choice of the function F. For example, a Coulomb field of force is known to influence the motion of a particle, even at large distance, in such a way that the function F will have to differ from the corresponding function for, say, a force field of finite range. The precise form of F will depend very much on the nature of the interaction, as is evident also for many particle problems, in which the possibilities for free evolution are both more numerous and necessarily more complicated. Consider a classical particle moving in one dimension, the displacement being given by $x=t+\log t$. Calculation of the acceleration as a function of position shows that the particle is moving in a potential $V(x)$, where $V\sim -m/x$ as $x\to\infty$. Note that, although the momentum of the particle tends to a limit, the tail of the potential still has an effect on the motion at large distances, in that $x-t$ does not approach a limit. In other words, the trajectory in space–time is not asymptotically linear. So even in classical mechanics, the asymptotic motion of a particle, even for quite familiar problems, must be described with some care.

The following theorem allows us to interpret the asymptotic condition in a mathematically much more convenient form.

Theorem 3.2

Let $\{g_t\}$ and $\{h_t\}$ be normalized one-parameter families of vectors in a Hilbert space, such that h_t converges weakly to zero as $t\to\infty$. Then $\{g_t\}$ is asymptotically indistinguishable from $\{h_t\}$ as $t\to\infty$ if and only if there exists a phase $e^{i\beta(t)}$ such that

$$\lim_{t\to\infty} \|g_t - e^{i\beta(t)} h_t\| = 0. \tag{3.1.16}$$

Proof

One half of the proof is easy. Suppose that (3.1.16) holds. Then, for any sequence $\{t_n\}$ satisfying (3.1.12), (3.1.16) implies $\lim_{n\to\infty} \langle g_{t_n}, Eg_{t_n}\rangle = \lim_{n\to\infty} \langle h_{t_n}, Eh_{t_n}\rangle = 1$, which contradicts (3.1.13). Hence there are no sequences that distinguish asymptotically between $\{g_t\}$ and $\{h_t\}$.

The converse is not quite so obvious. Suppose that $\{g_t\}$ is asymptotically indistinguishable from $\{h_t\}$ and that $h_t\to 0$ weakly. We first prove that g_t also converges weakly to zero.

We suppose that this is not so. Since any bounded sequence has a weakly convergent subsequence, we can certainly find a sequence $\{t_n\}$, tending to infinity, for which g_{t_n} converges weakly to some limit ϕ, say, where $\phi\neq 0$. We let E_N be the projection onto the subspace spanned by the h_{t_k} satisfying

$1 \le k \le N$. If $E_N h_{t_{N+1}} = 0$, so that $h_{t_{N+1}}$ is orthogonal to this subspace, we should have $E_{N+1} = E_N + |h_{t_{N+1}}\rangle\langle h_{t_{N+1}}|$. More generally, we have

$$E_{N+1} = E_N + \frac{|(1-E_N)h_{t_{N+1}}\rangle\langle(1-E_N)h_{t_{N+1}}|}{\|(1-E_N)h_{t_{N+1}}\|^2}. \tag{3.1.17}$$

Now, if $\psi_k (1 \le k \le N)$ is an orthonormal basis for the range of E_N then we have

$$\|E_N h_{t_{N+1}}\|^2 = \sum_{k=1}^{N} |\langle \psi_k, h_{t_{N+1}}\rangle|^2,$$

which will be very small provided that the inner product of $h_{t_{N+1}}$ with each of the ψ's is sufficiently small. Since $h_{t_{N+k}}$ converges weakly to zero as $k \to \infty$, this can in any case be arranged by replacing $h_{t_{N+1}}$ by $h_{t_{N+k}}$ in the series, for some k sufficiently large. The right-hand side of (3.1.17) is then closely approximated by $E_N + |h_{t_{N+1}}\rangle\langle h_{t_{N+1}}|$, with redefinition of the $(N+1)$th term of the series. We can now look at the difference between E_{N+2} and E_{N+1}. Provided that we redefine $h_{t_{N+2}}$ in a suitable way, this difference will be closely approximated by $|h_{t_{N+2}}\rangle\langle h_{t_{N+2}}|$. Proceeding in this way, it is possible to arrange that

$$\sum_{N=1}^{\infty} \|E_{N+1} - E_N - |h_{t_{N+1}}\rangle\langle h_{t_{N+1}}|\| < \infty, \tag{3.1.18}$$

where $\{t_n\}$ here is (possibly) a subsequence of the original sequence. Exactly the same estimates apply to the projection $E_N^{(n)}$ onto the subspace spanned by the vectors h_{t_k} for $n \le k < n+N$, so that

$$\sum_{N=1}^{\infty} \|E_{N+1}^{(n)} - E_N^{(n)} - |h_{t_{N+n}}\rangle\langle h_{t_{N+n}}|\| < \infty.$$

It follows that the projection $E_\infty^{(n)}$ onto the subspace spanned by all h_{t_k} for $k \ge n$ is closely approximated by $\Sigma_{k=n}^{\infty} |h_{t_k}\rangle\langle h_{t_k}|$, in the sense that

$$\lim_{n\to\infty} \left\| E_\infty^{(n)} - \sum_{k=n}^{\infty} |h_{t_k}\rangle\langle h_{t_k}| \right\| = 0. \tag{3.1.19}$$

Since h_{t_k} lies in the range of $E_\infty^{(n)}$ for $k \ge n$, the indistinguishability property allows us to assert that $\langle g_{t_k}, E_\infty^{(n)} g_{t_k}\rangle \to 1$ as $k \to \infty$, i.e. g_{t_k} is asymptotically in the range of $E_\infty^{(n)}$. Hence

$$\phi = \text{w-}\lim_{k\to\infty} g_{t_k} = \text{w-}\lim_{k\to\infty} E_\infty^{(n)} g_{t_k} = E_\infty^{(n)} \phi.$$

But this implies $\phi = \text{s-lim}_{n\to\infty} E_\infty^{(n)}\phi = 0$, since as $n \to \infty$ the $E_\infty^{(n)}$ converge strongly to zero. This contradicts our original hypothesis that $\phi \ne 0$, and we can conclude that indeed g_t converges weakly to zero.

Next we prove that

$$\lim_{t\to\infty} |\langle h_t, g_t\rangle| = 1. \tag{3.1.20}$$

We suppose that this is not so. Then we choose a sequence $\{t_n\}$ such that

(i) (3.1.19) holds,
(ii) $\lim_{n\to\infty}|\langle h_{t_n}, g_{t_n}\rangle| = \beta < 1$,
(iii) $\Sigma_{j,k;j\neq k} |\langle h_{t_k}, g_{t_j}\rangle| < \infty$.

To verify that (iii) can be satisfied, we suppose inductively that $t_1, t_2, \ldots, t_n$ have been chosen so that

$$\sum_{\substack{1\le j,k\le n\\ j\neq k}} |\langle h_{t_k}, g_{t_j}\rangle| < C_n. \tag{3.1.21}$$

Then on replacing n by $n+1$ on the left-hand side, we have to add to the left-hand side the further contribution

$$\sum_{j=1}^{n} \{|\langle h_{t_{n+1}}, g_{t_j}\rangle| + |\langle h_{t_j}, g_{t_{n+1}}\rangle|\},$$

which, by the weak convergence to zero of g_t and h_t, becomes arbitrarily small as $t_{n+1}\to\infty$. Hence, by suitable choice of t_{n+1}, (3.1.21) holds for $n\to n+1$, with C_{n+1} arbitrarily close to C_n. To verify (iii), the sequence $\{t_n\}$ has only to be chosen such that C_n remains bounded as n increases, and this is evidently possible.

We now consider $\lim_{j\to\infty, n\to\infty} \|E_\infty^{(n)} g_{t_j}\|$. Since h_{t_j} lies in the range of $E_\infty^{(n)}$ for $j\ge n$, and we cannot allow the sequence $\{t_n\}$ to distinguish between g_t and h_t, it follows that $\lim_{j\to\infty} \|E_\infty^{(n)} g_{t_j}\| = 1$, so that certainly

$$\lim_{\substack{j\to\infty\\ n\to\infty}} \|E_\infty^{(n)} g_{t_j}\| = 1. \tag{3.1.22}$$

On the other hand, according to (3.1.19), this limit is exactly

$$\lim_{\substack{j\to\infty\\ n\to\infty}} \left\| \sum_{k=n}^{\infty} \langle h_{t_k}, g_{t_j}\rangle h_{t_k} \right\| \le \lim_{\substack{j\to\infty\\ n\to\infty}} \sum_{k=n}^{\infty} |\langle h_{t_k}, g_{t_j}\rangle|.$$

From (iii), the contribution to the sum from $j\neq k$ is vanishingly small in the limit $j\to\infty$, so we need consider only the single term $j=k$, whose contribution, according to (ii), cannot exceed β in the limit $n\to\infty$. Thus $\beta<1$ contradicts (3.1.22), and we can only conclude that (3.1.20) holds.

We now let $\beta(t)$ be real and satisfy

$$\lim_{t\to\infty} e^{i\beta(t)}\langle g_t, h_t\rangle = 1. \tag{3.1.23}$$

Then

$$\|g_t - e^{i\beta} h_t\|^2 = \|g_t\|^2 + \|h_t\|^2 - 2\mathrm{Re}\{e^{i\beta}\langle g_t, h_t\rangle\}$$
$$= 2\,\mathrm{Re}\{1 - e^{i\beta}\langle g_t, h_t\rangle\} \to 0 \quad \text{as } t \to \infty,$$

so that (3.1.16) has been verified. ■

This argument was composed of little else but technical embellishments on the idea that, if $\{h_n\}$ is a normalized sequence converging rapidly to zero in the weak sense then the h_n can be approximated by an *orthonormal* sequence, and the projection onto the h_n-subspace will then look like $|h_n\rangle\langle h_n|$. The same idea is already needed in the proof of Theorem 3.1 to show that if $e^{-iHt}f$ goes weakly to zero then a projection E may be found such that $\langle e^{-iHt}f, Ee^{-iHt}f\rangle$ does *not* converge to a limit. In that case E could be defined, for a rapidly increasing sequence $\{t_n\}$, to be the projection onto the subspace spanned by the $e^{-iHt_n}f$ for even n. Then $\|Ee^{-iHt_n}f\|^2$ tends to 1 as $n \to \infty$ through even values, and to zero as $n \to \infty$ through odd values, so that a limit independent of the sequence $\{t_n\}$ cannot be found.

Remark

The proof of Theorem 3.2 relies very much on the weak convergence of h_t to zero. This condition cannot be removed altogether, but may be weakened by requiring only that there be no sequence $\{t_n\}$ such that $t_n \to \infty$ and such that h_{t_n} converge *strongly* to a limit. We omit the proof, which is similar to but technically more involved than the proof of Theorem 3.2. Since the existence of such sequences would lead us to think in terms of asymptotically stationary states, which we have discussed already, rather than free-particle states, the conclusion of Theorem 3.2 holds, in the present context, with great generality. Weak convergence of h_t to zero holds, in any case, for all of the examples of free evolution that we shall wish to consider.

We can now apply Theorem 3.2 to the asymptotic condition. The proof of the following result is an immediate consequence of Definition 3.3.

Theorem 3.3

Suppose that $e^{-iHt}g$ satisfies the asymptotic condition as $t \to \infty$, for some $\psi \in \mathscr{H}^{(0)}$, and that either $F(\boldsymbol{P}^{(0)}, t)$ converges weakly to zero or (more generally), there is no sequence $\{t_n\}$ such that $t_n \to \infty$ and such that $F(\boldsymbol{P}^{(0)}, t_n)\psi$ converges strongly to a limit. Then there exists a phase $e^{i\beta(t)}$ such that, for some $\psi \in \mathscr{H}^{(0)}$,

$$\lim_{t \to \infty} \|e^{-iHt} g - U_0\, e^{i\beta(t)}\, F(\boldsymbol{P}^{(0)}, t)\psi\| = 0. \tag{3.1.24}$$

A similar equation will hold for states satisfying the asymptotic condition as $t\to-\infty$. Equation (3.1.24), as we shall presently see, takes us to the very heart of scattering theory.

3.2 WAVE OPERATORS—THE GENERAL THEORY

Equation (3.1.24) is a consequence of the asymptotic condition, and expresses the fact that, for large positive times, evolution looks like free evolution. Since, as has been emphasised already, it will be an experimentally measurable property, for the quantum system in a given initial state g, whether or not evolution becomes asymptotically free, it is a consequence of the fundamental postulates of quantum mechanics that the states g satisfying the asymptotic condition for $t\to+\infty$ should define a *subspace* of the Hilbert space $\mathscr{H}$. We shall denote this subspace by $\mathscr{D}_0^+$ ($\mathscr{D}_0^-$ for $t\to-\infty$). Therefore (3.1.24) will hold for all $g\in\mathscr{D}_0^+$ and for all corresponding $\psi\in\mathscr{H}^{(0)}$, and we need no longer require that g and ψ be normalized.

The phase $e^{i\beta(t)}$, as is already implicit in our notation, may be taken, without loss of generality, to be independent of the vectors g and ψ. This is because, if g_1 corresponds to ψ_1, g_2 to ψ_2, and ag_1+bg_2 to ψ_{12}, the consistency of (3.1.24) for the three possible choices of g requires that

$$ae^{i\beta_1}\psi_1+be^{i\beta_2}\psi_2 \quad \text{approach} \quad e^{i\beta_{12}}\psi_{12}$$

in norm as $t\to\infty$, where $\|\psi_{12}\|=\|a\psi_1+b\psi_2\|$.

If $\langle\psi_1,\psi_2\rangle\neq0$ then this can only happen, for all possible choices of the constants a and b, if $e^{i(\beta_1-\beta_2)}\to1$, in which case, without loss of generality, we may take $\beta_1=\beta_2$ for *all* t. (If $\langle\psi_1,\psi_2\rangle=0$ then we simply find a third vector that is not orthogonal to ψ_1 or to ψ_2, and the same conclusion follows.) Therefore $\beta_1=\beta_2=\beta$ for all possible g and ψ.

There is now nothing to prevent us, in (3.1.24), from absorbing the phase $e^{i\beta(t)}$ into the definition of the function F, so that we have, without loss of generality,

$$\lim_{t\to\infty}\|e^{-iHt}g-U_0F(\boldsymbol{P}^{(0)},t)\psi\|=0, \tag{3.2.1}$$

where $\psi\in\mathscr{H}^{(0)}$ and $g\in\mathscr{D}_0^+$. The correspondence between ψ and g is linear, and we may define a *wave operator* $\Omega_+^{(0)}$ by the equation $g=\Omega_+^{(0)}\psi$. Since e^{iHt} is unitary, this relation may be expressed as follows.

Definition 3.4

The wave operator $\Omega_+^{(0)}$ is defined by

$$\Omega_+^{(0)}\psi=\text{s-}\lim_{t\to\infty}e^{iHt}\,U_0F(\boldsymbol{P}^{(0)},t)\psi \tag{3.2.2}$$

for $\psi \in \mathscr{H}^{(0)}$. Similarly, if we are concerned with evolution as $t \to -\infty$, so that $F(\boldsymbol{P}^{(0)}, t)$ is defined for $t < 0$, the wave operator $\Omega_{-}^{(0)}$ will be defined by

$$\Omega_{-}^{(0)} = \underset{t \to -\infty}{\text{s-lim}}\ e^{iHt}\, U_0 F(\boldsymbol{P}^{(0)}, t)\psi. \tag{3.2.2}$$

These two equations for the wave operators $\Omega_{\pm}^{(0)}$ are fundamental to the whole of time-dependent scattering theory. Let us explore some of their consequences.

Since all three operators e^{iHt}, U_0 and $F(\boldsymbol{P}^{(0)}, t)$ are unitary, each wave operator will be a strong limit of unitary operators. Hence $\Omega_{\pm}^{(0)}$ are both linear and isometric. There is a slight distinction to be made here. $\Omega_{\pm}^{(0)}$ are *unitary* regarded as mappings from $\mathscr{H}^{(0)}$ to $\mathscr{D}_0^{\pm}$ (since the range is the whole of $\mathscr{D}_0^{\pm}$), but *isometric* regarded as mappings from $\mathscr{H}^{(0)}$ to $\mathscr{H}$, since the range of $\Omega_{\pm}^{(0)}$ will not in general be the whole of $\mathscr{H}$. Thus

$$D(\Omega_{\pm}^{(0)}) = \mathscr{H}^{(0)}, \quad \text{range } (\Omega_{\pm}^{(0)}) = \mathscr{D}_0^{\pm} \tag{3.2.3}$$

The subspace $\mathscr{D}_0^{+}$ bears a very special relation to the Hamiltonian H. If $g \in \mathscr{D}_0^{+}$ is any vector satisfying the asymptotic condition then $e^{-iH\tau}g$ must also satisfy the asymptotic condition, i.e. belong to $\mathscr{D}_0^{+}$. This is because the evolution $e^{-iHt}(e^{-iH\tau}g)$ differs from $e^{-iHt}g$ only in the choice of the zero of time (t becomes $t+\tau$), so that if the initial state g becomes asymptotically free then so will the initial state $e^{-iH\tau}g$ (but τ units of time earlier!). Hence $\mathscr{D}_0^{+}$ is invariant under the operation $e^{-iH\tau}$ for any $\tau \in \mathbb{R}$, or, in other words, $\mathscr{D}_0^{+}$ is a reducing subspace for H.

For $g \in \mathscr{D}_0^{+}$, let us write $e^{-iH\tau}g = \Omega_{+}^{(0)}\psi_\tau$, so that ψ_τ is a one-parameter family of normalized vectors in $\mathscr{H}^{(0)}$, satisfying $\psi_0 = \psi$. Then

$$e^{-iH\tau}g = \underset{t \to \infty}{\text{s-lim}}\ e^{iHt}\, U_0 F(\boldsymbol{P}^{(0)}, t)\psi_\tau. \tag{3.2.4}$$

On the other hand, substituting $t \to t+\tau$ in (3.2.2), we find

$$e^{-iH\tau}g = e^{-iH\tau}\Omega_{+}^{(0)}\psi = \underset{t \to \infty}{\text{s-lim}}\ e^{iHt}\, U_0 F(\boldsymbol{P}^{(0)}, t+\tau)\psi. \tag{3.2.4$'$}$$

Comparing (3.2.4) and (3.2.4)$'$, we see that

$$\lim_{t \to \infty} \|F(\boldsymbol{P}^{(0)}, t)\psi_\tau - F(\boldsymbol{P}^{(0)}, t+\tau)\psi\| = 0. \tag{3.2.5}$$

Applying within the norm the unitary operator $F^*(\boldsymbol{P}^{(0)}, t)$, we may deduce the existence of the limit $U(\tau)$, defined on $\mathscr{H}^{(0)}$ by

$$U(\tau)\psi = \underset{t \to \infty}{\text{s-lim}}\ F^*(\boldsymbol{P}^{(0)}, t) F(\boldsymbol{P}^{(0)}, t+\tau). \tag{3.2.6}$$

Comparison with (3.2.5) shows that

$$U(\tau)\psi = \psi_\tau.$$

Hence $\Omega_+^{(0)}U(\tau)\psi=\Omega_+^{(0)}\psi_\tau=\mathrm{e}^{-\mathrm{i}H\tau}g=\mathrm{e}^{-\mathrm{i}H\tau}\Omega_+^{(0)}\psi$, so that

$$\Omega_+^{(0)}U(\tau)=\mathrm{e}^{-\mathrm{i}H\tau}\Omega_+^{(0)}. \tag{3.2.7}$$

It follows, by repeated application of this formula, that

$$\begin{aligned}\Omega_+^{(0)}U(\tau_1+\tau_2)&=\mathrm{e}^{-\mathrm{i}H(\tau_1+\tau_2)}\Omega_+^{(0)}=\mathrm{e}^{-\mathrm{i}H\tau_1}(\mathrm{e}^{-\mathrm{i}H\tau_2}\Omega_+^{(0)})\\&=\mathrm{e}^{-\mathrm{i}H\tau_1}\Omega_+^{(0)}U(\tau_2)=\Omega_+^{(0)}U(\tau_1)U(\tau_2).\end{aligned}$$

Since $\Omega_+^{(0)}$ is isometric and

$$\Omega_+^{(0)}\{U(\tau_1+\tau_2)-U(\tau_1)U(\tau_2)\}=0,$$

we can only have

$$U(\tau_1+\tau_2)=U(\tau_1)U(\tau_2) \tag{3.2.8}$$

Since $\mathrm{e}^{-\mathrm{i}H\tau}$ is strongly continuous in τ, (3.2.7) implies that the same is true of the one-parameter family of operators $U(\tau)$. Hence (3.2.8) allows us to infer, by Stone's Theorem, the existence of a self-adjoint operator that generates this unitary group of operators. We shall refer to this operator as the *free Hamiltonian.*

Definition 3.5

The free Hamiltonian $H^{(0)}$ is the self-adjoint operator in $\mathscr{H}^{(0)}$ defined implicitly by the equation

$$U(\tau)=\mathrm{e}^{-\mathrm{i}H^{(0)}\tau} \tag{3.2.9}$$

where $U(\tau)$ is the one-parameter unitary group of operators given by (3.2.6).

We can see, from (3.2.6) and (3.2.9), that $H^{(0)}$ is some function of the free-momentum operator $\boldsymbol{P}^{(0)}$.

We shall also suppose that (3.2.6) holds with the limit $t\to-\infty$ on the right-hand side. This establishes a correspondence between the family of operators $F(\boldsymbol{P}^{(0)}, t)$ for $t>0$ and the "same" family for $t<0$. Free evolutions for positive and for negative times should live in the same Hilbert space $\mathscr{H}^{(0)}$ and should derive from the same free Hamiltonian $H^{(0)}$. This is what we mean by the identification of a particular asymptotic type of evolution as $t\to+\infty$ with the corresponding evolution as $t\to-\infty$. The correspondence between F at $t>0$ and F at $t<0$ will generally be given by the operation of time reversal, i.e. in this case

$$F(\boldsymbol{P}^{(0)}, -t)=F^*(-\boldsymbol{P}^{(0)}, t).$$

Since evolution in $\mathscr{H}^{(0)}$ should reflect the symmetries of the Hamiltonian H, the free Hamiltonian will normally be invariant under time reversal. In the following paragraphs, we shall usually deal with the limit $t \to +\infty$, but it will always be clear that equivalent results for the limit $t \to -\infty$ can be written down.

Equation (3.2.7) immediately gives

$$\Omega_+^{(0)} e^{-iH^{(0)}\tau} = e^{-iH\tau} \Omega_+^{(0)}. \tag{3.2.10}$$

This is the so called *intertwining relation* between the free Hamiltonian $H^{(0)}$ and the (total) Hamiltonian H. Differentiating with respect to τ at $\tau = 0$, we can express the intertwining relation as

$$\Omega_+^{(0)} H^{(0)} = H \Omega_+^{(0)}. \qquad (3.2.10)'$$

We see that $H^{(0)}$ becomes H on commuting through the wave operator. It follows easily from (3.2.10)′ that the resolvent operator satisfies the intertwining relation, i.e.

$$\Omega_+^{(0)} (H^{(0)} - z)^{-1} = (H - z)^{-1} \Omega_+^{(0)}. \qquad (3.2.10)''$$

Via the resolvent operator, we can proceed to spectral projections (2.6.10), and thence to functions of the Hamiltonian. This more general statement of intertwining takes the form

$$\Omega_+^{(0)} Q(H^{(0)}) = Q(H) \Omega_+^{(0)}, \qquad (3.2.10)'''$$

The wave operator, being isometric, also satisfies

$$(\Omega_+^{(0)})^* \Omega_+^{(0)} = I^{(0)} \tag{3.2.11}$$

($I^{(0)}$ is the identity operator in $\mathscr{H}^{(0)}$), so that from (3.2.10)′ we obtain the expression

$$H^{(0)} = (\Omega_+^{(0)})^* H \Omega_+^{(0)} \tag{3.2.12}$$

for the free Hamiltonian in closed from. Since $\Omega_+^{(0)}$, regarded as an operator from $\mathscr{H}^{(0)}$ to $\mathscr{D}_0^+$, is unitary, (3.2.12) shows that H, *restricted to* $\mathscr{D}_0^+$ (that is, to the range of $\Omega_+^{(0)}$), is unitarily equivalent to $H^{(0)}$. Or $H^{(0)}$ is unitarily equivalent to the restriction of H to a reducing subspace. This means, for example, that the spectral properties of $H^{(0)}$ have important consequences for the spectral properties of H. Since it is often the case that a spectral analysis of the free Hamiltonian may be carried out in a relatively straightforward manner, we are here furnished with a powerful tool in the spectral theory of self-adjoint operators.

Equation (3.2.6) provides further characterization of the unitary operators $F(\boldsymbol{P}^{(0)}, t)$, namely

$$\operatorname*{s-lim}_{t \to \infty} \{F(\boldsymbol{P}^{(0)}, t + \tau) - e^{-iH^{(0)}\tau} F(\boldsymbol{P}^{(0)}, t)\} = 0. \tag{3.2.13}$$

The interpretation of this equation is that, for large t, the evolution of states in the free Hilbert space $\mathscr{H}^{(0)}$ between times t and $t+\tau$ looks like the Hamiltonian evolution generated, over the time interval τ, by $H^{(0)}$. Therefore as $t\to\infty$ there will be increasingly long periods of time over which evolution in $\mathscr{H}^{(0)}$ looks like free-Hamiltonian evolution. Since we have now met two distinct types of evolution in $\mathscr{H}^{(0)}$, it will be convenient to introduce a new terminology which differs somewhat from our previous usage, and at the same time to formalize the concept of free evolution.

Definition 3.6

A *modified free evolution* in the free Hilbert space $\mathscr{H}^{(0)}$ is a one-parameter family of unitary operators $F(\boldsymbol{P}^{(0)}, t)$ satisfying (3.2.13). We shall refer to the special case of the one-parameter family $e^{-iH^{(0)}t}$ simply as the *free evolution* in $\mathscr{H}^{(0)}$.

If, in (3.2.13), we write

$$F(\boldsymbol{P}^{(0)}, t)=e^{iH^{(0)}t}Z^{(0)}(\boldsymbol{P}^{(0)}, t) \tag{3.2.14}$$

then this equation becomes

$$\operatorname*{s-lim}_{t\to\infty}\{Z^{(0)}(t+\tau)-Z^{(0)}(t)\}=0, \tag{3.2.15}$$

expressing the fact that $Z^{(0)}(t)$ varies more and more slowly with t as t increases.

The simplest "solution" of (3.2.15) is that in which $Z^{(0)}(t)$ is assumed to be independent of t. In this case, $Z^{(0)}(t)\equiv Z^{(0)}$ is a unitary operator commuting with $H^{(0)}$, and $F(\boldsymbol{P}^{(0)}, t)=Z^{(0)}e^{-iH^{(0)}t}$. Redefining the unitary operator U_0 in (3.2.2) to include the factor $Z^{(0)}$, the wave operators then become

$$\Omega_{\pm}^{(0)}\psi=\operatorname*{s-lim}_{t\to\pm\infty} e^{iHt}U_0e^{-iH^{(0)}t}\psi. \tag{3.2.16}$$

In cases in which the Hilbert space $\mathscr{H}^{(0)}$ can be identified with a subspace $\mathscr{D}_0$ of $\mathscr{H}$, so that U_0 becomes an identification operator (i.e. the identity operator from $\mathscr{D}_0$ to $\mathscr{D}_0$), (3.2.16) is simply

$$\Omega_{\pm}^{(0)}\psi=\operatorname*{s-lim}_{t\to\pm\infty} e^{iHt}e^{-iH^{(0)}t}\psi. \tag{3.2.16$'$}$$

Equation (3.2.16)′ is the definition of the wave operators that has, to date, received the greatest attention in scattering theory, despite the fact it does not apply to Coulomb scattering, where the more general formula (3.2.2) has to be used.

Equations (3.2.16) and (3.2.16)′ describe the situation in which states look asymptotically like states evolving according to the time development generated by the free Hamiltonian $H^{(0)}$. If (3.2.16)′ holds then let us suppose that $\psi \in D(H^{(0)})$ and that $\|H^{(0)}e^{-iHt}g\|$ is bounded, where $g = \Omega_{+}^{(0)}\psi$. ($H^{(0)}$ is defined *ab initio* only on $\mathscr{D}_0$, but may be extended by taking, for example, $H^{(0)} = 0$ on $\mathscr{D}_0^{\perp}$.) Then

$$\begin{aligned}
&\lim_{t\to\infty} \langle e^{-iHt}g, H^{(0)}e^{-iHt}g\rangle \\
&\qquad = \lim_{t\to\infty} \langle H^{(0)}e^{-iHt}g, e^{-iHt}g\rangle \\
&\qquad = \lim_{t\to\infty} \langle H^{(0)}e^{-iHt}g, e^{-iH^{(0)}t}\psi\rangle
\end{aligned}$$

(on using the fact that $e^{-iHt}g$ approaches $e^{-iH^{(0)}t}\psi$ in norm)

$$\begin{aligned}
&= \lim_{t\to\infty} \langle g, e^{iHt}e^{-iH^{(0)}t}H^{(0)}\psi\rangle \\
&= \langle g, \Omega_{+}^{(0)}H^{(0)}\psi\rangle \\
&= \langle g, Hg\rangle
\end{aligned}$$

(on using the intertwining relation (3.2.10)′). Hence

$$\lim_{t\to\infty} \langle e^{-iHt}g, (H - H^{(0)})e^{-iHt}g\rangle = 0. \tag{3.2.17}$$

Equation (3.2.17) is one of a number of equations that could be written down expressing the fact that, for a state satisfying the asymptotic condition, the free Hamiltonian $H^{(0)}$ and the total Hamiltonian H are asymptotically the same. Another way of expressing this is to define an *interaction Hamiltonian* $V^{(0)}$ by $H = H^{(0)} + V^{(0)}$, in which case, as $t \to \infty$, $V^{(0)}$ will asymptotically approach zero. In fact this argument, to be made formally correct, requires some care owing to the fact that domain questions may have to be taken into account. ($D(H)$ and $D(H^{(0)})$ might intersect only in the zero vector, so that the right-hand side of (3.2.17) is then ill-defined.) Nevertheless we can obtain an intuitive picture of what is happening by these considerations. See also Section 3.3, where it will be found quite generally that $H_0 = U_0 H^{(0)} U_0^{-1}$ is asymptotically the same as H.

Equation (3.2.17) also holds in the more general case (3.2.2), whenever U_0 is an identification operator. The difference between the two cases lies in the rate at which the interaction Hamiltonian approaches zero at large times. To anticipate a little some of the ideas of the following chapters, the difference between free evolution and modified evolution can be expressed more explicitly as follows. (W shall, however, in the present heuristic treatment, proceed without due attention to questions of domains.)

We suppose that, for the initial state $g = \Omega_{+}^{(0)}\psi$, $V^{(0)}$ tends to zero sufficiently rapidly that

$$\int_0^\infty \|V^{(0)} e^{-iHt} g\| \, dt < \infty.$$

This means, in particular, that $\langle g, e^{-iHt} g \rangle$ cannot decrease to zero less rapidly than t^{-1}. Then

$$\frac{d}{dt} e^{iH(0)t} e^{-iHt} g = -i \, e^{iH(0)t} V^{(0)} e^{-iHt} g,$$

which is integrable in norm from 0 to ∞. This implies the existence of the limit

$$Wg = \operatorname*{s-lim}_{t \to \infty} e^{iH(0)t} e^{-iHt} g.$$

Hence

$$\begin{aligned} &\operatorname*{s-lim}_{t \to \infty} e^{iH(0)t} F(\boldsymbol{P}^{(0)}, t)\psi \\ &\quad = \operatorname*{s-lim}_{t \to \infty} e^{iH(0)t} e^{-iHt} (e^{iHt} F(\boldsymbol{P}^{(0)}, t)\psi) \\ &\quad = W\Omega_+^{(0)}\psi = Z\psi, \end{aligned}$$

using (3.2.2) with $U_0 = I$. In the notation of (3.2.14), we then have

$$\operatorname*{s-lim}_{t \to \infty} Z^{(0)}(\boldsymbol{P}^{(0)}, t)\psi = Z\psi,$$

so that, in (3.2.2), $F(\boldsymbol{P}^{(0)}, t)$ becomes $Z \, e^{-iH(0)t}$ in the limit. Hence (3.2.16) holds, with redefinition of U_0 to be the unitary mapping from $\mathscr{D}_0$ to the range of Z.

Therefore if $V^{(0)}$, acting on $e^{-iHt} g$, where g satisfies the asymptotic condition, converges to zero sufficiently rapidly, there is no need to modify the free evolution $e^{-iH(0)t}$. On the other hand, if $V^{(0)}$ converges to zero more slowly then it may be necessary to modify the free evolution to take account of the residual effect of the interaction at large times. This will be necessary, for example, in Coulomb scattering, about which we shall have more to say in Chapters 8 and 9. In Coulomb scattering, it turns out that the decay of $V^{(0)}$ is like t^{-1}, and that $F(\boldsymbol{P}^{(0)}, t)$ has to be defined very precisely in order to accurately reflect the influence of the potential at large distances from the scattering centre.

3.3 ASYMPTOTIC LIMITS OF MOMENTUM OBSERVABLES

One of the original motivations for the study of wave operators was to find a mathematical expression of the idea that for certain initial states (i.e. those satisfying the asymptotic condition), momentum observables approach limits asymptotically as $t \to \infty$, or as $t \to -\infty$. It will throw further light on the nature of wave operators to re-examine this idea in the light of what we have discovered so far. In doing so, we shall find that momentum observables converge to limits, even in the strong sense.

We suppose that the wave operator $\Omega_+^{(0)}$ is defined, by (3.2.2), as a mapping from the free Hilbert space $\mathscr{H}^{(0)}$ into the Hilbert space $\mathscr{H}$. As in (3.1.10), we define

$$\boldsymbol{P}_0 = U_0 \boldsymbol{P}^{(0)} U_0^{-1}.$$

The components of $\boldsymbol{P}_0$ are self-adjoint operators in the subspace $\mathscr{D}_0$, which we extend to self-adjoint operators in $\mathscr{H}$. It will be unimportant for the moment how precisely this extension is carried out. The following result allows us to conclude that $\boldsymbol{P}_0$, defined as an operator-valued function of t in the Heisenberg picture, converges strongly to a limit on states satisfying the asymptotic condition. To avoid domain questions, we shall deal with functions of $\boldsymbol{P}_0$, rather than $\boldsymbol{P}_0$ directly.

Theorem 3.4

Suppose that $g \in \mathscr{H}$ satisfies the asymptotic condition, and let $\Phi(\boldsymbol{P}_0)$ be a bounded operator function of $\boldsymbol{P}_0$, acting in $\mathscr{H}$. Then

$$\operatorname*{s-lim}_{t \to \infty} e^{iHt} \Phi(\boldsymbol{P}_0) e^{-iHt} g = \Omega_+^{(0)} \Phi(\boldsymbol{P}^{(0)}) (\Omega_+^{(0)})^* g. \tag{3.3.1}$$

Proof

We let $g = \Omega_+^{(0)} \psi$ for $\psi \in \mathscr{H}^{(0)}$. Then, as $t \to \infty$, $e^{-iHt} g$ approaches $U_0 F(\boldsymbol{P}^{(0)}, t)\psi$ in norm, so the limit on the left-hand side of (3.3.1) is

$$\begin{aligned}
&\operatorname*{s-lim}_{t \to \infty} e^{iHt} \Phi(\boldsymbol{P}_0) U_0 F(\boldsymbol{P}^{(0)}, t)\psi \\
&\quad = \operatorname*{s-lim}_{t \to \infty} e^{iHt} (U_0 \Phi(\boldsymbol{P}^{(0)}) U_0^{-1}) U_0 F(\boldsymbol{P}^{(0)}, t)\psi \\
&\quad = \operatorname*{s-lim}_{t \to \infty} (e^{iHt} U_0 F(\boldsymbol{P}^{(0)}, t)) \Phi(\boldsymbol{P}^{(0)})\psi \\
&\quad = \Omega_+^{(0)} \Phi(\boldsymbol{P}^{(0)})\psi,
\end{aligned}$$

on using the definition of the wave operator. Equation (3.3.1) now follows on noting, from (3.2.11), that

$$(\Omega_+^{(0)})^* g = (\Omega_+^{(0)})^* \Omega_+^{(0)} \psi = \psi.$$

∎

It is instructive to seek a possible converse of this result. In particular, we should ask the following question. Given momentum operators $\boldsymbol{P}_0$ that are known to converge strongly in the Heisenberg picture as $t \to \infty$, to what extent can we reconstruct the modified free evolution, wave operator and so on?

Theorem 3.5

Let $\boldsymbol{P}^{(0)}$ be the k-component momentum operator acting in the free Hilbert space $\mathscr{H}^{(0)} = L^2(\mathbb{R}^k)$. Let U_0 be a unitary map from $\mathscr{H}^{(0)}$ onto the subspace $\mathscr{D}_0$ of a Hilbert space $\mathscr{H}$. Define $\boldsymbol{P}_0$, acting in $\mathscr{H}$, to be some self-adjoint extension of $U_0 \boldsymbol{P}^{(0)} U_0^{-1}$, acting in $\mathscr{D}_0$. Let $x \in \mathscr{H}^{(0)}$ be any unit vector such that the set $\{\Phi(\boldsymbol{P}^{(0)})x\}$, as the bounded function Φ is varied, is dense in $\mathscr{H}^{(0)}$, and let $y \in \mathscr{H}$ be a unit vector such that

(a) $e^{-iHt}y$ belongs asymptotically to $\mathscr{D}_0$ as $t \to \infty$; in other words $P_{\mathscr{D}_0}^{\perp} e^{-iHt} y$ converges strongly to zero, where $P_{\mathscr{D}_0}$ is the orthogonal projection onto the subspace $\mathscr{D}_0$,
(b) $e^{iHt}\Phi(\boldsymbol{P}_0)e^{-iHt}y$ converges strongly to a limit as $t \to \infty$, for bounded functions Φ.

Then there exists a linear operator $W_+^{(0)}$ from $\mathscr{H}^{(0)}$ to $\mathscr{H}$, together with a one-parameter family $\{F_t\}$ of linear operators from $\mathscr{H}^{(0)}$ to $\mathscr{H}^{(0)}$, such that

(i) both $W_+^{(0)}$ and $\{F_t\}$ are densely defined in $\mathscr{H}^{(0)}$,
(ii) F_t commutes with functions of $\boldsymbol{P}^{(0)}$,
(iii) $W_+^{(0)} x = y$,
(iv)
$$\operatorname*{s-lim}_{t \to \infty} e^{iHt} U_0 F_t \psi = W_+^{(0)} \psi \tag{3.3.2}$$
for a dense set of ψ in $\mathscr{H}^{(0)}$.

Proof

For Φ bounded, we define F_t by

$$F_t \Phi(\boldsymbol{P}^{(0)})x = \Phi(\boldsymbol{P}^{(0)}) U_0^{-1} X_0(t) e^{-iHt} y, \tag{3.3.3}$$

where $\{X_0(t)\}$ is a one-parameter family of bounded operators from $\mathscr{H}$ to $\mathscr{D}_0$ satisfying

$$\operatorname*{s-lim}_{t \to \infty} e^{iHt} X_0(t) e^{-iHt} y = y. \tag{3.3.4}$$

That such a family exists is a consequence of assumption (a) of the theorem, from which it follows that a possible choice is $X_0(t) = \boldsymbol{P}_{\mathscr{D}_0}$, independently of t. On the other hand, *if* the wave operator $\Omega_+^{(0)}$, defined in the usual sense, were known to exist, and *if* $y = \Omega_+^{(0)}x$ (in which case certainly all hypotheses of the theorem are satisfied) then a possible choice would be

$$X_0(t) = U_0 F(\boldsymbol{P}^{(0)}, t)(\Omega_+^{(0)})^* e^{iHt}, \tag{3.3.5}$$

so that

$$F_t \equiv F(\boldsymbol{P}^{(0)}, t).$$

Note that if $\Phi_1(\boldsymbol{P}^{(0)})x = \Phi_2(\boldsymbol{P}^{(0)})x$ then $(\Phi_1 - \Phi_2)\Phi x = 0$, so that the property of the unit vector x that $\{\Phi x\}$ is dense in $\mathscr{H}^{(0)}$ implies that $\Phi_1 = \Phi_2$. Hence (3.3.3) defines the linear operator F_t uniquely. If Ψ is any other bounded function of $\boldsymbol{P}^{(0)}$ then we also have

$$F_t \Psi\Phi x = \Psi\Phi U_0^{-1} X_0 e^{-iHt} y = \Psi F_t \Phi x.$$

Hence $[\Psi, F_t] = 0$, which confirms (ii). The F_t are defined on a dense subset of $\mathscr{H}^{(0)}$. Moreover,

$$\begin{aligned} W_+^{(0)}\Phi x &\equiv \operatorname*{s-lim}_{t\to\infty} e^{iHt} U_0 F_t \Phi x \\ &= \operatorname*{s-lim}_{t\to\infty} e^{iHt} U_0 \Phi U_0^{-1} X_0 e^{-iHt} y \\ &= \operatorname*{s-lim}_{t\to\infty} e^{iHt} \Phi(\boldsymbol{P}_0) e^{-iHt} e^{iHt} X_0 e^{-iHt} y \\ &= \operatorname*{s-lim}_{t\to\infty} e^{iHt} \Phi(\boldsymbol{P}_0) e^{-iHt} y. \end{aligned} \tag{3.3.6}$$

(using (3.3.4)).

And this limit is known to exist, from (b) of the theorem. Since $\{\Phi x\}$ is dense in $\mathscr{H}^{(0)}$, so that (i) is now completely verified, it remains only to prove (iii). But this follows immediately from our evaluation of $W_+^{(0)}\Phi x$, on taking the special case $\Phi \equiv 1$, and the proof of the theorem is finished. ∎

Theorem 3.5 enables us to construct "wave operators" $W_+^{(0)}$, starting from the hypothesis of strong convergence of momentum operators $\boldsymbol{P}_0$ in the Heisenberg picture. The $W_+^{(0)}$ lack, however, one essential property of wave operators. Although $\|W_+^{(0)}x\| = \|y\| = \|x\|$, there is no reason in general to suppose that $W_+^{(0)}$ is isometric, and indeed $W_+^{(0)}$ may fail to be isometric under the hypotheses of the theorem. In order to remedy this defect, it is necessary to look at the norm of the right-hand side of (3.3.6).

We define a linear functional $\mathscr{L}$ on positive bounded functions Φ^2 from $\mathbb{R}^k$ to $\mathbb{R}$ by

$$\begin{aligned} \mathscr{L}(\Phi^2) &= \lim_{t\to\infty} \|\Phi(\boldsymbol{P}_0)\, e^{-iHt} y\|^2 \\ &= \lim_{t\to\infty} \langle e^{-iHt} y, \Phi^2(\boldsymbol{P}_0)\, e^{-iHt} y\rangle. \end{aligned} \tag{3.3.7}$$

The most that can be said on general grounds is that (3.3.7) defines a *measure*, according to the formula

$$\mathscr{L}(\Phi^2) = \int \Phi^2 \, \mathrm{d}\mu(\boldsymbol{p}).$$

Let us make, at this point, the additional hypothesis that (3.3.7) defines a *function*, according to the formula

$$\mathscr{L}(\Phi^2) = \int [\Phi(\boldsymbol{p})]^2 u^2(\boldsymbol{p}) \, \mathrm{d}^k \boldsymbol{p}, \tag{3.3.8}$$

where u is some non-negative function from $\mathbb{R}^k$ to $\mathbb{R}$. Setting $\Phi \equiv 1$ in (3.3.7), we see that in fact u is square-integrable and has unit norm, regarded as an element of $\mathscr{H}^{(0)} = L^2(\mathbb{R}^k)$ (momentum space). The vector x in (3.3.6) could be an arbitrary element of $\mathscr{H}^{(0)}$. Now we *choose* x in such a way that $|x(\boldsymbol{p})| = u(\boldsymbol{p})$, where $x(\boldsymbol{p})$ denotes the representative function of x in $\mathscr{H}^{(0)}$. Assuming that $u(\boldsymbol{p}) > 0$ for almost all $\boldsymbol{p} \in \mathbb{R}^k$, $\{\Phi x\}$ will then be dense in $\mathscr{H}^{(0)}$, so that this choice of x satisfies the conditions of the theorem. In this case, we have

$$\begin{aligned} &\|W_+^{(0)} \Phi x\|^2 \\ &= \mathscr{L}(|\Phi|^2) = \int |\Phi(\boldsymbol{p})|^2 u^2(\boldsymbol{p}) \, \mathrm{d}^k \boldsymbol{p} \\ &= \int |\Phi(\boldsymbol{p}) x(\boldsymbol{p})|^2 \, \mathrm{d}^k \boldsymbol{p} = \|\Phi x\|^2, \end{aligned}$$

so that $W_+^{(0)}$ is now isometric.

The additional hypothesis (3.3.8), which makes $W_+^{(0)}$ isometric, may be interpreted as asserting that the probability distribution, in momentum space, of the state $\mathrm{e}^{-\mathrm{i}Ht} y$, approaches a limit as $t \to \infty$, and that this limiting distribution is given by the density function $u^2(\boldsymbol{p})$. This additional hypothesis may fail if, for example, the limiting momentum is precisely concentrated in a particular direction, in which case $u^2(\boldsymbol{p})$ corresponds to a generalized function that is singular along a line. So the physical interpretation of (3.3.8), and of the conditions under which it might not apply, may be understood in terms of the asymptotic behaviour of momentum variables.

It should be noted that $|x(\boldsymbol{p})| = u(\boldsymbol{p})$ is the *only* possible choice of the amplitude of the wave function $x(\boldsymbol{p})$ that will guarantee the isometry of $W_+^{(0)}$, since the density function of the distribution is determined almost everywhere. Hence the only ambiguity in the definition of $W_+^{(0)}$ by (3.3.6) lies in the choice of the phase of the wave function $x(\boldsymbol{p})$ for varying $\boldsymbol{p}$ in $\mathbb{R}^k$.

Since $W_+^{(0)}$ depends on x only to the extent of this phase, we can see that $W_+^{(0)}$ is determined up to multiplication by a phase $U(\boldsymbol{P}^{(0)})$. In other words, two isometric operators $W_+^{(0)}$ and $(W_+^{(0)})'$, defined with different choices of the unit vector x in $\mathscr{H}^{(0)}$, will be related by an equation of the form

$$(W_+^{(0)})' = W_+^{(0)} U(\boldsymbol{P}^{(0)}).$$

Even the additional hypothesis (3.3.8) does not guarantee the existence of a true wave operator $\Omega_+^{(0)}$. This is because the isometry of $W_+^{(0)}$ allows us only to deduce the *asymptotic* isometry (in an obvious sense) of the F_t as $t \to \infty$. To illustrate this point, we consider the one-parameter family $\{f_t\}$ of vectors in $L^2(0, 1)$ defined by

$$f_t(p) = \sin pt + \cos pt.$$

Then $\lim_{t\to\infty} \|f_t\|^2 = 1$, and, moreover, for Φ bounded,

$$\lim_{t\to\infty} \|\Phi(p) f_t(p)\|^2 = \int_0^1 |\Phi(p) f(p)|^2 \, \mathrm{d}p,$$

with $f(p) \equiv 1$. On the other hand, f_t by no means approaches $|f|$ in norm as $t \to \infty$. So we would be unable to represent the asymptotic evolution described by the family $\{f_t\}$ in terms of wave operators with F_t unitary. This illustrative example shows that more has to be assumed in order to make the F_t unitary.

Suppose that it is given, in addition to the assumption of theorem 3.5, that

$$\lim_{t\to\infty} \|\mathrm{e}^{-iHt} y - U_0 u_t\| = 0, \tag{3.3.9}$$

for some one-parameter family of unit vectors in $\mathscr{H}^{(0)}$, such that, in $L^2(\mathbb{R}^k)$, $|u_t(\boldsymbol{p})| = u(\boldsymbol{p})$, where the right-hand side is strictly positive and independent of t. Then certainly (3.3.8) is satisfied, with $\mathscr{L}(\Phi^2)$ defined by (3.3.7). The one-parameter family $X_0(t)$ may now be defined by

$$X_0(t) = X(\boldsymbol{P}_0, t) P_{\mathscr{D}_0},$$

where

$$X(\boldsymbol{p}, t) = \frac{u_t(\boldsymbol{p})}{(U_0^{-1} P_{\mathscr{D}_0} \mathrm{e}^{-iHt} y)(\boldsymbol{p})},$$

assuming that the denominator does not vanish, which it is reasonable to suppose, for suitable y. With this choice of $X_0(t)$, (3.3.4) is satisfied, and $W_+^{(0)}$ may be defined by (3.3.6), again with $|x(\boldsymbol{p})| = u(\boldsymbol{p})$. It is then not difficult to verify that $\|F_t\| = 1$ for all t. Since F_t has been shown to commute with functions of $\boldsymbol{P}^{(0)}$, it follows in this case that

$$F_t = F(\boldsymbol{P}^{(0)}, t)$$

for some F satisfying $|F(\boldsymbol{p}, t)| \equiv 1$.

Hence (3.3.9) represents the extra assumption that is needed to assert that the operator $W_+^{(0)}$ defined by (3.3.6) is a true wave operator. In that case only are we entitled to write $\Omega_+^{(0)} = W_+^{(0)}$. Given the free Hilbert space $\mathscr{H}^{(0)}$, the unitary map U_0 from $\mathscr{H}^{(0)}$ into $\mathscr{H}$ and a vector y on which functions of $\boldsymbol{P}_0$ converge in the Heisenberg picture, we can use (3.3.6) to construct a wave operator $\Omega_+^{(0)}$. Moreover, as we have seen, *any* wave operator will be given by $\Omega_+^{(0)} = W_+^{(0)}$ as a special case of (3.3.6), for example with $X_0(t)$ given by (3.3.5), with y any unit vector satisfying the asymptotic condition, and $F_t \equiv F(\boldsymbol{P}^{(0)}, t)$. The wave operator will not be unique, since different choices for the family $F(\boldsymbol{P}^{(0)}, t)$ are possible. However, as we have seen in connection with the $W_+^{(0)}$, any two possibilities for $\Omega_+^{(0)}$ are necessarily related by

$$(\Omega_+^{(0)})' = \Omega_+^{(0)} U(\boldsymbol{P}^{(0)})$$

for some unitary function U of the free momentum. We let $(H^{(0)})'$ be the free Hamiltonian corresponding to $(\Omega_+^{(0)})'$, and $H^{(0)}$ corresponding to $\Omega_+^{(0)}$. Then by the intertwining relation (3.2.10)′ applied successively to each of the possible wave operators, we have

$$H(\Omega_+^{(0)})' = (\Omega_+^{(0)})'(H^{(0)})' = \Omega_+^{(0)} U(\boldsymbol{P}^{(0)})(H^{(0)})'.$$

whereas also

$$H(\Omega_+^{(0)})' = H\Omega_+^{(0)} U(\boldsymbol{P}^{(0)}) = \Omega_+^{(0)} H^{(0)} U(\boldsymbol{P}^{(0)}).$$

Since $\Omega_+^{(0)}$ is isometric, these two results may be combined to give

$$U(\boldsymbol{P}^{(0)})(H^{(0)})' = H^{(0)} U(\boldsymbol{P}^{(0)}).$$

But $H^{(0)}$ commutes with $U(\boldsymbol{P}^{(0)})$, so that we may infer that $(H^{(0)})' = H^{(0)}$.

In other words, the free Hamiltonian $H^{(0)}$ is completely determined by a specification of $\mathscr{H}^{(0)}$, U_0 and a single vector y satisfying the asymptotic condition. The uniqueness of the free Hamiltonian may also be deduced from (3.2.6) and (3.2.9), by showing that $(H^{(0)})'$ is independent of the unitary operator U. This also means that, in any case where unmodified free evolution may be used, $F(\boldsymbol{P}^{(0)}, t) = e^{-iH^{(0)}t}$ is the unique canonical choice for F, and the resulting wave operator will also be unique.

It is often convenient to define the free-energy operator

$$H_0 = U_0 H^{(0)} U_0^{-1}, \tag{3.3.10}$$

acting in the subspace $\mathscr{D}_0$ of $\mathscr{H}$. The following characterization of this operator is sometimes useful.

Theorem 3.6

The free energy H_0 is that (unique) function of the momentum operators $\boldsymbol{P}_0$ satisfying

$$\operatorname*{s-lim}_{t\to\infty} e^{iHt}\{(H-\lambda)^{-1}-(H_0-\lambda)^{-1}\}e^{-iHt}g=0 \tag{3.3.11}$$

for all g satisfying the asymptotic condition and for all $\lambda\in\mathbb{C}$ with $\operatorname{Im}(\lambda)\neq 0$.

Proof

We let $(H^{(0)}-\lambda)^{-1}=\Phi(\boldsymbol{P}^{(0)})$ in (3.3.1), so that $(H_0-\lambda)^{-1}=\Phi(\boldsymbol{P}_0)$. By using the intertwining relation, the right-hand side of (3.3.1) is $\Omega_+^{(0)}(H^{(0)}-\lambda)^{-1}(\Omega_+^{(0)})^* g=(H-\lambda)^{-1}g$, where we have written $\Omega_+^{(0)}\psi=g$ and $(\Omega_+^{(0)})^*g=\psi$. Hence (3.3.11) is satisfied. On the other hand,

$$(H-\lambda)^{-1}g=\Omega_+^{(0)}\Phi(\boldsymbol{P}^{(0)})(\Omega_+^{(0)})^*g$$

for all g in the range of $\Omega_+^{(0)}$ has the unique solution $\Phi(\boldsymbol{P}^{(0)})=(H^{(0)}-\lambda)^{-1}$, implying $\Phi(\boldsymbol{P}_0)=H_0-\lambda)^{-1}$. ■

This result also allows us to determine the free Hamiltonian, via the formula

$$H^{(0)}=U_0^{-1}H_0U_0.$$

What we have been concerned with here is the so-called algebraic theory of scattering, in which one tries to construct wave operators, starting from asymptotic limits, in the Heisenberg picture, of functions of momentum. Bounded functions $\Phi(\boldsymbol{P}^{(0)})$ are elements of a commutative or Abelian algebra of operators in $\mathscr{H}^{(0)}$. Sometimes in applications it is convenient to restrict attention to a smaller algebra, for example the algebra of bounded *continuous* functions of momentum, and sometimes one looks at an enlarged algebra (typically the commutant of $\mathscr{H}^{(0)}$, consisting of all bounded linear operators in $\mathscr{H}^{(0)}$ that commute with $H^{(0)}$). In each case, one starts with limits of the form $\omega(A)=\text{s-lim}_{t\to\infty} e^{iHt}U_0AU_0^{-1}e^{-iHt}g$, for elements A of the algebra and for some suitable set of vectors g, and tries to represent the mapping $A\to\omega(A)$ in the form $\omega(A)=W_+^{(0)}A(W_+^{(0)})^*$ for some isometric operator $W_+^{(0)}$ from $\mathscr{H}^{(0)}$ to $\mathscr{H}$ (cf. (3.3.1)). If this is possible then the mapping is referred to as a *spatial isomorphism.* As we have seen, in order to define true wave operators, the mere existence of the limits $\omega(A)$ is not enough, and additional hypotheses are required.

3.4 FURTHER EXTENSIONS AND DEVELOPMENTS

The ideas of wave operators in this chapter are of a very wide applicability. Here we shall note a few directions in which some of these results may be taken.

We have assumed so far that states in the free Hilbert space, evolving according to some free evolution, or modified free evolution, are completely described by some set $\boldsymbol{P}^{(0)}$ of momentum observables. In practice, this takes no account of the fact that $\mathscr{H}^{(0)}$ may have to be enlarged in order to accomodate other observables that commute with $\boldsymbol{P}^{(0)}$. One example of this is the possible need to describe asymptotic evolution of particles with spin. For a single particle, asymptotically free and of spin $\frac{1}{2}$, the free Hilbert space $\mathscr{H}^{(0)}$ would have to be not $L^2(\mathbb{R}^3)$ but rather the tensor product of $L^2(\mathbb{R}^3)$ with the appropriate two-dimensional spin space. The theory that we have developed goes through in that case with minor modifications. The modified free evolution might take the more general form $F(\boldsymbol{P}^{(0)}, \boldsymbol{S}, t)$, where $\boldsymbol{S}$ is the two-component spin operator.

Even in the spinless case, it may be possible to label free-particle states by quantum numbers corresponding to observables that are not functions of $\boldsymbol{P}^{(0)}$. This will be the case, for example, if there is some self-adjoint operator T having a discrete spectrum and commuting with $F(\boldsymbol{P}^{(0)}, t)$ for all t. If F is a function of $|\boldsymbol{P}^{(0)}|^2$ and t then the angular–momentum operators in $\mathscr{H}^{(0)}$ will have this property, and free-particle states may be labelled by quantum numbers of angular momentum. This has the important consequence that, in this case, wave operators may be defined and treated on each angular-momentum subspace separately. (That is, in the case of a single particle in three dimensions, on each partial-wave subspace corresponding to angular-momentum quantum numbers l and m; this results in the simplification that, on each such subspace, F is a function of $H^{(0)}$ and t alone.)

It is often of interest to compare, asymptotically as $t \to \infty$, the evolutions generated by two self-adjoint operators H_1 and H_2, acting in a Hilbert space $\mathscr{H}$. A natural generalization of the arguments of previous sections leads us to consider the limits

$$\Omega_{\pm}(H_1, H_2) \equiv \underset{t \to \pm\infty}{\text{s-lim}}\; e^{iH_1 t} e^{-iH_2 t}, \tag{3.4.1}$$

defined on some suitable set of vectors in $\mathscr{H}$. More generally, H_1 and H_2 may act in different Hilbert spaces $\mathscr{H}_1$ and $\mathscr{H}_2$, and, just as we used the unitary mapping U_0, we may be interested in wave operators of the form

$$\Omega_{\pm}(H_1, H_2 : J) \equiv \underset{t \to \pm\infty}{\text{s-lim}}\; e^{iH_1 t} J e^{-iH_2 t}, \tag{3.4.1$'$}$$

where J is some linear operator from $\mathscr{H}_2$ to $\mathscr{H}_1$. As in (3.2.10)′, these wave operators satisfy an intertwining relation of the form

$$H_1\Omega_\pm=\Omega_\pm H_2. \tag{3.4.2}$$

It is useful here to mention also the so-called *transitivity property* of the wave operators. Assuming that the range of $\Omega_\pm(H_1, H_2)$ is contained in the domain of $\Omega_\pm(H, H_1)$, where H is a further self-adjoint operator in $\mathscr{H}$, we find

$$\Omega_\pm(H,H_2)=\text{s-}\lim_{t\to\pm\infty} (e^{iHt}e^{-iH_1t})(e^{iH_1t}e^{-iH_2t}),$$

giving the identity

$$\Omega_\pm(H,H_2)=\Omega_\pm(H,H_1)\Omega_\pm(H_1,H_2) \tag{3.4.3}$$

A similar identity, where J-operators have to be taken into account, holds in the more general case where wave operators are given by (3.4.2).

For a scattering system in quantum mechanics, there may be a variety of possible modes of asymptotic evolution. Each such mode, for a given scattering system, is referred to as a *channel* of the system, and each scattering channel will be described by a corresponding free Hilbert space in which the asymptotic free evolution will act. Consider, for example, a system for which there are N channels. We shall then have, correspondingly, N free Hilbert spaces $\mathscr{H}^{(0)}, \mathscr{H}^{(1)}, \ldots, \mathscr{H}^{(N-1)}$.

In each free Hilbert space $\mathscr{H}^{(j)}$ there will be a modified free evolution $F_j(\boldsymbol{P}^{(j)}, t)$, where $\boldsymbol{P}^{(j)}$, acting in $\mathscr{H}^{(j)}$, will be the associated free momentum operator. Equation (3.2.9) defines a free Hamiltonian $H^{(j)}$ for each channel.

For each value of j there will be a unitary mapping U_j from $\mathscr{H}^{(j)}$ on to a subspace $\mathscr{D}_j$ of the underlying Hilbert space $\mathscr{H}$. Given any vector $\psi^{(j)}\in\mathscr{H}^{(j)}$, a wave operator $\Omega_+^{(j)}$ for the jth channel is given, as in (3.2.2), by the formula

$$\Omega_+^{(j)}\psi^{(j)}=\text{s-}\lim_{t\to\infty} e^{iHt}U_jF_j(\boldsymbol{P}^{(j)},t)\psi^{(j)}, \tag{3.4.4}$$

Then $\Omega_+^{(j)}$ is isometric as an operator from $\mathscr{H}^{(j)}$ to $\mathscr{H}$. The range of $\Omega_+^{(j)}$ will be denoted by $\mathscr{D}_j^+$, and the channel wave operator is unitary as an operator from $\mathscr{H}^{(j)}$ on to $\mathscr{D}_j^+$. Wave operators $\Omega_-^{(j)}$ may be defined in a similar way.

A vector $g^{(j)}=\Omega_+^{(j)}\psi^{(j)}$, in the range of $\Omega_+^{(j)}$, is said to satisfy the asymptotic condition for the jth scattering channel, in the limit $t\to\infty$. Such an initial state, at time $t=0$, will evolve asymptotically at large positive times according to the modified free evolution for channel j. It is fundamental to the mathematical description of many-channel scattering systems that it should be observable, in the asymptotic limit $t\to\infty$, in which of the various scattering channels the evolving state lies, and that, moreover, any two distinct channels should be mutually exclusive. This implies that $\langle g^{(i)}, g^{(j)}\rangle=0$ for $i\neq j$, so that the ranges of distinct channel wave operators are mutually orthogonal. (Similarly, range $(\Omega_-^{(i)})\perp$ range $(\Omega_-^{(j)})$ for $i\neq j$, but it is not in general true that range $(\Omega_+^{(i)}) \perp$ range $(\Omega_-^{(j)})$. This property is known as the orthogonality of channels. Note, however, that the subspaces $\mathscr{D}_j$ of $\mathscr{H}$ will *not*, generally, be mutually orthogonal. Indeed, these subspaces frequently overlap, and there may well be a scattering channel for which the corresponding subspace is the whole of $\mathscr{H}$. If such a subspace exists, it may conveniently have attached the label $j=0$, in which case U_0 maps from $\mathscr{H}^{(0)}$ onto the whole of $\mathscr{H}$. It need not then follow that the range of $\Omega_+^{(0)}$ is the whole of $\mathscr{H}$; this would, in fact, imply that the system had only a single channel.

All of the properties of wave operators that we have obtained so far apply equally to many-channel wave operators. Thus the intertwining property (3.2.10)′ gives, in general,

$$\Omega_+^{(j)} H^{(j)} = H \Omega_+^{(j)}, \tag{3.4.5}$$

and (3.2.11) may be extended, using orthogonality of channels, to give

$$(\Omega_+^{(i)})^* (\Omega_+^{(j)}) = \delta_{ij} I^{(j)}, \tag{3.4.6}$$

where $I^{(j)}$ is the identity operator on $\mathscr{H}^{(j)}$.

It is possible, in some applications of scattering theory to many-channel systems, to deal with operators in the single Hilbert space $\mathscr{H}$ rather than introduce a number of free Hilbert spaces $\mathscr{H}^{(j)}$. To do this, we write, in (3.4.4), $f^{(j)} = U_j \psi^{(j)}$ and $\boldsymbol{P}_j = U_j \boldsymbol{P}^{(j)} U_j^{-1}$ as a (vector) operator in $\mathscr{D}_j$. Channel Hamiltonians can also be defined as operators in $\mathscr{D}_j$ by the formula $H_j = U_j H^{(j)} U_j^{-1}$. If $U_j = U_0$ maps onto the whole of $\mathscr{H}$ then this defines H_0 as a self-adjoint operator in $\mathscr{H}$. Otherwise, H_j, and also $\boldsymbol{P}_j$, may be extended, in some convenient way, to act as self-adjoint operators in $\mathscr{H}$. (For example, one could define $H_j h = 0$ for $h \perp \mathscr{D}_j$; in practice, however, alternative extensions of these operators may be more useful.)

The right-hand side of (3.4.4) now becomes

$$\operatorname*{s-lim}_{t\to\infty} e^{iHt} F_j(\boldsymbol{P}_j, t) f^{(j)}.$$

In the simplest case, in which wave operators exist corresponding to an unmodified free evolution, we have

$$\operatorname*{s-lim}_{t\to\infty} e^{iHt} e^{-iH_j t} f^{(j)}.$$

In either case, the right-hand side is closer to the kind of formula that might be written down if we effectively *identify* each free Hilbert space with a subspace of $\mathscr{H}$ (with the whole of $\mathscr{H}$ in the case $\mathscr{H}^{(j)} = \mathscr{H}^{(0)}$). We shall find, however, that it is not possible to dispense entirely with the Hilbert spaces $\mathscr{H}^{(j)}$, which will play an important role (Chapter 4) in the definition of the many-channel scattering operator.

Chapter 4

The Structure of Wave Operators

4.1 WAVE OPERATORS AND ASYMPTOTIC COMPLETENESS

Scattering theory is a study of the asymptotic evolution of states in the limits as $t \to \pm\infty$. The idea of an evolution, generated by the total Hamiltonian H, and asymptotically indistinguishable from that of a freely evolving system, leads necessarily (as we saw in the last chapter) to the existence of wave operators in the associated Hilbert space. Asymptotically free evolution also entails the existence of a free-Hamiltonian operator H_0. In the simplest situation, scattering states will approach, in the limits $t \to \pm\infty$, states that evolve according to the free evolution generated by the unitary group $e^{-iH_0 t}$. More generally, comparison must be made with a modified evolution that is generated by a family of operators $F(\boldsymbol{P}, t)$ in the limit $t \to \infty$, and by $F^*(-\boldsymbol{P}, -t)$ in the limit $t \to -\infty$.

The developments of Chapter 3 are important to the foundations of scattering theory, in that they establish a *raison d'être* for the free-Hamiltonian and wave operators in any mathematical description of a quantum-scattering system. For a concrete system, the free and total Hamiltonian are usually known in advance, though their precise definitions as self-adjoint operators in a Hilbert space may need further analysis. The question then presents itself, how to define wave operators and to prove their existence as genuine mathematical objects. Having answered this question, we should like to know what kind of mathematical objects wave operators are, and if possible to obtain explicit formulae for them in terms of H and H_0. Finally we should like to relate wave operators to those physical quantities that are capable of direct physical observation, such as scattering cross-sections.

Each of these questions will be dealt with in subsequent chapters. We shall treat in particular detail the canonical example of scattering in non-relativistic quantum mechanics, that of a particle scattered by a potential V in one or three dimensions. In Chapter 5 we shall consider how the free and total Hamiltonians are to be defined as self-adjoint operators, with particular attention to the characterization of their domains. Contrary to what might be supposed, a study of domains in this context is not a pedantic mathematical exercise but will have an important bearing on key areas of physical interpretation. Chapter 6 will present a detailed analysis of the Schrödinger operator in one dimension, leading to the spectral analysis of such operators in Chapter 7. Only then, in the following and subsequent chapters, shall we be ready to establish the existence of wave operators in potential-scattering theory. In the one-dimensional problem, for both short- and long-range potentials, wave operators can be defined explicitly in terms of the solution of the Schrödinger equation at energy λ, which is an ordinary differential equation; asymptotic analysis of this equation leads to the definition of phase shifts, which, although not themselves directly observable, are closely related to experimental quantities. We shall also deal with potential-scattering theory in three-dimensional space. Here we are concerned mathematically with the Schrödinger operator as a partial differential operator, and with the Schrödinger equation as a PDE. We shall be concerned particularly to show how mathematical developments go hand in hand with an analysis of the physics of asymptotic evolution.

In the present chapter we shall consider the following question: *given* the existence of wave operators in a Hilbert space $\mathscr{H}$, what can we deduce about their general structure and properties? We shall find, in Section 4.2, that wave operators can, under fairly general conditions, be written down as a special kind of integral with respect to the spectral family of the free Hamiltonian.

Consider, then, a quantum system for which the free Hamiltonian H_0 and total Hamiltonian H are represented by self-adjoint operators in a Hilbert space $\mathscr{H}$. (More generally, H_0 could act in a subspace of $\mathscr{H}$; however, we take first of all the case in which the space $\mathscr{H}^{(0)}$, on which free evolution is defined, is unitarily equivalent to the whole of $\mathscr{H}$, so that we can *identify* $\mathscr{D}_0$ and $\mathscr{H}$.)

We shall suppose that wave operators may be defined on the whole space either by

$$\Omega_{\pm} = \operatorname*{s-lim}_{t \to \pm\infty} e^{iHt} e^{-iH_0 t} \tag{4.1.1}$$

or by

$$\Omega_{\pm} = \operatorname*{s-lim}_{t \to \pm\infty} e^{iHt} F(\boldsymbol{P}, t). \tag{4.1.2}$$

Equation (4.1.1) defines so-called unmodified wave operators, which we shall find to be appropriate in describing scattering by short-range potentials. Equation (4.2.2) defines wave operators by comparison with a modified free evolution, where the one-parameter family of operators $F(\boldsymbol{P},t)$ is defined by functions of the momentum operator $\boldsymbol{P}$. Since the free Hamiltonian H_0 is also a function of momentum, (4.1.1) is a special case of the general formula (4.1.2).

The operator family $F(\boldsymbol{P}, t)$ is not arbitrary. According to (3.2.14), we assume that

$$F(\boldsymbol{P}, t) = \mathrm{e}^{-\mathrm{i}H_0 t} Z(\boldsymbol{P}, t), \tag{4.1.3}$$

where $Z(\boldsymbol{P}, t)$ is a slowly varying function of t, in the sense that

$$\operatorname*{s-lim}_{t \to \pm\infty} \{Z(\boldsymbol{P}, t+s) - Z(\boldsymbol{P}, t)\} = 0. \tag{4.1.4}$$

In any concrete situation it is necessary either to prove the existence of the limits on the right-hand side of (4.1.1), or to prove the existence of the limits (4.1.2) for some suitable operator family $\{F(\boldsymbol{P}, t)\}$. In the case of a modified free evolution, we should also verify that (4.1.4) is satisfied. Here we shall be content to explore some of the *consequences* of (4.1.1)–(4.1.4).

We have already proved, in (3.4.3), the intertwining property of wave operators. In the present context, this stems from the fact that in the limits (4.1.1) and (4.1.2) we can replace t by $t+s$, making use of (4.1.4) in the case of modified wave operators, to obtain

$$\mathrm{e}^{\mathrm{i}Hs} \Omega_{\pm} \mathrm{e}^{-\mathrm{i}H_0 s} = \Omega_{\pm},$$

or

$$\Omega_{\pm} \mathrm{e}^{-\mathrm{i}H_0 s} = \mathrm{e}^{-\mathrm{i}Hs} \Omega_{\pm} \tag{4.1.5}$$

It then follows, on differentiating with respect to s, that

$$H\Omega_{\pm} = \Omega_{\pm} H_0. \tag{4.1.6}$$

Equation (4.1.6) also implies the corresponding result for the resolvent operator; thus

$$(H-z)^{-1}\Omega_{\pm} = \Omega_{\pm}(H_0 - z)^{-1}, \tag{4.1.7}$$

where z is any complex number such that $\mathrm{Im}(z) \neq 0$. An alternative proof of (4.1.7) is to multiply (4.1.5) by e^{izs} and to integrate with respect to s from 0 to ∞ ($\mathrm{Im}(z) > 0$) or from $-\infty$ to 0 ($\mathrm{Im}(z) < 0$). From the intertwining property for resolvent operators, we may deduce the corresponding result for the respective spectral families of H and H_0. To do this, we first of all note that, for an arbitrary self-adjoint operator T, the operator $T^3(T^2+\beta^2)^{-1}\,(T^2+c)^{-1}$ may be expressed, for β, $c > 0$ and $c \neq \beta^2$, as a linear combination of resolvent operators for T. The matrix element of this operator between vectors $f, g \in \mathscr{H}$ may be written

$$\int_{-\infty}^{\infty} \lambda^3(\lambda^2+\beta^2)^{-1}(\lambda^2+c)^{-1}\, d\langle g, E_\lambda f\rangle,$$

where $\{E_\lambda\}$ is the spectral family of T. On integrating with respect to β, and noting the result

$$\frac{1}{\pi}\int_{-\infty}^{\infty} \frac{\lambda^3\, d\beta}{(\lambda^2+\beta^2)(\lambda^2+c)} = \frac{\lambda|\lambda|}{\lambda^2+c},$$

we have the identity

$$\frac{1}{\pi}\int_{-\infty}^{\infty} T^3(T^2+\beta^2)^{-1}(T^2+c)^{-1}\, d\beta = T|T|(T^2+c)^{-1}.$$

By applying appropriate integration to both sides of (4.1.7), the intertwining relation is obtained in the form

$$H|H|(H^2+c)^{-1}\Omega_\pm = \Omega_\pm H_0|H_0|(H_0^2+c)^{-1}.$$

Hence the intertwining property also holds with

$$(H^2-H|H|+2c)(2H^2+2c)^{-1} \tag{4.1.8}$$

on the left-hand side.

Now $\lim_{c\to 0+}(\lambda^2-\lambda|\lambda|+2c)(2\lambda^2+2c)^{-1}$ is just the characteristic function of the interval $(-\infty, 0]$. It follows from the Spectral Theorem and the Lebesgue Dominated-Convergence Theorem that the expression (4.1.8) converges strongly, in the limit $c\to 0+$, to the spectral projection of H for the interval $(-\infty, 0]$. This argument may be repeated and applied to the spectral projection of H for the general interval $(-\infty, \lambda]$, from which we may deduce the identity

$$E_\lambda(H)\Omega_\pm = \Omega_\pm E_\lambda(H_0) \tag{4.1.9}$$

satisfied by the respective projections of H and H_0. Once (4.1.9) has been verified, the result may be generalized to spectral projections associated with arbitrary Borel subsets of $\mathbb{R}$. We have seen, in Chapter 2, that one such projection is that operator projecting onto the absolutely continuous subspace. Denoting these operators, for H and H_0 respectively, by $P_{ac}(H)$ and $P_{ac}(H_0)$, we then have

$$P_{ac}(H)\Omega_\pm = \Omega_\pm P_{ac}(H_0). \tag{4.1.10}$$

With $H_0 = F(\boldsymbol{P})$, we suppose that $F^{-1}(\mathscr{S})$ has Lebesgue measure zero in $\mathbb{R}^k$ ($\boldsymbol{P} = (P_1, P_2, \dots P_k)$), for any subset $\mathscr{S}$ of $\mathbb{R}$ such that the Lebesgue measure of $\mathscr{S}$ is zero. Since the spectral projection, for H_0, associated with the subset $\mathscr{S}$ is multiplication in momentum space by the characteristic function of the set $F^{-1}(\mathscr{S})$, it follows that this projection is always zero. Hence H_0, with the above assumption, has a purely absolutely continuous spectrum, so that $P_{ac}(H_0) = 1$. In this case (4.1.10) becomes

$$P_{ac}(H)\Omega_{\pm} = \Omega_{\pm}. \tag{4.1.10$'$}$$

In non-relativistic potential-scattering theory $H_0 = \boldsymbol{P}^2$ in suitable units, and (4.1.10)′ holds. More generally, (4.1.10)′ will remain valid for any free Hamiltonian H_0 that has a purely absolutely continuous spectrum.

Equation (4.1.10)′ leads immediately to the important result that

$$\text{range}\,(\Omega_{\pm}) \subseteq \mathscr{M}_{ac}(H). \tag{4.1.11}$$

For a vector $g = \Omega_+ f$ in the range of the wave operator Ω_+, we have, from (4.1.2),

$$\lim_{t \to \infty} \|e^{-iHt} g - F(\boldsymbol{P}, t) f\| = 0,$$

with $F(\boldsymbol{P}, t) = \mathrm{e}^{-iH_0 t}$ if unmodified wave operators are defined. Such vectors therefore describe those initial states (i.e. states at time $t = 0$) that become asymptotically free in the limit as $t \to \infty$. These states are called *scattering states* for the limit $t \to \infty$, and the range of Ω_+ is a scattering subspace. A scattering subspace may also be defined for the limit $t \to -\infty$, and is identical with the range of Ω_-. Equation (4.1.11) now tells us that scattering states belong to the absolutely continuous subspace for the total Hamiltonian.

Let us construct projection operators onto the respective scattering subspaces. We note first of all that each wave operator is an isometry. For example,

$$\begin{aligned}\langle \Omega_+ g, \Omega_+ f \rangle &= \operatorname*{s-lim}_{t \to \infty} \langle \mathrm{e}^{iHt} F(\boldsymbol{P}, t) g, \mathrm{e}^{iHt} F(\boldsymbol{P}, t) f \rangle \\ &= \langle g, f \rangle,\end{aligned}$$

since the operators $F(\boldsymbol{P}, t)$ are unitary, for fixed t.

Since $\langle \Omega_+ g, \Omega_+ f \rangle = \langle g, \Omega_+^* \Omega_+ f \rangle$, it follows that

$$\Omega_+^* \Omega_+ = \Omega_-^* \Omega_- = 1. \tag{4.1.12}$$

For $g = \Omega_+ f$ belonging to the range of Ω_+, we have

$$\Omega_+ \Omega_+^* g = \Omega_+ (\Omega_+^* \Omega_+ f) = \Omega_+ f = g.$$

On the other hand, if a vector h is orthogonal to the range of Ω_+, so that

$\langle h, \Omega_+ f\rangle = \langle \Omega_+^* h, f\rangle = 0$ for all $f \in \mathscr{H}$, it follows that $\Omega_+^* h = 0$. In this case

$$\Omega_+ \Omega_+^* h = \Omega_+ (\Omega_+^* h) = 0.$$

Hence we have shown that $\Omega_+ \Omega_+^*$ is the projection onto the range of Ω_+, and similarly $\Omega_- \Omega_-^*$ projects onto the range of Ω_-.

We now consider an initial state $g = \Omega_- f$ that is asymptotically free in the limit as $t \to -\infty$; thus

$$\lim_{t \to -\infty} \|e^{-iHt} g - F(\boldsymbol{P}, t) f\| = 0.$$

The initial state g may be expressed as the sum of a part that is in the range of Ω_+ and a part that is orthogonal to the range of Ω_+. The former contribution is $\Omega_+ \Omega_+^* \Omega_- f = \Omega_+ S f$, where $S = \Omega_+^* \Omega_-$. Hence

$$g = \Omega_+ S f + g^\perp = g_+ + g^\perp,$$

where $g^\perp$ is orthogonal to the range of Ω_+, and hence satisfies $\Omega_+^* g = 0$. By the adjoint of the intertwining relation (4.1.5), we note that

$$\Omega_+^* e^{-iHt} g^\perp = e^{-iH_0 t} \Omega_+^* g^\perp = 0,$$

so that, as t increases, the state $e^{-iHt} g^\perp$ remains orthogonal to the $t \to +\infty$ scattering subspace.

In the limit as $t \to \infty$, the part of the state $e^{-iHt} g$ that is asymptotically free is $e^{-iHt} g_+$, where with $g_+ = \Omega_+ S f$ we have

$$\lim_{t \to \infty} \|e^{-iHt} g_+ - F(\boldsymbol{P}, t) S f\| = 0 \tag{4.1.13}$$

The linear operator S, relating the free-particle state $F(\boldsymbol{P}, t) f$ in the limit $t \to -\infty$ to that part of the state $F(\boldsymbol{P}, t) S f$ that is asymptotically free in the limit $t \to +\infty$, is called the *scattering operator* for the quantum system. We have thus derived the formula

$$S = \Omega_+^* \Omega_- \tag{4.1.14}$$

for the scattering operator in terms of the wave operators $\Omega_\pm$. By the intertwining equation (4.1.6) and its adjoint, we have

$$H_0 S = H_0 \Omega_+^* \Omega_- = \Omega_+^* H \Omega_- = \Omega_+^* \Omega_- H_0 = S H_0,$$

so that $[H_0, S] = 0$. The following lemma demonstrates a connection between the ranges of $\Omega_\pm$ and properties of the scattering operator.

Lemma 4.1

$$\text{range}(\Omega_-) \subseteq \text{range}(\Omega_+) \Leftrightarrow S \quad \text{is isometric;}$$
$$\text{range}(\Omega_+) \subseteq \text{range}(\Omega_-) \Leftrightarrow S^* \text{ is isometric;}$$
$$\text{range}(\Omega_+) = \text{range}(\Omega_-) \Leftrightarrow S \quad \text{is unitary.}$$

Proof

We suppose first that range $(\Omega_-) \subseteq$ range (Ω_+). For any $f \in \mathscr{H}$ we then have $\Omega_- f \in$ range (Ω_+). Since $\Omega_+ \Omega_+^*$ projects onto the range of Ω_+, this implies that $\Omega_+ \Omega_+^* \Omega_- f = \Omega_- f$, and we have the chain of equalities

$$\|Sf\| = \|\Omega_+ Sf\| = \|\Omega_+ \Omega_+^* \Omega_- f\| = \|\Omega_- f\| = \|f\|.$$

Therefore S is isometric.

If, conversely, S is assumed to be isometric then, the same chain of equalities tells us that the projection of $\Omega_- f$ onto the range of Ω_+ has norm identical with $\|\Omega_- f\|$. Hence $\Omega_- f \in$ range (Ω_+), and the first statement of the lemma is proved.

To prove the second statement of the lemma, we observe that $S^* = \Omega_-^* \Omega_+$, and carry through the same argument as before, with plus and minus exchanged.

Finally, we note that S is unitary if and only if $SS^* = S^*S = 1$, i.e. if and only if both S^* and S are isometric. Hence the third statement of the lemma follows immediately from the first two. ■

The condition range $(\Omega_-) \subseteq$ range (Ω_+) may be given a physical interpretation. As we showed in the proof of the lemma, this condition implies, for any $g = \Omega_- f$, that $g = \Omega_+ Sf$. Hence $g^\perp = g - \Omega_+ Sf = 0$. so that (4.1.13) holds with $g = g_+$. In other words, the condition range $(\Omega_-) \subseteq$ range (Ω_+) implies that states that are asymptotically free in the limit $t \to -\infty$ become asymptotically free at time $t = +\infty$. The scattering operator S then implements the transformation between free-particle states at time $t = -\infty$ and free-particle states at time $t = +\infty$. A similar interpretation applies to the condition range $(\Omega_+) \subseteq$ range (Ω_-), and S will be unitary whenever the subspaces of free-particle states, respectively at $t = \pm\infty$, may be identified. In this case there is a unitary one-to-one correspondence between the asymptotic states for large positive and negative times. In the case of unmodified wave operators, this correspondence is defined by the property that a given initial state g, evolving to $e^{-iHt} g$ at general time t, will approach $e^{-iH_0 t} f$ as $t \to -\infty$ and $e^{-iH_0 t} Sf$ as $t \to +\infty$.

Definition 4.1

The wave operators $\Omega_\pm$ defined by (4.1.1) or (4.1.2) are said to be asymptotically complete if range $(\Omega_+) =$ range (Ω_-).

Asymptotic completeness is thus equivalent to unitarity of the scattering operator, and holds whenever "incoming" freely evolving states correspond to "outgoing" freely evolving states, generated in each case by an asymptotically free evolution $F(\boldsymbol{P}, t)$. Asymptotic completeness, as we are defining it here, is appropriate to a single scattering channel, and cannot be expected to hold, for example, for many particle systems where we have a variety of different asymptotic modes of evolution or scattering channels. In the case of a single particle moving non-relativistically in a conservative field of force, we shall see that, depending on the potential V, (asymptotic) completeness may or may not hold, and it can be a difficult technical problem to decide.

In considering the question of completeness for a scattering system, it is necessary to characterize the respective ranges of the wave operators. If $g=\Omega_+ f$ is any vector in the range of Ω_+ then we have

$$\operatorname*{s-lim}_{t\to\infty} F^*(\boldsymbol{P},t)e^{-iHt}g=\operatorname*{s-lim}_{t\to\infty} F^*(\boldsymbol{P},t)e^{-iHt}(e^{iHt}F(\boldsymbol{P},t)f)=f,$$

since both the evolution operator e^{-iHt} and the free evolution $F(\boldsymbol{P},t)$ are unitary. Conversely, let g be any vector in $\mathscr{H}$ such that the limit $\text{s-lim}_{t\to\infty} F^*(\boldsymbol{P},t)e^{-iHt}g$ exists. Since

$$\begin{aligned}\lim_{t\to\infty}\langle h, F^*(\boldsymbol{P},t)e^{-iHt}g\rangle &= \lim_{t\to\infty}\langle e^{iHt}F(\boldsymbol{P},t)h, g\rangle\\ &=\langle\Omega_+h,g\rangle=\langle h,\Omega_+^*g\rangle,\end{aligned}$$

it follows necessarily that indeed

$$\operatorname*{s-lim}_{t\to\infty} F^*(\boldsymbol{P},t)e^{-iHt}g=\Omega_+^*g. \tag{4.1.15}$$

Again, because both evolution operators are unitary, we have, in this case $\|\Omega_+^*g\|=\|g\|$. Since Ω_+ is isometric, projecting onto the range of Ω_+ now gives $\|(\Omega_+\Omega_+^*)g\|=\|g\|$, from which it follows that $g\in\text{range}(\Omega_+)$. Hence $g\in\text{range}(\Omega_+)$ is *equivalent* to the existence of the limit on the left-hand side of (4.1.15), and we may characterize the range of this wave operator by the condition that this limit exists. A similar result holds for the range of Ω_-. Hence, to prove asymptotic completeness, we need to prove the existence of a single subspace $\mathscr{M}$, defined both as the space of vectors for which the limit (4.1.15) exists and as the space of vectors for which the limit

$$\operatorname*{s-lim}_{t\to-\infty} F^*(\boldsymbol{P},t)e^{-iHt}g=\Omega_-^*g \tag{4.1.15$'$}$$

exists. We already know that $\text{range}(\Omega_\pm)\subseteq\mathscr{M}_{ac}(H)$, so that, in particular, asymptotic completeness will hold provided that both limits (4.1.15) and (4.1.15)$'$ hold for all $g\in\mathscr{M}_{ac}(H)$. In this case we have $\text{range}(\Omega_+)=\text{range}(\Omega_-)=\mathscr{M}_{ac}(H)$, a result that is sometimes known as *strong asymptotic completeness*. Although in the vast majority of cases for which completeness has been proved one also has the stronger result (see, e.g. Chapters 11 and 12), it is quite possible to have completeness with $\text{range}(\Omega_\pm)\neq\mathscr{M}_{ac}(H)$. In Chapter 13 we shall see that the absolutely continuous subspace for H has particular mathematical and physical importance in the theory of potential scattering.

What if asymptotic completeness fails? Of course, one reason may be that we have failed to take into account more than one possible scattering channel. However, we shall see later in this chapter how an appropriate modification of the statement of asymptotic completeness can often be applied in that case. The more serious question arises, for example in potential scattering of a single particle, if completeness fails in a context in which there is no apparent reason for having introduced further channels. We may then have incoming free-particle states at time $t=-\infty$ that are no longer asymptotically free in the limit $t\to+\infty$. The scattering operator will then be non-unitary, and there will be states in $\mathscr{M}_{ac}(H)$ that are orthogonal to the range of Ω_+, and hence that satisfy $\Omega_+^* f=0$. How do such states evolve in the limit $t\to\infty$? They are not scattering states, nor are they "bound states" in the normally accepted sense of being eigenvectors of the total Hamiltonian. We may also ask the question, how should we describe the asymptotic evolution of states lying in $\mathscr{M}_{sc}(H)$? These states are also orthogonal to the range of Ω_+—indeed to the range of both wave operators—and again cannot be described as scattering states, although like scattering states they belong to the spectrally continuous subspace for H.

It turns out that these questions cannot be answered in the degree of generality and abstraction in which they are here presented. They need to be discussed in the framework of a more detailed analysis of the evolution of states in both position and momentum space. Part of the solution will be found in the pages that follow. A continuing study of such problems is, to the present author, one of the main areas of interest in a field that is throwing new light on the foundations of scattering theory.

A common theme that underlies a number of possible approaches to the completeness question is to seek operators J such that $(1-J)e^{-iH_0t}$ converges strongly to zero, and such that both of the limits

$$\Omega_+ = \operatorname*{s-lim}_{t\to\infty} e^{iHt} J e^{-iH_0 t},$$

$$\Omega_+^* = \operatorname*{s-lim}_{t\to\infty} e^{iH_0 t} J^* e^{-iHt},$$

exist on the entire space $\mathscr{H}$. Appropriate changes to these formulae can be made to accommodate possible modified free evolutions. Vectors g that are orthogonal to the $t\to\infty$ scattering subspace will then satisfy $\Omega_+^* g=0$, so that $\operatorname{s-lim}_{t\to\infty} J^* e^{-iHt} g=0$. This equation can then be used, for suitable J, to characterize the time evolution of non-scattering states. We shall leave further development and applications of these ideas to subsequent chapters.

4.2 THE WAVE OPERATOR AS A SPECTRAL INTEGRAL

Given the existence of wave operators defined by (4.1.1) or (4.1.2), how are the strong limits as $t \to \pm\infty$ to be evaluated? A general answer to this question cannot be given without further analysis of the operators H and H_0. Here we shall make one or two simplifying domain assumptions and show, under these conditions, how the wave operators $\Omega_{\pm}$ may be expressed, in each case, as a spectral integral.

We have already met the idea of a spectral integral in a more limited context. If $T = \int \lambda \, \mathrm{d}E_\lambda$ is a self-adjoint operator then we can define functions of T by the formula

$$F(T) = \int F(\lambda)\, \mathrm{d}E_\lambda. \tag{4.2.1}$$

The spectral integral on the right-hand side of this equation is the integral of a complex-valued function $F(\lambda)$ with respect to the spectral family $\{E_\lambda\}$ of a self-adjoint operator. In scattering theory, however, we need a more general notion of spectral integral, in which the function $F(\lambda)$ is replaced by an *operator-valued* function $L(\lambda)$. That is, we have to consider integrals of the form $\int L(\lambda)\, \mathrm{d}E_\lambda$ where, for every (or almost every) $\lambda \in \mathbb{R}$ a linear operator $L(\lambda)$, acting in some Hilbert space $\mathscr{H}$, is defined. We shall not use the full power of the Lebesgue theory to define such integrals. It will be sufficient for our purposes to consider approximating Riemann sums. The following definition provides a theoretical framework for considering the integral of a one-parameter family $\{L(\lambda)\}$ of linear operators with respect to the spectral family of a self-adjoint operator.

Definition 4.2

Let $\{E_\lambda\}$ be the spectral family of a self-adjoint operator T in a Hilbert space $\mathscr{H}$; thus $T = \int \lambda \, \mathrm{d}E_\lambda$. Denote by $E(\mathscr{I})$ the spectral projection of T for the interval $\mathscr{I}$. Let $L(\lambda)$, $\lambda \in \mathbb{R}$, be a one-parameter family of linear operators in $\mathscr{H}$. For some $f \in \mathscr{H}$ suppose that $E(\mathscr{I})f \in D(L(\lambda))$ for every $\lambda \in \mathbb{R}$ and for every *finite* interval $\mathscr{I}$.

Partition the real line $\mathbb{R}$ into the union of a countable set of finite intervals $\mathscr{I}_k = (\lambda_{k-1}, \lambda_k]$, $k = 0, \pm 1, \pm 2, \ldots$, with $\lambda_{k-1} < \lambda_k$ and such that $\sigma \equiv \sup_k (\lambda_k - \lambda_{k-1}) < \infty$. Let $\tilde{\lambda}_k$, $k = 0, \pm 1, \pm 2, \ldots$, satisfy $\tilde{\lambda}_k \in \mathscr{I}_k$ for each k, and suppose that, for every possible choice of the $\tilde{\lambda}_k$, the limit

$$\sum_{k=-\infty}^{\infty} L(\tilde{\lambda}_k) E(\mathscr{I}_k) f = \operatorname*{s-lim}_{\substack{M \to -\infty \\ N \to +\infty}} \sum_{k=M}^{N} L(\tilde{\lambda}_k) E(\mathscr{I}_k) f \tag{4.2.2}$$

exists. Now consider a sequence $\mathscr{S}_n = \{\mathscr{I}_k^{(n)}\}$ of such partitions for which $\lim_{n\to\infty} \sigma_n = 0$, where $\sigma_n = \sup_k (\lambda_k^{(n)} - \lambda_k^{(n-1)})$, is the maximum length of interval in the nth partition of the sequence. We define the spectral integral $\int L(\lambda)\,\mathrm{d}E_\lambda$ by

$$\int_{-\infty}^{\infty} L(\lambda)\,\mathrm{d}E_\lambda f = \operatorname*{s-lim}_{n\to\infty} \sum_{k=-\infty}^{\infty} L(\tilde{\lambda}_k^{(n)})E(\mathscr{I}_k^{(n)})f, \tag{4.2.3}$$

provided that this limit exists for every possible such sequence $\mathscr{S}_n$ of partitions, and for every choice of the $\lambda_k^{(n)}$ within the nth partition of the sequence.

Equation (4.2.3) is a natural way to define the spectral integral of an operator family. By demanding the existence of the strong limit for every sequence of partitions and for every choice of the $\lambda_k^{(n)}$, we guarantee the existence of a limit that is independent of the choice of $\{\mathscr{S}_n\}$ and of $\lambda_k^{(n)}$. Although it is easy to set up such a definition, it is not so clear under what conditions integrals such as the left-hand side of (4.2.3) will exist. Certainly it is insufficient, for the existence of the spectral integral, that $L(\lambda)$ be (strongly) continuous as a function of λ. One possible approach is to attempt to justify (4.2.3) by means of integration by parts on the left-hand side. However, the method that we shall use is particularly adapted to the requirements of scattering theory, and has the further advantage that we are able to reduce the problem to that of evaluating more standard integrals.

Lemma 4.2

Let $\phi(s, \lambda)$ be some continuous complex-valued function of s and λ for $s \in (a, b)$ and $\lambda \in \mathbb{R}$, where the interval (a, b) may be finite or infinite. Let $\Phi(s)$ be a non-negative bounded function such that

$$|\phi(s, \lambda)| \leq \Phi(s) \tag{4.2.4}$$

for all $s \in (a, b)$ and all $\lambda \in \mathbb{R}$, and suppose further that

$$\int_a^b \Phi(s)\,\mathrm{d}s < \infty. \tag{4.2.5}$$

Let $T = \int \lambda\,\mathrm{d}E_\lambda$ be a self-adjoint operator and let $K(s)$, $s \in (a, b)$ be a family of linear operators such that, for any $g \in D(T)$, $K(s)g$ is strongly continuous in s and satisfies

$$\|K(s)g\| \leq \alpha\|Tg\| + \beta\|g\|, \tag{4.2.6}$$

where α and β are independent of s and g. Define on $D(T)$ the operator family $L(\lambda)$ by

$$L(\lambda)g = \int_a^b ds\, \phi(s, \lambda)K(s)g. \tag{4.2.7}$$

(Such integrals may be defined weakly; i.e. consider

$$\langle h, L(\lambda)g\rangle = \int_a^b ds\, \phi(s, \lambda)\langle h, K(s)g\rangle.)$$

Then the spectral integral defined by (4.2.3) exists, and is given by

$$\int_{-\infty}^{\infty} L(\lambda)\, dE_\lambda f = \int_a^b ds\, K(s)\phi(s, T)f \tag{4.2.8}$$

for any $f \in D(T)$.

Proof

In (4.2.8), $\phi(s, T)$ is defined by means of (4.2.1); thus $\phi(s, T) = \int \phi(s, \lambda)\, dE_\lambda$. Then $\phi(s, T)f$ is strongly continuous in s, and so is $\phi(s, T)Tf$. Using the strong continuity of $K(s)$, it follows that the integrand on the right-hand side of (4.2.8) is strongly continuous, and also bounded in norm by const $\Phi(s)$, on using (4.2.4) and (4.2.6). Since $\Phi(s)$ is integrable by (4.2.5), it follows that the right-hand side of (4.2.8) exists as a vector in the Hilbert space $\mathscr{H}$, for any $f \in D(T)$.

We now consider a partition $\{\mathscr{I}_k\}$ of $\mathbb{R}$, with associated choice of the sequence $\{\lambda_k\}$. For $s \in (a, b)$ and $\lambda \in \mathbb{R}$ we define a complex-valued function $\tilde{\phi}(s, \lambda)$ by

$$\tilde{\phi}(s, \lambda) = \phi(s, \tilde{\lambda}_k) \tag{4.2.9}$$

whenever $\lambda \in \mathscr{I}_k$. For fixed s, $\tilde{\phi}(s, \lambda)$ is a step function of λ, whereas for fixed λ, $\tilde{\phi}(s, \lambda)$ is a continuous function of s. In view of the continuity in s, we can repeat our previous argument and deduce the existence of the integrals

$$\int_a^b ds\, K(s)\tilde{\phi}(s, T)f \quad \text{and} \quad \int_a^b ds K(s)\tilde{\phi}(s, T)Tf.$$

For any $h \in \mathscr{H}$ we have

$$\tilde{\phi}(s, T)h = \int_{-\infty}^{\infty} \tilde{\phi}(s, \lambda)\, \mathrm{d}E_{\lambda} h$$

$$= \sum_{k=-\infty}^{\infty} \phi(s, \tilde{\lambda}_k) E(\mathscr{I}_k) h, \tag{4.2.10}$$

where the sum on the right-hand side is defined as a strong limit. We also have

$$\left\| \sum_{k=M}^{N} \phi(s, \tilde{\lambda}_k) E(\mathscr{I}_k) f - \sum_{k=-\infty}^{\infty} \phi(s, \tilde{\lambda}_k) E(\mathscr{I}_k) f \right\|$$

$$\leqq \Phi(s) \left\| f - \sum_{k=M}^{N} E(\mathscr{I}_k) f \right\|,$$

and the same inequality holds with f replaced by Tf. (We have used here the result $E(\mathscr{I}_k)E(\mathscr{I}_l) = \delta_{kl} E(\mathscr{I}_k)$.) From these two inequalities, together with (4.2.6) and the Lebesgue Dominated-Convergence Theorem, we have

$$\lim_{\substack{M \to -\infty \\ N \to \infty}} \int_a^b \mathrm{d}s \left\| K(s) \left\{ \sum_{k=M}^{N} \phi(s, \tilde{\lambda}_k) E(\mathscr{I}_k) f - \sum_{k=-\infty}^{\infty} \phi(s, \tilde{\lambda}_k) E(\mathscr{I}_k) f \right\} \right\| = 0.$$

Using (4.2.7) and (4.2.10), we now have

$$k = \sum_{-\infty}^{\infty} L(\tilde{\lambda}_k) E(\mathscr{I}_k) f$$

$$= \operatorname*{s-lim}_{\substack{M \to -\infty \\ N \to \infty}} \sum_{k=M}^{N} L(\tilde{\lambda}_k) E(\mathscr{I}_k) f = \int_a^b \mathrm{d}s\, K(s) \tilde{\phi}(s, T) f. \tag{4.2.11}$$

We now take a sequence $\{\mathscr{S}_n\}$ of partitions, as in the statement of Definition 4.2, with $\lim_{n \to \infty} \sigma_n = 0$, and for each partition we define, as in (4.2.9), a corresponding function $\tilde{\phi}_n(s, \lambda)$. By (4.2.11), it will be sufficient, in proving the lemma, to show that

$$\operatorname*{s-lim}_{n \to \infty} \int_a^b \mathrm{d}s\, K(s) \tilde{\phi}_n(s, T) f = \int_a^b \mathrm{d}s\, K(s) \phi(s, T) f. \tag{4.2.12}$$

Now

$$\lim_{n\to\infty} \|\tilde{\phi}_n(s, T)f - \phi(s, T)f\|^2 = \lim_{n\to\infty} \int_{-\infty}^{\infty} |\tilde{\phi}_n(s, \lambda) - \phi(s, \lambda)|^2 \, \mathrm{d}\langle f, E_\lambda f\rangle$$

$$= 0,$$

by the Lebesgue Dominated-Convergence Theorem, since $\phi(s, \lambda)$ is continuous so that $\tilde{\phi}_n(s, \lambda) \to \phi(s, \lambda)$. The same result holds with f replaced by Tf. Using (4.2.6), this implies that

$$\lim_{n\to\infty} \|K(s)(\tilde{\phi}_n(s, T) - \phi(s, T))f\| = 0.$$

Hence

$$\lim_{n\to\infty} \int_a^b \mathrm{d}s \, \|K(s)(\tilde{\phi}_n(s, T) - \phi(s, T))f\| = 0,$$

by the Lebesgue Dominated-Convergence Theorem, the integrand being dominated by const $\Phi(s)$, which is integrable. We have therefore verified (4.2.12), and the proof of the lemma is complete. ■

In order to apply Lemma 4.2 to evaluate wave operators, we shall make appropriate assumptions, for the operators H and H_0, that will allow us to satisfy the inequality (4.2.6). We assume that $H = H_0 + V$, where $D(H) = D(H_0)$. Then the interaction V is also defined on the common domains of H and H_0. We may assume without loss of generality that V is a closed operator (otherwise we take the closure of V). The operator $V(H_0 - z)^{-1}$, for $\mathrm{Im}\,(z) \neq 0$, is then closed and defined on the entire Hilbert space. Hence, by the Closed-Graph Theorem, this operator is bounded. Allowing $V(H_0 - z)^{-1}$ to act on a vector $(H_0 - z)f$, we readily deduce the inequality

$$\|Vf\| \leq \alpha \|H_0 f\| + \beta \|f\|, \tag{4.2.13}$$

for all $f \in D(H_0)$, which may be compared with assumption (4.2.6) of the lemma. A similar inequality to (4.2.13) holds with H_0 replaced by H.

We shall suppose, for simplicity, that unmodified wave operators exist, defined by (4.1.1). This allows us to replace the limit $t \to \infty$ by a limit as $\varepsilon \to 0$, by writing

$$\Omega_+ f = \operatorname*{s-lim}_{\varepsilon \to 0+} \int_0^{\infty} \mathrm{d}s \, \varepsilon \, \mathrm{e}^{-\varepsilon s} \, \mathrm{e}^{\mathrm{i}Hs} \, \mathrm{e}^{-\mathrm{i}H_0 s} f. \tag{4.2.14}$$

In order to verify (4.2.14), given any $\delta>0$, we first choose s_0 sufficiently large that $\|e^{iHs}e^{-iH_0s}f-\Omega_+f\|<\delta$ provided that $s>s_0$. Then

$$\int_0^\infty ds\,\varepsilon e^{-\varepsilon s}e^{iHs}e^{-iH_0s}f-\Omega_+f$$

$$=\int_0^\infty ds\,\varepsilon e^{-\varepsilon s}(e^{iHs}e^{-iH_0s}f-\Omega_+f),$$

which is bounded in norm by

$$2\int_0^{s_0} ds\,\varepsilon e^{-\varepsilon s}\|f\|$$

$$+\int_{s_0}^\infty ds\,\varepsilon e^{-\varepsilon s}\|e^{iHs}e^{-iH_0s}f-\Omega_+f\|.$$

The first integral, for fixed s_0, converges to zero in the limit $\varepsilon\to 0$, whereas, by the choice of s_0, the second integral is bounded by $\int_0^\infty ds\varepsilon e^{-\varepsilon s}\delta=\delta$. By letting δ approach zero, (4.2.14) is verified. A somewhat more useful version of (4.2.14) is obtained by assuming $f\in D(H_0)$, in which case $e^{iHs}e^{-iH_0s}f$ is strongly differentiable with respect to s. We can then integrate by parts in (4.2.14) to obtain

$$\Omega_+f=f+\operatorname*{s-lim}_{\varepsilon\to 0+}i\int_0^\infty ds\,e^{-\varepsilon s}e^{iHs}Ve^{-iH_0s}f. \tag{4.2.15}$$

The integrand is again strongly continuous in s. Let us now make use of Lemma 4.2 by making an appropriate choice of the function $\phi(s,\lambda)$ and the operator family $\{K(s)\}$. We allow s to range over the interval $(0,\infty)$, and set $T=H_0=\int\lambda\,dE_\lambda$.

As a first application, we consider

$$\phi(s,\lambda)=\varepsilon\exp[-i(\lambda-i\varepsilon)s],\qquad K(s)=\exp(iHs).$$

Note that $|\phi(s,\lambda)|\le\varepsilon e^{-\varepsilon s}$, which is integrable over the interval $(0,\infty)$. Using (4.2.7) and (4.2.8), together with (4.2.14), we find $L(\lambda)=-i\varepsilon(\lambda-H-i\varepsilon)^{-1}$, and

$$\Omega_+f=\operatorname*{s-lim}_{\varepsilon\to 0+}\int i\varepsilon(H-\lambda+i\varepsilon)^{-1}\,dE_\lambda f. \tag{4.2.16}$$

Here $K(s)$ is bounded, and since, in the lemma, (4.2.6) holds with $\alpha=0$, the formula (4.2.16) for the wave operator Ω_+ will be valid for all $f\in\mathscr{H}$, not simply for $f\in D(H_0)$.

Alternatively, we can use (4.2.15) by taking

$$\phi(s,\lambda)=\exp[-\mathrm{i}(\lambda-\mathrm{i}\varepsilon)s], \qquad K(s)=[\exp(\mathrm{i}Hs)]V.$$

This gives, for $f\in D(H_0)$, the identity

$$\Omega_+ f=f-\operatorname*{s\text{-}lim}_{\varepsilon\to 0+}\int(H-\lambda+\mathrm{i}\varepsilon)^{-1}V\,\mathrm{d}E_\lambda f. \tag{4.2.17}$$

Note how the resolvent operator $R(z)=(H-z)^{-1}$ enters into both of the formulae (4.2.16) and (4.2.17) for the wave operator. What is involved here is the resolvent evaluated for values of z very close, in the complex plane, to the real axis. The fact that $z=\lambda-\mathrm{i}\varepsilon$, in (4.2.16) and (4.2.17), lies in the *lower* half-plane, has led to the use, by some authors, of an alternative notation in which the right-hand side of (4.2.16) is the wave opeator Ω_- rather than Ω_+. We prefer to retain our stated convention, in view of the limit $t\to+\infty$ by which Ω_+ is defined.

For any z such that $\operatorname{Im}(z)\neq 0$, we may define on $D(V)$ an operator $T(z)$ by

$$T(z)=V-V(H-z)^{-1}V \tag{4.2.18}$$

Then

$$\begin{aligned}(H_0-z)^{-1}T(z)&\\ &=(H_0-z)^{-1}V-(H_0-z)^{-1}[(H-z)-(H_0-z)](H-z)^{-1}V\\ &=(H-z)^{-1}V,\end{aligned} \tag{4.2.19}$$

and similarly

$$T(z)(H_0-z)^{-1}=V(H-z)^{-1}. \tag{4.2.19$'$}$$

Hence (4.2.17) may be written as

$$\Omega_+ f=f-\operatorname*{s\text{-}lim}_{\varepsilon\to 0+}\int(H_0-\lambda+\mathrm{i}\varepsilon)^{-1}T(\lambda-\mathrm{i}\varepsilon)\,\mathrm{d}E_\lambda f. \tag{4.2.20}$$

Substituting from (4.2.19) into (4.2.18), the operator $T(z)$ satisfies

$$T(z)=V-V(H_0-z)^{-1}T(z), \tag{4.2.21}$$

so that we can write

$$T(z)=(1+V(H_0-z)^{-1})^{-1}V.$$

The linear operator $V(H_0-z)^{-1}$ is known to be bounded, and if $\|V(H_0-z)^{-1}\|<1$ then we can use this formula to write a convergent series expansion for $T(z)$; thus

$$T(z)=V-V(H_0-z)^{-1}V+V(H_0-z)^{-1}V(H_0-z)^{-1}V\ldots.$$

Such a series is sometimes used, on substituting the first few terms into the formula (4.2.20) for Ω_+, to obtain an expansion of the wave operator to some finite order of the "potential" V.

There is a rather simple and elegant representation for $V\Omega_+$ as a spectral integral of the operator family $T(\lambda-\mathrm{i}\varepsilon)$. To obtain this result, we shall make use of the identity, valid for any $f\in D(H_0)$

$$\int_{-\infty}^{\infty} L(\lambda)(H_0-\lambda)\,\mathrm{d}E_\lambda f=0, \tag{4.2.22}$$

where $L(\lambda)$ is an operator family such that $\|L(\lambda)\|\leq$const, uniformly in λ. To prove (4.2.22), we return to the definition (4.2.3) of the spectral integral.

Let $\mathscr{S}$ be a partition of $\mathbb{R}$ into subintervals $\mathscr{I}_k=(\lambda_{k-1},\lambda_k]$. Since $\|L\|(\lambda)\|\leq$const, it will be sufficient, by (4.2.3), to obtain an estimate for $\sum_{k=-\infty}^{\infty}\|(H_0-\tilde{\lambda}_k)E(\mathscr{I}_k)f$ is in the domain of H_0, we can write $f=(H_0+\mathrm{i})^{-1}g$. Moreover, as may be verified by the spectral theorem, the restriction of $(H_0-\tilde{\lambda}_k)(H_0+\mathrm{i})^{-1}$ to the range of the projection $E(\mathscr{I}_k)$ has a norm that cannot exceed $\sup_{\lambda\in(\lambda_{k-1},\lambda_k]}|(\lambda-\tilde{\lambda}_k)(\lambda+\mathrm{i})^{-1}|$. Since $\tilde{\lambda}_k\in\mathscr{I}_k$, this gives an estimate for the norm that is of order $|\mathscr{I}_k|/(1+|\lambda_k|)$, where $|\mathscr{I}_k|$ denotes the length of the interval $\mathscr{I}_k$. Hence

$$\begin{aligned}
&\sum_{k=-\infty}^{\infty}\|H_0-\tilde{\lambda}_k)E(\mathscr{I}_k)f\|\\
&\leq C\sum_{k=-\infty}^{\infty}|\mathscr{I}_k|(|\lambda_k|+1)^{-1}\|E(\mathscr{I}_k)g\|\\
&\leq\text{const}\left(\sum_{k=-\infty}^{\infty}|\mathscr{I}_k|^2(|\lambda_k|+1)^{-2}\right)^{1/2}\left(\sum_{k=-\infty}^{\infty}\|E(\mathscr{I}_k)g\|^2\right)^{1/2},
\end{aligned}$$

where

$$\sup_k|\mathscr{I}_k|\leq\sigma,$$

$$\sum_{k=-\infty}^{\infty}\|E(\mathscr{I}_k)g\|^2=\|g\|^2,$$

and we have used the Schwarz inequality.

It is not difficult to verify that $\Sigma_{k=-\infty}^{\infty}|\mathscr{I}_k|(|\lambda_k|+1)^{-2}$, in the limit $\sigma\to0$, approaches the Riemann integral $\int_{-\infty}^{\infty} d\lambda(|\lambda|+1)^{-2}$; hence $\Sigma_{k=-\infty}^{\infty} \|(H_0-\tilde{\lambda}_k)E(\mathscr{I}_k)f\|$ is bounded by const. $\sigma^{1/2}$ in the limit as $\sigma \to 0$, and converges to zero. We have thus verified (4.2.22). For $f \in D(H_0)$, we now substitute $H_0 f$ for f into (4.2.16) and use the intertwining property of the wave operators to write

$$H\Omega_+ f = \Omega_+ H_0 f$$

$$= \operatorname*{s-lim}_{\varepsilon \to 0+} \int i\varepsilon(H-\lambda+i\varepsilon)^{-1} H_0 \, dE_\lambda f. \tag{4.2.23}$$

Writing $H_0 = H - V$, this gives

$$H\Omega_+ f = \operatorname*{s-lim}_{\varepsilon \to 0+} \int i\varepsilon H(H-\lambda+i\varepsilon)^{-1} \, dE_\lambda f, \tag{4.2.24}$$

where we have used (4.2.17) to show that

$$\operatorname*{s-lim}_{\varepsilon \to 0+} \int i\varepsilon(H-\lambda+i\varepsilon)^{-1} V \, dE_\lambda f = 0.$$

Again, multiplying (4.2.16) by z, with $\operatorname{Im}(z) \neq 0$, and substituting from (4.2.24), we have

$$(H-z)\Omega_+ f = \operatorname*{s-lim}_{\varepsilon \to 0+} i\varepsilon \int (H-z)(H-\lambda+i\varepsilon)^{-1} \, dE_\lambda f.$$

Since, by hypothesis, the operator $V(H-z)^{-1}$ is bounded, we can now operate on both sides by this operator to obtain, for $f \in D(H_0)$,

$$V\Omega_+ f = \operatorname*{s-lim}_{\varepsilon \to 0+} \int i\varepsilon V(H-\lambda+i\varepsilon)^{-1} \, dE_\lambda f. \tag{4.2.25}$$

(Since V may be unbounded, we could not have obtained (4.2.25) directly from (4.2.16) on operating throughout by V.) We now write in (4.2.25), to the *right* of the resolvent operator,

$$i\varepsilon = (H-\lambda+i\varepsilon) - V - (H_0-\lambda).$$

The integral involving the last term $H_0-\lambda$ will vanish, because of (4.2.22). The first term will remove the resolvent operator and contribute $\int V \, dE_\lambda f = Vf$, and with the second contribution to the right-hand side of (4.2.25) we have, finally,

$$V\Omega_+ f = Vf - \operatorname*{s-lim}_{\varepsilon \to 0+} \int V(H-\lambda+i\varepsilon)^{-1} V \, dE_\lambda f, \tag{4.2.26}$$

which holds for all $f \in D(H_0)$. Equation (4.2.26), satisfied by the wave operator Ω_+, is of particular interest, because it is often the case that, although the resolvent $R(z)$ itself does not approach a limit as z tends to the real axis, an operator such as $V(H-z)^{-1}V$ may, under suitable conditions, have both upper and lower boundary values, which will be different (see Chapter 12). Using (4.2.18), we may rewrite (4.2.26) as

$$V\Omega_+ = \operatorname*{s-lim}_{\varepsilon\to 0+} \int T(\lambda - i\varepsilon)\, dE_\lambda f. \tag{4.2.27}$$

Although, in a sense, we have only replaced one limit $t\to\infty$ in (4.1.1) by another, $\varepsilon\to 0$, the limit $\varepsilon\to 0$ is in general easier to handle in that we may expect to be able to take this limit into the integrand of (4.2.27). In order to derive analogous expressions for the scattering operator S as a spectral integral, we shall need a slight generalization of Lemma 4.2, which allows the single integration variable s to be replaced by a pair (s_1, s_2) of integration variables. The proof of this extension of the lemma goes through as before, except that we now write $ds = ds_1\, ds_2$, where s_1 ranges over an interval (a_1, b_1) and s_2 over an interval (a_2, b_2).

Let us take each of these intervals to be $(0, \infty)$, and write

$$\phi(s,\lambda) = (\varepsilon_1+\varepsilon_2)\exp[-i(\lambda - i\varepsilon_1)s_1]\exp[i(\lambda + i\varepsilon_2)s_2],$$
$$K(s) = \exp(iH_0 s_1)\exp(-iHs_2)V.$$

From (4.2.7), we then have

$$L(\lambda) = (\varepsilon_1+\varepsilon_2)(H_0-\lambda+i\varepsilon_1)^{-1}(H-\lambda-i\varepsilon_2)^{-1}V,$$

which, in view of (4.2.19), becomes

$$L(\lambda) = i\{(H_0-\lambda+i\varepsilon_1)^{-1} - (H_0-\lambda-i\varepsilon_2)^{-1}\}T(\lambda+i\varepsilon_2). \tag{4.2.28}$$

We can then use (4.2.8) to evaluate the spectral integral of the operator family $L(\lambda)$, giving

$$\int L(\lambda)\, dE_\lambda f = \int_0^\infty ds_1\,(\varepsilon_1+\varepsilon_2)e^{-\varepsilon_1 s_1}e^{iH_0 s_1}\left\{\int_0^\infty ds_2\, e^{-\varepsilon_2 s_2}e^{-iHs_2}Ve^{iH_0 s_2}\right\} e^{-iH_0 s_1}f.$$

If we denote the right-hand side by $I(\varepsilon_1,\varepsilon_2)f$, we have

$$\left\| I(\varepsilon_1,\varepsilon_2)f - i\int_0^\infty ds_1\,(\varepsilon_1+\varepsilon_2)e^{-\varepsilon_1 s_1}e^{iH_0 s_1}(\Omega_- - 1)e^{-iH_0 s_1}f \right\|$$
$$\le \int_0^\infty ds_1\,(\varepsilon_1+\varepsilon_2)e^{-\varepsilon_1 s_1}$$
$$\left\| \left\{\int_0^\infty ds_2\, e^{-\varepsilon_2 s_2}e^{-iHs_2}Ve^{iH_0 s_2} - i(\Omega_- - 1)\right\}e^{-iH_0 s_1}f \right\|. \tag{4.2.29}$$

Now the analogous result to (4.2.15), for the wave operator Ω_-, is

$$\Omega_- g = g - \operatorname*{s\text{-}lim}_{\varepsilon \to 0+} \mathrm{i} \int_0^\infty \mathrm{d}s\, \mathrm{e}^{-\varepsilon s}\, \mathrm{e}^{-\mathrm{i}Hs} V \mathrm{e}^{\mathrm{i}H_0 s} g.$$

Writing $\varepsilon = \varepsilon_2$, $g = \mathrm{e}^{-\mathrm{i}H_0 s_1} f$ and $s = s_2$, the right-hand side is uniformly bounded in norm. The expression within the norm on the right-hand side of (4.2.29) is therefore uniformly bounded and converges strongly to zero. Hence, by the Lebesgue Dominated-Convergence Theorem, it follows from (4.2.29) that

$$\operatorname*{s\text{-}lim}_{\varepsilon_2 \to 0+} I(\varepsilon_1, \varepsilon_2) f = \mathrm{i} \int_0^\infty \mathrm{d}s_1\, \varepsilon_1 \mathrm{e}^{-\varepsilon_1 s_1} \mathrm{e}^{\mathrm{i}H_0 s_1} (\Omega_- - 1) \mathrm{e}^{-\mathrm{i}H_0 s_1} f.$$

Assuming that asymptotic completeness holds, or more generally that range $(\Omega_-) \subseteq \operatorname{range}(\Omega_+)$, we know that

$$\operatorname*{s\text{-}lim}_{t \to \infty} \mathrm{e}^{\mathrm{i}H_0 t} \Omega_- \mathrm{e}^{-\mathrm{i}H_0 t} = \operatorname*{s\text{-}lim}_{t \to \infty} \mathrm{e}^{\mathrm{i}H_0 t} \mathrm{e}^{-\mathrm{i}Ht} \Omega_- = \Omega_+^* \Omega_- = S.$$

Using the analogous result to (4.2.14) for the adjoint of Ω_+ then gives

$$\operatorname*{s\text{-}lim}_{\substack{\varepsilon_2 \to 0+ \\ \varepsilon_1 \to 0+}} I(\varepsilon_1, \varepsilon_2) = \mathrm{i}(\Omega_+^* \Omega_- - 1) f = \mathrm{i}(S - 1) f,$$

which by (4.2.8) is the spectral integral, in the respective ε limits, of the operator family $L(\lambda)$ given by (4.2.28). We then have

$$Sf = f + \operatorname*{s\text{-}lim}_{\substack{\varepsilon_2 \to 0+ \\ \varepsilon_1 \to 0+}} \int_{-\infty}^{\infty} [(H_0 - \lambda + \mathrm{i}\varepsilon_1)^{-1} - (H_0 - \lambda - \mathrm{i}\varepsilon_2)^{-1}] T(\lambda + \mathrm{i}\varepsilon_2)\, dE_\lambda. \quad (4.2.30)$$

A similar calculation may be carried out to evaluate $(\Omega_-^* \Omega_- - 1) f$. Here, of course, $\Omega_-^* \Omega_- = 1$, since Ω_- is isometric, and we obtain in this case, replacing (4.2.30),

$$0 = \operatorname*{s\text{-}lim}_{\substack{\varepsilon_2 \to 0+ \\ \varepsilon_1 \to 0+}} \int_{-\infty}^{\infty} [(H_0 - \lambda - \mathrm{i}\varepsilon_1)^{-1} - (H_0 - \lambda - \mathrm{i}\varepsilon_2)^{-1}] T(\lambda + \mathrm{i}\varepsilon_2)\, \mathrm{d}E_\lambda f.$$

Comparing with (4.2.30), we may rewrite this equation as

$$Sf = f + \operatorname*{s\text{-}lim}_{\substack{\varepsilon_2 \to 0+ \\ \varepsilon_1 \to 0+}} \int_{-\infty}^{\infty} [(H_0 - \lambda + \mathrm{i}\varepsilon_1)^{-1} - (H_0 - \lambda - \mathrm{i}\varepsilon_1)^{-1}] T(\lambda + \mathrm{i}\varepsilon_2)\, \mathrm{d}E_\lambda. \quad (4.2.30)'$$

Equation (4.2.30), which holds for all $f \in D(H_0)$, is the most useful integral representation that can be obtained for the scattering operator, in a general abstract setting, i.e. before we take into account a more detailed analysis of the free and total Hamiltonian in the case of particular scattering systems. Note that, in (4.2.30) and (4.2.30)′, the two limits $\varepsilon_2 \to 0+$, $\varepsilon_1 \to 0+$ have to be taken in that order. If $T(z)$ has a boundary value $T_+(z)$ as $z = \lambda + \mathrm{i}\varepsilon$ approaches the real axis from above, we may also take the formal limit $\varepsilon_1 \to 0+$ in (4.2.30)′, by making use of the identity

$$\lim_{\varepsilon \to 0+} [(x+\mathrm{i}\varepsilon)^{-1} - (x-\mathrm{i}\varepsilon)^{-1}] = -2\pi\mathrm{i}\delta(x).$$

We then obtain

$$Sf = f - 2\pi\mathrm{i} \int_{-\infty}^{\infty} \delta(H_0 - \lambda) T_+(\lambda)\, \mathrm{d}E_\lambda, \tag{4.2.31}$$

which is an expression that many theoretical physicists will recognize!

4.3 THE WAVE AND SCATTERING OPERATORS FOR MANY-CHANNEL SYSTEMS

We have dealt so far, in this chapter, with wave and scattering operators in the case of a system described by a total Hamiltonian H and a single free-Hamiltonian operator H_0 acting as self-adjoint operators in a Hilbert space $\mathscr{H}$. To describe many-channel systems, we have to deal not only with the underlying Hilbert space $\mathscr{H}$, in which the total Hamiltonian H acts, but also with a number of free Hilbert spaces $\mathscr{H}^{(0)}$, $\mathscr{H}^{(1)}, \ldots$, one for each possible scattering channel. For channel j there will then be corresponding wave operators $\Omega_\pm^{(j)}$, defined as in Section 3.4, and we refer the reader to Chapter 3 for the definitions of the operators U_j, and for clarification of the relationship between the modified free evolutions $F_j(\boldsymbol{P}^{(j)}, t)$ and the corresponding free Hamiltonians $H^{(j)}$. We shall denote by $\mathscr{D}_j^\pm$ the respective ranges of the channel wave operators $\Omega_\pm^{(j)}$. The subspaces $\mathscr{D}_j^+$, as j is varied, are mutually orthogonal. The operator $H^{(j)}$ may be represented as the operator of multiplication by a real function of momentum in some L^2 (momentum) space. Provided the inverse image, for this function, of any subset of $\mathbb{R}$ having Lebesgue measure zero also has measure zero as a set of momenta, the operator $H^{(j)}$ will have a purely absolutely continuous spectrum. In that, case, arguing as in Section 4.1 from the intertwining property

$$H\Omega_\pm^{(j)} = \Omega_\pm^{(j)} H^{(j)},$$

it follows that the range of $\Omega_{\pm}^{(j)}$ is contained in $\mathcal{M}_{\rm ac}(H)$. This is the usual situation for the wave operators defined in many-channel scattering. On the other hand, it may be possible to define operators $H^{(j)}$ that have singular spectra, in which case H will also have a non-trivial singular subspace. In particular, corresponding to any eigenvector f_i of H, a one-dimensional space $\mathscr{H}^{(i)}$ may be defined, in which $H^{(i)}$ is the operator of multiplication by the relevant eigenvalue. Since the asymptotic evolution of eigenstates of H may be described extremely simply, these one-dimensional free Hilbert spaces are usually omitted in the description of channels, though strictly if we are to consider *all* modes of asymptotic evolution then they should be included. Wave operators associated with singular continuous spectra for H and $H^{(j)}$ have received no attention, though we shall have more to say in later chapters concerning the evolution of states in the singular continuous subspace.

How is the formula (4.1.14) for the scattering operator to be generalized to the many-channel situation? Consider a system for which there are N channels, corresponding to $j=0, 1, \ldots, N-1$, and define a Hilbert space $\mathscr{H}_N$ by

$$\mathscr{H}_N = \mathscr{H}^{(0)} \oplus \mathscr{H}^{(1)} \oplus \ldots \oplus \mathscr{H}^{(N-1)} \tag{4.3.1}$$

The Hilbert space $\mathscr{H}_N$ is formed from the free Hilbert spaces by making each $\mathscr{H}^{(i)}$ orthogonal to $\mathscr{H}^{(j)}$, for $i \neq j$. Thus an element $\psi \in \mathscr{H}_N$ will be of the form

$$\psi = \psi^{(0)} \oplus \psi^{(1)} \oplus \ldots \psi^{(N-1)},$$

where $\psi^{(j)} \in \mathscr{H}^{(j)}$ for each j. The inner product is given by

$$\begin{aligned}\langle \psi_1, \psi_2 \rangle &= \langle \psi_1^{(0)} \oplus \psi_1^{(1)} \oplus \ldots \oplus \psi_1^{(N-1)}, \psi_2^{(0)} \oplus \psi_2^{(1)} \oplus \ldots \oplus \psi_2^{(N-1)} \rangle \\ &= \sum_{j=0}^{N-1} \langle \psi_1^{(j)}, \psi_2^{(j)} \rangle. \end{aligned} \tag{4.3.2}$$

We now define, for each pair i, j, the operator

$$S_{ij} = (\Omega_{+}^{(i)})^* (\Omega_{-}^{(j)}). \tag{4.3.3}$$

Now $\Omega^{(j)}$ maps from $\mathscr{H}^{(j)}$ into $\mathscr{H}$, and $\Omega_{+}^{(i)}$ from $\mathscr{H}^{(i)}$ into $\mathscr{H}$, so that the adjoint of $\Omega_{+}^{(i)}$ is defined as an operator from $\mathscr{H}$ into $\mathscr{H}^{(i)}$. It follows that the S_{ij} maps from $\mathscr{H}^{(j)}$ to $\mathscr{H}^{(i)}$. We now define the many-channel scattering operator S, as an operator from $\mathscr{H}_N$ to $\mathscr{H}_N$, by

$$\begin{aligned} S\psi = \left(\sum_{j=0}^{N-1} S_{0j} \psi^{(j)} \right) \left(\sum_{j=1}^{N-1} S_{1j} \psi^{(j)} \right) \oplus \ldots \\ \ldots \oplus \left(\sum_{j=0}^{N-1} S_{N-1,j} \psi^{(j)} \right) \end{aligned} \tag{4.3.4}$$

Thus, if we think of $\psi \in \mathscr{H}_N$ as being represented by an N-component column vector, of which the jth component is $\psi^{(j)} \in \mathscr{H}^{(j)}$ ($j=0, 1, \ldots, N-1$), then the scattering operator may be represented as multiplication by the matrix having (operator-valued) entries S_{ij}.

Equation (4.3.4) is the extension of our previous definition of the scattering operator to the many-channel situation. The notion of asymptotic completeness also extends in a natural way, as follows.

Definition 4.3

The many-channel wave operators $\Omega_{\pm}^{(j)}$ ($j=0, 1, \ldots, N-1$) are said to be *asymptotically complete* if

$$\bigoplus_{J=0}^{N-1} \text{range}\,(\Omega_{+}^{(j)}) = \bigoplus_{j=0}^{N-1} \text{range}\,(\Omega_{-}^{(j)}).$$

Equality here means that the subspace $\mathscr{M}$ of $\mathscr{H}$ defined by $\Sigma_{j=0}^{N-1} \Omega_{+}^{(j)} \psi_{-}^{(j)}$, as the $\psi^{(j)}$ vary over $\mathscr{H}^{(j)}$, may be identified with the corresponding subspace with $\Omega_{+}^{(j)}$ replaced by $\Omega_{-}^{(j)}$. The idea is that any vector f belonging to $\mathscr{M}$ may be decomposed as a sum of N mutually orthogonal components, each corresponding to a particular type of asymptotic evolution in the limit as $t \to +\infty$, and that a similar decomposition holds for the limit $t \to -\infty$. Asymptotic completeness is a guarantee that an incoming state, in a particular channel at time $t=-\infty$, will with probability 1 evolve at time $t=+\infty$ into one of a set of N mutually exclusive outgoing channels. In such a situation, our list of possible modes of asymptotic evolution for the outgoing state may be regarded as complete, and the matrix element of S between an incoming state in channel i and an outgoing state in channel j gives a transition amplitude between these two states.

It is an instructive exercise for the reader to carry out a proof of the analogue to Lemma 4.1, in the case of many-channel scattering. (It will be helpful, in generalizing the proof of this lemma, to regard $\Omega_{\pm}$ as replaced in the many-channel case by a *row vector* $\{\Omega_{\pm}^{(j)}\}$, and $\Omega_{\pm}^{*}$ by the corresponding *column* vector of adjoints. With this interpretation, an operator such as $\Omega_{+}\Omega_{+}^{*}$ will look like $\Sigma_{j=1}^{N-1}\Omega_{+}^{(j)}(\Omega_{+}^{(j)})^{*}$, which is the projection onto the sum of the ranges of the $\Omega_{+}^{(j)}$.) A particular consequence of the generalization of Lemma 4.1 is that, as before, asymptotic completeness holds if and only if S is unitary.

If asymptotic completeness holds, and if the subspace defined above is identical with $\mathscr{M}_{\text{ac}}(H)$, then we again speak of strong asymptotic completeness. (This requires the omission of any possible channels corresponding to eigenstates of H. If these channels are included, and if H has no singular continuous spectrum, then we should have $\mathscr{M}=\mathscr{H}$.)

The proof of asymptotic completeness for concrete systems in scattering theory is one of the most challenging problems confronting research workers in this field, and one that has yet to be solved in many situations of interest to the physicist.

Chapter 5

The Free Hamiltonian and its Perturbations

5.1 THE FREE HAMILTONIAN IN ONE DIMENSION

In most elementary treatments of quantum mechanics, considerable attention is given to one-dimensional problems. Usually this involves an account of the harmonic oscillator, of the hydrogen atom treated by ordinary-differential-equation methods, and of scattering by square-well and step potentials, and so on. Such problems do, of course, serve as excellent introductions to the quantum mechanics of full three-dimensional systems, in that they illustrate a number of the basic principles, while avoiding the need to enter the theory of partial differential operators and other analytic tools in higher dimensions. At the same time, a one-dimensional analysis can sometimes be carried out exactly, or at least taken further than in the corresponding three-dimensional problem.

We begin this chapter by establishing the framework within which a reduction to a one-dimensional representation of the free Hamiltonian as an ordinary, rather than a partial, differential operator, may be carried out. We start from the free Hamiltonian $H_0 = -\Delta$ as a differential operator in $\mathscr{H} = L^2(\mathbb{R}^3)$ (units $\hbar = 2m = 1$).

Since the Laplacian operator commutes with each component of the operator $\boldsymbol{L}$ for orbital angular momentum, any sensible definition of H_0 as a self-adjoint operator must have as reducing subspace each angular momentum subspace $\mathscr{H}_{l,m}$. The quantum numbers l and m here correspond to eigenvalues $l(l+1)$ and m of $\boldsymbol{L}^2$ and L_z respectively. Any $g \in \mathscr{H}_{l,m}$ will be represented by a wave function of the form

$$g(|\boldsymbol{r}|)Y_{lm}(\theta, \varphi),$$

where (r, θ, φ) are spherical polar coordinates and Y_{lm} is a spherical-harmonic function. The functions Y_{lm} form an orthogonal set, and we adopt the normalization

$$\int Y_{lm}(\theta,\varphi)\bar{Y}_{l'm'}(\theta,\varphi)\,\mathrm{d}\omega=\delta_{ll'}\delta_{mm'},$$

where $\mathrm{d}\omega=\sin\theta\ \mathrm{d}\theta\mathrm{d}\varphi$ is an element of solid angle. Corresponding to any $g\in\mathscr{H}_{l,m}$, we can define a function $g_{lm}\in L^2(0,\infty)$ by the formula

$$g_{lm}(r)=rg(r).$$

Since in that case $\int_0^\infty|g_{lm}(r)|^2\,\mathrm{d}r=\int|g(\mathbf{r})^2|\,\mathrm{d}^3\mathbf{r}$, we have here a one-to-one mapping from $\mathscr{H}_{l,m}$ onto $L^2(0,\infty)$ that preserves norms and inner products. Any element of $\mathscr{H}_{l,m}$ may be represented in this way by a wave function in $L^2(0,\infty)$, and any linear operator, such as H_0, for which $\mathscr{H}_{l,m}$ is a reducing subspace, may be implemented as a linear operator in each $\mathscr{H}_{l,m}$. Viewed in this way, we have decomposed the entire Hilbert space into infinitely many copies of $L^2(0,\infty)$.

Since

$$-\Delta=-\frac{1}{r}\frac{\partial^2}{\partial r^2}(r\cdot)+\frac{\mathbf{L}^2}{r^2},$$

the free Hamiltonian, acting in the copy of $L^2(0,\infty)$ labelled by quantum numbers l and m of angular momentum, may be implemented as the ordinary differential operator

$$-\frac{\mathrm{d}^2}{\mathrm{d}r^2}+\frac{l(l+1)}{r^2}.$$

If we are restricting our attention to a single partial-wave subspace, then, we may write without too much abuse of notation,

$$H_0=-\frac{\mathrm{d}^2}{\mathrm{d}r^2}+\frac{l(l+1)}{r^2}. \tag{5.1.1}$$

What kind of operator is defined by (5.1.1)? To start with, if we are to define H_0 as a *self-adjoint* operator in $L^2(0,\infty)$, we have to be more precise about domains; i.e. about the set of wave functions $f(r)$ in $L^2(0,\infty)$ on which H_0 acts. The standard way to specify a domain for H_0 is to define first of all a linear operator $\hat{H}_0=-\mathrm{d}^2/\mathrm{d}r^2+l(l+1)/r^2$ having rather a small domain—we shall take $D(\hat{H}_0)=\hat{C}_0^\infty(0,\infty)$, the linear space of functions ϕ that are infinitely differentiable and that vanish near $r=0$ and for sufficiently large values of r. Though symmetric, the linear operator $\hat{H}_0$ will not be self-adjoint, since $D(\hat{H}_0^*)$ is strictly larger than $D(\hat{H}_0)$. We can then define H_0 as a self-adjoint extension of $\hat{H}_0$. The domain of H_0 will be larger than the domain of $\hat{H}_0$, but smaller than or equal to the domain of $\hat{H}_0^*$. (See Section 2.5 on operator extensions.)

The condition for f to belong to the domain of $\hat{H}_0^*$ may be written

$$\langle \hat{H}_0 \phi, f \rangle = \langle \phi, f^* \rangle \tag{5.1.2}$$

for all $\phi \in C_0^\infty(0, \infty)$, where $f^* = \hat{H}_0^* f$. In the terminology of Section 2.2, we may say that $f(r)$ satisfies the differential equation

$$-\frac{d^2 f}{dr^2} + \frac{l(l+1)}{r^2} f = f^*$$

in the sense of distributions.

Chapter 6 will be concerned with an analysis of the general one-dimensional Schrödinger operator on an interval. Rather than entering here into a detailed discussion of the analysis of ordinary differential operators and their spectra, we shall summarize some of the results as they apply to the free Hamiltonian, in this section, and to the free Hamiltonian plus potential in the following section. In this way it should be possible for the reader to obtain an overview of what is known about these operators, before seeing how these results can be obtained.

For $f \in D(\hat{H}^*)$, the differential equation relating f and f^*, which we wrote down as a consequence of (5.1.2), may be interpreted in the following sense. A function $f \in L^2(0, \infty)$ will belong to $D(\hat{H}^*)$ if and only if there is a function $g^* \in L^2(0, \infty)$ such that

$$-\frac{d^2 f}{dr^2} + \frac{l(l+1)}{r^2} f = f^* \quad \text{for all } r > 0,$$

where $f(r)$ and $df(r)/dr$ are absolutely continuous functions of r for $r > 0$. Moreover, $f^* = \hat{H}_0^* f$.

As we saw in Chapter 2, the possibility of finding self-adjoint extensions of $\hat{H}_0$ depends on the *deficiency indices* of the operator $\hat{H}_0$. We are already assured that one self-adjoint extension at least must exist, since $\hat{H}_0$ is a real operator. (For further discussion of real extensions of the general Schrödinger operator see Chapter 6.) The question here is the number of such extensions. To determine the deficiency indices of $\hat{H}_0$, we need to consider solutions g of the equations $(\hat{H}_0^* - z)g = 0$, $(\hat{H}_0^* - \bar{z})g = 0$. The situation here is different depending on whether $l = 0$ or $l \neq 0$. For $l = 0$, and using our characterization of $D(\hat{H}_0^*)$, $g \in L^2(0, \infty)$ must satisfy the differential equation

$$-\frac{d^2 g}{dr^2} = zg, \qquad \text{or } -\frac{d^2 g}{dr^2} = \bar{z} g,$$

with $\mathrm{Im}(z) \neq 0$.

Apart from a multiplicative constant, this gives, respectively, $g = \exp(iz^{1/2} r)$ or $g = \exp(i\bar{z}^{1/2} r)$, where in both instances the square root is taken to have positive rather than negative imaginary part. Thus we obtain, for $l=0$, just one linearly independent solution for g, with the result that the deficiency indices are (1,1). We therefore have a one-parameter family of self-adjoint extensions. Each extension may be defined by means of a boundary condition at $r=0$, of the form $\cos\beta\, f(0)+\sin\beta\, f'(0)=0$. The boundary condition $f(0)=0$, corresponding to taking $\beta=0$, is the natural one to take on several counts, including the following.

(i) A function $f(r)\in L^2\ (0,\infty)$ corresponds to the wave function $f(|\mathbf{r}|)/r$ in $L^2(\mathbb{R}^3)$, apart from a normalization constant $(4\pi)^{-1/2}$. We want $f(|\boldsymbol{r}|)/r$ to be in the domain of the three-dimensional Laplacian Δ, which is an operator invariant under translations. Thus the singularity at $r=0$ must be only apparent, since by a different choice of origin we could have shifted this point elsewhere. It follows that we must have $f(|\boldsymbol{r}|)=0$ in any sensible definition of H_0 in the $l=0$ subspace.

ii) If $\beta\neq 0\ (\operatorname{mod}\pi)$ one has an eigenfunction e^{-kr} with $k=\cot\beta$. These spurious eigenfunctions have no physical significance, and we expect on physical grounds that the kinetic energy of a free particle will have a purely continuous spectrum.

(iii) An alternative approach is to define H_0 by methods based on sesquilinear forms (see Chapter 2). A form extension of H_0 leads without ambiguity to the boundary condition $f(0)=0$.

For these reasons, we define the free Hamiltonian $H_0=-\mathrm{d}^2/\mathrm{d}r^2$ in $L^2(0,\infty)$ to have domain $D(H_0)$ consisting of those $f\in L^2(0,\infty)$ for which $f(r)$, $\mathrm{d}f(r)/\mathrm{d}r$ are absolutely continuous $(r>0)$, and for which $\mathrm{d}^2f/\mathrm{d}r^2\in L^2(0,\infty)$ and $f(0)=0$. With this domain, H_0 is self-adjoint as an operator in the $l=0$ version of $L^2(0,\infty)$.

The case $l\neq 0$ must be considered separately. We then have to consider solutions of the differential equations

$$-\frac{\mathrm{d}^2g}{\mathrm{d}r^2}+\frac{l(l+1)}{r^2}g=zg,\qquad -\frac{\mathrm{d}^2g}{\mathrm{d}r^2}+\frac{l(l+1)}{r^2}g=\bar{z}g.$$

Following the discussion of Section 2.8, of which we have here a very special case, we know, for the first of these equations, that we have as possible asymptotic behaviour r^{-l}, r^{l+1} as $r\to 0$, and $\exp(\mathrm{i}z^{1/2}r)$, $\exp(-\mathrm{i}z^{1/2}r)$ as $r\to\infty$. The only possibility compatible with $g\in L^2(0,\infty)$ in this case is $g(r)\sim\mathrm{const}\, r^{l+1}$ as $r\to 0$, with $g(r)\sim\mathrm{const}\exp(\mathrm{i}z^{1/2}r)$ as $r\to\infty$, again with $\operatorname{Im}(z^{1/2})>0$. It is, however, a consequence of a general identity for Wronskians ((6.2.5) of Chapter 6) that no single function $g(r)$ can combine these two modes of asymptotic behaviour, apart from the trivial function $g(r)\equiv 0$. For $l\neq 0$ we have the same conclusion with $\bar{z}$ instead of z, and the deficiency indices in this case must be (0, 0). This means that the operator $\hat{H}_0$ for $l \neq 0$ is essentially self-adjoint, and we may define H_0 to be the unique self-adjoint extension. There is no need for a boundary condition at $r=0$, and the domain of H_0, for $l \neq 0$, consists of those $f \in L^2(0, \infty)$ for which $f(r)$ and $\mathrm{d}f(r)/\mathrm{d}r$ are absolutely continuous $(r>0)$, and for which $-\mathrm{d}^2 f/\mathrm{d}r^2 + l(l+1) f/r^2 \in L^2(0, \infty)$. Note here that $\mathrm{d}^2 f/\mathrm{d}r^2$ and f/r^2 need *not* separately belong to $L^2(0, \infty)$, since there may, in general, be cancellations between the two contributions to $H_0 f$.

We may also approach the free Hamiltonian in each partial-wave subspace starting from the momentum-space representation of the Hilbert space. In momentum space, any $g \in \mathscr{H}_{l,m}$ will be represented by a wave function of the form

$$\hat{g}(|\boldsymbol{k}|)\bar{Y}_{lm}(\theta', \varphi'),$$

where $(|\boldsymbol{k}|, \theta', \varphi')$ are spherical polar coordinates for the momentum vector $\boldsymbol{k}$. Complex-conjugation of the spherical harmonic function enters owing to the fact that the z-component of angular momentum is represented in position space by the operator $-\mathrm{i}\partial/\partial\varphi$, and in momentum space by the operator $+\mathrm{i}\partial/\partial\varphi'$. Corresponding to any $\hat{g} \in \mathscr{H}_{l,m}$, we can define $\hat{g}_{lm} \in L^2(0, \infty)$ by the formula

$$\hat{g}_{lm}(k) = k\hat{g}(k).$$

This effects, as before, a unitary transformation from $\mathscr{H}_{l,m}$ to $L^2(0, \infty)$, the advantage in the present case being that the free Hamiltonian H_0 will be implemented as a multiplication operator, rather than a differential operator, on one-dimensional wave functions.

From given $f \in \mathscr{H}_{l,m}$, we may arrive at a representation either $f(r) \in L^2(0, \infty)$, or $\hat{f}(k) \in L^2(0, \infty)$. As an operator in the position-space representation, we have already seen how to define

$$H_0 = -\frac{\mathrm{d}^2}{\mathrm{d}r^2} + \frac{l(l+1)}{r^2}$$

as a self-adjoint operator in each partial-wave subspace. As an operator in the momentum-space representation, we have, simply,

$$(H_0 \hat{f})(k) = k^2 \hat{f}(k) \tag{5.1.3}$$

The domain of an operator of multiplication is generally simpler to define than that of a differential operator. In (5.1.3), the domain of H_0 consists of those $\hat{f} \in L^2(0, \infty)$ such that $k^2 \hat{f} \in L^2(0, \infty)$. The two versions, position-space and momentum-space, are of course unitarily equivalent, since we started from two unitarily equivalent representations of wave functions in $L^2(\mathbb{R}^3)$. Indeed, a stricter insistence on formality of notation would require a different symbol for each possible implementation of the free-Hamiltonian operator. We shall adopt the less purist approach, which, with due care, results in greater clarity and ease of application.

The unitary transformation that allows us to pass from position space to momentum space, in a single (l, m) partial wave, is called the Fourier–Hankel transform, and is defined by the formula

$$\hat{f}(k) = \left(\frac{2}{\pi}\right)^{1/2} \int_0^\infty \mathrm{d}r\, f(r) kr j_l(kr), \tag{5.1.4}$$

where j_l is a modified Bessel function. If we define the Bessel function $J_p(x)$ through the standard power-series expansion, with $J_p(x) \sim (\frac{1}{2}x)^p / \Gamma(p+1)$ in the limit as $x \to 0$, the modified Bessel function $j_l(x)$ is given by

$$j_l(x) = \left(\frac{\pi}{2x}\right)^{1/2} J_{l+1/2}(x) \tag{5.1.5}$$

This results in the asymptotic behaviour

$$j_l(x) \sim \frac{2^l(l!)}{(2l+1)!} x^l \quad \text{as } x \to 0 \tag{5.1.6}$$

$$j_l(x) = \frac{1}{x} \sin(x - \tfrac{1}{2}l\pi + o(1)) \quad \text{as } x \to \infty. \tag{5.1.7}$$

If p is half an odd integer then the Bessel differential equation may be solved exactly. This allows us to express the modified Bessel function, in closed form, in terms of trigonometric functions, through the formula

$$j_l(x) = x^l \left(-\frac{1}{x}\frac{\mathrm{d}}{\mathrm{d}x}\right)^l \frac{\sin x}{x}. \tag{5.1.8}$$

The special case $l = 0$ gives $j_0(x) = (\sin x)/x$, in which case (5.1.4) is the standard formula for the Fourier sine transform:

$$\hat{f}(k) = \left(\frac{2}{\pi}\right)^{1/2} \int_0^\infty \mathrm{d}r\, f(r) \sin kr. \tag{5.1.9}$$

The factor $(2/\pi)^{1/2}$ is a normalization constant ensuring that the transformation is unitary. The inverse transformation to (5.1.9) is given by

$$f(r) = \left(\frac{2}{\pi}\right)^{1/2} \int_0^\infty \mathrm{d}k\, \hat{f}(k) \sin kr, \tag{5.1.10}$$

and, more generally, the inverse of (5.1.4) is

$$f(r)=\left(\frac{2}{\pi}\right)^{1/2}\int_0^\infty \mathrm{d}k\,\hat{f}(k)krj_l(kr), \tag{5.1.11}$$

Just as for the standard Fourier transform, the right-hand sides of these equations should be interpreted as strong limits, in $L^2(0, \infty)$, of integrals over the interval $[0, n]$ in the limit as $n\to\infty$. If $f\in D(H_0)$, so that $H_0\hat{f}=k^2\hat{f}\in L^2(0, \infty)$, application of the Schwarz inequality shows that $\hat{f}\in L^1(0, \infty)$. In this case, the right-hand sides of (5.1.10) and (5.1.11) exist as Lebesgue integrals, and one may verify from (5.1.10) by direct substitution the boundary condition $f(0)=0$ that applies to $D(H_0)$ in the $l=0$ partial wave. Although differential-operator techniques will be employed extensively in Chapter 6, spectral properties for the free Hamiltonian may be obtained most directly from the momentum-space representation. In the partial-wave subspace $\mathcal{H}_{l,m}$ the resolvent operator $(H_0-z)^{-1}$ is the operator of multiplication by $(k^2-z)^{-1}$, and is bounded for all $z\in\mathbb{C}$, apart from z lying on the negative real axis. Hence the spectrum of H_0 is the positive real axis $[0, \infty)$, as we should expect for an operator representing kinetic energy.

For $\lambda\le 0$, the spectral projection E_λ for H_0 is the zero operator. For $\lambda>0$, E_λ is given by

$$(E_\lambda\hat{f})(k)=\begin{cases}\hat{f}(k) & (k^2<\lambda)\\ 0 & (k^2\ge\lambda)\end{cases}$$

Hence

$$\langle\hat{g}, E_\lambda\hat{f}\rangle=\int_0^{\lambda^{1/2}}\bar{\hat{g}}(k)\hat{f}(k)\,\mathrm{d}k$$

and

$$\frac{\mathrm{d}}{\mathrm{d}\lambda}\langle\hat{g}, E_\lambda\hat{f}\rangle=\frac{1}{2\lambda^{1/2}}\bar{\hat{g}}(\lambda^{1/2})\hat{f}(\lambda^{1/2}) \quad \text{for } \lambda>0.$$

It follows that the spectrum of H_0 is purely absolutely continuous.

The momentum-space representation for H_0 is also a useful starting point for studying the free time evolution of states in a particular partial-wave subspace. For an initial momentum-space wave function $\hat{f}(k)$, the state at times t is described by the wave function $\mathrm{e}^{-\mathrm{i}k^2t}\hat{f}(k)$. According to (5.1.11), the time evolution of a free-particle state in position space is given by

$$f(r, t)=\left(\frac{2}{\pi}\right)^{1/2}\int_0^\infty \mathrm{d}k\,\mathrm{e}^{-\mathrm{i}k^2t}\hat{f}(k)krj_l(kr). \tag{5.1.12}$$

For suitable $\hat{f}$, say for $\hat{f}\in C_0^\infty(0,\infty)$, the right-hand side of (5.1.12) may be integrated by parts by means of the substitution $e^{-ik^2t}=i(2kt)^{-1}(d/dk)(e^{-ik^2t})$, to show that $f(r,t)\to 0$ as $t\to\infty$, uniformly in r over bounded intervals. Hence

$$\lim_{t\to\infty}\int_0^a dr\,|(e^{-iH_0t}f)(r)|^2=0,$$

showing that, in the limit $t\to\infty$, the probability of finding a freely moving particle in the finite region $0<r<a$ decreases to zero. Since $C_0^\infty(0,\infty)$ is dense in $L^2(0,\infty)$, this result may be extended to arbitrary $f\in L^2(0,\infty)$. A freely moving particle, at large times, is located far from the origin $r=0$.

In position space one of the most important features of the free Hamiltonian H_0 is that H_0 is a local operator. That is, $(H_0 f)(r)$ depends only on the derivatives of f *at the point* r. For more precise definitions, and some consequences of locality of differential operators, see Chapter 6. We shall note, in the following section, that locally the domains of H_0 and H_0+V look the same, a fact that will have far-reaching consequences for the time evolution of states. Our discussion so far has centred on the free Hamiltonian as an operator in $L^2(0,\infty)$, arising from the decomposition of the full space $L^2(\mathbb{R}^3)$ into its partial-wave subspaces. Let us also deal briefly with the free Hamiltonian $H_0=-d^2/dx^2$ for a particle moving along the entire real axis $\mathbb{R}$. Such a description of a freely moving particle may, of course, be criticized as being physically unrealistic. In quantum mechanics, particles do not confine themselves to motion along a straight line. Nevertheless, a study of $H_0=-d^2/dx^2$ in $L^2(\mathbb{R})$ may be regarded as a useful adjunct to a treatment of the Laplacian in $L^2(\mathbb{R}^3)$, since $\Delta=\partial^2/\partial x^2+\partial^2/\partial y^2+\partial^2/\partial z^2$ is the sum of three commuting "copies" of d^2/dx^2. More precisely, in tensor-product notation, we may write

$$\Delta=\frac{d^2}{dx^2}\otimes I\otimes I+I\otimes\frac{d^2}{dy^2}\otimes I+I\otimes I\otimes\frac{d^2}{dz^2},$$

where I is the identity operator in $L^2(\mathbb{R})$. Thus, for example, spectral properties of the one-dimensional kinetic-energy operator will be reflected in corresponding properties of the total kinetic energy of a particle in three dimensions.

The reader should have no difficulty in verifying that $\hat{P}=-i\,d/dx$ and $\hat{H}_0=-d^2/dx^2$, with $D(\hat{P})=D(\hat{H}_0)=C_0^\infty(\mathbb{R})$, are essentially self-adjoint. We define the free Hamiltonian H_0, in $L^2(\mathbb{R})$, to be the unique self-adjoint extension of $\hat{H}_0$. The domain of H_0 consists of those $f\in L^2(\mathbb{R})$ such that f and df/dx are absolutely continuous and $H_0 f\equiv -d^2f/dx^2\in L^2(\mathbb{R})$.

For $f \in D(H_0)$, we can see also that $\mathrm{d}f/\mathrm{d}x \in L^2(\mathbb{R})$. For $f \in D(H_0)$, let us suppose that $\int_0^\infty |\mathrm{d}f/\mathrm{d}x|^2 \,\mathrm{d}x = \infty$. Since

$$\int_0^n \left|\frac{\mathrm{d}f}{\mathrm{d}x}\right|^2 \mathrm{d}x = \int_0^n \frac{\mathrm{d}\bar{f}}{\mathrm{d}x}\frac{\mathrm{d}f}{\mathrm{d}x}\,\mathrm{d}x = \left[\frac{\mathrm{d}\bar{f}}{\mathrm{d}x} f\right]_0^n - \int_0^n \frac{\mathrm{d}^2\bar{f}}{\mathrm{d}x^2} f \,\mathrm{d}x,$$

where $\int_0^\infty (\mathrm{d}^2\bar{f}/\mathrm{d}x^2) f \,\mathrm{d}x$ is convergent by the Schwarz inequality, it follows that $\mathrm{Re}\,((\mathrm{d}\bar{f}/\mathrm{d}x) f) \to \infty$ as $x \to \infty$. Hence

$$\frac{1}{2}\left(\frac{\mathrm{d}\bar{f}}{\mathrm{d}x} f + \frac{\mathrm{d}f}{\mathrm{d}x}\bar{f}\right) = \frac{1}{2}\frac{\mathrm{d}}{\mathrm{d}x}|f|^2 \to \infty \quad \text{as } x \to \infty,$$

so that $|f|^2 \to \infty$. It follows that $f \notin L^2(0, \infty)$, which contradicts our original assumption. Hence $\mathrm{d}f/\mathrm{d}x \in L^2(0, \infty)$. Similarly, $\mathrm{d}f/\mathrm{d}x \in L^2(-\infty, 0)$, from which the result follows.

A further consequence of this result, for $f \in D(H_0)$, is that $f(x) \to 0$ as $|x| \to \infty$, which follows from the fact that $(\mathrm{d}/\mathrm{d}x)|f|^2 \in L^1(\mathbb{R})$, where $f \in L^2(\mathbb{R})$. Thus functions in the domain of H_0 decay to zero at infinity.

For $H_0 = -\mathrm{d}^2/\mathrm{d}x^2$ in one dimension, the transformation to a representation of H_0 as a multiplication operator is effected through the Fourier transform. We then have

$$\hat{f}(k) = (2\pi)^{-1/2} \int_{-\infty}^{\infty} \mathrm{e}^{-ikx} f(x)\,\mathrm{d}x, \tag{5.1.13}$$

and $(H_0 \hat{f})(k) = k^2 \hat{f}(k)$, the domain of H_0 in momentum space consisting of those $\hat{f} \in L^2(\mathbb{R})$ for which $k^2 \hat{f} \in L^2(\mathbb{R})$.

Again, H_0 has a purely absolutely continuous spectrum along the positive axis $[0, \infty)$, and formulae for the spectral projections E_λ may be written down without difficulty.

There is, however, an important difference between H_0 as a self-adjoint operator in $L^2(\mathbb{R})$ and H_0 as a self-adjoint operator in $L^2(0, \infty)$. The difference lies in the fact that $-\mathrm{d}^2/\mathrm{d}x^2$ in $L^2(\mathbb{R})$ has a *degenerate spectrum*, whereas $H_0 = -\mathrm{d}^2/\mathrm{d}r^2$ in $L^2(0, \infty)$ has a *simple spectrum*. A self-adjoint operator T in a Hilbert space $\mathscr{H}$ is said to have simple spectrum if there exists a vector $g \in \mathscr{H}$, a so-called *cyclic vector*, such that any $f \in \mathscr{H}$ may be represented in the form $f = F(T)g$. Since $\|f\|^2 = \int_{-\infty}^{\infty} |F(\lambda)|^2 \,\mathrm{d}\langle g, E_\lambda g\rangle$, this requires in each case that $F(\lambda)$ be square-integrable with respect to the Lebesgue–Stieltjes measure μ generated by $\langle g, E_\lambda g\rangle$. A self-adjoint operator for which there is no cyclic vector is said to have a degenerate spectrum. The reader should be able to appreciate the idea behind this terminology by verifying that, for an operator T having a *discrete* spectrum, T will have a simple spectrum if and only if the eigenspace corresponding to each eigenvalue is one-dimensional, i.e. non-degenerate.

In the momentum-space representation, a cyclic vector $\hat{g}(k)$ for H_0 in $L^2(0, \infty)$ requires, for any $\hat{f} \in L^2(0, \infty)$, a representation $\hat{f}(k) = F(k^2)\hat{g}(k)$, for some F. Such a cyclic vector is furnished, for example, by $\hat{g}(k) = \mathrm{e}^{-k}$. However, a cyclic vector cannot be found for H_0 in $L^2(\mathbb{R})$. For example, with $\hat{g}(k) = \mathrm{e}^{-|k|}$, we find that a representation exists only for those $\hat{f}(k)$ that are *even*. More generally, if $\hat{f}_1$ and $\hat{f}_2$ are two momentum-space wave functions that may each be represented in the form $F(k^2)\hat{g}(k)$, we must have $\hat{f}_1(-k)\hat{f}_2(k) = \hat{f}_2(-k)\hat{f}_1(k)$. This condition does not apply to all $\hat{f} \in L^2(\mathbb{R})$, and we conclude that the free Hamiltonian in $L^2(\mathbb{R})$ has a degenerate absolutely continuous spectrum. The Hilbert space may, however, be decomposed into two mutually orthogonal subspaces (even and odd functions respectively), on each of which H_0 has a simple spectrum. It is for this reason that we describe the spectrum of H_0 in $L^2(\mathbb{R})$ as being *doubly degenerate*.

A particularly simple representation may be obtained for the free evolution of a wave function in $L^2(\mathbb{R})$. We let $f(x)$ denote the state at time $t=0$ in position space. From (5.1.13) we see that the state at time t, in momentum space, is described by the wave function

$$\hat{f}(k, t) = (2\pi)^{-1/2} \int_{-\infty}^{\infty} \mathrm{e}^{-\mathrm{i}(ky + k^2 t)} f(y)\, \mathrm{d}y.$$

Returning to position space through the inverse Fourier transform, we have

$$f(x, t) = \frac{1}{2\pi} \int_{-\infty}^{\infty} K(x, y; t) f(y)\, \mathrm{d}y, \tag{5.1.14}$$

where

$$K(x, y; t) = \int_{-\infty}^{\infty} \exp\left[\mathrm{i}(k(x-y) - k^2 t)\right] \mathrm{d}k.$$

Despite appearances, this integral *does* converge, although not absolutely, and we obtain

$$K(x, y; t) = \left(\frac{\pi}{\mathrm{i}t}\right)^{1/2} \exp\left(\frac{\mathrm{i}(x-y)^2}{4t}\right).$$

Hence

$$(e^{-iH_0t}f)(x)=\left(\frac{1}{4\pi it}\right)^{1/2}\int_{-\infty}^{\infty}\exp\left(\frac{i(x-y)^2}{4t}\right)f(y)\,dy. \tag{5.1.15}$$

Equation (5.1.15) may again be used as a starting point for the study of evolution of wave functions in position space. Both (5.1.15), and the extension that we shall obtain for a free particle in three-dimensional space, are important ingredients in the definition of wave operators in scattering theory.

5.2 PERTURBATIONS OF THE FREE HAMILTONIAN IN ONE DIMENSION

For a particle moving in three dimensions in a field of force described by a spherically symmetric potential $V(|\mathbf{r}|)$, the total Hamiltonian $H=H_0+V$ commutes with each component of the orbital-angular-momentum operator $\mathbf{L}$, so that each partial-wave subspace $\mathscr{H}_{l,m}$ is a reducing subspace of H. In position space, acting in the copy of $L^2(0, \infty)$ labelled by quantum numbers l and m of angular momentum, H may be implemented as an ordinary differential operator

$$H=-\frac{d^2}{dr^2}+V(r)+\frac{l(l+1)}{r^2}. \tag{5.2.1}$$

In order to define H as a self-adjoint operator in $L^2(0, \infty)$, we start by defining the linear operator

$$\hat{H}=-\frac{d^2}{dr^2}+V(r)+\frac{l(l+1)}{r^2},$$

with $D(\hat{H})=C_0^\infty(0, \infty)$. In order that $\hat{H}$ be well-defined with this domain, it is necessary for $V(r)$ to be square-integrable locally in $(0, \infty)$. This means that V must be square-integrable over any closed bounded subset of $(0, \infty)$, but does not impose any further restriction on the behaviour of V near $r=0$, and for large values of r. The Hamiltonian H will then be a self-adjoint extension of $\hat{H}$. Possible deficiency indices for $\hat{H}$ are $(0, 0)$, $(1, 1)$ and $(2, 2)$. In the first case, $\hat{H}$ is essentially self-adjoint, and H may be defined to be the closure of $\hat{H}$. In the second case, a single boundary condition must be imposed, either at $r=0$ or at $r=\infty$. In Chapter 6 we shall examine the nature of these boundary conditions, in the light of a deeper analysis of the Schrödinger operator. If $\hat{H}$ has deficiency indices $(2, 2)$ then a boundary condition must be imposed *both* at $r=0$ and at $r=\infty$ in order to define H as a self-adjoint operator.

Whether or not boundary conditions are required will depend on the nature of the potential, and in particular on the asymptotic behaviour of solutions of the Schrödinger equation. We shall discuss this question in greater detail in Chapter 6, when we shall deal with the limit-point/limit-circle criterion for an ordinary differential operator. Almost universally in applications to scattering theory, no boundary conditions are required at infinity, and we are in the situation of deficiency indices either $(0,0)$ or $(1,1)$ for $\hat{H}$.

If $V(r)$ is non-singular at $r=0$ (roughly $V(r)$ is bounded or diverges less rapidly than const/r^2), a boundary condition will be necessary only if $l=0$, in which case we apply the same condition $f(0)=0$ as for the free Hamiltonian. If $V(r)$ is singular at $r=0$, say $V(r)=g/r^n$ for $n>2$, the deficiency indices will depend on whether V is attractive or repulsive. For a repulsive singular potential, $g>0$, $\hat{H}$ is essentially self-adjoint, whereas in the attractive case, $g<0$, a boundary condition must be imposed. Since the spectrum of an attractive singular potential is unbounded from below, such potentials are considered generally to be of little physical significance.

The distinction between attractive and repulsive singular potentials is not a clear-cut one. Potentials can be defined that are oscillatory near $r=0$ and that can be considered neither attractive nor repulsive. In some cases, such potentials can lead to unusual spectral properties and associated phenomena including asymptotic localization of states; we shall consider instances of this behaviour in Chapter 14.

In Chapter 6 we shall consider the analogue of the Fourier–Hankel transform (5.1.4) for the total Hamiltonian H. The aim will be to effect a unitary transformation from $L^2(0,\infty)$ onto an L^2 space in which H is represented by a multiplication operator. Just as in the case of H_0, a spectral analysis of the Hamiltonian will follow more directly from this representation than from the formula (5.1.1). A case of particular interest to scattering theory will be considered in Chapter 7, for which a generalization of (5.1.4) may be obtained in which the modified Bessel function is replaced by some suitable solution of the Schrödinger equation at energy k^2.

A wide variety of spectral properties is possible for the Schrödinger operator in $L^2(0,\infty)$. For $V(r)$ decaying sufficiently rapidly at infinity (say $V\in L^1(c,\infty)$, but extending also to a large class of long-range potentials, including the Coulomb), H will have an absolutely continuous component, often along the positive real line. However, a singular continuous spectrum is possible for certain slowly decaying potentials. Potentials singular at $r=0$ may be found that lead to a degeneracy of the spectrum at positive energies, or to an absolutely continuous spectrum at negative energies. Spectral analysis of the total Hamiltonian will be the subject of Chapter 7, and it will be left to consider in subsequent chapters the profound influence of these results on the time evolution of states, and on scattering theory in particular.

Finally, reference should again be made to the total Hamiltonian H as a *local* operator. As long as V is square-integrable locally in $(0, \infty)$, the domains of H and H_0 are locally the same. For if $f \in D(H)$ we find $\phi f \in D(H_0)$ for any $\phi \in C_0^\infty(0, \infty)$. Thus we can convert functions in the domain of H into functions in the domain of H_0 by multiplying by a factor ϕ that is infinitely differentiable and that vanishes near $r=0$ and for all sufficiently large r. Conversely, we also have $\phi f \in D(H)$ for all $f \in D(H_0)$, $\phi \in C_0^\infty(0, \infty)$. The fact that H is a local operator has an important bearing on the degree to which states can be localized in position space; the theory of localization will be taken up in Chapter 10.

This section has briefly surveyed some of the more fundamental ideas relating to perturbations of H_0 in one dimension. We refer to *perturbations* of H_0, motivated by the expectation that, in a number of important respects such as the construction of self-adjoint extensions, domains, spectral properties, time evolution and so on, the analysis of H_0+V is to a greater or lesser degree similar to that of H_0. Part of this book may be regarded as a study of the extent to which this expectation is fulfilled in practice. And a great deal of the fascination of the subject, at least to the present author, lies in exploring the boundaries of the standard theory, at which the effect of a so-called "perturbation" may be very dramatic indeed.

5.3 THE FREE HAMILTONIAN IN THREE DIMENSIONS

The Hamiltonian for a particle moving freely in three-dimensional space is $\boldsymbol{P}^2/2m$, where m is the mass of the particle and $\boldsymbol{P}$ is its momentum. As operators in $L^2(\mathbb{R}^3)$, we have $\boldsymbol{P}=(P_1, P_2, P_3)$, where $P_i = -i\hbar\partial/\partial x_i$, so that the Hamiltonian operator is $-(\hbar^2/2m)\Delta$, where Δ is the Laplacian operator.

Since we shall not here be making explicit calculations, we may employ units in which $\hbar = 2m = 1$, in which case the components of momentum are $P_i = -i\,\partial/\partial x_i$ and the free Hamiltonian is just $-\Delta$. We should first of all like to define both momentum and Hamiltonian operators as self-adjoint operators in the Hilbert Space $\mathscr{H}=L^2(\mathbb{R}^3)$. Let us start by defining each operator on $C_0^\infty(\mathbb{R}^3)$, the space of infinitely differentiable functions having compact support in $\mathbb{R}^3$. We shall denote by $\hat{P}_i$ and $\hat{H}_0$ the differential operators with domains restricted in this way.

Thus $\hat{P}_i = -i\,\partial/\partial x_i$ and $\hat{H}_0 = -\Delta$, with $D(\hat{P}_i)=D(\hat{H}_0)=C_0^\infty(\mathbb{R}^3)$. It is a simple exercise to verify that $\hat{P}_i$ and $\hat{H}_0$ are symmetric linear operators. In fact, both $\hat{P}_i$ and $\hat{H}_0$ are essentially self-adjoint. To see this for $\hat{P}_i$, we have to show that there is no vector g, apart from $g=0$, that is orthogonal to the range of $\hat{P}_i+\mathrm{i}$, and similarly for the range of $\hat{P}_i-\mathrm{i}$. Now any vector g orthogonal to the range of $\hat{P}_1+\mathrm{i}$ (to take $i=1$) will be orthogonal to $(\hat{P}_1+\mathrm{i})\phi$ for every ϕ of the form

$$\phi(x_1,x_2,x_3)=\pi_1(x_1)\pi_2(x_2,x_3).$$

where $\pi_1\in C_0^\infty(\mathbb{R})$ and $\pi_2\in C_0^\infty(\mathbb{R}^2)$.

Setting $\bar{g}_1(x_1)=\int\bar{g}(x_1,x_2,x_3)\pi_2(x_2,x_3)\,\mathrm{d}x_2\,\mathrm{d}x_3$, we have $\langle g,(\hat{P}_1+\mathrm{i})\phi\rangle=\langle g_1,(\hat{P}+\mathrm{i})\pi_1\rangle=0$ for all $\pi_1\in C_0^\infty(\mathbb{R})$, where $\hat{P}=-\mathrm{i}\partial/\partial x_1$ acting on $C_0^\infty(\mathbb{R})$. Since we saw in Section 5.1 that $\hat{P}$ is essentially self-adjoint, it follows that $g_1=0$ (for almost all x_1). Regarding $g(x_1,x_2,x_3)$, for fixed x_1, as belonging to $L^2(\mathbb{R}^2)$ as a function of x_2 and x_3, we have shown that g is orthogonal to $C_0^\infty(\mathbb{R}^2)$. Hence $g=0$, since $C_0^\infty(\mathbb{R}^2)$ is dense in $L^2(\mathbb{R}^2)$. Similarly, only the zero vector is orthogonal to the range of $\hat{P}_i+\mathrm{i}$, for $i=2$, 3, and to the range of $\hat{P}_i-\mathrm{i}$, for $i=1$, 2, 3, so that $\hat{P}_i$ is indeed essentially self-adjoint.

We shall denote by P_i the (unique) self-adjoint operator that is the closure of $\hat{P}_i$.

To show that $\hat{H}_0$ is essentially self-adjoint, we suppose that $\langle h,(\hat{H}_0+\mathrm{i})\phi\rangle=0$ for all $\phi\in C_0^\infty(\mathbb{R}^3)$. In particular, we suppose that $\phi(\boldsymbol{r})=r^{-1}\psi_{lm}(r)Y_{lm}(\theta,\varphi)$ in spherical polar coordinates, where $\psi_{lm}\in C_0^\infty(0,\infty)$ and Y_{lm} is the spherical harmonic corresponding to angular-momentum eigenvalues $l(l+1)$ and m for $\boldsymbol{L}^2$ and L_z respectively ($\boldsymbol{L}$ is orbital angular momentum).

For such ϕ, $\langle h,(\hat{H}_0+\mathrm{i})\phi\rangle=\langle E_{lm}h,(\hat{H}_0+\mathrm{i})\phi\rangle$, where E_{lm} is the projection onto the angular-momentum eigenspace corresponding to quantum numbers l and m. Writing $(E_{lm}h)(\boldsymbol{r})=r^{-1}h_{lm}(r)\;Y_{lm}(\theta,\varphi)$, and using the representation

$$\Delta=\frac{1}{r}\frac{\mathrm{d}^2}{\mathrm{d}r^2}(r\cdot)-\frac{\boldsymbol{L}^2}{r^2}$$

for the Laplacian, we have

$$\left\langle h_{lm},\left(-\frac{\mathrm{d}^2}{\mathrm{d}r^2}+\frac{l(l+1)}{r^2}+\mathrm{i}\right)\psi_{lm}\right\rangle=0\quad\text{for }\psi_{lm}\in C_0^\infty(0,\infty),$$

where the inner product is now in the space $L^2(0,\infty)$. Since (Section 5.1) for $l\neq0$ the operator $-\mathrm{d}^2/\mathrm{d}r^2+l(l+1)/r^2$, acting on $C_0^\infty(0,\infty)$, is essentially self-adjoint, we may deduce that $h_{lm}=0$ for $l\neq0$, so that the vector h must lie entirely in the $l=m=0$ subspace. For $l=0$ we also have, now with $h(\boldsymbol{r})=h_{00}(r)/r$,

$$\left\langle h_{00},\left(-\frac{\mathrm{d}^2}{\mathrm{d}r^2}+\mathrm{i}\right)\psi_{00}\right\rangle=0\quad\text{for }\psi_{00}\in C_0^\infty(0,\infty),$$

so that from Section 5.1 we have $h_{00}(r)=Ae^{-(1-i)r/\sqrt{2}}$, which is the only $L^2(0,\infty)$ solution of $h''+ih=0$.

Again, we have $\langle h_{00},(-d^2/dr^2+i)\psi_{00}\rangle=0$ for any infinitely differentiable function $\psi_{00}(r)$ having compact support in $[0,\infty)$, and such that $\psi_{00}(r)\equiv r$ near $r=0$. In that case, we have, on integrating by parts,

$$\begin{aligned}0&=\langle +h''_{00}+ih_{00},\psi_{00}\rangle+\langle h_{00},-\psi''_{00}+i\psi_{00}\rangle\\&=[\bar{h}'_{00}\psi_{00}-\bar{h}_{00}\psi'_{00}]=\bar{h}_{00}(0)=\bar{A}.\end{aligned}$$

Hence $A=0$, so that $h(\boldsymbol{r})=0$, and we have shown $\hat{H}_0$ to be essentially self-adjoint. We shall denote by H_0 the (unique) self-adjoint operator that is the closure of $\hat{H}_0$. Since no-one, presumably, would wish to deny membership of $D(P_i)$ and $D(H_0)$ to any function belonging to the space $C_0^\infty(\mathbb{R}^3)$, the self-adjoint operators P_i and H_0 are uniquely defined. It is important to observe that each of these operators are *local* operators. That is, if for example $f\in D(P_i)$ then we can operate by P_i on f locally, in the sense that $(P_i f)(\boldsymbol{r})$, in the neighbourhood of any point $\boldsymbol{r}_0$, depends only on the function $f(\boldsymbol{r})$ in the same neighbourhood. A more concrete expression of this property of locality for the operator P_i is the identity

$$P_i\rho f=\rho P_i f-i\frac{\partial\rho}{\partial x_i}f, \tag{5.3.1}$$

which holds for all $\rho\in C_0^\infty(\mathbb{R}^3)$ and $f\in D(P_i)$. We take the liberty here of using the same symbol ρ to denote the operator of multiplication by $\rho(\boldsymbol{r})$ (that is, the bounded operator on $L^2(\mathbb{R}^3)$ defined by $f(\cdot)\to\rho(\cdot)f(\cdot)$) and the function ρ itself. In the same way, $(\partial\rho/\partial x_i)f$ stands for the element of $L^2(\mathbb{R}^3)$ whose value, for almost all $\boldsymbol{r}$, is $(\partial\rho(\boldsymbol{r}/\partial x_i)f(\boldsymbol{r})$.

That (5.3.1) holds for all $f\in C_0^\infty(\mathbb{R}^3)$ and $\rho\in C_0^\infty(\mathbb{R}^3)$ is a trivial exercise in differentiation. More generally, for arbitrary $f\in D(P_i)$, we know that f is in the domain of the closure of $\hat{P}_i$, so that a sequence $\{\phi_n\}$ of functions in the space $C_0^\infty(\mathbb{R}^3)$ exists satisfying

$$\phi_n\to f,\qquad P_i\phi_n\to P_i f\qquad \text{.(strong limits)}.$$

Taking the limit $n\to\infty$ in the identity

$$P_i\rho\phi_n=\rho P_i\phi_n-i\frac{\partial\phi}{\partial x_i}\phi_n, \tag{5.3.1$'$}$$

we see that $P_i\rho\phi_n$ converges strongly to a limit. Since $\text{s-lim}_{n\to\infty}\rho\phi_n=\rho f$, we may deduce, from the property that P_i is closed, that $\rho f\in D(P_i)$ and that $\text{s-lim}_{n\to\infty}P_i\rho\phi_n=P_i\rho f$. Equation (5.3.1) then follows immediately from (5.3.1)$'$ on taking the limit $n\to\infty$. (This is a simple example of *proof by closure*, where a result is extended by using the closure property of an operator in an essential way. It is often sufficient for proof by extension (Section 2.2) that an operator is merely closed rather than bounded.)

A similar result to (5.3.1) may be obtained for the operator H_0, which is also a local operator.

We suppose that $g \in D(H_0)$, so that a sequence $\{\psi_n\}$ of $C_0^\infty(\mathbb{R}^3)$ functions exists satisfying

$$\psi_n \to g, \qquad H_0\psi_n \to H_0 g \qquad \text{(strong limits)}.$$

For $\phi \in C_0^\infty(\mathbb{R}^3)$ we have, on integration by parts, the identity

$$\langle \phi, H_0\phi \rangle = \sum_k \|P_k\phi\|^2. \tag{5.3.2}$$

Hence

$$\begin{aligned}\sum_k \|P_k(\psi_m - \psi_n)\|^2 &= \langle \psi_m - \psi_n, H_0(\psi_m - \psi_n)\rangle \\ &\leq \|\psi_m - \psi_n\| \, \|H_0\psi_m - H_0\psi_n\| \text{ (Schwarz inequality)} \\ &\to 0 \quad \text{as } m, n \to \infty.\end{aligned}$$

Hence $P_k\psi_n$ is a Cauchy sequence and converges strongly to a limit. Since P_k is closed, we have

$$\operatorname*{s-lim}_{n\to\infty} P_k\psi_n = P_k g.$$

Hence we have shown that $D(H_0) \subseteq D(P_i)$. Indeed, taking $\phi = \psi_n$ in (5.3.2) and proceeding to the limit, we have

$$\langle g, H_0 g \rangle = \sum_k \|P_k g\|^2 \tag{5.3.3}$$

for all $g \in D(H_0)$.

As for (5.3.1), we may start with $g \in C_0^\infty(\mathbb{R}^3)$ and extend to arbitrary $g \in D(H_0)$ to obtain the identities

$$\begin{aligned} H_0\rho g &= \rho H_0 g - (\Delta\rho)g - 2\mathrm{i}\sum_k \frac{\partial\rho}{\partial x_k} P_k g \\ &= \rho H_0 g + (\Delta\rho)g - 2\mathrm{i}\sum_k P_k\left(\frac{\partial\rho}{\partial x_k} g\right). \end{aligned} \tag{5.3.4}$$

Equation (5.3.4) holds for all $\rho \in C_0^\infty(\mathbb{R}^3)$ and for all $g \in D(H_0)$. We shall say that a function h is *locally square-integrable* in $\mathbb{R}^3$ if $\int_E |h|^2 \, \mathrm{d}^3\boldsymbol{r} < \infty$ for every closed bounded subset E of $\mathbb{R}^3$. Alternatively, we shall say in that case that h is *locally* L^2, and write $h \in L^2_{\text{loc}}(\mathbb{R}^3)$. The spaces $L^p_{\text{loc}}(\mathbb{R}^3)$ for $p \geq 1$ are defined in a similar way. Note that h is locally L^2 if and only if $\rho h \in L^2(\mathbb{R}^3)$ for every $\rho \in C_0^\infty(\mathbb{R}^3)$. This motivates the definition that a function h in $L^2_{\text{loc}}(\mathbb{R}^3)$ is said to be *locally in the domain of* H_0 if $\rho h \in D(H_0)$ for every

$\rho \in C_0^\infty(\mathbb{R}^3)$. We then write $h \in D_{\text{loc}}(H_0)$. The spaces $D_{\text{loc}}(P_i)$, and in general $D_{\text{loc}}(T)$ for an arbitrary self-adjoint operator T in $L^2(\mathbb{R}^3)$, are defined in a similar way.

Equation (5.3.4) tells us that $h \in D(H_0) \Rightarrow h \in D_{\text{loc}}(H_0)$, and similarly $h \in D(P_i) \Rightarrow h \in D_{\text{loc}}(P_i)$ from (5.3.1). In order to study the local domains of $\boldsymbol{P}$ and H_0 more closely, it will be useful to prove the following lemma.

Lemma 5.1

(i) Let $\phi \in C_0^\infty(\mathbb{R}^3)$. Then

$$\phi(\boldsymbol{r}) = -\frac{1}{4\pi}\int \frac{(\Delta\phi)(\boldsymbol{r}')\,\mathrm{d}^3\boldsymbol{r}'}{|\boldsymbol{r}-\boldsymbol{r}'|}, \tag{5.3.5}$$

(ii) Let the function h satisfy

$$\int \frac{|h(\boldsymbol{r}')|\,\mathrm{d}^3\boldsymbol{r}'}{1+|\boldsymbol{r}'|} < \infty,$$

and suppose that $\phi \in C_0^\infty(\mathbb{R}^3)$. Then

$$\left\langle \Delta\phi, \int \frac{h(\boldsymbol{r}')\,\mathrm{d}^3\boldsymbol{r}'}{|\boldsymbol{r}-\boldsymbol{r}'|} \right\rangle = -4\pi\langle \phi, h\rangle. \tag{5.3.6}$$

Proof

(i) The proof of (5.3.5) is a standard exercise in elementary potential theory. On the right-hand side we integrate over the region $|\boldsymbol{r}-\boldsymbol{r}'|>\varepsilon$, and finally take the limit $\varepsilon \to 0+$ (this is permitted by the Lebesgue Dominated-Convergence Theorem). Writing, for a scalar field $\psi(\boldsymbol{r}')$ and vector field $\boldsymbol{u}(\boldsymbol{r}')$,

$$\operatorname{grad}\psi(\boldsymbol{r}') = \left(\frac{\partial\psi}{\partial x_1'}, \frac{\partial\psi}{\partial x_2'}, \frac{\partial\psi}{\partial x_3'}\right), \qquad \operatorname{div}\boldsymbol{u}(\boldsymbol{r}') = \sum_k \frac{\partial u_k}{\partial x_k'}$$

($\boldsymbol{u} = u_1, u_2, u_3$), we have the identity

$$-\frac{1}{4\pi}\frac{\Delta\phi(\boldsymbol{r}')}{|\boldsymbol{r}-\boldsymbol{r}'|} = \operatorname{div}\left(\frac{\phi}{4\pi}\operatorname{grad}\frac{1}{|\boldsymbol{r}-\boldsymbol{r}'|} - \frac{1}{4\pi|\boldsymbol{r}-\boldsymbol{r}'|}\operatorname{grad}\phi\right).$$

Integrating over the region $|\boldsymbol{r}-\boldsymbol{r}'|>\varepsilon$ and applying the divergence theorem ($\int_{\mathscr{V}} \operatorname{div}\boldsymbol{u}\,\mathrm{d}V = \int_S \boldsymbol{u}\cdot\boldsymbol{n}\,\mathrm{d}S$, where $\boldsymbol{n}$ is the outward normal to the boundary S of $\mathscr{V}$), the right-hand side of (5.3.5) reduces to

$$\int_{S_\varepsilon}\left(\frac{\phi(\boldsymbol{r}')}{4\pi}\frac{(\boldsymbol{r}'-\boldsymbol{r})\cdot\boldsymbol{n}}{|\boldsymbol{r}-\boldsymbol{r}'|^3} + \frac{1}{4\pi|\boldsymbol{r}-\boldsymbol{r}'|}(\operatorname{grad}\phi)\cdot\boldsymbol{n}\right)\mathrm{d}S,$$

where now $\boldsymbol{n}=(\boldsymbol{r}'-\boldsymbol{r})/|\boldsymbol{r}'-\boldsymbol{r}|$ and S_ε is the sphere $|\boldsymbol{r}-\boldsymbol{r}'|=\varepsilon$. There is no other boundary contribution to the integral, since ϕ has compact support. Now $\phi(\boldsymbol{r}')-\phi(\boldsymbol{r})$, for $\boldsymbol{r}'$ on S_ε, vanishes in the limit $\varepsilon\to 0$. Using spherical polar coordinates centred on $\boldsymbol{r}$, we obtain $\phi(\boldsymbol{r})$ in the limit, and (5.3.5) is proved. A similar method of proof, removing the volume $|\boldsymbol{r}-\boldsymbol{r}'|<\varepsilon$ from the integration region, yields the identity

$$\int \frac{\phi(\boldsymbol{r}')(x_i-x_i')}{|\boldsymbol{r}-\boldsymbol{r}'|^3}\,\mathrm{d}^3\boldsymbol{r}' = -\int \frac{\partial\phi(\boldsymbol{r}')}{\partial x_i'}\frac{\mathrm{d}^3\boldsymbol{r}'}{|\boldsymbol{r}-\boldsymbol{r}'|}, \tag{5.3.7}$$

valid for all $\phi\in C_0^\infty(\mathbb{R}^3)$.

(ii) We use the notation $\langle g_1, g_2\rangle=\int \bar{g}_1 g_2\,\mathrm{d}^3\boldsymbol{r}$ irrespective of whether g_1 and g_2 belong to $L^2(\mathbb{R}^3)$. The left-hand side of (5.3.6) is $\iint \mathrm{d}^3\boldsymbol{r}\,\mathrm{d}^3\boldsymbol{r}'\,\Delta\bar{\phi}(\boldsymbol{r})h(\boldsymbol{r}')/|\boldsymbol{r}-\boldsymbol{r}'|$. Since the double integral is absolutely convergent ($\int \mathrm{d}^3\boldsymbol{r}|\Delta\bar{\phi}(\boldsymbol{r})/|\boldsymbol{r}-\boldsymbol{r}'|<\text{const}/(1+|\boldsymbol{r}'|)$) the order of integration does not matter, Integrating first over $\boldsymbol{r}$ and using (5.3.5) with $\boldsymbol{r}$ and $\boldsymbol{r}'$ interchanged, (5.3.6) follows. A similar identity for $\partial\phi/\partial x_1$ follows from (5.3.7), namely

$$\left\langle \frac{\partial\phi(\boldsymbol{r})}{\partial x_i}, \int\frac{h(\boldsymbol{r}')\,\mathrm{d}^3\boldsymbol{r}'}{|\boldsymbol{r}-\boldsymbol{r}'|}\right\rangle = \left\langle \phi, \int\frac{h(\boldsymbol{r}')(x_i-x_i')}{|\boldsymbol{r}-\boldsymbol{r}'|^3}\,\mathrm{d}^3\boldsymbol{r}'\right\rangle. \tag{5.3.7'}$$

■

In the language of distribution theory (Section 2.2), (5.3.6) may be interpreted as asserting

$$\Delta\int\frac{h(\boldsymbol{r}')\mathrm{d}^3\boldsymbol{r}'}{|\boldsymbol{r}-\boldsymbol{r}'|} = -4\pi h(\boldsymbol{r})$$

in the sense of distributions. (Actually the result may be generalized still further, in the direction of replacing h by a measure over $\mathbb{R}^3$.) In electrostatics h would be interpreted as the charge density and $\int \mathrm{d}^3\boldsymbol{r}'h(\boldsymbol{r}')/|\boldsymbol{r}-\boldsymbol{r}'|$ as the electrostatic potential. Equation (5.3.6) then says that the electrostatic potential satisfies Poisson's equation (in the sense of distributions). In the present context, we may use (5.3.6) to construct functions belonging to $D_{\mathrm{loc}}(H_0)$.

Corollary

Let h be any locally square-integrable function satisfying $\int \mathrm{d}^3\boldsymbol{r}'|h(\boldsymbol{r}')|/(1+|\boldsymbol{r}'|)<\infty$. Then

$$\int\frac{h(\boldsymbol{r}')\,\mathrm{d}^3\boldsymbol{r}'}{|\boldsymbol{r}-\boldsymbol{r}'|}\in D_{\mathrm{loc}}(H_0).$$

Moreover, if in addition $\int \mathrm{d}^3\boldsymbol{r}'\,h(\boldsymbol{r}')/|\boldsymbol{r}-\boldsymbol{r}'|\in L^2(\mathbb{R}^3)$ and $h\in L^2(\mathbb{R}^3)$ then

$$\int \frac{h(\boldsymbol{r}')\,\mathrm{d}^3\boldsymbol{r}'}{|\boldsymbol{r}-\boldsymbol{r}'|} \in D(H_0)$$

and

$$H_0 \int \frac{h(\boldsymbol{r}')\,\mathrm{d}^3\boldsymbol{r}'}{|\boldsymbol{r}-\boldsymbol{r}'|} = 4\pi h(\boldsymbol{r}). \tag{5.3.8}$$

Proof

Given $R>0$, $\int \mathrm{d}^3\boldsymbol{r}'h(\boldsymbol{r}')/|\boldsymbol{r}-\boldsymbol{r}'|$ is bounded for $|\boldsymbol{r}|<R$. For, with $|\boldsymbol{r}|<R$, we have $|\boldsymbol{r}-\boldsymbol{r}'|>\text{const}\,(1+|\boldsymbol{r}'|)$ for $|\boldsymbol{r}'|>2R$, so that

$$\begin{aligned}
\left|\int \frac{h(\boldsymbol{r}')\,\mathrm{d}^3\boldsymbol{r}'}{|\boldsymbol{r}-\boldsymbol{r}'|}\right| &\le \int\limits_{|\boldsymbol{r}'|>2R} \frac{|h(\boldsymbol{r}')|\,\mathrm{d}^3\boldsymbol{r}'}{|\boldsymbol{r}-\boldsymbol{r}'|} + \int\limits_{|\boldsymbol{r}'|<2R} \frac{|h(\boldsymbol{r}')|\,\mathrm{d}^3\boldsymbol{r}'}{|\boldsymbol{r}-\boldsymbol{r}'|} \\
&\le \text{const} \int\limits_{|\boldsymbol{r}'|>2R} \frac{|h(\boldsymbol{r}')|\,\mathrm{d}^3\boldsymbol{r}'}{1+|\boldsymbol{r}'|} \\
&\quad + \left(\int\limits_{|\boldsymbol{r}'|<2R} |h^2(\boldsymbol{r}')|\,\mathrm{d}^3\boldsymbol{r}'\right)^{1/2} \left(\int\limits_{|\boldsymbol{r}'|<2R} \frac{1}{|\boldsymbol{r}-\boldsymbol{r}'|^2}\,\mathrm{d}^3\boldsymbol{r}'\right)^{1/2} \\
&< \text{const},
\end{aligned}$$

where we have used the Schwarz inequality for the second contribution to the integral.

Hence certainly $\int \mathrm{d}^3\boldsymbol{r}'\,h(\boldsymbol{r}')/|\boldsymbol{r}-\boldsymbol{r}'|$ is locally square-integrable. It is also true, although not quite so obvious, that $\int \mathrm{d}^3\boldsymbol{r}'\,h(\boldsymbol{r}')/|\boldsymbol{r}-\boldsymbol{r}'|^2$ is locally square-integrable. To show this, as in the previous argument, it is necessary only to show that $\int_{|\boldsymbol{r}'|<2R} \mathrm{d}^3\boldsymbol{r}'\,h(\boldsymbol{r}')/|\boldsymbol{r}-\boldsymbol{r}'|^2$ is square-integrable over the region $|\boldsymbol{r}|<R$. If f is any function that is square-integrable over the region $|\boldsymbol{r}|<R$ then we have, integrating only over this region,

$$\begin{aligned}
&\left|\left\langle f, \int\limits_{|\boldsymbol{r}'|<2R} \frac{h(\boldsymbol{r}')\,\mathrm{d}^3\boldsymbol{r}'}{|\boldsymbol{r}-\boldsymbol{r}'|^2}\right\rangle\right| \\
&\le \iint\limits_{\substack{|\boldsymbol{r}|<R \\ |\boldsymbol{r}'|<2R}} \mathrm{d}^3\boldsymbol{r}\,\mathrm{d}^3\boldsymbol{r}'\,\frac{|f(\boldsymbol{r})|\,|h(\boldsymbol{r}')|}{|\boldsymbol{r}-\boldsymbol{r}'|^2} \\
&\le \left\{\left(\iint\limits_{\substack{|\boldsymbol{r}|<R \\ |\boldsymbol{r}'|<2R}} \frac{|h^2(\boldsymbol{r}')|}{|\boldsymbol{r}-\boldsymbol{r}'|^2}\,\mathrm{d}^3\boldsymbol{r}\,\mathrm{d}^3\boldsymbol{r}'\right)\left(\iint\limits_{\substack{|\boldsymbol{r}|<R \\ |\boldsymbol{r}'|<2R}} \frac{|f^2(\boldsymbol{r})|}{|\boldsymbol{r}-\boldsymbol{r}'|^2}\,\mathrm{d}^3\boldsymbol{r}\,\mathrm{d}^3\boldsymbol{r}'\right)\right\}^{1/2}
\end{aligned}$$

(by the Schwarz inequality). For the first factor we integrate initially with respect to $\boldsymbol{r}$, and for the second with respect to $\boldsymbol{r}'$, where $\int_{|\boldsymbol{r}|<R} \mathrm{d}^3\boldsymbol{r}/|\boldsymbol{r}-\boldsymbol{r}'|^2$ and $\int_{|\boldsymbol{r}'|<2R} \mathrm{d}^3\boldsymbol{r}'/|\boldsymbol{r}-\boldsymbol{r}'|^2$ are each bounded. We then have

$$\left|\left\langle f, \int\limits_{|\boldsymbol{r}'|<2R} \frac{|h(\boldsymbol{r}')|\,\mathrm{d}^3\boldsymbol{r}'}{|\boldsymbol{r}-\boldsymbol{r}'|^2}\right\rangle\right| \leq \text{const}\,\|f\|,$$

the constant being proportional to the L^2 norm of h in the region $|\boldsymbol{r}'|<2R$.

Since $I(\boldsymbol{r}) \equiv \int_{|\boldsymbol{r}'|<2R} \mathrm{d}^3\boldsymbol{r}'|h(\boldsymbol{r}')|/|\boldsymbol{r}-\boldsymbol{r}'|^2$ is locally L^1, we know that $I(\boldsymbol{r})<\infty$ for almost all $\boldsymbol{r}$. Setting

$$f = f_M(\boldsymbol{r}) = \begin{cases} I(\boldsymbol{r}) & \text{if } I(\boldsymbol{r}) \leq M, \\ 0 & \text{otherwise,} \end{cases}$$

we have

$$|\langle f_M, I\rangle| = \int\limits_{I(\boldsymbol{r})\leq M} I^2(\boldsymbol{r})\,\mathrm{d}^3\boldsymbol{r} \leq \text{const}\,\|f_M\|.$$

Therefore $\|f_M\| \leq \text{const}$ and $\|I\| = \lim_{M\to\infty}\|f_M\| < \infty$. Hence we have shown that $\int \mathrm{d}^3\boldsymbol{r}'\,h(\boldsymbol{r}')/|\boldsymbol{r}-\boldsymbol{r}'|^2 \in L^2_{\text{loc}}(\mathbb{R}^3)$. Since $|x_i - x_i'| \leq |\boldsymbol{r}-\boldsymbol{r}'|$, we also have

$$\int \frac{h(\boldsymbol{r}')(x_i - x_i')}{|\boldsymbol{r}-\boldsymbol{r}'|^3}\,\mathrm{d}^3\boldsymbol{r}' \in L^2_{\text{loc}}(\mathbb{R}^3). \tag{5.3.9}$$

Now, for $\phi \in C_0^\infty(\mathbb{R}^3)$, $\rho \in C_0^\infty(\mathbb{R}^3)$, we have

$$\begin{aligned}\left\langle -\mathrm{i}\frac{\partial\phi}{\partial x_i}, \rho\int\frac{h(\boldsymbol{r}')\,\mathrm{d}^3\boldsymbol{r}'}{|\boldsymbol{r}-\boldsymbol{r}'|}\right\rangle &= \left\langle -\mathrm{i}\bar{\rho}\frac{\partial\phi}{\partial x_i}, \int\frac{h(\boldsymbol{r}')\,\mathrm{d}^3\boldsymbol{r}'}{|\boldsymbol{r}-\boldsymbol{r}'|}\right\rangle \\ &= \left\langle -\mathrm{i}\frac{\partial}{\partial x_i}(\bar{\rho}\phi), \int\frac{h(\boldsymbol{r}')\,\mathrm{d}^3\boldsymbol{r}'}{|\boldsymbol{r}-\boldsymbol{r}'|}\right\rangle \\ &\quad + \left\langle \mathrm{i}\phi, \frac{\partial\rho}{\partial x_i}\int\frac{h(\boldsymbol{r}')\,\mathrm{d}^3\boldsymbol{r}'}{|\boldsymbol{r}-\boldsymbol{r}'|}\right\rangle.\end{aligned}$$

From (5.3.9) and (5.3.7)′ with $\bar{\rho}\phi$ for ϕ, we see that

$$\left\langle -\mathrm{i}\frac{\partial}{\partial x_i}(\bar{\rho}\phi), \int\frac{h(\boldsymbol{r}')\,\mathrm{d}^3\boldsymbol{r}'}{|\boldsymbol{r}-\boldsymbol{r}'|}\right\rangle = \langle\phi, h_1^*\rangle,$$

where

$$h_1^* = \mathrm{i}\rho\int\frac{h(\boldsymbol{r}')(x_i - x_i')\,\mathrm{d}^3\boldsymbol{r}'}{|\boldsymbol{r}-\boldsymbol{r}'|^3} \in L^2(\mathbb{R}^3).$$

Hence

$$\left\langle -\mathrm{i}\frac{\partial\phi}{\partial x_i}, \rho\int\frac{h(\boldsymbol{r}')\,\mathrm{d}^3\boldsymbol{r}'}{|\boldsymbol{r}-\boldsymbol{r}'|}\right\rangle = \langle\phi, h^*\rangle, \tag{5.3.10}$$

where

$$h^* = h_1^* - \mathrm{i}\frac{\partial\rho}{\partial x_i}\int\frac{h(\boldsymbol{r}')\,\mathrm{d}^3\boldsymbol{r}'}{|\boldsymbol{r}-\boldsymbol{r}'|}\in L^2(\mathbb{R}^3).$$

Since $\hat{P}_i = -\mathrm{i}\partial/\partial x_i$ on $C_0^\infty(\mathbb{R}^3)$, and P_i is the closure of $\hat{P}_i$, it follows that

$$\rho\int\frac{h(\boldsymbol{r}')\,\mathrm{d}^3\boldsymbol{r}'}{|\boldsymbol{r}-\boldsymbol{r}'|}\in D(P_i)$$

and

$$P_i\rho\int\frac{h(\boldsymbol{r}')\,\mathrm{d}^3\boldsymbol{r}'}{|\boldsymbol{r}-\boldsymbol{r}'|} = h^*.$$

Writing

$$\bar{\rho}\Delta\phi = \Delta(\bar{\rho}\phi) - 2\sum_k\frac{\partial\bar{\rho}}{\partial x_k}\frac{\partial\phi}{\partial x_k} - \phi\Delta\bar{\rho},$$

we also have, for ϕ, $\rho\in C_0^\infty(\mathbb{R}^3)$,

$$\begin{aligned}\left\langle\Delta\phi, \rho\int\frac{h(\boldsymbol{r}')\,\mathrm{d}^3\boldsymbol{r}'}{|\boldsymbol{r}-\boldsymbol{r}'|}\right\rangle &= \left\langle\Delta(\bar{\rho}\phi), \int\frac{h(\boldsymbol{r}')\,\mathrm{d}^3\boldsymbol{r}'}{|\boldsymbol{r}-\boldsymbol{r}'|}\right\rangle \\ &\quad -2\sum_k\left\langle\frac{\partial\phi}{\partial x_k}, \frac{\partial\rho}{\partial x_k}\int\frac{h(\boldsymbol{r}')\,\mathrm{d}^3\boldsymbol{r}'}{|\boldsymbol{r}-\boldsymbol{r}'|}\right\rangle \\ &\quad -\left\langle\phi, (\Delta\rho)\int\frac{h(\boldsymbol{r}')\,\mathrm{d}^3\boldsymbol{r}'}{|\boldsymbol{r}-\boldsymbol{r}'|}\right\rangle.\end{aligned}$$

Using (5.3.6), with $\bar{\rho}\phi$ for ϕ, together with (5.3.10) with $\partial\rho/\partial x_k$ for ρ, we have

$$\left\langle -\Delta\phi, \rho\int\frac{h(\boldsymbol{r}')\,\mathrm{d}^3\boldsymbol{r}'}{|\boldsymbol{r}-\boldsymbol{r}'|}\right\rangle = \langle\phi, k^*\rangle \quad \text{for some } k^*\in L^2(\mathbb{R}^3).$$

Since $\hat{H}_0 = -\Delta$ on $C_0^\infty(\mathbb{R}^3)$, and H_0 =closure of $\hat{H}_0$, $=\hat{H}_0^*$, it follows that $\rho\int\mathrm{d}^3\boldsymbol{r}'\,h(\boldsymbol{r}')/|\boldsymbol{r}-\boldsymbol{r}'|\in D(\hat{H}_0^*)$, so that we have verified the first part of the corollary. Equation (5.3.8) is seen as a straightforward consequence of (5.3.6) under the additional hypotheses. ∎

Equation (5.3.8) can be applied to functions of compact support that are in the domain of H_0. We let g be such a function, and set $h=H_0 g$. We let $\rho \in C_0^\infty(\mathbb{R}^3)$ be such that $\operatorname{supp}\rho$ and $\operatorname{supp} g$ do not intersect. Then $\rho g=(\Delta\rho)g=(\partial\rho/\partial x_k)g=0$, so that (5.3.4) implies that $\rho H_0 g=0$. It then follows that $H_0 g$ has compact support, and that $\operatorname{supp} H_0 g \subseteq \operatorname{supp} g$. We now suppose that $\boldsymbol{r}_0 \notin \operatorname{supp} g$, and let $\psi \in C_0^\infty(\mathbb{R}^3)$ be real such that $\psi(\boldsymbol{r})\equiv 1$ on $\operatorname{supp} g$ and $\psi\equiv 0$ in a neighbourhood of $\boldsymbol{r}_0$. Then $\psi(\boldsymbol{r})/\boldsymbol{r}-\boldsymbol{r}_0| \in C_0^\infty(\mathbb{R}^3)$, and we have

$$\begin{aligned}\int \frac{h(\boldsymbol{r}')\,\mathrm{d}^3\boldsymbol{r}'}{|\boldsymbol{r}_0-\boldsymbol{r}'|} &= \int \frac{\psi(\boldsymbol{r}')h(\boldsymbol{r}')\,\mathrm{d}^3\boldsymbol{r}'}{|\boldsymbol{r}_0-\boldsymbol{r}'|}\\ &= \left\langle \frac{\psi}{|\boldsymbol{r}-\boldsymbol{r}_0|}, H_0 g\right\rangle = \left\langle H_0\left(\frac{\psi}{|\boldsymbol{r}-\boldsymbol{r}_0|}\right), g\right\rangle\\ &=0\end{aligned}$$

since

$$\Delta\left(\frac{\psi}{|\boldsymbol{r}-\boldsymbol{r}_0|}\right)=\Delta\left(\frac{1}{|\boldsymbol{r}-\boldsymbol{r}_0|}\right)=0 \quad \text{on supp } g.$$

It then follows that $\int \mathrm{d}^3\boldsymbol{r}'\,h(\boldsymbol{r}')/|\boldsymbol{r}-\boldsymbol{r}'|$ is a locally square-integrable function having compact support, and hence belongs to $L^2(\mathbb{R}^3)$. Therefore (5.3.8) holds in this case, and for g of compact support we have

$$\left.\begin{aligned} &H_0 g=h,\\ &H_0\int \frac{h(\boldsymbol{r}')\,\mathrm{d}^3\boldsymbol{r}'}{|\boldsymbol{r}-\boldsymbol{r}'|}=4\pi h(\boldsymbol{r}).\end{aligned}\right\} \tag{5.3.11}$$

We should like to conclude from (5.3.11) that, in this case,

$$g(\boldsymbol{r})=\frac{1}{4\pi}\int \frac{h(\boldsymbol{r}')\,\mathrm{d}^3\boldsymbol{r}'}{|\boldsymbol{r}-\boldsymbol{r}'|},$$

but in order to justify this step we have first to show that the solution for g of $H_0 g=h$ is unique. Equivalently, we must show that $H_0 f=0$ implies $f=0$. Now $H_0 f=0$ means that $\langle H_0\phi, f\rangle=0$ for all $\phi \subseteq C_0^\infty(\mathbb{R}^3)$, which is the same as saying that f satisfies Laplace's equation in the sense of distributions. The following result implies that any distributional solution of Laplace's equation is infinitely differentiable, and applies to local solutions as well as to global solutions.

Lemma 5.2

Let $\mathscr{B}$ denote the region $|\boldsymbol{r}-\boldsymbol{r}_0|<R$. For some $f\in L^2(\mathscr{B})$, suppose that

$$\langle \Delta\phi, f \rangle = 0 \tag{5.3.12}$$

for all $\phi \in C_0^\infty(\mathscr{B})$. Then f is (almost everywhere equal to) an infinitely differentiable function in $\mathscr{B}$, and satisfies $\Delta f = 0$.

Proof

Since Δ is translation-invariant, it will be sufficient to treat the case $\boldsymbol{r}_0 = \mathbf{0}$, for which $\mathscr{B}$ is the region $|\boldsymbol{r}| < R$. The main idea of the proof is to take for $\phi(\boldsymbol{r})$ in (5.3.12) a C_0^∞ function that is as close as we can get to $|\boldsymbol{r} - \boldsymbol{r}'|^{-1}$ for some fixed $\boldsymbol{r}' \in \mathscr{B}$. However, such a function does not belong to $C_0^\infty(\mathscr{B})$, and we have to make two smooth cutoffs, one near the singularity and the other near the surface of the sphere $|\boldsymbol{r}| = R$.

We let $\mathscr{B}_0$ denote the region $|\boldsymbol{r}| \le R_0$ for some R_0 such that $0 < R_0 < R$, and for $\boldsymbol{r} \in \mathscr{B}_0$, $\boldsymbol{r}' \in \mathscr{B}$, we define $\phi(\boldsymbol{r}, \boldsymbol{r}')$ by

$$\phi(\boldsymbol{r}, \boldsymbol{r}'; r_0) = \frac{1}{|\boldsymbol{r} - \boldsymbol{r}'|} \Phi_1\left(\frac{|\boldsymbol{r} - \boldsymbol{r}'|}{r_0}\right) \Phi_2(|\boldsymbol{r} - \boldsymbol{r}'|), \tag{5.3.13}$$

where $r_0 > 0$, $\Phi_1(r)$ is a smooth non-decreasing function satisfying

$$\Phi_1(r) = \begin{cases} 0 & (0 \le r \le 1), \\ 1 & (r \ge 2), \end{cases} \tag{5.3.14}$$

and $\Phi_2(r)$ is a smooth non-increasing function satisfying, for some δ in the range $0 < \delta < R - R_0$,

$$\Phi_2(r) = \begin{cases} 1 & (0 \le r \le \delta), \\ 0 & (r \ge R - R_0). \end{cases} \tag{5.3.15}$$

For fixed $\boldsymbol{r}' \in \mathscr{B}_0$, $\phi(\boldsymbol{r}, \boldsymbol{r}'; r_0)$ is an infinitely differentiable function of $\boldsymbol{r}$ having compact support in $\mathscr{B}$, and we shall denote simply by $\Delta\phi(\boldsymbol{r}, \boldsymbol{r}'; r_0)$ the Laplacian Δ applied to this function. Equivalently, for fixed $\boldsymbol{r} \in \mathscr{B}_0$, $\Delta\phi(\boldsymbol{r}, \boldsymbol{r}'; r_0)$ may be obtained by applying the Laplacian with respect to $\boldsymbol{r}'$ to give a function in $C_0^\infty(\mathscr{B})$.

We now define an integral operator T_{r_0} from $L^2(\mathscr{B})$ to $L^2(\mathscr{B}_0)$ by

$$(T_{r_0} g)(\boldsymbol{r}) = \int_{\mathscr{B}} \Delta\phi(\boldsymbol{r}, \boldsymbol{r}'; r_0) g(\boldsymbol{r}') \, \mathrm{d}^3\boldsymbol{r}'. \tag{5.3.16}$$

From (5.3.13), using the fact that $\Delta(|\boldsymbol{r} - \boldsymbol{r}'|^{-1}) = 0$, we see that, on the support of $\Delta\phi(\boldsymbol{r}, \boldsymbol{r}'; r_0)$, either $r_0 \le |\boldsymbol{r} - \boldsymbol{r}'| \le 2r_0$ or $|\boldsymbol{r} - \boldsymbol{r}'| \ge \delta$. We shall fix δ, and consider what happens to T_{r_0} in the limit as $r_0 \to 0$. On the support of

$\Delta\phi(\boldsymbol{r},\boldsymbol{r}';r_0)$ we have, in this limit,

$$\frac{1}{|\boldsymbol{r}-\boldsymbol{r}'|}=O\left(\frac{1}{r_0}\right),\qquad \Delta\Phi_1\left(\frac{|\boldsymbol{r}-\boldsymbol{r}'|}{r_0}\right)=O\left(\frac{1}{r_0^2}\right),\qquad \text{etc.,}$$

so that, uniformly for $\boldsymbol{r}\in\mathscr{B}_0$, $\boldsymbol{r}'\in\mathscr{B}$,

$$\Delta\phi(\boldsymbol{r},\boldsymbol{r}';r_0)\le\frac{\text{const}}{r_0^3}. \tag{5.3.17}$$

By the Schwarz inequality applied to (5.3.16),

$$|(T_{r_0}g)(\boldsymbol{r})|^2\le\left[\int_{\mathscr{B}}|\Delta\phi(\boldsymbol{r},\boldsymbol{r}';r_0)|^{1/2}|\Delta\phi(\boldsymbol{r},\boldsymbol{r}';r_0)|^{1/2}|g(\boldsymbol{r}')|\,\mathrm{d}^3\boldsymbol{r}'\right]^2$$
$$\le\int_{\mathscr{B}}|\Delta\phi(\boldsymbol{r},\boldsymbol{r}';r_0)|\,\mathrm{d}^3\boldsymbol{r}'\int_{\mathscr{B}}|\Delta\phi(\boldsymbol{r},\boldsymbol{r}';r_0)||g(\boldsymbol{r}')|^2\,\mathrm{d}^3\boldsymbol{r}'.$$

By (5.3.17), for $\boldsymbol{r}\in\mathscr{B}_0$ the contribution to $\int_{\mathscr{B}}|\Delta\phi(\boldsymbol{r},\boldsymbol{r}';r_0)|\,\mathrm{d}^3\boldsymbol{r}'$ from integrating over the region $r_0\le|\boldsymbol{r}-\boldsymbol{r}'|\le 2r_0$ is bounded uniformly in $\boldsymbol{r}$ and r_0 (this region having volume of order r_0^3). And the contribution from integrating over the region $|\boldsymbol{r}-\boldsymbol{r}'|\ge\delta$ is also bounded uniformly since $|\Delta\phi(\boldsymbol{r},\boldsymbol{r}';r_0)|\le$const. for $|\boldsymbol{r}-\boldsymbol{r}'|\ge\delta$. It follows that

$$\int_{\mathscr{B}}|\Delta\phi(\boldsymbol{r},\boldsymbol{r}';r_0)|\,\mathrm{d}^3\boldsymbol{r}'\le\text{const.}$$

Similarly

$$\int_{\mathscr{B}_0}|\Delta\phi(\boldsymbol{r},\boldsymbol{r}';r_0)|\,\mathrm{d}^3\boldsymbol{r}\le\text{const},$$

and hence

$$\int_{\mathscr{B}_0}|(T_{r_0}g)(\boldsymbol{r})|^2\,\mathrm{d}^3\boldsymbol{r}\le\text{const}\int_{\mathscr{B}}|g(\boldsymbol{r}')|^2\,\mathrm{d}^3\boldsymbol{r}'.$$

We have verified that T_{r_0} is a bounded linear operator from $L^2(\mathscr{B})$ to $L^2(\mathscr{B}_0)$, and moreover that $\|T_{r_0}\|$ remains bounded in the limit as r_0 approaches zero. We shall obtain a formula for $\text{s-lim}_{r_0\to 0}T_{r_0}u$ in the case $u\in C_0^\infty(\mathscr{B})$ and then use the fact that $\|T_{r_0}\|$ is bounded to extend this formula to the whole of $L^2(\mathscr{B})$, thus obtaining $\text{s-lim}_{r_0\to 0}T_{r_0}g$ for arbitrary

$g \in L^2(\mathscr{B})$. This proof by extension (Section 2.2) works because $C_0^\infty(\mathscr{B})$ is dense in $L^2(\mathscr{B})$.

For $u \in C_0^\infty(\mathscr{B})$, we have, from (5.3.16),

$$\begin{aligned}(T_{r_0}u)(\boldsymbol{r}) &= \langle \Delta\phi(\boldsymbol{r},\cdot\,;r_0), u\rangle \\ &= \langle \phi(\boldsymbol{r},\cdot\,;r_0), \Delta u\rangle. \end{aligned} \tag{5.3.18}$$

Using the explicit form (5.3.13) for ϕ, the inner product in (5.3.18), in the limit $r_0 \to 0$, converges uniformly (over $\boldsymbol{r} \in \mathscr{B}_0$) and hence also in $L^2(\mathscr{B}_0)$ norm to

$$\int_{\mathscr{B}} \frac{\Phi_2(|\boldsymbol{r}-\boldsymbol{r}'|)}{|\boldsymbol{r}-\boldsymbol{r}'|} \Delta u(\boldsymbol{r}')\,\mathrm{d}^3\boldsymbol{r}'.$$

Applying the same technique as for the proof of Lemma 5.1(i), i.e. removing the region of integration $|\boldsymbol{r}-\boldsymbol{r}'| < \varepsilon$ and then taking the limit $\varepsilon \to 0$, we deduce that, for $u \in C_0^\infty(\mathscr{B})$,

$$\operatorname*{s-lim}_{r_0 \to 0} T_{r_0}u = \int_{\mathscr{B}} K(\boldsymbol{r}-\boldsymbol{r}')u(\boldsymbol{r}')\,\mathrm{d}^3\boldsymbol{r}' - 4\pi u(\boldsymbol{r}),$$

where

$$K(\boldsymbol{r}) = \Delta\left(\frac{\Phi_2(|\boldsymbol{r}|)}{|\boldsymbol{r}|}\right) \qquad (K \equiv 0 \quad \text{near } \boldsymbol{r} = 0).$$

By taking strong limits of such u, we have, for all $g \in L^2(\mathscr{B})$,

$$\operatorname*{s-lim}_{r_0 \to 0} T_{r_0}g = \int_{\mathscr{B}} K(\boldsymbol{r}-\boldsymbol{r}')g(\boldsymbol{r}')\,\mathrm{d}^3\boldsymbol{r}' - 4\pi g(\boldsymbol{r}). \tag{5.3.19}$$

However, (5.3.12) with $\phi = \phi(\boldsymbol{r},\cdot\,;r_0)$ implies that $T_{r_0}f = 0$ for all $r_0 > 0$, in which case (5.3.19) with $g = f$ gives

$$f(\boldsymbol{r}) = \frac{1}{4\pi}\int_{\mathscr{B}} K(\boldsymbol{r}-\boldsymbol{r}')f(\boldsymbol{r}')\,\mathrm{d}^3\boldsymbol{r}'. \tag{5.3.20}$$

It follows by differentiation under the integral sign that f is infinitely differentiable at each $\boldsymbol{r} \in \mathscr{B}_0$, since the kernel K is infinitely differentiable. Moreover, each $\boldsymbol{r} \in \mathscr{B}$ is in $\mathscr{B}_0$ for R_0 sufficiently close to R, so that $f(\boldsymbol{r})$ is infinitely differentiable at each $\boldsymbol{r} \in \mathscr{B}$. Hence we can write, for arbitrary $\psi \in C_0^\infty(\mathscr{B})$, $\langle \Delta\psi, f\rangle = \langle \psi, \Delta f\rangle = 0$, and because such ψ are dense in $L^2(\mathscr{B})$ we can deduce that $\Delta f = 0$. ■

Corollary

$$H_0 f = 0 \Rightarrow f = 0.$$

Proof

$H_0 f = 0 \Rightarrow \langle \Delta\phi, f \rangle = 0$ for all $\phi \in C_0^\infty(\mathbb{R}^3)$, and the lemma shows that such f must be infinitely differentiable and satisfy Laplace's equation. We let the projection of f onto the (l, m) partial wave be $r^{-1} f_{lm}(r) Y_{lm}(\theta, \varphi)$. Then $f_{lm}(\cdot) \in L^2(0, \infty)$ and satisfies

$$-\frac{d^2 f_{lm}}{dr^2} + \frac{l(l+1)}{r^2} f_{lm} = 0,$$

whose general solution is $f_{lm} = Ar^{l+1} + B/r^l$. The only L^2 solution has $A = B = 0$. Hence $f_{lm} = 0$ for every pair (l, m), and it follows that $f = 0$. ■

Equipped with this result, we may conclude from (5.3.11) for g of compact support that

$$H_0 g = h \Rightarrow g(\boldsymbol{r}) = \frac{1}{4\pi} \int \frac{h(\boldsymbol{r}')\mathrm{d}^3\boldsymbol{r}'}{|\boldsymbol{r} - \boldsymbol{r}'|}.$$

This result can be generalized as follows. We define a sequence $\{\rho_n(r)\}$ of smooth non-increasing functions such that

$$\rho_n(r) = \begin{cases} 1 & (0 \le r \le n), \\ 0 & (r \ge (n+1)), \end{cases}$$

and allow ρ_n also to denote the operator of multiplication by $\rho_n(|\boldsymbol{r}|)$ in $L^2(\mathbb{R}^3)$. For any $f \in L^2(\mathbb{R}^3)$, a straightforward application of the Schwarz inequality shows that

$$\lim_{n \to \infty} \int_{n \le |\boldsymbol{r}'| \le n+1} \frac{|f(\boldsymbol{r}')| \, \mathrm{d}^3\boldsymbol{r}'}{|\boldsymbol{r} - \boldsymbol{r}'|} = 0.$$

We are thinking here of the pointwise limit. However, convergence is uniform over bounded subsets of $\mathbb{R}^3$, so that we also have L^2 convergence over such subsets.

For arbitrary $g \in D(H_0)$, we set $h = H_0 g$, $g_n = \rho_n g$ and $h_n = \rho_n h$. Then g_n and h_n are smoothly cutoff versions of g and h respectively, and the cutoff is removed in the limit as $n \to \infty$. Since $\Delta\rho_n(|\boldsymbol{r}|)$ and $\partial\rho_n(|\boldsymbol{r}|)/\partial x_k$ are bounded uniformly in both n and $\boldsymbol{r}$, we can use the identity (5.3.4), with ρ_n for ρ, to deduce that

$$\lim_{n \to \infty} \int_{|\boldsymbol{r}'| \le n+1} \frac{(H_0 g_n - h_n)(\boldsymbol{r}') \, \mathrm{d}^3\boldsymbol{r}'}{|\boldsymbol{r} - \boldsymbol{r}'|} = 0, \tag{5.3.21}$$

the only contribution to the integral coming from the region $n \le |\mathbf{r}'| \le n+1$. Since the support of g_n is contained in $|\mathbf{r}'| \le n+1$, we also have

$$g_n = \frac{1}{4\pi}\int \frac{(H_0 g_n)(\mathbf{r}')\,\mathrm{d}^3\mathbf{r}'}{|\mathbf{r}-\mathbf{r}'|} = \frac{1}{4\pi}\int_{|\mathbf{r}'|\le n+1} \frac{H_0 g_n(\mathbf{r}')\,\mathrm{d}^3\mathbf{r}'}{|\mathbf{r}-\mathbf{r}'|}$$

Hence

$$g = \operatorname*{s-lim}_{n\to\infty} g_n = \lim_{n\to\infty} \int_{|\mathbf{r}'|\le n+1} \frac{h_n(\mathbf{r}')\,\mathrm{d}^3\mathbf{r}'}{|\mathbf{r}-\mathbf{r}'|}.$$

But we know that the contribution to the integral from the region $n \le |\mathbf{r}'| \le n+1$ is vanishingly small as $n \to \infty$, so that we might just as well integrate over $|\mathbf{r}'| \le n$, in which case $\rho_n = 1$ and $h_n = h$.

We have, finally, for arbitrary $g \in D(H_0)$,

$$g \in D(H_0) \Rightarrow g(\mathbf{r}) = \lim_{n\to\infty} \frac{1}{4\pi} \int_{|\mathbf{r}'|\le n} \frac{(H_0 g)(\mathbf{r}')\,\mathrm{d}^3\mathbf{r}'}{|\mathbf{r}-\mathbf{r}'|}. \tag{5.3.22}$$

As before, the limit may be viewed either as a pointwise limit (almost everywhere), or as an L^2 limit over bounded subsets of $\mathbb{R}^3$.

Equation (5.3.22) provides us with an integral representation of the (unbounded) operator H_0^{-1}, which may be thought of as an integral operator having kernel $1/4\pi|\mathbf{r}-\mathbf{r}'|$. Thus, for $h \in D(H_0^{-1})$.

$$(H_0^{-1} h)(\mathbf{r}) = \lim_{n\to\infty} \frac{1}{4\pi} \int_{|\mathbf{r}'|\le n} \frac{h(\mathbf{r}')\,d^3\mathbf{r}'}{|\mathbf{r}-\mathbf{r}'|}. \tag{5.3.22$'$}$$

An entirely analogous formula, of which (5.3.22)$'$ is the limiting case, $z=0$, may be derived for the resolvent operator $R(z) = (H_0 - z)^{-1}$. We let K be any complex number of which $\operatorname{Re}(K) > 0$, and set $K^2 = -z$. Following the proof of Lemma 5.1 closely, and using the identity

$$-\frac{1}{4\pi}\left(\frac{\Delta\phi(\mathbf{r}') + z\phi(\mathbf{r}')}{|\mathbf{r}-\mathbf{r}'|}\right)\mathrm{e}^{-K|\mathbf{r}-\mathbf{r}'|}$$

$$= \operatorname{div}\left\{\frac{\phi}{4\pi}\operatorname{grad}\frac{\mathrm{e}^{-K|\mathbf{r}-\mathbf{r}'|}}{|\mathbf{r}-\mathbf{r}'|} - \frac{\mathrm{e}^{-K|\mathbf{r}-\mathbf{r}'|}}{|\mathbf{r}-\mathbf{r}'|}\operatorname{grad}\frac{\phi}{4\pi}\right\},$$

we arrive at the result

$$\phi(\mathbf{r}) = \frac{-1}{4\pi}\int \frac{\mathrm{e}^{-K|\mathbf{r}-\mathbf{r}'|}}{|\mathbf{r}-\mathbf{r}'|}(\Delta\phi(\mathbf{r}') + z\phi(\mathbf{r}'))\mathrm{d}^3\mathbf{r}',$$

which is the analogue of (5.3.5) for $z \neq 0$, and holds for all $\phi \in C_0^\infty(\mathbb{R}^3)$. Just as before, we may deduce, for $\phi \in C_0^\infty(\mathbb{R}^3)$, that

$$\left\langle \Delta\phi + z\phi, \int \frac{e^{-K|\boldsymbol{r}-\boldsymbol{r}'|}}{|\boldsymbol{r}-\boldsymbol{r}'|} h(\boldsymbol{r}')\, d^3\boldsymbol{r}' \right\rangle = -4\pi\langle \phi, h\rangle, \tag{5.3.23}$$

which is the analogue of (5.3.6), but now holds for all $h \in L^2(\mathbb{R}^3)$. Note that the Fourier transform of $e^{-K|\boldsymbol{r}|}/4\pi|\boldsymbol{r}|$ is $(2\pi)^{-3/2}(|\boldsymbol{k}|^2 - z)^{-1}$ which is bounded, so that the integral operator having kernel $e^{-K|\boldsymbol{r}-\boldsymbol{r}'|}/|\boldsymbol{r}-\boldsymbol{r}'|$ is bounded and defined on the whole of $L^2(\mathbb{R}^3)$. The equation corresponding to (5.3.8) now becomes

$$(H_0 - z)\int \frac{e^{-K|\boldsymbol{r}-\boldsymbol{r}'|}}{|\boldsymbol{r}-\boldsymbol{r}'|} h(\boldsymbol{r}')\, d^3\boldsymbol{r}' = 4\pi h(\boldsymbol{r}). \tag{5.3.24}$$

(It follows that z cannot be an eigenvalue of H_0, even for real K and $z<0$, since the corresponding eigenvector f, by (5.3.24), would have to satisfy

$$\langle f, h\rangle = \left\langle (H_0 - z)f, \frac{1}{4\pi}\int \frac{e^{-K|\boldsymbol{r}-\boldsymbol{r}'|}}{|\boldsymbol{r}-\boldsymbol{r}'|} h(\boldsymbol{r}')\, d^3\boldsymbol{r}' \right\rangle = 0$$

for all $h \in L^2(\mathbb{R}^3)$.) We are therefore able to extend (5.7.22)′ to all z off the positive real axis ($z = -K^2$ with $\mathrm{Re}(K) > 0$), to obtain

$$((H_0 - z)^{-1}h)(\boldsymbol{r}) = \frac{1}{4\pi}\int \frac{e^{-K|\boldsymbol{r}-\boldsymbol{r}'|}}{|\boldsymbol{r}-\boldsymbol{r}'|} h(\boldsymbol{r}')\, d^3\boldsymbol{r}' \tag{5.3.24$'$}$$

Equation (5.3.24)′ holds for all $h \in L^2(\mathbb{R}^3)$, and tells us that the resolvent operator $(H_0 - z)^{-1}$ is a bounded integral operator, having kernel $(4\pi|\boldsymbol{r}-\boldsymbol{r}'|)^{-1}\exp[-(-z)^{1/2}|\boldsymbol{r}-\boldsymbol{r}'|]$, where $w^{1/2}$, for complex w, denotes the square root having positive real part.

Equation (5.3.24)′ may be written as

$$(H_0 - z)^{-1} = \mathscr{F}^{-1}(K_0 - z)^{-1}\mathscr{F}, \tag{5.3.25}$$

where K_0 is the multiplication operator (in momentum space) defined by $(K_0\hat{g})(\boldsymbol{k}) = |\boldsymbol{k}|^2 g(\boldsymbol{k})$. Equation (5.3.25) expresses the fact that the kinetic-energy operator, through Fourier transformation, is unitarily equivalent to multiplication by $|\boldsymbol{k}|^2$. Indeed, many properties of H_0 may be deduced directly from the corresponding properties of K_0. The corresponding spectral families $\{E_\lambda(H_0)\}$ and $\{E_\lambda(K_0)\}$ are related by

$$E_\lambda(H_0) = \mathscr{F}^{-1}E_\lambda(K_0)\mathscr{F},$$

so that the spectral properties of H_0 may be deduced from those of K_0. As may readily be verified, we have

$$\langle \hat{f}, E_\lambda(K_0)\hat{g}\rangle = \int\limits_{|\boldsymbol{k}|^2 \leq \lambda} \bar{\hat{f}}(\boldsymbol{k})\hat{g}(\boldsymbol{k})\, \mathrm{d}^3\boldsymbol{k}.$$

Writing $\boldsymbol{k} = k\boldsymbol{\omega}$. $\mathrm{d}^3\boldsymbol{k} = k^2\,\mathrm{d}k\ \ \mathrm{d}\omega$, with $\mathrm{d}\omega = \sin\theta\,\mathrm{d}\theta\,\mathrm{d}\varphi$ in spherical polar coordinates, we have, on making the change of integration variable $k = \surd(\lambda')^{1/2}$, and taking $\lambda \geq 0$,

$$\langle \hat{f}, E_\lambda(K_0)\hat{g}\rangle = \int_0^\lambda \mathrm{d}\lambda \tfrac{1}{2}(\lambda')^{1/2} \int \bar{\hat{f}}((\lambda')^{1/2}\boldsymbol{\omega})\hat{g}((\lambda')^{1/2}\boldsymbol{\omega})\,\mathrm{d}\omega.$$

Hence the measure defined by $\langle \hat{f}, E_\lambda(K_0)\hat{g}\rangle$ is absolutely continuous, with density function given by

$$F(\lambda) = \begin{cases} \tfrac{1}{2}\lambda^{1/2}\int \bar{\hat{f}}(\lambda^{1/2}\boldsymbol{\omega})\hat{g}(\lambda^{1/2}\boldsymbol{\omega})\,\mathrm{d}\omega & (\lambda \geq 0), \\ 0 & (\lambda < 0). \end{cases} \tag{5.3.26}$$

It follows that K_0 and hence also H_0 have absolutely continuous spectra, along the positive real line.

It is sometimes convenient to carry out the diagonalization of the free Hamiltonian H_0 through the *direct integral representation*. For $f \in L^2(\mathbb{R}^3)$, we let $\hat{f}$ denote the Fourier transform of f, so that $\hat{f}(\boldsymbol{k})$ describes the state of a particle in momentum space. We now define the function $f(\lambda, \boldsymbol{\omega})$ for $\lambda \geq 0$ and $|\boldsymbol{\omega}| = 1$ by the formula

$$f(\lambda, \boldsymbol{\omega}) = \frac{1}{\sqrt{2}}\lambda^{1/4}\hat{f}(\lambda^{1/2}\boldsymbol{\omega}). \tag{5.3.27}$$

It is common to regard $f(\lambda, \boldsymbol{\omega})$ as defining a one-parameter family of functions f_λ, according to the formula $f_\lambda(\boldsymbol{\omega}) = f(\lambda, \boldsymbol{\omega})$. Then

$$\|f\|^2 = \int_0^\infty \mathrm{d}\lambda \int_{S^{(2)}} |f_\lambda(\boldsymbol{\omega})|^2\,\mathrm{d}\omega. \tag{5.3.28}$$

We let $L^2(S^{(2)})$ denote the space of square-integrable functions on the unit sphere $|\boldsymbol{\omega}| = 1$, i.e. with inner product $\langle u, v\rangle_0 = \int_{S^{(2)}} \bar{u}(\boldsymbol{\omega})v(\boldsymbol{\omega})\,\mathrm{d}\omega$. Then (5.3.28) may be written as

$$\|f\|^2 = \int_0^\infty \mathrm{d}\lambda\, \|f_\lambda\|_0^2. \tag{5.3.29}$$

The representation of f by a one-parameter family in this way is called the direct integral representation; the inner product is given by

$$\langle f, g\rangle = \int_0^\infty \mathrm{d}\lambda \, \langle f_\lambda, g_\lambda\rangle_0. \tag{5.3.30}$$

Equation (5.3.30) actually defines the inner product in the space $L^2(0, \infty; L^2(S^{(2)}))$, in which each element may be represented by a norm-square-integrable function from $(0, \infty)$ to $L^2(S^{(2)})$. We say that this space is a *direct integral*, as λ runs from 0 to ∞, of (infinitely many) copies of the space $L^2(S^{(2)})$.

In the direct integral representation, H_0 is represented very simply by the operator of multiplication by λ. Thus

$$H_0 : \{f_\lambda(\boldsymbol{\omega})\} \to \{\lambda f_\lambda(\boldsymbol{\omega})\}. \tag{5.3.31}$$

(More exactly, H_0 is *unitarily equivalent* to this operator.) Of course, what is gained in a simple representation for H_0 may be lost in a more complicated representation for other operators. As an example, we consider the operator $|\boldsymbol{X}|^2$ of multiplication by $|\boldsymbol{r}|^2$: in position space, $f(\boldsymbol{r}) \to |\boldsymbol{r}|^2 f(\boldsymbol{r})$. In momentum space, this operator is represented by $-\Delta$ (negative Laplacian with respect to $\boldsymbol{k}$). Hence

$$|\boldsymbol{X}|^2 = -\frac{\partial^2}{\partial k^2} - \frac{2}{k}\frac{\partial}{\partial k} - \frac{1}{k^2}\left\{\frac{1}{\sin\theta}\frac{\partial}{\partial\theta}\left(\sin\theta\frac{\partial}{\partial\theta}\right) + \frac{1}{\sin^2\theta}\frac{\partial^2}{\partial\phi^2}\right\},$$

which becomes, in the direct integral representation,

$$|\boldsymbol{X}|^2 = -4\frac{\partial}{\partial\lambda}\left(\lambda\frac{\partial}{\partial\lambda}\right) + \frac{1}{4\lambda} - \frac{1}{\lambda}\left\{\frac{1}{\sin\theta}\frac{\partial}{\partial\theta}\left(\sin\theta\frac{\partial}{\partial\theta}\right) + \frac{1}{\sin^2\theta}\frac{\partial^2}{\partial\phi^2}\right\}.$$

There is a useful relationship between derivatives with respect to λ and certain polynomials of momentum observables. The operator $\partial/\partial\lambda$, in the direct integral representation, corresponds to the operator

$$\frac{1}{2k}\frac{\partial}{\partial k} + \frac{1}{4k^2}$$

acting on $\hat{f}(k\boldsymbol{\omega})$. Now

$$\frac{\partial}{\partial k} = \sum_{i=1}^{3} \frac{k_i}{k}\frac{\partial}{\partial k_i},$$

so that $\partial/\partial\lambda$ corresponds to the operator

$$\sum_{i=1}^{3} \frac{k_i}{2k^2}\frac{\partial}{\partial k_i} + \frac{1}{4k^2}$$

on $\hat{g}$. Thus $\partial/\partial\lambda$ may be represented by a polynomial in $\partial/\partial k_i$ of degree 1, whose coefficients are functions of $\boldsymbol{k}$, or equivalently of λ and $\boldsymbol{\omega}$. Repeated application of $\partial/\partial\lambda$ now allows us to express $\partial^n/\partial\lambda^n$, acting in the direct integral representation, as a polynomial of degree n in $X_i = \mathrm{i}\,\partial/\partial k_i$. For example, we have

$$\frac{\partial^2 f_\lambda}{\partial\lambda^2} = \left\{\left(\sum_{l,m} \beta_{lm} X_l X_m + \sum_j \beta_j X_j + \beta_0\right) f\right\}_\lambda, \tag{5.3.32}$$

where all the β's may be regarded as functions of λ and $\boldsymbol{\omega}$, and are, moreover, uniformly bounded over closed subsets of the interval $0 < \lambda < \infty$.

Using (5.3.29), we obtain from (5.3.32) the useful estimate

$$\begin{aligned}\int_{\lambda_1}^{\lambda_2} \mathrm{d}\lambda \left\| \frac{\partial^2 f_\lambda}{\partial\lambda^2} \right\|^2 &\le \text{const} \int_{\lambda_1}^{\lambda_2} \mathrm{d}\lambda \left\{ \sum_{l,m} \|(X_l X_m f)_\lambda\|_0^2 + \sum_j \|(X_j f)_\lambda\|_0^2 + \|f_\lambda\|_0^2 \right\} \\ &\le \text{const}\, \|(|\boldsymbol{X}|^2 + 1) f\|^2,\end{aligned}$$

and more generally we have

$$\int_{\lambda_1}^{\lambda_2} \mathrm{d}\lambda \left\| \frac{\partial^n f_\lambda}{\partial\lambda^n} \right\|_0^2 \le \text{const}\, \|(|\boldsymbol{X}|^2 + 1)^{n/2} f\|^2, \tag{5.3.33}$$

an estimate that will be needed later in Chapter 12.

For a particle moving freely in three dimensions, the *evolution operator* U_t is given by $U_t = \mathrm{e}^{-\mathrm{i}H_0 t}$. We then have $U_t = \mathscr{F}^{-1}\mathrm{e}^{-\mathrm{i}K_0 t}\mathscr{F}$, where $\mathrm{e}^{-\mathrm{i}K_0 t}$ is the operator of multiplication by $\mathrm{e}^{-\mathrm{i}|\boldsymbol{k}|^2 t}$. In order to derive an integral representation for U_t, it is convenient first of all to obtain a formula for $\mathrm{e}^{-\mathrm{i}H_0(t-\mathrm{i}\varepsilon)} = \mathscr{F}^{-1}\mathrm{e}^{-iK_0(t-\mathrm{i}\varepsilon)}\mathscr{F}$, thereby escaping the problems of convergence of integrals that would otherwise arise. We shall take $\varepsilon > 0$ and finally take the limit as $\varepsilon \to 0$.

Using the identity

$$\int \mathrm{d}^3\boldsymbol{k}\, \exp(-a|\boldsymbol{k}|^2 + 2\boldsymbol{b}\cdot\boldsymbol{k}) = \left(\frac{\pi}{a}\right)^{3/2} \exp\left(\frac{\boldsymbol{b}^2}{a}\right)$$

($\mathrm{Re}(a) > 0$), we find that the inverse Fourier transform of $\mathrm{e}^{-\mathrm{i}|\boldsymbol{k}|^2(t-\mathrm{i}\varepsilon)}$ is

$$\left(\frac{1}{2\pi}\right)^{3/2} \left(\frac{\pi}{\mathrm{i}(t-\mathrm{i}\varepsilon)}\right)^{3/2} \exp\left(\frac{\mathrm{i}|\boldsymbol{r}|^2}{4(t-\mathrm{i}\varepsilon)}\right),$$

where $w^{3/2}$, for complex w, is to be interpreted as $(w^{1/2})^3$, and $w^{1/2}$ is the square root having positive real part. Hence

$$(\mathrm{e}^{-\mathrm{i}H_0(t-\mathrm{i}\varepsilon)}f)(\boldsymbol{r})$$
$$=\left(\frac{1}{2\pi}\right)^3\left(\frac{\pi}{\mathrm{i}(t-\mathrm{i}\varepsilon)}\right)^{3/2}\int\exp\left(\frac{\mathrm{i}|\boldsymbol{r}-\boldsymbol{r}'|^2}{4(t-\mathrm{i}\varepsilon)}\right)f(\boldsymbol{r}')\,\mathrm{d}^3\boldsymbol{r}'.$$

By the spectral theorem, s-$\lim_{\varepsilon\to 0+}\mathrm{e}^{-\mathrm{i}H_0(t-\mathrm{i}\varepsilon)}=\mathrm{e}^{-\mathrm{i}H_0 t}$, so we can take the limit $\varepsilon\to 0+$ in (5.3.26) to obtain

$$(U_t f)(\boldsymbol{r})=(\mathrm{e}^{-\mathrm{i}H_0 t}f)(\boldsymbol{r})=\left(\frac{1}{4\pi\mathrm{i}t}\right)^{3/2}\int\exp\left(\frac{\mathrm{i}|\boldsymbol{r}-\boldsymbol{r}'|^2}{4t}\right)f(\boldsymbol{r}')\,\mathrm{d}^3\boldsymbol{r}',\quad(5.3.34)$$

which may be compared with the corresponding result (5.1.14) in one dimension. For f of compact support, the integral on the right-hand side is uniformly (in $\boldsymbol{r}$) and absolutely convergent. Arbitrary $f\in L^2(\mathbb{R}^3)$ may be expressed as strong limits of such f. For general $f\in L^2(\mathbb{R}^3)$ the integral in (5.3.34) must be understood, just as for the Fourier transform, as a strong limit as $n\to\infty$ of the integral over $|\boldsymbol{r}|<n$.

We therefore have that the evolution operator U_t is an integral operator with kernel

$$\left(\frac{1}{4\pi\mathrm{i}t}\right)^{3/2}\exp\left(\frac{\mathrm{i}|\boldsymbol{r}-\boldsymbol{r}'|^2}{4t}\right).\qquad(5.3.35)$$

(Although (5.3.34) does not make sense for $t=0$, using the change of integration variable $\boldsymbol{r}'=\boldsymbol{r}+(4|t|)^{1/2}\boldsymbol{u}$, it may be verified that U_t converges strongly to the identity operator as $t\to 0$. Another way of expressing this is to say that, in the limit as $t\to 0$, the kernel approaches $\delta(\boldsymbol{r}-\boldsymbol{r}')$, where δ is the Dirac δ-function. Indeed this statement is literally true, in the sense of distributions.)

As for the Fourier-transform operator, the kernel (5.3.35) defines a bounded, in fact unitary, operator, but is not square-integrable. We may even relate the kernel (5.3.35) to that of the Fourier transform by writing $|\boldsymbol{r}-\boldsymbol{r}'|^2=|\boldsymbol{r}|^2-2\boldsymbol{r}\cdot\boldsymbol{r}'+|\boldsymbol{r}'|^2$. To do this, we define on $L^2(\mathbb{R}^3)$ a one-parameter family of operators by

$$(\mathcal{U}_t f)(\boldsymbol{r})=\left(\frac{1}{2\mathrm{i}t}\right)^{3/2}\hat{f}\left(\frac{\boldsymbol{r}}{2t}\right),\qquad(5.3.36)$$

where $\hat{f}$ is the Fourier transform of f. $\mathcal{U}_t$ is unitary since the Fourier transform is unitary, and with the change of variable $\boldsymbol{r}=2|t|\boldsymbol{r}'$ we have $|2t|^{-3}\,\mathrm{d}^3\boldsymbol{r}=\mathrm{d}\boldsymbol{r}'$. Explicitly,

$$(\mathcal{U}_t f)(\boldsymbol{r})=\left(\frac{1}{4\pi\mathrm{i}t}\right)^{3/2}\int\exp\left(\frac{-\mathrm{i}\boldsymbol{r}\cdot\boldsymbol{r}'}{2t}\right)f(\boldsymbol{r}')\,\mathrm{d}^3r'.$$

If we now define a one-parameter family $\{W_t\}$ of unitary multiplication operators by

$$(W_t f)(\boldsymbol{r}) = \exp\left(\frac{\mathrm{i}r^2}{4t}\right) f(\boldsymbol{r}); \tag{5.3.37}$$

then from (5.3.34) we have, for $t \neq 0$,

$$U_t = W_t \mathscr{U}_t W_t. \tag{5.3.38}$$

The identity (5.3.38) is particularly useful in allowing an analysis of the evolution of states in position space, and lends itself especially to a study of what happens for large positive and negative times. Since, for large times, the behaviour of the evolution e^{-iHt}, where H is the total Hamiltonian, may often be compared, through the use of wave operators, with the corresponding behaviour of the free evolution, this information will be of value in describing the scattering of particles in the presence of interactions.

Suppose, for example, for a particle evolving freely, that we are interested in the asymptotic probability, as $t \to \infty$, of the particle being located within a region $\mathscr{R}$ of $\mathbb{R}^3$. Using (5.3.38), this probability is

$$\begin{aligned} &\mathrm{prob}_t(\boldsymbol{r} \in \mathscr{R}) \\ &= \|E_{\boldsymbol{r}\in\mathscr{R}} \mathrm{e}^{-\mathrm{i}H_0 t} f\|^2 \\ &= \|E_{\boldsymbol{r}\in\mathscr{R}} \mathscr{U}_t W_t f\|^2, \end{aligned}$$

where f is the state at time $t=0$, and $E_{\boldsymbol{r}\in\mathscr{R}}$ is the operator of multiplication, in position space, by the characteristic function of the region $\mathscr{R}$. From (5.3.37), it is easy to prove, using the Lebesgue Dominated-Convergence Theorem, that $W_t f$ converges strongly to f in the limit as $t \to \infty$. Hence asymptotically prob_t $(\boldsymbol{r} \in \mathscr{R})$ approaches $\|E_{\boldsymbol{r}\in\mathscr{R}} \mathscr{U}_t f\|^2$.

Now

$$\begin{aligned} \|E_{\boldsymbol{r}\in\mathscr{R}} \mathscr{U}_t f\|^2 &= \left(\frac{1}{2t}\right)^3 \int_{\mathscr{R}} \left|\hat{f}\left(\frac{\boldsymbol{r}}{2t}\right)\right|^2 \mathrm{d}^3\boldsymbol{r} \\ &= \int_{2t\boldsymbol{k}\in\mathscr{R}} |\hat{f}(\boldsymbol{k})|^2 \,\mathrm{d}^3\boldsymbol{k}, \end{aligned}$$

on making the change of variable $\boldsymbol{r} = 2t\boldsymbol{k}$. It is more suggestive, remembering that we have chosen units such that $2m=1$, to express this result in the form

$$\lim_{t\to\infty} \left\{ \mathrm{prob}_t\,(\boldsymbol{r}\in\mathscr{R}) - \int_{\boldsymbol{k}t/m\in\mathscr{R}} |\hat{f}(\boldsymbol{k})|^2 \,\mathrm{d}^3\boldsymbol{k} \right\} = 0. \tag{5.3.39}$$

Now $\boldsymbol{k}t/m$ is the position vector, at time t, of a classical particle that starts at the origin and moves with constant momentum $\boldsymbol{k}$. Equation (5.3.39) tells us that, asymptotically as $t \to \infty$, the probability of finding the particle within $\mathscr{R}$ is the same as the probability of finding a classical particle within $\mathscr{R}$, if the classical particle starts at the origin and moves with constant momentum, where $|\hat{f}(\boldsymbol{k})|^2$ is the probability density for this momentum. (In this sense, $|\hat{f}(\boldsymbol{k})|^2$ retains its quantum-mechanical interpretation.) Of course, the choice of origin here is arbitrary. The classical particle may be regarded as starting at any other point $\boldsymbol{r} = \boldsymbol{a}$.

Of particular importance, in (5.3.39), is the case in which the region $\mathscr{R}$ is a cone. Taking the vertex of the cone to be at $\boldsymbol{r} = 0$, we have in this case $\boldsymbol{k}t/m \in \mathscr{R}$ if and only if $\boldsymbol{k} \in \mathscr{R}$ (for $t > 0$!). This gives, for such a cone $\mathscr{C}$,

$$\lim_{t \to \infty} \{\text{prob}_t(\boldsymbol{r} \in \mathscr{C})\} = \int_{\mathscr{C}} |\hat{f}(\boldsymbol{k})|^2 \, \mathrm{d}^3\boldsymbol{k}, \tag{5.3.39$'$}$$

Thus for large t the probability of finding the particle within the cone $\mathscr{C}$ approaches a limit, and this limit is identical with the probability of the particle momentum lying within the same cone. It is this fact that allows us to speak of the asymptotic probability, as $t \to \infty$, of finding the particle within a cone. A corresponding result holds in the limit as $t \to -\infty$, namely

$$\lim_{t \to -\infty} \{\text{prob}_t(\boldsymbol{r} \in \mathscr{C}_-)\} = \int_{\mathscr{C}} |\hat{f}(\boldsymbol{k})|^2 \, \mathrm{d}^3\boldsymbol{k}, \tag{5.3.39$''$}$$

where $\mathscr{C}_-$ is the cone obtained from $\mathscr{C}$ through reflection in the origin, $\boldsymbol{r} \to -\boldsymbol{r}$. All of this agrees very well with our intuition concerning the motion of freely moving particles. A particle is to be found in $\mathscr{C}_-$, for large negative times, and in $\mathscr{C}$, for large positive times, if the direction of its momentum lies in the cone $\mathscr{C}$.

5.4 L² PERTURBATIONS OF THE FREE HAMILTONIAN

The Hamiltonian for a particle moving in three-dimensional space, in a conservative field of force described by a potential $V(\boldsymbol{r})$, is given by $H = H_0 + V(\boldsymbol{r})$, where $H_0 = \boldsymbol{P}^2/2m$ is the free Hamiltonian with which we dealt in the last section. In units such that $\hbar = 2m = 1$, H is represented as a linear operator in $L^2(\mathbb{R}^3)$ by

$$H = -\Delta + V(\boldsymbol{r}).$$

There may be technical difficulties to be overcome in defining H in this way. This is because the domains of the linear operators H_0 and V need not be the same, in which case it is not entirely a straightforward matter to add two such operators and to arrive, possibly by extension of the domain, at a self-adjoint operator for H. There also may be problems of defining H uniquely, if we arrive first of all at an operator that is not essentially self-adjoint.

In the case of the free Hamiltonian H_0, we started by defining an operator $\hat{H}_0$, with domain $C_0^\infty(\mathbb{R}^3)$. The operator H_0 was then defined as the unique self-adjoint extension of $\hat{H}_0$. In a similar way, let us define an operator $\hat{H}$ acting on C_0^∞ functions. The definition of $\hat{H}$ in this way only makes sense if the potential is at least locally square-integrable in some subset of $\mathbb{R}^3$. Let us therefore assume that we are dealing with a potential $V(\boldsymbol{r})$ that is square-integrable over some neighbourhood of a point $\boldsymbol{r}_0$. To be precise, we let $\mathscr{B}$ denote the region $|\boldsymbol{r}-\boldsymbol{r}_0|<R$, and suppose that $V\in L^2(\mathscr{B})$. The operator $\hat{H}$ will act on functions belonging to $C_0^\infty(\mathscr{B})$ according to the formula

$$\hat{H}\phi = -\Delta\phi + V\phi, \qquad \phi\in C_0^\infty(\mathscr{B}). \tag{5.4.1}$$

We do not assume here that $C_0^\infty(\mathscr{B})$ is the *whole* of the domain of $\hat{H}$. In most cases of interest the potential is locally square-integrable in the complement of some set Σ having Lebesgue measure zero. In this case we can cover the complement of Σ with a collection of neighbourhoods $\mathscr{B}$, in each of which $\hat{H}$ is defined by (5.4.1), and the resulting operator, extended by linearity, will have domain dense in $L^2(\mathbb{R}^3)$. Our present intention, however, is to explore the consequences of (5.4.1) for just one such neighbourhood.

The linear operator $\hat{H}$ defined on $C_0^\infty(\mathscr{B})$ by (5.4.1) is easily verified to be symmetric. A vector f in the domain of $\hat{H}^*$ will satisfy

$$\langle -\Delta\phi + V\phi, f\rangle = \langle \phi, f^*\rangle \tag{5.4.2}$$

for all $\phi\in C_0^\infty(\mathscr{B})$, where $f^* = \hat{H}^* f$. On the left-hand side of (5.4.2) we have

$$\langle V\phi, f\rangle = \int \bar{\phi}(\boldsymbol{r}')V(\boldsymbol{r}')f(\boldsymbol{r}')\,\mathrm{d}^3\boldsymbol{r}'.$$

The inner product is independent of the values of $f(\boldsymbol{r}')$ for $|\boldsymbol{r}'-\boldsymbol{r}_0|>R$, so that, in evaluating the integral, we may replace f by $\chi_{\mathscr{B}}f$, where $\chi_{\mathscr{B}}$ is the characteristic function of $\mathscr{B}$. Since $Vf\in L^1(\mathscr{B})$, we can apply Lemma 5.1, with $h(\boldsymbol{r}')=V(\boldsymbol{r}')f(\boldsymbol{r}')$, and (5.3.6) gives

$$\langle V\phi, f\rangle = -\frac{1}{4\pi}\left\langle \Delta\phi, \int_{\mathscr{B}} \frac{V(\boldsymbol{r}')f(\boldsymbol{r}')\,\mathrm{d}^3\boldsymbol{r}'}{|\boldsymbol{r}-\boldsymbol{r}'|}\right\rangle. \tag{5.4.3}$$

In the same way, we have

$$\langle \phi, f^*\rangle = -\frac{1}{4\pi}\left\langle \Delta\phi, \int_{\mathscr{B}} \frac{f^*(\boldsymbol{r}')\,\mathrm{d}^3\boldsymbol{r}'}{|\boldsymbol{r}-\boldsymbol{r}'|}\right\rangle. \tag{5.4.4}$$

Combining (5.4.2)–(5.4.4), we obtain

$$\left\langle \Delta\phi, f + \frac{1}{4\pi}\int_{\mathscr{B}} \frac{V(\boldsymbol{r}')f(\boldsymbol{r}')\,\mathrm{d}^3\boldsymbol{r}'}{|\boldsymbol{r}-\boldsymbol{r}'|} - \frac{1}{4\pi}\int_{\mathscr{B}} \frac{f^*(\boldsymbol{r}')\,\mathrm{d}^3\boldsymbol{r}'}{|\boldsymbol{r}-\boldsymbol{r}'|}\right\rangle = 0 \tag{5.4.5}$$

for all $\phi \in C_0^\infty(\mathscr{B})$.

By Lemma 5.2 the right-hand factor in this inner product is an infinitely differentiable function of $\boldsymbol{r}$ and a solution of Laplace's equation for $\boldsymbol{r} \in \mathscr{B}$. In particular, it is a *bounded* function of $\boldsymbol{r}$ in any region $|\boldsymbol{r}-\boldsymbol{r}_0| < R_1 < R$. Moreover, an application of the Schwarz inequality shows that $\int_{\mathscr{B}} \mathrm{d}^3\boldsymbol{r}'\, f^*(\boldsymbol{r}')/|\boldsymbol{r}-\boldsymbol{r}'|$ is bounded for $\boldsymbol{r} \in \mathscr{B}$, so that we can write, for $\boldsymbol{r} \in \mathscr{B}$,

$$f(\boldsymbol{r}) = g_1(\boldsymbol{r}) - \frac{1}{4\pi}\int_{\mathscr{B}} \frac{V(\boldsymbol{r}')f(\boldsymbol{r}')\,\mathrm{d}^3\boldsymbol{r}'}{|\boldsymbol{r}-\boldsymbol{r}'|}, \tag{5.4.6}$$

where $g_1(\boldsymbol{r})$ is bounded in the region $|\boldsymbol{r}-\boldsymbol{r}_0| < R_1 < R$. Certainly $f \in L^2(\mathscr{B})$, and, by the Schwarz inequality,

$$\begin{aligned}
&\int_{\mathscr{B}} \mathrm{d}^3\boldsymbol{r} \left| \int_{\mathscr{B}} \frac{V(\boldsymbol{r}')f(\boldsymbol{r}')\,\mathrm{d}^3\boldsymbol{r}'}{|\boldsymbol{r}-\boldsymbol{r}'|} \right|^2 \\
&\quad \le \text{const} \int_{\mathscr{B}} \mathrm{d}^3\boldsymbol{r} \int_{\mathscr{B}} \mathrm{d}^3\boldsymbol{r}' \frac{|V(\boldsymbol{r}')|^2}{|\boldsymbol{r}-\boldsymbol{r}'|^2} \\
&\quad = \text{const} \int_{\mathscr{B}} \mathrm{d}^3\boldsymbol{r}' |V(\boldsymbol{r}')|^2 \int_{\mathscr{B}} \frac{\mathrm{d}^3\boldsymbol{r}}{|\boldsymbol{r}-\boldsymbol{r}'|^2}.
\end{aligned}$$

Since $\int_{\mathscr{B}} \mathrm{d}^3\boldsymbol{r}/|\boldsymbol{r}-\boldsymbol{r}'|^2 \le \text{const}$ fot $\boldsymbol{r}' \in \mathscr{B}$, and since $V \in L^2(\mathscr{B})$, we see that the integral on the right-hand side of (5.4.6) defines a function of $\boldsymbol{r}$ that is square-integrable over $\mathscr{B}$. It follows from (5.4.6) that $g_1 \in L^2(\mathscr{B})$.

Now (5.4.6) may be "iterated" to obtain information on the local behaviour of $f(\boldsymbol{r})$. At the first iteration, we replace $f(\boldsymbol{r}')$ in the integrand on the right-hand side of (5.4.6) by

$$g_1(\boldsymbol{r}') - \frac{1}{4\pi}\int_{\mathscr{B}} \frac{V(\boldsymbol{r}'')f(\boldsymbol{r}'')\,\mathrm{d}^3\boldsymbol{r}''}{|\boldsymbol{r}'-\boldsymbol{r}''|}.$$

We then have

$$f(\boldsymbol{r}) = g_2(\boldsymbol{r}) + \left(\frac{1}{4\pi}\right)^2 \int_{\mathscr{B}}\int_{\mathscr{B}} \frac{V(\boldsymbol{r}')V(\boldsymbol{r}'')f(\boldsymbol{r}'')\,\mathrm{d}^3\boldsymbol{r}'\,\mathrm{d}^3\boldsymbol{r}''}{|\boldsymbol{r}-\boldsymbol{r}'|\,|\boldsymbol{r}'-\boldsymbol{r}''|}, \tag{5.4.7}$$

where

$$g_2(\boldsymbol{r}) = g_1(\boldsymbol{r}) - \frac{1}{4\pi} \int_{\mathscr{B}} \frac{V(\boldsymbol{r}')g_1(\boldsymbol{r}')\,\mathrm{d}^3\boldsymbol{r}'}{|\boldsymbol{r}-\boldsymbol{r}'|} \tag{5.4.8}$$

To estimate the integral on the right-hand side of (5.4.8), we divide the integration region into two parts. In the first subregion, defined by the inequality $|\boldsymbol{r}'-\boldsymbol{r}_0| < R_1$, we know that $|g_1(\boldsymbol{r}')| \leq$const, and the integral is bounded in absolute value by const $\int \mathrm{d}^3\boldsymbol{r}' |V(\boldsymbol{r}')|/|\boldsymbol{r}-\boldsymbol{r}'|$, which is bounded for $\boldsymbol{r} \in \mathscr{B}$. In the second subregion we have $R_1 < |\boldsymbol{r}'-\boldsymbol{r}_0| < R$. If we restrict $\boldsymbol{r}$ to lie in the region $|\boldsymbol{r}-\boldsymbol{r}_0| < R_2 < R_1$, then we have $1/|\boldsymbol{r}-\boldsymbol{r}'| \leq 1/(R_1-R_2)$. In this case, the integral on the right-hand side of (5.4.8) over this subregion is again bounded, since V and g_1 both belong to $L^2(\mathscr{B})$. Therefore $g_2(\boldsymbol{r})$ in (5.4.8) is bounded for $|\boldsymbol{r}-\boldsymbol{r}_0| < R_2 < R_1$, and again we can show that $g_2 \in L^2(\mathscr{B})$.

After three successive iterations of (5.4.6), we obtain, with a change of notation for integration variables, the following estimate for $f(\boldsymbol{r})$, valid in the region $|\boldsymbol{r}-\boldsymbol{r}_0| < R_4 < R_3 < R_2 < R_1 < R$:

$$|f(\boldsymbol{r})| \leq \text{const.} + \int_{\mathscr{B}}\int_{\mathscr{B}}\int_{\mathscr{B}}\int_{\mathscr{B}} \frac{\mathrm{d}^3\boldsymbol{r}_1\,\mathrm{d}^3\boldsymbol{r}_2\,\mathrm{d}^3\boldsymbol{r}_3\,\mathrm{d}^3\boldsymbol{r}_4\,|V(\boldsymbol{r}_1)V(\boldsymbol{r}_2)V(\boldsymbol{r}_3)V(\boldsymbol{r}_4)f(\boldsymbol{r}_4)|}{|\boldsymbol{r}-\boldsymbol{r}_1|\,|\boldsymbol{r}_1-\boldsymbol{r}_2|\,|\boldsymbol{r}_2-\boldsymbol{r}_3|\,|\boldsymbol{r}_3-\boldsymbol{r}_4|}.$$

Since $\int_{\mathscr{B}}\int_{\mathscr{B}}\int_{\mathscr{B}}\int_{\mathscr{B}} |V(\boldsymbol{r}_1)V(\boldsymbol{r}_2)V(\boldsymbol{r}_3)V(\boldsymbol{r}_4)|^2\,\mathrm{d}^3\boldsymbol{r}_1\,\mathrm{d}^3\boldsymbol{r}_2\,\mathrm{d}^3\boldsymbol{r}_3\,\mathrm{d}^3\boldsymbol{r}_4 < \infty$, the Schwarz inequality gives the estimate

$$\text{const.}\left(\int_{\mathscr{B}}\int_{\mathscr{B}}\int_{\mathscr{B}}\int_{\mathscr{B}} \frac{\mathrm{d}^3\boldsymbol{r}_1\,\mathrm{d}^3\boldsymbol{r}_2\,\mathrm{d}^3\boldsymbol{r}_3\,\mathrm{d}^3\boldsymbol{r}_4}{|\boldsymbol{r}-\boldsymbol{r}_1|^2\,|\boldsymbol{r}_1-\boldsymbol{r}_2|^2\,|\boldsymbol{r}_2-\boldsymbol{r}_3|^2\,|\boldsymbol{r}_3-\boldsymbol{r}_4|^2}\,|f(\boldsymbol{r}_4)|^2\right)^{1/2}$$

for the multiple integral. Now the substitution $\boldsymbol{r}_1 = \boldsymbol{r} + |\boldsymbol{r}-\boldsymbol{r}_2|\mathbf{z}$ allows us to deduce that

$$\int_{\mathscr{B}} \frac{\mathrm{d}^3\boldsymbol{r}_1}{|\boldsymbol{r}-\boldsymbol{r}_1|^2\,|\boldsymbol{r}_1-\boldsymbol{r}_2|^2} \leq \int_{\mathbb{R}^3} \frac{\mathrm{d}^3\boldsymbol{r}_1}{|\boldsymbol{r}-\boldsymbol{r}_1|^2\,|\boldsymbol{r}_1-\boldsymbol{r}_2|^2} = \frac{\text{const.}}{|\boldsymbol{r}-\boldsymbol{r}_2|},$$

and a similar estimate holds for the integral with respect to $\boldsymbol{r}_3$. Moreover, the bound $(|\boldsymbol{r}-\boldsymbol{r}_2|\,|\boldsymbol{r}_2-\boldsymbol{r}_4|)^{-1} \leq |\boldsymbol{r}-\boldsymbol{r}_2|^{-2} + |\boldsymbol{r}_2-\boldsymbol{r}_4|^{-2}$ implies that $\int_{\mathscr{B}} \mathrm{d}^3\boldsymbol{r}_2/|\boldsymbol{r}-\boldsymbol{r}_2|\,|\boldsymbol{r}_2-\boldsymbol{r}_4| \leq$const. Finally we have $\int_{\mathscr{B}} \mathrm{d}^3\boldsymbol{r}_4 |f(\boldsymbol{r}_4)|^2 < \infty$, so that the multiple integral is bounded for $\boldsymbol{r} \in \mathscr{B}$.

Hence $|f(\boldsymbol{r})| \leq$const. provided that $|\boldsymbol{r}-\boldsymbol{r}_0| < R_4$. This immediately leads us to the following lemma, which states that, if V is *locally* square-integrable then vectors f in the domain of $\hat{H}^*$ are *locally* in the domain of H_0.

Lemma 5.3

Let $\mathscr{B}$ be any open subset of $\mathbb{R}^3$ such that $V \in L^2(\mathscr{B})$. Define a symmetric linear operator $\hat{H}$ in $L^2(\mathbb{R}^3)$ that satisfies

$$\hat{H}\phi = -\Delta\phi + V\phi \qquad \text{for } \phi \in C_0^\infty(\mathscr{B}),$$

and let $\hat{H}^*$ be the adjoint of $\hat{H}$. Then any f in the domain of $\hat{H}^*$ is locally, in $\mathscr{B}$, in the domain of H_0. That is,

$$f \in D(\hat{H}^*) \Rightarrow \rho f \in D(H_0) \quad \text{for any } \rho \in C_0^\infty(\mathscr{B}).$$

If the total Hamiltonian H is any self-adjoint extension of $\hat{H}$ then

$$f \in D(H) \Rightarrow \rho f \in D(H_0) \quad \text{for any } \rho \in C_0^\infty(\mathscr{B}).$$

Proof

We shall verify the lemma in the case for which $\mathscr{B}$ is a ball $|\boldsymbol{r} - \boldsymbol{r}_0| < R$. The general case in which $\mathscr{B}$ is an arbitrary open subset of $\mathbb{R}^3$ follows similarly. The fact that we allow $\mathscr{B}$ to be unbounded in the general case does not cause difficulty, since the support of ρ is bounded. (For fixed ρ, $\mathscr{B}$ in the hypothesis of the lemma may be replaced by the intersection of $\mathscr{B}$ with any open ball containing the support of ρ.)

We therefore suppose that $\mathscr{B}$ is the ball $|\boldsymbol{r} - \boldsymbol{r}_0| < R$ and let $\mathscr{B}_1$ be the ball $|\boldsymbol{r} - \boldsymbol{r}_0| < R_1 < R$, where, for given $\rho \in C_0^\infty(\mathscr{B})$, R_1 is chosen such that the support of ρ is contained in $\mathscr{B}_1$. Now (5.4.5) will hold, with $\mathscr{B}$ replaced by $\mathscr{B}_1$, for any $\phi \in C_0^\infty(\mathscr{B})$. Proceeding as above, we may deduce that the function $g(\boldsymbol{r})$, defined for $\boldsymbol{r} \in \mathscr{B}_1$ by

$$g(\boldsymbol{r}) = f(\boldsymbol{r}) + \frac{1}{4\pi}\int_{\mathscr{B}_1} \frac{V(\boldsymbol{r}')f(\boldsymbol{r}')\,d^3\boldsymbol{r}'}{|\boldsymbol{r} - \boldsymbol{r}'|} - \frac{1}{4\pi}\int_{\mathscr{B}_1} \frac{f^*(\boldsymbol{r}')\,d^3\boldsymbol{r}'}{|\boldsymbol{r} - \boldsymbol{r}'|}, \tag{5.4.9}$$

is infinitely differentiable, for $\boldsymbol{r} \in \mathscr{B}_1$, and a solution of Laplace's equation. Hence $\rho g \in C_0^\infty(\mathscr{B}_1) \subset C_0^\infty(\mathbb{R}^3)$, and we have $H_0 \rho g = -\Delta \rho g$.

From (5.4.9) we write, for $\boldsymbol{r} \in \mathscr{B}_1$,

$$g(\boldsymbol{r}) = f(\boldsymbol{r}) + \int_{\mathscr{B}_1} \frac{h(\boldsymbol{r}')\,d^3\boldsymbol{r}'}{|\boldsymbol{r} - \boldsymbol{r}'|}, \tag{5.4.10}$$

with $h(\boldsymbol{r}') = (4\pi)^{-1}(V(\boldsymbol{r}')f(\boldsymbol{r}') - f^*(\boldsymbol{r}'))$. We have seen already that $f(\boldsymbol{r}')$ is bounded for $\boldsymbol{r}' \in \mathscr{B}_1$. Hence $h \in L^2(\mathscr{B}_1)$. We may now apply the Corollary to Lemma 5.1. to deduce that $\rho\int_{\mathscr{B}_1} d^3\boldsymbol{r}' h(\boldsymbol{r}')/|\boldsymbol{r} - \boldsymbol{r}'|$ is in the domain of H_0. Since we already have $\rho g \in D(H_0)$, (5.4.10) implies $\rho f \in D(H_0)$, as required. The rest of the lemma follows from the fact that $\hat{H}^*$ is an extension of H, so that $f \in D(H) \Rightarrow f \in D(\hat{H}^*)$. ■

The following lemma applies and extends this result to an important special case.

Lemma 5.4

Suppose that $V \in L^2_{\text{loc}}(\mathbb{R}^3 \setminus \{0\})$, meaning that V is square-integrable over any closed bounded subset of $\mathbb{R}^3$ that does not contain the origin. Define $\hat{H}$, with $D(\hat{H}) = C_0^\infty(\mathbb{R}^3 \setminus \{0\})$ by

$$\hat{H}\phi = -\Delta\phi + V\phi, \qquad \phi \in C_0^\infty(\mathbb{R}^3 \setminus \{0\}).$$

Suppose that either

(i) $\rho \in C_0^\infty(\mathbb{R}^3 \setminus \{0\})$, or
(ii) ρ is infinitely differentiable, with $\rho(\boldsymbol{r}) \equiv 1$ for $|\boldsymbol{r}|$ sufficiently large, and $\rho(\boldsymbol{r}) \equiv 0$ for $|\boldsymbol{r}|$ sufficiently small.

Then $f \in D(\hat{H}^*) \Rightarrow \rho f \in D(\hat{H}^*)$.

Proof

If ρ satisfies (i), and $f \in D(\hat{H}^*)$, then we have already shown that $\rho f \in D(H_0)$. We have also shown that f is locally bounded. Hence ρf is bounded with compact support not containing $\boldsymbol{r} = 0$, and it follows that $\rho f \in D(V)$. Thus $\rho f \in D(H_0) \cap D(V)$. For $\phi \in D(\hat{H})$, we have

$$\begin{aligned}\langle \hat{H}\phi, \rho f \rangle &= \langle -\Delta\phi + V\phi, \rho f \rangle \\ &= \langle \phi, H_0 \rho f + V \rho f \rangle.\end{aligned}$$

It follows, in this case, that

$$\hat{H}^* \rho f = H_0 \rho f + V \rho f. \tag{5.4.11}$$

Now we suppose that ρ satisfies (ii). For $f \in D(\hat{H}^*)$ and $\phi \in C_0^\infty(\mathbb{R}^3 \setminus \{0\})$, we have

$$\begin{aligned}\langle \hat{H}\phi, \rho f \rangle &= \langle -\Delta\phi + V\phi, \rho f \rangle \\ &= \langle (-\Delta + V)\rho\phi, f \rangle + \langle \phi\Delta\rho + 2(\nabla\rho)\cdot(\nabla\phi), f \rangle \\ &= \langle \hat{H}\rho\phi, f \rangle + \langle \phi\Delta\rho, f \rangle + \left\langle 2\mathrm{i}\sum_j \frac{\partial\rho}{\partial x_j} P_j\phi, f \right\rangle\end{aligned} \tag{5.4.12}$$

where $P_j = -\mathrm{i}\,\partial/\partial x_j$ are the components of the momentum operator. Now $\partial\rho/\partial x_j \in C_0^\infty(\mathbb{R}^3 \setminus \{0\})$, and f is locally in the domain of P_j (even H_0). Hence

$$\left| \left\langle \frac{\partial\rho}{\partial x_j} P_j \phi, f \right\rangle \right| = \left| \left\langle \phi, P_j \frac{\partial\rho}{\partial x_j} f \right\rangle \right| \leq \text{const.} \|\phi\|.$$

Also, on the right-hand side of (5.4.12), $|\langle\hat{H}\rho\phi, f\rangle| = |\langle\phi, \rho\hat{H}^* f\rangle| \leq \text{const}\|\phi\|$, and $|\langle\phi\Delta\rho, f\rangle| \leq \text{const}\|\phi\|$. Hence we have shown that $|\langle\hat{H}\phi, \rho f\rangle| \leq \text{const}\|\phi\|$, and it follows for ρ satisfying (ii), that $\rho f \in D(\hat{H}^*)$. ■

Corollary

Define $\hat{H}$ as in the lemma, with the same condition on V. Let H be any self-adjoint extension of $\hat{H}$. Suppose that either

(i) ρ satisfies condition (i) of the lemma, or
(ii) ρ satisfies condition (ii) of the lemma and $V = V_1 + V_2$, where $\int_{|\boldsymbol{r}|>1} |V_1(\boldsymbol{r})|^2 \, d^3\boldsymbol{r} < \infty$ and $\sup_{|\boldsymbol{r}|>1} |V_2(\boldsymbol{r})| < \infty$ (i.e. for $|\boldsymbol{r}| > 1$, V is the sum of a square-integrable potential and a bounded potential).

Then $f \in D(H) \Rightarrow \rho f \in D(H) \cap D(H_0) \cap D(V)$, and

$$H\rho f = H_0 \rho f + V\rho f. \tag{5.4.13}$$

Proof

We suppose that $f \in D(H)$. Since $\hat{H}^* \supseteq H$, we then have $f \in D(\hat{H}^*)$, from which the lemma implies $\rho f \in D(\hat{H}^*)$. If (i) above is satisfied then (5.4.11) follows, and in the proof of Lemma 5.5 we shall show that (5.4.11) also holds in case (ii). We let ρ_1 satisfy the same condition as ρ, with, in addition, $\rho_1(\boldsymbol{r}) \equiv 1$ on the support of ρ. For $h \in D(\hat{H}^*)$ it is then easy to see that $(\hat{H}^* h)(\boldsymbol{r}) \equiv (\hat{H}^* \rho_1 h)(\boldsymbol{r})$ provided that $\boldsymbol{r}$ lies in the support of ρ. We then have, for $h \in D(\hat{H}^*)$,

$$\begin{aligned}\langle\hat{H}^* h, \rho f\rangle &= \langle\hat{H}^* \rho_1 h, \rho f\rangle = \langle H_0 \rho_1 h + V\rho_1 h, \rho f\rangle \\ &= \langle\rho_1 h, H_0 \rho f + V\rho f\rangle = \langle h, H_0 \rho f + V\rho f\rangle.\end{aligned}$$

It follows that $\rho f \in D(\hat{H}^{**})$, and that

$$\hat{H}^{**} \rho f = H_0 \rho f + V\rho f.$$

Since $\hat{H}^{**}$ is the closure of $\hat{H}$, and H is an extension of $\hat{H}^{**}$, (5.4.13) follows, and we have proved the corollary. ■

The final result of this section deals with a class of regular potentials for which the domain of the Hamiltonian has a particularly simple description.

Lemma 5.5

Suppose that $V \in L^2(\mathbb{R}^3) + L^\infty(\mathbb{R}^3)$ (i.e. $V = V_1 + V_2$, where V_1 is square-integrable and V_2 is bounded). Define $\hat{H}$ on $C_0^\infty(\mathbb{R}^3)$ by

$$\hat{H}\phi = -\Delta\phi + V\phi, \qquad \phi \in C_0^\infty(\mathbb{R}^3).$$

Then $\hat{H}$ is essentially self-adjoint, and the unique self-adjoint extension H of $\hat{H}$ has $D(H)=D(H_0)$. Moreover, for $f\in D(H)$ we have $Hf=H_0f+Vf$.

Proof

We assume that V satisfies the conditions of the lemma. The key to the proof is the observation that $D(H_0)\subset D(V)$. To obtain this result, we use (5.3.24)′ to show that $V(H_0-z)^{-1}$, for Im (z)$\neq 0$, is an integral operator with kernel

$$K(\mathbf{r},\mathbf{r}')=\frac{1}{4\pi}\frac{V(\mathbf{r})e^{-K|\mathbf{r}-\mathbf{r}'|}}{|\mathbf{r}-\mathbf{r}'|},$$

where $z=-K^2$ and $\mathrm{Re}(K)>0$. If $V=V_1\in L^2(\mathbb{R}^3)$ then the kernel is Hilbert–Schmidt, whereas if $V=V_2\in L^\infty(\mathbb{R}^3)$ then $V_2(H_0-z)^{-1}$ is bounded. Hence $V(H_0-z)^{-1}$ is a bounded linear operator. Since the domain of H_0 is the range of $(H_0-z)^{-1}$, our assertion that $D(H_0)\subset D(V)$ is verified. In Section 5.3 we saw that $\hat{H}_0$, with the same domain as $\hat{H}$, is essentially self-adjoint. For any $g\in D(H_0)$ there exists a sequence $\{g_n\}$, converging strongly to g, such that $H_0g_n\to H_0g$ strongly. In this case, $Vg_n=\{V(H_0-z)^{-1}\}(H_0-z)g_n$ converges strongly to $V(H_0-z)^{-1}(H_0-z)g=Vg$, since $V(H_0-z)^{-1}$ is bounded. Hence Hg_n converges strongly to H_0g+Vg. Since H is a closed operator, we have shown that $g\in D(H_0)\Rightarrow g\in D(H)$ and $Hg=H_0g+Vg$.

To show that $\hat{H}$ is essentially self-adjoint, we suppose that, for some $x\in L^2(\mathbb{R}^3)$, and for Im $(z)\neq 0$,

$$\langle x,(\hat{H}-z)\phi\rangle=0 \quad \text{for all } \phi\in D(\hat{H}).$$

Since $(\hat{H}-z)\phi=(1+V(H_0-z)^{-1})(\hat{H}_0-z)\phi$, we may write this equation in the form

$$\langle(1+V(H_0-z)^{-1})^*x,(\hat{H}_0-z)\phi\rangle=0.$$

Since $\hat{H}_0$ is essentially self-adjoint, this implies that

$$(1+V(H_0-z)^{-1})^*x=0. \tag{5.4.14}$$

However, an estimate of the Hilbert–Schmidt norm shows that $\|V(H_0-z)^{-1}\|<1$ for suitable z, in which case we can invert the operator $(1+V(H_0-z)^{-1})$ (or its adjoint) by means of a power series, and deduce from (5.4.14) that $x=0$. Hence $\hat{H}$ has deficiency indices (0,0), and is essentially self-adjoint. We can also verify that any vector in the domain of H is in the domain of H_0. Indeed, we now have

$$(H_0-z)(H-z)^{-1}=(1+V(H_0-z)^{-1})^{-1}, \tag{5.4.15}$$

as may be shown by a little elementary manipulation. Hence $D(H)=D(H_0)$, and the lemma is proved. ■

These results also allow us to complete the proof of (5.4.11) for $f \in D(\hat{H}^*)$ in case (ii) of Lemma 5.4. For in that case, with $\phi \in C_0^\infty(\mathbb{R}^3 \setminus \{0\})$ and ρ satisfying the condition (ii), $\langle \hat{H}\phi, \rho f \rangle$ will be unchanged if we alter $V(\boldsymbol{r})$ near $\boldsymbol{r}=0$ (i.e. *outside* the support of ρ) to obtain a new potential belonging to $L^2(\mathbb{R}^3)+L^\infty(\mathbb{R}^3)$. We can then use the present lemma to obtain

$$\langle \hat{H}\phi, \rho f \rangle = \langle \phi, H_0 \rho f + V \rho f \rangle,$$

from which (5.4.11) follows.

As we have already noted, the local domain properties of the total Hamiltonian, which we have considered in this section, will be found to be of considerable value in our treatment of the time evolution of states. We can summarize some of our conclusions qualitatively by the observation that, away from the singularities of the potential, the domains of H_0 and H are locally the same. This fact will influence the asymptotic localization of states in position space, for a particle moving quantum-mechanically in a potential field of force.

5.5 L^1 PERTURBATIONS OF THE FREE HAMILTONIAN

So far, we have defined the total Hamiltonian H to be some self-adjoint extension of the operator $\hat{H} = -\Delta + V(\boldsymbol{r})$, acting on some suitable class of infinitely differentiable functions. If $V(\boldsymbol{r})$ is singular at $\boldsymbol{r}=0$, say $V(\boldsymbol{r}) = |\boldsymbol{r}|^{-3}$ or $V(r, \theta) = \cos\theta/r^2$ in spherical polar coordinates then a suitable domain for $\hat{H}$ is $C_0^\infty(\mathbb{R}^3 \setminus \{0\})$. If, however, $V(\boldsymbol{r})$ is regular at $\boldsymbol{r}=0$ or not *too* singular, say $V(\boldsymbol{r}) = e^{-|\boldsymbol{r}|}/|\boldsymbol{r}|$ which is an $L^2(\mathbb{R}^3)$ potential, we can take $D(\hat{H}) = C_0^\infty(\mathbb{R}^3)$. Singularities of the potential at more general sets of points can be dealt with in a similar way. For example, if $V(\boldsymbol{r})$ is singular on some closed bounded subset $\mathscr{S}$ of $\mathbb{R}^3$, but is square-integrable locally in the complement of $\mathscr{S}$, then a natural way of defining the domain of $\hat{H}$ is to take $D(\hat{H}) = C_0^\infty(\mathbb{R}^3 \setminus \mathscr{S})$.

It is sometimes better to use form extensions, rather than operator extensions, to define the total Hamiltonian. One advantage of using forms is that the domain of the sesquilinear form associated with a self-adjoint operator will in general be larger than the operator domain. (Thus if $T \geq 0$ we have $Q(T) \equiv D(T^{1/2}) \supseteq D(T)$.) In dealing with Hamiltonians defined by forms, we are naturally led to consider potentials $V(\boldsymbol{r})$ that are locally L^1 rather than locally L^2. Note again that $L^1_{\mathrm{loc}}(\mathbb{R}^3) \supseteq L^2_{\mathrm{loc}}(\mathbb{R}^3)$ by the Schwarz inequality.

As we saw in Section 2.7, the semiboundedness property plays a central role in the theory of forms.

When is the Hamiltonian $-\Delta + V(\boldsymbol{r})$ semibounded, as a form? The canonical example of a Hamiltonian that is semibounded is provided by

$-\Delta-\frac{1}{4}|\boldsymbol{r}|^{-2}$. The importance of the potential $V(\boldsymbol{r})=-\frac{1}{4}|\boldsymbol{r}|^{-2}$ is that, with this potential, we are at the boundary between semibounded and not-semibounded. Thus it turns out that $-\Delta+g/|\boldsymbol{r}|^2$ is semibounded for $g\geq-\frac{1}{4}$, and is not bounded below for $g<-\frac{1}{4}$. (Note, however, that, as is true in many areas of mathematics—for example in establishing conditions for the convergence/divergence of infinite series–here the boundary between semibounded and not-semibounded cannot be drawn in any very precise sense. One can always change these borderline cases either way by including powers of log $|\boldsymbol{r}|$, or log log $|\boldsymbol{r}|,\ldots$). We can say, roughly, that a negative potential V will give rise to a Hamiltonian that is bounded below provided that V is no more singular at $\boldsymbol{r}=\mathbf{0}$ than $-\frac{1}{4}|\boldsymbol{r}|^{-2}$. Thus $-\Delta-|\boldsymbol{r}|^{-3}$ is unbounded below, whereas $-\Delta-|\boldsymbol{r}|^{-1}$ (a Coulomb Hamiltonian) *is* bounded below. To justify these results, we turn now to a closer examination of the case $V=-\frac{1}{4}|\boldsymbol{r}|^{-2}$.

The differential expression $-\Delta-\frac{1}{4}|\boldsymbol{r}|^{-2}$ defines a sesquilinear form τ on $C_0^\infty(\mathbb{R}^3\setminus\{0\})\times C_0^\infty(\mathbb{R}^3\setminus\{0\})$, according to the formula

$$\tau(\psi,\phi)=\langle\psi,-\Delta\phi\rangle-\frac{1}{4}\int\frac{\bar{\psi}(\boldsymbol{r})\phi(\boldsymbol{r})}{|\boldsymbol{r}|^2}\,d^3\boldsymbol{r} \tag{5.5.1}$$

We have thus taken $Q(\tau)=C_0(\mathbb{R}^3\setminus\{0\})$. We could alternatively have set $Q(\tau)=C_0^\infty(\mathbb{R}^3)$, since $V\in L^1_{\text{loc}}(\mathbb{R}^3)$ in this case. However, in removing $\boldsymbol{r}=\mathbf{0}$ from the support of ϕ and ψ, we avoid from the start the need to cope with a singularity in the integrand. The basic tool in proving semiboundedness is the identity

$$(\text{grad}\,\phi)^2-\frac{\phi^2}{4|\boldsymbol{r}|^2}=-\frac{1}{2}\,\text{div}\left(\frac{\phi^2\boldsymbol{r}}{|\boldsymbol{r}|^2}\right)+\frac{(\text{grad}\,(|\boldsymbol{r}|^{1/2}\phi))^2}{|\boldsymbol{r}|}. \tag{5.5.2}$$

On the right-hand side of (5.5.2) we have, using elementary vector identities.

$$\begin{aligned}\text{div}\left(\frac{\phi^2\boldsymbol{r}}{r^2}\right)&=\phi^2\,\text{div}\left(\frac{\boldsymbol{r}}{r^2}\right)+\frac{\boldsymbol{r}}{r^2}\cdot\text{grad}\,\phi^2\\&=\phi^2\left(\frac{1}{r^2}\,\text{div}\,\boldsymbol{r}-\frac{\boldsymbol{r}\cdot 2\boldsymbol{r}}{r^4}\right)+\frac{2\phi\boldsymbol{r}\cdot\text{grad}\,\phi}{r^2}\\&=\frac{\phi^2}{r^2}+\frac{2\phi\boldsymbol{r}\cdot\text{grad}\,\phi}{r^2}\end{aligned}$$

and

$$\begin{aligned}&(\text{grad}\,(r^{1/2}\phi))^2\\&=(r^{1/2}\text{grad}\,\phi+\tfrac{1}{2}r^{-3/2}\phi\boldsymbol{r})^2\\&=r(\text{grad}\,\phi)^2+\frac{\phi\boldsymbol{r}\cdot\text{grad}\,\phi}{r}+\frac{\phi^2}{4r}.\end{aligned}$$

Substituting these two expressions into the right-hand side of (5.5.2), the identity is verified. For complex ϕ, (5.5.2) may be applied successively to the real and imaginary parts of ϕ, and we obtain

$$(\operatorname{grad}\bar{\phi})\cdot(\operatorname{grad}\phi)-\frac{|\phi|^2}{4|\boldsymbol{r}|^2} = -\frac{1}{2}\operatorname{div}\left(\frac{|\phi|^2\boldsymbol{r}}{|\boldsymbol{r}|^2}\right)+\frac{|\operatorname{grad}(|\boldsymbol{r}|^{1/2}\phi)|^2}{|\boldsymbol{r}|}. \tag{5.5.2$'$}$$

From (5.3.2) we have, with $\tau(\phi,\phi)$ given by (5.5.1),

$$\tau(\phi,\phi)=\int(\operatorname{grad}\bar{\phi})\cdot(\operatorname{grad}\phi)\,\mathrm{d}^2\boldsymbol{r}-\frac{1}{4}\int\frac{|\phi|^2}{|\boldsymbol{r}|^2}\,\mathrm{d}^3\boldsymbol{r}.$$

Hence we can use (5.5.2)′ to evaluate $\tau(\phi,\phi)$. By the divergence theorem, there is no contribution from the integral of the first term on the right-hand side of (5.5.2)′. (The boundary contribution vanishes because of the support properties of ϕ.) We now have,

$$\tau(\phi,\phi)=\int\frac{|\operatorname{grad}(|\boldsymbol{r}|^{1/2}\phi)|^2}{|\boldsymbol{r}|}\,\mathrm{d}^3\boldsymbol{r}\geq 0$$

and we have proved that the form defined by $-\Delta\ -\frac{1}{4}|\boldsymbol{r}|^{-2}$ is indeed bounded below; in fact $-\Delta-\frac{1}{4}|\boldsymbol{r}|^{-2}\geq 0$. We know, from Lemma 2.11, that τ is a closable form. If τ' is the closure of τ then we can define a self-adjoint operator H such that $Q(\tau')=D(H^{1/2})$, with

$$\tau(\psi,\phi)=\langle H^{1/2}\psi,H^{1/2}\phi\rangle,$$

for all $\phi,\psi\in\mathscr{C}_0^\infty(\mathbb{R}^3\setminus\{0\})$. Moreover, H is an extension of the operator $\hat{H}=-\Delta-\frac{1}{4}|\boldsymbol{r}|^{-2}$, with $D(\hat{H})=C_0^\infty(\mathbb{R}^3\setminus\{0\})$. H is not the only self-adjoint extension of $\hat{H}$, since it turns out that $\hat{H}$ is not essentially self-adjoint. However, looked at from the point of view of sesquilinear forms, H is the most natural extension to take, being defined uniquely in terms of closures. Thus the closure of τ defines a self-adjoint operator, whereas, in this case, the closure of $\hat{H}$ does not.

We can proceed in exactly the same way to define the free Hamiltonian H_0 by means of forms. We start with a form τ_0 defined by

$$\tau_0(\psi,\phi)=-\langle\psi,\Delta\phi\rangle, \tag{5.5.3}$$

with $Q(\tau_0)=C_0^\infty(\mathbb{R}^3\setminus\{0\})$. Clearly τ_0 is semibounded, since $-\Delta\geq 0$. We denote the closure of τ_0 by τ_0'. Then we can define a self-adjoint operator H_0 such that $Q(\tau_0')=D(H_0^{1/2})$, with

$$\tau_0(\psi,\phi)=\langle H_0^{1/2}\psi,H_0^{1/2}\phi\rangle$$

for all $\phi,\ \psi\in C_0^\infty(\mathbb{R}^3\setminus\{0\})$. H_0 is an extension of the operator $\hat{H}_0=-\Delta$, with $D(\hat{H}_0)=C_0^\infty(\mathbb{R}^3\setminus\{0\})$. Since we know that $\hat{H}_0$ is not essentially self-adjoint, it is necessary to check that the self-adjoint operator H_0 obtained in this way is the same as the free Hamiltonian H_0 defined in Section 5.3. To do this, we first note that each component P_k of the momentum operator $\boldsymbol{P}$ is essentially self-adjoint on $C_0^\infty(\mathbb{R}^3\setminus\{0\})$. (For $\phi\in C_0^\infty(\mathbb{R})$ and $\phi(y,z)\in C_0^\infty(\mathbb{R}^2\setminus(0,0))$, we have $\phi(x)\phi(y,z)\in C_0^\infty(\mathbb{R}^3\setminus\{0\})$. We let h be a vector orthogonal to $(P_1+i)\phi(x)\phi(y,z)$ for all such ϕs. Since $-\mathrm{i}\,d/dx$ is essentially self-adjoint on $C_0^\infty(\mathbb{R})$, the range of $\hat{P}_1+\mathrm{i}$, acting on $C_0^\infty(\mathbb{R})$, is dense in $L^2(\mathbb{R})$. Also, the functions $\phi(y,z)$ are dense in $L^2(\mathbb{R}^2)$. Since $L^2(\mathbb{R}^3)$ is the tensor product of $L^2(\mathbb{R})$ with $L^2(\mathbb{R}^2)$, the linear span of $\{f(x)f(y,z);\ f(x)\in L^2(\mathbb{R}),\ f(y,z)\in L^2(\mathbb{R}^2)\}$ is dense in $L^2(\mathbb{R}^3)$. We have therefore shown that the range of $P_1+\mathrm{i}$, acting on $C_0^\infty(\mathbb{R}^3\setminus\{0\})$ is everywhere dense, so that $h=0$. Similarly for $P_1-\mathrm{i}$, and for the other components P_2 and P_3 of the momentum operator. It follows that each component of $\boldsymbol{P}$ has deficiency indices $(0,0)$, and is essentially self-adjoint.

For $f\in D(H_0^{1/2})$, we have

$$\|H_0^{1/2}f\|^2=\sum_{k=1}^{3}\|P_kf\|^2, \tag{5.5.4}$$

so that, for each k, we have

$$\|P_kf\|\le\|H_0^{1/2}f\|. \tag{5.5.5}$$

Using (5.5.4) and (5.5.5), it is straightforward to show that, for a given sequence $\{f_n\}$ of elements of $L^2(\mathbb{R}^3)$, $H_0^{1/2}f_n$ converges strongly to a limit (i.e. $\{H_0^{1/2}f_n\}$ is a Cauchy sequence) if and only if P_kf_n converges strongly to a limit for each k. In particular, if we start by defining $H_0^{1/2}$ on $C_0^\infty(\mathbb{R}^3\setminus\{0\})$ then the closure of $H_0^{1/2}$ will have as its domain the intersection of the domains of each of the components of the (self-adjoint) momentum operator. This characterization of the domain of $H_0^{1/2}$ is independent of whether we define H_0 by forms, or by operators as in Section 5.3. From (2.7.17), the form extension H_0 will have

$$D(H_0)=D(H_0^{1/2})\cap D(\hat{H}_0^*), \tag{5.5.6}$$

where $D(\hat{H}_0)=C_0^\infty(\mathbb{R}^3\setminus\{0\})$. On the other hand, for the operator extension considered previously, we have

$$D(H_0)\subseteq D(H_0^{1/2})\cap D(\hat{H}_0^*).$$

Since the form version of H_0 cannot be a proper extension of the operator version, as both are self-adjoint, it follows that (5.5.6) must hold in both cases. Hence the self-adjoint operator H_0, obtained from the form closure with $Q(\tau)=C_0^\infty(\mathbb{R}^3\setminus\{0\})$, is indeed the free Hamiltonian previously defined.

The same argument would lead to the same result if we had started with $Q(\tau)=C_0^\infty(\mathbb{R}^3)$. However, when dealing with singular potentials, it is often convenient to start from the same form domain for the free Hamiltonian as for the total Hamiltonian.

The following result tells us that, for potentials that are locally L^2 in $\mathbb{R}^3$, semiboundedness is a sufficient condition for essential self-adjointness of the total Hamiltonian.

Lemma 5.6

Suppose that $V\in L^2_{\text{loc}}(\mathbb{R}^3)$, and that $-\Delta+V$, defined as a form with $Q(\tau)=C_0^\infty(\mathbb{R}^3)$, is bounded below. Then $\hat{H}=-\Delta+V$, defined as an *operator* with $D(\hat{H})=C_0^\infty(\mathbb{R}^3)$, is essentially self-adjoint.

Proof

We let H be any self-adjoint extension of $\hat{H}$, and let $\rho=\rho(|\mathbf{r}|)\in C_0^\infty(\mathbb{R}^3)$. We know from Section 5.4 that both ρf and $\rho^2 f$ are in the domain of H provided that f is. From (5.3.4), with $H\rho^2 f=H_0\rho^2 f+V\rho^2 f$, we have, for $f\in D(H)$, and ρ real,

$$\begin{aligned}\langle\rho^2 f,(H+c)f\rangle&=\langle(H+c)\rho^2 f, f\rangle\\&=\langle\rho^2(H+c)f, f\rangle-\langle(\Delta\rho^2)f, f\rangle\\&\quad-2\sum_k\left\langle\left(\frac{\partial}{\partial x_k}\rho^2\right)\frac{\partial f}{\partial x_k}, f\right\rangle.\end{aligned}\tag{5.5.7}$$

Similarly,

$$\begin{aligned}\langle\rho f,(H+c)\rho f\rangle&=\langle\rho f,\rho(H+c)f\rangle-\langle\rho f,(\Delta\rho)f\rangle-2\sum_k\left\langle\rho f,\frac{\partial\rho}{\partial x_k}\frac{\partial f}{\partial x_k}\right\rangle\\&=\langle\rho f,\rho(H+c)f\rangle-\langle\rho f,(\Delta\rho)f\rangle-\tfrac{1}{2}\overline{\langle\rho^2(H+c)f, f\rangle}\\&\quad+\tfrac{1}{2}\overline{\langle(\Delta\rho^2)f, f\rangle}+\tfrac{1}{2}\overline{\langle\rho^2 f,(H+c)f\rangle},\end{aligned}\tag{5.5.8}$$

on substituting from (5.5.7). This gives

$$\begin{aligned}\langle\rho f,(H+c)\rho f\rangle&=\tfrac{1}{2}(\langle\rho^2 f,(H+c)f\rangle+\langle(H+c)f,\rho^2 f\rangle)\\&\quad-\langle\rho f,(\Delta\rho)f\rangle+\tfrac{1}{2}\langle f,(\Delta\rho^2)f\rangle.\end{aligned}\tag{5.5.9}$$

Moreover, since $\rho=\rho(r)$, we have

$$\tfrac{1}{2}\Delta\rho^2-\rho\Delta\rho=\frac{1}{2r^2}\frac{\mathrm{d}}{\mathrm{d}r}\left(2r^2\rho\frac{\mathrm{d}\rho}{\mathrm{d}r}\right)-\frac{\rho}{r^2}\frac{\mathrm{d}}{\mathrm{d}r}\left(r^2\frac{\mathrm{d}\rho}{\mathrm{d}r}\right)$$
$$=\left(\frac{\mathrm{d}\rho}{\mathrm{d}r}\right)^2,$$

so that (5.5.9), for $f\in D(H)$, leads to the identity

$$\langle\rho f,(H+c)\rho f\rangle=\mathrm{Re}\,(\langle\rho^2 f,(H+c)f\rangle)+\left\langle f,\left(\frac{\mathrm{d}\rho}{\mathrm{d}r}\right)^2 f\right\rangle. \qquad (5.5.10)$$

We now let $\rho(r)\equiv 1$ for $r\leq 1$ and $\rho(r)\equiv 0$ for $r\geq R$, say, and define a sequence $\{\rho_n\}$ by $\rho_n(r)=\rho(r/n)$. Replacing ρ in (5.5.10) by $\rho_m-\rho_n$, we take the limit as m, $n\to\infty$. It is straightforward to verify, for example by the Lebesgue Dominated-Convergence Theorem, that both $(\rho_m-\rho_n)^2 f$ and $[(\mathrm{d}/\mathrm{d}r)(\rho_m-\rho_n)]^2 f$ converge strongly to zero. Hence from (5.5.10) we have

$$\lim_{m,n\to\infty}\langle(\rho_m-\rho_n)f,(H+c)(\rho_m-\rho_n)f\rangle=0, \qquad (5.5.11)$$

where c may be chosen such that $H+c\geq 0$. We let H_τ denote the Friedrichs extension of $\hat{H}$, defined in the proof of Lemma 2.14. Any self-adjoint extension must contain $(\rho_m-\rho_n)f$ in its domain, since such an operator is a restriction of $\hat{H}^*$. Moreover, we can apply the same arguments as before to obtain (5.5.11), with H replaced by H_τ. We may conclude that both $(H+c)^{1/2}\rho_n f$ and $(H_\tau+c)^{1/2}\rho_n f$ converge strongly to a limit. Since $\rho_n f\in D(H)\cap D(H_\tau)$ and $(H+c)\rho_n f=(H_\tau+c)\rho_n f=(H_0+V+c)\rho_n f$, we may use the fact that $(H+c)^{1/2}$ and $(H_\tau+c)^{1/2}$ are closed operators to deduce that, in the limit $n\to\infty$ and with $\rho=\rho_n$ in (5.5.10),

$$\langle(H+c)^{1/2}f,(H+c)^{1/2}f\rangle=\langle(H_\tau+c)^{1/2}f,(H_\tau+c)^{1/2}f\rangle \qquad (5.5.12)$$

for all $f\in D(H)$. Now any vector f in $D((H+c)^{1/2})$ is a strong limit of a sequence $\{f_n\}$ such that $f_n\in D(H)$ and $(H+c)^{1/2}f_n$ converges strongly to $(H+c)^{1/2}f$. For example, f_n could be obtained from f by applying the spectral projection of H for a finite interval $(-n,n]$. For such a sequence $\{f_n\}$, (5.5.12) implies that $(H_\tau+c)^{1/2}(f_m-f_n)\to 0$ strongly. Hence $(H_\tau+c)^{1/2}f_n$ is a Cauchy sequence. Since $(H_\tau+c)^{1/2}$ is a closed operator, we have $\text{s-lim}_{n\to\infty}(H_\tau+c)^{1/2}f_n=(H_\tau+c)^{1/2}f$, so that, setting $f=f_n$ in (5.5.12) and proceeding to a limit we find that (5.5.12) holds for all $f\in D((H+c)^{1/2})$. In particular, we have

$$D((H+c)^{1/2})\subseteq D((H_\tau+c)^{1/2}).$$

However, from the construction of H_τ in terms of the form closure of τ with $Q(\tau)=C_0^\infty(\mathbb{R}^3)$, we see also that

$$D((H_\tau+c)^{1/2})\subseteq D((H+c)^{1/2}).$$

Hence $(H_\tau+c)^{1/2}$ and $(H+c)^{1/2}$ have the same domain, and (5.5.12) holds for all vectors in this common domain. If we now follow the same argument as succeeded (2.7.22), we may deduce that $H=H_\tau$.

Since $\hat{H}$ has only a single self-adjoint extension, it follows that $\hat{H}$ is essentially self-adjoint. ∎

The following two results, while allowing $V(\mathbf{r})$ to be locally singular, require somewhat stronger conditions than the semiboundedness of H. Under these conditions, we are able to obtain some rather useful properties of the domain of H_τ.

Lemma 5.7

Let τ be the sesquilinear form defined by $-\Delta+V$, with $Q(\tau)=C_0^\infty(\mathbb{R}^3\setminus\{0\})$, and let H_τ be the self-adjoint operator defined as in the proof of Lemma 2.14. Suppose further that $-\Delta+gV$, defined as a form with form domain $C_0^\infty(\mathbb{R}^3\setminus\{0\})$, is semibounded for some $g>1$. Then $Q(H_\tau)\subseteq Q(H_0)$, where $Q(T)$ denotes the domain of $(T+c)^{1/2}$. If, in addition, $V\leq 0$ then $Q(H_\tau)=Q(H_0)$.

Proof

We suppose $-\Delta+gV\geq -c$, as a form on $C_0^\infty(\mathbb{R}^3\setminus\{0\})$. Given any $f\in Q(H_\tau)$, there will be a sequence $\{\phi_n\}$, with $\phi_n\to f$ strongly, and

$$\langle\phi_m-\phi_n,(-\Delta+V)(\phi_m-\phi_n)\rangle\to 0 \quad \text{as } m, n\to\infty. \tag{5.5.13}$$

(Strictly, here $\langle\phi_m-\phi_n, V(\phi_m-\phi_n)\rangle$ means $\int d^3\mathbf{r}V(\mathbf{r})|\phi_m(\mathbf{r})-\phi_n(\mathbf{r})|^2$, since we assume only $V\in L^1_{\text{loc}}(\mathbb{R}^3\setminus\{0\})$ rather than that V is locally square-integrable in general.) Writing

$$-\Delta+V=\frac{-\Delta+gV}{g}-\frac{(g-1)\Delta}{g},$$

$$-\Delta+V+\frac{c}{g}\geq -\frac{(g-1)\Delta}{g}$$

From (5.5.13), we now have,

$$\lim_{m,n\to\infty} \langle \phi_m - \phi_n, -\Delta(\phi_m - \phi_n\rangle$$

$$\leq g(g-1)^{-1} \lim_{m,n\to\infty} \left\langle \phi_m - \phi_n, \left(-\Delta + V + \frac{c}{g}\right)(\phi_m - \phi_n)\right\rangle,$$

so that the first limit also converges to zero. It follows that $f \in Q(H_0)$, and we have shown that $Q(H_\tau) \subseteq Q(H_0)$.

We now suppose that $V \leq 0$. Then $-V \leq -\Delta/g + c/g$, and the argument used above shows that $Q(H_0) \subseteq Q(-V)$. Clearly, then, $Q(H_0) \subseteq Q(-\Delta + V)$, so that $Q(H_0) \subseteq Q(H_\tau)$ in this case.

We therefore have $Q(H_\tau) = Q(H_0)$, provided that $V \leq 0$, and we should also note the useful result $Q(H_\tau) \subseteq D(|V|^{1/2})$ that then holds. A similar result is obtained, under slightly different conditions, for potentials that may be of arbitrary sign. Thus if we suppose that $-\Delta + gV_- \geq -c$, for some $g > 1$, where $V_\pm(\boldsymbol{r})$ are defined by

$$V = \begin{cases} V_+(\boldsymbol{r}) & \text{wherever } V(\boldsymbol{r}) \geq 0, \\ V_-(\boldsymbol{r}) & \text{wherever } V(\boldsymbol{r}) < 0, \end{cases}$$

then the above arguments may be used to justify the results $Q(H_\tau) \subseteq Q(H_0)$ and $Q(H_\tau) \subseteq Q(V_\pm)$. This suggests, in particular, that if V_+ is highly singular at $\boldsymbol{r} = 0$ then vectors belonging to $Q(H_\tau)$ must have a zero of high order at the origin. ■

We showed at the start of this section that $-\Delta - \frac{1}{4}|\boldsymbol{r}|^{-2} \geq 0$. Hence $r^{-2} \leq -4\Delta$, and we may deduce that $Q(H_0) \subseteq Q(|\boldsymbol{r}|^{-2})$. In other words, $D(H_0^{1/2}) \subseteq D(|\boldsymbol{r}|^{-1})$, which is a result widely used in applications.

The following result, a corollary to the last lemma, shows that, as for operator extensions, form extensions of Hamiltonian operators are local, in a sense to be made precise.

Corollary

Under the same hypothesis as in the lemma, with $V \leq 0$, $f \in Q(H_\tau) \Rightarrow \rho f \in Q(H_\tau)$, for any C^∞ bounded function $\rho(\boldsymbol{r})$ having bounded first derivatives.

Proof

In the notation of the proof of the lemma, it is only necessary to show that

$$\lim_{m,n\to\infty} \langle \rho(\phi_m - \phi_n), -\Delta\rho(\phi_m - \phi_n)\rangle = 0. \tag{5.5.14}$$

since $f \in Q(-V) \Rightarrow \rho f \in Q(-V)$. The left-hand side of (5.5.14) is $\Sigma_{k=1}^{3} \| P_k \rho(\phi_m - \phi_n) \|^2$, where $P_k \rho(\phi_m - \phi_n) = \rho P_k(\phi_m - \phi_n) - \mathrm{i}(\partial \rho / \partial x_k)(\phi_m - \phi_n)$. Equation (5.5.14) now follows, since both $P_k(\phi_m - \phi_n)$ and $(\partial \rho / \partial x_k)(\phi_m - \phi_n)$ converge strongly to zero. ■

Remark

The conclusion of the corollary again follows, for non-negative potentials, under the alternative hypothesis $-\Delta + gV_- \geq -c$, for some $g > 1$. One can also deduce the standard condition for a local operator, namely

$$f \in D(H_\tau) \Rightarrow \rho f \in D(H_\tau),$$

where the C^∞ bounded function is required only to have bounded first and second derivatives.

Chapter 6

The Schrödinger Operator

6.1 THE MINIMAL OPERATOR AND ITS ADJOINT

This chapter will deal entirely with differential operators of the form $-d^2/dr^2+V(r)$, acting on functions belonging to the Hilbert space $L^2(I)$, where I is an interval (a, b). The case of an infinite or semi-infinite interval will be included, by taking $a=-\infty$ or $b=+\infty$ as appropriate. We shall always suppose V to be real.

The case $I=\mathbb{R}$ applies to the one-dimensional motion of a quantum-mechanical particle in a potential V (units $\hbar=2m=1$.) With $I=(0, \infty)$, we may absorb the centrifugal term $l(l+1)/r^2$ into the definition of V, in order to treat (in a single partial-wave subspace) three-dimensional motion in a spherical potential $V=V(|\mathbf{r}|)$. These two applications alone would justify a study of the scattering and spectral theory of ordinary differential operators. Moreover, there is no doubt that many of the most fruitful and central ideas of the general theory find their first and most concrete expression in the treatment of ordinary Schrödinger operators.

Two key words will be *domain* and *spectrum*. It is clear that no differential operator will be defined on the whole of $\mathscr{H}=L^2(I)$, since "most" functions in this space come nowhere near to being differentiable. Nevertheless, a dense subset of $\mathscr{H}$ will lie in the domain of $-d^2/dr^2+V$, and the question is how to extend the operator in such a way that $H=-d^2/dr^2+V$ becomes a *self-adjoint* operator in $\mathscr{H}$. This leads to the imposition of *boundary conditions* at the endpoints a and b of the interval, and H will be self-adjoint subject to these boundary conditions. It may be that $-d^2/dr^2+V$, defined on the original dense subset, is already essentially self-adjoint. In this case no boundary conditions are necessary.

Having defined a self-adjoint operator H, and only then, we are able to carry out a spectral analysis. A complete analysis can be achieved by representing H as a multiplication operator in some L^2 space, and we shall be able to carry out this representation rather explicitly. The L^2 space is with respect to some measure. Since the spectrum of the second-order differential operator is at most doubly degenerate, it turns out that we arrive at a set of measures $d\rho_{ij}$, where each of i and j can take on two possible values. Again we shall be able to construct ρ_{ij}, called the *spectral matrix* of the differential operator. A knowledge of the spectral matrix, which reduces in the non-degenerate case to a *spectral function*, is equivalent to a complete spectral analysis of H.

We shall avoid, in the first instance, imposing special conditions on the potential that simplify the analysis. In doing so the value of this investigation would be greatly reduced, since it will be part of our purpose to exhibit domain and spectral properties in the greatest possible generality in order to show what is common to virtually *all* potentials. If our task is ultimately, say, to classify potentials with respect to the resulting spectral properties of H then we should not wish to start by removing a considerable set of potentials from this classification. Therefore we shall assume that V is reasonably regular in the *interior* of the interval I, while allowing quite arbitrary behaviour at each of the endpoints a and b.

More precisely, we shall assume only that V is locally square-integrable in I. (The use of the word *local* will be taken as usual to refer to the *interior* of the interval I. In other words, V is assumed to be square-integrable over any closed subinterval of the open interval I; V will not, in general, be square integrable across the interval I itself.) We shall write

$$V \in L^2_{\mathrm{loc}}(I).$$

This is the minimal condition necessary to guarantee that $-\mathrm{d}^2/\mathrm{d}r^2 + V$ will be defined as a linear operator on $C_0^\infty(I)$, the set of infinitely differentiable functions vanishing identically in a neighbourhood of $r=a$ and $r=b$. (The support of such functions will be closed bounded subsets of I.) Two examples are

$$-\frac{\mathrm{d}^2}{\mathrm{d}r^2} + \exp(r^2) \quad \text{in } L^2(\mathbb{R})$$

and

$$-\frac{\mathrm{d}^2}{\mathrm{d}r^2} - \frac{1}{r^4} \quad \text{in } L^2(0, 1).$$

We should like to reserve the symbol H for fully fledged self-adjoint operators in the Hilbert space $\mathscr{H} = L^2(I)$, so let us denote by $\hat{H}$ the differential operator acting on functions in $C_0^\infty(I)$. We should first check that $C_0^\infty(I)$ is dense in $\mathscr{H}$. To do this, it will be sufficient to verify that

$$\langle \phi, f \rangle = 0 \text{ for all } \phi \in C_0^\infty(I) \text{ implies } f = 0.$$

Now, if $\int_a^b \bar{\phi} f \,\mathrm{d}r = 0$ then it is not difficult to find a sequence $\{\phi_n\}$ of functions in $C_0^\infty(I)$ such that, for given ξ, $\eta \in I$,

$$\begin{aligned} \phi_n(r) &\to 1 \quad (\xi \leq r \leq \eta) \\ &\to 0 \quad \text{otherwise,} \end{aligned}$$

and for which

$$\lim_{n\to\infty} \int_a^b \bar{\phi}_n(r) f(r)\,\mathrm{d}r = \int_\xi^\eta f(r)\,\mathrm{d}r.$$

Since the integral on the left-hand side is always zero by assumption, we then have $\int_\xi^\eta f(r)\,\mathrm{d}r = 0$ for arbitrary ξ, $\eta \in I$, from which it follows by differentiation that $f = 0$. Hence $C_0^\infty(I)$ is indeed dense in $\mathscr{H}$, and setting

$$\hat{H}\phi = -\frac{\mathrm{d}^2\phi}{\mathrm{d}r^2} + V\phi, \qquad \phi \in C_0^\infty(I),$$

it is a simple matter to verify (integration by parts) that $\hat{H}$ is symmetric with this domain.

Definition 6.1

$$\hat{H} = -\frac{\mathrm{d}^2}{\mathrm{d}r^2} + V, \qquad \text{with } D(\hat{H}) = C_0^\infty(I). \tag{6.1.1}$$

Here $C_0^\infty(I)$ may be regarded as a *minimal domain*, in the sense that it would not be appropriate or justifiable to refer to the differential operator $-\mathrm{d}^2/\mathrm{d}r^2 + V$ at all if we did not include at least these functions in the domain. We also know that there is a maximal domain as well, in the sense that any self-adjoint extension H of $\hat{H}$ cannot have a larger domain than $D(\hat{H}^*)$. In fact H must be a restriction of $\hat{H}^*$.

As a first step to defining H, then, let us look at $\hat{H}^*$, the adjoint of $\hat{H}$. A vector f of $\mathscr{H}$ will lie in the domain of $\hat{H}^*$ if and only if

$$\langle \hat{H}\phi, f\rangle = \langle -\phi'' + V\phi, f\rangle = \langle \phi, f^*\rangle \tag{6.1.2}$$

for some $f^* \in L^2(I)$ and for all $\phi \in D(\hat{H})$. In this case, $\hat{H}^* f = f^*$. Since $f \in L^2(I)$ and $V \in L^2_{\mathrm{loc}}(I)$, we have, on integrating over any closed subinterval of I,

$$\left(\int |Vf|\,\mathrm{d}r\right)^2 \leq \left(\int |V|^2\,\mathrm{d}r\right)\left(\int |f|^2\,\mathrm{d}r\right) < \infty,$$

so that $Vf \in L^1_{\mathrm{loc}}(I)$. Moreover, $f^* \in L^2(I) \subseteq L^1_{\mathrm{loc}}(I)$, so that we can define, for any $\xi \in I$, the function h by

$$h(r) = -\int_{\xi}^{r} \mathrm{d}s \left\{ \int_{\xi}^{s} \mathrm{d}t (f^*(t) - V(t) f(t)) \right\}.$$

Both h and $\mathrm{d}h/\mathrm{d}r$ are locally absolutely continuous in I, and we have

$$-\frac{\mathrm{d}^2 h}{\mathrm{d}r^2} = f^* - Vf \tag{6.1.3}$$

for almost all r in I.

From (6.1.2), on integrating by parts, we have

$$\begin{aligned} \langle \phi'', f \rangle &= -\langle \phi, f^* - Vf \rangle = \langle \phi, h'' \rangle \\ &= \langle \phi'', h \rangle = \int_I \bar{\phi}'' h \, \mathrm{d}r. \end{aligned} \tag{6.1.4}$$

(We have taken the liberty here of writing down inner products with elements h *and* h'' that are not necessarily in $\mathscr{H}$, but the use of this notation should be transparent. Note that there are no boundary contributions since $\phi \equiv 0$ near the boundary to I.)

For $\phi \in C_0^\infty(I)$, we define an equivalence relation $\phi_1 \sim \phi_2$ (ϕ_1 equivalent to ϕ_2) by $\phi_1 \sim \phi_2$ whenever there exists $\psi \in C_0^\infty(I)$ such that $\phi_1 - \phi_2 = \psi''$. Clearly, if $\phi_1 \sim \phi_2$ then

$$\int_a^b (\phi_1 - \phi_2) \, \mathrm{d}r = [\psi']_a^b = 0,$$

and

$$\int_a^b r(\phi_1 - \phi_2) \, \mathrm{d}r = [r\psi' - \psi]_a^b = 0,$$

so that in this case

$$\left. \begin{aligned} \int_a^b \phi_1 \, \mathrm{d}r &= \int_a^b \phi_2 \, \mathrm{d}r, \\ \int_a^b r\phi_1 \, \mathrm{d}r &= \int_a^b r\phi_2 \, \mathrm{d}r. \end{aligned} \right\} \tag{6.1.5}$$

Conversely, if we suppose that (6.1.5) holds, and define

$$\psi(r)=\int_a^r \mathrm{d}s\left\{\int_a^s \mathrm{d}t\,(\phi_1(t)-\phi_2(t))\right\},$$

then ψ is identically zero near $r=a$, and near $r=b$ we have, on integrating by parts,

$$\psi(r)=-\int_a^r \mathrm{d}s\,(\phi_1(s)-\phi_2(s))s$$

$$=-\int_a^b \mathrm{d}s\,(\phi_1(s)-\phi_2(s))s,$$

so that $\psi\equiv 0$ there also. Hence $\psi\in C_0^\infty(I)$, and also $\phi_1-\phi_2=\psi''$, so that we can write $\phi_1\sim\phi_2$.

It follows that (6.1.5) is both necessary and sufficient for $\phi_1\sim\phi_2$. Since (6.1.4) holds with ϕ replaced by ψ, we also have

$$\langle\psi'',(f-h)\rangle=\langle\phi_1-\phi_2,(f-h)\rangle=0$$

whenever $\phi_1\sim\phi_2$. So two functions ϕ_1 and ϕ_2 in the same equivalence class have the same value of $\langle\phi,f-h\rangle$. This means that $\langle\phi,f-h\rangle$, for $\phi\in C_0^\infty(I)$, depends solely on the equivalence class to which ϕ belongs, which, by (6.1.5), is determined solely by the values of the integrals $\int_a^b\phi\,\mathrm{d}r$ and $\int_a^b r\phi\,\mathrm{d}r$. Moreover, $\overline{\langle\phi,(f-h)\rangle}$ depends *linearly* on ϕ, so we can only have, for some A and B depending on $f-h$,

$$\langle\phi,f-h\rangle=A\int_a^b\bar{\phi}\,\mathrm{d}r+B\int_a^b r\bar{\phi}\,\mathrm{d}r.$$

Thus $\langle\phi,f-h-A-Br\rangle=0$ for all $\phi\in C_0^\infty(I)$. From the proof that $C_0^\infty(I)$ is dense in $\mathscr{H}$, this allows us to assert that $f-h-A-Br=0$. Note that this does not require the *a priori* estimate that $f-h-A-Br\in\mathscr{H}=L^2(I)$, but rather in the proof we needed only to use $f-h-A-Br\in L^1_{\text{loc}}(I)$, which is certainly satisfied in this case. So we now have

$$f(r)=h(r)+A+Br. \tag{6.1.6}$$

From the known properties of h, together with (6.1.3), (6.1.6) tells us that if $f\in D(\hat{H}^*)$ and $\hat{H}^*f=f^*$ then both f and $\mathrm{d}f/\mathrm{d}r$ are locally absolutely continuous in I, and $-\mathrm{d}^2f/\mathrm{d}r^2+Vf=-\mathrm{d}^2h/\mathrm{d}r^2+Vf=f^*$ almost everywhere in I. Conversely, it is easy to verify (integration by parts) that for any $f \in \mathscr{H}$ such that f and $\mathrm{d}f/\mathrm{d}r$ are locally absolutely continuous in I, and such that $f^* = -\mathrm{d}^2 f/\mathrm{d}r^2 + Vf \in L^2(I)$, (6.1.2) holds, implying that $f \in D(\hat{H}^*)$ and $\hat{H}^* f = f^*$. All of this may be stated formally as follows.

Theorem 6.1

For $V \in L^2_{\mathrm{loc}}(I)$, define $\hat{H}$ by (6.1.1), and let f belong to $\mathscr{H} = L^2(I)$. Then $f \in D(\hat{H}^*)$ if and only if

(i) f and $\mathrm{d}f/\mathrm{d}r$ are locally absolutely continuous in I, and
(ii) $\hat{H}^* f = -\mathrm{d}^2 f/\mathrm{d}r^2 + Vf \in L^2(I)$.

Theorem 6.1 specifies a maximal domain for any self-adjoint extension of H. It may happen that $\hat{H}^*$ is itself self-adjoint. In this case $\hat{H}^* = (\hat{H}^*)^*$, which is just the closure of $\hat{H}$, and $\hat{H}^*$ is then the only self-adjoint extension of $\hat{H}$. $\hat{H}$, defined on $D(\hat{H}) = C_0^\infty(I)$ is then essentially self-adjoint. In all other cases the domain of any self-adjoint extension H of $\hat{H}$ will be both strictly larger than $D(\hat{H})$ and strictly smaller than $D(\hat{H}^*)$. In fact, $\hat{H} \subset H \subset \hat{H}^*$. We can see from (ii) above that in all cases Hf will be obtained, as we should expect, by direct application of the differential operator $-\mathrm{d}^2/\mathrm{d}r^2 + V$ to the wave function $f(r)$. That part (i) of the specification of the domain of $\hat{H}^*$ can by no means be ignored will be made clearer by the following considerations.

We let $f(r)$ be any function twice differentiable in I, except possibly at a finite number of exceptional points. Then $-\mathrm{d}^2 f/\mathrm{d}r^2 + Vf$ is defined, except at these points, and we suppose that $g^* \equiv -\mathrm{d}^2 f/\mathrm{d}r^2 + Vf \in L^2(I)$. Then absolute continuity of f and $\mathrm{d}f/\mathrm{d}r$ means that both f and $\mathrm{d}f/\mathrm{d}r$ should be continuous at these exceptional points. It is possible to infer that $f \in D(\hat{H}^*)$ only if the continuity of f and $\mathrm{d}f/\mathrm{d}r$ at these points has first of all been verified. It is not, however, necessary to have continuity at the endpoints of the interval. The equations of continuity of f and $\mathrm{d}f/\mathrm{d}r$ will, no doubt, be already familiar to the reader. They are to be found, for example, in the standard treatment of the square-well potential. Here we see, from a more fundamental point of view, that they reflect a regularity property of functions in the domain of $\hat{H}^*$, and hence of functions in the domain of any self-adjoint extension of the minimal differential operator $\hat{H}$.

Example 1

We take $\hat{H} = -\mathrm{d}^2/\mathrm{d}r^2$ in $L^2(0, \infty)$, with $D(\hat{H}) = C_0^\infty(0, \infty)$, and define $f(r) = \exp[-(1+\mathrm{i})r]$. Then $f \in D(\hat{H}^*)$ and $\hat{H}^* f = -2\mathrm{i}f$. This shows us that $\hat{H}$ in this case cannot be essentially self-adjoint, since $\hat{H}^*$ would then be self-adjoint and could not have complex eigenvalue $-2\mathrm{i}$.

Example 2

We take $\hat{H}$ as above, and define

$$f(r)=\begin{cases}\exp[-(r-1)] & (r>1),\\ 2-r & (0<r<1).\end{cases}$$

Then f and df/dr are both continuous at $r=1$. So $f\in D(\hat{H}^*)$, with

$$\hat{H}^*f=\begin{cases}-\exp[-(r-1)] & (r>1)\\ 0 & (0<r<1).\end{cases}$$

Example 3

We take $\hat{H}=-d^2/dr^2-1/4r^2$ in $L^2(0,\infty)$, with $D(\hat{H})=C_0^\infty(0,\infty)$, and define $f(r)=r^{1/2}(1+r^2)^{-1}$. Then $f\in D(\hat{H}^*)$. In this example, df/dr is unbounded near $r=0$, and neither d^2f/dr^2 nor $f/4r^2$ belong to the Hilbert space. Nevertheless $\hat{H}^*f\in L^2(0,\infty)$ because of cancellation between the two contributions.

6.2 LIMIT POINT AND LIMIT CIRCLE

A spectral analysis of the Hamiltonian H will be closely tied up with the study of solutions of the time-independent Schrödinger equation at complex energy z. If H has a purely discrete spectrum then it is already clear that we are interested in solutions with $\text{Im}(z)=0$, since these (normalizable) solutions will be the eigenfunctions of the problem. It is also not unreasonable to expect that even the non-normalizable solutions will play a crucial role in describing the continuous part of the spectrum, as z runs over the support of the continuous spectrum.

But why complex z? One way of understanding the need to have complex z is to consider the resolvent operator $R(z)=(H-z)^{-1}$, which is not defined, in general, for real values of z. The importance of this operator in spectral and domain analysis is twofold. First, $R(z)$ gives information on the spectrum of H (complement of the set of z for which $R(z)$ is bounded). And, secondly, the domain of H is exactly the range of $R(z)$ for $\text{Im}(z)\neq 0$, so that the resolvent allows us to characterize the domain of H in a most convenient way.

Now if $g=R(z)f$, we have

$$(H-z)g=f.$$

If H is an extension of the operator $\hat{H}$ defined in (6.1.1) then Theorem 6.1 (ii) allows us to infer that g satisfies the inhomogenous differential equation

$$-\frac{d^2 g}{dr^2}+Vg-zg=f.$$

The method of variation of constants may then be used to solve for g in terms of f, and the solution will depend on the solutions of the corresponding homogeneous differential equation. This equation is precisely the time-independent Schrödinger equation at complex energy z, and in fact two independent solutions of this equation form the basis of the construction of the kernel of $R(z)$ as an integral operator (see Section 6.4). Hence solutions of the time-independent Schrödinger equation play a central role. We can approach the resolvent operator through these solutions, and are thereby brought closer to a more detailed understanding of the Hamiltonian H.

We fix some $\xi \in (a,b)$, and define two solutions u_1 and u_2 of the differential equation

$$-\frac{d^2 u}{dr^2}+Vu=zu, \tag{6.2.1}$$

subject to the initial conditions

$$\left.\begin{aligned} u_1(\xi)&=0, \qquad u_1'(\xi)=1,\\ u_2(\xi)&=1, \qquad u_2'(\xi)=0. \end{aligned}\right\} \tag{6.2.2}$$

u_1 and u_2 will of course depend on z, but we shall not yet need to make this dependence explicit.

We define the *Wronskian* of two functions v and w by

$$W(v,w)=vw'-v'w, \tag{6.2.3}$$

where $v'\equiv dv/dr$ and $w'\equiv dw/dr$. The Wronskian is a function of r, and we shall denote by W_η the value of the Wronskian at $r=\eta$. The Wronskian has the following properties.

(i)
$$\int_\xi^\eta [-\bar{f}''+V\bar{f})g+\bar{f}(g''-Vg)]\,dr$$
$$=[W(\bar{f},g)]_\xi^\eta \equiv W_\eta(\bar{f},g)-W_\xi(\bar{f},g). \tag{6.2.4}$$

A special case of (i), applying whenever two functions u and v each satisfy (6.2.1), is

(ii)
$$2i\,\mathrm{Im}\,(z)\int_\xi^\eta \bar{v}u\,dr=[W(u,\bar{v})]_\xi^\eta \tag{6.2.5}$$

(set $f=\bar{u}, g=\bar{v}$ in (i)).

Again for any two solutions of (6.2.1), we have

(iii) $$W(u,v)=\text{const} \qquad (6.2.6)$$

(set $f=\bar{u}$, $g=v$ in (i)).

A *simple boundary condition* at $r=\eta$ is an equation of the form

$$(\alpha u+\beta u')|_{r=\eta}=0, \qquad (6.2.7)$$

where α and β are real numbers not both zero. Now the general solution of (6.2.1) will be a constant multiple of u_2+mu_1 (except for solutions proportional to u_1, which may be regarded as corresponding to $m=\infty$). A key question is: for which (complex) values of m does u_2+mu_1 satisfy a simple boundary condition at $r=\eta$, for some α and β? We take here Im $(z)\neq 0$, and $\eta>\xi$. In fact one can characterize geometrically the possible values of m. For a simple boundary condition to hold at $r=\eta$, it requires only that the ratios $u{:}u'$ and $\bar{u}{:}\bar{u}'$ should agree at this point. This leads to the equation

$$W_\eta(u_2+mu_1,\bar{u}_2+\bar{m}\bar{u}_1)=0. \qquad (6.2.8)$$

We now define the real numbers A, B, C and D by

$$W_\eta(u_1,\bar{u}_1)=\mathrm{i}A,$$
$$W_\eta(u_2,\bar{u}_2)=\mathrm{i}D$$

and

$$W_\eta(u_1,\bar{u}_2)=\mathrm{i}(B-\mathrm{i}C),$$
$$W_\eta(u_2,\bar{u}_1)=\mathrm{i}(B+\mathrm{i}C).$$

To interpret (6.2.8) geometrically in the complex plane, we set $m=x+\mathrm{i}y(x,y$ real), to obtain

$$A(x^2+y^2)+2Bx+2Cy+D=0, \qquad (6.2.9)$$

which is the equation of a circle. We have therefore shown that all values of m such that u_2+mu_1 satisfies a simple boundary condition at $r=\eta$ lie on a circle C_η. The radius of C_η is given by

$$R_\eta^2=\frac{B^2+C^2-AD}{A^2}.$$

Note that $A\neq 0$, in fact from (6.2.5) with $u=v=u_1$ we have

$$A=2\,\mathrm{Im}\,(z)\int_\xi^\eta |u_1|^2\,\mathrm{d}r. \qquad (6.2.10)$$

We also have, at $r=\eta$,

$$B^2+C^2-AD=-W(u_1,\bar{u}_2)W(u_2,\bar{u}_1)+W(u_1,\bar{u}_1)W(u_2,\bar{u}_2),$$

which, on writing out the Wronskians explicitly and simplifying, becomes

$$\begin{aligned}B^2+C^2-AD&=(u_1u_2'-u_1'u_2)(\bar{u}_1\bar{u}_2'-\bar{u}_1'\bar{u}_2)\\&=|W_\eta(u_1,u_2)|^2.\end{aligned}$$

Moreover, since u_1 and u_2 are both solutions of (6.2.1), (6.2.6) applies and $|W_\eta(u_1,u_2)|^2=|W_\xi(u_1,u_2)|^2=1$ in view of the initial conditions (6.2.2).

Using (6.2.10), we now have

$$R_\eta=\frac{1}{2|\mathrm{Im}\,(z)|\int_\xi^\eta|u_1|^2\,dr}. \tag{6.2.11}$$

With $u=v=u_2+mu_1$, (6.2.5) also implies

$$\begin{aligned}2\,\mathrm{Im}\,(z)&\int_\xi^\eta|u_2+mu_1|^2\,dr\\&=A(x^2+y^2)+2Bx+2Cy+D+2\,\mathrm{Im}\,(m),\end{aligned} \tag{6.2.12}$$

in this case for *arbitrary* $m=x+iy$. Hence the equation (6.2.9) for the circle C_η may also be written as

$$\mathrm{Im}\,(z)\int_\xi^\eta|u_2+mu_1|^2\,dr=\mathrm{Im}\,(m), \tag{6.2.13}$$

showing that C_η lies entirely in the same half-plane, upper or lower, as that in which the point z is to be found. The exterior of the circle C_η in the complex m-plane will be defined by the inequality

$$|\mathrm{Im}\,(z)|\int_\xi^\eta|u_2+mu_1|^2\,dr>|\mathrm{Im}\,(m)|, \tag{6.2.14}$$

provided again that m lies in the same half-plane as z.

What happens to the circle C_η as the point η is moved further to the right, ultimately approaching $r=b$? A comparison of (6.2.13) and (6.2.14) reveals that, for $\eta'>\eta$, any point m lying on the circle C_η will be *exterior* to the circle $C_{\eta'}$, since the integral in (6.2.13) can only increase with η. In other words, for $\eta'>\eta$, $C_{\eta'}$ will lie entirely in the interior of C_η. Equation (6.2.11) confirms that the radius R_η decreases with η.

We therefore have just two distinct possibilities, characterized by the following definition.

Definition 6.2

Let u_1 be the solution of (6.2.1) subject to the initial condition (6.2.2), and let R_η, the radius of the circle C_η, be given by (6.2.11). We shall say that we have the *limit-point* case at $r=b$ if $\lim_{r\to b} R_\eta=0$, and the *limit-circle* case at $r=b$ if $\lim_{r\to b} R_\eta \equiv R_b>0$.

In the limit-point case, then, the circle C_η shrinks down to a limiting point as η approaches b. We shall denote this limiting point by m_b. Thus the point m_b is interior to each of the circles C_η for $\xi \le \eta < b$, and as $\eta \to b$ the radius of C_η tends to zero. We shall write $m_b(z)$ whenever we wish to refer explicitly to the dependence on the complex number z.

In the limit-circle case we have a family of circles approaching a limiting circle as $\eta \to b$. We shall denote this limiting circle by C_b and note that again C_b is interior to each of the circles C_η for $\xi \le \eta < b$. We shall denote by m_b in this case a typical point on C_b.

Since in both the limit-point and limit-circle cases m_b is interior to C_η and lies in the same half-plane as z, we have from the converse of (6.2.14) that

$$\int_\xi^\eta |u_2+m_b u_1|^2 \, \mathrm{d}r < \mathrm{Im}\,(m_b)/\mathrm{Im}\,(z).$$

Certainly, then $\int_\xi^b |u_2+m_b u_1|^2 \, \mathrm{d}r < \infty$, which implies that there is at least one non-trivial solution of (6.2.1) that is square-integrable near b. In the limit-point case $R_\eta \to 0$ in (6.2.11), so that u_1 cannot lie in $L^2(\xi, b)$. There is then *exactly* one linearly independent solution of (6.2.1) that is square-integrable near b. On the other hand, in the limit-circle case $u_2+m_b u_1 \in L^2(\xi, b)$ for all m_b on C_b, implying that *every* solution of (6.2.1) is square-integrable near b in this case. In fact, this applies even to solutions of (6.2.1) with any other complex value of z; that is, one has the limit-circle case at b independently of the value of z.

To confirm this result, we let z' be an arbitrary complex number, and let $g \in L^2(\xi, b)$. Assuming the limit-circle case at b, we let f be the solution of the differential equation

$$-\frac{\mathrm{d}^2 f}{\mathrm{d}r^2}+Vf-zf=g,$$

subject to the initial conditions

$$f(\eta)=u_1(\eta), \qquad f'(\eta)=u_1'(\eta).$$

Then

$$f(r)=u_1(r)+u_2(r)\int_\eta^r u_1(t)g(t)\,\mathrm{d}t$$

$$-u_1(r)\int_\eta^r u_2(t)g(t)\,\mathrm{d}t.$$

Since b is limit-circle, both u_1 and u_2 belong to the Hilbert space $L^2(\eta,b)$ and hence so also does f. In fact, we may write

$$f=u_1+Lg,$$

where L is a bounded linear operator on the same Hilbert space, and $\|L\|$ becomes arbitrarily small as η approaches b.

Now we let f be the solution, subject to the same initial conditions, of the differential equation

$$-\frac{\mathrm{d}^2 f}{\mathrm{d}r^2}+Vf-z'f=0. \tag{6.2.15}$$

In this case, writing $g=(z'-z)f$, we have

$$(1+(z-z')L)f=u_1. \tag{6.2.16}$$

Choosing η sufficiently close to b that

$$\|(z-z')L\|<1,$$

the operator $1+(z-z')L$ in (6.2.16) is invertible, and (6.2.15) may be solved for f to give

$$f=(1+(z-z')L)^{-1}u_1. \tag{6.2.16$'$}$$

Certainly $f\in L^2(\eta,b)$, and a second linearly independent solution of (6.2.15), also in $L^2(\eta,b)$, may be constructed subject to the initial conditions

$$f(\eta)=u_2(\eta),\qquad f'(\eta)=u_2'(\eta).$$

Therefore in the limit-circle case at b every solution of (6.2.15), i.e. of (6.2.1) with z replaced by z', will be square-integrable in the neighbourhood of $r=b$. This means in particular that the limit-point/limit-circle criterion is independent of the particular choice, in (6.2.1), of the complex number z, and also of the point ξ of the interval (a,b). There is, of course a similar classification at $r=a$, and there will be no difficulty in writing down in that case analogues of all the results that we have obtained. Whether a particular endpoint a is limit point or limit circle will depend entirely on the behaviour of the potential in the neighbourhood of that endpoint. It follows that there is no connection between the classifications at a and b; any of the four combinations limit-point/circle at a with limit-point/circle at b are possible. In the limit-point case at b, let us set $m = m_b$ in (6.2.12) and take the limit $\eta \to b$. On the right-hand side we can write

$$\begin{aligned} &|A(x^2+y^2)+2Bx+2Cy+D| \\ &\quad = |A[(x-x_\eta)^2+(y-y_\eta)^2-R_\eta^2]|, \end{aligned}$$

where (x_η, y_η) and R_η are respectively the centre and radius of C_η, and $x+iy=m_b$. Since m_b is interior to C_η and $A=1/R_\eta$, the expression within the modulus signs is bounded by $|AR_\eta^2|=R_\eta$, and therefore tends to zero as $\eta \to b$. Hence from (6.2.12) we find

$$\operatorname{Im}(z) \int_\xi^b |u_2+m_b u_1|^2 \, dr = \operatorname{Im}(m_b). \tag{6.2.17}$$

With $u=v=u_2+m_b u_1$ in (6.2.5), the left-hand side of (6.2.17) may be expressed in terms of Wronskians to give

$$W_b(u_2+m_b u_1, \bar{u}_2+\bar{m}_b \bar{u}_1)=0, \tag{6.2.17$'$}$$

where the Wronskian at $r=b$ is defined by

$$W_b = \lim_{\eta \to b} W_\eta.$$

Equation (6.2.17) may be used as an alternative characterization of the limit point m_b. We note, however, that the left-hand side cannot be expanded, since for example $W(u_2, \bar{u}_2)$ will not exist.

In the limit-circle case at b, we again set $m=m_b$ in (6.2.12) and take the limit $\eta \to b$. Since in this case A is bounded, and (x_η, y_η) and R_η approach respectively the centre and radius of the limit circle, we have once more $|A(x^2+y^2)+2Bx+2Cy+D| \to 0$, where $x+iy=m_b$ is this time an arbitrary point on the limit circle. Hence (6.2.17) and (6.2.17)′ also hold, with a different interpretation, in the limit-circle case, and may then be regarded as alternative representations of the limit circle. The left-hand side of (6.2.17)′ *can* then be expanded as a quadratic polynomial in m_b amd $\bar{m}_b$.

The following theorem adds a further result to our conclusions so far.

Theorem 6.2

The endpoint b of the interval I is in the limit-point case if and only if, for any z such that $\operatorname{Im}(z) \neq 0$, there is exactly one linearly independent solution of (6.2.1) that is square-integrable in the neighbourhood of b.

The endpoint b is in the limit-circle case if, for any z (allowing $\mathrm{Im}(z)=0$), every solution of (6.2.1) is square-integrable in the neighbourhood of b.

If b is limit-point then the unique value m_b (the limit point) for a given value of z, ($\mathrm{Im}(z)\neq 0$) such that $u_2+m_b u_1\in L^2(\xi,b)$, satisfies (6.2.17) and (6.2.17)′, where u_1 and u_2 are solutions of (6.2.1) subject to the initial conditions (6.2.2). If b is limit-circle then there is a unique circle in the complex plane (the limit circle) for a given value of z ($\mathrm{Im}(z)\neq 0$), and any point m_b on this circle satisfies (6.2.17) and (6.2.17)′.

Analogous results hold for the endpoint a, and the limit points/limit circles appropriate to a and b respectively do not intersect; i.e. $m_a\neq m_b$.

Proof

It remains only to prove the final statement $m_a\neq m_b$. We set $u=v=u_2+mu_1$ in (6.2.5), and take limits $\xi\to a$, $\eta\to b$. If $m=m_a=m_b$, this gives

$$2\mathrm{i}\,\mathrm{Im}(z)\int_a^b |u_2+mu_1|^2\,\mathrm{d}r$$

$$=W_b(u_2+m_b u_1,\bar{u}_2+\bar{m}_b\bar{u}_1)-W_a(u_2+m_a u_1,\bar{u}_2+\bar{m}_a\bar{u}_1)$$

$$=0,$$

on using (6.2.17)′ and the corresponding equation at $r=a$. But the integral is strictly positive, and we have a contradiction. So indeed $m_a\neq m_b$.

This means that a limit point at a could not lie on a limit circle at b, two limit circles cannot intersect, and so on. ■

The following concluding result of this section illustrates the importance of limit-point/limit-circle in considering self-adjoint extensions of $\hat{H}=-\mathrm{d}^2/\mathrm{d}r^2+V$.

Theorem 6.3

The operator $\hat{H}=-\mathrm{d}^2/\mathrm{d}r^2+V$ defined on $C_0^\infty(I)$ is essentially self-adjoint if and only if both endpoints of the interval I are limit-point.

Proof

We suppose that both endpoints are limit-point. Then, according to the analysis of Section 2.6, $\hat{H}$ will be essentially self-adjoint provided that neither of the equations

$$(\hat{H}^*-\bar{z})f=0,\qquad(\hat{H}^*-z)f=0,$$

for Im $(z) \neq 0$, have non-trivial solutions in the Hilbert space $L^2(I)$. In fact, it will be sufficient to consider the second equation only, since solutions of the two equations are related by complex-conjugation. Now Theorem 6.1 (ii) shows that f must satisfy the differential equation

$$-\frac{d^2 f}{dr^2} + Vf - zf = 0.$$

Theorem 6.2 then shows that, in order to be square-integrable near $r = b$, f must be a constant multiple of $u_2 + m_b u_1$. Similarly, since $r = a$ is also limit-point, f has to be a multiple of $u_2 + m_a u_1$. Since $m_a \neq m_b$, this leaves only the possibility $f = 0$. Conversely, if b, say, were limit-circle then $f = u_2 + m_a u_1$ would satisfy $(\hat{H}^* - z)f = 0$ and be square-integrable at both endpoints. In this case $f \in L^2(I)$, and $\hat{H}$ cannot be essentially self-adjoint. ■

6.3 BOUNDARY CONDITIONS AND SELF-ADJOINT EXTENSIONS

The reason that $\hat{H} = -d^2/dr^2 + V$ on $C_0^\infty(I)$ fails to be essentially self-adjoint if either a or b is limit-circle lies in the multiplicity of L^2 solutions of the differential equation $-d^2 f/dr^2 + Vf = zf$. Now the obvious way to restrict the solution set of a differential equation is by the application of boundary conditions. We have already met, in (6.2.7), simple boundary conditions of the form

$$(\alpha u + \beta u')|_{r=\eta} = 0,$$

where α and β are real numbers not both zero. This boundary condition, at $r = \eta$, may equivalently be written as

$$W_\eta(u, v) = 0,$$

where W_η refers to the Wronskian at $r = \eta$, and v is related to the solutions u_1 and u_2 of (6.2.1) and (6.2.2) by

$$v = \alpha u_1 - \beta u_2.$$

Expressed in this way, the idea of boundary condition may more readily be generalized and applied to boundary conditions at the endpoints a and b. These (generalized) boundary conditions need not take such a simple form as (6.2.7).

Definition 6.3

A (generalized) boundary condition at $r = b$ is an equation of the form

$$W_b(u, h) = 0, \tag{6.3.1}$$

where W_b refers to the Wronskian W_η in the limit $\eta \to b$ and h is a (fixed) function such that both h and dh/dr are locally absolutely continuous in the interval I.

Boundary conditions at $r=a$ may be defined in the obvious way.

Now we can already see, from (6.2.17)′, that if m_b is either the limit point at b or any point on the limit circle at b then the solution $u = u_2 + m_b u_1$ satisfies at $r=b$ the boundary condition

$$W_b(u, \bar{u}_2 + \bar{m}_b \bar{u}_1) = 0. \tag{6.3.2}$$

Condition (6.3.2) fills exactly the required role, as the following result shows.

Theorem 6.4

Every solution that is square-integrable near $r=b$ of the differential equation $-d^2u/dr^2 + Vu = zu$, subject to boundary condition (6.3.2), is a constant multiple of $u_2 + m_b u_1$.

Proof

If b is limit-point then, according to Theorem 6.2, $u_2 + m_b u_1$ and constant multiples are the *only* solutions that are square-integrable in a neighbourhood of b. We therefore consider the limit-circle case at b, and suppose that there are two linearly independent solutions, both satisfying (6.3.2). Then

$$W_b(u_1, \bar{u}_2 + \bar{m}_b \bar{u}_1) = W_b(u_2, \bar{u}_2 + \bar{m}_b \bar{u}_1) = 0.$$

These equations enable us to deduce that

$$\begin{aligned} W_b(u_2, \bar{u}_2) &= -\bar{m}_b W_b(u_2, \bar{u}_1) = -\bar{m}_b \bar{W}_b(\bar{u}_2, u_1) \\ &= \bar{m}_b \bar{W}_b(u_1, \bar{u}_2) = -m_b \bar{m}_b \bar{W}_b(u_1, \bar{u}_1) \\ &= m_b \bar{m}_b W_b(u_1, \bar{u}_1). \end{aligned}$$

In a similar way, each of the remaining expressions $W_b(u_i, \bar{u}_j)$ may be expressed as some multiple of $W_b(u_1, \bar{u}_1)$.

From Theorem 6.2 and (6.2.17)′, the limit circle on which m_b lies has the equation

$$W_b(u_2 + mu_1, \bar{u}_2 + \bar{m}\bar{u}_1) = 0,$$

which, on substituting for $W_b(u_i, \bar{u}_j)$ and dividing by $W_b(u_1, \bar{u}_1)$ (note that $|W_b(u_1, \bar{u}_1)| = 1/R_b \neq 0$), becomes

$$m\bar{m} - m\bar{m}_b - \bar{m}m_b + m_b \bar{m}_b = 0,$$

or simply

$$|m-m_b|^2=0.$$

But this equation characterizes just the single point $m=m_b$, contradicting the original supposition that b is limit circle. Hence in this case also the solution set of the differential equation, subject to (6.3.2), is a one-dimensional vector space of functions. ■

It might be objected that, as opposed to (6.2.7), (6.3.2) may not define a *real* boundary condition, in that the complex conjugate of a function satisfying (6.3.2) need not itself satisfy it. Real boundary conditions will have a special importance in view of the fact that $\hat{H}=-\mathrm{d}^2/\mathrm{d}r^2+V$ on $C_0^\infty(I)$ is a real differential operator. The obvious way to replace (6.3.2) by a real boundary condition is to substitute its real or imaginary part. We shall not usually have to carry this out explicitly, since it will turn out that solutions subject to this modified boundary condition corresponds exactly to solutions subject to (6.3.2). We then have the following corollary to the theorem.

Corollary

For all except at most one exceptional value of ξ in (6.2.2), every solution of the differential equation $-\mathrm{d}^2u/\mathrm{d}r^2+Vu=zu$, subject to the real boundary condition

$$W_b(u, \operatorname{Re}(u_2+m_b u_1))=0, \qquad (6.3.2)'$$

is a constant multiple of $u_2+m_b u_1$.

Proof

Certainly $u=u_2+m_b u_1$ satisfies (6.3.2), and also, trivially, the boundary condition

$$W_b(u, u_2+m_b u_1)=0.$$

By combining these two results, (6.3.2)′ is satisfied. We suppose on the other hand that there are two linearly independent solutions, both satisfying (6.3.2)′. Then both u_1 and u_2 satisfy (6.3.2)′, and hence also, by complex-conjugation, so do $\bar{u}_1$ and $\bar{u}_2$. If the point $\xi\in I$ had been chosen differently, say $\tilde{\xi}$ instead of ξ, then $\tilde{u}_2$ and $\tilde{u}_1$, defined by (6.2.1) and (6.2.2) with $\tilde{\xi}$ instead of ξ could be used to construct a solution $\tilde{u}_2+\tilde{m}_b\tilde{u}_1$, where $\tilde{m}_b$ is a point on the new limit circle, or the new limit point. If b is limit-point then $\tilde{u}=\tilde{u}_2+\tilde{m}_b\tilde{u}_1$ is the unique solution of the differential equation, square-integrable near b, and satisfying $\tilde{u}(\tilde{\xi})=1$. Hence in this case $\tilde{u}=(u_2+m_b u_1)/$

$(u_2+m_b u_1)|_{r=\xi}$. Even if b is limit-circle, this may be taken as the defining equation for $\tilde{u}$, thus establishing a correspondence between points m_b on the limit circle with ξ and points $\tilde{m}_b$ on the limit circle with $\bar{\xi}$. Now $\mathrm{Re}(u_1)$ and $\mathrm{Im}(u_1)$, $\mathrm{Re}(u_2)$ and $\mathrm{Im}(u_2)$, by hypothesis, all satisfy (6.3.2)′; they cannot all satisfy the corresponding equation with $\bar{\xi}$, namely

$$W_b(u, \mathrm{Re}(\tilde{u}))=0$$

for in this case, noting that $\tilde{u}=\gamma(u_2+m_b u_1)$ with $\mathrm{Im}(\gamma)\neq 0$, and taking various linear combinations, we could deduce $W_b(u_1, u_2+m_b u_1)=0$, which contradicts $W_b(u_1, u_2)=W_\xi(u_1, u_2)=-1$.

(We cannot have $1/\gamma=u_2+m_b u_1$ real at $r=\bar{\xi}$ since this would imply, by (6.2.8) with $m=m_b$ and $\eta=\bar{\xi}$, that m_b lies on the circle $C_{\bar{\xi}}$, whereas in fact we know that m_b is strictly interior to all such circles.)

The conclusion is that, with either ξ or $\bar{\xi}$, (6.3.2)′ is sufficient to define, up to constant multiples, the solution $u=u_2+m_b u_1$ of $-\mathrm{d}^2u/\mathrm{d}r^2+Vu=zu$. ■

We are now ready to use generalized boundary conditions to define self-adjoint extensions H of the operator $\hat{H}=-\mathrm{d}^2/\mathrm{d}r^2+V$ on $C_0^\infty(I)$. For this purpose, we define smooth functions ρ_a and ρ_b such that

$$\rho_a(r)\equiv 1 \quad \text{near } r=a, \qquad \rho_a(r)\equiv 0 \quad \text{near } r=b,$$

and

$$\rho_b(r)\equiv 1 \quad \text{near } r=b, \qquad \rho_b(r)\equiv 0 \quad \text{near } r=a.$$

It will also be convenient to choose ρ_a and ρ_b such that their respective supports do not overlap; in other words such that $\rho_a(r)=0$ wherever $\rho_b(r)\neq 0$, and $\rho_b(r)=0$ wherever $\rho_a(r)\neq 0$.

We define u_a and $u_b \in L^2(I)$ by

$$\left.\begin{aligned} u_a(r)&=\rho_a(r)(u_2(r)+m_a u_1(r)),\\ u_b(r)&=\rho_b(r)(u_2(r)+m_b u_1(r)),\end{aligned}\right\} \tag{6.3.3}$$

where m_a is the limit point if a is limit-point and some chosen point on the limit circle otherwise, and similarly for m_b. In order to define H, we need first of all to enlarge the domain $D(\hat{H})$ by including two additional functions u_a and u_b. (The precise manner in which ρ_a and ρ_b are chosen will be immaterial, since with any two possible choices, say ρ_a and ρ_a^*, the difference $u_a-u_a^*$ will already lie in $D(\hat{H})$, so that the linear span of the enlarged domain will not depend on this choice.) Then we define $H_1=-\mathrm{d}^2/\mathrm{d}r^2+V$ on this enlarged domain $D(H_1)$. H_1 will be essentially self-adjoint, and the unique self-adjoint extension H of H_1 will be the required self-adjoint operator in $L^2(I)$. The following theorem guarantees all of the above results, and at the same time characterizes the domain of H by means of generalized boundary conditions at the respective endpoints $r=a$, $r=b$ of the interval $I=(a,b)$.

Theorem 6.5

Define in $L^2(I)$ the differential operator

$$H_1 = -\frac{d^2}{dr^2}+V, \qquad \text{with } D(H_1)=C_0^\infty(I)+\{u_a\}+\{u_b\}, \tag{6.3.4}$$

where u_a and u_b are given by (6.3.3). (In other words, the domain of H_1 consists of $C_0^\infty(I)$ together with u_a, u_b, and all linear combinations of finitely many such functions.) Then

(i) H_1 is a symmetric extension of the operator $\hat{H}$ defined by (6.1.1); in fact,

(ii) H_1 is essentially self-adjoint.

Let H, the closure of H_1, denote the unique self-adjoint extension of H_1, i.e. $H=H_1^*$. Then

(iii) the domain of H consists of all $f\in L^2(I)$ such that

$$\left.\begin{aligned} &f \text{ and } \frac{df}{dr} \text{ are locally absolutely continuous in } I,\\ &Hf\equiv -\frac{d^2 f}{dr^2}+Vf\in L^2(I), \end{aligned}\right\} \tag{6.3.5}$$

and such that f satisfies the boundary conditions

$$\left.\begin{aligned} W_a(f,\bar{u}_a)=0,\\ W_b(f,\bar{u}_b)=0. \end{aligned}\right\} \tag{6.3.6}$$

(iv) If either a or b is limit-point then the corresponding boundary condition in (6.3.6) may be omitted. That is, for example, if $r=b$ is limit-point then the conditions (6.3.5) automatically guarantee that the second condition in (6.3.6) is satisfied. (If *both* a and b are limit-point then (6.3.6) may be omitted altogether, in which case comparison with Theorem 6.1 shows that $H=\hat{H}^*$, the closure of $\hat{H}$. Thus Theorem 6.3 may be regarded as a special case of the present result.)

Proof

(i) Clearly H_1 is an extension of $\hat{H}$. To show that H_1 is symmetric, we note first of all that trivially

$$\langle H_1 u_a, u_b \rangle = \langle u_a, H_1 u_b \rangle,$$

since the respective supports of u_a and u_b are disjoint. Also, for $\phi \in C_0^\infty(I)$,

$$\langle H_1 \phi, u_a \rangle = \langle \phi, H_1 u_a \rangle$$

is a consequence of integration by parts, there being no boundary contributions as ϕ has compact support. Similarly

$$\langle H_1 \phi, u_b \rangle = \langle \phi, H_1 u_b \rangle.$$

It now needs only to be verified, that

$$\langle H_1 u_b, u_b \rangle = \langle u_b, H_1 u_b \rangle, \tag{6.3.7}$$

and similarly for u_a. From (6.2.4) with $f = g = u_b$, and taking limits $\xi \to a$, $\eta \to b$, we have

$$\langle H_1 u_b, u_b \rangle - \langle u_b, H_1 u_b \rangle = [-W(u_b, \bar{u}_b)]_a^b.$$

But $W_a(u_b, \bar{u}_b) = 0$ since $u_b \equiv 0$ near $r = a$, and (6.2.17)′, with u_b given by (6.3.3), is nothing but the statement that $W_b(u_b, \bar{u}_b) = 0$. Hence (6.3.7) is satisfied, and similarly for u_a, and it follows easily that H_1 is symmetric.

(ii) From Section 2.6, we have to show that if $\langle (H_1 - \bar{z})\psi, u \rangle = 0$ for all $\psi \in D(H_1)$ then $u = 0$; and similarly with z instead of $\bar{z}$. Suppose then that $\langle (H_1 - \bar{z})\psi, u \rangle = 0$ for all $\psi \in D(H_1)$, and first take $\psi \in C_0^\infty(I)$, which is part of $D(H_1)$. Then $\psi \in D(\hat{H})$, and equivalently we may write $\langle (\hat{H} - \bar{z})\psi, u \rangle = 0$ for all $\psi \in D(\hat{H})$. This means that $(\hat{H}^* - z)u = 0$, so that, by Theorem 6.1 (ii), u must satisfy the differential equation $-\mathrm{d}^2u/\mathrm{d}r^2 + Vu - zu = 0$. We can also take $\psi = u_b$, giving $\langle (H_1 - \bar{z})u_b, u \rangle = 0$. But from (6.2.4) with $f = u_b$, $g = u$, and taking limits $\xi \to a$, $\eta \to b$, we have, on taking account of the differential equation for u,

$$\begin{aligned} 0 &= \langle (H_1 - \bar{z})u_b, u \rangle - \langle u_b, 0 \rangle \\ &= [-W(u, \bar{u}_b)]_a^b = -W_b(u, \bar{u}_b), \end{aligned}$$

since u_b vanishes identically near $r = a$. So u satisfies the second of the two boundary conditions (6.3.6), or equivalently the condition (6.3.2), so that by Theorem 6.4, u is a constant multiple of $u_2 + m_b u_1$. Taking $\psi = u_a \in D(H_1)$, we may similarly deduce that u is a constant multiple of $u_2 + m_a u_1$. Hence, as in the proof of Theorem 6.3, $m_a \neq m_b$, and we can have only $u = 0$.

A similar argument goes through with $\bar{z}$ replaced by z. Following the proof through in this case, the function u satisfying the boundary condition $W_b(u, \bar{u}_2 + \bar{m}_b \bar{u}_1) = 0$ now also satisfies the same differential equation as that for $\bar{u}_2 + \bar{m}_b \bar{u}_1$, namely $-\mathrm{d}^2u/\mathrm{d}r^2 + Vu - \bar{z}u = 0$. Hence $W_\eta(u, \bar{u}_2 + \bar{m}_b \bar{u}_1) = 0$, the Wronskian being independent of η, so that again u can only be a constant multiple of $\bar{u}_2 + \bar{m}_b \bar{u}_1$. From this point the proof proceeds as before.

(iii) We suppose that $f \in D(H)$; this means that

$$\langle H_1 \psi, f \rangle = \langle \psi, H_1 f \rangle \quad \text{for all } \psi \in D(H_1).$$

As in the proof of (ii) we first take $\psi \in C_0^\infty(I)$ and deduce that $H_1 f = \hat{H}^* f$. Theorem 6.1 may then be used to obtain (6.3.5). The boundary conditions (6.3.6) follow by setting successively $\psi = u_b$ and u_a, and again using (6.2.4) as in (ii).

(iv) We suppose that b is limit-point. Note that the proof in (ii) of the essential self-adjointness of H_1 is valid even if u_b is omitted from $D(H_1)$. This is because, for each of the differential equations $-\mathrm{d}^2 u/\mathrm{d}r^2 + Vu - zu = 0$ and $-\mathrm{d}^2 u/\mathrm{d}r^2 + Vu - \bar{z}u = 0$ there is just one linearly independent solution ($u_2 + m_b u_1$ and $\bar{u}_2 + \bar{m}_b \bar{u}_1$ respectively) that is square-integrable near $r = b$. It then becomes unnecessary to use $u_b \in D(H_1)$ in order to derive the boundary condition that determines u, up to a multiplicative constant, in each case. We then define

$$H_2 = -\frac{\mathrm{d}^2}{\mathrm{d}r^2} + V \qquad \text{with } D(H_2) = C_0^\infty(I) + \{u_a\},$$

so that H_2 is essentially self-adjoint. Clearly H_2 is a restriction of H_1, so that the closure of H_2 is a restriction of the closure of H_1; i.e. $H_2^* \subseteq H_1^* = H$. Since H_2^* and H are both self-adjoint, we can have only $H = H_2^*$. So if b is limit-point then H may alternatively be defined as the closure of H_2. The proof of (iii) may now be followed through as before, but with H_1 replaced by H_2; and we find that the second boundary condition in (6.3.6) is omitted. This completes the final part of the Theorem. ■

Theorem 6.5 shows how to construct a self-adjoint extension H of the differential operator $-\mathrm{d}^2/\mathrm{d}r^2 + V$ in $L^2(I)$. A single boundary condition, having one of the two forms (6.3.6), is imposed at each endpoint (if any) that is limit-circle. In order to write down this boundary condition, say at $r = b$, a point m_b on the limit circle has to be selected, and u_b is defined by (6.3.3). There is thus a one-to-one correspondence between possible boundary conditions and points on the limit circle at b. Equivalently, boundary conditions may be parametrized, modulo 2π, by an angle θ. This was already true of the simple boundary condition (6.2.7), where the angle θ might be defined, for example, by $\theta = \tan^{-1}(\beta/\alpha)$.

If either a or b is limit-circle then the domain of the self-adjoint operator H defined by Theorem 6.5 is larger than the domain of $\hat{H}$, but smaller than the domain of $\hat{H}^*$. The domain of H is also larger than the domain of $\hat{H}^{**}$, the closure of $\hat{H}$. This may be clearly seen from the following Theorem, which characterizes $D(\hat{H}^{**})$ by means of boundary conditions at a and b. We can see that elements in $D(\hat{H}^{**})$ satisfy automatically the boundary conditions (6.3.6) for elements in $D(H)$.

Theorem 6.6

Let $\hat{H}^{**}$ be the closure of the operator $\hat{H}=-\mathrm{d}^2/\mathrm{d}r^2+V$, with $D(\hat{H})=C_0^\infty(I)$. Then $D(\hat{H}^{**})$ consists of all $f\in L^2(I)$ such that f and $\mathrm{d}f/\mathrm{d}r$ are locally absolutely continuous in I,

$$\hat{H}^{**}f\equiv-\frac{\mathrm{d}^2 f}{\mathrm{d}r^2}+Vf\in L^2(I), \tag{6.3.8}$$

and such that f satisfies the boundary conditions

$$\left.\begin{aligned} &W_a(f,\bar{u}_1)=W_a(f,\bar{u}_2)=0 \qquad \text{(if } a \text{ is limit-circle)}\\ &\text{and}\\ &W_b(f,\bar{u}_1)=W_b(f,\bar{u}_2)=0 \qquad \text{(if } b \text{ is limit-circle)}\end{aligned}\right\} \tag{6.3.9}$$

Proof

We note that there are four boundary conditions if both a and b are limit-circle, two if a is limit-circle and b limit-point, and so on.

Equation (6.3.8) is simply a consequence of the fact that $\hat{H}^{**}$ is a restriction of $\hat{H}^*$. We let $T=-\mathrm{d}^2/\mathrm{d}r^2+V$, restricted to the above domain. First let us verify that T is closed. To show this, we let $\{g_n\}\in\mathscr{H}$ be a sequence converging strongly to some element g, and such that $Tg_n\to h$ strongly. We have to prove that $h=Tg$. This requires us to derive boundary conditions (6.3.9) for g, given that each g_n satisfies the same boundary conditions. As in (6.3.3), we let ρ_b be a smooth function such that

$$\rho_b(r)\equiv 1 \quad \text{near } r=b, \qquad \rho_b(r)\equiv 0 \quad \text{near } r=a.$$

We assume the limit-circle case at $r=b$. From (6.2.4) with $f=\rho_b u_i$ $(i=1,2)$ and $\xi\to a$, $\eta\to b$, we have

$$\begin{aligned} W_b(g,\bar{u}_i)&=W_b(g,\rho_b\bar{u}_i)\\ &=\langle\rho_b u_i,\hat{H}^*g\rangle-\langle H^*\rho_b u_i,g\rangle\\ &=\lim_{n\to\infty}\{\langle\rho_b u_i,\hat{H}^*g_n\rangle-\langle\hat{H}^*\rho_b u_i,g_n\rangle\}\\ &=\lim_{n\to\infty}W_b(g_n,\rho_b\bar{u}_i)\\ &=\lim_{n\to\infty}(g_n,\bar{u}_i)\\ &=0.\end{aligned}$$

We have used here the fact that the closed operator $\hat{H}^*$ is an extension of T, so that we may write throughout $Tg_n = \hat{H}^* g_n$, which converges strongly to $h = \hat{H}^* g$. Hence g satisfies the boundary conditions (6.3.9) at $r = b$, and similarly at $r = a$ if a is limit-circle. So T is closed.

Let u be any solution of $(\hat{H}^* - z)u = 0$. Then, for $f \in D(T)$, $\rho_b f \in D(T)$ and

$$\begin{aligned}\langle (T - \bar{z})\rho_b f, u\rangle &= \langle (\hat{H}^* - \bar{z})\rho_b f, u\rangle \\ &= \langle \rho_b f, (\hat{H}^* - z)u\rangle - W_b(u, \bar{f}) \\ &= \bar{W}_b(f, \bar{u}) \\ &= 0,\end{aligned}$$

since u is a linear combination of u_1 and u_2, on using (6.3.9). Similarly, with $\rho_a = 1 - \rho_b$,

$$\langle (T - \bar{z})\rho_a f, u\rangle = 0.$$

Adding these two results, we have

$$\langle (T - \bar{z})f, u\rangle = 0 \quad \text{for all } f \in D(T).$$

What has been shown is that $(\hat{H}^* - z)u = 0 \Rightarrow (T^* - z)u = 0$. We have assumed here that b and a are limit-circle, but the argument may readily be extended to other cases. For example, if we suppose that b is limit-point and that u satisfies $(\hat{H}^* - z)u = 0$ then $u = \text{const } (u_2 + m_b u_1)$. Then for $f \in D(T)$, $\rho_b f \in D(T)$, so that $\rho_b f$ will lie in the domain of some self-adjoint extension H of T. From (6.3.6), this implies that $W_b(f, \bar{u}_2 + \bar{m}_b \bar{u}_1) = 0$, so that $W_b(f, \bar{u}) = 0$, and the same argument as before allows us to infer again that $\langle (T - \bar{z})\rho_b f, u\rangle = 0$, and similarly if a is limit-point.

So, in all cases, $(\hat{H}^* - z)u = 0 \Rightarrow (T^* - z)u = 0$. Conversely, $(T^* - z)u = 0 \Rightarrow (\hat{H}^* - z)u = 0$, since $\hat{H}^*$ is an extension of T^*. Therefore the respective solution sets of $(T^* - z)u = 0$ and $(\hat{H}^* - z)u = 0$ are identical and hence the range of $T - \bar{z}$ is identical with the range of $\hat{H}^{**} - \bar{z}$, since these are respectively the orthogonal subspaces to the above solution sets. So $D(T) = D(\hat{H}^{**})$, since any proper extension T of $\hat{H}^{**}$ would lead to an enlarged range for $T - \bar{z}$ in comparison with the range of $\hat{H}^{**} - \bar{z}$. We have now proved that $T = \hat{H}^{**}$, so that (6.3.8) and (6.3.9) do indeed characterize the closure $\hat{H}^{**}$ of $\hat{H}$. ■

With Theorem 6.6, the characterization of the respective domains of the operators $\hat{H}, \hat{H}^{**}$, $H, \hat{H}^*$ is now complete, and it may be helpful to summarize some of our conclusions. Each member of the chain $\hat{H} \subseteq \hat{H}^{**} \subseteq H \subseteq \hat{H}^*$ is an extension of all preceeding operators in the chain.

We start with $\hat{H} = -\mathrm{d}^2/\mathrm{d}r^2 + V$ defined on $C_0^\infty(I)$, the minimal domain. On taking the closure of $\hat{H}$, we arrive at $\hat{H}^{**}$; the domain of this operator is given in Theorem 6.6. If a and b are both limit-point then the closure of $\hat{H}$ is already self-adjoint, and we need proceed no further (in this case, $\hat{H}^{**}=H=\hat{H}^{*}$). If either a or b (or both) is limit-circle then (6.3.9) shows that elements in the closure of $\hat{H}$ satisfy, for example at a, every conceivable boundary condition $W_a(f,\bar{u})=0$, where u is any solution of $-\mathrm{d}^2u/\mathrm{d}r^2+Vu-zu=0$. In order to define a self-adjoint extension H, one of these boundary conditions $W_a(f,\bar{u}_2+\bar{m}_a\bar{u}_1)=0$ is selected, corresponding to some chosen point m_a on the limit circle; see Theorem 6.5 and (6.3.6). The operator $\hat{H}^*$, where domain is given in Theorem 6.1, represents a maximal operator from which all self-adjoint H may be derived by the application of boundary conditions.

The following table gives in each possible case the *deficiency indices* (n_1,n_2) of the closure of $\hat{H}$, together with a basis, where appropriate, of the solution space of the equation $(\hat{H}^*-z)u=0$. The solution space of $(\hat{H}^*-\bar{z})u=0$ may be derived by complex conjugation.

	Limit point at b	Limit circle at b
Limit point at a	$(0,0)$	$(1,1)$; $u_2+m_au_1$
Limit circle at a	$(1,1)$; $u_2+m_bu_1$	$(2,2)$; u_1,u_2

6.4 LOCALITY, REALITY AND ANALYTICITY

Definition 6.4

A linear operator T acting in $\mathscr{H}=L^2(I)$ is said to be a *local operator* whenever

(i) $f\in D(T)\Rightarrow \rho f\in D(T)$ for any smooth function $\rho(r)$ such that either

(a) $\rho(r)\equiv 1$ near $r=a$ and $\rho(r)\equiv 0$ near $r=b$, or
(b) $\rho(r)\equiv 1$ near $r=b$ and $\rho(r)\equiv 0$ near $r=a$, or
(c) $\rho(r)\equiv 0$ near $r=a$ *and* near $r=b$; that is $\rho\in C_0^\infty(I)$;

(ii) $f\in D(T)$ and $f(r)\equiv 0$ for $\xi\le r\le\eta$ together imply $(Tf)(r)\equiv 0$ for $\xi\le r\le\eta$.

If T is a local operator, whether or not an element f belongs to $D(T)$ will depend on the local behaviour of $f(r)$ in the interior of the interval I, and on the behaviour of $f(r)$ in the neighbourhood of each endpoint a,b. Moreover, (ii) implies that $(Tf)(r)$ may be determined *locally*; that is, to evaluate $(Tf)(r)$ for r in a subinterval $[\xi,\eta]$ of I, it is necessary only to know the function $f(r)$ in the same subinterval. We shall see in Chapter 10 that the idea of locality may be generalized, and carries with it important physical consequences for quantum mechanics. The relevance of locality in the present context is that each of the operators $\hat{H}, \hat{H}^{**}$, H and $\hat{H}^*$ considered in Section 6.3 is a local operator. This may be verified directly from Theorems 6.1, 6.5 and 6.6, where the respective differential operators and their domains are specified. (In the case of $\hat{H}$ with $D(\hat{H}) = C_0^\infty(I)$, verification is immediate.)

It should be realized at this point that if both a and b are limit-circle then there will be self-adjoint extensions of $\hat{H}$, other than those defined by Theorem 6.5, which will be non-local in the above sense. These will be extensions defined by boundary conditions that "mix" the endpoints a and b. An example familiar from Fourier series is that of so-called periodic boundary conditions (which are not boundary conditions at all in our sense!). A self-adjoint extension of $\hat{H} = -\mathrm{d}^2/\mathrm{d}r^2$ in $L^2(-\pi, \pi)$ may be defined subject to the periodic conditions $\phi(-\pi) = \phi(\pi)$, $\phi'(-\pi) = \phi'(\pi)$. The eigenfunctions of this extension then become the basis functions appropriate to Fourier-series expansions over this interval.

Non-local self-adjoint extensions of this type will exist only if both a and b are limit-circle. Even in this case, there are usually sound physical reasons to prefer local extensions in quantum mechanics, and what Theorem 6.5 does is to classify and to describe completely all local self-adjoint extensions of $-\mathrm{d}^2/\mathrm{d}r^2 + V$.

Let us now turn to the subject of *reality*.

Definition 6.5

A linear operator T acting in $\mathscr{H} = L^2(I)$ is said to be a real operator if $f \in D(T) \Rightarrow \bar{f} \in D(T)$, where

$$T\bar{f} = \overline{(Tf)}.$$

It is immediately evident that $\hat{H}$ and therefore also the closure $\hat{H}^{**}$ and the adjoint $\hat{H}^*$ of $\hat{H}$ are real operators. It is not quite so clear, although also true, that the self-adjoint extension H defined by Theorem 6.5 is a real operator. To see this, we note first that

$$f, g \in D(H) \Rightarrow W_b(f, \bar{g}) = 0. \tag{6.4.1}$$

This comes from the fact that if $\rho_b(r) \equiv 1$ near $r = b$ and $\rho_b(\mathrm{r}) \equiv 0$ near $r = a$ then $\rho_b f$, $\rho_b g \in D(H)$ by locality of H, so that (6.4.1) follows from (6.2.4) in the usual way, using

$$\langle H\rho_b f, \rho_b g \rangle = \langle \rho_b f, H\rho_b g \rangle.$$

Certainly $\bar{u}_b \in D(H)$, the function $\bar{u}_b$ satisfying trivially the boundary conditions (6.3.6). Hence (6.4.1) with $g=\bar{u}_b$ implies $W_b(f,u_b)=0$. Hence also $W_b(\bar{f},\bar{u}_b)=0$, so that $f\in D(H)$ is such that $\bar{f}$ satisfies the second boundary condition (6.3.6). The first condition (6.3.6) follows similarly, and we have

$$f\in D(H)\Rightarrow \bar{f}\in D(H),$$

where $H\bar{f}=\overline{Hf}$ is easily verified.

Therefore H in Theorem 6.5 is a real operator. Using the Corollary to Theorem 6.4, the boundary conditions (6.3.6) may be rewritten in a form that explicitly exhibits the reality of H. In order to do this, we follow through the proof of (i)–(iii) of Theorem 6.5, except that in (6.3.4) u_a and u_b are to be replaced respectively by $\mathrm{Re}\,(u_a)$ and $\mathrm{Re}\,(u_b)$ in $D(H_1)$. The self-adjoint operator H, defined as the closure of H_1, will be unaffected by this replacement, since $\mathrm{Re}\,(u_a)$ and $\mathrm{Re}\,(u_b)$ satisfy the boundary conditions (6.3.6) and hence belong already to $D(H)$. On carrying through the proof, we find that (6.3.6) may be rewritten in the alternative form

$$\left.\begin{aligned} W_a(f,\mathrm{Re}\,(u_a))=0,\\ W_b(f,\mathrm{Re}\,(u_b))=0.\end{aligned}\right\} \tag{6.4.2}$$

More generally, we can carry out the proof of (i)–(iii) of Theorem 6.5, replacing u_a and u_b in (6.3.4) by f_a and f_b, where f_a, $f_b\in D(H)$ with $f_a(r)\equiv 0$ near $r=b$ and $f_b(r)\equiv 0$ near $r=a$. We suppose also that $W_b(f_b,\bar{u}_1)$ and $W_b(f_b,\bar{u}_2)$ are not both zero, and that $W_a(f_a,\bar{u}_1)$, $W_a(f_a,\bar{u}_2)$ are not both zero. By (6.3.9), this means, in the limit-circle case, that neither f_a nor f_b should belong to $D(\hat{H}^{**})$. By reality, it follows that $\bar{f}_a$ and $\bar{f}_b$ cannot belong to this domain either. In other words, we cannot have, for example, $W_b(\bar{f}_b,\bar{u}_1)=W_b(\bar{f}_b,\bar{u}_2)=0$, so that, by complex-conjugation, $W_b(f_b,u_1)$ and $W_b(f_b,u_2)$ cannot both vanish. This means, in the proof of (i)–(iii) of Theorem 6.5 with f_a,f_b instead of u_a,u_b, that we can continue to rely on the implication

$$W_b(u,\bar{f}_b)=0\Rightarrow u=\mathrm{const}\,(u_2+m_b u_1)$$

and

$$W_b(u,f_b)=0\Rightarrow u=\mathrm{const}\,(\bar{u}_2+\bar{m}_b\bar{u}_1),$$

for solutions u respectively of $-\mathrm{d}^2u/\mathrm{d}r^2+Vu-zu=0$ and $-\mathrm{d}^2u/\mathrm{d}r^2+Vu-\bar{z}u=0$. We then find that (6.3.6) may be rewritten in the alternative more general form

$$\left.\begin{aligned} W_a(f,\bar{f}_a)=0,\\ W_b(f,\bar{f}_b)=0.\end{aligned}\right\} \tag{6.4.3}$$

Thus the boundary condition at $r=b$, applied to an element f of $D(H)$, is nothing but a restatement of (6.4.1) for some fixed f_b, vanishing near $r=a$, such that $f_b \in D(H)$ but $f_b \notin D(\hat{H}^{**})$. Equation (6.3.6) corresponds to the special case $f_b = u_b$, and (6.4.2) to the special case $f_b = \mathrm{Re}\,(u_b)$. Of course, at any endpoint that is limit-point, no boundary condition is necessary.

The third notion that we shall want to consider is that of analyticity. So far we have ignored the fact that the solutions u_1 and u_2 of (6.2.1) and (6.2.2) depend on the complex parameter z. Let us now explicitly take account of this dependence by writing respectively $u_1(r,z)$ and $u_2(r,z)$. It is a consequence of standard arguments in the theory of differential equations that the solutions $u_i(r,z)$ of the equation $-\mathrm{d}^2u/\mathrm{d}r^2 + Vu - zu = 0$, with fixed initial conditions at the regular point $r = \xi$, depend *analytically* on z. In other words, for each $r \in I$, $u_1(r,z)$ and $u_2(r,z)$ are analytic functions of z in the entire complex plane.

We are interested particularly in what happens at the endpoints $r = a$, $r = b$. There is no reason to expect that u_1 and u_2 are defined, in general, at these endpoints. Instead, it will be consistent with our approach so far to consider the dependence on z of $m_a(z)$ and $m_b(z)$, which are defined for $\mathrm{Im}\,(z) \neq 0$.

How is the z-dependence to be defined? We consider for example the endpoint $r = b$. If b is limit-point then we know that this will be so independently of z, for $\mathrm{Im}\,(z) \neq 0$. The limit point m_b then depends on the complex parameter z in (6.2.1). On the other hand, let us suppose that b is limit-circle. We define a self-adjoint operator H, through Theorem 6.5, by means of the boundary condition

$$W_b(f, \bar{u}_2(\cdot, z_0) + m_b(z_0)u_1(\cdot, z_0)) = 0,$$

where $m_b(z_0)$ is some selected point on the limit circle, for solutions of $-\mathrm{d}^2u/\mathrm{d}r^2 + Vu - z_0u = 0$, and z_0 $(\mathrm{Im}(z_0) \neq 0)$ is some point in the complex plane. If z is any other point in the complex plane $(\mathrm{Im}\,(z) \neq 0)$ then the *same* self-adjoint operator H could be defined by means of a boundary condition

$$W_b(f, \bar{u}_2(\cdot, z) + m_b(z)u_1(\cdot, z)) = 0,$$

where $m_b(z)$ is a point on the limit circle for solutions of $-\mathrm{d}^2u/\mathrm{d}r^2 + Vu - zu = 0$. The point $m_b(z)$, for given z, is uniquely determined once the self-adjoint extension H has been defined through choice of the point $m_b(z_0)$. The choice of the single point $m_b(z_0)$ thus induces a correspondence between points on the various limit circles for different complex values of z. We may therefore enquire, as in the limit-point case, what is the nature of dependence of $m_b(z)$, and similarly $m_a(z)$, on the complex parameter z. We cannot expect here that $m_b(z)$ will be analytic in the entire complex plane, since points on the real axis lie outside the domain of definition. The most we can expect, in general, is that $m_b(z)$ be analytic in the complex plane cut along the real axis.

Lemma 6.1

Let Σ be a closed bounded subset of the complex plane, such that Σ does not intersect the real axis. Then

(i) in the limit-point case at b, $\int_\xi^\eta |u_1(r,z)|^2\,dr$ tends to infinity as $\eta \to b$, and the convergence is uniform in z for $z \in \Sigma$;
(ii) in the limit-circle case at b, $\int_\eta^b |u_i(r,z)|^2\,dr$ $(i=1,2)$ tends to zero as $\eta \to b$, and the convergence is uniform in z for $z \in \Sigma$.

Proof

The existence of these limits comes from the result that in the limit-point case u_1 is not square-integrable near $r=b$, whereas in the limit-circle case both u_1 and u_2 are square-integrable near $r=b$ (cf. Theorem 6.2). Hence what is in question here is *uniformity* of convergence.

(i) We consider first the limit-point case. Given $N>0$, for each $z\in\Sigma$ we can find η', depending on z, such that $\int_\xi^{\eta'} |u_1(r,z)|^2\,dr > N$. Through the arguments leading to (6.2.16)′, with the operator L now acting in $L^2(\xi,\eta')$ and with initial conditions imposed at ξ instead of η, we prove that, as z' approaches z, $u_1(r,z')$ approaches $u_1(r,z)$ in norm. It follows that, by choice of η', the inequality $\int_\xi^\eta |u_1(r,z)|^2\,dr > N$ will hold in some *neighbourhood* of the point z. In this case all these neighbourhoods, for $z\in\Sigma$, will cover the closed bounded set Σ, and, by the Heine–Borel Theorem, Σ is covered by a finite collection of these neighbourhoods. Choosing the largest of the values of η' corresponding to these neighbourhoods, we now have $\int_\xi^{\eta'} |u_1(r,z)|^2\,dr > N$ for all $z\in\Sigma$. Part (i) of the lemma now follows, since N is arbitrary.

(ii) We suppose that b is limit-circle and apply the same argument as above, using the fact that $u_i(r,z')$ approaches $u_i(r,z)$ as $z'\to z$, in norm in the Hilbert space $L^2(\eta,b)$. ■

Theorem 6.7

$m_b(z)$ is an analytic function of z for $\operatorname{Im}(z)\neq 0$.

Proof

(i) We consider first the limit-point case at b. From (6.2.11) and (i) of the lemma, the radius R_η of the circle C_η described by (6.2.8) or (6.2.13) converges uniformly to zero for $z\in\Sigma$. This implies that, as η approaches b, points on C_η converge uniformly to the limit point $m_b(z)$. We choose on C_η the representative point $-u_2(\eta,z)/u_1(\eta,z)$. This corresponds in (6.2.7) to the simple boundary condition $u(\eta)=0$. Certainly $u_1(\eta,z)\neq 0$, since a zero value would imply $R_\eta=\infty$. Thus for $z\in\Sigma$

$$m_b(z)=-\lim_{\eta\to b}\frac{u_2(\eta,z)}{u_1(\eta,z)},$$

which is a uniform limit of functions analytic in z. So $m_b(z)$ is analytic in z, for $\operatorname{Im}(z)\neq 0$.

(ii) We now consider the limit-circle case at b, and use Theorem 6.5 to define a self-adjoint operator H corresponding to some point $m_b(z_0)$ on the limit circle for z_0. By (6.3.3) and (6.3.4), $m_b(z)$ may be characterized by the boundary condition (6.3.6) satisfied at $r=b$ by the function $u_2(r,z)+m_b(z)u_1(r,z)$. More explicitly, this boundary condition takes the form

$$W_b(u_2(\cdot,z)+m_b(z)u_1(\cdot,z),\bar{u}_2(\cdot,z_0)+\bar{m}_b(z_0)\bar{u}_1(\cdot,z_0))=0. \tag{6.4.4}$$

Solving for $m_b(z)$, we have

$$m_b(z)=-\frac{\lim\limits_{\eta\to b} W_\eta(u_2(\cdot,z),\bar{u}_2(\cdot,z_0)+\bar{m}_b(z_0)\bar{u}_1(\cdot,z_0))}{\lim\limits_{\eta\to b} W_\eta(u_1(\cdot,z),\bar{u}_2(\cdot,z_0)+\bar{m}_b(z_0)\bar{u}_1(\cdot,z_0))}. \tag{6.4.5}$$

For fixed η, both $u_i(\eta,z)$ and $u_i'(\eta,z)$ $(i=1,2)$ are analytic in z in the entire complex plane, and certainly in (6.4.5) the denominator cannot vanish. Once again, it remains to prove uniformity of convergence, for $z\in\Sigma$.

Arguing once again from (6.2.4), with $f=u_2(\cdot,z_0)+m_b(z_0)u_1(\cdot,z_0)$ and $g=u_2(\cdot,z)$, the Wronskian in the numerator of (6.4.5) satisfies

$$W_b-W_\eta=(z-\bar{z}_0)\int_\eta^b(\bar{u}_2(r,z_0)+\bar{m}_b(z_0)\bar{u}_1(r,z_0))u_2(r,z)\,\mathrm{d}r,$$

which by the Schwarz inequality is bounded in absolute value by const $(\int_\eta^b|u_2(r,z)|^2\,\mathrm{d}r)^{1/2}$. Then, by (ii) of the lemma, the convergence of W_η to W_b is *uniform* for $z\in\Sigma$. A similar argument applies to the limit in the denominator in (6.4.5). We again have a uniform limit of analytic functions, so that $m_b(z)$ is analytic in z, for $\operatorname{Im}(z)\neq 0$, also in the limit-circle case. ∎

An original motivation for studying solutions of the time-independent Schrödinger equation at complex rather than real energies z was to furnish information concerning the resolvent operator $R(z)=(H-z)^{-1}$. The following theorem shows how to construct $R(z)$ as an integral operator, whose kernel may be expressed simply in terms of $u_i(r,z)$ and $m_a(z)$, $m_b(z)$. Here, as usual, $m_b(z)$ (respectively $m_a(z)$) will denote either the limit point or the point on the limit circle that is used to define H, by means of the boundary condition at b.

Theorem 6.8

The resolvent $R(z)=(H-z)^{-1}$ is an integral operator, with kernel

$$K(r,r';z)=\begin{cases} \dfrac{[u_2(r,z)+m_a(z)u_1(r,z)][u_2(r',z)+m_b(z)u_1(r',z)]}{m_a(z)-m_b(z)} & (r\leq r'), \\ \\ \dfrac{[u_2(r,z)+m_b(z)u_1(r,z)][u_2(r',z)+m_a(z)u_1(r',z)]}{m_a(z)-m_b(z)} & (r\geq r'). \end{cases} \tag{6.4.6}$$

The interpretation of this statement is that, for $f\in L^2(I)$,

$$((H-z)^{-1}f)(r)=\int_a^b \mathrm{d}r'\,K(r,r';z)f(r') \tag{6.4.7}$$

Proof

We note first that the kernel is well defined, since $m_a\neq m_b$ by Theorem 6.2. Indeed, Theorem 6.7 implies that $K(r,r';z)$ is analytic in z ($\mathrm{Im}\,(z)\neq 0$) for fixed $r,r'\in(a,b)$. Since $u_2+m_a u_1$ is square-integrable near $r=a$, and $u_2+m_b u_1$ is square-integrable near $r=b$, it follows easily from (6.4.6) that, for $\mathrm{Im}(z)\neq 0$ and $r\in(a,b)$,

$$\int_a^b \mathrm{d}r'\,|K(r,r';z)|^2<\infty. \tag{6.4.8}$$

The right-hand side of (6.4.7) is therefore a convergent integral for $r\in(a,b)$. Since $R(z)$ is a bounded operator and may be extended by continuity, it will be sufficient to prove (6.4.7) for $f\in C_0^\infty(I)$. We therefore suppose that this is so, and that the support of f is contained in the subinterval $[\alpha,\beta]$ of I. From the differential equation $-\mathrm{d}^2u/\mathrm{d}r^2+Vu-zu=0$ satisfied by u_i $(i=1,2)$, together with local absolute continuity in I of u_i and u_i' as functions of r, it is a straightforward exercise to verify that the right-hand side of (6.4.7) and its derivative with respect to r are locally absolutely continuous in I, and that the right-hand side satisfies the differential equation

$$-\frac{\mathrm{d}^2u}{\mathrm{d}r^2}+Vu-zu=\frac{W(u_2+m_bu_1,u_2+m_au_1)}{m_a-m_b}f(r).$$

As in (6.2.6), the Wronskian is independent of r, and so may conveniently be evaluated at $r=\xi$. By (6.2.2), this gives just

$$W(u_2+m_bu_1,u_2+m_au_1)=m_a-m_b,$$

so that the differential equation satisfied by the right-hand side of (6.4.7) is exactly

$$-\frac{d^2 u}{dr^2}+Vu-zu=f.$$

From the characterization, in Theorem 6.5, of the domain of the Hamiltonian H, we have only to verify the boundary conditions (6.3.6) in order to be permitted to write

$$(H-z)u=f, \qquad \text{or } u=(H-z)^{-1} f \text{ as required.}$$

Using (6.3.6), the right-hand side of (6.4.7) is given by

$$u(r)=\begin{cases} C_a(u_2(r,z)+m_b(z)u_1(r,z)) & (r>\beta), \\ C_b(u_2(r,z)+m_a(z)u_1(r,z)) & (r<\alpha), \end{cases}$$

where

$$(m_a(z)-m_b(z))C_a=\int_a^b dr'\,(u_2(r',z)+m_a(z)u_1(r',z))\,f(r'),$$

$$(m_a(z)-m_b(z))C_b=\int_a^b dr'\,(u_2(r',z)+m_b(z)u_1(r',z))\,f(r').$$

According to (6.3.3), $u(r)$ is then a constant multiple of u_a near $r=a$, and a constant multiple of u_b near $r=b$. Since u_a and u_b, by (6.3.4), satisfy the respective boundary conditions (6.3.6) at $r=a$ and $r=b$, it follows that $u(r)$ satisfies both boundary conditions, and hence that $u\in D(H)$. We are now able to write down (6.4.7), and the proof is complete. ■

There is a simple rule, derived from (6.4.6), for writing down the integral kernel of the resolvent $R(z)$ for an arbitrary Schrödinger operator in the interval (a,b). First we observe, by explicit calculation of the Wronskian at $r=\xi$, that

$$W(u_2+m_a(z)u_1, u_2+m_b(z)u_1)=m_b(z)-m_a(z).$$

Now, for $r\le r'$, the kernel $K(r,r';z)=u(r,z)v(r',z)$ is a product of two functions u and v, say

$$u(r,z)=\frac{u_2(r,z)+m_a(z)u_1(r,z)}{m_a(z)-m_b(z)}$$

$$v(r',z)=u_2(r',z)+m_b(z)u_1(r',z),$$

satisfying the following properties:

(a) u and v each satisfy the Schrödinger equation at (complex) energy z;
(b) $u \in L^2(a,\xi)$ in the limit-point case at a, u satisfies the boundary condition at a in the limit-circle case; $v \in L^2(\xi, b)$ in the limit-point case at b, v satisfies the boundary condition at b in the limit-circle case;
(c) $W(u,v) = -1$.

Conditions (a)–(b) define u and v, each up to a multiplicative function of z. If condition (c) is taken into account, we see that the product $u(r,z)v(r',z)$ is uniquely defined, since any function of z that multiplies u must divide v.

As a simple application, we consider the case in which v is integrable over $[a,c]$, for $a<c<b$. As we saw in Chapter 2, with $l=0$, we can then find solutions $f_1(r,z)$ and $f_2(r,z)$ of the Schrödinger equation, such that

$$f_1(a,z)=0, \qquad f'_1(a,z)=1$$
$$f_2(a,z)=1, \qquad f'_2(a,z)=0.$$

For fixed $m \in \mathbb{R} \cup \{\infty\}$, we define at $r=a$ a boundary condition by $f'(a)/f(a)=m$, with $m=\infty$ corresponding to the boundary condition $f(a)=0$. It is straightforward to verify that $u_2(\cdot,z)+m_a(z)u_1(\cdot,z)$ will satisfy this condition provided that

$$m_a(z) = \frac{u'_2(a,z) - mu_2(a,z)}{mu_1(a,z) - u'_1(a,z)}, \tag{6.4.9}$$

with $m_a(z) = -u_2(a,z)/u_1(a,z)$ for $m=\infty$. This defines $m_a(z)$ as a point on the limit circle, for each value of z, and we may use (6.4.4), for example, with a instead of b, to check that this definition of $m_a(z)$ is consistent with our theory developed above. For the definition of $m_b(z)$, it is convenient in this case to specify boundary conditions at $r=a$ instead of the general point $r=\xi$; i.e. we define $m_b(z)$ by the condition

$$f_2(\cdot,z)+m_b(z)f_1(\cdot,z) \in L^2(c,b).$$

We may then verify, by explicit calculation of the Wronskian at $r=a$, that functions u and v satisfying properties (a)–(c) may be given by

$$u(r,z) = \frac{f_2(r,z)+mf_1(r,z)}{m-m_b(z)},$$

$$v(r,z) = f_2(r,z)+m_b(z)f_1(r,z).$$

The resolvent in this case then has a kernel

$$K(r,r';z) = \begin{cases} \dfrac{[f_2(r,z)+mf_1(r,z)][f_2(r',z)+m_b(z)f_1(r',z)]}{m-m_b(z)} & (r \le r') \\[2ex] \dfrac{[f_2(r,z)+m_b(z)f_1(r,z)][f_2(r',z)+mf_1(r',z)]}{m-m_b(z)} & (r \ge r'). \end{cases} \tag{6.4.10}$$

Note that (6.4.10) may be identified with (6.4.6), where, through (6.4.9) with f_1, f_2 for u_1, u_2, we have $m_a(z)$ replaced by m. Equation (6.4.10) is the standard formula for the resolvent kernel if solutions $f_1(r,z)$ and $f_2(r,z)$ can be found that are differentiable at $r=a$, but does not apply, for example, if $V(r)$ is singular at $r=a$, in which case the more general result (6.4.6) may be used.

6.5 EIGENFUNCTION EXPANSIONS AND THE SPECTRAL MATRIX

One of the most important techniques of spectral analysis is the method of eigenfunction expansions. In its simplest form, for an operator H having a purely discrete spectrum, this involves forming an orthonormal basis of eigenvectors of H, and expressing an arbitrary element f of the Hilbert space as a linear combination of these basis elements.

The situation is a little different if H is allowed to have a continuous spectrum. In this case one sets out to express f as an integral rather than a sum, over a family of so-called improper eigenvectors. If H is a differential operator acting in a function space, then these improper eigenfunctions will be formal solutions of the differential equation $(H-\lambda)u=0$, but will not, in general, be square-integrable functions in the Hilbert space. In many applications, they will belong instead to some associated Banach space. The following definition gives a precise characterization of eigenfunction expansions for the differential operator $-\mathrm{d}^2/\mathrm{d}r^2+V(r)$ in $L^2(I)$.

Definition 6.6

Two linearly independent families of solutions $u_1(r,\lambda)$ and $u_2(r,\lambda)$ of the differential equation

$$-\frac{\mathrm{d}^2 u}{\mathrm{d}r^2}+Vu=\lambda u \quad (\lambda\in\mathbb{R},\ r\in I=(a,b)) \tag{6.5.1}$$

are said to define an eigenfunction expansion if every $f\in L^2(I)$ may be expressed, for some Borel measures $\mathrm{d}f_1$ and $\mathrm{d}f_2$, depending linearly on f, in the form

$$f(r)=\int_{-\infty}^{\infty} u_1(r,\lambda)\,\mathrm{d}f_1(\lambda)+\int_{-\infty}^{\infty} u_2(r,\lambda)\,\mathrm{d}f_2(\lambda). \tag{6.5.2}$$

(The integrals on the right-hand side are to be defined as strong limits, as $N \to \infty$, of the corresponding integrals between $-N$ and $+N$.)

Equation (6.5.2) includes discrete expansions as a special case, if f_1 and f_2 are step functions.

The following theorem establishes the existence of eigenfunction expansions in this sense, and shows how to construct an eigenfunction expansion corresponding to any self-adjoint operator $H = -d^2/dr^2 + V$ in $L^2(I)$.

Theorem 6.9

Let $u_1(r,\lambda)$ and $u_2(r,\lambda)$ be the solutions of (6.5.1), subject to the initial conditions

$$\left.\begin{aligned} u_1(\xi,\lambda)&=0, & u_1'(\xi,\lambda)&=1,\\ u_2(\xi,\lambda)&=1, & u_2'(\xi,\lambda)&=0, \end{aligned}\right\} \tag{6.5.3}$$

where a prime denotes differentiation with respect to r. Then these solutions define an eigenfunction expansion.

Moreover, if $\{E_\lambda\}$ denotes the spectral family associated with some self-adjoint operator $H = -d^2/dr^2 + V$ defined as in Theorem 6.5, and if $f(r,\lambda)$ is defined by

$$f(r,\lambda) = ((E_\lambda - E_0)f)(r), \tag{6.5.4}$$

then, for all $\lambda \in \mathbb{R}$,

$$f(r,\lambda) = \int_0^\lambda u_1(r,\mu)\,df_1(\mu) + \int_0^\lambda u_2(r,\mu)\,df_2(\mu), \tag{6.5.5}$$

where the non-decreasing functions f_1 and f_2 may be given in terms of one-parameter families $\{\Gamma_i(\lambda)\}$ of vectors in $L^2(I)$ by

$$f_i(\lambda) = \langle \Gamma_i(\lambda), f\rangle \tag{6.5.6}$$

Proof

We notice first that $(E_\lambda - E_0)f$ belongs to $D(H)$, since, for example, with $\lambda > 0$ we have

$$\|H(E_\lambda - E_0)f\|^2 = \int_0^\lambda \mu^2\, d\langle f, E_\mu f\rangle < \infty.$$

According to (6.3.5), $(E_\lambda - E_0)f$ may be represented in $L^2(I)$, for fixed λ, by a wave function $f(r,\lambda)$ such that both f and f' are locally absolutely continuous in I.

Since $(E_\lambda - E_0)f = (H-z)^{-1}(H-z)(E_\lambda - E_0)f$, we may use the representation (6.4.6) of the kernel $K(r,r')$ of the resolvent (we drop for the moment explicit reference to the complex number z) to write

$$f(r,\lambda) = \int_a^b \mathrm{d}r'\, K(r,r')((H-z)(E_\lambda - E_0)f)(r'). \tag{6.5.7}$$

Since $K(r,\cdot) \in L^2(I)$ (meaning that $\int_a^b \mathrm{d}r'\,|K(r,r')|^2 < \infty$), (6.5.7) may be rewritten as

$$\begin{aligned} f(r,\lambda) &= \langle \bar{K}(r,\cdot),\, (H-z)(E_\lambda - E_0)f \rangle \\ &= \langle (H-\bar{z})(E_\lambda - E_0)\bar{K}(r,\cdot), f \rangle. \end{aligned} \tag{6.5.7$'$}$$

Hence for $\lambda \geq 0$ we have

$$\begin{aligned} -f''(r,\lambda) + V(r)f(r,\lambda) &= (H(E_\lambda - E_0)f)(r) \\ &= \langle (H-\bar{z})(E_\lambda - E_0)\bar{K}(r,\cdot), H(E_\lambda - E_0)f \rangle \\ &= \int_0^\lambda \mu\, \mathrm{d}_\mu \langle (H-\bar{z})(E_\lambda - E_0)\bar{K}(r,\cdot), E_\mu f \rangle \\ &= \int_0^\lambda \mu\, \mathrm{d}_\mu \langle (H-\bar{z})(E_\mu - E_0)\bar{K}(r,\cdot), f \rangle, \end{aligned}$$

where we have used

$$E_\lambda E_\mu = E_\mu, \qquad E_0 E_\mu = E_0, \qquad [E_\mu, H] = 0.$$

Comparison with (6.5.7)$'$ now gives

$$-f''(r,\lambda) + V(r) f(r,\lambda) = \int_0^\lambda \mu\, \mathrm{d}_\mu f(r,\mu), \tag{6.5.8}$$

and the same equation also holds for $\lambda < 0$.

This is an integro-differential equation for $f(r,\lambda)$. Note that, on integrating by parts, the right-hand side could also be written as $\lambda f(r,\lambda) - \int_0^\lambda \mathrm{d}\mu\, f(r,\mu)$. Regarding the right-hand side as the inhomogeneous term of a second-order differential equation, the method of variation of constants gives

$$f(r,\lambda)=f_1(\lambda)u_1(r,\lambda)+f_2(\lambda)u_2(r,\lambda)$$
$$+\int_\xi^r \mathrm{d}r'[u_2(r,\lambda)u_1(r',\lambda)-u_1(r,\lambda)u_2(r',\lambda)]$$
$$\times\left[-\int_0^\lambda \mu \mathrm{d}_\mu f(r',\mu)\right], \tag{6.5.9}$$

where

$$f_1(\lambda)=f'(\xi,\lambda), \qquad f_2(\lambda)=f(\xi,\lambda) \tag{6.5.10}$$

Equation (6.5.9) is an integral equation for $f(r,\lambda)$. We need this equation only to observe that the solution, and hence also the solution of (6.5.8), may be obtained by standard iterative methods, and is unique for given $f_1(\lambda)$ and $f_2(\lambda)$. (That is, $f_1(\lambda)$ and $f_2(\lambda)$ must be known for *all* real λ, in order to determine $f(r,\lambda)$ for all $r\in I, \lambda\in\mathbb{R}$.)

We next observe that the right-hand side of (6.5.5), with *arbitrary* choice of f_1 and f_2, is a solution of (6.5.8), since, by differentiation under the integral sign, (6.5.5) implies

$$-\frac{\mathrm{d}^2 f(r,\lambda)}{\mathrm{d}r^2}+V(r)f(r,\lambda)$$
$$=\int_0^\lambda \{\mu u_1(r,\mu)\,\mathrm{d}f_1(\mu)+\mu u_2(r,\mu)\,\mathrm{d}f_2(\mu)\}$$
$$=\int_0^\lambda \mu\,\mathrm{d}_\mu\left\{\int_0^\mu (u_1(r,\mu')\mathrm{d}f_1(\mu')+u_2(r,\mu')\mathrm{d}f_2(\mu'))\right\}.$$

At $r=\xi$ the right-hand side of (6.5.5), using (6.5.3), is

$$\int_0^\lambda \{0\,\mathrm{d}f_1(\mu)+1\,\mathrm{d}f_2(\mu)\}=f_2(\lambda)-f_2(0),$$

and similarly the derivative with respect to ξ, at $r=\xi$ is $f_1(\lambda)-f_1(0)$. Hence, with the choice (6.5.10) of the functions f_1 and f_2, so that $f_1(0)=f_2(0)=0$, the right-hand side of (6.5.5) satisfies both the integro-differential equation (6.5.8) and the initial conditions at $r=\xi$. It follows at once, by uniqueness, that the function $f(r,\lambda)$ defined in (6.5.4) is indeed given by (6.5.5) with measures df_i given by (6.5.10).

Equation (6.5.2) may now be deduced from (6.5.5), using the fact that $f(r)$ is the strong limit, as $N\to\infty$, of

$$f(r,N)-f(r,-N).$$

We may now use (6.5.10), together with (6.5.7)′, to evaluate $f_1(\lambda)$ and $f_2(\lambda)$. Writing

$$\gamma_1=\bar{K}'(\xi,\cdot), \qquad \gamma_2=\bar{K}(\xi,\cdot), \tag{6.5.11}$$

we see that (6.5.6) is satisfied with

$$\Gamma_i(\lambda)=(H-\bar{z})(E_\lambda-E_0)\gamma_i. \tag{6.5.12}$$

■

Notice that $f_i(\lambda)$, by (6.5.10), is independent of z, so that we know in (6.5.6), which determines $\Gamma_i(\lambda)$ once the $f_i(\lambda)$ are known for each f, that $\Gamma_i(\lambda)$ is also independent of z (despite the fact that γ_i depends on z in (6.5.12)). Note also that the zero of energy $\lambda=0$ plays no essential role in the theory, and we could just as easily have considered, say, $(E_\lambda-E_\beta)f$ instead of $(E_\lambda-E_0)f$, with arbitrary β.

If in (6.5.12) γ_i lies in $D(H)$, we write

$$\Gamma_i(\lambda)=(E_\lambda-E_0)\Gamma_i, \tag{6.5.13}$$

where $\Gamma_i=(H-\bar{z})\gamma_i$ is again independent of z. For notational convenience we shall continue to write (6.5.13) even if γ_i does not lie in $D(H)$. We may then think of Γ_i as some non-normalized limit of $\Gamma_i(N)-\Gamma_i(-N)$ as $N\to\infty$ in some larger space than $\mathscr{H}=L^2(I)$. It will be unnecessary to consider this limit in more detail, since we shall take care that vectors that we write down will be expressible in terms of the $\Gamma_i(\lambda)$.

Since the spectral family $\{E_\lambda\}$ of a real operator H will itself be a family of real operators, it follows from (6.5.4) that if $f\in L^2(I)$ is represented by a real wave function $f(r)$ then $f(r,\lambda)$ will also be real for each λ. From (6.5.6), each $\Gamma_i(\lambda)\in L^2(I)$ must then be represented by a real wave function.

Let us now define a one-parameter family $\{\rho_{ij}(\lambda)\}$ of 2×2 matrices by

$$\rho_{ij}(\lambda)=\langle\Gamma_i,(E_\lambda-E_0)\Gamma_j\rangle. \tag{6.5.14}$$

In other words, remembering that

$$(E_\lambda-E_0)^2=\begin{cases}(E_\lambda-E_0) & (\lambda\geq 0),\\ -(E_\lambda-E_0) & (\lambda<0),\end{cases}$$

we have, from (6.5.13),

$$\rho_{ij}(\lambda)=\begin{cases}\langle\Gamma_i(\lambda),\Gamma_j(\lambda)\rangle & (\lambda\geq 0),\\ -\langle\Gamma_i(\lambda),\Gamma_j(\lambda)\rangle & (\lambda<0).\end{cases}\tag{6.5.14$'$}$$

Since the $\Gamma_i(\lambda)$ are real, $\rho_{ij}(\lambda)$, for each λ, will be a real symmetric matrix.

Definition 6.7

The matrix-values function $\rho_{ij}(\lambda)$, defined by (6.5.14)$'$, is called a *spectral matrix* associated with the self-adjoint operator H. (There will be a one-parameter family of spectral matrices, corresponding to the choice of the point $\xi\in(a,b)$.) The Hilbert space $L^2(\mathbb{R};\mathrm{d}\rho_{ij})$ consists of ordered pairs of functions $h=\{h_1(\lambda),h_2(\lambda)\}$ such that

$$\|h\|^2\equiv\sum_{i,j}\int_{-\infty}^{\infty}\bar{h}_i(\lambda)h_j(\lambda)\,\mathrm{d}\rho_{ij}(\lambda)<\infty.\tag{6.5.15}$$

(More precisely, $L^2(\mathbb{R};\mathrm{d}\rho_{ij})$ consists of equivalence classes of ordered pairs in the usual way. Thus h and h^* will belong to the same equivalence class provided that $\|h-h^*\|=0$, and we adopt the usual practice of *identifying* the functions h and h^* in this case, as representing the same element of the Hilbert space. We can see that $\|h\|^2\geq 0$ in (6.5.15), since, from (6.5.14),

$$\|h\|^2=\|\sum_i h_i(H)\Gamma_i\|^2,\tag{6.5.16}$$

whenever the right-hand side is finite. Although Γ_i in this equation need not belong to $\mathscr{H}$, $h_i(H)\Gamma_i$ may be defined as the strong limit, as $N\to\infty$, of

$$\int_{-N}^{N}h_i(\lambda)\,\mathrm{d}\Gamma_i(\lambda).$$

The inner product in $L^2(\mathbb{R};\mathrm{d}\rho_{ij})$, having specified the norm, is defined in an obvious way.)

The importance of spectral matrices and their Hilbert spaces comes from the following result.

Theorem 6.10

For $f\in L^2(I)\equiv L^2(a,b)$, let

$$h_i(\lambda)=\int_a^b \mathrm{d}r\,f(r)u_i(r,\lambda),\tag{6.5.17}$$

the right-hand side being defined as a strong limit as $\xi \to a$, $\eta \to b$, in $L^2(\mathbb{R}, d\rho_{ij})$, of the integral between ξ and η. Then (6.5.17) defines a unitary mapping from the whole of $L^2(I)$ onto $L^2(\mathbb{R}, d\rho_{ij})$, the inverse transformation being

$$f(r) = \sum_{i,j} \int_{-\infty}^{\infty} h_i(\lambda) u_j(r, \lambda)\, d\rho_{ij}(\lambda) \tag{6.5.17$'$}$$

(strong limit, in $L^2(I)$, of the integral from $-N$ to N). Equations (6.5.17) and (6.5.17)$'$ establish the unitary equivalence between the self-adjoint operator $H = -d^2/dr^2 + V$ in $L^2(I)$ and the operator of multiplication by λ in $L^2(\mathbb{R}, d\rho_{ij})$. The solution $u_i(r, \lambda)$ satisfies the relation

$$\sum_{i,j} \int_{-\infty}^{\infty} u_i(r, \lambda) u_j(r', \lambda)\, d\rho_{ij}(\lambda) = \delta(r - r'), \tag{6.5.18}$$

in the sense described below.

Proof

We define $\Gamma_j(r, \lambda)$ by the equation corresponding to (6.5.4) with Γ_j instead of f. Comparison with (6.5.13) shows that

$$\Gamma_j(\cdot, \lambda) = \begin{cases} (E_\lambda - E_0)\Gamma_j(\lambda) & (\lambda \geq 0), \\ -(E_\lambda - E_0)\Gamma_j(\lambda) & (\lambda < 0). \end{cases}$$

In (6.5.5) with Γ_j instead of f and with $f_i(\lambda) = \langle \Gamma_i(\lambda), \Gamma_j \rangle = \rho_{ij}(\lambda)$ from (6.5.6) and (6.5.14), we have

$$\Gamma_j(r, \lambda) = \sum_i \int_0^{\lambda} u_i(r, \lambda)\, d\rho_{ij}(\lambda). \tag{6.5.19}$$

We now take f to be any element of $L^2(I)$ having compact support in I, in which case (6.5.19) applied to (6.5.5) and (6.5.6) implies

$$\begin{aligned} f(r, \lambda) &= \sum_j \int_0^{\lambda} u_j(r, \mu)\, d_\mu \left\{ \sum_i \int_0^{\mu} \langle u_i(\cdot, \mu'), f \rangle\, d\rho_{ij}(\mu') \right\} \\ &= \sum_{i,j} \int_0^{\lambda} u_j(r, \mu) h_i(\mu)\, d\rho_{ij}(\mu), \end{aligned} \tag{6.5.20}$$

where $h_i(\mu)=\langle u_i(\cdot,\mu),f\rangle$ as in (6.5.17). Multiplying (6.5.20) by $\bar{f}(r)$ and integrating over r, we have

$$\langle f,(E_\lambda - E_0)f\rangle = \int_I \bar{f}(r)f(r,\lambda)\,\mathrm{d}r$$

$$=\sum_{i,j}\int_0^\lambda h_i(\mu)\bar{h}_j(\mu)\,\mathrm{d}\rho_{ij}(\mu),$$

from which it easily follows, on taking limits $\lambda\to\pm\infty$, that

$$\|f\|^2=\sum_{i,j}\int_{-\infty}^{\infty} h_i(\lambda)\bar{h}_j(\lambda)\,\mathrm{d}\rho_{ij}(\lambda).$$

Comparison with (6.5.15) now shows that the mapping defined by (6.5.17) is norm-preserving. Equations (6.5.17) and (6.5.20) may immediately be extended to arbitrary $f\in L^2(I)$ by taking strong limits, and (6.5.17)′ is a consequence of the fact that $f(r)$ is the strong limit in $L^2(I)$, as $N\to\infty$, of $f(r,N)-f(r,-N)$.

Using (6.5.19), (6.5.17)′ may be written as

$$f=\sum_i \int_{-\infty}^{\infty} h_i(\lambda)\,\mathrm{d}\Gamma_i(\lambda)=\sum_i h_i(H)\Gamma_i,$$

which means, by (6.5.16), that in (6.5.17)′ $f\in L^2(I)$ may be defined for arbitrary $\{h_1,h_2\}$ in $L^2(\mathbb{R},\mathrm{d}\rho_{ij})$. In other words, the mapping from $L^2(I)$ to $L^2(\mathbb{R},\mathrm{d}\rho_{ij})$ is *onto*, and hence also unitary.

From (6.5.8), we have, for $f\in D(H)$,

$$(Hf)(r,\lambda)=\int_0^\lambda \mu\,\mathrm{d}f(r,\mu),$$

and, on substituting from (6.5.20) for $f(r,\mu)$, we find that, on operating by H on $f, h_i(\mu)$ on the right-hand side of (6.5.20) is to be replaced by $\mu h_i(\mu)$. In other words, the operation of H corresponds, in $L^2(\mathbb{R},\mathrm{d}\rho_{ij})$, to the operator of multiplication by λ.

From (6.5.20), we are able to infer that, for any given λ, the operator

$E_\lambda - E_0$ is an integral operator acting in $L^2(I)$, with kernel given by

$$k(r,r',\lambda)=\sum_{i,j}\int_0^\lambda u_i(r,\lambda)u_j(r',\lambda)\,\mathrm{d}\rho_{ij}(\lambda).$$

Using (6.5.19), the kernel may also be written $\Sigma_{i,j}\ (u_i(r,H)(E_\lambda - E_0)\Gamma_i)(r')$, showing that, as for the (not unrelated) kernel $K(r,r',\lambda)$ of Theorem 6.8, we have, for each $r\in I$,

$$\int_I \mathrm{d}r'|k(r,r',\lambda)|^2<\infty.$$

If Λ is an arbitrary bounded Borel subset of $\mathbb{R}$ then the spectral projection of H associated with the set Λ will again be an integral operator, with kernel

$$k(r,r',\Lambda)=\sum_{i,j}\int_\Lambda u_i(r,\lambda)u_j(r',\lambda)\,\mathrm{d}\rho_{ij}(\lambda).$$

Equation (6.5.18) may now be regarded as the limiting case in which the subset Λ expands to the entire real line. In this case, we have

$$\lim_{\Lambda\to R} k(r,r',\Lambda)=\delta(r-r'),$$

in the sense that, for arbitrary $f\in L^2(I)$,

$$\operatorname*{s-lim}_{\Lambda\to \mathrm{R}}\int_\Lambda \mathrm{d}r'\,k(r,r',\Lambda)f(r')=f(r).$$

In this sense, the kernel in the limit may be regarded as approaching a δ-distribution. ■

Equations (6.5.14) and (6.5.14)′ do not really furnish us with the means to carry out a direct evaluation of the spectral matrix in particular cases, since neither the spectral projections nor the Γ_i are usually susceptible to determination in any simple way. It is, however, possible to express $\rho_{ij}(\lambda)$ in terms of the $m_a(z)$ and $m_b(z)$ that have been used, in Theorem 6.5, to define the self-adjoint operator $H=-\mathrm{d}^2/\mathrm{d}r^2+V$ in $L^2(I)$. As usual, z is the complex parameter in the differential equation $-\mathrm{d}^2u/\mathrm{d}r^2+Vu-zu=0$, and $m_a(z)$ (similarly $m_b(z)$) is the limit point if $r=a$ is limit-point and otherwise the point on the limit circle at $r=a$, which is used to define the boundary condition at a.

We already know (Theorem 6.7) that $m_a(z)$ and $m_b(z)$ are analytic functions of z for $\mathrm{Im}\,(z)\neq 0$, and that $m_a(z)\neq m_b(z)$ for $\mathrm{Im}\,(z)\neq 0$ (Theorem 6.2). These two results allow us to define a 2×2 matrix $M_{ij}(z)$, having elements analytic in z for $\mathrm{Im}\,(z)\neq 0$, by

$$\left.\begin{aligned} M_{11}(z)&=\frac{m_a(z)m_b(z)}{m_a(z)-m_b(z)},\\ M_{12}(z)=M_{21}(z)&=\frac{m_a(z)+m_b(z)}{2(m_a(z)-m_b(z))},\\ M_{22}(z)&=\frac{1}{m_a(z)-m_b(z)}.\end{aligned}\right\} \tag{6.5.21}$$

We also know (Section 6.4) that H is a real operator, so that, in the notation of that section,

$$\rho_b(u_2(r,z)+m_b(z)u_1(r,z))\in D(H)$$

implies also

$$\rho_b(\bar{u}_2(r,z)+\bar{m}_b(z)\bar{u}_1(r,z))\in D(H).$$

Since $\bar{u}_i(r,z)=u_i(r,\bar{z})$ $(i=1,2)$, we then have

$$\rho_b(u_2(r,\bar{z})+\bar{m}_b(z)u_1(r,\bar{z})\in D(H),$$

which implies that

$$\bar{m}_b(z)=m_b(\bar{z}), \tag{6.5.22}$$

and similarly for m_a. From (6.5.21), we now have

$$\bar{M}_{ij}(z)=M_{ij}(\bar{z}). \tag{6.5.22$'$}$$

A function $f(z)$ that is analytic for $\mathrm{Im}\,(z)\neq 0$ and that satisfies $\bar{f}(z)=f(\bar{z})$ for $\mathrm{Im}\,(z)\neq 0$ is said to be *real analytic*. Thus we have shown that $m_a(z)$, $m_b(z)$ and $M_{ij}(z)$ are all real analytic in the complex plane for $\mathrm{Im}\,(z)\neq 0$.

The following theorem expresses the spectral matrix ρ_{ij}, up to an additive constant matrix, in terms of boundary values, as z approaches the real axis, of the matrix $M_{ij}(z)$. More precisely, the spectral measure $\rho_{ij}(\beta)-\rho_{ij}(\alpha)$ of an interval is expressed as an integral between α and β of the *discontinuity* between $M_{ij}(z)$ for values of z respectively just above and just below the axis $\mathrm{Im}\,(z)=0$. Since from (6.5.14), and the corresponding property of the spectral family $\{E_\lambda\}$, $\rho_{ij}(\lambda)$ will be right-continuous, it will be sufficient to evaluate $\rho_{ij}(\beta)-\rho_{ij}(\alpha)$ for points $\lambda=\alpha,\beta$ at which $\rho_{ij}(\lambda)$ is continuous. The points of discontinuity, being eigenvalues of H, are countable in number, and the spectral matrix at these points may be obtained by taking limits from the right.

We then have the following theorem.

Theorem 6.11

The spectral matrix $\rho_{ij}(\lambda)$ (Definition 6.7) is given, at points of continuity α, β, by

$$\rho_{ij}(\beta)-\rho_{ij}(\alpha)=\lim_{\varepsilon\to 0+}\frac{1}{\pi}\int_{\alpha}^{\beta}\mathrm{d}x\,\mathrm{Im}\,(M_{ij}(x+\mathrm{i}\varepsilon)) \tag{6.5.23}$$

$$=\lim_{\varepsilon\to 0+}\frac{1}{2\pi\mathrm{i}}\int_{\alpha}^{\beta}\mathrm{d}x\,\{M_{ij}(x+\mathrm{i}\varepsilon)-M_{ij}(x-\mathrm{i}\varepsilon)\}. \tag{6.5.23$'$}$$

Inversely, the imaginary part of the matrix $M_{ij}(z)$ is given, in terms of the spectral matrix, by

$$\mathrm{Im}\,(M_{ij}(z))=\int_{-\infty}^{\infty}\frac{\mathrm{Im}\,(z)}{|\lambda-z|^2}\,\mathrm{d}\rho_{ij}(\lambda), \tag{6.5.24}$$

where in all formulae the matrix M_{ij} is given by (6.5.21).

Proof

From (6.5.12) and (6.5.14)$'$, by means of the spectral theorem, we have (always for $\mathrm{Im}\,(z)\neq 0$)

$$\rho_{ij}(\lambda)=\int_{0}^{\lambda}|\lambda-z|^2\,\mathrm{d}\langle\gamma_i, E_\lambda\gamma_j\rangle, \tag{6.5.25}$$

and we may also write down an inverse to this equation, namely

$$\langle\gamma_i,\gamma_j\rangle=\int_{-\infty}^{\infty}\mathrm{d}\langle\gamma_i, E_\lambda\gamma_j\rangle$$

$$=\int_{-\infty}^{\infty}|\lambda-z|^{-2}\,\mathrm{d}\rho_{ij}(\lambda). \tag{6.5.25$'$}$$

The strategy will be to evaluate the left-hand side of this equation, which for each value of z may be expressed explicitly in terms of m_a and m_b. For example, from (6.5.11) for γ_1 and γ_2 we have

$$\langle\gamma_1,\gamma_2\rangle=\int_a^b \mathrm{d}r'\, \bar{K}(\xi,r')K'(\xi,r'). \tag{6.5.26}$$

Using Theorem 6.8 to substitute in (6.5.26) for the kernel K, and dividing the integration region $I=(a,b)$ into two subintervals, (6.5.26) becomes

$$\langle\gamma_1,\gamma_2\rangle=(m_a-m_b)^{-1}(\bar{m}_a-\bar{m}_b)^{-1}\Big\{(\bar{u}_2+\bar{m}_a\bar{u}_1)_{r=\xi}(u_2'+m_a u_1')_{r=\xi}$$

$$\times\int_\xi^b \mathrm{d}r'\,(\bar{u}_2+\bar{m}_b\bar{u}_1)(u_2+m_b u_1)+(\bar{u}_2+\bar{m}_b\bar{u}_1)_{r=\xi}(u_2'+m_b u_1')_{r=\xi}$$

$$\times\int_a^\xi \mathrm{d}r'\,(\bar{u}_2+\bar{m}_a\bar{u}_1)(u_2+m_a u_1)\Big\} \tag{6.5.27}$$

At $r=\xi$, the initial conditions (6.5.3) for u_1 and u_2 imply

$$(\bar{u}_2+\bar{m}_a\bar{u}_1)(u_2'+m_a u_1')=m_a,$$

$$(\bar{u}_2+\bar{m}_b\bar{u}_1)(u_2'+m_b u_1')=m_b.$$

By (6.2.5), we also have

$$2\mathrm{i}\,\mathrm{Im}\,(z)\int_\xi^b \mathrm{d}r'(\bar{u}_2+\bar{m}_b\bar{u}_1)(u_2+m_b u_1)$$

$$=[W(u_2+m_b u_1,\bar{u}_2+\bar{m}_b\bar{u}_1)]_\xi^b$$

$$=-W_\xi(u_2+m_b u_1,\bar{u}_2+\bar{m}_b\bar{u}_1),$$

the Wronskian vanishing, by Theorem 6.2, at $r=b$. Again substituting for u_i and u_i' at $r=\xi$ from the initial conditions, we obtain

$$\int_\xi^b \mathrm{d}r'\,(\bar{u}_2+\bar{m}_b\bar{u}_1)(u_2+m_b u_1)=\frac{1}{2\mathrm{i}\,\mathrm{Im}\,(z)}(m_b-\bar{m}_b). \tag{6.5.28}$$

The second integral on the right-hand side of (6.5.27) may be evaluated in the same way, and (6.5.27) becomes

$$\begin{aligned}\langle\gamma_1,\gamma_2\rangle &= \frac{1}{2\mathrm{i}\,\mathrm{Im}\,(z)}\,\frac{m_b\bar{m}_a - m_a\bar{m}_b}{(m_a-m_b)(\bar{m}_a-\bar{m}_b)} \\ &= \frac{1}{2\mathrm{i}\,\mathrm{Im}\,(z)}\left\{\frac{m_b}{m_a-m_b} - \frac{\bar{m}_b}{\bar{m}_a-\bar{m}_b}\right\} \\ &= \frac{1}{\mathrm{Im}\,(z)}\,\mathrm{Im}\left(\frac{m_b}{m_a-m_b}\right) \\ &= \frac{1}{\mathrm{Im}\,(z)}\,\mathrm{Im}\left(\frac{m_a+m_b}{2(m_a-m_b)}\right).\end{aligned}$$

Comparing with (6.5.21), we then have

$$\langle\gamma_1,\gamma_2\rangle = \frac{\mathrm{Im}\,(M_{12}(z))}{\mathrm{Im}\,(z)}.$$

The remaining inner products $\langle\gamma_i,\gamma_j\rangle$ may be evaluated in a similar way, and we find quite generally

$$\langle\gamma_i,\gamma_j\rangle = \frac{\mathrm{Im}\,(M_{ij}(z))}{\mathrm{Im}\,(z)}.$$

Comparison with (6.5.25)′ now leads to (6.5.24). Thus the integral on the right-hand side of (6.5.24) is seen to be the imaginary part of a function analytic in the entire complex plane apart from the line $\mathrm{Im}\,(z)=0$. We shall say that $M_{ij}(z)$ is analytic in the *cut plane*, the cut lying along the real axis. This relationship between the integral and an analytic function may be brought out more clearly by writing

$$\frac{\mathrm{Im}\,(z)}{|\lambda-z|^2} = \mathrm{Im}\left(\frac{1}{\lambda-z}\right),$$

in which case (6.5.24) becomes

$$\mathrm{Im}\,(M_{ij}(z)) = \int_{-\infty}^{\infty} \mathrm{Im}\left(\frac{1}{\lambda-z}\right) \mathrm{d}\rho_{ij}(\lambda). \qquad (6.5.29)$$

Whereas $\mathrm{Im}\,((\lambda-z)^{-1})$ decreases for large $|\lambda|$ like λ^{-2}, $\mathrm{Re}((\lambda-z)^{-1})$ decreases only like λ^{-1}. Hence problems of convergence may prevent us from writing the right-hand side of (6.5.29) as $\mathrm{Im}\int_{-\infty}^{\infty}(\lambda-z)^{-1}\,\mathrm{d}\rho_{ij}(\lambda)$. Instead of doing this, we consider first $M_{ij}(z)-M_{ij}(z_0)$, where z_0 is an arbitrary fixed complex number ($\mathrm{Im}\,(z_0)\neq 0$). Since $(\lambda-z)^{-1}-(\lambda-z_0)^{-1}$ decreases for large $|\lambda|$ like λ^{-2}, (6.5.29) allows us to identify the imaginary parts of the functions $M_{ij}(z)-M_{ij}(z_0)$ and $\int_{-\infty}^{\infty}\{(\lambda-z)^{-1}-(\lambda-z_0)^{-1}\}\,\mathrm{d}\rho_{ij}(\lambda)$, for $\mathrm{Im}\,(z)\neq 0$.

Note that this integral is already analytic in the cut plane, being a uniformly convergent integral of an analytic function of z in any closed and bounded region not intersecting the real axis. It is also an elementary consequence of the Cauchy–Riemann equations that any two analytic functions having the same imaginary part can differ only by a constant. Since each function in this case vanishes at $z=z_0$, the constant is identically zero, so that

$$M_{ij}(z)-M_{ij}(z_0)=\int_{-\infty}^{\infty}\{(\lambda-z)^{-1}-(\lambda-z_0)^{-1}\}\,\mathrm{d}\rho_{ij}(\lambda). \qquad (6.5.30)$$

Equation (6.5.30) provides us with a representation of $M_{ij}(z)$ as an integral with respect to the spectral matrix, with *one subtraction*. In order to express the spectral matrix in terms of M_{ij}, we need to invert this equation. To do so, we set $z=x+\mathrm{i}\varepsilon$ in (6.5.24) and integrate from α to β with respect to x. There is no problem exchanging orders of integration on the right-hand side

Writing

$$I_\varepsilon=\int_\alpha^\beta \frac{\varepsilon}{(\lambda-x)^2+\varepsilon^2}\,\mathrm{d}x=\left[\tan^{-1}\left(\frac{x-\lambda}{\varepsilon}\right)\right]_\alpha^\beta,$$

we have the limits

$$\lim_{\varepsilon\to 0+} I_\varepsilon=\begin{cases}0 & (\lambda\notin[\alpha,\beta]),\\ \pi & (\lambda\in(\alpha,\beta)),\\ \tfrac{1}{2}\pi & (\lambda=\alpha \text{ or } \lambda=\beta).\end{cases}$$

Since, for fixed α and β, I_ε is both bounded uniformly in ε as $\varepsilon\to 0$ through positive values, and bounded uniformly in ε for large $|\lambda|$ by const$/\lambda^2$, the Lebesgue Dominated-Convergence Theorem may be used to exchange the limit $\varepsilon\to 0+$ with integration with respect to λ. Assuming that α and β are not points of discontinuity of the spectral matrix as a function of λ, we arrive at (6.5.23). If α and β *are* points of discontinuity then there will be further contributions coming from the endpoint of the integration in λ. Equation (6.5.23)′ now follows on identifying $M_{ij}(x-\mathrm{i}\varepsilon)$ with the complex conjugate of $M_{ij}(x+\mathrm{i}\varepsilon)$. ∎

6.6 REGIONS OF MEROMORPHY

We have seen, in Theorem 6.7, that $m_b(z)$, and similarly of course $m_a(z)$, are analytic functions of z as long as we keep away from the real axis. Whereas $m_b(z)$ has positive imaginary part for z in the upper half-plane, $m_a(z)$ will have negative imaginary part in the same region.

The region of analyticity for $m_b(z)$ does not extend to the real axis. Indeed, the nature of the singularities on the real axis of the functions $M_{ij}(z)$ defined by (6.5.21) play a crucial role in determining the spectral properties of the Schrödinger Hamiltonian, as we shall see in the next chapter. There are, however, two important cases for which we may find a domain of *meromorphy*, which includes at least part of the real axis. A function $F(z)$ is said to be meromorphic, as a function of the complex variable z in some domain, if $F(z)$ is analytic in that domain, apart from isolated poles of finite order. The following theorem gives sufficient conditions for $m_a(z)$ or $m_b(z)$ to be meromorphic in a domain including a subinterval of $\mathbb{R}$.

Theorem 6.12

For a range of energies $\lambda_1 < \lambda < \lambda_2$, suppose that there is a non-trivial solution $f(r, \lambda)$ of the Schrödinger equation

$$-\frac{d^2 f}{dr^2} + V(r) f = \lambda f \tag{6.6.1}$$

that is square-integrable at $r = a$ (that is, $f(\cdot, \lambda) \in L^2(a, c)$ for any c such that $a < c < b$). Then for each ξ in the interval (a, b) *either* $f'(\xi, \lambda)/f(\xi, \lambda)$, regarded as a function of λ for fixed ξ, assumes every real value in the interval (λ_1, λ_2), *or* $m_a(z)$ is meromorphic in the entire complex plane apart from the subset $(-\infty, \lambda_1] \cup [\lambda_2, \infty)$ of the real axis. The same result holds with the endpoint b replacing the endpoint a throughout.

If for $\lambda \in (\lambda_1, \lambda_2)$ there are *two* linearly independent solutions of (6.6.1), each square-integrable at $r = a$ (respectively $r = b$) then $m_a(z)$ (respectively $m_b(z)$) is meromorphic in the entire complex plane.

Proof

We shall prove the first part of the theorem for the endpoint $r = b$. The endpoint $r = a$ may be treated similarly. We define, as previously, solutions $u_1(r, \lambda)$ and $u_2(r, \lambda)$ of (6.6.1.), subject to the conditions

$$u_1(\xi, \lambda) = 0, \qquad u_1'(\xi, \lambda) = 1,$$

$$u_2(\xi, \lambda) = 1, \qquad u_2'(\xi, \lambda) = 0.$$

For $\lambda\in(\lambda_1,\lambda_2)$, we define $m_b(\lambda)$ by the condition

$$u_2(\cdot,\lambda)+m_b(\lambda)u_1(\cdot,\lambda)\in L^2(\xi,b),$$

with $m_b(\lambda)=\infty$ if u_1 is square-integrable. We assume in the first instance that $r=b$ is in the limit-point case.

We now suppose that there is some real m such that $f'(\xi,\lambda)/f(\xi,\lambda)$ does *not* assume the value m, for $\lambda\in(\lambda_1,\lambda_2)$. We define in $L^2(\xi,b)$ the Schrödinger operator $H_b=-\mathrm{d}^2/\mathrm{d}r^2+V$, with boundary condition $f'/f=m$ at $r=\xi$. Referring to (6.4.10), the resolvent operator of H_b will have kernel

$$K_b(r,r';z)=\begin{cases}\dfrac{[u_2(r,z)+mu_1(r,z)][u_2(r',z)+m_b(z)u_1(r',z)]}{m-m_b(z)} & (r\le r'),\\[2ex] \dfrac{[u_2(r,z)+m_b(z)u_1(r,z)][u_2(r',z)+mu_1(r',z)]}{m-m_b(z)} & (r\ge r').\end{cases}\tag{6.6.2}$$

Since $f'(\xi,\lambda)/f(\xi,\lambda)=m_b(\lambda)\neq m$, K_b is defined even for $z=\lambda$ with $\lambda\in(\lambda_1,\lambda_2)$. For $\phi\in C_0^\infty(\xi,b)$, we define $\psi(r,\lambda)$ by

$$\psi(r,\lambda)=\int_\xi^b \mathrm{d}r'\,K_b(r,r';\lambda)\phi(r').$$

Then $(H_b-\lambda)\psi=\phi$, or $\psi=(H_b-\lambda)^{-1}\phi$. Thus $\phi\in D((H_b-\lambda)^{-1})$ for every $\lambda\in(\lambda_1,\lambda_2)$. Writing $H_b=\int\lambda\,\mathrm{d}E_\lambda$, this implies that

$$\|(H_b-\lambda)^{-1}\phi\|^2=\int(\lambda-\mu)^{-2}\,\mathrm{d}\langle\phi,E_\mu\phi\rangle<\infty.$$

In particular, by the Lebesgue Dominated-Convergence Theorem, we have

$$\lim_{\varepsilon\to0}\int\frac{\varepsilon}{(\lambda-\mu)^2+\varepsilon^2}\,\mathrm{d}\langle\phi,E_\mu\phi\rangle=0.\tag{6.6.3}$$

From the results of Section 2.3, it follows that the Lebesgue–Stieltjes measure generated by $\langle\phi,E_\mu\phi\rangle$ assigns zero measure to the interval (λ_1,λ_2). Since this is true for all ϕ belonging to $C_0^\infty(\xi,b)$, which is dense in $L^2(\xi,b)$, the spectrum of H_b cannot intersect the interval (λ_1,λ_2). So $\langle\phi,(H_b-z)^{-1}\phi\rangle$ is analytic in $\mathbb{C}$, apart from the subset $(-\infty,\lambda_1]\cup[\lambda_2,\infty)$ of $\mathbb{R}$. Since

$$\langle\phi,(H_b-z)^{-1}\phi\rangle=\int_\xi^b\mathrm{d}r\int_\xi^b\mathrm{d}r'\,K_b(r,r';z)\bar{\phi}(r)\phi(r'),$$

we can use (6.6.2) to evaluate this matrix element, to obtain an expression of the form

$$\frac{A(z)+B(z)m_b(z)}{m-m_b(z)},$$

where $A(z)$ and $B(z)$ are analytic in the given region. Solving for $m_b(z)$ as a rational function of $\langle\phi,(\mathrm{H}_b-z)^{-1}\phi\rangle$, we deduce the meromorphy of $m_b(z)$ as required.

What if there are two linearly independent solutions of (6.6.1), each square-integrable at $r=b$? We are then in the limit-circle case at $r=b$. The reader should follow through the proof of Theorem 6.7 in that case to confirm the meromorphy of $m_b(z)$ in the entire complex plane. ■

The proof of the theorem may easily be modified to cover the situation in which the function $m_b(\lambda)$ assumes a given value m only finitely many times, for λ in the interval (λ_1,λ_2). Those λ for which $m_b(\lambda)=m$ are eigenvalues of the operator H_b, which has an empty spectrum in the interval, apart from these values. Again $m_b(z)$ is meromorphic in the entire complex plane, apart from the set $(-\infty,\lambda_1]\cup[\lambda_2,\infty)$. Even if $m_b(\lambda)$ assumes the value m infinitely many times for $\lambda\in(\lambda_1,\lambda_2)$, the spectrum of H_b in this interval will be purely discrete, with eigenvalues at the (countable many) values of λ for which $m_b(\lambda)=m$. The operator H_b then necessarily has a limit point of eigenvalues at some point in the closed interval $[\lambda_1,\lambda_2]$. Assuming, then, that for each $\lambda\in(\lambda_1,\lambda_2)$ there is at least one linearly independent non-trivial solution of (6.6.1) that is square-integrable at $r=b$, there are only two possibilities. Either subintervals of (λ_1,λ_2) can be found on which $m_b(z)$ is meromorphic, or the operator H_b has discrete spectrum that is dense in (λ_1,λ_2). The second possibility for H_b must apply for all possible boundary conditions at $r=\xi$. This can only occur if $m_b(\lambda)$ assumes *every* real value m in every subinterval of (λ_1,λ_2), in which case $m_b(\lambda)$ is a very pathological function indeed!

What consequences follow from the meromorphy of $m_a(z)$ or $m_b(z)$ in some region that includes a subinterval (λ_1,λ_2) of the real axis? Meromorphy allows us to analytically continue, say, $m_a(z)$ from the upper half-plane to the lower half-plane. Since, for $\lambda\in(\lambda_1,\lambda_2)$, $m_a(\lambda)$ is real, apart from isolated poles, we also know in this case that $m_a(z)$ has a real boundary value as z approaches the real axis from above or below. In the following chapter, we shall apply the results of Section 2.3, dealing with the analysis of measures, and introduce a new notion of subordinacy, to extend some of these results. Our goal will be the ambitious one of a complete spectral analysis of the arbitrary Schrödinger operator in an interval (a,b), which may be finite or infinite.

Chapter 7

Spectral Analysis of the Schrödinger Operator

7.1 THE STURM–LIOUVILLE OPERATOR

In Chapter 6 we saw how to define the Schrödinger operator $H=-\mathrm{d}^2/\mathrm{d}r^2+V(r)$, in an interval $a<r<b$, as a self-adjoint operator in the Hilbert space $L^2(a,b)$. A spectral analysis of the Schrödinger operator will be closely related to the behaviour of the potential $V(r)$ near the endpoints $r=a$ and $r=b$. For convenience, we shall restate here some of the relevant notation. Let $u_1(r,z)$ and $u_2(r,z)$ satisfy the Schrödinger equation at complex energy z, subject to the conditions, at some interior point ξ of the interval (a,b).

$$\left.\begin{aligned} u_1(\xi,z)=0, \quad u_1'(\xi,z)=1,\\ u_2(\xi,z)=1, \quad u_2'(\xi,z)=0.\end{aligned}\right\} \tag{7.1.1}$$

We define, for $\mathrm{Im}\,(z)\neq 0$, the functions $m_a(z)$ and $m_b(z)$ by the requirements that $u_2(\cdot,z)+m_a(z)u_1(\cdot,z)\in L^2(a,\xi)$, $u_2(\cdot,z)+m_b(z)u_1(\cdot,z)\in L^2(\xi,b)$. If, at either endpoint, there are two linearly independent solutions, both of which are square-integrable at that endpoint, we define $m_a(z)$ or $m_b(z)$ such that an appropriate boundary condition is satisfied. The functions m_a and m_b are analytic for $\mathrm{Im}\,(z)\neq 0$, with, in the upper half-plane, $\mathrm{Im}\,(m_a)<0$, $\mathrm{Im}\,(m_b)>0$. A matrix-valued analytic function of z, $\mathrm{Im}\,(z)\neq 0$, is defined by

$$\left.\begin{aligned} M_{11}(z)&=\frac{m_a(z)m_b(z)}{m_a(z)-m_b(z)},\\ M_{12}(z)=M_{21}(z)&=\frac{m_a(z)+m_b(z)}{2(m_a(z)-m_b(z))},\\ M_{22}(z)&=\frac{1}{m_a(z)-m_b(z)},\end{aligned}\right\} \tag{7.1.2}$$

and a corresponding family $\rho_{ij}(\lambda)$ may be defined up to an additive constant, at points of continuity, by

$$\rho_{ij}(\beta)-\rho_{ij}(\alpha)=\lim_{\varepsilon\to 0+}\frac{1}{2\pi\mathrm{i}}$$

$$\int_{\alpha}^{\beta}\mathrm{d}x\,\{M_{ij}(x+\mathrm{i}\varepsilon)-M_{ij}(x-\mathrm{i}\varepsilon)\}. \tag{7.1.3}$$

Using the analytic properties of $M_{ij}(z)$, the right-hand side of (7.1.3) may be replaced by an integral round a suitable contour.

The Schrödinger operator H is unitarily equivalent to the operator of multiplication by λ in the Hilbert space $L^2(\mathbb{R};\mathrm{d}\rho_{ij})$. This unitary equivalence is implemented by the transformation

$$h_i(\lambda)=\int_a^b f(r)u_i(r,\lambda)\,\mathrm{d}r, \tag{7.1.4}$$

where $f\in L^2(a,b)$, $\{h_i\}\in L^2(\mathbb{R},\mathrm{d}\rho_{ij})$ and $u_i(r,\lambda)$ is the solution of the Schrödinger equation satisfying the conditions (7.1.1) at real energy $z=\lambda$.

Equation (7.1.3) may be inverted to express $f(r)$ in terms of the function pair $\{h_1(\lambda),\, h_2(\lambda)\}$, to give

$$f(r)=\sum_{i,j}\int h_i(\lambda)u_j(r,\lambda)\,\mathrm{d}\rho_{ij}(\lambda). \tag{7.1.4$'$}$$

Spectral properties of the Hamiltonian H may be deduced from spectral properties of the measures defined by the $\rho_{ij}(\lambda)$ ($\rho_{12}=\rho_{21}$ defines a charge rather than a measure, the charge being absolutely continuous with respect to the measures defined by ρ_{11} and ρ_{22}). For example, if ρ_{11} and ρ_{22} define measures that are absolutely continuous (with respect to Lebesgue measure) then ρ_{12} will give rise to an absolutely continuous charge, and the spectrum of H will be purely absolutely continuous. Although this is the case of most interest in scattering theory, it is instructive first of all to consider a number of situations leading to a discrete measure. This case corresponds to the classical Sturm–Liouville theory, and arises for example if both $r=a$ and $r=b$ are limit-circle. We then know (Chapter 6) that each of the functions $m_a(z)$ and $m_b(z)$ is meromorphic in the entire complex plane, including the real axis. From (7.1.2), the only singularities of the function $M_{22}(z)$ occur at points for which $m_a(z)=m_b(z)$. Such points cannot be in the upper half-plane, where $\mathrm{Im}\,(m_b(z))>0$, and $\mathrm{Im}\,(m_a(z))<0$, and similarly there are no singularities of $M_{22}(z)$ in the lower half-plane. Hence $M_{22}(z)$ is an analytic function of z, apart from isolated poles, of finite multiplicity, on the real axis. Since $\operatorname{Im}(M_{22}(z))$ has the same sign as $\operatorname{Im}(z)$, we can go further and argue that each pole must be of multiplicity 1, and have negative residue. (If we consider, for example, a pole of order 2 at $z=0$, having the asymptotic behaviour c/z^2 as $|z|\to 0$, where $c=|c|e^{i\alpha}$ and $z=|z|e^{i\theta}$, then $c/z^2 \sim |c/z^2|e^{i(\alpha-2\theta)}$, which cannot have positive imaginary part for all θ in the range $0<\theta<\pi$.)

Let λ_0 be such a pole of $M_{22}(z)$ with residue $-c_0$. Then $m_a(\lambda_0)=m_b(\lambda_0)$; from (7.1.2) we see that the residues at $z=\lambda_0$ of the functions $M_{12}(z)$ and $M_{11}(z)$ are given respectively by $-c_0 m_a(\lambda_0)$ and $-c_0[m_a(\lambda_0)]^2$. For an interval (α,β) containing the point λ_0 but no other pole, the value of $\rho_{ij}(\beta)-\rho_{ij}(\alpha)$ may be determined from (7.1.3) by converting the right-hand side to an integral round a closed contour, to give

$$\rho_{ij}(\beta)-\rho_{ij}(\alpha)=b_i b_j, \tag{7.1.5}$$

where $b_1=c_0^{1/2}m_a(\lambda_0)$ and $b_2=c_0^{1/2}$. Note that the ρ_{11} and ρ_{22} measures of the point λ_0 are positive, whereas ρ_{12} will give a positive or negative contribution, depending on the sign of $m_a(\lambda_0)$. The right-hand side of (7.1.5) is seen to factorize, a result sometimes known as the *factorization of residues.*

Using (7.1.5) and inserting the values of b_1 and b_2, we find that the contribution of the single point $\lambda=\lambda_0$ to the right-hand side of (7.1.4)′ is just

$$c_0 h(\lambda_0)(u_2(r,\lambda_0)+m_a(\lambda_0)u_1(r,\lambda_0)),$$

where

$$h(\lambda)=h_2(\lambda)+m_a(\lambda)h_1(\lambda). \tag{7.1.6}$$

Taking a linear combination of (7.1.4) with $i=1$ and $i=2$ respectively, we also have

$$h(\lambda)=\int_a^b f(r)u(r,\lambda)\,dr, \tag{7.1.7}$$

where

$$u(r,\lambda)=u_2(r,\lambda)+m_a(\lambda)u_1(r,\lambda).$$

For every eigenvalue λ_j of H ($j=0, 1, 2,\ldots$), $u(r,\lambda_j)$ is an eigenfunction, satisfying appropriate boundary condition at each endpoint of the interval (a,b). Equation (7.1.7) defines $h(\lambda_j)$ as an inner product of a given function $f\in L^2(a,b)$ with the jth eigenfunction of H. And, summing over all of the eigenvalues of H, (7.1.4)′ becomes

$$f(r)=\sum_j c_j h(\lambda_j)u(r,\lambda_j), \tag{7.1.8}$$

which expresses f as a linear combination of the (orthogonal) basis of eigenfunctions. This $h(\lambda_j)$ is a generalized Fourier coefficient for the function f, and (7.1.8) defines a generalized Fourier series.

For consistency between (7.1.7) and (7.1.8), we need to take account of the fact that the eigenfunctions $u(r,\lambda_j)$ are not normalized, and it follows that

$$c_j = \frac{1}{\|u(\cdot,\lambda_j)\|^2}, \tag{7.1.9}$$

where

$$\|u(\cdot,\lambda_j)\|^2 = \int_a^b |u(r,\lambda_j)|^2 \, \mathrm{d}r.$$

Since $-c_j$ is the residue at $\lambda = \lambda_j$ of the function $M_{22}(z)$, we know, from (7.1.2), that

$$c_j = (m_b'(\lambda_j) - m_a'(\lambda_j))^{-1}. \tag{7.1.10}$$

which, in combination with (7.1.9), leads to

$$\|u(\cdot,\lambda_j)\|^2 = m_b'(\lambda_j) - m_a'(\lambda_j). \tag{7.1.11}$$

Equation (7.1.11) may be verified directly, by dividing the interval (a,b) into two integration regions (a,ξ) and (ξ,b), with $m(\lambda)=m_a(\lambda)$ and $m(\lambda)=m_b(\lambda)$ respectively in the two regions. It should be noted that, away from poles, $m_b(\lambda)$ is an increasing function, whereas $m_a(\lambda)$ is decreasing. For this reason, the right-hand side of (7.1.11) is automatically positive, as required.

The above construction shows, for a Schrödinger Sturm–Liouville operator with limit-circle endpoints, that the Hamiltonian H has purely discrete spectrum. Equations (7.1.4) and (7.1.4)′ correspond in this case to a discrete-eigenfunction expansion in terms of an orthogonal basis of $L^2(a,b)$ consisting of (non-degenerate) eigenfunctions. There is a unitary transformation,

$$\{h_1(\lambda), h_2(\lambda)\} \to h_2(\lambda) + m_a(\lambda) h_1(\lambda)$$

from

$$L^2(\mathbb{R}; \mathrm{d}\rho_{ij}(\lambda)) \text{ onto } L^2(\mathbb{R}; \mathrm{d}\rho(\lambda)),$$

where $\mathrm{d}\rho(\lambda)$ defines a point measure concentrated at the eigenvalues of H, the eigenvalue λ_j having ρ-measure c_j given by (7.1.10).

Since

$$\|f\|^2 = \sum_{i,j} \int h_i(\lambda) h_j(\lambda) \, \mathrm{d}\rho_{ij}(\lambda) = \int |h(\lambda)|^2 \, \mathrm{d}\rho(\lambda),$$

where $h(\lambda)$ is given by (7.1.7), we have established a unitary transformation between $L^2(a,b)$ and $L^2(\mathbb{R}; \mathrm{d}\rho(\lambda))$, where in the second space the operator H is represented by the operator of multiplication by λ.

The above analysis applies to any situation in which $m_a(z)$ and $m_b(z)$ are meromophic in the entire complex plane, and Theorem 6.12 may be used to extend the results to a much wider class of potentials than those for which $r=a$ and $r=b$ are limit-circle. If we suppose, for example, that the potential $V(r)$ at $r=a$ is singular and positive then the asymptotic analysis of solutions of the Schrödinger equation in Section 2.8 allows us to conclude that there is just one linearly independent solution, for each energy, that is square-integrable at $r=a$. We can then verify that the hypotheses of Theorem 6.12 apply, leading to the conclusion that $m_a(z)$ is again meromorphic. If $r=b$ is then limit-circle, or if $V(r)$ is singular and positive at $r=b$, it follows that H has a purely discrete spectrum, with the same kind of eigenfunction expansion as before.

Again, we may use Theorem 6.12 in certain cases to show that the spectrum of H is discrete *in an interval* (λ_1, λ_2), whereas there may be a continuous spectrum outside this interval. Let us suppose, for example, that we have $H = -\mathrm{d}^2/\mathrm{d}r^2 + V(r)$ as a self-adjoint operator in $L^2(0, \infty)$, and let us consider the case in which V is short-range (i.e. $V \in L^1(c, \infty)$ for $c > 0$) and either non-singular ($V \in L^1(0, c)$) or singular and positive at $r=0$ (e.g. $V = g/r^{2+\varepsilon}$ with $g, \varepsilon > 0$). Under such hypotheses, $m_a(z)$ is meromorphic in the entire complex plane. On the other hand, $m_b(z)$ is *not* meromorphic everywhere; there is a branch cut along the positive real axis. However, for $\lambda < 0$, the analysis of Section 2.8 shows that there is just one linearly independent solution $f(r, \lambda)$ of the Schrödinger equation that is square-integrable in (c, ∞), this solution decaying exponentially for large r. Moreover, this asymptotic analysis shows that $f'(c,\lambda)/f(c,\lambda)$ is a continuous function of λ, unless $f(c,\lambda)=0$. We can therefore apply Theorem 6.12 to deduce that $m_b(z)$ is meromorphic in the entire complex plane apart from the positive axis $[0, \infty)$. Negative eigenvalues of H are then solutions λ_j of the equation $m_a(\lambda_j) = m_b(\lambda_j)$, and the discrete-eigenfunction expansion to which we were led in (7.1.7) and (7.1.8) applies to any $f \in L^2(0, \infty)$ that lies in the subspace corresponding to the negative part of the spectrum of H. The negative spectrum of H is thus purely discrete, and $\lambda = 0$ is the only possible point of accumulation of these eigenvalues.

This analysis accounts, as one special example, for the negative discrete spectrum of hydrogen in the $l=0$ partial wave, and extends easily to higher partial waves. Indeed, the area of applicability of these ideas is so wide that one is tempted to conclude that, for *any* Schrödinger operator H in $L^2(0, \infty)$ that is of short range, one has a purely discrete spectrum for $\lambda < 0$. However, both absolutely continuous and singular continuous spectra are possible for potentials behaving rather wildly at $r=0$, and we shall see examples of each kind of behaviour in Chapter 14. It remains true, in the great majority of applications to scattering theory, that one has to deal with a purely discrete negative spectrum.

The reader may have observed, in our analysis of the Sturm–Liouville Schrödinger operator in an interval, and the negative spectrum of a wider class of Schrödinger operators, the crucial role played by an asymptotic analysis, at each endpoint, of solutions of the time-independent Schrödinger equation. In particular, special importance is attached to those solutions $f(r, \lambda)$ that have more regular asymptotic behaviour in the sense of being square-integrable at one endpoint or the other, or at both endpoints. In the following section we shall see how a closely related notion, that of *subordinacy*, leads to a spectral analysis that deals with the continuous part of the spectrum as well as the discrete. This will allow us to unify, by means of a few basic ideas and techniques, the entire treatment of the spectrum of the one-dimensional Schrödinger operator.

7.2 SUBORDINACY

The following definition makes precise the notion that a particular solution of the Schrödinger equation is asymptotically smaller than all other solutions, at an endpoint, say $r=b$.

Definition 7.1

Let $f(r, \lambda)$ be a non-trivial solution of the Schrödinger equation $-f''+Vf=\lambda f$ in the interval $a<r<b$, for some fixed real value of λ. Suppose that for any solution $g(r, \lambda)$, of the Schrödinger equation that is not a constant multiple of $f(r, \lambda)$ we have

$$\lim_{N\to b} \frac{\|f(\cdot, \lambda)\|_N}{\|g(\cdot, \lambda)\|_N}=0 \tag{7.2.1}$$

where $\|\cdot\|_N$ denotes the norm, for $\xi<N<b$, in the Hilbert space $L^2(\xi, N)$. Then we shall say that $f(r, \lambda)$ is a *subordinate solution* of the Schrödinger equation, at energy λ, at the endpoint $r=b$.

Subordinacy is a quantification of the idea that $f(r, \lambda)$ has asymptotically vanishing norm, compared with that of any solution linearly independent of $f(r, \lambda)$, near $r=b$. A similar definition applies to solutions subordinate at $r=a$. Note that any (non-zero) constant multiple of a subordinate solution is subordinate, and also that there can be at most one linearly independent subordinate solution for given λ. To prove f subordinate in (7.2.1), it is sufficient to verify (7.2.1) for just one solution $g(r,\lambda)$ that is not a constant multiple of f. Since λ is real, we may assume without loss of generality that $f(r,\lambda)$ is also real. Note that subordinacy is independent of the value of ξ.

Of course, one realization of (7.2.1) occurs if $f(r,\lambda)$ is the only solution (apart from constant multiples) that is square-integrable at $r=b$. However, subordinate solutions may also exist even if no solution is square-integrable at b. In this case, in (7.2.1), $\|f(\cdot,\lambda)\|_N$ diverges in the limit as $N\to b$, but $\|g(\cdot,\lambda)\|_N$ diverges more rapidly. Another special case, assuming $r=b$ to be limit-point, occurs if a real solution $f_0(r,\lambda)$ exists that is non-vanishing near $r=b$. We then have, for any real solution g that is independent of f_0,

$$\frac{\mathrm{d}}{\mathrm{d}r}\left(\frac{g}{f_0}\right)=\frac{f_0 g'-g f'_0}{f_0^2}=\frac{\text{const}}{f_0^2},$$

since the Wronskian is constant. Hence g/f_0 is monotonic, and must therefore approach a limit, finite or infinite, as $r\to b$. If $g/f_0\to\infty$ then f_0 is itself a subordinate solution, being asymptotically vanishingly small compared with g. On the other hand, if $g/f_0\to c$ then $(g-cf_0)/f_0\to 0$, and the solution $g-cf_0$ is subordinate. In either case, a solution that is non-oscillatory near $r=b$ is sufficient to guarantee the existence of a subordinate solution, provided that $r=b$ is limit-point. The notion of subordinacy is, however, of much wider applicability than this special case would suggest, and is particularly useful in analysing the Schrödinger operator in cases where oscillatory solutions exist.

Rather than taking a continuous limit as N approaches b, we can also consider sequences. This leads to the following definition, which is closely related to Definition 7.1.

Definition 7.2

Suppose, instead of (7.2.1), that there exists a *sequence* $N_1, N_2, N_3, \ldots$ such that $N_j\to b$ and

$$\lim_{j\to\infty}\frac{\|f(\cdot,\lambda)\|_{N_j}}{\|g(\cdot,\lambda)\|_{N_j}}=0 \tag{7.2.2}$$

for any solution $g(r,\lambda)$ that is not a constant multiple of $f(r,\lambda)$. Then we shall say that $f(r,\lambda)$ is a *sequentially subordinate solution* of the Schrödinger equation, at energy λ, at the endpoint $r=b$.

Note that every subordinate solution is sequentially subordinate, although the converse is, in general, false. It should also be noted that different sequences $\{N_j\}$ may give rise to different sequentially subordinate solutions.

The following result establishes an important link between subordinate solutions at $r=b$ and the behaviour of the function $m_b(z)$ as z approaches the real axis.

Theorem 7.1

Let $r=b$ be limit-point. For some fixed $\lambda \in \mathbb{R}$ suppose that $m_b(\lambda+i\varepsilon)$ has a real limit as ε tends to zero. (This means that $m_b(z)$ has a real limit as z approaches the real axis, perpendicularly, at λ.) Denote this limit by $m_b(\lambda)$. Then the solution $u_2(r,\lambda)+m_b(\lambda)u_1(r,\lambda)$ is subordinate at $r=b$. If, on the other hand, $|m_b(\lambda+i\varepsilon)|\to\infty$ as $\varepsilon\to 0$ then the solution $u_1(r,\lambda)$ is subordinate.

Proof

We shall rely on the identity, recalled from (6.2.17)

$$\frac{\mathrm{Im}\,(m_b(z))}{\mathrm{Im}\,(z)} = \int_{\xi}^{b} \mathrm{d}r\, |u_2(r,z)+m_b(z)u_1(r,z)|^2 \tag{7.2.3}$$

For $z=\lambda+i\varepsilon$, we define the function $K(\varepsilon)$ by

$$\|u_2(\cdot,z)+m_b(z)u_1(\cdot,z)\|_b = K(\varepsilon), \tag{7.2.4}$$

where $\|\cdot\|_b$ denotes a norm in $L^2(\xi,b)$; and suppose that $m_b(z)$ approaches the real limit $m_b(\lambda)$ as $\varepsilon\to 0$. We may assume that $K(\varepsilon)\to\infty$ as $\varepsilon\to 0$. (For if it does not then, a sequence of ε's, converging to zero, can be found such that $K(\varepsilon)\le K<\infty$. Since $\|\cdot\|_N\le\|\cdot\|_b$ for any N such that $\xi<N<b$, and $u_j(\cdot,z)$ converges in norm to $u_j(\cdot,\lambda)$ in the space $L^2(\xi,N)$, we can take the limit $\varepsilon\to 0$ in (7.2.4) to deduce that $\|u_2(\cdot,\lambda)+m_b(\lambda)u_1(\cdot,\lambda)\|_N\le K$ for all such N. Letting N approach b, it then follows that $\|u_2(\cdot,\lambda)+m_b(\lambda)u_1(\cdot,\lambda)\|_b\le K$, so that $u_2(\cdot,\lambda)+m_b(\lambda)u_1(\cdot,\lambda)\in L^2(\xi,b)$. By hypothesis, $r=b$ is in the limit-point case, so that $u_2(\cdot,\lambda)+m_b(\lambda)u_1(\cdot,\lambda)$ is the *only* linearly independent solution that is square-integrable at b, and hence is subordinate at $r=b$.)

Equation (7.2.3), together with the fact that $m_b(z)$ is bounded as $\varepsilon\to 0$, tells us that $K(\varepsilon)$ cannot diverge too rapidly, since

$$\lim_{\varepsilon\to 0} \varepsilon^{1/2}K(\varepsilon)=0. \tag{7.2.5}$$

Since $u_1(r,\lambda)$ and $u_2(r,\lambda)$ cannot both be square-integrable at b, we know that, as N increases from ξ to b, $\sup_j \|u_j(\cdot,\lambda)\|_N$ increases continuously from 0 to ∞. Given any positive integer n, we can find $N(\varepsilon)$, depending continuously on ε for $\varepsilon>0$, such that

$$\sup_j \|u_j(\cdot, \lambda)\|_{N(\varepsilon)} = nK(\varepsilon). \tag{7.2.6}$$

From (7.2.5), this implies that

$$\lim_{\varepsilon \to 0} \varepsilon \|u_1(\cdot, \lambda)\|_{N(\varepsilon)} \|u_2(\cdot, \lambda)\|_{N(\varepsilon)} = 0. \tag{7.2.7}$$

Equation (7.2.7) allows us to obtain an iterative solution, in the space $L^2(\xi, N(\varepsilon))$, of the integral equation

$$\begin{aligned} u(r, z) &= u_2(r, \lambda) + m_b(z) u_1(r, \lambda) \\ &\quad + u_2(r, \lambda) \int_\xi^r i\varepsilon u_1(t, \lambda) u(t, z)\, dt \\ &\quad - u_1(r, \lambda) \int_\xi^r i\varepsilon u_2(t, \lambda) u(t, z)\, dt. \end{aligned} \tag{7.2.8}$$

Using (7.2.7), we can prove convergence of the iteration, for ε small enough, to a solution $u(r, z)$ for which, in the limit $\varepsilon \to 0$, we have

$$\|u(\cdot, z)\|_{N(\varepsilon)} = \|u_2(\cdot, \lambda) + m_b(z) u_1(\cdot, \lambda)\|_{N(\varepsilon)} (1 + o(1)) \tag{7.2.9}$$

By direct verification, from (7.2.8), of the differential equation, and initial conditions at $r=\xi$, satisfied by the solution $u(r, z)$ of the integral equation, we find

$$u(r, z) = u_2(r, z) + m_b(z) u_1(r, z).$$

From (7.2.9), we can therefore deduce that

$$\|u_2(\cdot, \lambda) + m_b(z) u_1(\cdot, \lambda)\|_{N(\varepsilon)} = \|u_2(\cdot, z) + m_b(z) u_1(\cdot, z)\|_{N(\varepsilon)} (1 + o(1)).$$

From (7.2.4), with the knowledge that $\|\cdot\|_{N(\varepsilon)} \leq \|\cdot\|_b$, this gives

$$\limsup_{\varepsilon \to 0} \frac{\|u_2(\cdot, \lambda) + m_b(z) u_1(\cdot, \lambda)\|_{N(\varepsilon)}}{K(\varepsilon)} \leq 1. \tag{7.2.10}$$

Since $m_b(z) = m_b(\lambda)(1 + o(1))$, and by (7.2.6), $\|u_1(\cdot, \lambda)\|_{N(\varepsilon)} \leq nK(\varepsilon)$, (7.2.10) implies that

$$\limsup_{\varepsilon \to 0} \frac{\|u_2(\cdot, \lambda) + m_b(\lambda) u_1(\cdot, \lambda)\|_{N(\varepsilon)}}{K(\varepsilon)} \leq 1. \tag{7.2.11}$$

Equation (7.2.6) also tells us that $\|u_2(\cdot,\lambda)\|_{N(\varepsilon)}+\|u_1(\cdot,\lambda)\|_{N(\varepsilon)}\geq nK(\varepsilon)$, so that (7.2.11) implies that

$$\limsup_{\varepsilon\to 0}\frac{\|u_2(\cdot,\lambda)+m_b(\lambda)u_1(\cdot,\lambda)\|_{N(\varepsilon)}}{\|u_1(\cdot,\lambda)\|_{N(\varepsilon)}+\|u_2(\cdot,\lambda)\|_{N(\varepsilon)}}\leq\frac{1}{n}. \tag{7.2.12}$$

As $\varepsilon\to 0$, $K(\varepsilon)\to\infty$ on the right-hand side of (7.2.6), so that $N(\varepsilon)\to\infty$ continuously in ε. Given any $\varepsilon'>0$, we have for all sufficiently large N that $N(\varepsilon)=N$ for some $\varepsilon<\varepsilon'$. It then follows, from (7.2.12), that

$$\limsup_{N\to\infty}\frac{\|u_2(\cdot,\lambda)+m_b(\lambda)u_1(\cdot,\lambda)\|_N}{\|u_1(\cdot,\lambda)\|_N+\|u_2(\cdot,\lambda)\|_N}\leq\frac{1}{n},$$

and, since the positive integer n on the right-hand side is arbitrary, we have simply

$$\lim_{N\to\infty}\frac{\|u_2(\cdot,\lambda)+m_b(\lambda)u_1(\cdot,\lambda)\|_N}{\|u_1(\cdot,\lambda)\|_N+\|u_2(\cdot,\lambda)\|_N}=0. \tag{7.2.13}$$

But

$$\frac{|\,\|u_2(\cdot,\lambda)\|_N-\|m_b(\lambda)u_1(\cdot,\lambda)\|_N|}{\|u_1(\cdot,\lambda)\|_N+\|u_2(\cdot,\lambda)\|_N}\to 0$$

also, this quotient being bounded by the left-hand side of (7.2.13). Writing $s=\|u_2(\cdot,\lambda)\|_N/\|u_1(\cdot,\lambda)\|_N$, this gives $|s-|m_b(\lambda)|\,|/(s+1)\to 0$ as $N\to\infty$, so that s converges to $|m_b(\lambda)|$ in this limit. Thus

$$\frac{\|u_2(\cdot,\lambda)+m_b(\lambda)u_1(\cdot,\lambda)\|_N}{\|u_1(\cdot,\lambda)\|_N}=\frac{(1+s)\|u_2(\cdot,\lambda)+m_b(\lambda)u_1(\cdot,\lambda)\|_N}{\|u_1(\cdot,\lambda)\|_N+\|u_2(\cdot,\lambda)\|_N}$$

$\to 0$ as $N\to\infty$ by (7.2.13).

Hence $u_2(r,\lambda)+m_b(\lambda)u_1(r,\lambda)$ is subordinate at $r=b$, as required.

The case in which $|m_b(\lambda+i\varepsilon)|\to\infty$ can be dealt with in a similar way. Again, we define $N(\varepsilon)$ by (7.2.6), with the definition of $K(\varepsilon)$ in (7.2.4) now replaced by

$$\|(m_b(z))^{-1}u_2(\cdot,z)+u_1(\cdot,z)\|_b=K(\varepsilon). \tag{7.2.14}$$

Equation (7.2.5) remains valid, in view of the fact that $\operatorname{Im}(m_b(z))/|m_b(z)|^2\to 0$ as $\varepsilon\to 0$. In place of (7.2.8), we can write down an integral equation for the function

$$u(r,z)=(m_b(z))^{-1}u_2(r,z)+u_1(r,z),$$

which may be iterated in the space $L^2(\xi,N(\varepsilon))$, to deduce that

$$\|(m_b(z)^{-1}u_2(\cdot,\lambda)+u_1(\cdot,\lambda)\|_{N(\varepsilon)}=\|(m_b(z))^{-1}u_2(\cdot,z)+u_1(\cdot,z)\|_{N(\varepsilon)}(1+o(1)).$$

We may now deduce, from (7.2.14) and the fact that $(m_b(z))^{-1}\to 0$, that $\limsup_{\varepsilon\to 0}\|u_1(\cdot,\lambda)\|_{N(\varepsilon)}/K(\varepsilon)\leq 1$. Proceeding as before, we find that $\|u_1(\cdot,\lambda)\|_N/\|u_2(\cdot,\lambda)\|_N\to 0$ as $N\to\infty$, so that $u_1(r,\lambda)$ is subordinate at $r=b$.

Corresponding results leading to subordinacy at $r=a$, given that either $m_a(z)$ has a real boundary value as z approaches λ, or that $|m_a(z)|\to\infty$ in this limit, follow by analogous arguments. ■

Corollary

Let $r=b$ be limit-point. For some fixed $\lambda\in\mathbb{R}$ suppose that a sequence $\{\varepsilon_i\}$ exists $(i=1,2,3,\ldots)$ such that $\varepsilon_i\to 0$ and $m_b(\lambda+i\varepsilon_i)$ has a real limit as $i\to\infty$. Denote this limit by $m_b(\lambda)$. Then $u_2(r,\lambda)+m_b(\lambda)u_1(r,\lambda)$ is *sequentially* subordinate at $r=b$. If, on the other hand, $|m_b(\lambda+i\varepsilon_i)|\to\infty$ as $i\to\infty$ then $u_1(r,\lambda)$ is sequentially subordinate.

Proof

We follow the steps of our previous proof, with the limit $\varepsilon\to 0$ replaced by a limiting *sequence* $\varepsilon_i\to 0$. Then $\{K(\varepsilon_i)\}$ and $\{N(\varepsilon_i)\}$ both define sequences that tend to infinity. With $N_i=N(\varepsilon_i)$, we arrive at a sequence of N values in (7.2.12), where

$$\limsup_{i\to\infty}\frac{\|u_2(\cdot,\lambda)+m_b(\lambda)u_1(\cdot,\lambda)\|_{N_i}}{\|u_1(\cdot,\lambda)\|_{N_i}+\|u_2(\cdot,\lambda)\|_{N_i}}\leq\frac{1}{n}. \tag{7.2.15}$$

The sequence $\{N_i\}$ depends on the value of n. By choosing successively larger values of n, and making, for each n, a suitable choice of N_i, we can construct a new sequence $\{N_i\}$ such that $N_i\to\infty$ and the quotient on the left-hand side of (7.2.15) converges to zero. This proves that $u_2(r,\lambda)+m_b(\lambda)u_1(r,\lambda)$ is sequentially subordinate, and a similar argument leads to sequential subordinacy for $u_1(r,\lambda)$ in the case that $|m_b(\lambda+i\varepsilon_i)|\to\infty$. ■

The following result provides a converse to Theorem 7.1.

Theorem 7.2

For some fixed $\lambda\in\mathbb{R}$, suppose that $m(\lambda)\in\mathbb{R}$ exists such that $u_2(r,\lambda)+m(\lambda)u_1(r,\lambda)$ is subordinate at $r=b$. Then $m_b(z)$, with $z=\lambda+i\varepsilon$, has a real boundary value, namely $\lim_{\varepsilon\to 0}m_b(\lambda+i\varepsilon)\equiv m_b(\lambda)=m(\lambda)$. If, on the other hand, $u_1(r,\lambda)$ is subordinate at $r=b$ then $|m_b(\lambda+i\varepsilon)|\to\infty$ as $\varepsilon\to 0$.

Proof

We first suppose that $u_2(r,\lambda)+m(\lambda)u_1(r,\lambda)$ is subordinate, and define $K(\varepsilon)$ as in (7.2.4). Note that $m_b(z)$ is bounded in the limit $\varepsilon\to 0$. (If not then there is a sequence ε_i such that $|m_b(\lambda+i\varepsilon_i)|\to\infty$. Hence, by the Corollary to Lemma 7.1, $u_1(r,\lambda)$ is sequentially subordinate, which contradicts the assumption that $u_2(r,\lambda)+m(\lambda)u_1(r,\lambda)$ is subordinate.) From (7.2.3), it follows that $K(\varepsilon)=O(\varepsilon^{-1/2})$. Given $\varepsilon>0$, we define $N(\varepsilon)$ such that $\|u_1(\,,\lambda)\|_{N(\varepsilon)}=nK(\varepsilon)$, where n is a positive integer. As $\varepsilon\to 0$, assume $K(\varepsilon)$, $N(\varepsilon)\to\infty$. (The case $K(\varepsilon)\not\to\infty$ may be treated more simply.) Since $u_2(r,\lambda)+m(\lambda)u_1(r,\lambda)$ is subordinate, it also follows that $\|u_2(\cdot,\lambda)+m(\lambda)u_1(\cdot,\lambda)\|_{N(\varepsilon)}=o(K(\varepsilon))$, so that, combining the various norm estimates, we have

$$\varepsilon\|u_1(\cdot,\lambda)\|_{N(\varepsilon)}\|u_2(\cdot,\lambda)+m(\lambda)u_1(\cdot,\lambda)\|_{N(\varepsilon)}\to 0$$

as $\varepsilon\to 0$. This allows us to write down integral equations for the functions $v_1(r,z)$ and $v_2(r,z)$ respectively, where $v_1(r,z)=u_1(r,z)$ and $v_2(r,z)=u_2(r,z)+m(\lambda)u_1(r,z)$, and to solve these integral equations iteratively, as in the proof of Theorem 7.1. The kernal of these integral equations may be expressed in terms of the solutions $u_1(r,\lambda)$, $u_2(r,\lambda)+m(\lambda)u_1(r,\lambda)$ of the Schrödinger equation at real energy λ. For the norms of v_1 and v_2, we have the estimates $\|v_1(\cdot,z)\|_{N(\varepsilon)}=nK(\varepsilon)(1+o(1))$, and $\|v_2(\cdot,z)\|_{N(\varepsilon)}=o(K(\varepsilon))$. Since

$$\begin{aligned}\|u_2(\cdot,z)+m_b(z)u_1(\cdot,z)\|_{N(\varepsilon)}&=\|v_2(\cdot,z)+(m_b(z)-m(\lambda))v_1(\cdot,z)\|_{N(\varepsilon)}\\&\le K(\varepsilon)\qquad\text{by (7.2.4),}\end{aligned}$$

the estimates for $\|v_1\|_{N(\varepsilon)}$ and $\|v_2\|_{N(\varepsilon)}$ can only hold provided that

$$\limsup_{\varepsilon\to 0}|m_b(z)-m(\lambda)|\le\frac{1}{n}.$$

But n is an arbitrary positive integer, and hence

$$m_b(z)\to m(\lambda)\quad\text{as }\varepsilon\to 0,$$

as required. Let us suppose, alternatively, that $u_1(r,\lambda)$ is the subordinate solution. In this case, we use (7.2.14) to define $K(\varepsilon)$. Note that $|m_b(z)|$ must be bounded away from zero. (If not then a sequence $\{\varepsilon_i\}$ may be found, converging to zero, such that $m_b(\lambda+i\varepsilon_i)\to 0$. In this case $u_2(r,\lambda)$ is sequentially subordinate, in contradiction with the hypothesis that $u_1(r,\lambda)$ is a subordinate solution.) The conclusion of the theorem now follows if we define $N(\varepsilon)$ by

$$\|u_2(\cdot,\lambda)\|_{N(\varepsilon)}=nK(\varepsilon),$$

with

$$\|u_1(\cdot,\lambda)\|_{N(\varepsilon)}=o(K(\varepsilon))$$

leading to

$$\limsup_{\varepsilon\to 0}\frac{1}{|m_b(z)|}\leq\frac{1}{n}.$$ ■

The results of this section, on subordinacy and its relation to boundary values of the functions $m_a(z)$ and $m_b(z)$, may, together with the results of Chapter 2 on the analysis of measures, be applied to an arbitrary Schrödinger operator in an interval (a,b). The following section will be devoted to a complete spectral analysis and eigenfunction expansion in the case of most interest in scattering theory. While leaving the interested reader to extend the applications of this theory to cases that may be important in other contexts, we close this section with two straightforward applications that illustrate the power of the method and the flavour of the results that can be obtained.

Theorem 7.3

Let $H=-d^2/dr^2+V(r)$ in $L^2(a,b)$. Suppose that $V(r)$ is bounded at $r=a$. Take boundary condition $f'(a)/f(a)=m\,(m=\infty$ corresponding to $f(a)=0)$. Assume that $r=b$ is limit-point.

(i) Suppose that, for almost all λ in the interval (λ_1,λ_2), a solution $u(r,\lambda)$ of the Schrödinger equation exists that is subordinate at $r=b$. Then H has a purely singular spectrum, in this interval.
(ii) Suppose that, for all λ in the interval (λ_1,λ_2), there is *no* solution of the Schrödinger equation that is even sequentially subordinate at $r=b$. Then H has a purely absolutely continuous spectrum in this interval.

Proof

(i) We relabel the endpoint a as $r=\xi$. We refer to (6.6.2), where a formula for the kernel of the resolvent operator of H is given. (The boundary condition corresponding to $m=\infty$ is a limiting case.) For almost all λ in the interval in question, a solution subordinate at $r=b$ exists. Hence for such λ either $m_b(z)$ has a real boundary value $m_b(\lambda)$ as z approaches the real axis, or $|m_b(z)|\to\infty$. By standard results in the theory of boundary values of analytic functions, we have, almost everywhere, $m_b(\lambda)\neq m$. ($m_b(\lambda)=m$ for a set of λ having positive measure would imply $m_b(z)\equiv m$, which would lead to an empty spectrum for H!) An examination of the kernel of $(H-z)^{-1}$ then shows that for almost all λ this kernel has a real and finite limit, for $z=\lambda+i\varepsilon$ in the limit $\varepsilon\to 0$. It follows, for any $\phi\in C_0^\infty(\xi,b)$, that the imaginary part of $\langle\phi,(H-z)^{-1}\phi\rangle$ converges to zero. In terms of the spectral family $\{E_u\}$ of H, we can now use the same argument as followed (6.6.3) to show the set of λ such that $m_b(z)$ has a real boundary value, or such that $|m_b(z)|\to\infty$, has zero H-measure (in the sense that the spectral projection of H, associated with this subset of $\mathbb{R}$, is the zero projection). Since in (λ_1,λ_2) the *complement* of this set has Lebesgue measure zero, we have proved that H is spectrally singular in this interval.

(ii) Under the given hypothesis, and using the Corollary to Theorem 7.1. we know for each $\lambda\in(\lambda_1,\lambda_2)$ that a sequence $\{\varepsilon_i\}$, converging to zero, cannot have a real limit for $m_b(\lambda+i\varepsilon_i)$, nor can $|m_b(\lambda+i\varepsilon_i)|$ tend to infinity. We may conclude from these two statements, first that $|m_b(z)|$ is bounded ($z=\lambda+i\varepsilon$, λ fixed), and secondly that constants C_1 and C_2 may be found for which

$$0<C_1\le \operatorname{Im}(m_b(z))\le C_2. \tag{7.2.16}$$

As for (i), we may now estimate the kernel of the resolvent operator and show that, for each $\lambda\in(\lambda_1,\lambda_2)$ and for $\phi\in C_0^\infty(\xi,b)$, $\operatorname{Im}(\langle\phi,(H-z)^{-1}\phi\rangle)$ is bounded as z approaches the real axis at λ. In terms of the spectral family of H, this tells us that

$$\int_{-\infty}^{\infty}\frac{\varepsilon}{(\lambda-\mu)^2+\varepsilon^2}\,d\langle\phi,E_\mu\phi\rangle$$

is bounded in the limit $\varepsilon\to 0$. An application of Lemmas 2.2 and 2.4 now allows us to conclude that H has an absolutely continuous spectrum in the interval (λ_1,λ_2). ■

Some comments are in order concerning applications of these results. Note first of all that we have left just one gap in the chain of results that relate subordinacy and sequential subordinacy for solutions of the Schrödinger equation to behaviour of $m_b(z)$ as z approaches the real axis. That is, we have omitted any proof that the existence of a *sequentially* subordinate solution implies that there is a sequence $\{\varepsilon_i\}$, converging to zero, for which either $m_b(\lambda+i\varepsilon_i)$ has a real limit, or $|m_b(\lambda+i\varepsilon_i)|\to\infty$. (The second possibility occurs if $u_1(r,\lambda)$ is sequentially subordinate.) The interested reader should remedy this omission by an appropriate modification of the proof of Theorem 7.2. Given this result, we have established necessary and sufficient conditions for both subordinacy and sequential subordinacy.

We may also conclude, in Theorem 7.3(i), that it is sufficient to assume sequential subordinacy for almost all $\lambda \in (\lambda_1, \lambda_2)$. For if $u(r, \lambda)$ is sequentially subordinate then we know that $m_b(z)$ has a *real* limit point as z approaches the real axis, or that $|m_b(z)|$ is unbounded. From complex-variable theory, we also know that $m_b(z)$ has a limit, for almost all λ, as z approaches the real axis. Combining these results, we infer that the boundary value of $m_b(z)$ must be real almost everywhere in (λ_1, λ_2), so that by Theorem 7.1 a subordinate solution exists for almost all λ in this interval. (Thus sequential subordinacy almost everywhere implies subordinacy almost everywhere, even though sequential subordinacy does not imply subordinacy!)

In Theorem 7.3(ii) it is equivalent to assume, for any two solutions $u(r, \lambda)$ and $v(r, \lambda)$ of the Schrödinger equation, with $\lambda \in (\lambda_1, \lambda_2)$, that the ratio $\|u(\cdot, \lambda)\|_N / \|v(\cdot, \lambda)\|_N$ remains bounded, both above and below, in the limit $N \to b$. Thus it is necessary to assume that any two solutions are comparable, in L^2 norm, over an interval (ξ, N) as N approaches the endpoint $r = b$. Any such behaviour of solutions then leads to an absolutely continuous spectrum. As simple applications, we consider respectively a potential $V(r)$ that is short-range as $r \to \infty$ and a potential with a Coulomb-like tail. In both cases we know (Chapter 2) that at positive energies solutions behave asymptotically like sine and cosine functions—modified in the Coulomb case by logarithmic corrections. In either case, it is straightforward to verify the required comparability in norm of any pair of solutions. The spectrum for $\lambda > 0$ is thus purely absolutely continuous in both cases, and it is necessary only to consult Section 2.8 to extend this spectral analysis to a wide variety of short- and long-range potentials.

The canonical spectral behaviour of most interest to scattering theory consists of purely discrete non-degenerate separated eigenvalues for $\lambda \le 0$, together with an absolutely continuous spectrum for $\lambda > 0$. In view of the importance of spectral analysis of such Schrödinger operators, we shall devote the following section to this and to the resulting eigenfunction expansion, within the context of the theory of subordinacy that we have developed.

7.3 SPECTRAL THEORY AND EIGENFUNCTION EXPANSIONS FOR THE ABSOLUTELY CONTINUOUS SPECTRUM

We consider the Schrödinger operator $H = -\mathrm{d}^2/\mathrm{d}r^2 + V(r)$ in the interval $0 < r < \infty$, under the following hypotheses.

(i) Either $r = 0$ is in the limit-circle case, or for almost all λ in some interval (λ_1, λ_2) a solution of the Schrödinger equation exists that is subordinate at $r=0$.

(ii) For all $\lambda \in (\lambda_1, \lambda_2)$ there is no solution of the Schrödinger equation that is sequentially subordinate at $r=\infty$, and $r=\infty$ is in the limit-point case.

Under these two hypotheses, we shall develop a spectral theory and eigenfunction expansion for H, appropriate to the range $\lambda_1 < \lambda < \lambda_2$ of the spectrum of H. For most applications to scattering theory, the energy range in question will be $0 < \lambda < \infty$. However, the analysis will cover other cases of interest as well. Indeed, the theory extends without difficulty to cover absolute continuity for the Schrödinger operator acting in more general Hilbert spaces, including $L^2(0, b)$, where b is finite. (That such operators may have absolutely continuous spectra will follow from examples to be considered in Chapter 14.)

If $r=0$ is in the limit-circle case then a boundary condition is needed in order to define H as a self-adjoint operator. We know in this case, from Theorem 6.12, that $m_0(z)$ is meromorphic in the entire complex plane. A consequence of this is that $m_0(z)$ will have a boundary value as z approaches the real axis at λ, apart from isolated simple poles, and we denote this boundary value by $m_0(\lambda)$. If, on the other hand, a solution subordinate at $r=0$ exists for almost all $\lambda \in (\lambda_1, \lambda_2)$, we know from Theorem 7.2 and the proof of Theorem 7.3 that $m_0(\lambda) \equiv \lim_{\varepsilon \to 0} m_0(\lambda + i\varepsilon)$ exists, apart from a set of λ having Lebesgue measure zero. So, in either case, we define $m_0(\lambda)$ to be the almost-everywhere boundary value of $m_0(z)$ for $\lambda \in (\lambda_1, \lambda_2)$.

At the endpoint $r=\infty$ the situation is quite different. There, from the proof of Theorem 7.3, and in particular from (7.2.16), we know that, for *fixed* $\lambda \in (\lambda_1, \lambda_2)$, $\operatorname{Im}(m_\infty(z))$ is bounded above and below by positive constants. By standard results in complex-variable theory, we know that $m_\infty(z)$ has an almost-everywhere boundary value as z approaches points λ on the real axis, and we denote this boundary value by $m_+(\lambda)$. Again, $m_+(\lambda)$ is defined, apart from a λ-set having Lebesgue measure zero, as an upper boundary value.

We refer to (6.4.6), where the kernel of the resolvent operator for H is given (we take $a=0$, $b=\infty$). For fixed λ, with $z=\lambda+i\varepsilon$, we have $\operatorname{Im}(m_0(z))<0$, $\operatorname{Im}(m_\infty(z))>0$, so that by the bounds on $\operatorname{Im}(m_\infty(z))$, together with the fact that $|m_\infty(z)|$ is also bounded, we may conclude that the functions $((m_0(z)-m_\infty(z))^{-1}$, $m_0(z)(m_0(z)-m_\infty(z))^{-1}$, $m_\infty(z)(m_0(z)-m_\infty(z))^{-1}$ and $m_0(z)m_\infty(z)(m_0(z)-m_\infty(z))^{-1}$ are all bounded, as ε approaches zero with $\lambda_1 < \lambda < \lambda_2$. As in the proof of Theorem 7.3(ii), we may conclude that H has a purely absolutely continuous spectrum in this interval. The functions $\rho_{ij}(\lambda)$, through (6.5.14), are expressible directly in terms of the spectral family of H, and lead evidently to absolutely continuous measures (charges in the case of ρ_{12} and ρ_{21}). Furthermore, we can apply Lemma 2.8(iii) to determine the density function of each of these measures. (Although Lemma 2.8 dealt explicitly with *finite* measures, it will be found that the measure outside the interval (λ_1, λ_2) makes, as we should expect, no contribution to the density function within that interval; so it is sufficient that the measure is finite over (λ_1, λ_2).) Comparing (2.4.6) with (6.5.24), we find that $\text{Im}\,(M_{ij}(z))$ is just the resolvent transform of the measure $d\rho_{ij}$. (See Section 2.4 for a general discussion of the resolvent transform.) Hence Lemma 2.8 leads to an expression for the density function, valid for almost all λ in the interval (λ_1, λ_2):

$$\frac{d\rho_{ij}(\lambda)}{d\lambda} = \frac{1}{\pi} \lim_{\varepsilon \to 0+} \text{Im}\,(M_{ij}(\lambda + i\varepsilon)). \tag{7.3.1}$$

We can now introduce on the right-hand side the explicit formulae (7.1.2) for the components of the matrix M_{ij}, and taking the limit as $\varepsilon \to 0$ gives, for almost all $\lambda \in (\lambda_1, \lambda_2)$,

$$\frac{d\rho_{11}(\lambda)}{d\lambda} = \frac{1}{\pi} \text{Im}\left(\frac{m_0(\lambda)m_+(\lambda)}{m_0(\lambda) - m_+(\lambda)}\right),$$

$$\frac{d\rho_{12}(\lambda)}{d\lambda} = \frac{d\rho_{21}(\lambda)}{d\lambda} = \frac{1}{\pi} \text{Im}\left(\frac{m_0(\lambda) + m_+(\lambda)}{2(m_0(\lambda) - m_+(\lambda))}\right), \tag{7.3.2}$$

$$\frac{d\rho_{22}(\lambda)}{d\lambda} = \frac{1}{\pi} \text{Im}\left(\frac{1}{m_0(\lambda) - m_+(\lambda)}\right).$$

The notation may be simplified by defining $\rho(\lambda) = \rho_{22}(\lambda)$. Noting that $m_0(\lambda)$ is real, we have

$$\frac{d\rho(\lambda)}{d\lambda} = \frac{1}{\pi} \frac{\text{Im}\,(m_+(\lambda))}{|m_0(\lambda) - m_+(\lambda)|^2}, \tag{7.3.3}$$

with

$$\frac{d\rho_{ij}(\lambda)}{d\lambda} = r_i(\lambda) r_j(\lambda) \frac{d\rho(\lambda)}{d\lambda} \tag{7.3.4}$$

where

$$r_1(\lambda) = m_0(\lambda), \qquad r_2(\lambda) = 1. \tag{7.3.5}$$

Comparison should be made here between (7.3.4) and the factorization of residues that is given in (7.1.5). The difference between the two cases arises from the fact that in a problem with a discrete spectrum the integral (7.1.3) may be evaluated, through Cauchy's Theorem, by evaluation of residues; whereas if the spectrum is absolutely continuous, we are led to an integral round a branch cut, corresponding roughly to the limiting case of a continuous distribution of poles along part of the real axis.

Given a function pair $\{h_1(\lambda), h_2(\lambda)\}$ belonging to the space $L^2(\mathbb{R}; d\rho_{ij}(\lambda))$, we have, by a linear combination of (7.1.4),

$$h_2(\lambda)+m_0(\lambda)h_1(\lambda)=\int_0^\infty dr f(r)\,(u_2(r,\lambda)+m_0(\lambda)u_1(r,\lambda)).$$

Moreover, the contribution, in (7.1.4)′, from integration over the interval (λ_1, λ_2), with (7.3.4) and (7.3.5), takes the form

$$\begin{aligned} f(r) &= \sum_{i,j}\int h_i(\lambda)u_j(r,\lambda)r_i(\lambda)r_j(\lambda)\,d\rho(\lambda) \\ &= \int (h_2(\lambda)+m_0(\lambda)h_1(\lambda))(u_2(r,\lambda)+m_0(\lambda)u_1(r,\lambda))\,d\rho(\lambda). \end{aligned}$$

In a similar way, an evaluation of the norm of the function pair $\{h_1, h_2\}$ in the space $L^2(\mathbb{R}; d\rho_{ij})$ leads to the contribution

$$\|\{h_1, h_2\}\|^2=\int |(h_2(\lambda)+m_0(\lambda)h_1(\lambda)|^2\,d\rho(\lambda).$$

For $\lambda_1<\lambda<\lambda_2$, we define the function $h\in L^2((\lambda_1, \lambda_2); d\rho(\lambda))$ by

$$h(\lambda)=h_2(\lambda)+m_0(\lambda)h_1(\lambda), \tag{7.3.6}$$

and the solution $u_s(r, \lambda)$ of the Schrödinger equation by

$$u_s(r,\lambda)=u_2(r,\lambda)+m_0(\lambda)u_1(r,\lambda). \tag{7.3.7}$$

Then for $f\in L^2(0, \infty)$, belonging to the range of the spectral projection of H associated with the interval (λ_1, λ_2), we can define the integral transform

$$h(\lambda)=\int_0^\infty f(r)u_s(r,\lambda)\,dr \tag{7.3.8}$$

and inverse transform

$$f(r)=\int_{(\lambda_1,\lambda_2)} h(\lambda)u_s(r,\lambda)\,d\rho(\lambda) \tag{7.3.9}$$

Equations (7.3.8) and (7.3.9) define a transformation from the subspace of $L^2(0, \infty)$ corresponding to the interval (λ_1, λ_2) of the spectrum of H onto the subspace $L^2((\lambda_1, \lambda_2); \mathrm{d}\rho)$. The transformation is unitary; i.e.

$$\int_0^\infty |f(r)|^2 \,\mathrm{d}r = \int_{(\lambda_1, \lambda_2)} |h(\lambda)|^2 \,\mathrm{d}\rho(\lambda). \tag{7.3.10}$$

In the space $L^2((\lambda_1, \lambda_2); \mathrm{d}\rho)$, H is implemented by the operator of multiplication by λ. As a special case, if the entire spectrum of H is contained in the interval (λ_1, λ_2) then H is unitarily equivalent to the operator of multiplication by λ in $L^2(\mathbb{R}; \mathrm{d}\rho(\lambda))$.

Note that the solution $u_s(r, \lambda)$ is subordinate at $r=0$—hence the suffix. It is thus the subordinate solution at $r=0$ that enters into the kernel of the integral transform that diagonalizes the total Hamiltonian H.

There are, however, two disadvantages in the use of the representation (7.3.8)–(7.3.10) in concrete applications. First, the kernel of the integral transform depends on the choice of an intermediate point ξ in the interval $(0, \infty)$. A different choice of ξ will give rise to a different normalization of the subordinate solution $u_s(r, \lambda)$. Since the value of ξ has no particular significance, we should like to free our integral transform from this arbitrary parameter. A second problem is that to determine, through (7.3.3), the measure $\mathrm{d}\rho(\lambda)$, we are faced first of all with the difficult task of evaluating the functions $m_0(\lambda)$ and $m_+(\lambda)$. Fortunately, both of these obstacles to the application of (7.3.8)–(7.3.10) can be overcome through the introduction of ideas that bring us close to the essence of spectral theory as it applies to scattering in quantum mechanics. It is first necessary to give some appropriate definitions.

Definition 7.3

A (complex-valued) solution $f_+(r, \lambda)$ of the Schrödinger equation, at real energies λ, is said to be an *upper solution* if $f_+(r, \lambda)$ is the limit, for almost all λ, as $z=\lambda+\mathrm{i}\varepsilon$ approaches the real axis from above, of solutions $f(r, z)$ at complex energy z that are square-integrable for large r. If, in addition, the condition

$$W(f_+, \bar{f}_+) \equiv f_+\bar{f}'_+ - \bar{f}_+ f'_+ = -\mathrm{i} \tag{7.3.11}$$

is satisfied then f_+ is said to be a normalized upper solution. (More precisely, these are *families* of solutions.)

Thus, with $V(r) \equiv 0$, $\mathrm{e}^{\mathrm{i}kr}$ with $k>0$ and $\lambda=k^2$ is an upper solution, being the upper boundary value of the square-integrable solution $\exp(iz^{1/2}r)$, where the square root having positive imaginary part is intended. However, e^{ikr} is not a normalized upper solution, since $W(e^{ikr}, e^{-ikr}) = -2ik$. A normalized upper solution in this case is $e^{ikr}/(2k)^{1/2}$. The reason for the factor $(2k)^{-1/2}$ in normalizing this solution is that our general eigenfunction expansion will be expressed in terms of integrals with respect to λ, where $d\lambda = 2k\,dk$. Note that the Wronskian on the left-hand side of (7.3.11) is pure imaginary, having, as we shall shortly verify, a negative imaginary part. Observe also that normalized upper solutions are determined up to a multiplicative phase factor, which may be a function of λ.

Definition 7.4

Given a normalized upper solution $f_+(r,\lambda)$, let u_+ and v_+ be real and imaginary parts respectively, so that

$$f_+(r,\lambda) = u_+(r,\lambda) + iv_+(r,\lambda). \tag{7.3.12}$$

For any *real* solution $u(r,\lambda)$ of the Schrödinger equation at real energy λ, write

$$u(r,\lambda) = A(\lambda)u_+(r,\lambda) + B(\lambda)v_+(r,\lambda). \tag{7.3.13}$$

Then the *spectral amplitude* of the solution $u(r,\lambda)$ is defined as $(A^2(\lambda)+B^2(\lambda))^{1/2}$.

From (7.3.11) and (7.3.12), it follows easily that

$$W(u_+, v_+) = \tfrac{1}{2}. \tag{7.3.14}$$

Hence

$$\begin{aligned} W(u, f_+) &= W(Au_+ + Bv_+, u_+ + + iv_+) \\ &= \tfrac{1}{2}(iA - B), \end{aligned}$$

so that the spectral amplitude of $u(r,\lambda)$ may be written in the form

$$\mathscr{A} = |2W(u, f_+)|. \tag{7.3.15}$$

Since f_+ is determined up to a (λ-dependent) phase factor, this proves that the spectral amplitude of a real solution $u(r,\lambda)$ is unique. For $V(r) \equiv 0$ and $u = (a_1 \cos kr + a_2 \sin kr)/(2k)^{1/2}$ the spectral amplitude is $(a_1^2 + a_2^2)^{1/2}$.

We are now ready to exhibit, in canonical form, the eigenfunction expansion for H in the interval (λ_1, λ_2).

Theorem 7.4

Suppose that conditions (i) and (ii) at the beginning of this section are satisfied. Let $\mathscr{H}_{(\lambda_1, \lambda_2)}$ *denote the subspace of* $L^2(0, \infty)$ corresponding to the interval (λ_1, λ_2) of the spectrum of H. For almost all $\lambda \in (\lambda_1, \lambda_2)$, let $u_0(r, \lambda)$ denote the solution of the Schrödinger equation at energy λ, subordinate at $r=0$ (or, in the limit-circle case at 0, satisfying the appropriate boundary condition at $r=0$), and normalized to have spectral amplitude 1. Then a unitary transformation from $\mathscr{H}_{(\lambda_1, \lambda_2)}$ onto $L^2(\lambda_1, \lambda_2)$ may be defined by the equations

$$F(\lambda) = \left(\frac{2}{\pi}\right)^{1/2} \int_0^\infty f(r) u_0(r, \lambda)\, \mathrm{d}r, \tag{7.3.16}$$

$$f(r) = \left(\frac{2}{\pi}\right)^{1/2} \int_{(\lambda_1, \lambda_2)} F(\lambda) u_0(r, \lambda)\, \mathrm{d}\lambda, \tag{7.3.17}$$

$$\int_0^\infty |f(r)|^2\, \mathrm{d}r = \int_{(\lambda_1, \lambda_2)} |F(\lambda)|^2\, \mathrm{d}\lambda, \tag{7.3.18}$$

such that H is implemented by the operator of multiplication by λ in $L^2(\lambda_1, \lambda_2)$.

Proof

In order to define the spectral amplitude, we must first find an upper solution of the Schrödinger equation. Since $u_2(r, z) + m_\infty(z) u_1(r, z)$ is square-integrable at infinity, an upper solution may be obtained by taking the limit $\varepsilon \to 0+$ with $z = \lambda + \mathrm{i}\varepsilon$. Thus $u_2(r, \lambda) + m_+(\lambda) u_1(r, \lambda)$ is an upper solution as required. Using the conditions (7.1.1) at $r = \xi$ to evaluate the left-hand side of (7.3.11), the Wronskian is given in this case by $W = \bar{m}_+(\lambda) - m_+(\lambda) = -2\mathrm{i}\, \mathrm{Im}\,(m_+(\lambda))$.

Hence a normalized upper solution is given by

$$f_+(r, \lambda) = \frac{u_2(r, \lambda) + m_+(\lambda) u_1(r, \lambda)}{[2 \mathrm{Im}(m_+(\lambda))]^{1/2}}. \tag{7.3.19}$$

A subordinate solution at $r=0$ is given, for almost all $\lambda \in (\lambda_1, \lambda_2)$, by

$$u_s(r, \lambda) = u_2(r, \lambda) + m_0(\lambda) u_1(r, \lambda),$$

where, again evaluating at $r = \xi$, we have

$$W(u_s, f_+) = \frac{m_+(\lambda) - m_0(\lambda)}{[2 \mathrm{Im}\,(m_+(\lambda))]^{1/2}}.$$

According to (7.3.15), we have, for the spectral amplitude of u_s.

$$\mathscr{A}=\left(\frac{2}{\pi}\right)^{1/2}\left|\frac{\pi(m_0(\lambda)-m_+(\lambda))^2}{\operatorname{Im}(m_+(\lambda))}\right|^{1/2},$$

which by (7.3.3) gives just

$$\mathscr{A}=\left(\frac{2}{\pi}\left(\frac{d\rho}{d\lambda}\right)^{-1}\right)^{1/2} \tag{7.3.20}$$

Normalizing $u_s(r,\lambda)$ to have spectral amplitude 1 now gives

$$u_0(r,\lambda)=\left(\frac{\pi}{2}\frac{d\rho}{d\lambda}\right)^{1/2}u_s(r,\lambda). \tag{7.3.21}$$

It is now a straightforward matter, starting from (7.3.8)–(7.3.10), with the substitutions $F(\lambda)=h(\lambda)(d\rho/d\lambda)^{1/2}$, $d\rho(\lambda)=(d\rho(\lambda)/d\lambda)\,d\lambda$, to verify (7.3.16)–(7.3.18). ∎

In (7.3.16)–(7.3.18) we have arrived at an eigenfunction expansion for H in the interval (λ_1, λ_2) in which all reference to the intermediate point ξ has been dropped. Instead, in defining the kernel $u_0(r,\lambda)$ of the integral transform, we have only to seek, at real energies λ, that solution of the Schrödinger equation which has appropriate asymptotic behaviour at $r=0$ and at $r=\infty$.

Let us consider, for example, a potential $V\in L^1(0,\infty)$. In this case, we can define f_+ to have asymptotic behaviour $e^{ikr}/(2k)^{1/2}$ as $r\to\infty$, and a suitably normalized solution for $u_0(r,\lambda)$ should behave asymptotically like $(2k)^{-1/2}\sin(kr+\delta)$. At $r=0$ we require $u_0(0,\lambda)=0$, corresponding to the boundary condition $f(0)=0$. If we wish to replace, in (7.3.17) and (7.3.18), the λ-integration by integration with respect to k, where $d\lambda=2k\,dk$, we need only to make the substitutions

$$v_0(r,k)=(2k)^{1/2}u_0(r,k^2), \qquad G(k)=(2k)^{1/2}F(k^2).$$

For $V(r)\equiv 0$, this reduces (7.3.16) and (7.3.17) to the Fourier sine transform and its inverse.

Of course, for general $V\in L^1(0,\infty)$ there will not only be an absolutely continuous spectrum for $\lambda>0$, but also possible zero- or negative-energy eigenvalues, in which case the full transform of a function $f\in L^2(0,\infty)$ should contain, in (7.3.16)–(7.3.18), additional contributions from the discrete part of the spectrum. The form that these extra terms should take has already been discussed in Section 7.1.

To deal with potentials that are of long range, such as the Coulomb potential, it is only necessary to use the analysis of Section 2.8. In particular,

Lemma 2.21 allows us to determine the appropriate asymptotic behaviour for $u_0(r,\lambda)$ in the large-r domain. In the case of the Coulomb potential, we have the asymptotic behaviour

$$\frac{1}{(2k)^{1/2}}\sin\left(kr-\frac{g}{2k}\log r+\delta\right)$$

for $u_0(r,\lambda)$ with $\lambda=k^2$.

We may also have to modify the behaviour of $u_0(r,\lambda)$ near the origin, in order to take account of any possible singularity of the potential. Thus, for a non-singular potential with centrifugal barrier, $u_0(r,\lambda)$ should behave at $r=0$ like const. r^{l+1}. For the appropriate modifications to take account of singular potentials, we again refer the reader to Section 2.8.

Chapter 8

Scattering by Central Potentials

8.1 SHORT- AND LONG-RANGE ASYMPTOTICS

For a particle moving in a central field of force, described by a potential $V(r)$, each partial-wave subspace, associated with quantum numbers l and m of angular momentum, is a reducing subspace for the total Hamiltonian H, which may be implemented in that subspace by

$$H=-\frac{\mathrm{d}^2}{\mathrm{d}r^2}+V(r)+\frac{l(l+1)}{r^2}, \tag{8.1.1}$$

as a self-adjoint operator in the space $L^2(0, \infty)$. In this chapter we shall consider scattering by such a potential, under each of the alternative hypotheses (a) $V \in L^1(c, \infty)$ for every $c>0$, and (b) $|V(r)| \leq \text{const}/r^\beta$, $|\mathrm{d}V(r)/\mathrm{d}r| \leq \text{const}/r^{\beta+1}$ for $r \geq c$, for some $\beta > \frac{1}{2}$. If $V \in L^1(c, \infty)$ then the potential is integrable at infinity, and thus of short range. On the other hand, the conditions $|V| \leq \text{const}/r^\beta$, $|\mathrm{d}V/\mathrm{d}r| \leq \text{const}/r^{\beta+1}$, with $\beta > \frac{1}{2}$, admit potentials that are of long range—in particular the Coulomb potential $V(r)=g/r$. There is no particular significance to the limiting value $\beta = \frac{1}{2}$; it is possible to extend the theory to allow β such that $\beta > 0$. If $\beta > 1$ then we are again in the short-range case.

Our free Hamiltonian H_0 is the self-adjoint operator

$$H_0=-\frac{\mathrm{d}^2}{\mathrm{d}r^2}+\frac{l(l+1)}{r^2}, \tag{8.1.2}$$

with boundary condition $f(0)=0$ imposed for $l=0$ at the origin. For the total Hamiltonian H, at $r=0$, we shall make the same assumption as in (i) at the start of Section 7.3. That is, we shall assume, for each l, that we have the limit-circle case at $r=0$ (when a suitable boundary condition must be imposed in order to define H as a self-adjoint operator), or else that for almost all positive energies λ there is a solution of the Schrödinger equation that is subordinate at $r=0$. As we saw in Chapter 7, these assumptions imply an absolutely continuous spectrum for H along the positive real line. The spectrum of H for negative energies will be of lesser importance in the present context, since scattering states belong to the positive-energy subspace. If $r=0$ is limit-circle, or if subordinate solutions exist for almost all $\lambda<0$, then the negative spectrum of H will be purely discrete, consisting of isolated eigenvalues having $\lambda=0$ as the only possible limit point. Our conditions allow a wide range of behaviour of the potential at $r=0$, including both non-singular potentials and singular potentials that may be either attractive or repulsive.

Scattering theory is a study of the large-time evolution of states. For central potentials, since each partial-wave subspace reduces the total Hamiltonian, time evolution may be considered separately in each of these subspaces. The evolution operator e^{-iHt} is best defined by means of an eigenfunction expansion. We have already established such expansions in Chapter 7, and in particular in Theorem 7.4. We set $\lambda=k^2$, with k as an integration variable. For arbitrary $f(r)\in L^2(0,\infty)$, we define $F(k)$ for $\lambda>0$ by

$$F(k)=\left(\frac{2}{\pi}\right)^{1/2}\int_0^\infty f(r)v_0(r,k)\,\mathrm{d}r. \tag{8.1.3}$$

If, in addition, as demanded by scattering theory, f belongs to the range of the projection associated with the positive part of the spectrum of H then (8.1.3) may be inverted to give

$$f(r)=\left(\frac{2}{\pi}\right)^{1/2}\int_0^\infty F(k)v_0(r,k)\,\mathrm{d}k. \tag{8.1.4}$$

We use here the notation $F(k)$ to refer to the function $(2k)^{1/2}F(k^2)$, where $F(\lambda)$, in our previous notation, is defined by Theorem 7.4.

In (8.1.3) and (8.1.4), $v_0(r,k)$ is that solution of the Schrödinger equation, at energy $\lambda=k^2$, which either satisfies the appropriate boundary condition at $r=0$ (if $r=0$ is in the limit-circle case), or is subordinate at $r=0$, and which in addition has the correct asymptotic behaviour in the limit as $r\to\infty$. This large-r behaviour of $v_0(r,k)$ depends on the nature of the potential at large r. If V is a short-range potential ($V\in L^1(c,\infty)$) then for some $\delta(k)$ we have the asymptotic behaviour

$$v_0(r,k)=\sin(kr-\tfrac{1}{2}l\pi+\delta(k))+o(1) \tag{8.1.5}$$

where $o(1)$ denotes a contribution that converges to zero for large r. The way in which we have written the argument of the sine function on the right-hand side of (8.1.5) is explained by the fact that $v_0(r,k)=krj_l(kr)$ in the case of zero potential. According to (5.1.7), we then have, for $V(r)\equiv 0$, the asymptotic behaviour $v_0(r,k)=\sin(kr-\frac{1}{2}l\pi)+o(1)$. Thus for a free particle, we have, in (8.1.5), $\delta(k)=0$.

The function $\delta(k)$ is called the *scattering phase shift* at energy $\lambda=k^2$. The magnitude of the phase shift is a measure of how the potential $V(r)$ affects the phase of the solution $v_0(r,k)$ of the Schrödinger equation. We suppress for the moment the l-dependence of the phase shift.

In the case of a long-range potential, the formula (8.1.5) no longer gives the correct large-r behaviour of $v_0(r)$. We must then use the asymptotic analysis of Section 2.8 (see, e.g. Lemma 2.21). Solutions of the Schrödinger equation may be defined having respectively the asymptotic behaviours

$$\cos\left[\int_c^r (k^2-V(t))^{1/2}\,dt\right] \quad \text{and} \quad \sin\left[\int_c^r (k^2-V(t))^{1/2}\,dt\right],$$

where, for fixed k^2, the positive constant c is taken sufficiently large that $|V(t)|<k^2$ for $t>c$. Since, by hypothesis, we have $|V(t)|=O(|t|^{-\beta})$ for some $\beta>\frac{1}{2}$, we can apply the binomial theorem to obtain

$$(k^2-V(t))^{1/2}=k-\frac{V(t)}{2k}+O(|t|^{-2\beta}).$$

Hence for large r we can write

$$\int_c^r (k^2-V(t))^{1/2}\,dt=kr-\frac{W(r)}{2k}+\theta(r,k),$$

where $V=dW/dr$, and the function θ has a finite limit as $r\to\infty$. Thus for long-range potentials we have, instead of (8.1.5),

$$v_0(r,k)=\sin\left(kr-\frac{l\pi}{2}-\frac{W(r)}{2k}+\delta(k)\right)+o(1). \tag{8.1.6}$$

We can again call $\delta(k)$ the phase shift for scattering by the long-range potential. There is, however, a fundamental ambiguity in defining a phase shift in this case. The function $W(r)$ in (8.1.6), with $V=dW/dr$, is determined only up to an additive constant. Any change in the value of this constant will add to the phase shift a function of k. We shall see another aspect of this ambiguity when we come to define wave operators. For the moment, we should note that (8.1.6) may be regarded as a generalization of (8.1.5), applicable to both short- and long-range potentials, in which we take respectively $W=0$ in the short-range and $V=\mathrm{d}W/\mathrm{d}r$ in the long-range case. It is also possible to see the difference between (8.1.5) and (8.1.6), which arises from the non-integrability of $V(r)$ at large distances in the long-range case, as resulting from a renormalization of phase through the subtraction of an infinite constant of integration; at any rate, a more or less unified treatment can be achieved by using (8.1.6) in both cases, with, however, alternative possible definitions of the function $W(r)$.

We now turn to a study of the time evolution of states, generated by the Hamiltonian H. We shall be particularly interested in a comparison with the evolution of free-particle states. Since free evolution leads a particle to move asymptotically to large distances from the scattering centre $r=0$, a crucial element of our analysis will be a treatment of large-t large-r asymptotics. We let f belong to the positive-energy subspace of H. Since, in the subspace derived by application of the integral transform (8.1.3), H is (unitarily equivalent to) multiplication by k^2, we can represent the state at time t in the transformed space by the function $\mathrm{e}^{-\mathrm{i}k^2t}F(k)$. Returning to position space, if $f(r,0)=f(r)$ is the wave function at time $t=0$, as an element of $L^2(0,\infty)$, then the wave function at general time t is given by

$$(\mathrm{e}^{-\mathrm{i}Ht}f)(r)=f(r,t)$$

$$=\left(\frac{2}{\pi}\right)^{1/2}\int_0^\infty \mathrm{e}^{-\mathrm{i}k^2t}F(k)v_0(r,k)\,\mathrm{d}k. \tag{8.1.7}$$

The following Lemma gives precise expression to the fact that, for large r, the function v_0 in the integrand of (8.1.7) may be replaced by its asymptotic expression.

Lemma 8.1

Let V be a short- or long-range potential such that the assumptions of the present section are satisfied, with $\beta>\frac{3}{4}$ for V of long range. Define

$$\tilde{v}_0(r,k)=\sin\left(kr-\frac{l\pi}{2}-\frac{W(r)}{2k}+\delta(k)\right), \tag{8.1.8}$$

with $V=\mathrm{d}W/\mathrm{d}r$ (long-range) or $V=0$ (short-range). Let

$$\tilde{f}(r,t)=\left(\frac{2}{\pi}\right)^{1/2}\int_0^\infty \mathrm{e}^{-\mathrm{i}k^2t}F(k)\tilde{v}_0(r,k)\,\mathrm{d}k. \tag{8.1.9}$$

Then, for $F \in C_0^\infty(0, \infty)$,

$$\lim_{R\to\infty} \int_R^\infty |f(r,t) - f(r,t)|^2 \, dr = 0, \tag{8.1.10}$$

convergence being uniform in t. (We shall indicate in the proof how the statement of the lemma may be modified to cover all $\beta > \frac{1}{2}$.)

Proof

The lemma states that f is close to $\tilde{f}$ in norm in the space $L^2(R, \infty)$, provided that R is sufficiently large. The proof proceeds a little differently, depending on whether V is short- or long-range. We first consider the short-range case, for which $V \in L^1(c, \infty)$. With $V_l(r) \equiv V(r) + l(l+1)/r^2$, the function $v_0(r,k)$ having the asymptotic behaviour (8.1.5) will be a solution of the integral equation

$$v_0(r,k) = \sin(kr - \tfrac{1}{2}l\pi + \delta(k)) - k^{-1} \int_r^\infty \sin k(r-s) V_l(s) v_0(s,k) \, ds. \tag{8.1.11}$$

In Chapter 2 we saw how to iterate such integral equations in order to extract information on the large-r asymptotic behaviour. For sufficiently large r, and $k > 0$, the iterative solution of (8.1.11) is an infinite series of the form

$$\begin{aligned} v_0(r,k) &= \tilde{v}_0(r,k) \\ &\quad - k^{-1} \int_r^\infty ds_1 \sin k(r-s_1) V_l(s_1) \sin(ks_1 - \tfrac{1}{2}l\pi + \delta(k)) \\ &\quad + k^{-2} \int_r^\infty ds_1 \left\{ \sin k(r-s_1) V_l(s_1) \int_{s_1}^\infty ds_2 \sin k(s_1 - s_2) V_l(s_2) \right. \\ &\qquad\qquad \left. \times \sin(ks_2 - \tfrac{1}{2}l\pi + \delta(k)) \right\} \end{aligned} \tag{8.1.12}$$

Now

$$f(r,t)-\tilde{f}(r,t)=\left(\frac{2}{\pi}\right)^{1/2}\int_0^\infty e^{-ik^2t}F(k)(v_0(r,k)-\tilde{v}_0(r,k))\,dk, \qquad (8.1.13)$$

and we can substitute into the integrand the series (8.1.12) for $v_0(r,k)$. Since $F(k)\in C_0^\infty(0,\infty)$, the integration is effectively over a finite interval $[k_1, k_2]$, with $0<k_1<k_2$. In this interval the right-hand side of (8.1.12) converges uniformly in k and r (for $r>R$ and R sufficiently large), so that integration of the series on the right-hand side of (8.1.13) may be carried out term by term. We then consider the contribution to this integral of the nth term on the right-hand side of (8.1.12); we have, necessarily, $n\geq 2$. This term, with factor $(-k)^{-(n-1)}$ will be an $(n-1)$-fold multiple integral with respect to variables $s_1, s_2, \ldots, s_{n-1}$, over the integration region $r<s_1<s_2\ldots<s_{n-1}$. The integrand will contain a product of n sine functions, of which the first is $\sin k(r-s_1)$. Using standard trigonometric formulae for products of sine and cosine functions, this product may be expressed as a sum and difference of 2^{n-1} functions, each of the form

$$\sin k(r-X(s_1, s_2, \ldots s_{n-1})\pm\delta(k))$$

or

$$\cos k(r-X(s_1,s_2,\ldots,s_{n-1})\pm\delta(k)),$$

with an overall factor of $2^{-(n-1)}$. Each of these 2^{n-1} functions may itself be expressed as a sum and difference of terms $\sin k(r-X)$ or $\cos k(r-X)$, multiplied by $\cos\delta(k)$ or $\sin\delta(k)$. Thus each integration with respect to k in (8.1.13) of a single term on the right-hand side of (8.1.12) results, for fixed s_1, $s_2, \ldots, s_{n-1}$, in an integrand that is a sum and difference of 2^n *sine or cosine transforms*, each of which looks like $\hat{h}(r-X)$ for some h having bounded norm in $L^2(0,\infty)$. There is again an overall numerical factor of $2^{-(n-1)}$. Applying the Schwarz inequality, integration with respect to $s_1, s_2, \ldots, s_{n-1}$ gives a bound

$$\left\{\int|\hat{h}(r-X(s_1, s_2\ldots s_{n-1})|^2\,|V_l(s_1)V_l(s_2)\ldots V_l(s_{n-1})|\,ds\right\}^{1/2}$$

$$\times\left\{\int|V_l(s_1)V_l(s_2)\ldots V_l(s_{n-1})|\,ds\right\}^{1/2},$$

where ds stands for $ds_1\,ds_2\ldots ds_{n-1}$, and, for $r>R$, each integral with respect to s_i is bounded by an integral from R to ∞. Squaring this estimate and carrying out the r-integration (taking as upper bound the integral with respect to r from $-\infty$ to ∞), we obtain a *norm* bound of the form const $\{\int_R^\infty |V_l(s)|\,ds\}^{n-1}$. There are 2^n such contributions coming from the nth term on the right-hand side of (8.1.12), each with a numerical factor $2^{-(n-1)}$. From (8.1.13) we then have

$$\left(\int_R^\infty |f(r,t)-\tilde{f}(r,t)|^2\,dr\right)^{1/2} \leq \text{const} \sum_{j=1}^{\infty} \left(\int_R^\infty |V_l(s)|\,ds\right)^j.$$

Since V_l is integrable at infinity, (8.1.10) now follows, for short-range potentials $V(r)$.

We turn now to consider long-range potentials satisfying $V(r)=O(r^{-\beta})$, $dV(r)/dr=O(r^{-(1+\beta)})$ as $r\to\infty$, for some $\beta>\frac{1}{2}$. Again, we need to estimate $|v_0(r,k)-\tilde{v}_0(r,k)|$ for large r, where now the function $\tilde{v}_0(r,k)$ is defined by (8.1.8). The required estimates in this case are provided by Lemma 2.18, where we have to obtain bounds, for large r, on the $o(1)$ contributions to (2.8.25). We refer to the differential equation (2.8.29) for a solution $f(x,k)$ of the Schrödinger equation, with $V_1(r)=V(r), V_2(r)=l(l+1)/r^2$, and where the new variable x is given in terms of r by (2.8.28). Note that $x\sim kr$ as $r\to\infty$. In the differential equation (2.8.30), it is easy to verify that $q_1(x)=O(x^{-(1+\beta)})$ and $q_2(x)=O(x^{-2})$ as $x\to\infty$. The resulting iterative solution for $f(x,k)$ and $f'(x,k)$ then gives rise to the estimate $f(x,k)=a\cos x+b\;\sin x+O(x^{-\beta})$. Returning to the original variable r, this tells us that $v_0(r,k)$ differs from its asymptotic limit $\sin[\int_c^r (k^2-V(t))^{1/2}\,dt\,\theta(k)]$ by at most $O(r^{-\beta})$. Since the norm of $r^{-\beta}$ in $L^2(R,\infty)$ approaches zero in the limit as $R\to\infty$, this allows us to write down (8.1.10), where $\tilde{f}(r,t)$ is given by (8.1.9), except that instead of (8.1.8) we have

$$\tilde{v}_0(r,k)=\sin\left(kr-\frac{l\pi}{2}-\frac{W(r)}{2k}+\delta(k)+\frac{1}{8k^3}\int_r^\infty V^2(s)\,ds+\ldots\right). \qquad (8.1.14)$$

The series inside the parentheses on the right-hand side is derived from integration of the binomial expansion of $(k^2-V)^{1/2}$. The series may be terminated by omitting all terms that go to zero as $r\to\infty$ more rapidly than $r^{-(1/2+\delta)}$ for some $\delta>0$. If $\beta>\frac{3}{4}$ then all terms of the series from $\int_R^\infty V^2(s)\,ds/8k^3$ onwards may be omitted. In this case, $\tilde{v}_0(r,k)$ may be defined by (8.1.8), and the proof is complete. More generally, for $\beta>\frac{1}{2}$, we may omit from (8.1.14) all integrals of the nth power of V for which $n\beta>\frac{3}{2}$. The modified function $\tilde{v}_0(r,k)$ obtained in this way replaces the definition (8.1.8), and the proof of (8.1.9) and (8.1.10) follows as before. ∎

The following lemma allows us to go a step further, in replacing the asymptotic expression $\tilde{v}_0(r,k)$ in (8.1.9) by the corresponding asymptotic expression $\tilde{v}_0^+(r,k)$ of an upper solution of the Schrödinger equation. Again we look at norms in $L^2(R,\infty)$, for large R; t is now positive.

Lemma 8.2

Let V be a short- or long-range potential as in Lemma 8.1, with $\beta>\frac{3}{4}$ in the long-range case. Define

$$\tilde{v}_0^+(r,k)=(2\mathrm{i})^{-1}\exp\left[\mathrm{i}\left(kr-\frac{l\pi}{2}-\frac{W(r)}{2k}+\delta(k)\right)\right], \tag{8.1.15}$$

and

$$\tilde{f}^+(r,t)=\left(\frac{2}{\pi}\right)^{1/2}\int_0^\infty \mathrm{e}^{-\mathrm{i}k^2t}F(k)\tilde{v}_0^+(r,k)\,\mathrm{d}k. \tag{8.1.16}$$

Then

$$\lim_{R\to\infty}\int_R^\infty |f(r,t)-\tilde{f}^+(r,t)|^2\,\mathrm{d}r=0, \tag{8.1.17}$$

convergence being uniform in t for $t>0$. A corresponding result, starting from (8.1.14) instead of (8.1.8), holds in the long-range case for all $\beta>\frac{1}{2}$.

Proof

The short-range case is readily dealt with. In (8.1.8) with $W(r)\equiv 0$, we express the sine function in terms of complex exponentials. We have only to show that

$$\int_0^\infty \mathrm{e}^{-\mathrm{i}(k^2t+kr+l\pi/2-\delta(k))}F(k)\,\mathrm{d}k$$

is negligible, in $L^2(R,\infty)$ norm, as $R\to\infty$, uniformly in $t>0$. Apart from numerical factors, this integral is the Fourier transform of a function $\mathrm{e}^{-\mathrm{i}(k^2t-\delta(k))}F(k)$. To verify the lemma for arbitrary $F\in L^2(k_1,k_2)$ having support in the interval $[k_1,k_2]$, it is sufficient to consider a dense subset of this L^2 space. It is therefore no loss of generality to suppose that

$$\mathrm{e}^{\mathrm{i}\delta(k)}F(k)\in C_0^\infty(k_1,k_2)\quad(0<k_1<k_2).$$

We can then write

$$\mathrm{e}^{-\mathrm{i}(k^2t+kr)}=\mathrm{i}(2kt+r)^{-1}\frac{\mathrm{d}}{\mathrm{d}k}\mathrm{e}^{-\mathrm{i}(k^2t+kr)}$$

and integrate by parts, with $t>0$, to obtain a suitable estimate. Equation (8.1.17) then follows, with little difficulty, from Lemma 8.1.

We now consider the case of a long-range potential. In the most general case, with $\beta>\frac{1}{2}$, the function $\tilde{v}_0(r,k)$ is given by (8.1.14), the series within the sine function terminating after a finite number of integrals of powers of V. We know that the integral operator having kernel $v_0(r,k)$ defines a bounded linear operator from $L^2(0,\infty)$ to $L^2(0,\infty)$, according to (8.1.4). By construction (that is, in $\tilde{v}_0$, omitting terms of order $r^{-(1/2+\delta)}$ with $\delta>0$), it is also apparent that the integral operator from $L^2(k_1,k_2)$ to $L^2(R,\infty)$ obtained by replacing v_0 by $\tilde{v}_0$ in (8.1.4) is also bounded for $R>0$. (We need to assume here that $F(k)$ has compact support away from $k=0$ in order that both k and k^{-1} are bounded in the resulting estimates.) Now $F\in L^2(k_1,k_2)$ is in the domain of multiplication by k^2, and hence corresponds to f in the domain of H. Such an f is *locally* in the domain of H_0, in the sense that $f\in D(H)\Rightarrow \rho f\in D(H_0)$ for any $\rho(r)$ that is infinitely differentiable, and such that $\rho\equiv 0$ for all $r>0$ sufficiently small, and $\rho\equiv 1$ for all r sufficiently large. (See, for example, Section 5.2 for local domain properties of H and H_0.) Now $\rho f\in D(H_0)\Rightarrow(\mathrm{d}/\mathrm{d}r)(\rho f)\in L^2(0,\infty)$, and it follows in particular that $f'(r)\in L^2(R,\infty)$. Differentiating (8.1.4) with respect to r, it follows that the linear operator from $F\in L^2(k_1,k_2)$ to $L^2(R,\infty)$ defined by

$$F(k)\rightarrow f'(r)=\left(\frac{2}{\pi}\right)^{1/2}\int_{k_1}^{k_2} F(k)v_0'(r,k)\,\mathrm{d}k \tag{8.1.18}$$

is bounded. Now Lemma 2.18, in addition to providing estimates of $v_0(r,k)$ for large r, also enables us to estimate derivatives with respect to r. Corresponding to (8.1.14), we have

$$\tilde{w}_0(r,k)=k\cos\left(kr-\frac{l\pi}{2}-\frac{W(r)}{2k}+\delta(k)+\frac{1}{8k^3}\int_r^\infty V^2(s)\,\mathrm{d}s+\dots\right),$$

where $\tilde{w}_0$ is an asymptotic expression for $v_0'(r,k)$ in the limit as $r\rightarrow\infty$, satisfying $|\tilde{w}_0(r,k)-v_0'(r,k)|=O(r^{-(1/2+\delta)})$ for some $\delta>0$. Comparing with (8.1.18), we see that

$$F(k)\rightarrow\left(\frac{2}{\pi}\right)^{1/2}\int_{k_1}^{k_2} F(k)\tilde{w}_0(r,k)\,\mathrm{d}k$$

defines a bounded linear operator from $L^2(k_1,k_2)$ to $L^2(R,\infty)$. Taking a linear combination of this with the corresponding result for $\tilde{v}_0$, it follows that

$$F(k)\to\int_{k_1}^{k_2} \mathrm{d}k F(k)\exp\left[-\mathrm{i}\left(k^2t+kr-\frac{l\pi}{2}\right.\right.$$

$$\left.\left.-\frac{W(r)}{2k}+\delta(k)+\frac{1}{8k^3}\int_r^\infty V^2(s)\,\mathrm{d}s+\dots\right)\right]$$

is a linear operator, bounded uniformly in t, between the same pair of spaces. As before, we must show that the $L^2(R,\infty)$ norm of the right-hand side approaches zero in the limit as $R\to\infty$, uniformly in $t>0$. We may again assume that $\mathrm{e}^{\mathrm{i}\delta(k)}F(k)\in C_0^\infty(k_1,k_2)$ and integrate by parts with

$$\mathrm{e}^{-\mathrm{i}(k^2t+kr)}=\mathrm{i}(2kt+r)^{-1}\frac{\mathrm{d}}{\mathrm{d}k}\mathrm{e}^{-\mathrm{i}(k^2t+kr)}.$$

The leading order term, for large r, has a factor

$$(2kt+r)^{-1}\frac{\mathrm{d}}{\mathrm{d}k}\left(\frac{W(r)}{2k}\right)$$

in the integrand, and is of order $r^{-\beta}$ ($(\log r)/r$ for $\beta=1$).

Hence the conclusion of the lemma is verified for all $\beta>\frac{1}{2}$. In the case $\beta>\frac{3}{4}$ it is unnecessary to retain terms of second or higher order in V in the complex exponential, and the simplified expression (8.1.15) results. ∎

Lemmas 8.1 and 8.2 furnish us with asymptotic formulae that characterize the large-r behaviour of the evolving wave function $f(r,t)$. Since, in each case, estimates are uniform in t, at least for $t>0$ in the case of Lemma 8.2, they are valid in the limit as $t\to\infty$. Before considering this limit, we need to consider the relation between large-t and large-r asymptotics. For a *classical* particle of mass m and momentum k, moving as $t\to\infty$ at large distances from the scattering centre, we expect the relation $r\approx kt/m$ to hold, though account must be taken of the effect on asymptotic time development of any long-range behaviour of the potentials. In our units, $\hbar=2m=1$, and $r\approx kt/m$ is equivalent to $r\approx 2kt$. The following lemma leads to a quantum analogue of this result. In integrals such as (8.1.16), the leading contribution for large r and t comes from the integration region for which $r\approx 2kt$.

Lemma 8.3

Given $\Phi\in C_0^\infty(0,\infty)$, define the integral $I(r,t)$ by

$$I(r,t)=\int_0^\infty dk\, e^{-i(k^2t-kr)}\Phi(k)\{W(r,k)-W(2kt,k)\}, \qquad (8.1.19)$$

where, for all k in the support of Φ, for $r \geq R_0 > 0$, and for some $\delta > 0$, the function $W(r,k)$ satisfies

(i) $|W(r,k)| \leq \text{const}, \qquad \left|\dfrac{\partial W(r,k)}{\partial k}\right| \leq \text{const}\, r^{(1/2-\delta)},$

$$\left|\frac{\partial W(r,k)}{\partial r}\right| \leq \text{const}\, r^{-(1/2+\delta)};$$

(ii) $\left|\dfrac{\partial^2 W(r,k)}{\partial r\,\partial k}\right| \leq \text{const}\, r^{-(1/2+\delta)},$

$$\left|\frac{\partial^2 W(r,k)}{\partial r^2}\right| \leq \text{const}\, r^{-(3/2+\delta)}.$$

Suppose that $W(r,k)$ is twice continuously differentiable. Then

$$\lim_{R\to\infty}\int_R^\infty dr\,|I(r,t)|^2=0, \qquad (8.1.20)$$

the limit being uniform in t for $t \geq T_0 = R_0/2k_1$, where k_1 is the infimum of the support of Φ.

Proof

In the integral (8.1.19), we write

$$e^{-i(k^2t-kr)}=i(2kt-r)^{-1}\frac{\partial}{\partial k}e^{-i(k^2t-kr)},$$

and integrate by parts. There are two contributions to the resulting integral, and we consider each in turn. The first is

$$i\int_0^\infty dk\,\Phi'(k)e^{-i(k^2t-kr)}\left\{\frac{W(r,k)-W(2kt,k)}{r-2kt}\right\}. \qquad (8.1.21)$$

Since W is bounded, the required estimate for (8.1.20) follows immediately if we take $2kt \leq \frac{1}{2}r$, since we have a bound for (8.1.21) of order r^{-1} in that case. We therefore assume that $2kt > \frac{1}{2}r$, and in the same way we may assume that $r > kt$. (Thus $kt < r < 4kt$, so that r and t are of the same order of magnitude.) By the Mean-Value Theorem, the expression within the curly brackets in

(8.1.21) is $[\partial W(x,k)/\partial x]_{x=\xi}$ for some ξ between r and $2kt$. Either $\xi > r$ or $\xi > 2kt$, and in either case $\xi > \frac{1}{2}r$. By condition (i) of the lemma, this gives $|\partial W(x,k)/\partial x|_{x=\xi} \leq \text{const}\, \xi^{-(1/2+\delta)} \leq \text{const}\, r^{-(1/2+\delta)}$, and the same bound applies to the integral (8.1.21), giving a contribution that is consistent with (8.1.20). The second contribution to the integral $I(r,t)$, on integrating by parts, is

$$\mathrm{i}\int_0^\infty dk\, \mathrm{e}^{-\mathrm{i}(k^2 t - kr)}\Phi(k)\frac{\partial}{\partial k}\left\{\frac{W(r,k)-W(2kt,k)}{r-2kt}\right\}. \tag{8.1.22}$$

The expression within curly brackets in (8.1.22), on differentiation with respect to k, gives rise to two contributions, of which the first is

$$(r-2kt)^{-1}\left\{\frac{\partial W(r,k)}{\partial k} - \left[\frac{\partial W(2kt,y)}{\partial y}\right]_{y=k}\right\}.$$

Here differentiation of $W(2kt,k)$ is carried out with respect to the second argument k only. Using the bound (i) on $\partial W(r,k)/\partial k$, we may restrict attention to the case for which $kt < r < 4kt$. We may then carry out the same kind of estimate as for the integrand of (8.1.21), except that now $W(r,k)$ is replaced by its derivative with respect to k. Instead of the bound on $\partial W(r,k)/\partial r$ in (i), we use the corresponding bound for $\partial^2 W(r,k)/\partial r\,\partial k$ in (ii), again leading to an estimate of order $r^{-(1/2+\delta)}$.

The remaining terms coming from differentiation of the curly bracket in (8.1.22) look like

$$-2t(r-2kt)^{-1}\left[\frac{\partial W(x,k)}{\partial x}\right]_{x=2kt} + 2t\frac{W(r,k)-W(2kt,k)}{(r-2kt)^2}. \tag{8.1.23}$$

For $2kt \leq \frac{1}{2}r$, on using (i) above, we obtain the bound $\text{const}\, t^{1/2-\delta}/r \leq \text{const}\, r^{-(1/2+\delta)}$ for the first term of (8.1.23), and the bound const/r for the second term. So it remains to consider (8.1.23) with $2kt > \frac{1}{2}r$. If $kt \geq r$ then we have $t^{1/2-\delta}(2kt-r)^{-1} \leq t^{-(1/2+\delta)}/k \leq \text{const}\, r^{-(1/2+\delta)}$, and the second term of (8.1.23) is bounded by $\text{const}\, r^{-1}$. So again in (8.1.23) we may restrict attention to the range of variables $kt < r < 4kt$. By Taylor expansion with second-order remainder, we have

$$W(r,k) = W(2kt,k) + (r-2kt)\left[\frac{\partial W(x,k)}{\partial x}\right]_{x=2kt} + \frac{(r-2kt)^2}{2}\left[\frac{\partial^2 W(x,k)}{\partial x^2}\right]_{x=\xi},$$

for some ξ lying between r and $2kt$. Substituting into (8.1.23), and noting the resulting cancellation, we obtain just $t[\partial^2 W(x,k)/\partial x^2]_{x=\xi}$, which, with hypothesis (ii) of the lemma, is bounded by $t\xi^{-(3/2+\delta)}$. But certainly $\xi > \frac{1}{2}r$, and also we are allowed to assume that $kt < r$, so that we have again a bound of order $r^{-(1/2+\delta)}$. We have now established bounds for each of the integrals (8.1.21) and (8.1.22) that follow from integrating $I(r,t)$ by parts. There are no boundary contributions, since Φ has compact support in $(0,\infty)$. Hence $I(r,t) \leq \text{const}\, r^{-(1/2+\delta)}$, from which (8.1.20) follows. ∎

Repeated application of Lemma 8.3 leads to a result that, in applications, represents a weakening of hypotheses (ii) of the lemma, while postulating further bounds on higher-order derivatives of $W(r,k)$. We shall state this result as a corollary to the lemma.

Corollary

Define $I(r,t)$ as in the lemma, but replace conditions (i) and (ii) by the bounds

$$\left|\frac{\partial^{i+j}}{\partial r^i\, \partial k^j} W(r,k)\right| \leq \text{const}\, r^{-\beta i + (1-\beta) j} \tag{8.1.24}$$

for some β satisfying $1 > \beta > \frac{1}{2}$, and for all $i, j = 0, 1, 2, \ldots$ (the constant on the right-hand side of (8.1.24) may depend on i and j). Then $I(r,t)$ may be expressed in the form

$$I(r,t) = I_1(r,t) + I_2(r,t), \tag{8.1.25}$$

where

$$\left.\begin{aligned} &\lim_{R\to\infty} \int_R^\infty \mathrm{d}r\, |I_1(r,t)|^2 = 0 \\ &\text{and} \\ &\lim_{t\to\infty} \int_{R_0}^\infty \mathrm{d}r\, |I_2(r,t)|^2 = 0. \end{aligned}\right\} \tag{8.1.26}$$

The first limit is again uniform in t for $t \geq T_0$.

Proof

If only derivatives up to first order are considered then (8.1.24) is equivalent to (i) of the lemma with $i, j = 0$ or 1. However, for (ii) of the lemma to apply, we need $\beta > \frac{3}{4}$. For $\beta > \frac{3}{4}$, (8.1.26) follows from the lemma with $I_2(r,t) \equiv 0$. We therefore consider the case $\frac{1}{2} < \beta \leq \frac{3}{4}$.

Integrating by parts as before, we arrive at the integral (8.1.22). All other contributions of $I(r,t)$ are of order $r^{-(1/2+\delta)}$, where $\beta = \delta + \frac{1}{2}$, and are absorbed into the function $I_1(r,t)$. Now (8.1.22) is of the form

$$i \int_0^\infty dk \, e^{-i(k^2 t - kr)} \Phi(k) W_1(r,k), \tag{8.1.27}$$

where

$$W_1(r,k) = \frac{\partial}{\partial k} \left\{ \frac{W(r,k) - W(2kt,k)}{r - 2kt} \right\}.$$

(For simplicity of notation, we suppress the t-dependence of W_1.) Defining W_1 to be continuous at $r = 2kt$, we have

$$W_1(2kt,k) = \left[\frac{\partial^2 W(x,k)}{\partial x \, \partial k} \right]_{x=2kt} + t \left[\frac{\partial^2 W(x,k)}{\partial x^2} \right]_{x=2kt},$$

which, by (8.1.24), gives

$$W_1(2kt,k) \leq \text{const}\, t^{-(2\beta-1)}.$$

Thus

$$i \int_0^\infty dk \, e^{-i(k^2 t - kr)} \Phi(k) W_1(2kt,k),$$

being the inverse Fourier transform with respect to k of the function $i(2\pi)^{1/2} e^{-ik^2 t} \Phi(k) W_1(2kt,k)$, has a norm in $L^2(R_0, \infty)$ that is bounded by $\text{const}\, \|\Phi\| t^{-(2\beta-1)}$. Since $\beta > \frac{1}{2}$, this integral may be absorbed into the function $I_2(r,t)$, and in place of (8.1.27) we may consider the integral

$$i \int_0^\infty dk e^{-i(k^2 t - kr)} \Phi(k) \{ W_1(r,k) - W_1(2kt,k) \}.$$

This integral is just (8.1.19) with W replaced by W_1. Repeating this process, we generate a sequence $W_1, W_2, W_3, \ldots$ of functions, defined iteratively by

$$W_{n+1}(r,k) = \frac{\partial}{\partial k} \left\{ \frac{W_n(r,k) - W_n(2kt,k)}{r - 2kt} \right\}.$$

At each stage, in passing from W_n to W_{n+1}, the large-r asymptotic behaviour is improved. If, for $\gamma \geq 0$, $W_n(r,k)$ satisfies the bounds

$$\left|\frac{\partial^{i+j} W_n(r,k)}{\partial r^i \, \partial k^j}\right| \leq \text{const}\, r^{-\gamma-\beta i+(1-\beta)j}$$

then

$$\left|\frac{\partial^{i+j} W_{n+1}(r,k)}{\partial r^i \, \partial k^j}\right|$$

satisfies the same bounds, except that γ is replaced by $\gamma+2\beta-1$. Since $\beta>\frac{1}{2}$, the value of γ increases at each stage, until a value of n is reached at which the function W_n satisfies each of the bounds (i) and (ii) of Lemma 8.3. Omitting contributions that are taken into the function $I_1(r,t)$ and $I_2(r,t)$, the remainder of the integral $I(r,t)$ then has the form (8.1.19) with W replaced by W_n and (i) and (ii) satisfied. The lemma then implies, for this final contribution to the integral, that (8.1.20) is satisfied, and we can add this term to the function $I_1(r,t)$. Thus $I(r,t)$ is indeed of the form (8.1.25), where (8.1.26) are satisfied. ■

(The reader with a mind to verify in detail the estimates that have been made will find the following elementary identities useful:

$$\frac{\partial}{\partial x}\frac{f(x)-f(y)}{x-y}=\frac{f'(x)-f'(y)}{x-y}-\frac{\partial}{\partial y}\frac{f(x)-f(y)}{x-y};$$

$$\frac{\partial^2}{\partial x^2}\frac{f(x)-f(y)}{x-y}=\frac{f''(x)-f''(y)}{x-y}-2\frac{\partial}{\partial y}\frac{f'(x)-f'(y)}{x-y}+\frac{\partial^2}{\partial y^2}\frac{f(x)-f(y)}{x-y}$$

(with extensions to higher-order derivatives with respect to x, the coefficients coming from the binomial expansion);

$$\frac{\partial^n}{\partial y^n}\frac{f(x)-f(y)}{x-y} = \frac{n!}{(x-y)^{n+1}}\left\{f(x)-f(y)-(x-y)f'(y)-\cdots-\frac{(x-y)^n}{n!}f^{(n)}(y)\right\},$$

where, by the Taylor Expansion Theorem, the right-hand side is of the form $(n+1)^{-1}f^{(n+1)}(\xi)$ for some ξ between x and y. In applying these identities to the evaluation and estimation of integrals related to $I(r,t)$, all integrands need to be considered for values in the range $kt<r<4kt$.)

Lemma 8.3 and its corollary allow us to take the asymptotic large-r and large-t limits of integrals such as (8.1.16), where under suitable hypotheses the function $W(r)$ may be replaced by $W(2kt)$. With these two results, together with the previous lemmas, we are provided with powerful tools of asymptotic analysis, which will lead us in the following section to wave and scattering operators for both short- and long-range potentials.

8.2. SHORT AND LONG-RANGE WAVE OPERATORS

Our study in the last section of large-r and large-t asymptotics for the wave function $f(r,t)=(e^{-iHt}f)(r)$ leads quite naturally to the definition of appropriate wave operators.

First of all, we consider scattering by a short-range potential $V(r)$. By Lemma 8.2, we know that $f(r,t)$ is closely approximated, in $L^2(R,\infty)$ norm and for $t>0$, by

$$\tilde{f}^{+}(r,t)=\frac{1}{2i}\left(\frac{2}{\pi}\right)^{1/2}\int_0^\infty e^{-i(k^2t-kr+l\pi/2-\delta(k))}F(k)\,dk \tag{8.2.1}$$

We have set $W(r)\equiv 0$ in (8.1.15), and taken F to have compact support in $L^2(0,\infty)$, where $F(k)$ is given in terms of the initial state $f(r,0)=f(r)$ by (8.1.3). We let U denote the unitary transformation from $L^2(0,\infty)$ to $L^2(0,\infty)$ defined by (8.1.3). Thus U sends $f(r)$ into $F(k)$. Let U_0 be the corresponding unitary transformation for zero potential. Then U is an integral operator with kernel $(2/\pi)^{1/2}v_0(r,k)$, and U_0 is an integral operator with kernel $(2/\pi)^{1/2}krj_l(kr)$.

We now define $g(r)\in L^2(0,\infty)$ by

then

$$\left.\begin{aligned} F&=Uf=U_0g;\\ g&=U_0^{-1}Uf. \end{aligned}\right\} \tag{8.2.2}$$

Motivated by (8.2.1), we consider the free evolution e^{-iH_0t} applied to the initial state $\exp(i\delta(H_0^{1/2}))g$. In momentum space, this state is given by $e^{i\delta(k)}F(k)$, and at time t the wave function in position space is

$$\left(\frac{2}{\pi}\right)^{1/2}\int_0^\infty e^{-i(k^2t-\delta(k))}F(k)krj_l(kr)\,dk.$$

Remembering that the free Hamiltonian gives rise to zero phase shift, the analysis leading to (8.2.1) tells us in this case that the freely evolving wave function is closely approximated, in $L^2(R, \infty)$ norm and for $t>0$, by precisely the function $\tilde{f}^{+}(r, t)$ defined by (8.2.1). It follows that the norm, in $L^2(R, \infty)$, of

$$e^{-iHt}f - e^{-iH_0t}\exp(i\delta(H_0^{1/2}))\,g$$

approaches zero in the limit as $R\to\infty$, and does so uniformly in t for $t>0$.

Since $v_0(r, k)$ is locally bounded as a function of r, it follows from (8.1.7) that $f(r,t)=(e^{-iHt}f)(r)$ is bounded on compact subsets of $(0, \infty)$, and converges to zero as $t\to\infty$, for fixed r, by the Riemann–Lebesgue Lemma. For $0<c<R$, the Lebesgue Dominated-Convergence Theorem implies that $\lim_{t\to\infty}\int_c^R|f(r,t)|^2\,dr=0$ and a similar result obtains with $f(r,t)$ replaced by $[e^{-iH_0t}\exp(i\delta(H_0^{1/2}))\,g](r)$.

Given $\varepsilon>0$, we fix R sufficiently large that, for $t>0$, the norm of $e^{-iHt}f - e^{-iH_0t}\exp(i\delta(H_0^{1/2}))\,g$ in the space $L^2(R, \infty)$ is smaller than $\frac{1}{2}\varepsilon$. For given $c>0$, we now take t sufficiently large that the $L^2(c, R)$ norm is also smaller than $\frac{1}{2}\varepsilon$. Then the norm in $L^2(c, \infty)$ is smaller than ε, for t sufficiently large, and we have shown that

$$\lim_{t\to\infty}\int_c^{\infty}|[e^{-iHt}f - e^{-iH_0t}\exp(i\delta(H_0^{1/2}))g](r)|^2\,dr=0. \tag{8.2.3}$$

Since $krj_l(kr)$ is bounded even in the limit as $r\to 0$, we can also show that

$$\lim_{t\to\infty}\int_0^{c}|[e^{-iH_0t}\exp(i\delta(H_0^{1/2}))g](r)|^2\,dr=0,$$

implying that, asymptotically for large t, the entire norm of $e^{-iH_0t}\exp(i\delta(H_0^{1/2}))g$ is concentrated in the interval $[c, \infty)$. Since

$$\|e^{-iH_0t}\exp(i\delta(H_0^{1/2}))g\| = \|g\| = \|f\| = \|e^{-iHt}f\|,$$

it follows that, asymptotically, the entire norm of $e^{-iHt}f$ is concentrated in the same interval. Thus we may set $c=0$ in (8.2.3), to deduce that

$$\lim_{t\to\infty}\|e^{-iHt}f - e^{-iH_0t}\exp(i\delta(H_0^{1/2}))g\|=0. \tag{8.2.4}$$

Defining the wave operator $\Omega_{+}=\text{s-lim}_{t\to\infty}\, e^{iHt}\,e^{-iH_0t}$, we now have

$$\Omega_{+}\exp(i\delta(H_0^{1/2}))\,g = f. \tag{8.2.5}$$

Equation (8.2.5) may be extended by continuity to arbitrary $g \in L^2(0, \infty)$, and to arbitrary f belonging to the range of the spectral projection for H corresponding to positive energies. The corresponding result for the limit $t \to -\infty$ is

$$\Omega_- \exp(-\mathrm{i}\delta(H_0^{1/2}))\, g = f. \tag{8.2.5$'$}$$

The wave operators $\Omega_\pm$ are asymptotically complete, the range in each case coinciding with the positive-energy subspace for H, and each wave operator is defined on the entire Hilbert space.

Substituting for g from (8.2.2), we can write

$$\Omega_\pm \exp(\pm\mathrm{i}\delta(H_0^{1/2}))\, U_0^{-1} U f = f,$$

so that we have the explicit formulae

$$\Omega_\pm = U^{-1} U_0 \exp(\mp\mathrm{i}\delta(H_0^{1/2})) = U^{-1} \mathrm{e}^{\mp\mathrm{i}\delta(k)} U_0, \tag{8.2.6}$$

where $\mathrm{e}^{\mp\mathrm{i}\delta(k)}$ stands for the operator of multiplication (in momentum space) by $\mathrm{e}^{\mp\mathrm{i}\delta(k)}$.

The scattering operator $S = \Omega_+^* \Omega_-$ may be written down immediately, using the unitary property of the transformation operators U_0 and U, to give

$$S = U_0^{-1}\, \mathrm{e}^{2\mathrm{i}\delta(k)}\, U_0. \tag{8.2.7}$$

Hence S is just the operator, in momentum space, of multiplication by $\mathrm{e}^{2\mathrm{i}\delta(k)}$.

We have dealt here with the case of a short-range potential, but all of these results have analogues in the case of long-range potentials. For a long-range potential with $|V(r)| \le \mathrm{const}\, r^{-\beta}$, $|\mathrm{d}V(r)/\mathrm{d}r| \le \mathrm{const}\, r^{-(1+\beta)}$ as $r \to \infty$, and $\beta > \frac{1}{2}$, (8.2.1) must be modified by using the appropriate expression for $\tilde{v}_0^+(r,k)$ in (8.1.16). For $\beta > \frac{3}{4}$, $\tilde{v}_0^+(r,k)$ may be defined by (8.1.15), where $V(r) = (\mathrm{d}/\mathrm{d}r) W(r)$. In this case, we can apply Lemma 8.3, with $W(r,k) = \mathrm{e}^{-\mathrm{i}W(r)/2k}$, and the reader should verify that the appropriate estimates apply. It follows, in (8.1.16) that, asymptotically (i.e. in $L^2(R, \infty)$ norm as $R \to \infty$), we may replace in the integrand $\mathrm{e}^{-\mathrm{i}W(r)/2k}$ by $\mathrm{e}^{-\mathrm{i}W(2kt)/2k}$. The evolving wave function $f(r,t) = (\mathrm{e}^{-\mathrm{i}Ht} f)(r)$ is then closely approximated, in $L^2(R, \infty)$ norm as $R \to \infty$, by

$$h(r,t) = \frac{1}{2\mathrm{i}} \left(\frac{2}{\pi}\right)^{1/2} \int_0^\infty \mathrm{e}^{-\mathrm{i}(k^2 t - kr + l\pi/2 + W(2kt)/2k - \delta(k))} F(k)\, \mathrm{d}k. \tag{8.2.8}$$

The term $W(2kt)/2k$ in the exponential prevents a direct comparison of (8.2.8) with a freely evolving state, as was possible for a short-range potential. Instead, we must consider a *modified* free evolution given by

$$e^{-iH_0t}\exp\left(-i\frac{W(2H_0^{1/2}t)}{2H_0^{1/2}}\right).$$

and define modified wave operators accordingly. This leads to wave operators

$$\Omega_\pm = \operatorname*{s-lim}_{t\to\pm\infty} e^{iHt}\,e^{-iH_0t}\exp\left(-i\frac{W(2H_0^{1/2}t)}{2H_0^{1/2}}\right), \tag{8.2.9}$$

with exactly the same formulae (8.2.6) and (8.2.7) for the wave and scattering operators, which are asymptotically complete as before.

For a long-range potential with $\beta>\frac{1}{2}$ but $\beta\leq\frac{3}{4}$, the Corollary to Lemma 8.3 may be applied, provided that suitable bounds hold for higher derivatives of the potential. In that case, as we saw in Section 8.1, additional terms will enter into the exponential in (8.2.8), and we must define $W(r,k)$ by

$$W(r,k)=\exp\left[-i\left(\frac{W(r)}{2k}-\frac{1}{8k^3}\int_r^\infty V^2(s)\,ds-\cdots\right)\right],$$

the series being complete after a finite number of terms. However, since all terms in the exponential following $W(r)/2k$ are of order $r^{-(2\beta-1)}$ with $\beta>\frac{1}{2}$, they will not contribute to the strong limit $t\to\infty$ on making the substitution $r=2kt$. For this reason, the definition (8.2.9) of the wave operators, and the subsequent formulae for the wave and scattering operators, remain valid for all long-range potentials with $\beta>\frac{1}{2}$.

We summarize the results of this section in the following theorem.

Theorem 8.1

Let $H_0=-d^2/dr^2+l(l+1)/r^2$ and $H=-d^2/dr^2+V(r)+l(l+1)/r^2$ be the free- and total-Hamiltonian operators, acting in a partial-wave subspace. At $r=0$, suppose either that the differential operator $-d^2/dr^2+V(r)+l(l+1)/r^2$ is in the limit-circle case, or that there is a subordinate solution $v_0(r,k)$, for almost all $k^2>0$, of the corresponding real-energy Schrödinger equation

$$-\frac{d^2v}{dr^2}+Vv+\frac{l(l+1)}{r^2}v=k^2v.$$

(If $r=0$ is limit-circle then define $v_0(r,k)$ to be a solution subject to a suitable boundary condition at $r=0$, and use this same boundary condition to define H as a self-adjoint operator.) As $r\to\infty$, suppose either that $V(r)$ is of short range (i.e. $\int_c^\infty |V(r)|\,dr < \infty$ for $c>0$), or that $V(r)$ is of long range, satisfying $|V(r)| = O(r^{-\beta})$, $|dV(r)/dr| = O(r^{-\beta'})$, $|d^nV/dr^n| = O(r^{-n\beta})$ for some $\beta > \frac{1}{2}$, $\beta' > \frac{3}{2}$, and for $n = 2, 3, 4, \ldots$. Let $V(r) = dW(r)/dr$ in the long-range case, and set $W(r) \equiv 0$ in the short-range case, and "normalize" the solution $v_0(r,k)$ to have the asymptotic form, as $r \to \infty$, $v_0(r,k) = \sin(kr - \frac{1}{2}l\pi - W(r)/2k + \delta(k)) + o(1)$.)Thus the phase shift $\delta(k)$ is defined up to an additive multiple of 2π.) Then wave operators

$$\Omega_{\pm} = \operatorname*{s-lim}_{t \to \pm\infty} e^{iHt}\, e^{-iH_0 t} \exp\left(-i\frac{W(2H_0^{1/2}t)}{2H_0^{1/2}}\right)$$

exist as strong limits on the entire Hilbert space, and are asymptotically complete. The range of each wave operator is the subspace of H associated with the positive spectrum. If, in addition, an $L^2(0,\infty)$ solution of the Schrödinger equation exists at all *negative* energies $-k^2$ then the spectrum of H is absolutely continuous for $\lambda > 0$ and purely discrete for $\lambda < 0$. In this case, the range of each wave operator is the absolutely continuous subspace for H (strong asymptotic completeness).

The scattering operator $S = \Omega_+^* \Omega_-$ is given by $S = U_0^{-1} e^{2i\delta(k)} U_0$, where U_0 is the integral transform from position space to momentum space in the given partial-wave subspace.

Proof

Sufficient details have been given, we hope, for the reader to complete any gaps in the argument. At the beginning of this chapter, we set $\beta' = \beta + 1$ in the long-range case, although this choice of β' is unnecessary. The reader should verify, in particular, that the analogue of Lemma 8.2 extends also to the long-range case, and follows as before on integration by parts. ■

Theorem 8.1 applies in great generality to short-range potentials and to a wide class of long-range potentials including $V(r) = g/r^\beta$ for any $\beta > \frac{1}{2}$. At $r = 0$ non-singular potentials, as well as highly singular potentials, both attractive and repulsive, may be considered. Indeed, just about any potential of interest to scattering theory, provided that it is spherically symmetric, is covered by the theorem. To find examples that cannot be treated by these methods will require the construction of Schrödinger operators H for which the wave operators are not asymptotically complete, or for which the spectrum may be singular continuous. We shall defer such bizarre possibilities to a later chapter, and return briefly to a more homely example—that of the Coulomb potential.

For the Coulomb Hamiltonian, the spectrum is absolutely continuous for $\lambda > 0$, and purely discrete for $\lambda < 0$. If $V(r) = g/r$ then we can take

$W(r) = g \log r$, in which case $W(2kt)/2k = g \log (2kt)/2k$. A modified free evolution can be defined by

$$e^{-iH_0 t} \exp\left(-i \frac{g \log (2H_0^{1/2} t)}{2H_0^{1/2}}\right).$$

It is more usual, in dealing with the Coulomb potential, to factor out the unitary operator $\exp(-ig \log (2H_0^{1/2})/2H_0^{1/2})$ and instead define the modified free evolution as

$$e^{-iH_0 t} \exp\left(-i \frac{g \log t}{2H_0^{1/2}}\right).$$

Wave operators may then be defined to be

$$\Omega_{\pm} = \operatorname*{s\text{-}lim}_{t \to \pm\infty} e^{iHt} e^{-iH_0 t} \exp\left(-i \frac{g \log t}{2H_0^{1/2}}\right),$$

which are the same as our previous wave operators, apart from right multiplication by a unitary function of H_0. The scattering operator S, which commutes with any function of H_0, is exactly as before.

Chapter 9

Scattering in Three Space Dimensions

9.1 SHORT- AND LONG-RANGE WAVE OPERATORS

In the last chapter, we considered scattering by a radial potential, in a given partial wave. The analysis may be regarded as complete, in the sense that explicit formulae were derived for both wave and scattering operators, for a class of potentials including short- and long-range, singular and non-singular. Since, in each case, the wave and scattering operators are expressible in terms of the scattering phase shift, their calculation is thereby reduced to that of the solution of a second-order linear differential equation, namely the time-independent Schrödinger equation at energy k^2. The behaviour of solutions of that equation, in the limits $r \to 0$ and $r \to \infty$, determines the phase shifts, and through them the wave and scattering operators.

This is not, however, the whole story. We may, as a matter of mathematical convenience in dealing with central potentials, restrict our attention to a single partial-wave subspace, characterized by angular-momentum quantum numbers l and m. However, a real scattering experiment is carried out in three-dimensional space, and can in no way be completely described by eigenstates of angular momentum. States with $l=0$, for example, are spherically symmetric, and clearly we cannot hope to surround our scattering target by an incoming spherical beam of particles. We know that, in a realistic quantum description, we have to deal with wave packets, and these are necessarily constructed from elements of the full space $L^2(\mathbb{R}^3)$. We must not, of course, replace one impossible idealization by another, and for this reason we shall avoid introducing such constructs as plane waves into the foundations of the theory. Plane waves, in scattering theory, do have their uses, as we shall see in the derivation of the Lippman–Schwinger equation, but they have no place in a Hilbert-space context, which is both mathematically convenient and physically necessary in discussing the time evolution of quantum systems.

This chapter, then, lays the framework for scattering theory in $L^2(\mathbb{R}^3)$. In going to three dimensions, we shall be able to deal, in addition, with non-central potentials $V(\boldsymbol{r})$, for which the treatment of the last chapter, based as it is on the use of ordinary rather than partial differential equations, is no longer appropriate. In the case of central potentials $V(r)$, we shall also see how our previous results in angular-momentum subspaces may be brought together and made consistent with the three-dimensional picture.

Mathematically, one of our principal aims will be to define wave operators, and to establish their existence as operators on $L^2(\mathbb{R}^3)$. Just as, in our one-dimensional Schrödinger equation, from which the kernel of an integral operator was constructed, so now we shall have to deal with integral operators based on suitable solutions of the full three-dimensional time-independent Schrödinger equation. The question of asymptotic completeness will be deferred until two later chapters, where it will be treated successively by trace methods, and by geometrical methods based on the use of the generator of the dilation group. The theory of localization, which provides the key to the understanding of the problem of asymptotic completeness for scattering by potentials that may be locally singular, will be developed in Chapter 10.

How, then, do we define wave operators for scattering by potentials $V(\boldsymbol{r})$ that are not necessarily spherically symmetric? We shall suppose, for simplicity, that $V(\boldsymbol{r})$ is locally square-integrable apart from a possible (arbitrary) singularity at $\boldsymbol{r}=\boldsymbol{0}$. (Further local singularities, and even singular surfaces, could be accommodated without difficulty.)

The free Hamiltonian H_0 is the unique self-adjoint extension of $\hat{H}_0=-\Delta$, with $D(\hat{H}_0)=C_0^\infty(\mathbb{R}^3)$. The total Hamiltonian H is a self-adjoint extension of $\hat{H}=-\Delta+V(\boldsymbol{r})$, with $D(\hat{H})=C_0^\infty(\mathbb{R}^3\setminus\{0\})$. If necessary, this will involve imposing boundary conditions at $\boldsymbol{r}=\boldsymbol{0}$; our conditions on the potential in the limit at $|\boldsymbol{r}|\to\infty$ will imply that no boundary conditions at infinity are required.

Experience with central potentials leads us to expect that wave operators may be defined, in the short-range case, by

$$\Omega_\pm = \operatorname*{s-lim}_{t\to\pm\infty} e^{iHt}\, e^{-iH_0t},$$

and that these limits will exist on the entire Hilbert space $\mathscr{H}=L^2(\mathbb{R}^3)$. For long-range potentials we expect that wave operators may be defined through some modified free evolution $F(\boldsymbol{P},t)$. The distinction between short- and long-range potentials, which even for central potentials imposes a somewhat arbitrary criterion, becomes even more ill-defined for general non-central potentials. Perhaps the most obvious analogue of the condition $V \in L^1(0, \infty)$ for $V = V(|\boldsymbol{r}|)$ would be the condition $V(\boldsymbol{r})/r^2 \in L^1(\mathbb{R}^3)$, which reduces in the central case to our former criterion for a potential to be short-range. (We are thinking here primarily of behaviour as $|\boldsymbol{r}| \to \infty$, so it would be better to suppose that $V(\boldsymbol{r})/r^2$ be integrable over any region $|\boldsymbol{r}| \geq c$, for $c > 0$.) However, $V(\boldsymbol{r})/r^2 \in L^1(\mathbb{R}^3)$ allows $V(\boldsymbol{r})$ to be locally as singular as $|\boldsymbol{r} - \boldsymbol{r}_0|^{-3+\delta}$ for some $\boldsymbol{r}_0 \neq \boldsymbol{0}$ and for $\delta > 0$. To admit such local singularities of the potential opens the door to a wide variety of pathologies (see, e.g. Chapter 14); indeed in these cases it becomes unclear, in general, how to define a Hamiltonian operator H. It is more in keeping with our approach, in requiring $V \in L^2_{\text{loc}}(\mathbb{R}^3 \setminus \{\boldsymbol{0}\})$, to define short-range potentials through some kind of L^2 condition. We cannot then expect complete accord with the L^1 criterion of Chapter 8. The following result demonstrates that (unmodified) wave operators exist for a class of short-range non-spherical potentials that is large enough to include all V satisfying $|V(\boldsymbol{r})| \leq \text{const}\, r^{-\beta}$ for some $\beta > 1$.

Theorem 9.1

Let $V(\boldsymbol{r})$ satisfy

$$\int_{|\boldsymbol{r}|>c} |\boldsymbol{r}|^{-1+\delta} |V(\boldsymbol{r})|^2 \, \mathrm{d}^3\boldsymbol{r} < \infty, \tag{9.1.1}$$

for all $c > 0$ and for some $\delta > 0$. Then wave operators $\Omega_{\pm}$, defined by

$$\Omega_{\pm} = \operatorname*{s-lim}_{t \to \pm\infty} \mathrm{e}^{\mathrm{i}Ht} \, \mathrm{e}^{-\mathrm{i}H_0 t}, \tag{9.1.2}$$

exist on the entire Hilbert space $\mathscr{H} = L^2(\mathbb{R}^3)$.

Proof

We suppose that (9.1.1) is satisfied, and let ρ denote multiplication in $L^2(\mathbb{R}^3)$, by some smooth non-decreasing function $\rho(|\boldsymbol{r}|)$, where $\rho(r) \equiv 1$ for all sufficiently large r, and $\rho(r) \equiv 0$ for all r in some neighbourhood of $r = 0$. Since, under the free evolution $\mathrm{e}^{-\mathrm{i}H_0 t}$, states are asymptotically concentrated, in norm, at large distances from $\boldsymbol{r} = \boldsymbol{0}$, we know that $(1 - \rho)\mathrm{e}^{-\mathrm{i}H_0 t}$ converges strongly to zero. Hence, for any $f \in \mathscr{H}$, we may replace (9.1.2) by

$$\Omega_{\pm} f = \operatorname*{s-lim}_{t \to \pm\infty} \mathrm{e}^{\mathrm{i}Ht} \rho \mathrm{e}^{-\mathrm{i}H_0 t} f, \tag{9.1.3}$$

provided that the limits on the right-hand side exist. Assuming $f \in D(H_0)$ and

$\rho e^{-iH_0t} f \in D(V)$, we have

$$\frac{d}{dt} e^{iHt} \rho e^{-iH_0t} f = i\, e^{iHt}(H\rho - \rho H_0) e^{-iH_0t} f, \tag{9.1.4}$$

where the derivative on the left-hand side is defined as a strong limit in the usual way. Now

$$H\rho - \rho H_0 = V\rho + [H_0, \rho],$$

where

$$[H_0, \rho] = [-\Delta, \rho] = -\Delta\rho - 2i\sum_j \frac{\partial \rho}{\partial x_j} P_j.$$

Here $\{P_j\}$ are components of the momentum operator $\boldsymbol{P}$. Let us suppose that we can prove, for some f, that

$$\int_0^\infty dt \left\| \frac{d}{dt} e^{iHt} \rho e^{-iH_0t} f \right\| < \infty. \tag{9.1.5}$$

Then the integral $\int_0^\infty dt\, (d/dt)\, e^{iHt} \rho e^{-iH_0t} f$ may be defined, for example, as a strong limit of approximating Riemann sums, and will be the strong limit, as $t \to \infty$, of $\int_0^t dt\, (d/dt) e^{iHt} \rho e^{-iH_0t} f$. Since this last integral is just $e^{iHt} \rho e^{-iH_0t} f - \rho f$, we shall then have proved the existence of the limit $t \to \infty$ on the right-hand side of (9.1.3). In the same way, we may deduce the existence of Ω_- as a strong limit provided that we can show that

$$\int_{-\infty}^0 dt \left\| \frac{d}{dt} e^{iHt} \rho e^{-iH_0t} f \right\| < \infty.$$

We first consider the wave operator Ω_+. Following the evaluation of $H\rho - \rho H_0$ on the right-hand side of (9.1.4), we have to consider three integrals, namely

$$I_1 = \int_0^\infty dt\, \|\rho V e^{-iH_0t} f\|, \qquad I_2 = \int_0^\infty dt\, \|(\Delta\rho) e^{-iH_0t} f\|,$$

and

$$I_3 = \int_0^\infty dt \left\| \frac{\partial \rho}{\partial x_j} P_j\, e^{-iH_0t} f \right\|.$$

We shall do so in the case where f is represented by a Gaussian wave function. We therefore suppose that for $a>0$

$$f(\boldsymbol{r})=\left(\frac{1}{2a}\right)^{3/2}\exp\left(-\frac{|\boldsymbol{r}|^2}{4a}\right). \tag{9.1.6}$$

In momentum space, the state f is represented by the wave function $\hat{f}(\boldsymbol{k})=\mathrm{e}^{-a|\boldsymbol{k}|^2}$. After time t the freely evolving state $\mathrm{e}^{-\mathrm{i}H_0t}f$ is represented by the wave function $\mathrm{e}^{(-\mathrm{i}|\boldsymbol{k}|^2t-a|\boldsymbol{k}|^2)}$. By the results preceeding (5.3.26), the wave function in position space is then, after time t,

$$(\mathrm{e}^{-\mathrm{i}H_0t}f)(\boldsymbol{r})=(2\mathrm{i}(t-\mathrm{i}a))^{-3/2}\exp\left(\frac{\mathrm{i}|\boldsymbol{r}|^2}{4(t-\mathrm{i}a)}\right). \tag{9.1.7}$$

Hence

$$|(\mathrm{e}^{-\mathrm{i}H_0t}f)(\boldsymbol{r})|\leq\text{const }(1+|t|)^{-3/2}\exp\left(-\frac{a|\boldsymbol{r}|^2}{4(t^2+a^2)}\right).$$

We first of all estimate the integral I_1. Since $\|\rho V\mathrm{e}^{-\mathrm{i}H_0t}f\|$ is a bounded function of t, for values of t in any finite interval $[0,t_0]$, we need obtain estimates only for $t>t_0$. We therefore assume that $t>a$, in which case $t^2+a^2<2t^2$ and

$$|(\mathrm{e}^{-\mathrm{i}H_0t}f)(\boldsymbol{r})|\leq\text{const }|t|^{-3/2}\exp\left(-\frac{a|\boldsymbol{r}|^2}{8t^2}\right). \tag{9.1.8}$$

In (9.1.1), we may assume without loss of generality that $\delta<1$, in which case the function $x^{(1-\delta)/2}\,\mathrm{e}^{-ax^2/8}$ is bounded for $x>0$. We then have

$$|\boldsymbol{r}|^{(1-\delta)/2}\,|t|^{-(1-\delta)/2}\exp\left(-\frac{a|\boldsymbol{r}|^2}{8t^2}\right)\leq\text{const},$$

so that, from (9.1.8),

$$|(\rho V\mathrm{e}^{-\mathrm{i}H_0t}f)(\boldsymbol{r})|\leq\text{const }|\boldsymbol{r}|^{-(1-\delta)/2}\,|(\rho V)(\boldsymbol{r})|\,|t|^{-1-\delta/2}.$$

Taking norms and using (9.1.1), it follows from the support properties of ρ that $\|\rho V\mathrm{e}^{-\mathrm{i}H_0t}f\|\leq\text{const }|t|^{-1-\delta/2}$. Since the norm is locally bounded, we have indeed

$$\|\rho V\mathrm{e}^{-\mathrm{i}H_0t}f\|\leq\text{const }(1+|t|)^{-1-\delta/2},$$

and the finiteness of I_1 follows on integrating this inequality with respect to t. The finiteness of the integral I_2 is now immediate, since (9.1.1) is satisfied with $\Delta\rho$ replacing V. It remains only to estimate the integral I_3. The only essential difference here is the presence of momentum operators P_j, which differentiate $(\mathrm{e}^{-\mathrm{i}H_0t}f)(\boldsymbol{r})$ with respect to the components x_j of $\boldsymbol{r}$. For large t, these derivatives introduce extra factors that are of order $|\boldsymbol{r}|/|t|$. Following (9.1.8), we need only observe that $x^{(3-\delta)/2}\,\mathrm{e}^{-ax^2/8}$ is bounded for $x>0$. The estimates then proceed as before, and we have $I_3<\infty$. Hence (9.1.5) holds, and we have shown that $\Omega_+ f$ exists, where $f(\boldsymbol{r})$ is given by (9.1.6). The existence of $\Omega_- f$, for f defined by (9.1.6), follows in the same way.

Of course, as a is varied, the family of wave functions given by (9.1.6) is insufficient to span the entire Hilbert space, since each f belongs to the $l=0$ partial-wave subspace. However, (9.1.1) implies, for any $\boldsymbol{r}_0\in\mathbb{R}^3$, that $|\boldsymbol{r}-\boldsymbol{r}_0|^{-(1-\delta)/2}V(\boldsymbol{r})$ is square-integrable over the region $|\boldsymbol{r}-\boldsymbol{r}_0|>c$, so that only slight modification of our previous estimates allows us to conclude the existence of each wave operator $\Omega_\pm$, acting on the wave function $f(\boldsymbol{r}-\boldsymbol{r}_0)$, where $f(\boldsymbol{r})$ is again given by (9.1.6). Now the Fourier transform of $f(\boldsymbol{r}-\boldsymbol{r}_0)$ is $\mathrm{e}^{-i\boldsymbol{k}\cdot\boldsymbol{r}_0}\hat{f}(\boldsymbol{k})$, and any $h\in L^2(\mathbb{R}^3)$ that is orthogonal to all $f(\boldsymbol{r}-\boldsymbol{r}_0)$, as $\boldsymbol{r}_0$ varies, must satisfy

$$\int \mathrm{d}^3\boldsymbol{k}\;\overline{\hat{h}(\boldsymbol{k})}\,\hat{f}(\boldsymbol{k})\mathrm{e}^{-i\boldsymbol{k}.\boldsymbol{r}_0}=0.$$

Since $\hat{f}$ is bounded, $\bar{\hat{h}}\hat{f}\in L^2(\mathbb{R}^3)$ and has zero Fourier transform. Moreover, $\hat{f}(\boldsymbol{k})\neq 0$, and it follows that $\hat{h}=0$. We have shown, then, that any h orthogonal to all $f(\boldsymbol{r}-\boldsymbol{r}_0)$ is identically zero. Thus the closed linear space spanned by $\{f(\boldsymbol{r}-\boldsymbol{r}_0)\}$, as $\boldsymbol{r}_0$ varies, is the whole of $L^2(\mathbb{R}^3)$. In proving the existence of wave operators acting on Gaussian wave functions, we have thereby defined $\Omega_\pm$ on a dense subset of the Hilbert space; hence wave operators, given by (9.1.2), exist on the entire Hilbert space, and the theorem is proved. ■

The condition (9.1.1) on the large-$|\boldsymbol{r}|$ behaviour of $V(\boldsymbol{r})$ is by no means optimal in proving the existence of short-range wave operators. In proving the theorem, we used the fact that $(1+|t|)^{-1-\delta/2}$ is integrable over $\mathbb{R}$. However, it is also true, for $\delta>0$, that the function $t^{-1}(\log t)^{-1-\delta/2}$ is integrable for large t. The proof can easily be modified to take this into account, leading, instead of (9.1.1), to the condition

$$\int_{|\boldsymbol{r}|>c} |\boldsymbol{r}|^{-1}(\log(2+|\boldsymbol{r}|))^{2+\delta}\,|V(\boldsymbol{r})|^2\,\mathrm{d}^3\boldsymbol{r}<\infty,$$

and we can introduce logarithms of logarithms to obtain an even stronger result! There is therefore no well-defined borderline between short- and long-range potentials that allows us to define wave operators, by (9.1.2), for some definite and canonical class of potentials. What is clear, once again, is that however the distinction between short- and long-range potentials is drawn, the Coulomb potential is very close to the borderline, but falls very definitely into the long-range camp.

What is the analogous result to Theorem 9.1, for a long-range potential $V(\boldsymbol{r})$? If $V(\boldsymbol{r})$ is long-range then we must first of all define a modified free evolution $\{F(\boldsymbol{P}, t)\}$. Unlike the short-range case, the family of unitary operators $\{F(\boldsymbol{P}, t)\}$ will here depend on the potential $V(\boldsymbol{r})$. Our experience in Chapter 8 with long-range *radial* potentials suggests that we define the modified free evolution in terms of a function $W(\boldsymbol{r})$ that is an integral of the potential. Since for a radial potential we wrote $V(r)=\mathrm{d}W(r)/\mathrm{d}r$, we shall define $W(\boldsymbol{r})$ in the three-dimensional problem by the equation

$$V(\boldsymbol{r})=\frac{\partial W(\boldsymbol{r})}{\partial r}, \tag{9.1.9}$$

where differentiation on the right-hand side is with respect to the radial coordinate r of spherical polars; thus $\boldsymbol{r}=r\boldsymbol{\omega}$, where $\boldsymbol{\omega}$ is a unit vector specifying the direction of $\boldsymbol{r}$, and is held constant for the partial derivative. Note that W in (9.1.9) is determined up to an additive function of the two remaining, angular, polar coordinates. Moreover, the definition of W, through (9.1.9), is dependent on a choice of the origin $\boldsymbol{r}=\boldsymbol{0}$. These ambiguities in the definition of W reflect the fact that, as we saw in Chapter 3, long-range wave operators are non-unique. Nevertheless, for a given potential $V(\boldsymbol{r})$, there may be a natural choice of origin, as for example in the case of the Coulomb potential, and a natural choice of the function $W(\boldsymbol{r})$, in (9.1.9).

Corresponding to a given $W(\boldsymbol{r})$, we now define a modified free evolution, for $t>0$, by

$$F(\boldsymbol{P}, t)=\mathrm{e}^{-\mathrm{i}H_0 t}\, Z(\boldsymbol{P}, t), \tag{9.1.10}$$

where

$$Z(\boldsymbol{P}, t)=\exp\left(-\mathrm{i}\frac{W(2\boldsymbol{P}t)}{|2\boldsymbol{P}|}\right), \tag{9.1.11}$$

where $\boldsymbol{P}$ is the momentum operator and $H_0=\boldsymbol{P}^2$ is the free Hamiltonian. The modified free evolution for $t<0$ is obtained from the identity $F(\boldsymbol{P}, -t)=F^*(-\boldsymbol{P}, t)$, and has the opposite sign to (9.1.11) in the argument of the complex exponential. For fixed $t>0$, the operator $Z(\boldsymbol{P}, t)$ given by (9.1.11) may be represented in momentum space as the operator of multiplication by

$$\exp\left(-\mathrm{i}\frac{W(2\boldsymbol{k}t)}{2|\boldsymbol{k}|}\right).$$

We can now define wave operators $\Omega_{\pm}$ as the strong limits of $\mathrm{e}^{\mathrm{i}Ht}F(\boldsymbol{P}, t)$ as $t\to\pm\infty$ respectively, and we shall find, as for the spherically symmetric case treated in Chapter 8, that such wave operators exist for a wide range of potentials $V(\boldsymbol{r})$ decaying at infinity as rapidly as $|\boldsymbol{r}|^{-\beta}$ for some $\beta>\frac{1}{2}$. As a preliminary to proving the existence of these modified wave operators, let us consider the modified free evolution for an initial state described by a wave function ϕ for which the Fourier transform $\hat{\phi}$ satisfies $\hat{\phi}\in C_0^\infty(\mathbb{R}^3)$. After time t, we have $\phi(\boldsymbol{r},t)=(e^{-iH_0t}Z(\boldsymbol{P},t)\phi)(\boldsymbol{r})$.

Let us write $\zeta(\boldsymbol{r},t)=(Z(\boldsymbol{P},t)\phi)(\boldsymbol{r})$. We shall suppose that the function $\zeta(\boldsymbol{r},t)$ varies sufficiently slowly, as a function of t, that the norm of $(e^{ir^2/4t}-1)\zeta(\boldsymbol{r},t)$, in $L^2(\mathbb{R}^3)$, tends to zero in the limit $t\to\infty$. Thus

$$\lim_{t\to\infty}\left\|\left[\exp\left(\frac{i\boldsymbol{r}^2}{4t}\right)-1\right]\zeta(\boldsymbol{r},t)\right\|=0. \tag{9.1.12}$$

Under what conditions is (9.1.12) satisfied? Since $|e^{ir^2/4t}-1|\le\text{const}\,\boldsymbol{r}^2/4t$, (9.1.12) will hold provided that $t^{-1}\|\boldsymbol{r}^2\zeta(\boldsymbol{r},t)\|\to0$ as $t\to\infty$. Moreover, the Fourier transform of $\boldsymbol{r}^2\zeta(\boldsymbol{r},t)$ is $-\Delta\{Z(\boldsymbol{k},t)\hat{\phi}(\boldsymbol{k})\}$, where Δ stands for $\Sigma_{j=1}^3\,\partial^2/\partial k_j^2$. Hence (9.1.12) will hold provided that

$$\lim_{t\to\infty}t^{-1}\Delta Z(\boldsymbol{k},t)=\lim_{t\to\infty}t^{-1}\frac{\partial Z(\boldsymbol{k},t)}{\partial k_j}=0, \tag{9.1.13}$$

where limits are uniform in k over the support of $\hat{\phi}$.

According to (9.1.11), we have $Z(\boldsymbol{k},t)=\exp(-iW(2\boldsymbol{k}t)/|2\boldsymbol{k}|)$. If ϕ is chosen such that $\boldsymbol{k}=\boldsymbol{0}$ does not lie in the support of $\hat{\phi}$ then we can relate bounds for the derivatives of $Z(\boldsymbol{k},t)$ with respect to $\boldsymbol{k}$ in the limit $t\to\infty$, to bounds for the derivatives of $W(\boldsymbol{r})$ with respect to $\boldsymbol{r}$ in the limit $|\boldsymbol{r}|\to\infty$. Provided that $W(\boldsymbol{r})$ satisfies the conditions

$$r^{-1/2}W(\boldsymbol{r})\to0,\quad r^{1/2}\frac{\partial W(\boldsymbol{r})}{\partial x_j}\to0,\quad r\frac{\partial^2W(\boldsymbol{r})}{\partial x_j^2}\to0\quad\text{as }|\boldsymbol{r}|\to\infty,$$

(9.1.13) will be satisfied. In terms of the potential $V(\boldsymbol{r})$, these conditions will hold (for suitable choice of $W(\boldsymbol{r})$) provided that

$$r^{1/2}V(\boldsymbol{r})\to0,\quad r^{3/2}\frac{\partial V(\boldsymbol{r})}{\partial x_j}\to0,\quad r^2\frac{\partial^2V(\boldsymbol{r})}{\partial x_i\,\partial x_j}\to0 \tag{9.1.14}$$

in the limit as $|\boldsymbol{r}|\to\infty$. Roughly, $V(\boldsymbol{r})$ should tend to zero as $|\boldsymbol{r}|\to\infty$ more rapidly than $|\boldsymbol{r}|^{-1/2}$, and each derivative with respect to some component of $\boldsymbol{r}$ should introduce into the estimate some further power of $|\boldsymbol{r}|^{-1/2}$. These properties of $V(\boldsymbol{r})$ are comparable with those demanded in Chapter 8 for central long-range potentials. It is also possible to handle potentials that are non-central and that decay at infinity like $|\boldsymbol{r}|^{-\beta}$ for $\beta>0$. In such cases, more careful asymptotic analysis leads to different definitions of the modified free evolution from that which we are following in (9.1.10) and (9.1.11).

Given (9.1.12), we can now introduce into the formula for $\phi(\boldsymbol{r},t)$ the expression (5.3.34) for the free evolution $e^{-iH_0 t}$ as an integral operator in position space, where, in the strong limit as $t\to\infty$, (9.1.12) tells us that $\zeta(\boldsymbol{r},t)$ may be replaced by $e^{-ir^2/4t}\zeta(\boldsymbol{r},t)$. Using (5.3.34), we have

$$\phi(\boldsymbol{r},t)=\left(\frac{1}{4\pi it}\right)^{3/2} e^{ir^2/4t}\int e^{-2i\boldsymbol{r}.\boldsymbol{r}'/4t}\zeta(\boldsymbol{r}',t)\,d^3\boldsymbol{r}'+\psi(\boldsymbol{r},t),$$

where $\|\psi\|$ converges to zero in the limit as $t\to\infty$. Hence

$$\phi(\boldsymbol{r},t)=\left(\frac{1}{2it}\right)^{3/2} e^{ir^2/4t}\hat{\xi}\left(\frac{\boldsymbol{r}}{2t},t\right)+\psi(\boldsymbol{r},t), \tag{9.1.15}$$

where $\hat{\zeta}(\boldsymbol{k},t)$ denotes the Fourier transform of $\zeta(\boldsymbol{r},t)$ with respect to $\boldsymbol{r}$. Since $\hat{\zeta}(\boldsymbol{k},t)=Z(\boldsymbol{k},t)\hat{\phi}(\boldsymbol{k})$, (9.1.15) implies that

$$\phi(\boldsymbol{r},t)=\left(\frac{1}{2it}\right)^{3/2} e^{ir^2/4t}J(\boldsymbol{r},t)\hat{\phi}\left(\frac{\boldsymbol{r}}{2t}\right)+\psi(\boldsymbol{r},t), \tag{9.1.16}$$

where

$$J(\boldsymbol{r},t)=Z(\boldsymbol{r}/2t,t). \tag{9.1.17}$$

The corresponding result with $V=0$ and $Z\equiv 1$ tells us that

$$(e^{-iH_0 t}\phi)(\boldsymbol{r})=\left(\frac{1}{2it}\right)^{3/2} e^{ir^2/4t}\hat{\phi}\left(\frac{\boldsymbol{r}}{2t}\right)+\psi_0(\boldsymbol{r},t), \tag{9.1.18}$$

where ψ_0 converges in norm to zero. Since $J(\boldsymbol{r},t)$ is bounded, a comparison of (9.1.16) with (9.1.18) shows that

$$\lim_{t\to\infty}\|F(\boldsymbol{P},t)\phi-J(\boldsymbol{r},t)e^{-iH_0 t}\phi\|=0, \tag{9.1.19}$$

where $J(\boldsymbol{r},t)$ now stands for the operator of multiplication by J in position space. Asymptotically in the limit as $t\to\infty$, we see that the state ϕ evolves, according to the modified free evolution, as a free-particle state $e^{-iH_0 t}$ multiplied by the complex phase function $J(\boldsymbol{r},t)$. In particular, it is interesting to note that the probability of finding the particle in any fixed region of position space approaches the corresponding probability for the free-particle state.

Note, from (9.1.16), that what matters is the value of $J(\boldsymbol{r},t)$ for values of $\boldsymbol{r}$ such that $\boldsymbol{r}/2t$ belongs to the support of $\hat{\phi}$, provided always that $\hat{\phi}$ has compact support that does not include the origin. It follows that (9.1.19) will remain valid if $J(\boldsymbol{r},t)$ is replaced by any other bounded function that agrees with J whenever $\boldsymbol{r}/2t\in\operatorname{supp}\hat{\phi}$. A convenient way of taking into account the support properties of $\hat{\phi}$ is to redefine $J(\boldsymbol{r},t)$, on multiplying by $\rho(\boldsymbol{r}/2t)$, where $\rho(\boldsymbol{r})$ is a function in $C_0^\infty(\mathbb{R}^3\setminus\{0\})$, such that $\rho(\boldsymbol{r})\equiv 1$ for all $\boldsymbol{r}\in\operatorname{supp}\hat{\phi}$. We may then assume, for example, without loss of generality, that the support of the function $J(\boldsymbol{r}, t)$ is contained in the region $2c_1 t < |\boldsymbol{r}| < 2c_2 t$ of $\mathbb{R}^3$, for some positive constants c_1 and c_2. (Note, however, that in doing this we lose the unitary property of J that previously followed from (9.1.17).)

Since $C_0^\infty(\mathbb{R}^3 \setminus \{0\})$ is dense in $L^2(\mathbb{R}^3)$, (9.1.19) may extended to apply to *any* $\phi \in \mathscr{H}$, with the original choice (9.1.17) of the function $J(\boldsymbol{r}, t)$. For the wave operator $\Omega_+ = \text{s-lim}_{t\to\infty} e^{iHt} F(\boldsymbol{P}, t)$, we can then write, for arbitrary $f \in L^2(\mathbb{R}^3)$,

$$\Omega_+ f = \underset{t\to\infty}{\text{s-lim}}\, e^{iHt} J(\boldsymbol{r}, t) e^{-iH_0 t} f. \tag{9.1.20}$$

(If the support of $\hat{f}(\boldsymbol{k})$ is contained in the region $c_1 < |\boldsymbol{k}| < c_2$ then we may again redefine $J(\boldsymbol{r}, t)$ in (9.1.20) to have support in the region $2c_1 t < |\boldsymbol{r}| < 2c_2 t$.)

Substituting for Z in (9.1.17), we have

$$J(\boldsymbol{r}, t) = \exp\left(-i\frac{tW(\boldsymbol{r})}{r}\right). \tag{9.1.21}$$

Noting, from (9.1.9), that the relation between V and W may be written in the form $V(\boldsymbol{r}) = (\boldsymbol{r}\cdot \text{grad}\, W(\boldsymbol{r}))/r$, it is straightforward to verify that $J(\boldsymbol{r}, t)$ satisfies the equation

$$i\frac{\partial J(\boldsymbol{r}, t)}{\partial t} = V(\boldsymbol{r}) J(\boldsymbol{r}, t) - i\frac{\boldsymbol{r}\cdot \text{grad}\, J(\boldsymbol{r}, t)}{t}. \tag{9.1.22}$$

Moreover, (9.1.22) remains satisfied if $J(\boldsymbol{r}, t)$ is multiplied by an arbitrary differentiable function of $\boldsymbol{r}/t$. We shall prove the existence of the limit (9.1.20), first of all for J satisfying (9.1.22) and having support in the region $2c_1 t < |\boldsymbol{r}/ = 2c_2 t$. To do this, we proceed as for short-range potentials, by taking f to be Gaussian. For such f, we have

$$\frac{d}{dt} e^{iHt} J(\boldsymbol{r}, t) e^{-iH_0 t} f = i\, e^{iHt}\left(HJ - JH_0 - i\frac{\partial J}{\partial t}\right) e^{-iH_0 t} f. \tag{9.1.23}$$

The terms within parentheses on the right-hand side become, on using the commutation relation for $[H_0, J]$, together with (9.1.22),

$$HJ - JH_0 - i\frac{\partial J}{\partial t}$$

$$= [H_0, J] + VJ - i\frac{\partial J}{\partial t}$$

$$= -\Delta J - 2\mathrm{i}\sum_j \frac{\partial J}{\partial x_j} P_j + \mathrm{i}\frac{\boldsymbol{r}\cdot \operatorname{grad} J}{t}$$

$$= -\Delta J + \mathrm{i}t^{-1}(\operatorname{grad} J)\cdot(\boldsymbol{r} - 2\boldsymbol{P}t). \tag{9.1.24}$$

Proceeding as in the proof of Theorem 9.1, we have to consider two integrals, namely

$$I_1 = \int_0^\infty \mathrm{d}t \, \|(\Delta J)\mathrm{e}^{-\mathrm{i}H_0 t} f\|$$

and

$$I_2 = \int_1^\infty \mathrm{d}t \, t^{-1} \|(\operatorname{grad} J)\cdot(\boldsymbol{r} - 2\boldsymbol{P}t)\mathrm{e}^{-\mathrm{i}H_0 t} f\|.$$

(We have changed the lower limit of integration here to avoid the apparent singularity at $t=0$.) In order to estimate the integrals I_1 and I_2, we shall make slightly stronger assumptions than (9.1.14), namely

$$|V(\boldsymbol{r})| \le \mathrm{const}\ r^{-\beta}, \qquad \left|\frac{\partial V(\boldsymbol{r})}{\partial x_j}\right| \le \mathrm{const}\, r^{-(1+\beta)},$$

$$\left|\frac{\partial^2 V(\boldsymbol{r})}{\partial x_i \partial x_j}\right| \le \mathrm{const}\, r^{-(3/2+\beta)}, \tag{9.1.25}$$

where β is a constant such that $1 > \beta > \frac{1}{2}$, and the estimates are to hold in the limit as $|\boldsymbol{r}| \to \infty$. Under these conditions, we obtain the following estimates for $W(\boldsymbol{r})$:

$$|W(\boldsymbol{r})| \le \mathrm{const}\, r^{1-\beta}, \qquad \left|\frac{\partial W(\boldsymbol{r})}{\partial x_j}\right| \le \mathrm{const}\ r^{-\beta},$$

$$\left|\frac{\partial^2 W(\boldsymbol{r})}{\partial x_i \partial x_j}\right| \le \mathrm{const}\ r^{-(1/2+\beta)},$$

again with $1 > \beta > \frac{1}{2}$.

These bounds may be deduced from (9.1.25) for a suitable choice of W; for example, we can take

$$W(\boldsymbol{r}) = \int_c^r V\left(\frac{s\boldsymbol{r}}{r}\right) \mathrm{d}s. \tag{9.1.26}$$

If we now replace the definition (9.1.21) of $J(\boldsymbol{r}, t)$ by

$$J(\boldsymbol{r}, t) = \rho\left(\frac{\boldsymbol{r}}{2t}\right) \exp\left(-\mathrm{i}\frac{tW(\boldsymbol{r})}{r}\right),$$

where the function ρ is in $C_0^\infty(\mathbb{R}^3 \setminus \{0\})$, then we can use our bounds on $W(\boldsymbol{r})$ and the derivatives of $W(\boldsymbol{r})$ to estimate $\Delta J(\boldsymbol{r}, t)$. Because of the support properties of J, we may treat $|\boldsymbol{r}|$ and t as being of the same order of magnitude (i.e. $|\boldsymbol{r}|/t$ is bounded above and below by positive constants). Thus power bounds in $|\boldsymbol{r}|$ may be converted into power bounds in t, and we obtain

$$|\Delta J(\boldsymbol{r}, t)| \leq \text{const } t^{-(\beta+1/2)} + \text{const } t^{-2\beta}. \tag{9.1.27}$$

With $\beta > \frac{1}{2}$, we have shown that the integral I_1 is convergent. Under the given hypothesis, $|\operatorname{grad} J|$ is of order $t^{-\beta}$, and to prove the convergence of the integral I_2 it remains only to derive a bound for $\|(x_j - 2P_j t)\mathrm{e}^{-\mathrm{i}H_0 t} f\|$. If $f(\boldsymbol{r})$ is the Gaussian wave function given by (9.1.6) then we have the explicit formula (9.1.7) for $(\mathrm{e}^{-\mathrm{i}H_0 t} f)(\boldsymbol{r})$, from which we may verify that

$$(\boldsymbol{r} - 2\boldsymbol{P}t)\mathrm{e}^{-\mathrm{i}H_0 t} f = -\mathrm{i}a\boldsymbol{r}(t - \mathrm{i}a)^{-1} \mathrm{e}^{-\mathrm{i}H_0 t} f. \tag{9.1.28}$$

The inequality (9.1.8) then gives

$$\|(x_j - 2P_j t)\mathrm{e}^{-\mathrm{i}H_0 t} f\|^2 \leq a^2 \int r^2 t^{-5} \mathrm{e}^{-ar^2/4t^2} \, \mathrm{d}^3\boldsymbol{r}.$$

Since the integral on the right-hand side is independent of t, we have

$$\|(x_j - 2P_j t)\mathrm{e}^{-\mathrm{i}H_0 t} f\| \leq \text{const}, \tag{9.1.29}$$

and the integral I_2 is convergent. We have thus shown that the right-hand side of (9.1.23) is integrable in norm with respect to t, from which we deduce the existence of the strong limit on the right-hand side of (9.1.20). This limit is not yet equal to $\Omega_+ f$ as in (9.1.20), owing to the presence of the factor $\rho(\boldsymbol{r}/2t)$ in the J-function. However, we can prove the existence of the limit for a sequence of J-functions such that

$$J(\boldsymbol{r}, t) \equiv 1 \quad \text{for } 2c_1 t < |\boldsymbol{r}| < 2c_2 t, \; |J(\boldsymbol{r}, t)| \leq 1,$$

and for which $c_1 \to 0$ and $c_2 \to \infty$. The L^2 norm of $\mathrm{e}^{-\mathrm{i}H_0 t} f$ taken over the region $|\boldsymbol{r}| < 2c_1 t$, by (9.1.8) cannot exceed

$$\left(\int_{|\boldsymbol{r}| < 2c_1 t} t^{-3} \mathrm{e}^{-ar^2/4t^2} \, \mathrm{d}^3\boldsymbol{r}\right)^{1/2}.$$

The integral here is independent of t, and converges to zero in the limit as $c_1 \to 0$. The same is true of the L^2 norm of $\mathrm{e}^{-\mathrm{i}H_0 t} f$ over the region $|\boldsymbol{r}| > 2c_2 t$. It follows that, in the limits $c_1 \to 0$, $c_2 \to \infty$, $J(\boldsymbol{r}, t)\mathrm{e}^{-\mathrm{i}H_0 t} f$ approaches in norm the corresponding expression with $J(\boldsymbol{r}, t)$ given by (9.1.21). This completes the proof of the existence of the wave operator Ω_+, given by (9.1.20) and acting on the Gaussian wave function $f(\boldsymbol{r})$ of (9.1.6).

In the same way, we can consider the operation of Ω_+ on the translated Gaussian wave function $f(\boldsymbol{r}-\boldsymbol{r}_0)$. The bound (9.1.29) applies in this case too, as may be verified either by an explicit estimation of the norm, or by an argument exploiting the translation invariance of the operators $\boldsymbol{P}$ and H_0. To estimate the norms of $e^{-iH_0t}f$ over the region $|\boldsymbol{r}|<2c_1t$, in the case $f=f(\boldsymbol{r}-\boldsymbol{r}_0)$ it is useful to note that this region is contained in the union of $|\boldsymbol{r}-\boldsymbol{r}_0|<4c_1t$ with the bounded region $|\boldsymbol{r}|<|\boldsymbol{r}_0|$. Since

$$\lim_{t\to\infty}\int_{|\boldsymbol{r}|<|\boldsymbol{r}_0|}|(e^{-iH_0t}f)(\boldsymbol{r})|^2\,d^3\boldsymbol{r}=0,$$

the norm of $e^{-iH_0t}f$ over the region $|\boldsymbol{r}|<2c_1t$ may be estimated for large t in terms of the corresponding norm over the region $|\boldsymbol{r}-\boldsymbol{r}_0|<4c_1t$, and this norm converges to zero as $c_1\to 0$, uniformly in t. The norm of $e^{-iH_0t}f$ over the region $|\boldsymbol{r}|>2c_2t$ also converges to zero as $c_2\to\infty$, uniformly in t for $t>1$, say. Taking the limits $c_1\to 0$, $c_2\to\infty$, we can again remove the factor $\rho(\boldsymbol{r}/2t)$ in the formula for $J(\boldsymbol{r}, t)$, and (9.1.20) is satisfied for $f=f(\boldsymbol{r}-\boldsymbol{r}_0)$, with $J(\boldsymbol{r}, t)$ given by (9.1.21).

As we have seen, the family of elements of $L^2(\mathbb{R}^3)$ defined by $f(\boldsymbol{r}-\boldsymbol{r}_0)$, as $\boldsymbol{r}_0$ is varied, span a dense subset of the Hilbert space, so that we can extend (9.1.20) by continuity to apply to any $f\in L^2(\mathbb{R}^3)$. We may summarize our conclusions in the following theorem.

Theorem 9.2

Let $V(\boldsymbol{r})$ be a (long-range) potential satisfying the bounds (9.1.25) on V and the derivatives of V, in the limit $|\boldsymbol{r}|\to\infty$. Let $V(\boldsymbol{r})=\partial W(\boldsymbol{r})/\partial r$, where $W(\boldsymbol{r})$ is given by (9.1.26). Define a modified free evolution by

$$F(\boldsymbol{P}, t)=\begin{cases} e^{-iH_0t}\,e^{-iW(2\boldsymbol{P}t)/2|\boldsymbol{P}|} & (t>0),\\ e^{-iH_0t}\cdot e^{iW(2\boldsymbol{P}t)/2|\boldsymbol{P}|} & (t<0).\end{cases}$$

Then wave operators $\Omega_\pm$, given by

$$\Omega_\pm f=\operatorname*{s-lim}_{t\to\pm\infty} e^{iHt}\,F(\boldsymbol{P}, t)f, \tag{9.1.30}$$

exist for all $f\in\mathscr{H}$. The wave operators are also given by

$$\Omega_\pm f=\operatorname*{s-lim}_{t\to\pm\infty} e^{iHt}J(\boldsymbol{r}, t)e^{-iH_0t}f,$$

where $J(\boldsymbol{r}, t)$ is the function expressed in terms of $W(\boldsymbol{r})$ by (9.1.21).

Proof

We have already proved the theorem for the wave operator Ω_+. The wave operator Ω_- may be treated in a similar way by taking the limit $t \to -\infty$. ■

In order to verify that the long-range wave operators $\Omega_\pm$, which we have defined, may be incorporated into the general framework of Chapter 4, we must show that the modified free evolution, given by (9.1.10) and (9.1.11), satisfies the asymptotic property

$$\operatorname*{s-lim}_{t \to \infty} \{Z(\boldsymbol{P}, t+s) - Z(\boldsymbol{P}, t)\} f = 0. \tag{9.1.31}$$

According to (9.1.11), $Z(\boldsymbol{P}, t)$ is the operator in momentum space of multiplication by $Z(\boldsymbol{k}, t)$, where, by (9.1.9) and the bound (9.1.25) on $V(\boldsymbol{r})$, we have

$$\left| \frac{\partial Z(\boldsymbol{k}, t)}{\partial t} \right| \leq \text{const } t^{-\beta}$$

as $t \to \infty$, provided that $\boldsymbol{k}$ lies in a compact subset of $\mathbb{R}^3 \setminus \{0\}$. By the Mean-Value Theorem, we then have

$$|Z(\boldsymbol{k}, t+s) - Z(\boldsymbol{k}, t)| \leq \text{const } t^{-\beta}.$$

Hence (9.1.31) is valid for $f = \phi$, provided that $\hat{\phi} \in C_0^\infty(\mathbb{R}^3 \setminus \{0\})$. Since such ϕ are dense in the Hilbert space, (9.1.31) is valid for all $f \in \mathscr{H}$. A similar result to (9.1.31) holds if we take the limit as $t \to -\infty$.

It is important to note that the existence of wave operators, both for short- and long-range potentials, depends on the asymptotic behaviour of the potential $V(\boldsymbol{r})$ for large $|\boldsymbol{r}|$, and is unaffected by local singularities of V. Thus Theorems 9.1 and 9.2 imply the existence of $\Omega_\pm$ as isometric operators on the entire Hilbert space for potentials that may have arbitrary local singularities in a bounded region.

Apart from conditions governing the behaviour of the potential in the limit as $|\boldsymbol{r}| \to \infty$, we have only to suppose that $V(\boldsymbol{r})$ is square-integrable locally, away from a closed bounded subset of $\mathbb{R}^3$ of Lebesgue measure zero. The total Hamiltonian H may then be defined, possibly with boundary conditions imposed at the singularities, and $\Omega_\pm(H, H_0)$ will exist for any such self-adjoint extension of $\hat{H}$.

In considering the question of asymptotic completeness, the situation is quite different. Local singularities of the potential *do* matter. It is not possible to decide whether $\Omega_\pm$ are asymptotically complete without further analysis of the time evolution of states in the neighbourhood of any singularities of the potential. This is already evident if we consider the corresponding classical problem. An attractive potential such as $V(\boldsymbol{r}) = -1/\boldsymbol{r}^2$ can lead, in classical mechanics, to the phenomenon of a particle spiralling in to the origin in a finite time. In the quantum domain the uncertainty principle acts against such extreme localization of a state, but nevertheless it is possible, for certain potentials V, to have asymptotically localizing states in the limit as $t \to \infty$ or as $t \to -\infty$. We shall turn to the important question of localization in the following chapter, and in Chapters 11 and 12 we shall deal with the problem of asymptotic completeness.

Many of the ideas and methods of this section will reappear in connection with asymptotic completeness. One may roughly characterize the proof of completeness as a repetition of the arguments of Theorems 9.1 and 9.2, but with the roles of H_0 and H reversed. For example, the important estimate (9.1.29), which shows that $\boldsymbol{r} \approx 2\boldsymbol{P}t$ $(=\boldsymbol{P}t/m)$ as $t \to \infty$, must be replaced, in Chapter 12, by a corresponding estimate in which the free evolution e^{-iH_0t} is replaced by the total evolution $e^{-iHt}f$. It is instructive to note that a satisfactory mathematical treatment of these questions emerged only when allied with physically motivated arguments based on the evolution of quantum states in position and momentum space.

9.2 A MATHEMATICAL DESCRIPTION OF A SCATTERING EXPERIMENT

In Section 5.3 we considered the asymptotic free evolution of states, and showed in (5.3.39) how the probability, for a freely evolving state, of finding the particle within a cone $\mathscr{C}$ approaches a constant value at large positive and negative times. The value of this limiting probability, for $t \to +\infty$, is given by the integral $\int_{\mathscr{C}} |\hat{f}(\boldsymbol{k})|^2 \, d^3\boldsymbol{k}$, where $\hat{f}(\boldsymbol{k})$ is the momentum-space wave function of the initial normalized state. In the limit $t \to -\infty$, we evaluate the same integral, with the cone $\mathscr{C}$ replaced by its reflection $\mathscr{C}_-$ through the origin. From (9.1.19) in the last section, it also follows that these asymptotic formulae for the probability of finding the particle within a cone $\mathscr{C}$ also hold if we replace free evolution by a modified free evolution.

We now consider a particle moving in a potential field $V(\boldsymbol{r})$, which may be of short or of long range, and let $\Omega_\pm$ be corresponding short- or long-range wave operators. We let $g = \Omega_- f$ be the normalized state at time $t = 0$, so that

$$\lim_{t \to -\infty} \| e^{-iHt} g - F(\boldsymbol{P}, t) f \| = 0, \tag{9.2.1}$$

where $F(\boldsymbol{P}, t) = e^{-iH_0t}$ in the short-range case.

The initial state g may be decomposed as the sum of a vector in the range of Ω_+ and a vector orthogonal to the range of Ω_+. Thus we can write

$$g=\Omega_+ h+g^\perp, \tag{9.2.2}$$

where $g^\perp$ is orthogonal to the range of Ω_+, implying that $\Omega_+^* g^\perp=0$. Since Ω_+ is isometric, we have $\Omega_+^*\Omega_+=1$, so that

$$h=\Omega_+^* g=\Omega_+^*\Omega_- f=Sf, \tag{9.2.3}$$

where S is the scattering operator.

The corresponding result to (9.2.1) in the limit $t\to+\infty$ gives

$$\lim_{t\to\infty}\|\mathrm{e}^{-\mathrm{i}Ht}\Omega_+ h-F(\boldsymbol{P},t)h\|=0. \tag{9.2.4}$$

In order to deduce from (9.2.4) an asymptotic formula for $\mathrm{e}^{-\mathrm{i}Ht}g$ in the limit as $t\to\infty$, we need to known about the asymptotic evolution of $\mathrm{e}^{-\mathrm{i}Ht}g^\perp$, for any vector $g^\perp$ orthogonal to the range of Ω_+. Since $g\in$ range $(\Omega_-)\subseteq \mathcal{M}_{\mathrm{ac}}(H)$, we also have $g^\perp\in\mathcal{M}_{\mathrm{ac}}(H)$, so that $g^\perp$ belongs to the orthogonal complement of range (Ω_+) in $\mathcal{M}_{\mathrm{ac}}(H)(g^\perp\in\mathcal{M}_{\mathrm{ac}}(H)\cap(\mathrm{range}(\Omega_+))^\perp)$. We shall defer detailed consideration of the evolution of such states until Chapter 11, where it will be shown that s-$\lim_{t\to\infty}E_{|\boldsymbol{r}|>R}\mathrm{e}^{-\mathrm{i}Ht}g^\perp=0$. Here $E_{|\boldsymbol{r}|>R}$ denotes the operator of multiplication, in position space, by the characteristic function of the region $|\boldsymbol{r}|>R$, for any $R>0$; thus $E_{|\boldsymbol{r}|>R}$ projects onto the complement of a ball of radius R. From (9.2.2) and (9.2.4), it therefore follows that

$$\lim_{t\to\infty}\|E_{|\boldsymbol{r}|>R}\mathrm{e}^{-\mathrm{i}Ht}g-E_{|\boldsymbol{r}|>R}F(\boldsymbol{P},t)h\|=0. \tag{9.2.5}$$

In the limit as $t\to\infty$, $F(\boldsymbol{P},t)h$ approaches $\mathrm{e}^{-\mathrm{i}H_0t}h$ in norm in position space, apart from a phase factor, and we have also $\lim_{t\to\infty}E_{|\boldsymbol{r}|<R}\mathrm{e}^{-\mathrm{i}H_0t}h=0$. These two facts, together with (9.2.5), imply that

$$\lim_{t\to\infty}\|E_{|\boldsymbol{r}|>R}\mathrm{e}^{-\mathrm{i}Ht}g\|^2=\lim_{t\to\infty}\|E_{|\boldsymbol{r}|>R}\mathrm{e}^{-\mathrm{i}H_0t}h\|^2=\|h\|^2, \tag{9.2.6}$$

and

$$\lim_{t\to\infty}\|E_{\boldsymbol{r}\in\mathscr{C}}E_{|\boldsymbol{r}|>R}\mathrm{e}^{-\mathrm{i}Ht}g\|^2=\lim_{t\to\infty}\|E_{\boldsymbol{r}\in\mathscr{C}}\mathrm{e}^{-\mathrm{i}H_0t}h\|^2, \tag{9.2.7}$$

where $E_{\boldsymbol{r}\in\mathscr{C}}$ projects in position space onto the cone $\mathscr{C}$ with vertex at the origin. Evaluating the limit on the right-hand side of (9.2.7), we have

$$\lim_{t\to\infty}\|E_{\boldsymbol{r}\in\mathscr{C}}E_{|\boldsymbol{r}|>R}\mathrm{e}^{-\mathrm{i}Ht}g\|^2=\int_{\mathscr{C}}|\hat{h}(\boldsymbol{k})|^2\,\mathrm{d}^3\boldsymbol{k}=\int_{\mathscr{C}}|(S\hat{f})(\boldsymbol{k})|^2\,\mathrm{d}^3\boldsymbol{k}. \tag{9.2.8}$$

The left-hand side of (9.2.8) gives the asymptotic probability, as $t \to \infty$, for the state $e^{-iHt}g$, to find the particle within the *truncated cone* that is the intersection of $\mathscr{C}$ with the region $|\boldsymbol{r}| > R$. If S is unitary (asymptotic completeness) then we shall have $\|h\| = \|Sf\| = \|f\| = \|g\| = 1$, and (9.2.6) shows in that case that the particle is asymptotically to be found within the region $|\boldsymbol{r}| > R$ with probability 1. Equation (9.2.8) then allows us to find the relative probabilities of lying in various subregions of $|\boldsymbol{r}| > R$, defined by cones. If S is non-unitary then with $\|h\| < 1$ there will be a non-zero probability of the particle remaining in the region $|\boldsymbol{r}| < R$ at large positive times; we shall analyse this situation in more detail in later chapters.

Since $g = \Omega_- f$, the corresponding result to (9.2.8) for the limit $t \to -\infty$ can be worked out without difficulty, and we have, for example, for a cone $\mathscr{C}'$ having vertex at the origin,

$$\lim_{t \to -\infty} \|E_{r \in \mathscr{C}'_-} e^{-iHt} g\|^2 = \int_{\mathscr{C}'} |\hat{f}(\boldsymbol{k})|^2 \, d^3\boldsymbol{k}. \tag{9.2.9}$$

There is no need here to truncate the cones $\mathscr{C}'_-$ or $\mathscr{C}'$, since for $g \in \text{range}(\Omega_-)$ the state $e^{-iHt}g$ is localized asymptotically as $t \to -\infty$ entirely in the region $|\boldsymbol{r}| > R$.

Now $|\hat{f}(\boldsymbol{k})|^2$ is the probability density for momentum of the particle at large negative times, whereas, from (9.2.8), $|(\hat{S}f)(\boldsymbol{k})|^2$ is the probability density for momentum at large positive times, provided that the particle is to be found at large distances from the origin. Equations (9.2.6) and (9.2.8) show how to relate these densities to the corresponding probabilities in position space. The case of most direct application to scattering theory is that in which the cone $\mathscr{C}'$ in (9.2.9) has small vertex angle, and the function $\hat{f}$ has support contained entirely within $\mathscr{C}'$. We choose the axis OZ to coincide with the axis of $\mathscr{C}'$, and let α be the semivertex angle. The cone $\mathscr{C}'_-$ then has the same semivertex angle and has axis along the negative Z-axis. If α is small then the state vector $e^{-iHt}g$ at large negative times describes an incoming particle having momentum close in direction to OZ, but with a small spread of direction through angle α. The incoming particle is also close to the negative Z-axis in the sense that the position vector $\boldsymbol{r}$ is asymptotically with probability 1 confined to $\mathscr{C}'_-$. If the cone $\mathscr{C}$ is also taken to have narrow vertex angle, but with axis in an arbitrary direction from O, then we can use (9.2.8) to evaluate the probability of the particle being scattered into $\mathscr{C}$, thus providing us with a probability distribution for scattering into various final directions. We can imagine an instrument, placed at large distance R from the scattering centre O, that will detect the arrival of the scattered particle, at large time t, into the truncated cone defined by $\mathscr{C}$. In this way the value of the left-hand side of (9.2.8), by repeated measurement for a given incoming state, is in principle capable of observation.

Of course this description of a scattering experiment, related as it is to the scattering of a single particle, is hardly realistic. In practice, we have to deal with an incoming beam of particles rather than a single particle. Another disadvantage of this application of (9.2.6) and (9.2.8) is that the transition probability into the cone $\mathscr{C}$ may depend crucially on how the function $\hat{f}(\boldsymbol{k})$, with support in $\mathscr{C}'$, is chosen. We shall remedy both of these defects with a more realistic mathematical model of a scattering process.

Let us again take $\mathscr{C}'$ with small vertex angle and with axis along OZ. We consider an incoming beam of particles, moving essentially parallel to the Z-axis, but with a small spread α of incoming momentum direction. We shall eventually consider the idealization of the limiting case $\alpha \to 0$.

An individual particle of the beam will be described not by a pure state, but rather by a statistical mixture of pure states, each pure state of the mixture being obtained from the others by a relative translation perpendicular to the Z-axis. Thus we let f be a normalized pure state vector such that $\hat{f}(\boldsymbol{k})$ has support in $\mathscr{C}'$, and we let $\boldsymbol{a} = (a_1, a_2)$ be a vector in the (X, Y) plane. In position space if $f = f(\boldsymbol{r})$ then the pure state obtained by relative translation through $\boldsymbol{a}$ is $f_{\boldsymbol{a}}(\boldsymbol{r}) = f(\boldsymbol{r} - \boldsymbol{a})$. In momentum space we have $\hat{f}_{\boldsymbol{a}}(\boldsymbol{k}) = \mathrm{e}^{-i\boldsymbol{k}\cdot\boldsymbol{a}} \hat{f}(\boldsymbol{k})$, so that $f_{\boldsymbol{a}} = \mathrm{e}^{-i\boldsymbol{P}\cdot\boldsymbol{a}} f$, and $\hat{f}_{\boldsymbol{a}}(\boldsymbol{k})$ again has support in $\mathscr{C}'$. We can think of f and $f_{\boldsymbol{a}}$ as describing respectively one possible pure state for a particle of the beam and another, moving parallel to it and displaced by a constant vector $\boldsymbol{a}$. A statistical mixture of such states, with probability density $p(\boldsymbol{a})$, will be represented by a density operator $\int p(\boldsymbol{a}) |f_{\boldsymbol{a}}\rangle\langle f_{\boldsymbol{a}}| \, \mathrm{d}^2\boldsymbol{a}$. Each particle of the beam will correspond to a wave function $f_{\boldsymbol{a}}(\boldsymbol{r})$, where there is a probability distribution for the value of $\boldsymbol{a}$. The probability of a value of $\boldsymbol{a}$ that lies in some subset $\mathscr{S}$ of $\mathbb{R}^2$ is then $\int_{\mathscr{S}} p(\boldsymbol{a}) \, \mathrm{d}^2\boldsymbol{a} = \int_{\mathscr{S}} p(a_1, a_2) \, \mathrm{d}a_1 \, \mathrm{d}a_2$, where naturally $\int_{\mathbb{R}^2} p(\boldsymbol{a}) \, \mathrm{d}^2\boldsymbol{a} = 1$.

Applying (9.2.1), the state at time $t = 0$ will be $g_{\boldsymbol{a}} = \Omega_- f_{\boldsymbol{a}}$ with probability density $p(\boldsymbol{a})$, where in the limit $t \to -\infty$, $\mathrm{e}^{-iHt} g_{\boldsymbol{a}}$ asymptotically approaches $F(\boldsymbol{P}, t) f_{\boldsymbol{a}}$ in norm. If the incident beam contains a large number of particles per unit time, projected towards the scattering centre and moving (approximately) parallel to the axis OZ, then particles of the beam can be found corresponding to many different values of $\boldsymbol{a}$. We can then speak of an *incident flux* of particles. If N particles are projected per unit time and if the support of $p(\boldsymbol{a})$ is the bounded region $\mathscr{S}$ of $\mathbb{R}^2$, with area $\mathscr{A}$, then the incident flux is defined to be $N/\mathscr{A}$. This is just the number of particles per unit area per unit time in the incident beam. Eventually we shall consider the idealized case in which $\mathscr{A}$ tends to infinity and becomes the whole of $\mathbb{R}^2$. For bounded $\mathscr{S}$, we could take $p(\boldsymbol{a}) = 1/\mathscr{A}$ for $\boldsymbol{a} \in \mathscr{S}$. Since this involves an abrupt jump of $p(\boldsymbol{a})$ from $1/\mathscr{A}$ to 0 at the boundary of $\mathscr{S}$, it is more realistic to suppose that $p(\boldsymbol{a})$ is approximately constant over most of the area of $\mathscr{S}$, but drops smoothly to zero as $\boldsymbol{a}$ approaches the boundary. In this case $p(\boldsymbol{a})=1/\mathscr{A}$ is an approximation that is satisfied very closely, except near to the boundary of $\mathscr{S}$.

We shall use (9.2.8), with $f=f_{\boldsymbol{a}}$, to calculate the probability of a particle of the beam being detected, at large positive times, in a given truncated cone. We shall make use of the direct integral representation of the Hilbert space to evaluate the right-hand side of (9.2.8). In the direct integral representation the vector $f\in L^2(\mathbb{R}^3)$ is represented by the one-parameter family $\{f_\lambda\}$, $f_\lambda\in L^2(\mathbb{S}^2)$, where

$$f_\lambda(\boldsymbol{\omega})=\frac{1}{\sqrt{2}}\lambda^{1/4}\hat{f}(\lambda^{1/2}\boldsymbol{\omega})=f(\lambda,\boldsymbol{\omega}).$$

Hence $f_{\boldsymbol{a}}=\mathrm{e}^{-\mathrm{i}\boldsymbol{P}\cdot\boldsymbol{a}}f$ has the representation

$$f_{\boldsymbol{a}}(\lambda,\boldsymbol{\omega})=\frac{1}{\sqrt{2}}\lambda^{1/4}\mathrm{e}^{-\mathrm{i}\lambda^{1/2}\boldsymbol{\omega}\cdot\boldsymbol{a}}\hat{f}(\lambda^{1/2}\boldsymbol{\omega}). \tag{9.2.10}$$

In the direct integral representation H_0 corresponds to the operator of multiplication by λ. Since S commutes with H_0, we can decompose the scattering operator into a family $\{S_\lambda\}$ of operators acting in $L^2(\mathbb{S}^2)$ with

$$(Sf)_\lambda=S_\lambda f_\lambda. \tag{9.2.11}$$

Here S_λ is defined for almost all $\lambda\geq 0$, and may be regarded as the scattering operator at energy λ, or the scattering operator on the energy shell. We shall refer to S_λ as the scattering matrix.

(To define S_λ a.e., we let S operate on vectors that in the direct integral representation are of the form $\{\phi(\lambda)h\}$ for some $h\in L^2(\mathbb{S}^2)$. In particular, if $\phi(\lambda)=\chi(\lambda)$ is the characteristic function of some subinterval $[a,b]$ of $[0,\infty)$ then S_λ can be defined by

$$S\{\chi(\lambda)h\}=\{\chi(\lambda)S_\lambda h\}. \tag{9.2.12}$$

If this formula is used to define, a.e., $S_\lambda h_n$ for each vector h_n of an orthonormal basis for $\mathscr{H}_0$ then the definition extends immediately to finite linear combinations of such h_n, and by continuity to the whole of $L^2(\mathbb{S}^2)$. The set of positive values of λ at which S_λ is *not* defined is then a countable union of such sets defined separately for each h_n, and hence is of Lebesgue measure zero. For more general $\phi(\lambda)$ we have, from (9.2.12),

$$\begin{aligned}S\{\phi(\lambda)\chi(\lambda)h\}&=S\phi(H_0)\{\chi(\lambda)h\}=\phi(H_0)S\{\chi(\lambda)h\}\\&=\{\phi(\lambda)\chi(\lambda)S_\lambda h\}\end{aligned}$$

and proceeding to the limit $a\to 0$, $b\to\infty$, we then have

$$S\{\phi(\lambda)h\}=\{S_\lambda\phi(\lambda)h\}, \tag{9.2.13}$$

provided that $\phi\in L^2(\mathbb{R})$. The formula (9.2.11) now follows if we remember that linear combinations of vectors of the form $\{\phi(\lambda)h\}$, as both ϕ and h are varied, are dense in the Hilbert space $\mathscr{H}$.)

As a linear operator on $L^2(\mathbb{S}^2)$ for almost all $\lambda\geq 0$, the scattering matrix has norm 1, and is unitary provided that S is unitary as an operator on $\mathscr{H}=L^2(\mathbb{R}^3)$.

(We leave these results as an exercise. However, to give a flavour of the kind of argument that allows us to deduce properties of S_λ from the corresponding properties of S, let us suppose that we start from $\|S\|\leq 1$. To prove $\|S_\lambda\|_0\leq 1$, a.e., we suppose on the contrary that there exists some subset $\mathscr{K}$ of $[0, \infty)$, with strictly positive Lebesgue measure, on which $\|S_\lambda\|_0>1$. Then for each $\lambda\in\mathscr{K}$ there exists f_λ with $\|S_\lambda f_\lambda\|_0>\|f_\lambda\|_0$. If we then define a vector in the direct integral representation for $\mathscr{H}$ to be $\{\chi(\lambda)f_\lambda\}$, where now χ is the characteristic function of $\mathscr{K}$, we have

$$\begin{aligned}\|S\{\chi(\lambda)f_\lambda\}\|^2&=\int_0^\infty \mathrm{d}\lambda\,\|\chi(\lambda)S_\lambda f_\lambda\|_0^2>\int_0^\infty \mathrm{d}\lambda\,\chi(\lambda)\|f_\lambda\|_0^2\\&=\|\{\chi(\lambda)f_\lambda\}\|^2.\end{aligned}$$

This contradicts the assumption that $\|S\|\leq 1$. Hence $\|S\|\leq 1\Rightarrow\|S_\lambda\|_0\leq 1$ for almost all $\lambda\geq 0$. Similar arguments may be used to show that S_λ is unitary provided that S is. If, on the other hand, S is non-unitary then there will be a subset of $[0, \infty)$ with positive Lebesgue measure at which S_λ is non-unitary.)

If the potential $V(\boldsymbol{r})$ is everywhere zero then we have $S=1$ (identity operator in $\mathscr{H}=L^2(\mathbb{R}^3)$), and $S_\lambda=1$ (identity operator in $\mathscr{H}_0=L^2(\mathbb{S}^2)$). In this case, a comparison of (9.2.8) and (9.2.9) shows that, for any cone $\mathscr{C}$ with vertex at the origin,

$$\lim_{t\to\infty}\text{prob}_t\,(\boldsymbol{r}\in\mathscr{C};|\boldsymbol{r}|>R)=\lim_{t\to-\infty}\text{prob}_t\,(\boldsymbol{r}\in\mathscr{C}_-),$$

where prob_t denotes a probability at time t. We also then have,

$$\lim_{t\to\infty}\text{prob}_t\,(\boldsymbol{P}\in\mathscr{C})=\lim_{t\to-\infty}\text{prob}_t\,(\boldsymbol{P}\in\mathscr{C}),$$

where $\boldsymbol{P}$ denotes momentum. These two equations state, in quantum-mechanical terms, that if $V(\boldsymbol{r})\equiv 0$ then every particle of the beam is unscattered. The momentum of the outgoing particle is that of the incoming particle, and a particle that is to be found within a cone $\mathscr{C}_-$ at large negative times will be found at large distances from the origin, and within the reflected cone $\mathscr{C}$, at large positive times. We can say that the beam is transmitted entirely into the *forward direction*, and the transmitted flux is the same as the incident flux.

Even for a potential that is not identically zero, we can expect that, for any particle of the beam corresponding to a large value of $|\boldsymbol{a}|$, there will be a high probability of the particle being scattered through a very small angle. Classically, we should say that such a particle followed an orbit that failed to bring it close to the scattering centre O. In this sense, we can then say that "most" particles hardly get scattered at all, in which case there will be a dominant contribution to the scattering amplitude that comes from particles being transmitted into the forward direction. We therefore separate out the forward contribution to the scattering matrix by writing

$$S_\lambda = 1 + T_\lambda, \tag{9.2.14}$$

where the linear operator T_λ on $L^2(\mathbb{S}^2)$, defined for almost all values of λ, is called the T-matrix. Later we shall see the connection between the T-matrix and the linear operator $T(z)$ defined in Chapter 4 (see (4.2.18) *et seq.*).

We shall assume for the moment that, for almost all λ, the T-matrix at energy λ may be represented as an integral operator with kernel $T(\lambda, \boldsymbol{\omega}, \boldsymbol{\omega}')$. Thus we assume a formula

$$(T_\lambda f_\lambda)(\boldsymbol{\omega}) = \int T(\lambda, \boldsymbol{\omega}, \boldsymbol{\omega}') f_\lambda(\boldsymbol{\omega}')\, d\omega' \tag{9.2.15}$$

An equation like (9.2.15) is in fact satisfied for a large class of potentials. Indeed, we shall find in the following section that if $V(\boldsymbol{r})$ decays at infinity more rapidly than $|\boldsymbol{r}|^{-(2+\varepsilon)}$ for some $\varepsilon > 0$ then T_λ is even Hilbert–Schmidt, so that $\iint |T(\lambda, \boldsymbol{\omega}, \boldsymbol{\omega}')|^2\, d\omega\, d\omega' < \infty$. More generally, it can often be proved that the kernel $T(\lambda, \boldsymbol{\omega}, \boldsymbol{\omega}')$ exists and is finite away from the forward direction; that is, the kernel is even continuous as a function of $\boldsymbol{\omega}$ and $\boldsymbol{\omega}'$, apart from a possible singularity at $\boldsymbol{\omega} = \boldsymbol{\omega}'$.

Assuming (9.2.15), we can now use the right-hand side of (9.2.8), in the direct integral representation, to calculate the limiting probability, as $t \to \infty$, of a particle of the beam being scattered into a given (truncated) cone $\mathscr{C}$. Remembering that each particle of the beam is described by a statistical mixture with probability density $p(\boldsymbol{a})$, where the pure state $f_{\boldsymbol{a}}$ is given in momentum space by $e^{-i\boldsymbol{k}\cdot\boldsymbol{a}}\hat{f}(\boldsymbol{k})$ and (by (5.3.27)) in the direct integral representation by

$$\frac{1}{\sqrt{2}} \lambda^{1/4} e^{-i\lambda^{1/2}\boldsymbol{\omega}\cdot\boldsymbol{a}} \hat{f}(\lambda^{1/2}\boldsymbol{\omega}) = e^{-i\lambda^{1/2}\boldsymbol{\omega}\cdot\boldsymbol{a}} f_\lambda(\boldsymbol{\omega}),$$

we find that

$$\lim_{t\to\infty} \text{prob}_t\,(\boldsymbol{r}\in\mathscr{C}, |\boldsymbol{r}|>R)$$
$$= \int_{\omega\in\mathscr{C}} \mathrm{d}\lambda\,\mathrm{d}\omega\,\mathrm{d}^2\boldsymbol{a}\,p(\boldsymbol{a}) \left| \mathrm{e}^{-\mathrm{i}\lambda^{1/2}\boldsymbol{\omega}\cdot\boldsymbol{a}} f_\lambda(\boldsymbol{\omega}) \right.$$
$$\left. + \int T(\lambda,\boldsymbol{\omega},\boldsymbol{\omega}')\mathrm{e}^{-\mathrm{i}\lambda^{1/2}\boldsymbol{\omega}'\cdot\boldsymbol{a}} f_\lambda(\boldsymbol{\omega}')\,\mathrm{d}\omega' \right|^2. \tag{9.2.16}$$

In (9.2.16) note that $\boldsymbol{\omega}\in\mathscr{C} \Rightarrow \lambda^{1/2}\boldsymbol{\omega}\in\mathscr{C}$. We can think of $\int \mathrm{d}\lambda\,\mathrm{d}^2\boldsymbol{a}\,p(\boldsymbol{a})|\cdot|^2$ as a probability density over the unit sphere, representing the probability, per unit solid angle, of a particle of the beam being scattered into the direction of the unit vector $\boldsymbol{\omega}$. The cone $\mathscr{C}$ is uniquely defined by its intersection with the unit sphere, so that $\mathscr{C}$ is characterized by a subset of $\mathbb{S}^2$. Our assumptions imply that the $\boldsymbol{\omega}'$-support of the function $f_\lambda(\boldsymbol{\omega}')$ is entirely contained within a cone $\mathscr{C}'$ of narrow vertex angle, with axis along the Z-axis. We can again regard $\mathscr{C}'$ as characterized by a subset of $\mathbb{S}^2$ with small solid angle. The right-hand side of (9.2.16) then becomes the transition probability density for scattering from an initial direction $\boldsymbol{\omega}'$, close to $\hat{z}$ ($\equiv$ unit vector in the Z-direction) into a final direction $\boldsymbol{\omega}$.

Let us consider first the evaluation of (9.2.16) in the case where the cones $\mathscr{C}$ and $\mathscr{C}'$ do not overlap. We then have a true *scattering* probability in the sense that $\boldsymbol{\omega}=\boldsymbol{\omega}'$ is not allowed. By the support properties of $f_\lambda(\boldsymbol{\omega})$, we then have $f_\lambda(\boldsymbol{\omega})=0$ for $\boldsymbol{\omega}\in\mathscr{C}$, so that the function $f_\lambda(\boldsymbol{\omega})$ within the modulus sign on the right-hand side of (9.2.16) may be omitted.

The integration with respect to $\boldsymbol{\omega}'$ looks very much like a two-dimensional Fourier transform. Indeed, if we take new integration variables (x', y'), defined to be the Cartesian coordinates of the projection of the vector $\lambda^{1/2}\boldsymbol{\omega}'$ onto the (X, Y) plane, then we have $\mathrm{d}\omega'=\lambda^{-1}\,\mathrm{d}x'\,\mathrm{d}y'/|\boldsymbol{\omega}'\cdot\hat{z}/$. (This can be verified by writing down the transformation from spherical polar coordinates to Cartesians and evaluating the Jacobian determinant.) In terms of the new integration variables, the integration in (9.2.16) with respect to $\boldsymbol{\omega}'$ takes exactly the form of a Fourier transform in two dimensions. The two-dimensional Fourier transform, defined with a numerical factor $(1/(2\pi)^{1/2})^2$, is unitary. We therefore have

$$\int \mathrm{d}^2\boldsymbol{a} \left| \int T(\lambda,\boldsymbol{\omega},\boldsymbol{\omega}')\mathrm{e}^{-\mathrm{i}\lambda^{1/2}\boldsymbol{\omega}'\cdot\boldsymbol{a}} f_\lambda(\boldsymbol{\omega}')\,\mathrm{d}\omega' \right|^2$$
$$=(2\pi)^2 \int \mathrm{d}^2\boldsymbol{a} \left| \int \left(\frac{1}{(2\pi)^{1/2}}\right)^2 \lambda^{-1}/\boldsymbol{\omega}'.\hat{z}|^{-1}\, T(\lambda,\boldsymbol{\omega},\boldsymbol{\omega}')\mathrm{e}^{-\mathrm{i}(a_1x'+a_2y')} f_\lambda(\boldsymbol{\omega}')\,\mathrm{d}x'\,\mathrm{d}y' \right|^2$$

$$=(2\pi)^2 \int |\lambda^{-1}|\boldsymbol{\omega}'\cdot\hat{z}|^{-1} T(\lambda,\boldsymbol{\omega},\boldsymbol{\omega}') f_\lambda(\boldsymbol{\omega}')|^2 \, dx' \, dy'$$

$$= (2\pi)^2 \int \lambda^{-1} |\boldsymbol{\omega}'\cdot\hat{z}|^{-1} |T(\lambda,\boldsymbol{\omega},\boldsymbol{\omega}')|^2 |f_\lambda(\boldsymbol{\omega}')|^2 \, d\omega'.$$

For non-overlapping cones $\mathscr{C}$ and $\mathscr{C}'$, so that $f_\lambda(\boldsymbol{\omega})$ is omitted in (9.2.16), we now proceed to the limiting case $\mathscr{A}\to\infty$, with $p(\boldsymbol{a})=1/\mathscr{A}$ over most of $\mathscr{S}$ and $p(\boldsymbol{a})=0$ outside $\mathscr{S}$. The right-hand side of (9.2.16) then becomes, asymptotically in the limit $\mathscr{A}\to\infty$.

$$\frac{1}{\mathscr{A}} \int_{\boldsymbol{\omega}\in\mathscr{C}} d\lambda \, d\omega \, d\omega' \, (2\pi)^2 \lambda^{-1} |\boldsymbol{\omega}'\cdot\hat{z}|^{-1} |T(\lambda,\boldsymbol{\omega},\boldsymbol{\omega}')|^2 |f_\lambda(\boldsymbol{\omega}')|^2.$$

If the incident beam is highly collimated, so that the incoming direction of each particle is almost parallel to the Z-axis, then we can take the limiting case $\boldsymbol{\omega}'\cdot\hat{z}=1$, since as α approaches zero $\boldsymbol{\omega}'\cdot\hat{z}$ tends to unity on the support of $f_\lambda(\boldsymbol{\omega}')$. Note also that $|f_\lambda(\boldsymbol{\omega}')|^2$ is the probability density function for the energy/direction of the particles of the incident beam, in the sense that if (λ_1, λ_2) is any energy interval and $\mathscr{C}''$ is a subset of $\mathbb{S}^2$ then the probability, for a particle of the incident beam, of $\lambda\in(\lambda_1,\lambda_2)$ and $\boldsymbol{\omega}\in\mathscr{C}''$ is

$$\int_{\boldsymbol{\omega}'\in\mathscr{C}''} d\omega' \left(\int_{\lambda_1}^{\lambda_2} d\lambda \, |f_\lambda(\boldsymbol{\omega}')|^2 \right).$$

Given this interpretation of $|f_\lambda(\boldsymbol{\omega}')|^2$ as a probability density, we may interpret

$$\frac{1}{\mathscr{A}} \int_{\boldsymbol{\omega}\in\mathscr{C}} d\omega \, (2\pi)^2 \lambda^{-1} |T(\lambda,\boldsymbol{\omega},\boldsymbol{\omega}')|^2$$

as a *conditional* probability, representing the probability of a particle of the incident beam, projected towards the target in direction $\boldsymbol{\omega}'$, being scattered into the cone $\mathscr{C}$. This is just the probability of a final direction $\boldsymbol{\omega}\in\mathscr{C}$, or equivalently the probability of detecting the particle, at large positive times, within a truncated cone $\mathscr{C}$, conditional on the incident particle direction being $\boldsymbol{\omega}'$.

Let us now suppose that $\boldsymbol{\omega}'$ is the incident-beam direction, and that N particles are projected per unit time. Of these N particles, the expectation value of the number of particles scattered into the cone $\mathscr{C}$ is

$(N/\mathscr{A})\int_{\omega\in\mathscr{C}} \mathrm{d}\omega\ (2\pi)^2\lambda^{-1}|T(\lambda,\boldsymbol{\omega},\boldsymbol{\omega}')|^2$. Here $N/\mathscr{A}$ is just the incident flux.

We define the *scattering cross-section* from incident beam direction $\boldsymbol{\omega}'$ into the cone $\mathscr{C}$, at energy λ, by the formula

$$\sigma(\lambda;\boldsymbol{\omega}'\to\mathscr{C}) = \int_{\omega\in\mathscr{C}} \mathrm{d}\omega\ (2\pi)^2\lambda^{-1}|T(\lambda,\boldsymbol{\omega},\boldsymbol{\omega}')|^2. \tag{9.2.17}$$

What we have shown is that $\sigma(\lambda;\boldsymbol{\omega}'\to\mathscr{C})$ is the ratio

$$\frac{\text{number of particles scattered per unit time into } \mathscr{C}}{\text{incident flux}}.$$

The *total cross-section* at energy λ is given in terms of the incident-beam direction $\boldsymbol{\omega}'$ by

$$\sigma(\lambda,\boldsymbol{\omega}') = \int_{\omega\in\mathbb{S}^2} \mathrm{d}\omega\ (2\pi)^2\lambda^{-1}|T(\lambda,\boldsymbol{\omega},\boldsymbol{\omega}')|^2. \tag{9.2.18}$$

Thus $\sigma(\lambda,\boldsymbol{\omega}')$ is the ratio

$$\frac{\text{total number of particles scattered per unit time}}{\text{incident flux}}.$$

In the classical theory of scattering by a potential $V(\boldsymbol{r})$, it is usually the case that *every* particle of the incident beam is scattered by the potential. Since we speak of the incident flux as a number of particles per unit time *per unit area*, and we have taken the limit $\mathscr{A}\to\infty$, this effectively means that, in the classical theory, the total number of particles scattered per unit time is infinite. Hence the classical total scattering cross-section is usually infinite. However, if $V(\boldsymbol{r})$ has compact support, or if we consider scattering by a finite obstacle, then the classical total scattering cross-section is finite. In the case of obstacle scattering, for example, a particle will be scattered provided that its trajectory meets the obstacle, with the result that the total cross-section is equal to the *cross-sectional area* presented by the obstacle to the incident beam. For this reason, we retain the term *cross-section*, even in quantum mechanics, where such an interpretation does not apply. In quantum mechanics we shall again find, in the following section, that $\sigma(\lambda,\boldsymbol{\omega}')$ is finite for potentials having compact support, and indeed more generally for potentials decaying sufficiently rapidly at infinity.

The differential cross-section is defined as

$$\frac{\mathrm{d}\sigma}{\mathrm{d}\omega}(\lambda;\boldsymbol{\omega}'\to\boldsymbol{\omega}) = (2\pi)^2\lambda^{-1}|T(\lambda,\boldsymbol{\omega},\boldsymbol{\omega}')|^2, \tag{9.2.19}$$

and is given by the ratio

$$\frac{\text{number of particles scattered per unit solid angle per unit time in direction } \boldsymbol{\omega}}{\text{incident flux}}$$

Note that

$$\sigma(\lambda, \boldsymbol{\omega}') = \int_{\mathbb{S}^2} \mathrm{d}\omega \, \frac{\mathrm{d}\sigma}{\mathrm{d}\omega}(\lambda; \boldsymbol{\omega}' \to \boldsymbol{\omega}). \tag{9.2.20}$$

Another important quantity is the averaged total cross-section, obtained by averaging over all possible directions of the incident beam. We then have

$$\begin{aligned} \sigma_{\text{ave}}(\lambda) &= \frac{1}{4\pi} \int_{\mathbb{S}2} \mathrm{d}\omega' \sigma(\lambda, \boldsymbol{\omega}') \\ &= \int_{\boldsymbol{\omega}, \boldsymbol{\omega}' \in \mathbb{S}2} \mathrm{d}\omega \, \mathrm{d}\omega' \, \pi\lambda^{-1} |T(\lambda, \boldsymbol{\omega}, \boldsymbol{\omega}')|^2. \end{aligned} \tag{9.2.21}$$

Since $T(\lambda, \boldsymbol{\omega}, \boldsymbol{\omega}')$ is the kernel of the integral operator T_λ acting in $L^2(\mathbb{S}^2)$, the right-hand side of (9.2.21) is proportional to the square of the Hilbert–Schmidt norm of T_λ, so that

$$\sigma_{\text{ave}}(\lambda) = \pi\lambda^{-1}(\|T_\lambda\|_2)^2. \tag{9.2.22}$$

Thus $\sigma_{\text{ave}}(\lambda)$ is finite provided that T_λ is Hilbert–Schmidt. We shall define the *scattering amplitude* $f(\lambda, \boldsymbol{\omega}' \to \boldsymbol{\omega})$ at energy λ, from incident direction $\boldsymbol{\omega}'$ into direction $\boldsymbol{\omega}$, by

$$f(\lambda; \boldsymbol{\omega}' \to \boldsymbol{\omega}) = -2\pi \mathrm{i} \lambda^{-1/2} T(\lambda, \boldsymbol{\omega}, \boldsymbol{\omega}'), \tag{9.2.23}$$

so that, for example,

$$\frac{\mathrm{d}\sigma}{\mathrm{d}\omega}(\lambda; \boldsymbol{\omega}' \to \boldsymbol{\omega}) = |f(\lambda; \boldsymbol{\omega}' \to \boldsymbol{\omega})|^2. \tag{9.2.24}$$

There is an important connection between the total scattering cross-section and the so-called *forward-scattering amplitude* $f(\lambda, \boldsymbol{\omega}' \to \boldsymbol{\omega}')$. We suppose that the scattering operator is unitary, so that $S_\lambda = 1 + T_\lambda$ is unitary for almost all λ. Then

$$S_\lambda^* S_\lambda = 1 \Rightarrow T_\lambda^* T_\lambda = -(T_\lambda + T_\lambda^*). \tag{9.2.25}$$

Now, if T_λ, as an integral operator, has kernel $T(\lambda, \boldsymbol{\omega}, \boldsymbol{\omega}')$ then T_λ^* has kernel $\overline{T(\lambda,\boldsymbol{\omega}',\boldsymbol{\omega})}$, and the product $T_\lambda^* T_\lambda$ has kernel given by the integral

$$\int_{\mathbb{S}^2} \mathrm{d}\omega'' \, \overline{T(\lambda, \boldsymbol{\omega}'', \boldsymbol{\omega})}\, T(\lambda, \boldsymbol{\omega}'', \boldsymbol{\omega}'),$$

which is the natural extension to integral operators of the law for matrix multiplication, and can be verified, for example, if T_λ is Hilbert–Schmidt.

If we now write $\boldsymbol{\omega} = \boldsymbol{\omega}'$ and compare the kernels of the respective integral operators in (9.2.25), we have

$$\int_{\mathbb{S}^2} \mathrm{d}\omega'' \, |T(\lambda, \boldsymbol{\omega}'', \boldsymbol{\omega}')|^2 = -2\,\mathrm{Re}\,(T(\lambda, \boldsymbol{\omega}', \boldsymbol{\omega}'))$$

$$= \pi^{-1}\lambda^{1/2}\,\mathrm{Im}\,(f(\lambda; \boldsymbol{\omega}' \to \boldsymbol{\omega}')). \qquad (9.2.26)$$

From (9.2.18), the left-hand side of (9.2.26) is $(2\pi)^{-2}\lambda\sigma\,(\lambda, \boldsymbol{\omega}')$, so that we have the identity

$$\sigma(\lambda, \boldsymbol{\omega}') = 4\pi\lambda^{-1/2}\,\mathrm{Im}\,(f(\lambda; \boldsymbol{\omega}' \to \boldsymbol{\omega}')). \qquad (9.2.27)$$

Equation (9.2.27) is a statement of the so-called *optical theorem*, and shows in particular that the forward-scattering amplitude has a positive imaginary part. The theorem can only be proved rigorously if a correct definition of $f(\lambda, \boldsymbol{\omega}' \to \boldsymbol{\omega}')$ can be given; for example, if $f(\lambda, \boldsymbol{\omega}' \to \boldsymbol{\omega})$ is a continuous function of $\boldsymbol{\omega}$ and $\boldsymbol{\omega}'$ in the neighbourhood of $\boldsymbol{\omega} = \boldsymbol{\omega}'$. It may happen that the imaginary part of the scattering amplitude contains a singularity as $\boldsymbol{\omega}'$ approaches $\boldsymbol{\omega}$, in which case the total cross-section is infinite.

We can also evaluate the expression on the right-hand side of (9.2.16) in the more general case where the cones $\mathscr{C}$ and $\mathscr{C}'$ are allowed to overlap. In this case there will be two additional contributions to the multiple integral. The first of these, arising from the term $|f_\lambda(\boldsymbol{\omega})|^2$ in the integrand, is identical with the evaluation of the right-hand side for $S = 1$, and has already been considered. This contribution arises from the fact that, with high probability, a particle of the incident beam will be transmitted without scattering. The second additional contribution to (9.2.16) comes from cross-terms in evaluating the factor $|\cdot|^2$ in the integrand, and takes the form

$$2\,\mathrm{Re}\left(\int_{\boldsymbol{\omega} \in \mathscr{C}} \mathrm{d}\lambda\, \mathrm{d}\omega\, \mathrm{d}\omega'\, \mathrm{d}^2\boldsymbol{a}\; p(\boldsymbol{a}) T(\lambda, \boldsymbol{\omega}, \boldsymbol{\omega}')\, \mathrm{e}^{\mathrm{i}\lambda^{1/2}(\boldsymbol{\omega} - \boldsymbol{\omega}')\cdot \boldsymbol{a}} \overline{f_\lambda(\boldsymbol{\omega})} f_\lambda(\boldsymbol{\omega}') \right).$$

Again, the integrations with respect to ω and ω' may be written as two-dimensional Fourier transforms. Setting $p(\boldsymbol{a})=1/\mathscr{A}$ in the limit $\mathscr{A}\to\infty$, let us take $\mathscr{C}=\mathscr{C}'$ and assume that the cone is of vanishingly small vertex angle. Integration with respect to $\boldsymbol{a}$ leads, under suitable assumptions concerning the continuity of $T(\lambda,\boldsymbol{\omega},\boldsymbol{\omega}')$ at $\boldsymbol{\omega}=\boldsymbol{\omega}'$, to the expression

$$-\frac{1}{\mathscr{A}}\int 4\pi\lambda^{-1/2}\,\mathrm{Re}\,(T(\lambda,\boldsymbol{\omega},\boldsymbol{\omega}'))|f_\lambda(\boldsymbol{\omega})|^2\,\mathrm{d}\omega,$$

where we have again proceeded to the limits $\boldsymbol{\omega}\cdot\hat{z}=\boldsymbol{\omega}'\cdot\hat{z}=1$. Using (9.2.23), together with the optical theorem, we arrive finally at $-\mathscr{A}^{-1}\sigma(\lambda,\boldsymbol{\omega}')$ in the limit $\alpha\to 0$, where $\boldsymbol{\omega}'$ is the incident-beam direction. This result confirms that the number of particles detected per unit time in the forward direction is just the number of particles per unit time in the incident beam, less the number of particles per unit time scattered away from the forward direction. Unitarity of the scattering operator thus provides us with a simple accounting system for determining what has become of all the particles of the incident beam. We cannot say for certain where any given incident particle will be detected in the limit as $t\to\infty$, since that will be determined by the laws of probability and quantum mechanics, but we can at least keep track of the overall distributions of particles in the scattered beam. If S is non-unitary then the "missing" particles are to be found in the neighbourhood of the scattering centre as $t\to\infty$—this will be a matter for further consideration.

9.3 THE SCATTERING CROSS-SECTION

Equations (9.2.21) and (9.2.22) of the last section establish a relationship between the averaged total scattering cross-section $\sigma_{\text{ave}}(\lambda)$ and the Hilbert–Schmidt norm of the operator T_λ. Assuming that T_λ is indeed Hilbert–Schmidt, for almost all (positive) energies λ, then $\sigma_{\text{ave}}(\lambda)$ will be finite. It also follows, in this case, that the total cross-section $\sigma(\lambda,\boldsymbol{\omega}')$ from incident-beam direction $\boldsymbol{\omega}'$ will be finite for almost all $\boldsymbol{\omega}'\in\mathbb{S}^2$. In addition, the scattering amplitude $f(\lambda;\boldsymbol{\omega}'\to\boldsymbol{\omega})$ will exist at almost all energies and for almost all directions $\boldsymbol{\omega}'$ and $\boldsymbol{\omega}$ of the incident and scattered beams. In view of these far-reaching consequences, it is important to establish conditions on the potential $V(\boldsymbol{r})$ under which $\|T_\lambda\|_2$ is finite for almost all λ.

The difficulty in proving T_λ to be Hilbert–Schmidt is that the most direct estimates are those for the scattering operator S. Although T_λ is derived from S in the direct integral representation, it is not always easy to pass from properties of S to properties of T_λ. Indeed, these two operators act in different spaces; S is an operator on $\mathscr{H}=L^2(\mathbb{R}^3)$, whereas T_λ acts on $L^2(\mathbb{S}^2)$. In order to clarify the link between these two operators, as far as Hilbert–

Schmidt norms are concerned, we start by defining, for any $\phi \in C_0^\infty(0, \infty)$, a corresponding linear operator $P(\phi)$ acting on $\mathscr{H}$. $P(\phi)$ is most conveniently defined in the direct integral representation, by the formula

$$(P(\phi)f)_\lambda(\boldsymbol{\omega}) = \phi(\lambda) \int_0^\infty \bar{\phi}(\mu) f_\mu(\boldsymbol{\omega}) \, \mathrm{d}\mu. \tag{9.3.1}$$

If $\phi(\lambda)$ is taken to be normalized, as an element of $L^2(0, \infty)$, then it is easy to verify that $P(\phi)$ is a projection operator. Indeed, $P(\phi)$ is then the projection onto the space of vectors $f = \{f_\lambda\}$ that are of the form $f_\lambda(\boldsymbol{\omega}) = \phi(\lambda) h(\boldsymbol{\omega})$ for some $h \in L^2(\mathbb{S}^2)$. In other words, in the direct integral representation, the energy dependence of vectors in the range of $P(\phi)$ is always in proportion to the given function $\phi(\lambda)$.

The operator $S - 1$ acting on $\mathscr{H}$ will not, in general, be Hilbert–Schmidt. Nevertheless, an operator may become Hilbert–Schmidt when projected onto a smaller subspace, so let us evaluate the Hilbert–Schmidt norm, in $\mathscr{H}$, of the operator $(S-1)P(\phi)$. For simplicity, we shall take ϕ to be both normalized and real. Let $\{h_i\}$ be an orthonormal basis of $L^2(\mathbb{S}^2)$. Then, in the direct integral representation, $\{\phi(\lambda) h_i\}$ is an orthonormal basis for the range of $P(\phi)$, so that

$$\begin{aligned}(\|(S-1)P(\phi)\|_2)^2 &= \sum_i \|(S-1)\{\phi(\lambda) h_i\}\|^2 \\ &= \sum_i \|\{T_\lambda \phi(\lambda) h_i\}\|^2 \\ &= \sum_i \int_0^\infty \mathrm{d}\lambda\, \phi^2(\lambda)(\|T_\lambda h_i\|_0^2) \\ &= \int_0^\infty \mathrm{d}\lambda\, \phi^2(\lambda) \left\{ \sum_i (\|T_\lambda h_i\|_0)^2 \right\},\end{aligned} \tag{9.3.2}$$

where $\|\cdot\|_0$ denotes a norm in $L^2(\mathbb{S}^2)$, and we have used the direct integral representation for $S = 1 + T$. The orders of summation/integration in (9.3.2) may be exchanged, provided that either side of the equation is absolutely convergent. Now, $\Sigma_i (\|T_\lambda h_i\|_0)^2$ is simply the square of the Hilbert–Schmidt norm of T_λ, as an operator in $L^2(\mathbb{S}^2)$ so that (9.3.2) becomes

$$(\|(S-1)P(\phi)\|_2)^2 = \int_0^\infty \mathrm{d}\lambda\, \phi^2(\lambda)(\|T_\lambda\|_2)^2. \tag{9.3.3}$$

Equation (9.3.3) tells us that, if $(S-1)P(\phi)$ is Hilbert–Schmidt as an operator in $\mathscr{H}$ then T_λ is Hilbert–Schmidt as an operator in $L^2(\mathbb{S}^2)$, for almost all λ in the support of ϕ. In particular, if $(S-1)P(\phi)$ is Hilbert–Schmidt for *all* $\phi \in C_0(0, \infty)$ then T_λ is Hilbert–Schmidt for almost all $\lambda > 0$. We can then infer the finiteness of total cross-sections at almost all positive energies.

How then may we estimate $\|(S-1)P(\phi)\|_2$? We let $V(\boldsymbol{r})$ be a short-range potential, with wave operators $\Omega_\pm$. Since Ω_+ is isometric, we have $\Omega_+^* \Omega_+ = 1$, and $\|\Omega_+^*\| = 1$, so that

$$\begin{aligned}\|(S-1)P(\phi)\|_2 &= \|(\Omega_+^*\Omega_- - 1)P(\phi)\|_2 \\ &= \|\Omega_+^*(\Omega_+ - \Omega_-)P(\phi)\|_2 \\ &\leq \|(\Omega_+ - \Omega_-)P(\phi)\|_2. \end{aligned} \tag{9.3.4}$$

Let us assume that V is bounded locally away from $\boldsymbol{r} = 0$, but make no special assumption concerning the behaviour of V in the neighbourhood of the origin. We introduce the operator ρ of multiplication by a smooth function $\rho(|\boldsymbol{r}|)$, where ρ is monotonic non-decreasing, with $\rho \equiv 0$ near to $\boldsymbol{r} = \boldsymbol{0}$, and $\rho \equiv 1$ for $|\boldsymbol{r}|$ sufficiently large. Then

$$\Omega_\pm = \operatorname*{s-lim}_{t \to \pm\infty} e^{iHt} \rho \, e^{-iH_0 t}$$

where

$$\frac{d}{dt} e^{iHt} \rho e^{-iH_0 t} = i \, e^{iHt} \{ -\Delta\rho - 2i \nabla\rho \cdot \boldsymbol{P} + V\rho \} e^{-iH_0 t}.$$

We can now write $\Omega_+ - \Omega_- = \int_{-\infty}^{\infty} (d/dt) \, e^{iHt} \rho e^{-iH_0 t} \, dt$, and (9.3.4) implies that

$$\begin{aligned}\|(S-1)P(\phi)\|_2 \leq \int_{-\infty}^{\infty} dt \, \{ &\|(V\rho - \Delta\rho)e^{-iH_0 t} P(\phi)\|_2 \\ &+ 2\|\nabla\rho \cdot \boldsymbol{P} e^{-iH_0 t} P(\phi)\|_2 \}. \end{aligned} \tag{9.3.5}$$

The bound (9.3.5) gives rise to an estimate for $\|(S-1)P(\phi)\|_2$ in terms of Hilbert–Schmidt norms of operators involving only V and the free evolution, to be evaluated on the range of the projection $P(\phi)$. Note, on the right-hand side of (9.3.5), that multiplication by the smooth cutoff function ρ effectively removes any possible singularities of the potential. It is a consequence of this that local singularities of the potential play no role in determining whether the total cross-section is finite or infinite. What matters is the behaviour of $V(\boldsymbol{r})$ at large distances.

In order to estimate the right-hand side of (9.3.5), we need to obtain bounds on integrals of the form

$$\int_{-\infty}^{\infty} dt \|q e^{-iH_0 t} P(\phi)\|_2$$

and

$$\int_{-\infty}^{\infty} dt \|q P_i e^{-iH_0 t} P(\phi)\|_2,$$

where q denotes the operator of multiplication by some real function $q(\mathbf{r})$. To obtain such bounds, we first of all consider the evaluation of related integrals in which $\|\cdot\|_2$ is replaced by $(\|\cdot\|_2)^2$.

Lemma 9.1

We have the following identities:

$$\int_{-\infty}^{\infty} dt\,(\|q e^{-iH_0 t} P(\phi)\|_2)^2 = (2\pi)^{-1} \|\lambda^{1/4} \phi\|^2 \, \|q\|^2, \tag{9.3.6}$$

$$\int_{-\infty}^{\infty} dt\,(\|q P_i e^{-iH_0 t} P(\phi)\|_2)^2 = (6\pi)^{-1} \|\lambda^{3/4} \phi\|^2 \, \|q\|^2, \tag{9.3.7}$$

$$\int_{-\infty}^{\infty} dt\, t^2 (\|q e^{-iH_0 t} P(\phi)\|_2)^2 = (2\pi)^{-1} \|(\lambda^{1/4} \phi)'\|^2 \|q\|^2 + (24\pi)^{-1} \|\lambda^{-1/4} \phi\|^2 \, \| |\mathbf{r}| q\|^2, \tag{9.3.8}$$

$$\int_{-\infty}^{\infty} dt\, t^2 (\|q P_i e^{-iH_0 t} P(\phi)\|_2)^2 = (6\pi)^{-1} \|(\lambda^{3/4} \phi)'\|^2 \, \|q\|^2 + (120\pi)^{-1} \|\lambda^{1/4} \phi\|^2 \| |\mathbf{r}| q\|^2 + (60\pi)^{-1} \|\lambda^{1/4} \phi\|^2 \, \|X_i q\|^2. \tag{9.3.9}$$

Proof

In these identities, $\|\lambda^{1/4}\phi\|$ stands for the norm in $L^2(0,\infty)$ of the function $\lambda^{1/4}\phi(\lambda)$, $\|q\|$ is the norm in $L^2(\mathbb{R}^3)$ of $q(\boldsymbol{r})$ and $(\lambda^{1/4}\phi)'$ denotes $(\mathrm{d}/\mathrm{d}\lambda)$ $(\lambda^{1/4}\phi(\lambda))$; X_i is the ith component of the position operator.

We consider the first of the four identities. The adjoint of the operator in the integrand is the operator $P(\phi)\mathrm{e}^{\mathrm{i}H_0 t}q$. Using the convolution property of Fourier transforms, we can evaluate $(\widehat{qf})(\boldsymbol{k})$ for any $f\in L^2(\mathbb{R}^3)$; thus

$$(\widehat{qf})(\boldsymbol{k})=(2\pi)^{-3/2}\int \hat{q}(\boldsymbol{k}-\boldsymbol{k}')\hat{f}(\boldsymbol{k}')\,\mathrm{d}^3\boldsymbol{k}'. \tag{9.3.10}$$

Equation (9.3.10) is an expression of the fact that q may be represented as an integral operator in momentum space, with kernel $(2\pi)^{-3/2}\hat{q}(\boldsymbol{k}-\boldsymbol{k}')$. Using (5.3.27), qf may be written in the direct integral representation as

$$(\widehat{qf})_\lambda=\frac{1}{\sqrt{2}}(2\pi)^{-3/2}\int \lambda^{1/4}\hat{q}(\lambda^{1/2}\boldsymbol{\omega}-\boldsymbol{k}')\hat{f}(\boldsymbol{k}')\,\mathrm{d}^3\boldsymbol{k}'.$$

Hence

$$(P(\phi)\mathrm{e}^{\mathrm{i}H_0 t}qf)_\lambda=\frac{1}{\sqrt{2}}(2\pi)^{-3/2}\phi(\lambda)$$

$$\int_0^\infty \mathrm{d}\mu\times\int \mathrm{d}^3\boldsymbol{k}'\,\mathrm{e}^{\mathrm{i}\mu t}\mu^{1/4}\phi(\mu)\hat{q}(\mu^{1/2}\boldsymbol{\omega}-\boldsymbol{k}')\hat{f}(\boldsymbol{k}'). \tag{9.3.11}$$

The right-hand side of (9.3.11) has the form

$$\int \mathrm{d}^3\boldsymbol{k}'\,K(\lambda,\boldsymbol{\omega},\boldsymbol{k}')\hat{f}(\boldsymbol{k}'),$$

for some kernel K. Using the unitary property of the Fourier transform, together with the standard expression for the Hilbert–Schmidt norm of an integral operator, we find

$$(\|P(\phi)\mathrm{e}^{-\mathrm{i}H_0 t}q\|_2)^2=\int \mathrm{d}\lambda\,\mathrm{d}\boldsymbol{\omega}\,\mathrm{d}^3\boldsymbol{k}'|K(\lambda,\boldsymbol{\omega},\boldsymbol{k}')|^2.$$

The Hilbert–Schmidt norm is unaffected by taking adjoints, so that, assuming ϕ to be normalized, we have

$$\int_{-\infty}^{\infty} \mathrm{d}t\,(\|q\mathrm{e}^{-\mathrm{i}H_0 t}P(\phi)\|_2)^2$$

$$= \int_{-\infty}^{\infty} \mathrm{d}t \int \mathrm{d}\omega\, \mathrm{d}^3\boldsymbol{k}' \left| \frac{1}{\sqrt{2}}(2\pi)^{-3/2} \int_0^{\infty} \mathrm{d}\mu\, \mathrm{e}^{\mathrm{i}\mu t}\mu^{1/4}\phi(\mu)\hat{q}(\mu^{1/2}\boldsymbol{\omega} - \boldsymbol{k}')\right|^2. \quad (9.3.12)$$

Since the μ-integration takes the form of a one-dimensional Fourier transform, we may use the unitary property of this transform to carry out the t-integration, obtaining

$$\int_0^{\infty} \mathrm{d}\mu \int \mathrm{d}\omega\, \mathrm{d}^3\boldsymbol{k}'\,(8\pi^2)^{-1}\mu^{1/2}\phi^2(\mu)|\hat{q}(\mu^{1/2}\boldsymbol{\omega} - \boldsymbol{k}')|^2.$$

On changing integration variable from $\boldsymbol{k}'$ to $\boldsymbol{p}' = \boldsymbol{k}' - \mu^{1/2}\boldsymbol{\omega}$, we can carry out the $\boldsymbol{p}'$ integration to obtain

$$\int_0^{\infty} \mathrm{d}\mu \int \mathrm{d}\omega\,(8\pi^2)^{-1}\mu^{1/2}\phi^2(\mu)\|q\|^2,$$

where this time we have used the unitary property of the three-dimensional Fourier transform to write $\|\hat{q}\| = \|q\|$. The integrand is now independent of $\boldsymbol{\omega}$. Since $\int \mathrm{d}\omega = 4\pi$, we arrive finally at (9.3.6).

In (9.3.7), note the presence on the left-hand side of P_i, the ith component of the momentum operator. In momentum space, P_i corresponds to the operator of multiplication by k_i, and in the direct integral representation P_i corresponds to multiplication by $\lambda^{1/2}\omega_i$. We therefore have to evaluate the right-hand side of (9.3.12), with a further factor $\mu^{1/2}\omega_i$ in the integrand of the μ-integral, and here we make use of the fact that $\int \omega_i^2\,\mathrm{d}\omega = \frac{4}{3}\pi$. The effect of these changes is to replace $\phi(\lambda)$ by $\lambda^{1/2}\phi(\lambda)$ in (9.3.6) and divide the resulting expression by 3, from which (9.3.7) follows.

To evaluate the integral in (9.3.8), we may take account of the factor t^2 by writing $(\mathrm{d}/\mathrm{d}\mu)\mathrm{e}^{\mathrm{i}\mu t}$ for $\mathrm{e}^{\mathrm{i}\mu t}$ in (9.3.12). On integrating by parts with respect to μ, the μ-derivative may be transferred to the other factors in the integrand. Note that

$$\frac{\mathrm{d}}{\mathrm{d}\mu}\hat{q}(\mu^{1/2}\boldsymbol{\omega} - \boldsymbol{k}') = -\tfrac{1}{2}\mathrm{i}\mu^{-1/2}\sum_{i=1}^{3}\omega_i\hat{q}_i(\mu^{1/2}\boldsymbol{\omega} - \boldsymbol{k}'),$$

where $\hat{q}_i$ denotes the three-dimensional Fourier transform of the function $q_i(\boldsymbol{r}) = x_i q(\boldsymbol{r})$ ($\boldsymbol{r} = (x_1, x_2, x_3)$). Proceeding as before, we obtain

$$\int_0^\infty \mathrm{d}\mu \int \mathrm{d}\omega\, \mathrm{d}^3\boldsymbol{k}'(8\pi^2)^{-1} |(\mu^{1/4}\phi(\mu))'\hat{q}(\mu^{1/2}\boldsymbol{\omega} - \boldsymbol{k}') - \tfrac{1}{2}\mathrm{i}\mu^{-1/4}\phi(\mu) \sum_{i=1}^{3} \omega_i \hat{q}_i(\mu^{1/2}\boldsymbol{\omega} - \boldsymbol{k}')|^2.$$

Here the ω-integration may be carried out by making use of the result $\int (a + \Sigma_{j=1}^3 b_j\omega_j)^2\, \mathrm{d}\omega = 4\pi(a^2 + \frac{1}{3}b^2)$. for a and $\boldsymbol{b} = (b_1, b_2, b_3)$ independent of $\boldsymbol{\omega}$. Again using the unitary property of the three-dimensional Fourier transform, we arrive at (9.3.8). The final integration (9.3.9) is slightly more complicated, but may be evaluated by the same methods as for the integrals (9.3.7) and (9.3.8). ■

In order to estimate the integrals of the form $\int_{-\infty}^{\infty} \mathrm{d}t \|q\mathrm{e}^{-\mathrm{i}H_0t} P(\phi)\|_2$, we have to *interpolate*, starting from (9.3.6) and (9.3.8), to powers of t in the integrand intermediate between 1 and t^2. We shall not seek precise estimates here, but shall be content to prove convergence of integrals, rather than obtain upper bounds for them. Suppose then that we have $q \in L^2(\mathbb{R}^3)$. Equation (9.3.6) shows that in this case we have

$$\int_{-\infty}^{\infty} \mathrm{d}t\, (\|q\mathrm{e}^{-\mathrm{i}H_0t} P(\phi)\|_2)^2 < \infty,$$

and (9.3.8), with $q(1 + |\boldsymbol{r}|)^{-1}$ for q, shows that

$$\int_{-\infty}^{\infty} \mathrm{d}t\, t^2(\|q(1 + |\boldsymbol{r}|)^{-1}\, \mathrm{e}^{-\mathrm{i}H_0t} P(\phi)\|_2)^2 < \infty,$$

For any orthonormal basis $\{e_i\}$,

$$\begin{aligned}
&(\|q(1 + |\boldsymbol{r}|)^{-1/2}\, \mathrm{e}^{-\mathrm{i}H_0t} P(\phi)\|_2)^2 \\
&= \sum_i \langle q\mathrm{e}^{-\mathrm{i}H_0t} P(\phi)e_i, q(1 + |\boldsymbol{r}|)^{-1}\, \mathrm{e}^{-\mathrm{i}H_0t} P(\phi)e_i\rangle \\
&\leq \sum_i \|q\mathrm{e}^{-\mathrm{i}H_0t} P(\phi)e_i\|\, \|q(1 + |\boldsymbol{r}|)^{-1}\, \mathrm{e}^{-\mathrm{i}H_0t} P(\phi)e_i\| \\
&\leq \left\{\left(\sum_i \|q\mathrm{e}^{-\mathrm{i}H_0t} P(\phi)e_i\|^2\right)\left(\sum_i \|q(1 + |\boldsymbol{r}|)^{-1}\, \mathrm{e}^{-\mathrm{i}H_0t} P(\phi)e_i\|^2\right)\right\}^{1/2} \\
&= \|q\mathrm{e}^{-\mathrm{i}H_0t} P(\phi)\|_2\, \|q(1 + |\boldsymbol{r}|)^{-1}\, \mathrm{e}^{-\mathrm{i}H_0t} P(\phi)\|_2,
\end{aligned}$$

where we have used the Schwarz inequality. Again applying the Schwarz inequality, we have

$$\int_{-\infty}^{\infty} dt\, |t| (\|q(1+|\boldsymbol{r}|)^{-1/2} e^{-iH_0 t} P(\phi)\|_2)^2$$

$$\leq \int_{-\infty}^{\infty} dt\, |t|\, \|q e^{-iH_0 t} P(\phi)\|_2 \, \|q(1+|\boldsymbol{r}|)^{-1} e^{-iH_0 t} P(\phi)\|_2$$

$$\leq \left\{ \int_{-\infty}^{\infty} dt (\|q e^{-iH_0 t} P(\phi)\|_2)^2 \times \int_{-\infty}^{\infty} dt\, t^2 (\|q(1+|\boldsymbol{r}|)^{-1} e^{-iH_0 t} P(\phi)\|_2)^2 \right\}^{1/2}$$

$$< \infty,$$

since we have seen that both integrals converge in the final expression. Roughly, we can say that the factor $|t|$ in the integrand is compensated by a factor $(1+|\boldsymbol{r}|)^{-1/2}$ multiplying q. This result extends in a similar way to the interpolation between powers $|t|$ and t^2, and we obtain

$$\int_{-\infty}^{\infty} dt\, |t|^{3/2} (\|q(1+|\boldsymbol{r}|)^{-3/4} e^{-iH_0 t} P(\phi)\|_2)^2 < \infty.$$

We can continue by interpolating between powers $|t|$ and $|t|^{3/2}$, $|t|$ and $|t|^{5/4}, \ldots$, obtaining, for $\delta = (\frac{1}{2})^n$ arbitrarily small and positive

$$\int_{-\infty}^{\infty} dt\, |t|^{1+2\delta} (\|q(1+|\boldsymbol{r}|)^{-(1/2+\delta)} e^{-iH_0 t} P(\phi)\|_2)^2 < \infty \qquad (9.3.13)$$

provided that $q \in L^2(\mathbb{R}^3)$. Equation (9.3.6), with q replaced by $q(1+|\boldsymbol{r}|)^{-(1/2+\delta)}$, shows that the same integral converges if the factor $|t|^{1+2\delta}$ is removed from the integrand. Hence we also have

$$\int_{-\infty}^{\infty} dt\, (1+|t|)^{1+2\delta} (\|q(1+|\boldsymbol{r}|)^{-(1/2+\delta)} e^{-iH_0 t} P(\phi)\|_2)^2 < \infty. \qquad (9.3.13)'$$

for $q \in L^2(\mathbb{R}^3)$ and $\delta > 0$ arbitrarily small. This implies that

$$\int_{-\infty}^{\infty} \mathrm{d}t \, \|q(1+|\boldsymbol{r}|)^{-(1/2+\delta)} \mathrm{e}^{-\mathrm{i}H_0 t} P(\phi)\|_2$$

$$= \int_{-\infty}^{\infty} \mathrm{d}t \, \{(1+|t|)^{-(1+2\delta)/2}\}\{(1+|t|)^{(1+2\delta)/2}$$

$$\times \|q(1+|\boldsymbol{r}|)^{-(1/2+\delta)} \mathrm{e}^{-\mathrm{i}H_0 t} P(\phi)\|_2\}$$

$$< \infty, \tag{9.3.14}$$

by the Schwarz inequality, using (9.3.13)′ and the convergence of the integral $\int_{-\infty}^{\infty} \mathrm{d}t \, (1+|t|)^{-(1+2\delta)}$. A similar result follows if we start from (9.3.7) and (9.3.9). Here we shall only need to use the fact that

$$\int_{-\infty}^{\infty} \mathrm{d}t \, \|qP_i \mathrm{e}^{-\mathrm{i}H_0 t} P(\phi)\|_2 < \infty \tag{9.3.15}$$

for any $q \in C_0^{\infty}(\mathbb{R}^3)$. The conclusions from the results (9.3.14) and (9.3.15) follow in the next theorem.

Theorem 9.3

Let $V(\boldsymbol{r})$ be any (short-range) potential such that, for some $\delta > 0$, $|\boldsymbol{r}|^{1/2+\delta} V(\boldsymbol{r})$ is square-integrable over the region $|\boldsymbol{r}| > R$, for any $R > 0$. Then, at almost all positive energies λ, the averaged total cross-section $\sigma_{\text{ave}}(\lambda)$ is finite; moreover, the total cross-section $\sigma(\lambda, \boldsymbol{\omega}')$ for scattering from incident-beam direction $\boldsymbol{\omega}'$ is also finite for almost all directions $\boldsymbol{\omega}'$.

Proof

Note that both $\nabla\rho$ and $\Delta\rho$ in (9.3.5) have compact support. From (9.3.14), with

$$q = (1+|\boldsymbol{r}|)^{(1/2+\delta)}(V\rho - \Delta\rho) \in L^2(\mathbb{R}^3),$$

and (9.3.15), with $q = \partial\rho/\partial x_i \in C_0^{\infty}(\mathbb{R}^3)$, we find that the right-hand side of (9.3.5) is a convergent integral. This proves that $(S-1)P(\phi)$ is Hilbert–Schmidt for any $\phi \in C_0^{\infty}(0, \infty)$. Our previous discussion, following (9.3.3), allows us to conclude that T_λ is Hilbert–Schmidt for almost all λ, and the finiteness of $\sigma_{ave}(\lambda)$ and $\sigma(\lambda, \boldsymbol{\omega}')$ then follow. ■

Theorem 9.3 implies finiteness of total cross-section for scattering by any potential, whether singular or non-singular, decaying at infinity more rapidly that $|\boldsymbol{r}|^{-(2+\delta)}$ for some $\delta>0$. Since the total cross-section may be explicitly shown to be infinite in the case $V(\boldsymbol{r})=|\boldsymbol{r}|^{-2}$, this result is close to the best possible. A particular case of interest is that in which $V=V(|\boldsymbol{r}|)$ is a radial potential. Then, by spherical symmetry, it is not difficult to show that the scattering amplitude is rotation-invariant. It follows that the dependence of $f(\lambda;\boldsymbol{\omega}',\boldsymbol{\omega})$ on $\boldsymbol{\omega}$ and $\boldsymbol{\omega}'$ can be only on the relative directions of $\boldsymbol{\omega}$ and $\boldsymbol{\omega}'$; that is, on $\boldsymbol{\omega}\cdot\boldsymbol{\omega}'$. Since $\sigma(\lambda,\boldsymbol{\omega}')$ is obtained from the transition probability $|f|^2$ by integration over $\boldsymbol{\omega}$, this integral must be independent of $\boldsymbol{\omega}'$. We then have $\sigma(\lambda,\boldsymbol{\omega}')=\sigma_{\text{ave}}(\lambda)$, which is finite at almost all energies.

It is possible to go further and use Lemma 9.1 to obtain precise upper bounds on the integral in (9.3.14). In this way, one arrives at an upper estimate for $\sigma_{\text{ave}}(\lambda)$, or rather for the averaged value of this cross-section, over a range of energies, with weight function $\phi^2(\lambda)$. By suitable choice of $\phi(\lambda)$, we can convert this estimate into a precise bound for the cross-section averaged (without weight function) over a finite range of energies. We can then study the asymptotic behaviour of cross-sections at high energies. In particular, for a potential of compact support within the sphere $|\boldsymbol{r}|<R$, a bound of order R^2 is obtained, in agreement with the result that we might anticipate by analogy with the cross-section for potential scattering in classical mechanics.

What can be said about cross-sections for potentials that decay at infinity less rapidly than $|\boldsymbol{r}|^{-2}$? We have already mentioned that in such cases the kernel $T(\lambda,\boldsymbol{\omega},\boldsymbol{\omega}')$ of the operator T_λ may be singular in the forward direction $\boldsymbol{\omega}=\boldsymbol{\omega}'$, and the nature of the singularity may be sufficient to prevent T_λ from being Hilbert–Schmidt. In this case, instead of considering the operator $T=S-1$, we can consider the commutator $[T,\boldsymbol{P}]$ of T with the components of the momentum operator $\boldsymbol{P}$. In the direct integral representation, $[T,\boldsymbol{P}]_\lambda$ will be implemented by a linear operator in $L^2(\mathbb{S}^2)$, at each energy λ, with kernel $\lambda^{1/2}T(\lambda,\boldsymbol{\omega},\boldsymbol{\omega}')\,(\boldsymbol{\omega}'-\boldsymbol{\omega})$. Thus $[T,\boldsymbol{P}]_\lambda$ may be expected to be less singular than T_λ at $\boldsymbol{\omega}=\boldsymbol{\omega}'$. If $V(\boldsymbol{r})$ is a short-range potential such that $|V(\boldsymbol{r})|=O(|\boldsymbol{r}|^{-(1+\delta)})$ as $|\boldsymbol{r}|\to\infty$ for some $\delta>0$ then one can show that $[T,\boldsymbol{P}]_\lambda$ is Hilbert–Schmidt for almost all λ. (Some differentiability of the potential is also required. In taking commutators with $\boldsymbol{P}=-\mathrm{i}\boldsymbol{\nabla}$, we need to have $|\operatorname{grad}V|=O(|\boldsymbol{r}|^{-(2+\delta)})$ as $|\boldsymbol{r}|\to\infty$.) Since we then have $\int d\omega\, d\omega'\, |\boldsymbol{\omega}-\boldsymbol{\omega}'|^2|T(\lambda,\boldsymbol{\omega},\boldsymbol{\omega}')|^2<\infty$, it follows that T_λ is square integrable with respect to $\boldsymbol{\omega}$ and $\boldsymbol{\omega}'$ over any region of $\mathbb{S}^2\times\mathbb{S}^2$ in which $|\boldsymbol{\omega}-\boldsymbol{\omega}'|\geq\text{const}>0$. In particular, the cross-section $\sigma(\lambda;\,\boldsymbol{\omega}'\to\mathscr{C})$ is finite at almost all energies and for almost all incident-beam directions $\boldsymbol{\omega}'$, provided that $\boldsymbol{\omega}'$ is not contained in the closure of $\mathscr{C}$. That is, cross-sections for short-range potentials are finite, provided that we keep away from the forward direction, and the differential cross-section $(\mathrm{d}\sigma/\mathrm{d}\omega)(\lambda;\boldsymbol{\omega}'\to\boldsymbol{\omega})$ exists for almost all $\lambda,\boldsymbol{\omega}$ and $\boldsymbol{\omega}'$. We shall not give proofs of these results here, nor of the extension to long-range potentials.

9.4 THE LIPPMANN–SCHWINGER EQUATION

In Chapter 4 we obtained formulae for the wave and scattering operators as spectral integrals. These formulae hold under very general conditions, for a pair of self-adjoint operators H_0,H for which the corresponding (unmodified) wave operators exist. In particular, the wave operator Ω_+ is given (4.2.17) by

$$\Omega_+ f = f - \operatorname*{s\text{-}lim}_{\varepsilon\to 0+}\int (H-\lambda+\mathrm{i}\varepsilon)^{-1} V\,\mathrm{d}E_\lambda f, \tag{9.4.1}$$

where f is an arbitrary vector belonging to $D(H_0)\equiv D(H)$. How does this result apply in the case of scattering of a particle in three dimensions by a short-range potential $V(\boldsymbol{r})$, with $H_0=-\Delta$ and $H=-\Delta+V(\boldsymbol{r})$? To see how (9.4.1) applies in this case, we first consider the operation of Ω_+ on the wave function $f(\boldsymbol{r})=\mathrm{e}^{\mathrm{i}\boldsymbol{k}\cdot\boldsymbol{r}}$, for some fixed $\boldsymbol{k}$. The function $\mathrm{e}^{\mathrm{i}\boldsymbol{k}\cdot\boldsymbol{r}}$ is a generalized, non-normalizable, "eigenfunction" of the momentum operator $\boldsymbol{P}$, representing a plane wave of momentum $\boldsymbol{k}$. We cannot really set $f=\mathrm{e}^{\mathrm{i}\boldsymbol{k}\cdot\boldsymbol{r}}$ in (9.4.1), since we have no reason to suppose that an expression such as $\Omega_+\mathrm{e}^{\mathrm{i}\boldsymbol{k}\cdot\boldsymbol{r}}$ makes sense. However, by considering formally what happens if we substitute a plane wave into (9.4.1), we shall see more clearly what results to aim at by a rigorous analysis. Since $\mathrm{e}^{\mathrm{i}\boldsymbol{k}\cdot\boldsymbol{r}}$ is also a non-normalizable "eigenfunction" of $H_0=\boldsymbol{P}^2$, with "eigenvalue" $\boldsymbol{k}^2$, we expect the entire contribution to the integral on the right-hand side of (9.4.1) to come from the single point $\lambda=\boldsymbol{k}^2$. In this case we can formally set $\lambda=\boldsymbol{k}^2$ in the integrand to obtain

$$\Omega_+\mathrm{e}^{\mathrm{i}\boldsymbol{k}\cdot\boldsymbol{r}} = \mathrm{e}^{\mathrm{i}\boldsymbol{k}\cdot\boldsymbol{r}} - \lim_{\varepsilon\to 0+}\int (H-\boldsymbol{k}^2+\mathrm{i}\varepsilon)^{-1} V\,\mathrm{d}E_\lambda\,\mathrm{e}^{\mathrm{i}\boldsymbol{k}\cdot\boldsymbol{r}}.$$

Since the integrand is now independent of λ, the integral can be carried out to yield the formal identity

$$\Omega_+\mathrm{e}^{\mathrm{i}\boldsymbol{k}\cdot\boldsymbol{r}} = \mathrm{e}^{\mathrm{i}\boldsymbol{k}\cdot\boldsymbol{r}} - \lim_{\varepsilon\to 0+}(H-\boldsymbol{k}^2+\mathrm{i}\varepsilon)^{-1} V\mathrm{e}^{\mathrm{i}\boldsymbol{k}\cdot\boldsymbol{r}}.$$

Of course, such manipulations can hardly be used to establish a rigorous result, since we require $f\in L^2(\mathbb{R}^3)$ in (9.4.1). However, through a three-dimensional Fourier transform, we can write any $f\in L^2(\mathbb{R}^3)$ as an integral of plane waves; thus $f(\boldsymbol{r})=(2\pi)^{-3/2}\int \mathrm{e}^{\mathrm{i}\boldsymbol{k}\cdot\boldsymbol{r}}\hat{f}(\boldsymbol{k})\,\mathrm{d}^3\boldsymbol{k}$. Applying this result to our formal identity leads to

$$\Omega_+ f = f - \left(\frac{1}{2\pi}\right)^{3/2} \lim_{\varepsilon \to 0+} \int \{(H - \boldsymbol{k}^2 + \mathrm{i}\varepsilon)^{-1} V \mathrm{e}^{\mathrm{i}\boldsymbol{k}\cdot\boldsymbol{r}}\} \hat{f}(\boldsymbol{k})\, \mathrm{d}^3\boldsymbol{k}.$$

The expression within the curly brackets may be thought of as a function of $\boldsymbol{r}$ and $\boldsymbol{k}$, representing part of the kernel of an integral operator that takes the wave function $\hat{f}(\boldsymbol{k})$ of a state in momentum space into the function $(\Omega_+ f)(\boldsymbol{r})$ that describes the state $\Omega_+ f$ in position space. The following theorem gives precise mathematical expression to this result, to which we have been led by heuristic arguments. We shall assume rather rapid decay of the potential at infinity. Such restrictions do not give the best possible result, and can be considerably weakened. However, by assuming more rapid decay than is actually needed, we are able to bypass a number of technical difficulties in the proof that would otherwise arise.

Theorem 9.4

Let $V(\boldsymbol{r})$ be a short-range potential satisfying $|V(\boldsymbol{r})| \leq \mathrm{const}/|\boldsymbol{r}|$ near $\boldsymbol{r}=0$ and

$$|V(\boldsymbol{r})| \leq \frac{\mathrm{const}}{|\boldsymbol{r}|^{3+\delta}} \quad \text{as } |\boldsymbol{r}| \to \infty \tag{9.4.2}$$

for some $\delta > 0$. Define a function $\psi_+(\boldsymbol{r},\boldsymbol{k})$ for $\boldsymbol{k} \neq 0$ by

$$\psi_+(\boldsymbol{r},\boldsymbol{k}) = \mathrm{e}^{\mathrm{i}\boldsymbol{k}\cdot\boldsymbol{r}} - \lim_{\varepsilon \to 0+} (H - \boldsymbol{k}^2 + \mathrm{i}\varepsilon)^{-1} V \mathrm{e}^{\mathrm{i}\boldsymbol{k}\cdot\boldsymbol{r}}. \tag{9.4.3}$$

On the right-hand side $V\mathrm{e}^{\mathrm{i}\boldsymbol{k}\cdot\boldsymbol{r}}$ denotes the element of $L^2(\mathbb{R}^3)$ (position space) with wave function $V(\boldsymbol{r})\mathrm{e}^{\mathrm{i}\boldsymbol{k}\cdot\boldsymbol{r}}$, for given fixed $\boldsymbol{k}$. The limit on the right-hand side is to be interpreted in the sense that

$$\operatorname*{s-lim}_{\varepsilon \to 0+} \frac{1}{1+|\boldsymbol{r}|} (H - \boldsymbol{k}^2 + \mathrm{i}\varepsilon)^{-1} V \mathrm{e}^{\mathrm{i}\boldsymbol{k}\cdot\boldsymbol{r}} = \frac{1}{1+|\boldsymbol{r}|} (\mathrm{e}^{\mathrm{i}\boldsymbol{k}\cdot\boldsymbol{r}} - \psi_+(\boldsymbol{r},\boldsymbol{k})). \tag{9.4.4}$$

Then

(i) $\psi_+(\boldsymbol{r},\boldsymbol{k})$, for fixed $\boldsymbol{k}$, is a solution of the three-dimensional time-independent Schrödinger equation at energy $\boldsymbol{k}^2$.

$$-\Delta\psi_+(\boldsymbol{r},\boldsymbol{k}) + V(\boldsymbol{r})\psi_+(\boldsymbol{r},\boldsymbol{k}) = \boldsymbol{k}^2\psi_+(\boldsymbol{r},\boldsymbol{k}), \tag{9.4.5}$$

subject to the boundary condition at infinity

$$\psi_+(\boldsymbol{r},\boldsymbol{k}) - \mathrm{e}^{\mathrm{i}\boldsymbol{k}\cdot\boldsymbol{r}} \sim \frac{\mathrm{e}^{-\mathrm{i}kr}}{r} f_+(\boldsymbol{\omega},\boldsymbol{k}) \tag{9.4.6}$$

for some $f_+(\boldsymbol{\omega},\boldsymbol{k})$ with $\boldsymbol{\omega} = \boldsymbol{r}/r$;

(ii) $\psi_+(\boldsymbol{r},\boldsymbol{k})$, for fixed $\boldsymbol{k}$, is a solution of the integral equation

$$\psi_+(\boldsymbol{r},\boldsymbol{k})=\mathrm{e}^{\mathrm{i}\boldsymbol{k}\cdot\boldsymbol{r}}-\frac{1}{4\pi}\int\frac{\mathrm{e}^{-\mathrm{i}k|\boldsymbol{r}-\boldsymbol{r}'|}}{|\boldsymbol{r}-\boldsymbol{r}'|}V(\boldsymbol{r}')\psi_+(\boldsymbol{r}',\boldsymbol{k})\,\mathrm{d}^3\boldsymbol{r}'. \qquad (9.4.7)$$

(iii) $(2\pi)^{-3/2}\psi_+(\boldsymbol{r},\boldsymbol{k})$ is a kernel for the wave operator Ω_+, represented as an integral operator from momentum space to position space; thus

$$(\Omega_+ f)(\boldsymbol{r})=\left(\frac{1}{2\pi}\right)^{3/2}\int\psi_+(\boldsymbol{r},\boldsymbol{k})\hat{f}(\boldsymbol{k})\,\mathrm{d}^3\boldsymbol{k} \qquad (9.4.8)$$

for all $\hat{f}\in C_0^\infty(\mathbb{R}^3)$.

Proof

Under the stated hypotheses, we have $V\in L^2(\mathbb{R}^3)$, so that $V(\boldsymbol{r})\mathrm{e}^{\mathrm{i}\boldsymbol{k}\cdot\boldsymbol{r}}$ is L^2 as a function of $\boldsymbol{r}$. Since the resolvent operator $R(z)=(H-z)^{-1}$ is bounded for $\mathrm{Im}\,(z)\neq 0$, $(H-\boldsymbol{k}^2+\mathrm{i}\varepsilon)^{-1}V\mathrm{e}^{\mathrm{i}\boldsymbol{k}\cdot\boldsymbol{r}}$ makes sense as an element of $L^2(\mathbb{R}^3)$, for fixed $\varepsilon>0$. In Section 12.2 we shall show that, for any short-range potential less singular at the origin than const/$|\boldsymbol{r}|$, and certainly for a potential satisfying the conditions of the present theorem, the limit $\lim_{\varepsilon\to 0+}$ $|\boldsymbol{r}|^{-1}(H-\lambda+\mathrm{i}\varepsilon)^{-1}|\boldsymbol{r}|^{-1}$ exists for $\lambda>0$ as a limit in the operator-norm topology. Since $(1+|\boldsymbol{r}|)^{-1}|\boldsymbol{r}|$ is bounded, this implies that s-$\lim_{\varepsilon\to 0+}$ $(1+|\boldsymbol{r}|)^{-1}$ $(H-\lambda+\mathrm{i}\varepsilon)^{-1}|\boldsymbol{r}|^{-1}g$ exists for arbitrary $g\in L^2(\mathbb{R}^3)$, and indeed that the limit is uniform for all $g\in\mathscr{H}$ such that $\|g\|=1$. Under the hypotheses of the theorem, we have $|\boldsymbol{r}|V(\boldsymbol{r})\mathrm{e}^{\mathrm{i}\boldsymbol{k}\cdot\boldsymbol{r}}\in L^2(\mathbb{R}^3)$ as a function of $\boldsymbol{r}$, implying the existence of the strong limit on the left-hand side of (9.4.4). Defining ψ_+ in this way, we have verified (9.4.3), provided that the limit is correctly interpreted.

To prove (i), we consider $\phi\in C_0^\infty(\mathbb{R}^3)$, and let ψ_ε denote the right-hand side of (9.4.3) before the limit is taken. Although ψ_ε will not in general belong to $L^2(\mathbb{R}^3)$, we know that ψ_ε is locally integrable, so that the inner product $\langle(H-\boldsymbol{k}^2-\mathrm{i}\varepsilon)\phi,\psi_\varepsilon\rangle$ is well defined. This inner product may be written as

$$\begin{aligned}&\langle(H_0-\boldsymbol{k}^2-\mathrm{i}\varepsilon)\phi,\mathrm{e}^{\mathrm{i}\boldsymbol{k}\cdot\boldsymbol{r}}\rangle+\langle V\phi,\mathrm{e}^{\mathrm{i}\boldsymbol{k}\cdot\boldsymbol{r}}\rangle\\&\qquad-\langle(H-\boldsymbol{k}^2-\mathrm{i}\varepsilon)\phi,\ (H-\boldsymbol{k}^2+\mathrm{i}\varepsilon)^{-1}V\mathrm{e}^{\mathrm{i}\boldsymbol{k}\cdot\boldsymbol{r}}\rangle\\&\quad=\langle(H_0-\boldsymbol{k}^2-\mathrm{i}\varepsilon)\phi,\mathrm{e}^{\mathrm{i}\boldsymbol{k}\cdot\boldsymbol{r}}\rangle,\end{aligned}$$

since the last inner product is just $\langle\phi,V\mathrm{e}^{\mathrm{i}\boldsymbol{k}\cdot\boldsymbol{r}}\rangle$. Moreover, $(\Delta+\boldsymbol{k}^2)\mathrm{e}^{\mathrm{i}\boldsymbol{k}\cdot\boldsymbol{r}}=0$, so that $\langle(H_0-\boldsymbol{k}^2)\phi,\ \mathrm{e}^{\mathrm{i}\boldsymbol{k}\cdot\boldsymbol{r}}\rangle=0$ on using the divergence theorem to transfer derivatives from ϕ to the plane-wave function. We then have

$$\langle(H-\boldsymbol{k}^2-\mathrm{i}\varepsilon)\phi,\psi_\varepsilon\rangle=\mathrm{i}\varepsilon\langle\phi,\mathrm{e}^{\mathrm{i}\boldsymbol{k}\cdot\boldsymbol{r}}\rangle \qquad (9.4.9)$$

The left-hand side may be written as $\langle(1+|\boldsymbol{r}|)(H-\boldsymbol{k}^2-\mathrm{i}\varepsilon)\phi,\ (1+|\boldsymbol{r}|)^{-1}\psi_\varepsilon\rangle$, and we take the limit $\varepsilon\to 0$, using (9.4.4), to give

$$\begin{aligned}&\langle(1+|\boldsymbol{r}|)(H-\boldsymbol{k}^2)\phi,\ (1+|\boldsymbol{r}|)^{-1}\psi_+\rangle\\&\quad=\langle(H-\boldsymbol{k}^2)\phi,\psi_+\rangle=0\quad\text{for all }\phi\in C_0^\infty(\mathbb{R}^3).\end{aligned}$$

Thus $\psi_+(\boldsymbol{r},\boldsymbol{k})$ satisfies $(-\Delta+V-\boldsymbol{k}^2)\psi_+=0$ in the sense of distributions, and we have seen already that this is sufficient for (9.4.5) to hold. We defer the proof of the boundary condition on ψ_+ until after that of (ii).

To prove (ii), we define ψ_ε as before, so that

$$\begin{aligned}&(H_0-\boldsymbol{k}^2+\mathrm{i}\varepsilon)^{-1}V\psi_\varepsilon\\&\quad=(H_0-\boldsymbol{k}^2+\mathrm{i}\varepsilon)^{-1}V\mathrm{e}^{\mathrm{i}\boldsymbol{k}\cdot\boldsymbol{r}}-(H_0-\boldsymbol{k}^2+\mathrm{i}\varepsilon)^{-1}V(H-\boldsymbol{k}^2+\mathrm{i}\varepsilon)^{-1}V\mathrm{e}^{\mathrm{i}\boldsymbol{k}\cdot\boldsymbol{r}}\end{aligned}$$

However,

$$\begin{aligned}&(H_0-\boldsymbol{k}^2+\mathrm{i}\varepsilon)^{-1}V(H-\boldsymbol{k}^2+\mathrm{i}\varepsilon)^{-1}\\&\quad=(H_0-\boldsymbol{k}^2+\mathrm{i}\varepsilon)^{-1}[(H-\boldsymbol{k}^2+\mathrm{i}\varepsilon)-(H_0-\boldsymbol{k}^2+\mathrm{i}\varepsilon)](H-\boldsymbol{k}^2+\mathrm{i}\varepsilon)^{-1}\\&\quad=(H_0-\boldsymbol{k}^2+\mathrm{i}\varepsilon)^{-1}-(H-\boldsymbol{k}^2+\mathrm{i}\varepsilon)^{-1},\end{aligned}$$

so that the above expression simplifies to

$$\begin{aligned}(H_0-\boldsymbol{k}^2+\mathrm{i}\varepsilon)^{-1}V\psi_\varepsilon&=(H-\boldsymbol{k}^2+\mathrm{i}\varepsilon)^{-1}V\mathrm{e}^{\mathrm{i}\boldsymbol{k}\cdot\boldsymbol{r}}\\&=\mathrm{e}^{\mathrm{i}\boldsymbol{k}\cdot\boldsymbol{r}}-\psi_\varepsilon,\end{aligned}$$

giving

$$\psi_\varepsilon=\mathrm{e}^{\mathrm{i}\boldsymbol{k}\cdot\boldsymbol{r}}-(H_0-\boldsymbol{k}^2+\mathrm{i}\varepsilon)^{-1}V\psi_\varepsilon. \tag{9.4.10}$$

We now multiply both sides by $(1+|\boldsymbol{r}|)^{-1}$ and proceed to the limit $\varepsilon\to 0$. We know in this case (see Section 12.2) that $|\boldsymbol{r}|^{-1}(H_0-\boldsymbol{k}^2+\mathrm{i}\varepsilon)^{-1}|\boldsymbol{r}|^{-1}$ converges to a limit in operator norm. The limit of $(1+|\boldsymbol{r}/)^{-1}\ (H_0-\boldsymbol{k}^2+\mathrm{i}\varepsilon)^{-1}V\psi_\varepsilon$ may be written as

$$\left(\lim_{\varepsilon\to 0+}(1+|\boldsymbol{r}|)^{-1}(H_0-\boldsymbol{k}^2+\mathrm{i}\varepsilon)^{-1}|\boldsymbol{r}|^{-1}\right)\left(\operatorname*{s-lim}_{\varepsilon\to 0+}|\boldsymbol{r}|V\psi_\varepsilon\right).$$

Since $|\boldsymbol{r}|(1+|\boldsymbol{r}|)V(\boldsymbol{r})$ is bounded and $(1+|\boldsymbol{r}|)^{-1}\ (\psi_\varepsilon-\mathrm{e}^{\mathrm{i}\boldsymbol{k}\cdot\boldsymbol{r}})$ converges strongly to $(1+|\boldsymbol{r}|)^{-1}\ (\psi_+-\mathrm{e}^{\mathrm{i}\boldsymbol{k}\cdot\boldsymbol{r}})$, we obtain finally, from (9.4.10),

$$\begin{aligned}(1+|\boldsymbol{r}|)^{-1}\psi_+&=(1+|\boldsymbol{r}|)^{-1}\mathrm{e}^{\mathrm{i}\boldsymbol{k}\cdot\boldsymbol{r}}\\&\quad-\operatorname*{s-lim}_{\varepsilon\to 0+}(1+|\boldsymbol{r}|)^{-1}(H_0-\boldsymbol{k}^2+\mathrm{i}\varepsilon)^{-1}V\psi_+.\end{aligned}\tag{9.4.11}$$

We can use (5.3.24)′ to write down an integral kernel for the resolvent operator of H_0. In this case,

$$((H_0-\boldsymbol{k}^2+\mathrm{i}\varepsilon)^{-1}V\psi_+)(\boldsymbol{r})=\frac{1}{4\pi}\int\frac{\mathrm{e}^{-K|\boldsymbol{r}-\boldsymbol{r}'|}}{|\boldsymbol{r}-\boldsymbol{r}'|}V(\boldsymbol{r}')\psi_+(\boldsymbol{r}',\boldsymbol{k})\,\mathrm{d}^3\boldsymbol{r}',$$

where $K^2=-\boldsymbol{k}^2+\mathrm{i}\varepsilon$ and K has positive real part. In the limit $\varepsilon\to 0+$, K converges to $\mathrm{i}k$, where $k=|\boldsymbol{k}|$. Even with $K=\mathrm{i}k$, the integrand is absolutely convergent, and we may use the Lebesgue Dominated-Convergence Theorem to take the limit $\varepsilon\to 0+$. By comparison with (9.4.11), (9.4.7) now follows.

Let us now estimate the integral on the right-hand side of (9.4.7). From the operator-norm convergence of $|\boldsymbol{r}|^{-1}\ (H-\boldsymbol{k}^2+\mathrm{i}\varepsilon)^{-1}|\boldsymbol{r}|^{-1}$ as $\varepsilon\to 0+$, we can replace $(1+|\boldsymbol{r}|)^{-1}$ by $|\boldsymbol{r}|^{-1}$ in (9.4.4) to deduce that $|\boldsymbol{r}|^{-1}(\mathrm{e}^{\mathrm{i}\boldsymbol{k}\cdot\boldsymbol{r}}-\psi_+(\boldsymbol{r},\boldsymbol{k}))\in L^2(\mathbb{R}^3)$ as a function of $\boldsymbol{r}$. Since $|V(\boldsymbol{r})|\le\text{const}\,|\boldsymbol{r}|^{-1}(1+|\boldsymbol{r}|)^{-(2+\delta)}$, it follows that we can write

$$|V(\boldsymbol{r})\psi_+(\boldsymbol{r},\boldsymbol{k})|\le\text{const}\,|\boldsymbol{r}|^{-1}(1+|\boldsymbol{r}|)^{-(2+\delta)} \\ +\text{const}\,(1+|\boldsymbol{r}|)^{-(2+\delta)}h(\boldsymbol{r}),$$

where $h(\boldsymbol{r})\in L^2(\mathbb{R}^3)$. Each of the integrals

$$I_1=\int\frac{\mathrm{d}^3\boldsymbol{r}'}{|\boldsymbol{r}'|(1+|\boldsymbol{r}'|)^{2+\delta}}\,\frac{1}{|\boldsymbol{r}-\boldsymbol{r}'|}$$

and

$$I_2=\int\frac{\mathrm{d}^3\boldsymbol{r}'}{(1+|\boldsymbol{r}'|)^{2+\delta}}\,\frac{h(\boldsymbol{r}')}{|\boldsymbol{r}-\boldsymbol{r}'|}$$

defines a bounded function of $\boldsymbol{r}$ that is bounded by $\text{const}/|\boldsymbol{r}|$ in the limit $|\boldsymbol{r}|\to\infty$. Hence in particular, the integral in (9.4.7) is bounded as a function of $\boldsymbol{r}$, so that $|\psi_+(\boldsymbol{r},\boldsymbol{k})|$ is bounded, for fixed $\boldsymbol{k}$. We subdivide the integration region with respect to $\boldsymbol{r}'$ into two parts. We define the first subregion $\mathscr{R}_1$ by the inequality $|\boldsymbol{r}-\boldsymbol{r}'|<\frac{1}{2}|\boldsymbol{r}|$. Since ψ_+ is bounded, the integral over this subregion is bounded in absolute value by

$$\text{const}\int_{\mathscr{R}_1}\frac{|V(\boldsymbol{r}')|\,\mathrm{d}^3\boldsymbol{r}'}{|\boldsymbol{r}-\boldsymbol{r}'|}.$$

In the limit as $|\boldsymbol{r}|\to\infty$, $|V(\boldsymbol{r}')|=O(|\boldsymbol{r}'|^{-(3+\delta)})=O(|\boldsymbol{r}|^{-(3+\delta)})$ in $\mathscr{R}_1$, since $|\boldsymbol{r}|$ and $|\boldsymbol{r}'|$ are of the same order for $|\boldsymbol{r}-\boldsymbol{r}'|<\frac{1}{2}|\boldsymbol{r}|$. Moreover, $\int_{\mathscr{R}_1}(\mathrm{d}^3\boldsymbol{r}'/|\boldsymbol{r}-\boldsymbol{r}'|)=\frac{1}{2}\pi|\boldsymbol{r}|^2$. Hence the integral over $\mathscr{R}_1$ is of order $|\boldsymbol{r}|^{-(1+\delta)}$ in the limit as $|\boldsymbol{r}|\to\infty$. It follows that

$$\lim_{|r|\to\infty}\int\frac{|\boldsymbol{r}|e^{ikr}e^{-ik|\boldsymbol{r}-\boldsymbol{r}'|}}{|\boldsymbol{r}-\boldsymbol{r}'|}V(\boldsymbol{r}')\psi_+(\boldsymbol{r}',\boldsymbol{k})\,d^3\boldsymbol{r}'$$

$$=\lim_{|r|\to\infty}\int\frac{\chi(\boldsymbol{r}')|\boldsymbol{r}|e^{ikr}e^{-ik|\boldsymbol{r}-\boldsymbol{r}'|}}{|\boldsymbol{r}-\boldsymbol{r}'|}V(\boldsymbol{r}')\psi_+(\boldsymbol{r}',\boldsymbol{k})\,d^3\boldsymbol{r}', \tag{9.4.12}$$

where χ is the characteristic function of the subregion $\mathscr{R}_2$ defined by the inequality $|\boldsymbol{r}-\boldsymbol{r}'|\geq\frac{1}{2}|\boldsymbol{r}|$. In taking the limit $r=|\boldsymbol{r}|\to\infty$, we write $\boldsymbol{r}=r\boldsymbol{\omega}$ and keep $\boldsymbol{\omega}$ fixed. Note that $\chi(\boldsymbol{r}')|\boldsymbol{r}|/|\boldsymbol{r}-\boldsymbol{r}'|\leq 2$, $V\in L^1(\mathbb{R}^3)$, and ψ_+ is bounded. Hence the integrand in (9.4.12) is bounded in absolute value by an integrable function of $\boldsymbol{r}'$ that is independent of $\boldsymbol{r}$. We can therefore apply the Lebesgue Dominated-Convergence Theorem to take the limit $r\to\infty$ inside the integral sign. For fixed $\boldsymbol{r}'$, we have $\lim_{r\to\infty}\chi(\boldsymbol{r}')|\boldsymbol{r}|/|\boldsymbol{r}-\boldsymbol{r}'|=1$, since asymptotically $|\boldsymbol{r}-\boldsymbol{r}'|>\frac{1}{2}|\boldsymbol{r}|$.

Moreover, $r-|\boldsymbol{r}-\boldsymbol{r}'|=r-r(1-2\boldsymbol{r}\cdot\boldsymbol{r}'/r^2+\boldsymbol{r}'^2/r^2)^{1/2}\to\boldsymbol{\omega}\cdot\boldsymbol{r}'$ by the Binomial Theorem. This allows us to take the limit in (9.4.12). Comparing with (9.4.7), where we have taken a factor $e^{ikr}|\boldsymbol{r}|$ into the integrand, we see that (9.4.6) holds, where

$$f_+(\boldsymbol{\omega},\boldsymbol{k})=-\frac{1}{4\pi}\int e^{ik\boldsymbol{\omega}\cdot\boldsymbol{r}'}V(\boldsymbol{r}')\psi_+(\boldsymbol{r}',\boldsymbol{k})\,d^3\boldsymbol{r}'. \tag{9.4.13}$$

To obtain the formula (9.4.8) for the wave operator Ω_+, we start from (4.2.15), which was the equation from which the spectral integral result (9.4.1) was derived. Using the inverse Fourier transform, we write $f(\boldsymbol{r})=(2\pi)^{-3/2}\int e^{i\boldsymbol{k}\cdot\boldsymbol{r}}\hat{f}(\boldsymbol{k})\,d^3\boldsymbol{k}$. Then $Ve^{-iH_0s}f$ has wave function in position space given by $(2\pi)^{-3/2}\int V(\boldsymbol{r})e^{i(\boldsymbol{k}\cdot\boldsymbol{r}-k^2s)}\hat{f}(\boldsymbol{k})\,d^3\boldsymbol{k}$. Note that $V(\boldsymbol{r})e^{i\boldsymbol{k}\cdot\boldsymbol{r}}\in L^2(\mathbb{R}^3)$ as a function of $\boldsymbol{r}$, and we can define a norm-continuous one-parameter family of vectors $g_{\boldsymbol{k}}\in L^2(\mathbb{R}^3)$ by $g_{\boldsymbol{k}}(\boldsymbol{r})=(2\pi)^{-3/2}V(\boldsymbol{r})e^{i\boldsymbol{k}\cdot\boldsymbol{r}}$; i.e. there is a single vector-valued parameter $\boldsymbol{k}$. We then have

$$Ve^{-iH_0s}f=\int d^3\boldsymbol{k}\,\hat{f}(\boldsymbol{k})e^{-ik^2s}g_{\boldsymbol{k}}. \tag{9.4.14}$$

For $\hat{f}\in C_0^\infty(\mathbb{R}^3)$, there is no difficulty in defining the right-hand side of (9.4.14), either as a strong limit of approximating Riemann sums, or weakly on taking inner products with a general element of $L^2(\mathbb{R}^3)$. From (4.2.15), the wave operator is given by

$$\Omega_+f=f+\operatorname*{s-lim}_{\varepsilon\to0+}i\int d^3\boldsymbol{k}\,\hat{f}(\boldsymbol{k})\int_0^\infty ds\,e^{-\varepsilon s}e^{-ik^2s}e^{iHs}g_{\boldsymbol{k}}.$$

Here we have taken the operator e^{iHs} into the integrand in (9.4.14), and inverted orders of integration. Since the double integral is absolutely convergent in norm, both steps are justified. We now carry out the integration with respect to s, to obtain

$$\Omega_+ f = f - \operatorname*{s-lim}_{\varepsilon \to 0+} \int d^3\boldsymbol{k}\, \hat{f}(\boldsymbol{k})(H - \boldsymbol{k}^2 + i\varepsilon)^{-1} g_{\boldsymbol{k}}.$$

Multiplying throughout by $(1+|\boldsymbol{r}|)^{-1}$ and using (9.4.4) gives

$$\begin{aligned}(1+|\boldsymbol{r}|)^{-1}&(\Omega_+ f - f)\\ &= -(2\pi)^{-3/2} \int d^3\boldsymbol{k} \hat{f}(\boldsymbol{k})\{(1+|\boldsymbol{r}|)^{-1}(e^{i\boldsymbol{k}\cdot\boldsymbol{r}} - \psi_+(\boldsymbol{r},\boldsymbol{k}))\}.\end{aligned}$$

On removing the factor $(1+|\boldsymbol{r}|)^{-1}$, use of the inverse Fourier transform to cancel $-f$ yields (9.4.8). ∎

Similar results may be obtained for the wave operator Ω_-, by replacing, in (9.4.3) and (9.4.4), ψ_+ and $+i\varepsilon$ by ψ_- and $-i\varepsilon$ respectively. The integral equation for ψ_- then becomes

$$\psi_-(\boldsymbol{r},\boldsymbol{k}) = e^{i\boldsymbol{k}\cdot\boldsymbol{r}} - \frac{1}{4\pi}\int \frac{e^{ik|\boldsymbol{r}-\boldsymbol{r}'|}}{|\boldsymbol{r}-\boldsymbol{r}'|} V(\boldsymbol{r}')\psi_-(\boldsymbol{r}',\boldsymbol{k})\, d^3\boldsymbol{r}'. \tag{9.4.15}$$

Asymptotically, we have in this case

$$\psi_-(\boldsymbol{r},\boldsymbol{k}) - e^{i\boldsymbol{k}\cdot\boldsymbol{r}} \sim \frac{e^{ikr}}{r} f_-(\boldsymbol{\omega},\boldsymbol{k}), \tag{9.4.16}$$

with

$$f_-(\boldsymbol{\omega},\boldsymbol{k}) = -\frac{1}{4\pi}\int e^{-ik\boldsymbol{\omega}\cdot\boldsymbol{r}'} V(\boldsymbol{r}')\psi_-(\boldsymbol{r}',\boldsymbol{k})\, d^3\boldsymbol{r}'. \tag{9.4.17}$$

Again, $(2\pi)^{-3/2}\psi_-(\boldsymbol{r},\boldsymbol{k})$ is a kernel for the integral operator Ω_-. The proof of the theorem and of the corresponding result for Ω_- depend on the existence of the norm limits, for $\lambda > 0$. $\lim_{\varepsilon\to 0+} |\boldsymbol{r}|^{-1}\,(H-\lambda \pm i\varepsilon)^{-1}|\boldsymbol{r}|^{-1}$. In Section 12.2 we shall prove not only that these limits exist, but also that the resulting linear operators are norm-continuous in λ, even Hölder-continuous. Since $|\boldsymbol{r}|V(\boldsymbol{r})e^{i\boldsymbol{k}\cdot\boldsymbol{r}}$, regarded as a one-parameter family of elements of $L^2(\mathbb{R}^3)$, is strongly continuous in $\boldsymbol{k}$ (use the Lebesgue Dominated-Convergence Theorem to show that $\lim_{\boldsymbol{k}'\to\boldsymbol{k}} \| \, |\boldsymbol{r}|(g_{\boldsymbol{k}} - g_{\boldsymbol{k}'})\| = 0$, where $g_{\boldsymbol{k}}$ is as in the proof of the theorem), we may deduce that the right-hand side of (9.4.4) is also strongly continuous in $\boldsymbol{k}$ for $\boldsymbol{k} \neq 0$. The integral on the right-hand side of

(9.4.13) may be written as the sum

$$\int e^{i(k\cdot r+k\omega r')}V(r')\,d^3r'$$

$$+\int e^{ik\omega\cdot r'}(1+|r'|)V(r')[(1+|r'|)^{-1}(\psi_+(r',k)-e^{ik\cdot r'})]\,d^3r'.$$

Using the Lebesgue Dominated-Convergence theorem, together with the facts that $V\in L^1(\mathbb{R}^3)$ and $(1+|r|)V\in L^2(\mathbb{R}^3)$, and norm continuity in k of $(1+|r|)^{-1}(\psi_+(r,k)-e^{ik\cdot r})$, we may deduce that $f_+(\omega,k)$ is a continuous function of ω and k. The same result applies to $f_-(\omega,k)$, defined by (9.4.17). Equation (9.4.7) and the corresponding equation for ψ_- are called the Lippmann–Schwinger equations for ψ_+ and ψ_- respectively. These equations are frequently used by physicists as a starting point for calculations of the scattering matrix, cross-sections and so on. We see in (9.4.8) how the solution of the Lippmann–Schwinger equation is intimately related to the wave operator. Indeed, following the heuristic analysis at the beginning of this section, we may regard, for example, $\psi_-(r,k)$ as a formal expression for $\Omega_- e^{ik\cdot r}$. Given the definition of Ω_- as a strong limit, we may interpret ψ_- as the (improper) wave function at time $t=0$ corresponding to an incoming plane-wave state $e^{ik\cdot r}$ at time $t=-\infty$. Of course neither ψ_- nor $e^{ik\cdot r}$ are normalizable states; both must be regarded in the sense of the generalized eigenfunctions that were discussed earlier in the context of the one-dimensional Schrödinger equation. From the asymptotic formula (9.4.16), it follows that, at large distances from the scattering centre, the incoming plane wave of momentum k gives rise to a transmitted wave $e^{ik\cdot r}$, together with an outgoing spherical wave. The outgoing spherical wave has kinetic energy k^2, as we should expect. The function f_- in (9.4.16) corresponds to the wave amplitude. The variation of f_- with ω, for fixed incoming momentum k, thus gives the amplitude for scattering into various outgoing directions. Of course, all these considerations, while strongly motivated physically, should not be thought of as replacing the more thorough analysis of three-dimensional scattering that we presented earlier in this chapter. The proper context for scattering theory, both mathematically and physically, is a study of the asymptotic evolution of wave packets in Hilbert space. The following theorem shows that, in (9.4.16), the amplitude of the outgoing spherical wave is identical with the scattering amplitude.

Theorem 9.5

Let $V(r)$ be as in Theorem 9.4, and define $f_-(\omega,k)$, as in (9.4.16), in terms of the asymptotic behaviour of the solution ψ_- of the Lippmann–Schwinger equation (9.4.15). Then the scattering amplitude $f(\lambda; \boldsymbol{\omega}' \to \boldsymbol{\omega})$ is given by

$$f(\lambda; \boldsymbol{\omega}' \to \boldsymbol{\omega}) = f_-(\boldsymbol{\omega}, \lambda^{1/2}\boldsymbol{\omega}'). \tag{9.4.18}$$

Proof

For $V \in L^1(\mathbb{R}^3) \cap L^2(\mathbb{R}^3)$, the wave operators are asymptotically complete. We shall prove this in Chapter 11 by trace-class methods, and completeness for an even wider class of potentials will follow in Chapter 12. Hence $\Omega_+^* = \text{s-lim}_{t\to\infty} e^{iH_0t} e^{-iHt}$ exists as a strong limit on the range of Ω_-. In fact, this limit exists on the whole of $\mathcal{M}_{ac}(H)$. We have, in this case

$$Sf = \Omega_+^* \Omega_- f = \underset{t\to\infty}{\text{s-lim}}\, e^{iH_0t}\, e^{-iHt}\, \Omega_- f.$$

For $f \in D(H_0)$, the intertwining property implies that $\Omega_- f \in D(H_0) = D(H)$. We then have, as a strong derivative,

$$\frac{d}{dt} e^{iH_0t}\, e^{-iHt}\, \Omega_- f = -i\, e^{iH_0t} V\, e^{-iHt}\, \Omega_- f$$

$$= -i\, e^{iH_0t} V \Omega_-\, e^{-iH_0t} f. \tag{9.4.19}$$

Since Ω_- is isometric, we have $\Omega_-^* \Omega_- = 1$. It follows on integrating (9.4.19) with respect to t from $-\infty$ to $+\infty$ that

$$(S-1)f = (\Omega_+^* \Omega_- - \Omega_-^* \Omega_-) f$$

$$= -i \int_{-\infty}^{\infty} dt\, e^{iH_0t} V \Omega_- e^{-iH_0t} f. \tag{9.4.20}$$

The integral on the right-hand side of (9.4.20) should be defined as a strong limit as $T \to \infty$ of the integral from $-T$ to $+T$. We shall evaluate this integral using the direct integral representation of the Hilbert space. For simplicity, we assume that $\hat{f} \in C_0^\infty(\mathbb{R}^3 \setminus \{0\})$. If $f = \{f_\lambda\}$, where $f_\lambda(\boldsymbol{\omega}) = f(\lambda, \boldsymbol{\omega})$ in the direct integral representation, then (5.3.27) implies that $\hat{f}(\boldsymbol{k}) = \sqrt{2}\, k^{-1/2} f(k^2, \boldsymbol{\omega})$, where we have written $\boldsymbol{k} = k\boldsymbol{\omega}$. Using the corresponding result to (9.4.8), for Ω_-, we have

$$(V\Omega_-\, e^{-iH_0t} f)(\boldsymbol{r})$$
$$= \left(\frac{1}{2\pi}\right)^{3/2} \int V(\boldsymbol{r}) \psi_-(\boldsymbol{r}, k'\boldsymbol{\omega}') \sqrt{2}\, (k')^{-1/2} e^{-ik'^2t} f(k'^2, \boldsymbol{\omega}') \frac{k'}{2}\, dk'^2\, d\omega',$$

where we have written $\boldsymbol{k}' = k'\boldsymbol{\omega}'$, and $d^3\boldsymbol{k}' = k'^2\, dk'\, d\omega' = \frac{1}{2} k'\, dk'^2\, d\omega'$. Taking a Fourier transform, the momentum-space wave function with variable $\boldsymbol{k}$ for $V\Omega_- \, \mathrm{e}^{-\mathrm{i}H_0 t} f$ is

$$\left(\frac{1}{2\pi}\right)^3 \int \mathrm{d}^3 \boldsymbol{r}' \, \mathrm{e}^{-\mathrm{i}\boldsymbol{k}\cdot\boldsymbol{r}} \left\{ \int V(\boldsymbol{r}')\psi_-(\boldsymbol{r}', k'\boldsymbol{\omega}') \right.$$

$$\left. \times \sqrt{2}\,(k')^{-1/2} \mathrm{e}^{-\mathrm{i}k'^2 t} f(k'^2, \boldsymbol{\omega}') \frac{k'}{2} \, \mathrm{d}k'^2 \, \mathrm{d}\omega' \right\}.$$

Again using (5.3.27), this becomes, in the direct integral representation as a function of λ and $\boldsymbol{\omega}$,

$$\left(\frac{1}{2\pi}\right)^3 \int \mathrm{d}^3 \boldsymbol{r}' \, \mathrm{e}^{-\mathrm{i}\lambda^{1/2}\boldsymbol{\omega}\cdot\boldsymbol{r}'} \left\{ \int V(\boldsymbol{r}')\psi_-(\boldsymbol{r}', k'\boldsymbol{\omega}') \right.$$

$$\left. \times \, (k')^{-1/2} \lambda^{1/4} \mathrm{e}^{-\mathrm{i}k'^2 t} f(k'^2, \boldsymbol{\omega}') \frac{k'}{2} \, \mathrm{d}k'^2 \, \mathrm{d}\omega' \right\}. \tag{9.4.21}$$

In this representation, we may operate by $-\mathrm{i}\,\mathrm{e}^{\mathrm{i}H_0 t}$, as in the integrand of (9.4.20), through multiplication by $-\mathrm{i}\,\mathrm{e}^{\mathrm{i}\lambda t}$. We then have to integrate with respect to t from $-T$ to $+T$, and proceed to the strong limit $T \to \infty$. The integrations that are to be carried out in (9.4.21), together with the final t-integral, represent an absolutely convergent multiple integral. After evaluating the integrals in (9.4.21) with respect to $\boldsymbol{r}'$ and ω', the result is an integral of the form $F(\lambda, \boldsymbol{\omega}, t) = \int_0^\infty \mathrm{e}^{-\mathrm{i}k'^2 t} h(k'^2, \lambda, \boldsymbol{\omega}) \, \mathrm{d}k'^2$, where by previous arguments the function h depends continuously on k', λ and $\boldsymbol{\omega}$. (Because of the support properties of f, the k'^2 support of h is compact in $(0, \infty)$.)

We let

$$I(\lambda, \boldsymbol{\omega}, T) = -\mathrm{i} \int_{-T}^{T} \mathrm{d}t \, \mathrm{e}^{\mathrm{i}\lambda t} F(\lambda, \boldsymbol{\omega}, t)$$

and

$$I(\lambda, \boldsymbol{\omega}) = -\mathrm{i} \int_{-\infty}^{\infty} \mathrm{d}t \, \mathrm{e}^{\mathrm{i}\lambda t} F(\lambda, \boldsymbol{\omega}, t),$$

in the sense of a strong limit. That is,

$$\int_0^\infty \mathrm{d}\lambda \int \mathrm{d}\omega \, |I(\lambda, \boldsymbol{\omega}, T) - I(\lambda, \boldsymbol{\omega})|^2 \to 0 \quad \text{as } T \to \infty.$$

We know that this strong limit exists; indeed, by (9.4.20), $I(\lambda, \boldsymbol{\omega})$ is just the direct integral representation for $(S-1)f$. Now $I(\lambda, \boldsymbol{\omega})$ is an inverse Fourier transform with respect to t of a function F that is itself a Fourier transform with respect to k'^2. We therefore have, by standard formulae for the transform and inverse transform,

$$I(\lambda, \boldsymbol{\omega}) = -2\pi \mathrm{i} h(\lambda, \lambda, \boldsymbol{\omega}).$$

(The explicit λ-dependence of $h(k'^2, \lambda, \boldsymbol{\omega})$ will not affect the argument; it may be verified that the t-integration in (9.4.21) may be carried out before integration with respect to $\boldsymbol{r}'$, in which case the dependence of the integrand on k' and on λ may be effectively separated.)

Returning to the explicit formula (9.4.21) for F, we now have

$$I(\lambda, \boldsymbol{\omega}) = \frac{-\mathrm{i}}{(2\pi)^2} \int \mathrm{d}^3\boldsymbol{r}' \, \mathrm{e}^{-\mathrm{i}\lambda^{1/2}\boldsymbol{\omega}\cdot\boldsymbol{r}'}$$

$$\times \left\{ \int V(\boldsymbol{r}')\psi_-(\boldsymbol{r}', \lambda^{1/2}\boldsymbol{\omega}') f(\lambda, \boldsymbol{\omega}') \frac{\lambda^{1/2}}{2} \mathrm{d}\omega' \right\}.$$

Again, the order of integration does not matter, and we may use (9.4.17) to carry out the integral with respect to $\boldsymbol{r}'$, giving

$$I(\lambda, \boldsymbol{\omega}) = \frac{\mathrm{i}}{2\pi} \int \lambda^{1/2} f_-(\boldsymbol{\omega}, \lambda^{1/2}\boldsymbol{\omega}') f(\lambda, \boldsymbol{\omega}') \, \mathrm{d}\omega' = ((S-1)f)(\lambda, \boldsymbol{\omega}).$$

Writing $S = 1 + T$, we find that, in the direct integral representation, T is an integral operator with kernel

$$T(\lambda, \boldsymbol{\omega}, \boldsymbol{\omega}') = \frac{\mathrm{i}}{2\pi} \lambda^{1/2} f_-(\boldsymbol{\omega}, \lambda^{1/2}\boldsymbol{\omega}'). \tag{9.4.22}$$

Comparison with (9.2.23) now gives (9.4.18). ∎

Theorem 9.5 is important in that it allows us to use the asymptotic behaviour of solutions of the Lippmann–Schwinger equation to evaluate scattering amplitudes and cross-sections. If, for example, the potential $V(\boldsymbol{r})$ is given as proportional to some small parameter g, or coupling constant, one may iterate (9.4.15) in increasing powers of g and substitute the resulting series for ψ_- into (9.4.17) to obtain a series for the scattering amplitude. To lowest order in g, (9.4.17) then becomes

$$f_-(\boldsymbol{\omega}, \boldsymbol{k}) \approx -\frac{1}{4\pi} \int \mathrm{e}^{\mathrm{i}(\boldsymbol{k}\cdot\boldsymbol{r}' - k\boldsymbol{\omega}\cdot\boldsymbol{r}')} V(\boldsymbol{r}') \, \mathrm{d}^3\boldsymbol{r}'.$$

With $\boldsymbol{k}=\lambda^{1/2}\boldsymbol{\omega}'$, this leads to

$$f(\lambda, \boldsymbol{\omega}' \to \boldsymbol{\omega}) \approx -\frac{1}{4\pi}\int \mathrm{e}^{-\mathrm{i}\lambda^{1/2}(\boldsymbol{\omega}-\boldsymbol{\omega}')\cdot\boldsymbol{r}'} V(\boldsymbol{r}')\,\mathrm{d}^3\boldsymbol{r}'$$

$$= -\left(\frac{\pi}{2}\right)^{1/2} \hat{V}(\lambda^{1/2}(\boldsymbol{\omega}-\boldsymbol{\omega}')), \tag{9.4.23}$$

where $\hat{V}$ is the Fourier transform of V. Equation (9.4.23) gives the *Born approximation* to the scattering amplitude.

In restricting the analysis to potentials satisfying the condition (9.4.2) at the start of this section, implying $V \in L^1(\mathbb{R}^3)$, we ensure the absolute convergence of integrals such as that in (9.4.17). We can then guarantee the continuity of both the Born approximation and the full scattering amplitude, as a function of λ, $\boldsymbol{\omega}$ and $\boldsymbol{\omega}'$. In particular, the scattering matrix will then be norm-continuous as a function of λ, and both differential and total cross-sections will vary continuously with energy. (Our analysis in Chapter 10 of singular points of the Hamiltonian will show that, in certain circumstances and for a less restricted class of potentials, it is possible for discontinuities as a function of energy to occur.)

These results can in part be extended to a wider class of potentials that decay more slowly at infinity. Solutions to the Lippmann–Schwinger equation may then exist such that integrals as in (9.4.17) are only conditionally convergent. We shall not explore these possibilities, nor the more delicate problem of writing down a generalized Lippmann–Schwinger equation for Coulomb and other long-range potentials.

The results of this section can be recast in terms of the operator $T(z)=V-V(H-z)^{-1}V$ introduced in (4.2.18). If we write $z=\lambda+\mathrm{i}\varepsilon$ then $T(z)$ has an upper boundary value, as z approaches the real axis from above, in the sense of operator norms, and we define

$$T_+(\lambda) = \lim_{\varepsilon\to 0+} T(\lambda+\mathrm{i}\varepsilon).$$

Using the corresponding results to (9.4.3) and (9.4.4) for ψ_- then gives the identity

$$V\psi_- = T_+(\boldsymbol{k}^2)\mathrm{e}^{\mathrm{i}\boldsymbol{k}\cdot\boldsymbol{r}}, \tag{9.4.24}$$

where the right-hand side can be understood as a strong limit, on multiplying by $(1+|\boldsymbol{r}|)^{-1}$, as in (9.4.4). The juxtaposition of + and − signs in (9.4.24) is unfortunate, but inevitable owing to the fact that *two* limits $t\to-\infty$ and $\varepsilon\to 0+$ are involved.

We now suppose that $T_+(\lambda)$ is given by an integral operator in momentum space with kernel $T_+(\lambda, \boldsymbol{k}', \boldsymbol{k}'')$. That is,

$$(\widehat{T_+(\lambda)f})(\boldsymbol{k}') = \int T_+(\lambda, \boldsymbol{k}', \boldsymbol{k}'')\hat{f}(\boldsymbol{k}'')\, \mathrm{d}^3\boldsymbol{k}''.$$

Formally, the Fourier transform of $e^{i\boldsymbol{k}\cdot\boldsymbol{r}'}$ with respect to $\boldsymbol{r}'$ is given as a function of $\boldsymbol{k}''$ by $(2\pi)^{3/2}\ \delta(\boldsymbol{k}''-\boldsymbol{k})$, and the right-hand side of (9.4.24) is represented in momentum space by the function of $\boldsymbol{k}'$

$$h(\boldsymbol{k}') = (2\pi)^{3/2} T_+(\boldsymbol{k}^2, \boldsymbol{k}', \boldsymbol{k}).$$

We may also interpret the integral in (9.4.17) as a Fourier transform of the left-hand side of (9.4.24), so that

$$f_-(\boldsymbol{\omega}, \boldsymbol{k}) = -\frac{(2\pi)^{3/2}}{4\pi} h(k\boldsymbol{\omega}) = -2\pi^2 T_+(k^2, k\boldsymbol{\omega}, \boldsymbol{k}) \tag{9.4.25}$$

Comparison with (9.4.22) now gives

$$T(\lambda, \boldsymbol{\omega}, \boldsymbol{\omega}') = -\mathrm{i}\pi\lambda^{1/2} T_+(\lambda, \lambda^{1/2}\boldsymbol{\omega}, \lambda^{1/2}\boldsymbol{\omega}'). \tag{9.4.26}$$

Equation (9.4.26) establishes the formal link between the respective kernels of T_λ as an operator in $L^2(\mathbb{S}^2)$ and $T_+(\lambda)$ as an operator in momentum space in three dimensions.

9.5 PARTIAL-WAVE ANALYSIS

The results of the last section, summarized in Theorems 9.4 and 9.5, apply to scattering by a large class of short-range potentials $V(\boldsymbol{r})$, whether central or non-central. Since many of the potentials with which physicists have to deal are of the form $V = V(|\boldsymbol{r}|)$, it is important to see what effect the further assumption of spherical symmetry has on our analysis. It is also instructive to see how to link up the results of Chapter 8 for spherical potentials in a single partial wave with the full three-dimensional treatment of scattering theory. In short, we want to see how the three-dimensional picture can be built up from its one-dimensional partial-wave components.

One way of doing this is to start from the Lippmann–Schwinger equation, with $V = V(r)$, and project the equation onto a partial-wave subspace characterized by quantum numbers l and m of angular momentum. An important identity here is the partial-wave decomposition of a plane wave $e^{i\boldsymbol{k}\cdot\boldsymbol{r}}$. With $\boldsymbol{k} = k\boldsymbol{\omega}'$, $\boldsymbol{r} = r\boldsymbol{\omega}$, we have

$$\begin{aligned} e^{\mathrm{i}\boldsymbol{k}\cdot\boldsymbol{r}} &= e^{\mathrm{i}kr\boldsymbol{\omega}\cdot\boldsymbol{\omega}'} \\ &= \frac{4\pi}{kr} \sum_{l,m} \mathrm{i}^l kr j_l(kr) Y_{lm}(\boldsymbol{\omega}) \bar{Y}_{lm}(\boldsymbol{\omega}'). \end{aligned} \tag{9.5.1}$$

The summation on the right-hand side of (9.5.1) is over $l = 0,1,2,\ldots$, with m taking integer values, for each l, in the range $-l \leq m \leq l$. Let us therefore make a similar expansion to (9.5.1) for ψ_-, namely

$$\psi_-(\boldsymbol{r}, \boldsymbol{k}) = \frac{4\pi}{kr} \sum_{l,m} \mathrm{i}^l \psi_l(r, k) Y_{lm}(\boldsymbol{\omega}) \bar{Y}_{lm}(\boldsymbol{\omega}'). \tag{9.5.2}$$

with again $\boldsymbol{k} = k\boldsymbol{\omega}'$, $\boldsymbol{r} = r\boldsymbol{\omega}$. Using the orthogonality property for the functions Y_{lm}, $\psi_l(r, k)$ may be determined from the identity

$$\int \psi_-(\boldsymbol{r}, \boldsymbol{k}) \bar{Y}_{lm}(\boldsymbol{\omega}) Y_{lm}(\boldsymbol{\omega}')\, \mathrm{d}\omega\, \mathrm{d}\omega' = \frac{4\pi}{kr} i^l \psi_l(r, k).$$

The fact that ψ_- has an expansion of the form (9.5.2) is a fundamental consequence of rotation invariance, given that $V = V(r)$, and follows directly by substituting the plane-wave expansion (9.5.1) into the left-hand side of (9.4.4). Since H commutes with angular momentum, the resolvent operator $(H - \boldsymbol{k}^2 + \mathrm{i}\varepsilon)^{-1}$ does not alter the angular dependence $Y_{lm}(\boldsymbol{\omega})$ of a wave function belonging to the (l, m) partial-wave subspace, and the same is true of the operator of multiplication by $V(r)$.

We now substitute (9.5.1) and (9.5.2) into the Lippmann–Schwinger equation (9.4.15), with $V(\boldsymbol{r}') = V(r')$. In the integral on the right-hand side of (9.4.15), the coefficient of $\bar{Y}_{lm}(\boldsymbol{\omega}')$ is

$$-\frac{\mathrm{i}^l}{kr} \int rr' \frac{\mathrm{e}^{\mathrm{i}k|\boldsymbol{r}-\boldsymbol{r}'|}}{|\boldsymbol{r} - \boldsymbol{r}'|} V(r') \psi_l(r', k) Y_{lm}(\boldsymbol{\omega}'')\, \mathrm{d}r'\, \mathrm{d}\omega'',$$

where we have written $\boldsymbol{r}' = r'\boldsymbol{\omega}''$ and $\mathrm{d}^3\boldsymbol{r}' = r'^2\, \mathrm{d}\omega''$. Multiplying by $\bar{Y}_{lm}(\boldsymbol{\omega})$ and integrating with respect to ω gives the coefficient, in this integral, of $Y_{lm}(\boldsymbol{\omega})\bar{Y}_{lm}(\boldsymbol{\omega}')$, and, comparing with the corresponding coefficient in the remaining terms of (9.4.15), we are led to the integral equation

$$\psi_l(r, k) = krj_l(kr) - \int_0^\infty \mathrm{d}r'\, G_l(k, r, r') V(r') \psi_l(r', k), \tag{9.5.3}$$

where the kernel of the equation is given by

$$G_l(k, r, r') = \frac{rr'}{4\pi} \int \frac{\mathrm{e}^{\mathrm{i}k|\boldsymbol{r}-\boldsymbol{r}'|}}{|\boldsymbol{r} - \boldsymbol{r}'|} \bar{Y}_{lm}(\boldsymbol{\omega}) Y_{lm}(\boldsymbol{\omega}'')\, \mathrm{d}\omega\, \mathrm{d}\omega''. \tag{9.5.4}$$

Equation (9.5.3) is just a one-dimensional version of the Lippmann-Schwinger equation, in a given partial wave. An alternative way of deriving (9.5.4) is to repeat the proof of Theorem 9.4, within a given partial wave, and, on replacing ψ_+ by ψ_-, (9.4.3) then becomes

$$\psi_l(r, k) = krj_l(kr) - \lim_{\varepsilon \to 0+} (H - k^2 - \mathrm{i}\varepsilon)^{-1} V krj_l(kr). \tag{9.5.5}$$

The kernel $G_l(k, r, r')$ in (9.5.4), as in the proof of Theorem 9.4, is then obtained as a limit as $\varepsilon \to 0+$ of the kernel of the resolvent operator, where here the resolvent is to be regarded as an integral operator in $L^2(0, \infty)$. We have already seen how to write down the kernel of such an operator in terms of solutions of the time-independent Schrödinger equation. The reader should verify that, in this case,

$$G_l(k, r, r') = \begin{cases} (-1)^l u(r, k) r' j_l(kr') & (r' < r), \\ (-1)^l r j_l(kr) u(r', k) & (r' > r), \end{cases} \tag{9.5.6}$$

where $u(r, k)$ is an upper solution of the Schrödinger equation at energy $\lambda = k^2$, with the asymptotic behaviour $u(r, k) \sim e^{i(kr + l\pi/2)}$ as $r \to \infty$. Adapting the rest of the proof of Theorem 9.4 to a single partial-wave subspace, we have, corresponding to (9.4.8),

$$(\Omega_- f)(r) = \left(\frac{2}{\pi}\right)^{1/2} \int_0^\infty \psi_l(r, k') \hat{f}(k') \, dk', \tag{9.5.7}$$

where $\hat{f}$ is the Fourier–Hankel transform of a function $f \in L^2(0, \infty)$. Given the asymptotic formula $\sin(kr - \frac{1}{2}l\pi)$ for the modified Bessel function, together with (9.5.6) for G_l and the asymptotic formula for the upper solution $u(r, k)$, we can use (9.5.3) to determine the asymptotic behaviour of $\psi_l(r, k)$ for large r. Thus we have, as $r \to \infty$,

$$\psi_l(r, k) = \sin(kr - \tfrac{1}{2}l\pi) - e^{i(kr + l\pi/2)} \int_0^\infty dr' \, (-1)^l r' V(r') j_l(kr') \psi_l(r', k) + o(1) \tag{9.5.8}$$

Starting from (9.4.20), we can proceed, as in the proof of Theorem 9.5, to obtain an expression for the scattering operator in a single partial wave. Thus, in momentum space, $V\Omega_- f$ is represented, on multiplying (9.5.7) by $V(r)$ and taking a Fourier–Hankel transform, by the function of k

$$\frac{2}{\pi} \int_0^\infty dr' \int_0^\infty dk'^2 \, (2k')^{-1} kr' j_l(kr') V(r') \psi_l(r', k') \hat{f}(k').$$

To evaluate $(S-1)f$ in momentum space, we introduce a factor $-i \, e^{i(k^2 - k'^2)t}$ and integrate from $-\infty$ to ∞ with respect to t. Using standard results for Fourier transform and inverse transform, we may deduce that the scattering operator S_l in a single partial wave is the operator defined in momentum space as multiplication by the function

$$S_l(k^2)=1-2i\int_0^\infty dr'\, r'V(r')j_l(kr')\psi_l(r',k). \tag{9.5.9}$$

Comparing with (9.5.8), we see that $\psi_l(r,k)$ has the asymptotic behaviour, as $r\to\infty$,

$$\begin{aligned}\psi_l(r,k)&=\sin(kr-\tfrac{1}{2}l\pi)+(2i)^{-1}(S_l-1)(-1)^l e^{i(kr+l\pi/2)}+o(1)\\ &=(2i)^{-1}\{S_l e^{i(kr-l\pi/2)}-e^{-i(kr-l\pi/2)}\}+o(1).\end{aligned} \tag{9.5.10}$$

Defining the partial-wave phase shift δ_l, up to multiples of π, by $S_l=\exp(2i\delta_l)$, (9.5.10) simplifies to

$$\psi_l(r,k)=e^{i\delta_l}\sin(kr-\tfrac{1}{2}l\pi+\delta_l)+o(1). \tag{9.5.11}$$

The reader should verify that this result, together with (9.5.7), is consistent with the formula (8.2.6) for the wave operator.

Let us now carry out a partial-wave analysis of the scattering amplitude $f(\lambda,\boldsymbol{\omega}'\to\boldsymbol{\omega})$. We start from (9.4.17) for $f_-(\boldsymbol{\omega},\boldsymbol{k})$ and use (9.5.1) and (9.5.2) to write

$$e^{-ik\boldsymbol{\omega}\cdot\boldsymbol{r}'}=\frac{4\pi}{kr'}\sum_{l,m}(-i)^l kr' j_l(kr')Y_{lm}(\boldsymbol{\omega})\bar{Y}_{lm}(\boldsymbol{\omega}''),$$

$$\psi_-(\boldsymbol{r}',\boldsymbol{k})=\frac{4\pi}{kr'}\sum_{l,m}i^l\psi_l(r',k)Y_{l,m}(\boldsymbol{\omega}'')\bar{Y}_{lm}(\boldsymbol{\omega}').$$

We have here $\boldsymbol{r}'=r'\boldsymbol{\omega}''$ and $\boldsymbol{k}=k\boldsymbol{\omega}'$. Multiplying these two expansions together, we can write $d^3\boldsymbol{r}'=r'^2\,d\omega''$ and carry out, on the right-hand side of (9.4.17), the integration with respect to ω''. Using the fact that the Y_{lm} are orthonormal in $L^2(\mathbb{S}^2)$, this gives

$$f_-(\boldsymbol{\omega},\boldsymbol{k})=-\frac{4\pi}{k}\sum_{l,m}\int_0^\infty dr'\, r'V(r')j_l(kr')\psi_l(r',k)Y_{lm}(\boldsymbol{\omega})\bar{Y}_{lm}(\boldsymbol{\omega}') \tag{9.5.12}$$

The integral on the right-hand side has already been written down in (9.5.9), and (9.5.12) becomes

$$f_-(\boldsymbol{\omega},\boldsymbol{k})=-2\pi i k^{-1}\sum_{l,m}(S_l(k^2)-1)Y_{l,m}(\boldsymbol{\omega})\bar{Y}_{lm}(\boldsymbol{\omega}'). \tag{9.5.12$'$}$$

Comparison with (9.4.18) and (9.4.22) now yields the formulae

$$f(\lambda,\boldsymbol{\omega}'\to\boldsymbol{\omega})=-2\pi i\lambda^{-1/2}\sum_{l,m}(S_l(\lambda)-1)Y_{lm}(\boldsymbol{\omega})\bar{Y}_{lm}(\boldsymbol{\omega}'), \tag{9.5.13}$$

$$T(\lambda, \boldsymbol{\omega}, \boldsymbol{\omega}') = \sum_{l,m} (S_l(\lambda) - 1) Y_{lm}(\boldsymbol{\omega}) \bar{Y}_{lm}(\boldsymbol{\omega}'). \tag{9.5.14}$$

Equation (9.5.13) gives the partial-wave decomposition of the scattering amplitude. We can use this equation to define a partial-wave scattering amplitude $a_l(\lambda)$ by the formula

$$a_l(\lambda) = -2\pi i \lambda^{-1/2} (S_l(\lambda) - 1). \tag{9.5.15}$$

In each of (9.5.12)–(9.5.14) the summation over m from $-l$ to $+l$ can be carried out explicitly by making use of the identity

$$4\pi \sum_{m=-l}^{l} Y_{lm}(\boldsymbol{\omega}) \bar{Y}_{lm}(\boldsymbol{\omega}') = (2l+1) P_l(\boldsymbol{\omega} \cdot \boldsymbol{\omega}'),$$

where P_l is a Legendre polynomial.

We can also use (9.5.14) to evaluate the Hilbert–Schmidt norm of the operator T_λ, giving

$$\begin{aligned} (\|T_\lambda\|_2)^2 &= \int d\omega \, d\omega' \, |T(\lambda, \omega, \omega')|^2 \\ &= \sum_{l=0}^{\infty} (2l+1) |S_l(\lambda) - 1|^2. \end{aligned} \tag{9.5.16}$$

In terms of the phase shifts δ_l, where $S_l = e^{2i\delta_l}$, we then have,

$$(\|T_\lambda\|_2)^2 = 4 \sum_{l=0}^{\infty} (2l+1) \sin^2 \delta_l. \tag{9.5.17}$$

According to (9.2.22), we then have the result

$$\sigma_{\text{ave}}(\lambda) = 4\pi \lambda^{-1} \sum_{l=0}^{\infty} (2l+1) \sin^2 \delta_l \tag{9.5.18}$$

for the total scattering cross-section at energy λ.

By specializing the results of the last section to the case of spherical symmetry, all of the results obtained above may be rigorously justified for a wide range of central potentials $V(r)$, including for example those potentials for which $V \in L^1(0, \infty)$. If we use (9.5.18) to define a partial-wave cross-section σ_l by

$$\sigma_l(\lambda) = 4\pi \lambda^{-1} \sin^2 \delta_l, \tag{9.5.19}$$

we see that the total cross-section will be finite whenever the phase shift decreases to zero sufficiently rapidly as a function of l. Indeed, the method of Section 9.3 can be applied to estimate the cross-section in each partial-wave subspace, and precise bounds on the behaviour of σ_l as a function of l

may be obtained. For a potential subject to the bound $|V(r)| \leq \text{const}/r^{2+\varepsilon}$ $(\varepsilon>0)$ as $r\to\infty$, δ_l decreases to zero more rapidly than $1/l$, and the series (9.5.18) is convergent. It is also interesting to note, from (9.5.19), that there is an absolute bound, independent of the potential, on the partial-wave cross-section as a function of energy.

To summarize, partial-wave analysis allows us, for central potentials, to treat the Lippmann–Schwinger equation as an integral equation in a single partial wave. The solution $\psi_l(r,k)$ of the Schrödinger equation at energy $\lambda=k^2$, subject to a regular boundary condition at the origin and the condition (9.5.10) at $r=\infty$, determines in principle the partial-wave scattering amplitude and partial-wave cross-section for each l-value. The complete picture of scattering in three-dimensional space may then be built up by summing over partial waves, and physical quantities such as scattering amplitudes and cross-sections may then be determined.

9.6 WAVE OPERATORS FOR TWO- AND THREE-PARTICLE SCATTERING

The simplest example of a two-particle system is that of a pair of particles, of masses m_1 and m_2 respectively, interacting through a potential $V(\boldsymbol{r}_1-\boldsymbol{r}_2)$, where $\boldsymbol{r}_1$ and $\boldsymbol{r}_2$ are the respective positions of the particles. The state of such a system is described by a wave function $f(\boldsymbol{r}_1,\boldsymbol{r}_2)$ belonging to the Hilbert space $L^2(\mathbb{R}^6)$, i.e. with norm given by $\|f\|^2=\int|f(\boldsymbol{r}_1,\boldsymbol{r}_2)|^2\,\mathrm{d}^3\boldsymbol{r}_1\,\mathrm{d}^3\boldsymbol{r}_2$. The momentum of particle 1 is given by the operator $\boldsymbol{P}_1=-\mathrm{i}\,\boldsymbol{\nabla}_{r_1}=-\mathrm{i}\,(\partial/\partial x_1,\partial/\partial y_1,\partial/\partial z_1)$, where $\boldsymbol{r}_1=(x_1,y_1,z_1)$ and we have chosen units such that $\hbar=1$. Similarly, for particle 2 we have $\boldsymbol{P}_2=-\mathrm{i}\,\boldsymbol{\nabla}_{r_2}$, and the total Hamiltonian for the system is

$$H=\frac{\boldsymbol{P}_1^2}{2m_1}+\frac{\boldsymbol{P}_2^2}{2m_2}+V(\boldsymbol{r}_1-\boldsymbol{r}_2). \tag{9.6.1}$$

As a differential operator in $L^2(\mathbb{R}^6)$, we then have,

$$H=-\frac{\Delta_{r_1}}{2m_1}-\frac{\Delta_{r_2}}{2m_2}+V(\boldsymbol{r}_1-\boldsymbol{r}_2), \tag{9.6.2}$$

where Δ_{r_i} denotes the Laplacian operator with respect to the coordinates of $\boldsymbol{r}_i$.

It is well known that for such a two-particle system the dynamics may be separated, as in the corresponding problem in classical mechanics, into centre-of-mass free motion together with motion relative to the centre of mass. This effectively reduces the mathematical description of the system to that of a single particle of mass $m = m_1 m_2/(m_1 + m_2)$ in a potential $V(\boldsymbol{r})$. To effect this reduction, we replace $\boldsymbol{r}_1$ and $\boldsymbol{r}_2$ by new coordinates $\boldsymbol{r}$ and $\boldsymbol{R}$, defined by

$$\boldsymbol{r} = \boldsymbol{r}_1 - \boldsymbol{r}_2, \qquad \boldsymbol{R} = \frac{m_1 \boldsymbol{r}_1 + m_2 \boldsymbol{r}_2}{m_1 + m_2}.$$

Note that $\int |f|^2 \, d^3\boldsymbol{r}_1 \, d^3\boldsymbol{r}_2 = \int |f|^2 \, d^3\boldsymbol{r} \, d^3\boldsymbol{R}$.

We also have, in terms of the components (x, y, z) and (X, Y, Z) of $\boldsymbol{r}$ and $\boldsymbol{R}$ respectively,

$$\frac{\partial}{\partial x_1} = \frac{\partial}{\partial x} + \frac{m_1}{m_1 + m_2} \frac{\partial}{\partial X},$$

$$\frac{\partial}{\partial x_2} = -\frac{\partial}{\partial x} + \frac{m_2}{m_1 + m_2} \frac{\partial}{\partial X},$$

together with similar equations for the derivatives with respect to the remaining components of $\boldsymbol{r}_1$ and $\boldsymbol{r}_2$. Making the change of variables in (9.6.2) now gives

$$H = -\frac{\Delta_r}{2m} + V(\boldsymbol{r}) - \frac{\Delta_R}{2(m_1 + m_2)}. \tag{9.6.3}$$

We can then write,

$$H = H_r + H_R, \tag{9.6.4}$$

where

$$H_r = -\frac{\Delta_r}{2m} + V(\boldsymbol{r}), \tag{9.6.5}$$

$$H_R = -\frac{\Delta_R}{2(m_1 + m_2)}. \tag{9.6.6}$$

Here H_r is now a one-particle Hamiltonian, with mass m and potential $V(\boldsymbol{r})$, and H_R is a Hamiltonian for a free particle of mass $m_1 + m_2$, describing the motion of the centre of mass. Since $m < m_1$, $m < m_2$, m is called the reduced mass of the two-particle system. Note that $[H_r, H_R] = 0$.

Because H_r and H_R commute, the evolution operator e^{-iHt} may be factorized as

$$e^{-iHt} = e^{-iH_r t} e^{-iH_R t}$$

and any factorizable wave function $f = f_1(\boldsymbol{r}) f_2(\boldsymbol{r})$ will lead to the time development

$$(e^{-iHt}f)(\boldsymbol{r}, \boldsymbol{R}) = f_1(\boldsymbol{r}, t) f_2(\boldsymbol{R}, t),$$

where $f_1(\boldsymbol{r}, t)$ satisfies the time-dependent Schrödinger equation with Hamiltonian H_r and $f_2(\boldsymbol{R}, t)$ with Hamiltonian $H_{\boldsymbol{R}}$. If we then define a free Hamiltonian H_0 by

$$H_0 = \frac{\boldsymbol{P}_1^2}{2m_1} + \frac{\boldsymbol{P}_2^2}{2m_2} = \frac{-\Delta_r}{2m} - \frac{\Delta_{\boldsymbol{R}}}{2(m_1+m_2)},$$

with

$$\Omega_\pm = \underset{t\to\pm\infty}{\text{s-lim}}\; e^{iHt}e^{-iH_0t}, \tag{9.6.7}$$

it is easy to see that

$$(\Omega_\pm f)(\boldsymbol{r}, \boldsymbol{R}) = (\Omega'_\pm f_1)(\boldsymbol{r}) f_2(\boldsymbol{R}), \tag{9.6.8}$$

where $\Omega'_\pm$ is the wave operator for a single particle of mass m in a potential $V(\boldsymbol{r})$. In the language of tensor products, (9.6.8) may be written as

$$\Omega_\pm = \Omega'_\pm \otimes I_{\boldsymbol{R}}. \tag{9.6.9}$$

where $I_{\boldsymbol{R}}$ is the identity operator in $\mathscr{H}_{\boldsymbol{R}}$ and we have set $\mathscr{H} = L^2(\mathbb{R}^6) = \mathscr{H}_r \otimes \mathscr{H}_{\boldsymbol{R}} = L^2(\mathbb{R}^3) \otimes L^2(\mathbb{R}^3)$. Since $L^2(\mathbb{R}^6)$ is spanned by factorizable wave functions, (9.6.9) is sufficient to define wave operators on the whole of $\mathscr{H}$. Equation (9.6.7) is appropriate if the potential $V(\boldsymbol{r})$ is of short range, so that we are able to define unmodified wave operators. If, on the other hand, $V(\boldsymbol{r})$ is of long range, we can define a modified free evolution $e^{-iH_0t}Z(\boldsymbol{P}_r, t)$, where $\boldsymbol{P}_r = -i\nabla_r$, and the form of the function Z is determined as in Section 9.1 by the large-$|\boldsymbol{r}|$ behaviour of $V(\boldsymbol{r})$. In this case too we obtain a (modified) wave operator (9.6.9), where $\Omega'_\pm$ is now a single-particle long-range wave operator appropriate to a particle of mass m moving in potential $V(\boldsymbol{r})$. (Note, however, that if units are taken with $\hbar = 1$ but $m \neq \frac{1}{2}$ then the right-hand side of (9.1.11) should read, in general, $\exp\{-imW(\boldsymbol{P}_r t/m)/|\boldsymbol{P}_r|\}_f$, where $W(\boldsymbol{r})$ satisfies (9.1.9).)

Scattering theory for a two-particle system interacting through a pair potential $V(\boldsymbol{r}_1 - \boldsymbol{r}_2)$ thus reduces to a single-particle problem with $V = V(\boldsymbol{r})$. Wave operators will exist if and only if they exist for the one-particle problem, and will be complete if and only if one-particle wave operators are complete. To treat a two-particle system in this way is equivalent to the classical approach of referring the motion to the centre-of-mass system. As in the classical theory, all other observable quantities such as two-particle scattering cross-sections can then be determined by a suitable change of frame of reference.

It is instructive to consider the dynamics corresponding to an eigenfunction of the one-particle Hamiltonian H_r. For such an eigenfunction $u(\boldsymbol{r})$, with eigenvalue λ_0, we let

$$f(\boldsymbol{r}, \boldsymbol{R}) = u(\boldsymbol{r}) f_2(\boldsymbol{R}).$$

Then

$$(Hf)(\boldsymbol{r}, \boldsymbol{R}) = u(\boldsymbol{r})\left(-\frac{\Delta_R}{2(m_1+m_2)} + \lambda_0\right) f_2(\boldsymbol{R}).$$

Thus the total energy of such a state is obtained by subtracting from the kinetic energy of the two-particle "bound state", of mass m_1+m_2, a "binding energy" $-\lambda_0$. Normally we expect $\lambda_0 < 0$, so that $-\lambda_0$ is positive. We also have, in this case,

$$(\mathrm{e}^{-\mathrm{i}Ht} f)(\boldsymbol{r}, \boldsymbol{R}) = \mathrm{e}^{-\mathrm{i}\lambda_0 t} u(\boldsymbol{r})(\mathrm{e}^{-\mathrm{i}H_R t} f_2)(\boldsymbol{R}),$$

showing that the motion of the bound state is that of a free particle located at the centre of mass. It is indeed possible to define a corresponding scattering channel, with free Hilbert space $\mathscr{H}^{(2)}$ and $f_2 \in \mathscr{H}^{(2)}$. The space $\mathscr{H}^{(2)}$ then contains functions $f_2(\boldsymbol{R}) \in L^2(\mathbb{R}^3)$, and $H^{(2)}$ may be defined as the operator $H_R + \lambda_0$, acting in this space. Let us define the isometry U_2 from $\mathscr{H}^{(2)}$ to $\mathscr{H}$ by

$$(U_2 f_2)(\boldsymbol{r}, \boldsymbol{R}) = u(\boldsymbol{r}) f_2(\boldsymbol{R}),$$

with $\|u\| = 1$; then the wave operators $\Omega_\pm^{(2)}$ are given by

$$\Omega_\pm^{(2)} = \operatorname*{s-lim}_{t \to \pm\infty} \mathrm{e}^{\mathrm{i}Ht} U_2 \, \mathrm{e}^{-\mathrm{i}H^{(2)}t},$$

in accordance with the theory of wave operators developed in Chapter 3. In this case we have simply $\Omega_\pm^{(2)} = U_2$, and the wave operators project on to the bound-state eigenspace. Scattering channels corresponding to two-particle bound states are not usually considered in discussing two-particle problems with pair potentials, and have no bearing on the question of asymptotic completeness. (The scattering operator between $\mathscr{H}^{(2)}$ and $\mathscr{H}^{(2)}$ in the above example acts as an identity operator on $\mathscr{H}^{(2)}$, reflecting the fact that a two-particle bound state is transmitted without scattering.) For this reason, the Hamiltonian given by (9.6.2) is usually regarded as defining a single-channel scattering system.

To consider a true many-channel problem with pair potentials $V(\boldsymbol{r}_i - \boldsymbol{r}_j)$, it is necessary to consider a system of at least three particles. Let us consider, for example, a system of three particles, described by wave functions

$f(\boldsymbol{r}_1, \boldsymbol{r}_2, \boldsymbol{r}_3) \in L^2(\mathbb{R}^9) = \mathscr{H}$, with total Hamiltonian

$$H = \frac{\boldsymbol{P}_1^2}{2m_1} + \frac{\boldsymbol{P}_2^2}{2m_2} + \frac{\boldsymbol{P}_3^2}{2m_3} + V_{12}(\boldsymbol{r}_1 - \boldsymbol{r}_2) + V_{23}(\boldsymbol{r}_2 - \boldsymbol{r}_3)$$
$$+ V_{31}(\boldsymbol{r}_3 - \boldsymbol{r}_1). \tag{9.6.10}$$

We assume first of all that $V_{ij}(\boldsymbol{r})$ are all short-range potentials. We define a channel, labelled by the index 0, corresponding to the asymptotic evolution of three free non-interacting particles. The free Hilbert space $\mathscr{H}^{(0)}$ is a copy of $L^2(\mathbb{R}^9)$, and U_0 is the identity mapping from $\mathscr{H}^{(0)}$ to $\mathscr{H}$. The free Hamiltonian in $\mathscr{H}^{(0)}$ is then

$$H^{(0)} = \frac{\boldsymbol{P}_1^2}{2m_1} + \frac{\boldsymbol{P}_2^2}{2m_2} + \frac{\boldsymbol{P}_3^2}{2m_3},$$

and $H_0 = U_0 H^{(0)} U_0^{-1}$ is the same operator acting in $\mathscr{H}$. Given $f^{(0)} \in \mathscr{H}^{(0)}$, with $f_0 = U_0 f^{(0)}$, the wave operators $\Omega_\pm^{(0)}$ for this channel are defined by

$$\Omega_\pm^{(0)} f^{(0)} = \operatorname*{s-lim}_{t \to \pm\infty} \mathrm{e}^{\mathrm{i}Ht}\, U_0\, \mathrm{e}^{-\mathrm{i}H^{(0)}t} f^{(0)}$$
$$= \operatorname*{s-lim}_{t \to \pm\infty} \mathrm{e}^{\mathrm{i}Ht}\, \mathrm{e}^{-\mathrm{i}H_0 t} f_0 \tag{9.6.11}$$

The existence of the strong limit on the right-hand side of (9.6.11) may be proved by essentially the same method as for the one-particle problem in Section 9.1. In this case, we replace the smooth multiplication operator $\rho(\boldsymbol{r})$ in (9.1.4) by the operator of multiplication by the function

$$\rho_0(\boldsymbol{r}_1, \boldsymbol{r}_2, \boldsymbol{r}_3) = \rho(\boldsymbol{r}_1 - \boldsymbol{r}_2)\, \rho(\boldsymbol{r}_2 - \boldsymbol{r}_3)\, \rho(\boldsymbol{r}_3 - \boldsymbol{r}_1).$$

In order to prove the existence of the limit for $\Omega_+^{(0)} f^{(0)}$, we have to estimate the integral

$$\int_0^\infty \mathrm{d}t \left\| \frac{\mathrm{d}}{\mathrm{d}t}\, \mathrm{e}^{\mathrm{i}Ht}\, \rho_0 \mathrm{e}^{-\mathrm{i}H_0 t} f_0 \right\|.$$

Noting that $H = H_0 + V_{12} + V_{23} + V_{31}$, one contribution to the bound on this integral is bounded by

$$\int_0^\infty \mathrm{d}t\, \|V_{23}(\boldsymbol{r}_2 - \boldsymbol{r}_3)\, \rho(\boldsymbol{r}_2 - \boldsymbol{r}_3)\, \mathrm{e}^{-\mathrm{i}H_0 t} f_0\|. \tag{9.6.12}$$

As before, it will be sufficient to represent f_0 by a Gaussian function, and assume that each pair potential satisfies the conditions of Theorem 9.1. We define $f_0(\boldsymbol{r}_1, \boldsymbol{r}_2, \boldsymbol{r}_3, t)$ by

$$f_0(\boldsymbol{r}_1, \boldsymbol{r}_2, \boldsymbol{r}_3, t) = (\mathrm{e}^{-\mathrm{i}H_0 t} f_0)(\boldsymbol{r}_1, \boldsymbol{r}_2, \boldsymbol{r}_3)$$

$$= \prod_{j=1}^{3} \left(2\mathrm{i}\left(\frac{t}{2m_j} - \mathrm{i}a_j\right)\right)^{-3/2} \exp\left(\mathrm{i}|\boldsymbol{r}_j|^2 \Big/ 4\left(\frac{t}{2m_j} - \mathrm{i}a_j\right)\right), \quad (9.6.13)$$

where $f_0(\boldsymbol{r}_1, \boldsymbol{r}_2, \boldsymbol{r}_3)$ is obtained by setting $t=0$ on the right-hand side. For $t \geq \text{const} > 0$, and some $K>0$, we have

$$|\mathrm{e}^{-\mathrm{i}H_0 t} f_0| \leq \text{const}\,(t^{-3/2})^3 \exp\left(\frac{-K(|\boldsymbol{r}_1|^2 + |\boldsymbol{r}_2|^2 + |\boldsymbol{r}_3|^2)}{t^2}\right).$$

Let us make the change of variables

$$\left.\begin{aligned} \boldsymbol{r} &= \boldsymbol{r}_2 - \boldsymbol{r}_3, \\ \boldsymbol{R} &= \frac{m_2\boldsymbol{r}_2 + m_3\boldsymbol{r}_3}{m_2 + m_3}. \end{aligned}\right\} \quad (9.6.14)$$

It may readily be verified that $|\boldsymbol{r}_2|^2 + |\boldsymbol{r}_3|^2 \geq C(|\boldsymbol{r}|^2 + |\boldsymbol{R}|^2)$, for some $C>0$, so that

$$|V_{23}(\boldsymbol{r}_2 - \boldsymbol{r}_3)\rho(\boldsymbol{r}_2 - \boldsymbol{r}_3)(\mathrm{e}^{-\mathrm{i}H_0 t} f_0)(\boldsymbol{r}_1, \boldsymbol{r}_2, \boldsymbol{r}_3)|$$
$$\leq \text{const}\,(t^{-3/2})^3 |V_{23}(\boldsymbol{r})\rho(\boldsymbol{r})|\, \mathrm{e}^{-KC|\boldsymbol{r}|^2/t^2}\, \mathrm{e}^{-K|\boldsymbol{r}_1|^2/t^2}\, \mathrm{e}^{-KC|\boldsymbol{R}|^2/t^2}.$$

Squaring and integrating with respect to $\boldsymbol{r}$, $\boldsymbol{R}$ and $\boldsymbol{r}_1$, we obtain a norm estimate for the integrand in (9.6.12), of the form

$$\|\cdot\|^2 \leq \text{const}\,|t|^{-3} \int \mathrm{d}^3\boldsymbol{r}\, |V_{23}(\boldsymbol{r})\rho(\boldsymbol{r})|^2 \exp\left(\frac{-2KC|\boldsymbol{r}|^2}{t^2}\right).$$

A comparison with (9.1.8) shows that this is exactly the estimate that was used in the single-particle problem, and, following the same arguments as before, we may deduce that the integral (9.6.12) is convergent. A similar result applies to the integrals involving V_{12} and V_{31}, and the terms involving the commutator of H_0 with ρ lead to similar bounds. We conclude that the wave operator $\Omega_+^{(0)}$ and similarly $\Omega_-^{(0)}$ exist as strong limits on the space $\mathscr{H}^{(0)}$. (We leave the reader to verify that $(1-\rho)\mathrm{e}^{-\mathrm{i}H_0 t}$ converges strongly to zero in this case.)

What other channels may be defined in the three-particle problem with Hamiltonian $\boldsymbol{H}$ given by (9.6.10)? Let us suppose that a "bound state" or eigenfunction exists for the two-particle subsystem, consisting of particle 2 and particle 3. More precisely, we let $u(\boldsymbol{r})$ be a normalized solution, with eigenvalue λ_1, of the equation

$$-\frac{\Delta_r u}{2m}+V_{23}(\boldsymbol{r})u=\lambda_1 u, \tag{9.6.15}$$

where now $m=m_2m_3/(m_2+m_3)$. We suppose, in addition, that $|u(\boldsymbol{r})|\leq\text{const}\,|\boldsymbol{r}|^{-(2-\delta/2)}$ and $|\partial u(\boldsymbol{r})/\partial x_j|\leq\text{const}\,|\boldsymbol{r}|^{-(2-\delta/2)}$, in the limit as $|\boldsymbol{r}|\to\infty$, where $V_{ij}(\boldsymbol{r})$ satisfies the conditions of Theorem 9.1. We can then define a scattering channel corresponding to the asymptotic evolution of *two* free particles, viz particle 1 and a (2,3) bound state. To do this, we define a free Hilbert space $\mathscr{H}^{(1)}=L^2(\mathbb{R}^6)$, consisting of square-integrable functions $f^{(1)}(\boldsymbol{r}_1, \boldsymbol{R})$, and define the isometric mapping U_1 from $\mathscr{H}^{(1)}$ into a subspace of $\mathscr{H}=L^2(\mathbb{R}^9)$ by

$$(U_1 f^{(1)})(\boldsymbol{r}_1, \boldsymbol{r}_2, \boldsymbol{r}_3)=f^{(1)}\left(\boldsymbol{r}_1, \frac{m_2\boldsymbol{r}_2+m_3\boldsymbol{r}_3}{m_2+m_3}\right)u(\boldsymbol{r}_2-\boldsymbol{r}_3). \tag{9.6.16}$$

The free Hamiltonian $H^{(1)}$, acting in $\mathscr{H}^{(1)}$, is defined by

$$H^{(1)}=-\frac{\Delta_{r_1}}{2m_1}-\frac{\Delta_R}{2(m_2+m_3)}+\lambda_1, \tag{9.6.17}$$

where $\boldsymbol{R}$ is the coordinate given by (9.6.14). We also define the operator H_1, acting in $\mathscr{H}$, by

$$H_1=\frac{\boldsymbol{P}_1^2}{2m_1}+\frac{\boldsymbol{P}_2^2}{2m_2}+\frac{\boldsymbol{P}_3^2}{2m_3}+V_{23}(\boldsymbol{r}_2-\boldsymbol{r}_3) \tag{9.6.18}$$

As a differential operator, in terms of coordinates $\boldsymbol{r}_1$, $\boldsymbol{r}$ and $\boldsymbol{R}$, where $\boldsymbol{r}$ and $\boldsymbol{R}$ are given by (9.6.14), we have

$$H_1=-\frac{\Delta_{r_1}}{2m_1}-\frac{\Delta_r}{2m}-\frac{\Delta_R}{2(m_2+m_3)}+V_{23}(\boldsymbol{r}) \tag{9.6.18$'$}$$

Allowing H_1 to operate on both sides of (9.6.16), we have, on using (9.6.15) and (9.6.18)′,

$$H_1U_1f^{(1)}=U_1H^{(1)}f^{(1)}.$$

It follows that H_1 is an extension of the operator defined on the range of U_1 to be $U_1H^{(1)}U_1^{-1}$. If we then define the wave operators $\Omega_\pm^{(1)}$ by

$$\Omega_\pm^{(1)}f^{(1)}=\underset{t\to\pm\infty}{\text{s-lim}}\ e^{iHt}\,U_1\,e^{-iH^{(1)}t}f^{(1)}, \tag{9.6.19}$$

we have

$$\Omega_\pm^{(1)}f^{(1)}=\underset{t\to\pm\infty}{\text{s-lim}}\ e^{iHt}\,e^{-iH_1t}f, \tag{9.6.20}$$

where $f = U_1 f^{(1)}$ is given by (9.6.16).

To prove the existence of the strong limit in (9.6.20), we can again without loss of generality represent f by a Gaussian wave function. We take, in this case,

$$(\mathrm{e}^{-\mathrm{i}H^{(1)}t} f^{(1)})(\boldsymbol{r}_1, \boldsymbol{R})$$
$$= \left(2\mathrm{i}\left(\frac{t}{2m_1} - \mathrm{i}a_1\right)\right)^{-3/2} \exp\left(\mathrm{i}|\boldsymbol{r}_1|^2 \Big/ 4\left(\frac{t}{2m_1} - \mathrm{i}a_1\right)\right)$$
$$\times \left(2\mathrm{i}\left(\frac{t}{2(m_2+m_3)} - \mathrm{i}a\right)\right)^{-3/2} \exp\left(\mathrm{i}|\boldsymbol{R}|^2 \Big/ 4\left(\frac{t}{2(m_2+m_3)} - \mathrm{i}a\right)\right)$$
$$\times \exp(-\mathrm{i}\lambda_0 t). \tag{9.6.21}$$

The function $(\mathrm{e}^{-\mathrm{i}H_1 t} f)(\boldsymbol{r}_1, \boldsymbol{r}_2, \boldsymbol{r}_3)$ is then obtained by multiplying on the right-hand side by $u(\boldsymbol{r}_2 - \boldsymbol{r}_3)$, and using (9.6.14) to substitute for $\boldsymbol{r}$ and $\boldsymbol{R}$ in terms of $\boldsymbol{r}_2$ and $\boldsymbol{r}_3$.

We can now proceed exactly as for $\Omega_\pm^{(0)}$ to prove the existence of the limits on the right-hand side of (9.6.20), noting in this case that $H - H_1 = V_{12}(\boldsymbol{r}_1 - \boldsymbol{r}_2) + V_{31}(\boldsymbol{r}_3 - \boldsymbol{r}_1)$.

Let us consider, for example, a norm estimate in $L^2(\mathbb{R}^6)$ of a function bounded by

$$(t^{-3/2})^2 |V_{12}(\boldsymbol{r}_1 - \boldsymbol{r}_2)\rho(\boldsymbol{r}_1 - \boldsymbol{r}_2)| \exp\left(\frac{-K(|\boldsymbol{r}_1|^2 + |\boldsymbol{R}|^2)}{t^2}\right) |u(\boldsymbol{r})|.$$

From the elementary inequality

$$|\boldsymbol{r}_1|^2 + |\boldsymbol{R}|^2 \geq \tfrac{1}{4}(|\boldsymbol{r}_1 - \boldsymbol{R}|^2 + |\boldsymbol{R}|^2),$$

the exponential is bounded by

$$\exp\left(\frac{-K(|\boldsymbol{r}_1 - \boldsymbol{R}|^2 + |\boldsymbol{R}|^2)}{4t^2}\right).$$

Since $\boldsymbol{r}_1 - \boldsymbol{r}_2 = \boldsymbol{r}_1 - \boldsymbol{R} - m_3\boldsymbol{r}/(m_2+m_3)$, we can change the integration variable from $\boldsymbol{r}_1$ to $\boldsymbol{r}' = \boldsymbol{r}_1 - \boldsymbol{R} - m_3\boldsymbol{r}/(m_2+m_3)$. The only dependence on $\boldsymbol{R}$ in the resulting integrand is then, on squaring, the factor $\exp(-K|\boldsymbol{R}|^2/2t^2)$. After carrying out the $\boldsymbol{R}$-integral, we arrive at the norm estimate

$$\|\cdot\|^2 \leq \text{const}\, t^{-3} \int \mathrm{d}^3\boldsymbol{r}'\, \mathrm{d}^3\boldsymbol{r}\, |V_{12}(\boldsymbol{r}')\rho(\boldsymbol{r}')|^2$$
$$\times \exp\left(-K\left|\boldsymbol{r}' + \frac{m_3\boldsymbol{r}}{m_2+m_3}\right|^2 \Big/ 2t^2\right) |u(\boldsymbol{r})|^2.$$

We now use the fact that, with $0 < \delta < 1$, $t^{-(1-\delta)/2}(1+|x|)^{(1-\delta)/2}\mathrm{e}^{-Kx^2/4t^2}$ is bounded uniformly in x and t for $t \geq \mathrm{const} > 0$. This gives the estimate

$$\|\cdot\|^2 \leq \mathrm{const}\, t^{-(2+\delta)} \int \mathrm{d}^3\boldsymbol{r}'\, |V_{12}(\boldsymbol{r}')\rho(\boldsymbol{r}')|^2 F(\boldsymbol{r}'), \tag{9.6.22}$$

where

$$F(\boldsymbol{r}') = \int \frac{|u(\boldsymbol{r})|^2\, \mathrm{d}^3\boldsymbol{r}}{(1 + |\boldsymbol{r}' + m_3\boldsymbol{r}/(m_2+m_3)|)^{1-\delta}}.$$

We divide the integration for $F(\boldsymbol{r}')$ into two regions $\mathscr{R}_1$ and $\mathscr{R}_2$, defined respectively by the conditions

$$\left| \boldsymbol{r}' + \frac{m_3\boldsymbol{r}}{m_2+m_3} \right| > \tfrac{1}{2}|\boldsymbol{r}'|$$

and

$$\left| \boldsymbol{r}' + \frac{m_3\boldsymbol{r}}{m_2+m_3} \right| \leq \tfrac{1}{2}|\boldsymbol{r}'|.$$

The integral over $\mathscr{R}_1$ is bounded for large $|\boldsymbol{r}'|$ by $\mathrm{const}/|\boldsymbol{r}'|^{1-\delta}$, since u is square-integrable. In the region $\mathscr{R}_2$, with volume of order $|\boldsymbol{r}'|^3$, $|\boldsymbol{r}|$ and $|\boldsymbol{r}'|$ are of the same order of magnitude. By hypothesis, $|u(\boldsymbol{r})| \leq \mathrm{const}/|\boldsymbol{r}'|^{2-\delta/2}$, so that the integral over $\mathscr{R}_2$ is bounded as $|\boldsymbol{r}'| \to \infty$ by $\mathrm{const}\, |\boldsymbol{r}'|^3\, |\boldsymbol{r}'|^{-(4-\delta)} = \mathrm{const}/|\boldsymbol{r}'|^{1-\delta}$. Hence $F(\boldsymbol{r}') \leq \mathrm{const}/|\boldsymbol{r}'|^{1-\delta}$, and the integral on the right-hand side of (9.6.22) is convergent, for V_{12} satisfying the hypothesis of Theorem 9.1. It follows that the norm in (9.6.22) is integrable with respect to t for large positive t. Replacing the strong limit on the right-hand side of (9.6.20) by $\text{s-lim}_{t\to\infty}\, e^{iHt}\rho_1 e^{-iH_1t}f$, with $\rho_1 = \rho(\boldsymbol{r}_1 - \boldsymbol{r}_2)\rho(\boldsymbol{r}_3 - \boldsymbol{r}_1)$, and estimating in norm the various contributions to $(\mathrm{d}/\mathrm{d}t)e^{iHt}\rho_1 e^{-iH_1t}$, we can confirm that this strong derivative is integrable in norm from 0 to ∞. (The additional hypothesis on the *derivatives* of $u(\boldsymbol{r})$ is demanded in order to accommodate derivatives arising from the commutator of H_1 and ρ_1.) We thus prove the existence of $\Omega_+^{(1)} f^{(1)}$, where $f(t)$ is given by the right-hand side of (9.6.21) with $t = 0$. (It may again be verified that $(1-\rho_1)e^{iH_1t}f \to 0$ strongly as $t \to \infty$.) Exactly the same kind of argument can be applied with $f^{(1)}(\boldsymbol{r}_1, \boldsymbol{R})$ replaced by $f^{(1)}(\boldsymbol{r}_1 - \boldsymbol{c}_1, \boldsymbol{R} - \boldsymbol{c})$. (It is useful to note here that, for example, $|\boldsymbol{r}_1 - \boldsymbol{c}|^2 \geq \tfrac{1}{2}|\boldsymbol{r}_1|^2 + \mathrm{const}$.) Proceeding as before, linear combinations of these translated functions are dense in $\mathscr{H}^{(1)}$, and it follows that $\Omega_+^{(1)}$ exists as a strong limit on the entire space $\mathscr{H}^{(1)}$. The same conclusion follows for $\Omega_-^{(1)}$.

We may summarize our conclusions for three-particle scattering in the following theorem.

Theorem 9.6

Suppose that the pair potentials V_{ij} for the three-particle Hamiltonian H defined by (9.6.10) satisfy the condition

$$\int_{|\boldsymbol{r}|>c} |\boldsymbol{r}|^{-1+\delta} |V_{ij}(\boldsymbol{r})|^2 \, d^3\boldsymbol{r} < \infty$$

for all $c > 0$ and for some $\delta > 0$. Then the wave operators $\Omega_{\pm}^{(0)}$ corresponding to the asymptotic three-free-particle evolution exist on the entire space $\mathscr{H}^{(0)}$.

Let λ_i be any eigenvalue for the (j, k) particle subsystem; that is, suppose that there is an eigenfunction $u(\boldsymbol{r})$ of the equation

$$-\frac{\Delta_r u}{2m} + V_{jk}(\boldsymbol{r})u = \lambda_i u$$

(where i, j, k is some even permutation of 1, 2, 3 and $m = m_j m_k/(m_j + m_k)$), and assume further that $|u(\boldsymbol{r})|$ and $|\nabla_r u(\boldsymbol{r})|$ are of order $|\boldsymbol{r}|^{-(2-\delta/2)}$ in the limit as $|\boldsymbol{r}| \to \infty$. Then the wave operators $\Omega_{\pm}^{(i)}$ corresponding to asymptotic free evolution of particle i and a bound state of particles j and k exist on the entire free-particle spaces $\mathscr{H}^{(i)}$.

Proof

We have already sketched the proof of this result for $i = 1$. The proof for $i = 2, 3$ follows in the same way. Apart from possible channels that may be defined corresponding to bound states of *all three* particles, and which give rise to contributions to the scattering operator that are described by the identity operator on the appropriate Hilbert spaces, the list of scattering channels detailed in the theorem is complete for most three-particle systems of interest to the physicist. The reader should have no difficulty in defining and proving the existence of the corresponding wave operators for pair potentials $V_{ij}(\boldsymbol{r})$ that are of long range. Here the relevant conditions on the pair potentials are those of Theorem 9.2, which allows a wide class of long-range potentials, including the Coulomb potential. The extension to systems of more than three particles presents no new difficulties, although the analysis becomes notationally more complex as the number of particles increases. ∎

Finally, it should be noted that a *two*-particle Hamiltonian of the form

$$H = \frac{\boldsymbol{P}_1^2}{2m_1} + \frac{\boldsymbol{P}_2^2}{2m_2} + V(\boldsymbol{r}_1 - \boldsymbol{r}_2) + V_1(\boldsymbol{r}_1) + V_2(\boldsymbol{r}_2)$$

defines in general a *many-channel* quantum system. It is a useful application of the ideas of this section to define channel Hamiltonians and wave operators for this Hamiltonian, which may also be regarded as a limiting, case of the three-particle Hamiltonian (9.6.10) as m_3 tends to infinity and $\boldsymbol{r}_3$ is held fixed.

9.7 BACK TO ONE DIMENSION

Let us now briefly return to the scattering problem in one dimension, in the light of the results of this chapter. We assume that $V=V(x)$, where $V\in L^{r}(\mathbb{R})$, with $H_0=-\mathrm{d}^2/\mathrm{d}x^2$ and $H=-\mathrm{d}^2/\mathrm{d}x^2+V(x)$. To simplify domain questions, let us make the stronger assumption for the moment that $|V(x)|\leq$ const $(1+|x|)^{-1}$. The analogue of (9.4.3) for ψ_- is

$$\psi_-(x,k)=\mathrm{e}^{ikx}-\lim_{\varepsilon\to 0+}(H-k^2-\mathrm{i}\varepsilon)^{-1}V\mathrm{e}^{ikx}. \tag{9.7.1}$$

Corresponding to (9.4.8), this leads to the formula

$$(\Omega_- f)(x)=\left(\frac{1}{2\pi}\right)^{1/2}\int_{-\infty}^{\infty}\psi_-(x,k')\hat{f}(k')\,\mathrm{d}k'. \tag{9.7.2}$$

Following the arguments of the proof of Theorem 9.4, (9.7.1) implies that ψ_- satisfies the equation

$$\psi_-(x,k)=\mathrm{e}^{ikx}-\lim_{\varepsilon\to 0+}(H_0-k^2-\mathrm{i}\varepsilon)^{-1}V\psi_-. \tag{9.7.3}$$

(The limits in (9.7.1) and (9.7.3) may be interpreted as strong limits, on multiplication by $(1+|x|)^{-1}$. Actually $(1+|x|)^{-1/2-\delta}$ for some $\delta>0$ will do.) The resolvent operator for the free Hamiltonian may be written down explicitly, and the right-hand side of (9.7.3) becomes, on taking the limit $\varepsilon\to 0+$, with $k>0$,

$$\psi_-(x,k)=\mathrm{e}^{ikx}+(2ik)^{-1}\left\{\mathrm{e}^{ikx}\int_{-\infty}^{x}\mathrm{e}^{-ikx'}V(x')\psi_-(x',k)\,\mathrm{d}x'\right.$$
$$\left.+\mathrm{e}^{-ikx}\int_{x}^{\infty}\mathrm{e}^{ikx'}V(x')\psi_-(x',k)\,\mathrm{d}x'\right\}. \tag{9.7.4}$$

It follows immediately that

$$\psi_-(x,k)\sim\begin{cases} e^{ikx}\left\{1+(2ik)^{-1}\displaystyle\int_{-\infty}^{\infty} e^{-ikx'}V(x')\psi_-(x',k)\,dx'\right\} & \text{as } x\to\infty,\\ e^{ikx}+(2ik)^{-1}e^{-ikx}\displaystyle\int_{-\infty}^{\infty} e^{ikx'}V(x')\psi_-(x',k)\,dx' & \text{as } x\to-\infty \end{cases}\tag{9.7.5}$$

Now using (9.7.2) for the wave operator Ω_- together with the formula (9.4.20) for the scattering operator S, we may evaluate the Fourier transform of Sf. If the support of $\hat{f}$ is contained in the interval $[0,\infty)$ then we find, proceeding as in Section 9.4, that

$$(\widehat{Tf})(k')=\begin{cases} \left(\dfrac{-i}{2k'}\displaystyle\int_{-\infty}^{\infty} e^{-ik'x'}V(x')\psi_-(x',k')\,dx'\right)\hat{f}(k') & (k'>0),\\ \left(\dfrac{i}{2k'}\displaystyle\int_{-\infty}^{\infty} e^{-ik'x'}V(x')\psi_-(x',-k')\,dx'\right)\hat{f}(-k') & (k'<0). \end{cases}\tag{9.7.6}$$

Comparing with (9.7.5), we may define transmission and reflection coefficients, for $k>0$, by the asymptotic formulae

$$\psi_-(x,k)\sim\begin{cases} T(k)e^{ikx} & \text{as } x\to\infty,\\ e^{ikx}+R(k)e^{-ikx} & \text{as } x\to-\infty. \end{cases}\tag{9.7.7}$$

Equation (9.7.7) defines transmission and reflection coefficients for the scattering of a particle of momentum $k>0$ incoming at $t=-\infty$ from the region $x=-\infty$. The functions $T(k)$ and $R(k)$ may be calculated from the solution of the time-independent Schrödinger equation at energy $\lambda=k^2$, subject to the prescribed conditions at $x=\pm\infty$. If, moreover, corresponding to any function $\hat{f}\in L^2(\mathbb{R})$ we define the function pair

$$\hat{f}\longleftrightarrow\begin{pmatrix} f_1(k)\\ f_2(k)\end{pmatrix},\qquad \text{for } k>0,$$

by $f_1(k)=\hat{f}(k)$, $f_2(k)=\hat{f}(-k)$ then we have shown in (9.7.6) that the scattering operator S corresponds to the transformation

$$S:\begin{pmatrix} f_1(k)\\ 0\end{pmatrix}\to\begin{pmatrix} T(k)f_1(k)\\ R(k)f_1(k)\end{pmatrix}.\tag{9.7.8}$$

A similar analysis may be carried out assuming that $\hat{f}$ has support in the region $(-\infty, 0]$, and we can define corresponding transmission and reflection coefficients $T'(k)$ and $R'(k)$ for a particle of momentum $-k$ incoming from $x = +\infty$. Combining these results, we have

$$S: \begin{pmatrix} f_1(k) \\ f_2(k) \end{pmatrix} \to \begin{pmatrix} T(k) & R'(k) \\ R(k) & T'(k) \end{pmatrix} \begin{pmatrix} f_1(k) \\ f_2(k) \end{pmatrix} \tag{9.7.9}$$

The scattering operator is here represented by a two-component matrix function of momentum. The fact that S is here a matrix rather than a function relates to the degeneracy of the continuous spectrum of $H_0 = -d^2/dx^2$ in $L^2(\mathbb{R})$. (Note that $H_0 = -d^2/dx^2$ in $L^2(0, \infty)$ has a simple spectrum.)

We thus confirm that the calculations of transmission and reflection probabilities $|T(k)|^2$ and $|R(k)|^2$ for one-dimensional scattering problems do indeed correspond to the correct mathematical structure of the scattering operator. The textbooks of quantum mechanics have it right!

Chapter 10

Localization

10.1 LOCALIZATION IN SCATTERING THEORY

In proving the existence of short- and long-range wave operators in the last chapter, we made essential use of the postulated asymptotic behaviour of the potential $V(\boldsymbol{r})$ at large distances ($|\boldsymbol{r}| \to \infty$), whereas possible local singularities of the potential appeared to play a minor role. Physically, the reason for this is that, at large positive times t, a freely evolving state $e^{-iH_0 t} f$ will be found asymptotically to be concentrated at large distances from $\boldsymbol{r} = \mathbf{0}$, and will therefore be outside the influence of any singularities of the potential. The same will be true, in the long-range case, if we replace free evolution by some modified free evolution.

We have also seen how the support properties of $\hat{f}(\boldsymbol{k})$ in momentum space control the subsequent large-time evolution in position space. For example, an initial state having kinetic energy in the interval $c_1^2 \leq \boldsymbol{k}^2 \leq c_2^2$ will, under free evolution or modified free evolution, be concentrated asymptotically in the region $2c_1 t \leq |\boldsymbol{r}| \leq 2c_2 t$. More precisely, the probability of finding the particle *outside* this region will approach zero in the limit $t \to \infty$. For such a state f, $g = \Omega_+ f$, through the intertwining property of wave operators, will have *total* energy λ in the interval $c_1^2 \leq \lambda \leq c_2^2$. In this case, $e^{-iHt} g$ at large positive times will be concentrated in the region $2c_1 t \leq |\boldsymbol{r}| \leq 2c_2 t$. We therefore find for scattering states g, an important link between the spectral support properties of g with respect to the total Hamiltonian and the large-time evolution $e^{-iHt} g$ in position space. Localization in total energy implies asymptotic large-distance localization in position space, at least for scattering states. Indeed, an important ingredient in the proof of asymptotic completeness for long-range potentials (Chapter 12) will be the verification that this connection between localization in total energy and large-distance localization in position space holds at all positive energies.

Possible local singularities of the potential, while playing a minor role in the proof of existence of wave operators, do however have an important bearing on the question of asymptotic completeness. In the presence of strong local singularities, we have to rule out the possibility of localization in total energy being coupled with localization in position space *in the neighbourhood of the singularity*, rather than at large distances. In other words, a potential $V(\boldsymbol{r})$ may behave locally in such a way that, asymptotically as $t \to \infty$, states can concentrate near $\boldsymbol{r}=\mathbf{0}$ rather than at infinity.

In considering this possibility, it will be convenient first of all to separate the two aspects of this question, namely localization in position and energy, and asymptotic time evolution. We consider the first aspect in the following section, and we shall find it to be closely related to an analysis of the spectrum of the Hamiltonian H.

10.2 LOCALIZATION IN POSITION AND ENERGY

For simplicity, we shall consider a potential $V(\boldsymbol{r})$ with at most one local singularity, which we shall take to be at $\boldsymbol{r}=\mathbf{0}$. We take $V(\boldsymbol{r})$ to be bounded in any region $|\boldsymbol{r}| \geq R$, for $R>0$, and suppose that $V(\boldsymbol{r}) \to 0$ in the limit as $|\boldsymbol{r}| \to \infty$. We make no further assumptions on the potential, which is allowed to be arbitrarily singular at $\boldsymbol{r}=\mathbf{0}$.

It will be convenient to introduce the notation $E_{|H-\lambda|<\varepsilon}$ to denote the spectral projection of the Hamiltonian H for the interval $(\lambda-\varepsilon, \lambda+\varepsilon)$. Here H will be some self-adjoint extension of the operator $\hat{H}=-\Delta+V(\boldsymbol{r})$, with $D(\hat{H})=C_0^\infty(\mathbb{R}^3 \setminus \{\mathbf{0}\})$. It is important to remember that *locally*, away from $\boldsymbol{r}=\mathbf{0}$, the domains of H and H_0 are the same, where H_0 is the unique self-adjoint extension of $\hat{H}_0=-\Delta$, with $D(\hat{H}_0)=C_0^\infty(\mathbb{R}^3)$. For example, if ρ denotes the operator of multiplication, in position space, by $\rho(|\boldsymbol{r}|)$, where $\rho(r)$ is non-decreasing and infinitely differentiable, with $\rho(r) \equiv 1$ for sufficiently large r and $\rho(r) \equiv 0$ for sufficiently small r, then $\rho D(H) \subseteq D(H_0)$. Similarly, $\rho D(H_0) \subseteq D(H)$, $\rho D(H_0) \subseteq D(H_0)$ and $\rho D(H) \subseteq D(H)$. One important consequence of the fact that H_0 and H are local operators is that $(H_0+1)\rho E_{|H-\lambda|<\varepsilon}$ is defined on the entire Hilbert space, and hence bounded, by an application of the Closed-Graph Theorem.

We can also make use of the identity

$$\begin{aligned}\rho(H_0-z)^{-1}-(H-z)^{-1}\rho &= (H-z)^{-1}([H_0,\rho]+V\rho)(H_0-z)^{-1} \\ &= (H-z)^{-1}(-\Delta\rho-2\mathrm{i}(\operatorname{grad}\rho)\cdot\boldsymbol{P})(H_0-z)^{-1} \\ &\quad +(H-z)^{-1}V\rho(H_0-z)^{-1}\end{aligned}$$

to show that this operator is compact. (Since ρ vanishes near $\boldsymbol{r}=0$, $V\rho$ is non-singular and vanishes at infinity. Hence $V\rho(H_0-z)^{-1}$ may be written as $\lim_{R\to\infty} V\rho E_{|\boldsymbol{r}|<R}(H_0-z)^{-1}$ (norm limit), where $E_{|\boldsymbol{r}|<R}(H_0-z)^{-1}$ is compact, even Hilbert–Schmidt. Here $E_{|\boldsymbol{r}|<R}$ denotes multiplication, in position space, by the characteristic function of the region $|\boldsymbol{r}|<R$. If $E_{H_0<M}$ is the spectral projection of H_0 for the interval $[0, M)$ then we also have $(\text{grad } \rho)\cdot\boldsymbol{P}(H_0-z)^{-1}=\lim_{M\to\infty} (\text{grad } \rho)\cdot\boldsymbol{P}(H_0-z)^{-1} E_{H_0<M}$, where again $(\text{grad}\,\rho)\cdot\boldsymbol{P}(H_0-z)^{-1} E_{H_0<M}$ is Hilbert–Schmidt. Finally, the operator $\Delta\rho(H_0-z)^{-1}$ is Hilbert–Schmidt.)

Starting from linear combinations of $(\lambda-z)^{-1}$ for various complex values of z, we can pass by uniform limits to C_0^∞ functions of λ. If $\{\phi_n(\lambda)\}$ converges, uniformly on $\mathbb{R}$, to $\phi(\lambda)$ then $\{\phi_n(H_0)\}$ and $\{\phi_n(H)\}$ converge in operator norm to $\phi(H_0)$ and $\phi(H)$ respectively, since the spectral theorem implies that $\|\phi_n(T)-\phi(T)\|\le\sup_{\lambda\in\mathbb{R}}|\phi_n(\lambda)-\phi(\lambda)|$ for any self-adjoint operator T. Moreover, limits in operator norm of compact operators are compact. Hence we may deduce the compactness of $\rho\phi(H_0)-\phi(H)\rho$, for any $\phi\in C_0^\infty(\mathbb{R})$, or more generally for any continuous $\phi(\lambda)$ that approaches zero in the limits $\lambda\to\pm\infty$.

By the same argument applied in the case $H=H_0$, we also have that $\rho\phi(H_0)-\phi(H_0)\rho$ is compact, so that by subtraction we now have the compactness of $\rho(\phi(H)-\phi(H_0))$. On taking adjoints, $(\phi(H)-\phi(H_0))\rho$ is compact. These two results give precise expression to the fact that, away from the singularity of the potential, functions of H behave like functions of H_0. We may summarize our conclusions in the following lemma, and add a further simple consequence.

Lemma 10.1

Suppose that $V(\boldsymbol{r})$ is bounded in the region $|\boldsymbol{r}|\ge R$ for any $R>0$, and that $V(\boldsymbol{r})\to 0$ as $|\boldsymbol{r}|\to\infty$. Let $\rho(|\boldsymbol{r}|)$ be infinitely differentiable and non-decreasing, with $\rho(r)\equiv 1$ for sufficiently large r and $\rho(r)\equiv 0$ for sufficiently small r, and let $\phi\in C_0^\infty(\mathbb{R})$. Then the operators $\rho\phi(H_0)-\phi(H)\rho$ and $E_{|\boldsymbol{r}|<1}\phi(H)E_{H_0<M}$ are compact.

Proof

It remains only to prove compactness of the second operator. If we choose ρ to have support in the interval $[1,\infty)$ then $E_{|\boldsymbol{r}|<1}\rho=0$, and

$$\begin{aligned} &E_{|\boldsymbol{r}|<1}\phi(H)E_{H_0<M} \\ &\quad=-E_{|\boldsymbol{r}|<1}\{\rho\phi(H_0)-\phi(H)\rho\}E_{H_0<M} \\ &\qquad+E_{|\boldsymbol{r}|<1}\phi(H)\{(1-\rho)E_{H_0<M}\}. \end{aligned}$$

Now $(1-\rho)E_{H_0<M}$ is Hilbert–Schmidt, since $1-\rho$ is of compact support. Hence both of the operators on the right-hand side within curly brackets are compact, and the result follows. ■

For $\lambda \in \mathbb{R}$, we wish to measure the degree to which a state may be localized to an arbitrarily small neighbourhood of the singularity $\boldsymbol{r}=\mathbf{0}$, while retaining a degree of localization in total energy. With this in mind, we define

$$\gamma_\varepsilon(\lambda) = \lim_{R\to 0} \|E_{|\boldsymbol{r}|<R} E_{|H-\lambda|<\varepsilon}\|. \tag{10.2.1}$$

The limit on the right-hand side will always exist, and we take the limit as $R \to 0$ through positive values. It will also be important to consider what happens as the range of energies is decreased to zero, and we define

$$\gamma(\lambda) = \lim_{\varepsilon\to 0} \gamma_\varepsilon(\lambda) = \lim_{\substack{R\to 0\\ \varepsilon\to 0}} \|E_{|\boldsymbol{r}|<R} E_{|H-\lambda|<\varepsilon}\|. \tag{10.2.2}$$

It is not difficult to see that the double limit on the right-hand side has a value independent of the manner in which R and ε approach zero. For example, we could let $\varepsilon \to 0$ followed by the limit $R \to 0$.

Since, by the uncertainty principle, a state having well-defined position cannot have a precise value of momentum, or kinetic energy, we expect such localized states to be of large kinetic energy, and the following lemma is an expression of this fact.

Lemma 10.2

Define $\gamma_\varepsilon(\lambda)$ and $\gamma(\lambda)$ by (10.2.1) and (10.2.2). Then

$$\gamma_\varepsilon(\lambda) = \lim_{M\to \infty} \|E_{H_0>M} E_{|H-\lambda|<\varepsilon}\|, \tag{10.2.3}$$

and

$$\gamma(\lambda) = \lim_{\substack{M\to \infty\\ \varepsilon\to 0}} \|E_{H_0>M} E_{|H-\lambda|<\varepsilon}\|. \tag{10.2.4}$$

Proof

It is only necessary to prove (10.2.3), since (10.2.4) is then an immediate consequence. We let $\gamma'_\varepsilon(\lambda)$ be the limit defined by the right-hand side of (10.2.3). Given $\delta > 0$, we fix M sufficiently large that

$$\|E_{H_0>M} E_{|H-\lambda|<\varepsilon}\| \le \lambda'_\varepsilon(\lambda) + \delta, \tag{10.2.5}$$

and write

$$E_{|r|<R}E_{|H-\lambda|<\varepsilon}=E_{|r|<R}E_{H_0>M}E_{|H-\lambda|<\varepsilon}$$
$$+E_{|r|<R}E_{H_0<M}E_{|H-\lambda|<\varepsilon}. \tag{10.2.6}$$

Now $E_{|r|<1}E_{H_0<M}$ is compact (Hilbert–Schmidt), and $E_{|r|<R}$ converges strongly to zero as $R\to 0$. Hence $E_{|r|<R}E_{H_0<M}=E_{|r|<R}\{E_{|r|<1}E_{H_0<M}\}$ converges to zero in operator norm. It follows from (10.2.5) and (10.2.6), on taking the limit $R\to 0$, that

$$\gamma_\varepsilon(\lambda)\leq\gamma'_\varepsilon(\lambda)+\delta.$$

Since δ was arbitrary, we then have $\gamma_\varepsilon(\lambda)\leq\gamma'_\varepsilon(\lambda)$. On the other hand, we can define the function $\rho(r)$ above to satisfy $\rho(r)\equiv 1$ for $r\geq R$, where by (10.2.1) we can take R sufficiently small that

$$\|(1-\rho)E_{|H-\lambda|<\varepsilon}\|\leq\|E_{|r|<R}E_{|H-\lambda|<\varepsilon}\|\leq\gamma_\varepsilon(\lambda)+\delta \tag{10.2.7}$$

In this case,

$$\|E_{H_0>M}E_{|H-\lambda|<\varepsilon}\|\leq\gamma_\varepsilon(\lambda)+\delta+\|E_{H_0>M}\rho E_{|H-\lambda|<\varepsilon}\|. \tag{10.2.8}$$

Moreover,

$$\|E_{H_0>M}\rho E_{|H-\lambda|<\varepsilon}\|=\|E_{H_0>M}(H_0+1)^{-1}(H_0+1)\rho E_{|H-\lambda|<\varepsilon}\|$$
$$\leq\|E_{H_0>M}(H_0+1)^{-1}\|\,\|(H_0+1)\rho E_{|H-\lambda|<\varepsilon}\|,$$

where

$$\|E_{H_0>M}(H_0+1)^{-1}\|\leq(M+1)^{-1}\to 0 \quad \text{as } M\to\infty.$$

Taking the limit $M\to\infty$ in (10.2.8) now gives, with (10.2.5), $\gamma'_\varepsilon(\lambda)\leq\gamma_\varepsilon(\lambda)+\delta$, from which it follows that $\gamma'_\varepsilon(\lambda)\leq\gamma_\varepsilon(\lambda)$. Coupled with our previous result $\gamma_\varepsilon(\lambda)\leq\gamma'_\varepsilon(\lambda)$, we now have $\gamma_\varepsilon(\lambda)=\gamma'_\varepsilon(\lambda)$, from which (10.2.3) follows. ∎

It will sometimes be convenient to replace the projection operator $E_{|H-\lambda|<\varepsilon}$ in (10.2.2) and (10.2.4) by a corresponding smooth function of H. We let $\phi_\varepsilon(x)$ be infinitely differentiable, increasing from $\lambda-2\varepsilon$ to $\lambda-\varepsilon$, decreasing from $\lambda+\varepsilon$ to $\lambda+2\varepsilon$, and such that $\phi_\varepsilon(x)\equiv 1$ for $|x-\lambda|\leq\varepsilon$, $\phi_\varepsilon(x)\equiv 0$ for $|x-\lambda|\geq 2\varepsilon$. If $\chi_\varepsilon(x)$ is the characteristic function of the interval $(\lambda-\varepsilon,\ \lambda+\varepsilon)$, we then have

$$\chi_\varepsilon(x)\leq\phi_\varepsilon(x)\leq\chi_{2\varepsilon}(x),$$

from which it follows that

$$\gamma(\lambda)=\lim_{\substack{R\to 0\\ \varepsilon\to 0}}\|E_{|r|<R}\phi_\varepsilon(H)\|=\lim_{\substack{M\to\infty\\ \varepsilon\to 0}}\|E_{H_0>M}\phi_\varepsilon(H)\|. \tag{10.2.9}$$

Equation (10.2.9) also holds with ϕ_ε replaced by ϕ_ε^2. We now have the basic result of the theory of localization.

Theorem 10.1

Define $\gamma(\lambda)$ by (10.2.2). Then, for given λ, $\gamma(\lambda)$ is either 0 or 1.

Proof

For $R \leq 1$, we write

$$\begin{aligned} E_{|r|<R}\phi_\varepsilon^2(H) = E_{|r|<R}\{E_{|r|<1}\phi_\varepsilon^2(H)E_{H_0<M}\} \\ + E_{|r|<R}\{E_{|r|<1}E_{H_0<M}\}\phi_\varepsilon^2(H)E_{H_0>M} \\ + E_{|r|<R}E_{H_0>M}\phi_\varepsilon^2(H)E_{H_0>M}. \end{aligned} \tag{10.2.10}$$

From Lemma 10.1 with $\phi = \phi_\varepsilon^2$, the first operator within the curly brackets is compact, as also is the second operator. Since $E_{|r|<R}$ converges strongly to zero, the first two terms on the right-hand side of (10.2.10) converge in norm to zero in the limit $R \to 0$. We therefore have, from (10.2.10),

$$\begin{aligned} \lim_{R\to 0} \|E_{|r|<R}\phi_\varepsilon^2(H)\| &\leq \|E_{H_0>M}\phi_\varepsilon^2(H)E_{H_0>M}\| \\ &= \|(E_{H_0>M}\phi_\varepsilon(H))(E_{H_0>M}\phi_\varepsilon(H))^*\| \\ &= \|E_{H_0>M}\phi_\varepsilon(H)\|^2. \end{aligned}$$

Taking the further limits $\varepsilon \to 0$, $M \to \infty$, and using (10.2.9) with ϕ_ε replaced by ϕ_ε^2, we now have,

$$\gamma(\lambda) \leq (\gamma(\lambda))^2.$$

However, from the definition (10.2.2) of $\gamma(\lambda)$, we have $0 \leq \gamma(\lambda) \leq 1$, implying that $\gamma(\lambda) \geq (\gamma(\lambda))^2$. It follows that $\gamma(\lambda) = (\gamma(\lambda))^2$, from which $\gamma(\lambda) = 0$ or 1. ■

Definition 10.1

We shall say that λ is a *regular* point of the Hamiltonian H if $\gamma(\lambda) = 0$, and a *singular point* if $\gamma(\lambda) = 1$.

We can obtain a similar result to that of the last theorem, with $\gamma(\lambda)$ replaced by $\gamma_\varepsilon(\lambda)$.

Theorem 10.2

Suppose that $\lambda \pm \varepsilon$ are regular points of H. Then, for given λ and ε, $\gamma_\varepsilon(\lambda)$ is either 0 or 1. Moreover, $\gamma_\varepsilon(\lambda)=1$ if and only if the interval $(\lambda-\varepsilon, \lambda+\varepsilon)$ contains at least one singular point.

Proof

We define a C_0^∞ function $\psi_\delta(x)$, increasing from $\lambda-\varepsilon-\delta$ to $\lambda-\varepsilon$ and decreasing from $\lambda+\varepsilon$ to $\lambda+\varepsilon+\delta$, with $\psi_\delta(x)\equiv 1$ for $x\in(\lambda-\varepsilon, \lambda+\varepsilon)$ and $\psi_\delta(x)\equiv 0$ for $x\notin(\lambda-\varepsilon-\delta, \lambda+\varepsilon+\delta)$. We can think of ψ_δ as a smooth approximation, for small δ, to the characteristic function of the interval $[\lambda-\varepsilon, \lambda+\varepsilon]$. Since ψ_δ differs from this characteristic function only within the intervals $(\lambda-\varepsilon-\delta, \lambda-\varepsilon)$ and $(\lambda+\varepsilon, \lambda+\varepsilon+\delta)$, we have

$$\begin{aligned} &\|E_{|r|<R}(E_{|H-\lambda|<\varepsilon}-\psi_\delta(H))\| \\ &\leq \|E_{|r|<R}E_{|H-\lambda-\varepsilon|<\delta}\|+\|E_{|r|<R}E_{|H-\lambda+\varepsilon|<\delta}\| \\ &\to 0 \quad \text{as } \delta\to 0, \end{aligned}$$

since $\lambda\pm\varepsilon$ are regular points. From (10.2.1), we then have,

$$\gamma_\varepsilon(\lambda)=\lim_{\substack{R\to 0\\ \delta\to 0}} \|E_{|r|<R}\psi_\delta(H)\|,$$

from which it follows as before that

$$\gamma_\varepsilon(\lambda)=\lim_{\substack{M\to\infty\\ \delta\to 0}} \|E_{H_0>M}\psi_\delta(H)\|.$$

We now repeat the proof of Theorem 10.1, with $\phi_\varepsilon(H)$ replaced by $\psi_\delta(H)$, and with the limit $\delta\to 0$ instead of $\varepsilon\to 0$. This proves that $\gamma_\varepsilon(\lambda)=0$ or 1.

If the interval $(\lambda-\varepsilon, \lambda+\varepsilon)$ contains at least one singular point, say λ', then

$$\gamma_\varepsilon(\lambda)\geq \lim_{\substack{R\to 0\\ \delta\to 0}} \|E_{|r|<R}E_{|H-\lambda'|<\delta}\|=1,$$

so that $\gamma_\varepsilon(\lambda)=1$. Conversely, let us suppose that $\gamma_\varepsilon(\lambda)=1$, and that the interval contains no singular point. Subdividing the interval into two subintervals $(\lambda-\varepsilon, \lambda)$ and $(\lambda, \lambda+\varepsilon)$, together with the single (regular) point λ, we must have either $\gamma_{\varepsilon/2}(\lambda-\frac{1}{2}\varepsilon)=1$, or $\gamma_{\varepsilon/2}(\lambda+\frac{1}{2}\varepsilon)=1$. (If both of these are zero then it follows easily that $\gamma_\varepsilon(\lambda)=0$.) Let us suppose, for example, that $\gamma_{\varepsilon/2}(\lambda-\frac{1}{2}\varepsilon)=1$. Then, again, either $\gamma_{\varepsilon/4}(\lambda-\frac{3}{4}\varepsilon)=1$ or $\gamma_{\varepsilon/4}(\lambda-\frac{1}{4}\varepsilon)=1$. Proceeding in this way, we arrive at a sequence of points $\{\lambda_n\}$ such that $\gamma_{\varepsilon/2^n}(\lambda_n)=1$, $|\lambda_n-\lambda_{n-1}|=\varepsilon/2^n$ and $\lambda_n\in(\lambda-\varepsilon, \lambda+\varepsilon)$. Let us suppose that $\lambda_n\to\lambda'$, say, where $\lambda'\in[\lambda-\varepsilon, \lambda+\varepsilon]$. Then certainly $\gamma(\lambda')=1$, since $\gamma_{\varepsilon/2^n}(\lambda_n)=1$ and $|\lambda_n-\lambda'|\leq\varepsilon/2^n$, where λ_n is a regular point, imply that $\gamma_{\varepsilon/2^n}(\lambda')=1$. But this contradicts the hypothesis that $[\lambda-\varepsilon, \lambda+\varepsilon]$ contains no singular point. We therefore conclude that $\gamma_\varepsilon(\lambda)=1$ implies the existence of at least one singular point in the interval. ■

Corollary

The set of singular points is closed. Moreover,

(i) if $\lambda \pm \varepsilon$ are regular points then $\gamma_\varepsilon(\lambda)=0$ if and only if the operator $E_{|\mathbf{r}|<1} E_{|H-\lambda|<\varepsilon}$ is compact;
(ii) $\gamma(\lambda)=0$ if and only if $E_{|\mathbf{r}|<1} E_{|H-\lambda|<\varepsilon}$ is compact for sufficiently small ε;
(iii) H has no singular points if and only if $E_{|\mathbf{r}|<1}(H-z)^{-1}$ is compact for $\operatorname{Im}(z) \neq 0$.

Proof

We let λ be the limit of a sequence of singular points. By the theorem, we then have $\gamma_\varepsilon(\lambda)=1$ for all ε. Hence $\gamma(\lambda)=1$. Hence the set of singular points is closed.

(i) We suppose that $E_{|\mathbf{r}|<1} E_{|H-\lambda|<\varepsilon}$ is compact. Then

$$\gamma_\varepsilon(\lambda)=\lim_{R \to 0} \| E_{|\mathbf{r}|<R}\{E_{|\mathbf{r}|<1} E_{|H-\lambda|<\varepsilon}\}\|=0,$$

since $E_{|\mathbf{r}|<R}$ converges strongly to zero. Conversely, let us suppose that $\gamma_\varepsilon(\lambda)=0$. For $0<R<1$, we let $\rho_1 \in C_0^\infty(\mathbb{R}^3 \setminus \{\mathbf{0}\})$, with $\rho_1(\mathbf{r}) \equiv 1$ for $R<|\mathbf{r}|<=1$. Then $\rho_1 E_{|H-\lambda|<\varepsilon}=(H_0+1)^{-1}\rho_2(H_0+1)\rho_1 E_{|H-\lambda|<\varepsilon}$ for any $\rho_2 \in C_0^\infty(\mathbb{R}^3)$ such that $\rho_2(\mathbf{r}) \equiv 1$ on the support of ρ_1. Now $(H_0+1)\rho_1 E_{|H-\lambda|<\varepsilon}$ is bounded, and $(H_0+1)^{-1}\rho_2$ is Hilbert–Schmidt. Hence $\rho_1 E_{|H-\lambda|<\varepsilon}$ is compact, from which it follows that

$$E_{R<|\mathbf{r}|<1} E_{|H-\lambda|<\varepsilon}=E_{R<|\mathbf{r}|<1}\rho_1 E_{|H-\lambda|<\varepsilon}$$

is compact. Now $\gamma_\varepsilon(\lambda)=\lim_{R\to 0}\|E_{|\mathbf{r}|<R}E_{|H-\lambda|<\varepsilon}\|=0$, so that

$$E_{|\mathbf{r}|<1} E_{|H-\lambda|<\varepsilon}=\lim_{R \to 0} E_{R<|\mathbf{r}|<1} E_{|H-\lambda|<\varepsilon}$$

is the norm limit of compact operators, and hence compact.

(ii) $\gamma(\lambda)=0 \Leftrightarrow [\lambda-\varepsilon, \lambda+\varepsilon]$ contains no singular points for small $\varepsilon \Leftrightarrow E_{|\mathbf{r}|<1} E_{|H-\lambda|<\varepsilon}$ is compact, by (i).

(iii) If $E_{|\mathbf{r}|<1}(H-z)^{-1}$ is compact then so is

$$E_{|\mathbf{r}|<1} E_{|H-\lambda|<\varepsilon}=E_{|\mathbf{r}|<1}(H-z)^{-1}[(H-z)E_{|H-\lambda|<\varepsilon}],$$

in which case, by (iii), there can be no singular points. Conversely, let us suppose that H has no singular points. Then $E_{|\mathbf{r}|<1} E_{|H|<M}$ is compact, since $[-M, M]$ contains no singular points. On the other hand,

$$\lim_{M \to \infty} E_{|H| \geq M}(H-z)^{-1}=0,$$

so that

$$E_{|\boldsymbol{r}|<1}(H-z)^{-1} = \lim_{M\to\infty} E_{|\boldsymbol{r}|<1} E_{|H|<M}(H-z)^{-1},$$

which is the norm limit of compact operators, and hence is compact. ■

Thus the classification of a point λ as a singular or regular point is reduced to a question of compactness of a product of projections that localize in position and total energy respectively. If λ is a regular point of the Hamiltonian and P_{d} is the projection onto the discrete subspace of H then we known that $E_{|H-\lambda|<\varepsilon}(1-P)$ converges strongly to zero in the limit $\varepsilon \to 0$, so that $E_{|\boldsymbol{r}|<1} E_{|H-\lambda|<\varepsilon}(1-P_{\mathrm{d}})$ converges in norm to zero. This means, for continuum states with energy concentrated within the interval $(\lambda-\varepsilon, \lambda+\varepsilon)$, that there is a high probability of finding the particle outside the region $|\boldsymbol{r}|<1$, and similarly for any other finite region $|\boldsymbol{r}|<R$. In other words, if λ is a regular point then continuum states that are highly localized in energy at λ will necessarily be found at large distances from $\boldsymbol{r}=\mathbf{0}$. Such states are effectively outside the influence of the potential, and are also highly localized in kinetic energy. They may be used to describe the scattering of an incident beam of particles that is prepared to have a small spread of energies.

If λ is a singular point of H then the situation is different. States can then be prepared that are both highly localized in total energy and at the same time concentrated in position space in a small neighbourhood of the singularity $\boldsymbol{r}=\mathbf{0}$. Such states can have a profound effect on scattering, and can in certain circumstances cause a violation of asymptotic completeness. We shall return to this question in later chapters. For the present, let us note, in the following theorem, that the presence of singular points is a rare phenomenon, which may be ruled out for a wide class of potentials.

Theorem 10.3

Suppose that one of the following conditions is satisfied:

(i) $D(H)\subseteq D(H_0^{\delta})$ or $D(H)\subseteq D(|\boldsymbol{r}|^{-\delta})$ for some $\delta>0$;

(ii) $H=H_0+V$ is defined by means of the Friedrichs extension, in terms of the sesquilinear form on $C_0^{\infty}(\mathbb{R}^3\setminus\{0\})$; either H_0+gV is bounded below for some $g>1$, or $H_0+V-g|\boldsymbol{r}|^{-\delta}$ is bounded below for some $g>0$, $\delta>0$.

Then H has no singular points.

Proof

(i) If $D(H)\subseteq D(H_0^\delta)$ then, by the Closed-Graph Theorem, $(H_0+1)^\delta E_{|H-\lambda|<\varepsilon}$ is bounded. Also $E_{|\boldsymbol{r}|<1}(H_0+1)^{-\delta}$ is compact. Hence $E_{|\boldsymbol{r}|<1}E_{|H-\lambda|<\varepsilon}$ is compact, so that λ cannot be a singular point.

If $D(H)\subseteq D(|\boldsymbol{r}|^{-\delta})$ then $|\boldsymbol{r}|^{-\delta}E_{|H-\lambda|<\varepsilon}$ is bounded. Also $\|E_{|\boldsymbol{r}|<R}|\boldsymbol{r}|^\delta\|\leq R^\delta\to 0$ as $R\to 0$. Hence $\lim_{R\to 0}\|E_{|\boldsymbol{r}|<R}E_{|H-\lambda|<\varepsilon}\|=0$, so that again λ cannot be a singular point.

(ii) If H_0+gV is bounded below with $g>1$, let us write $H_0+V=g^{-1}(H_0+gV)+(1-g^{-1})H_0$. From the definition of the Friedrichs extension, it follows that the domain of H is contained in the form domain of H_0+gV and of H_0. That is, $D(H)\subseteq D(H_0^{1/2})$, and the result follows from (i). Similarly, if $H_0+V-g|\boldsymbol{r}|^{-\delta}$ is bounded below with $g>0$, $\delta>0$ then it follows that $D(H)\subseteq D(|\boldsymbol{r}|^{-\delta/2})$. Again H can have no singular points. ■

The following theorem provides a more direct method of obtaining the singular points, if any, of a Schrödinger Hamiltonian.

Theorem 10.4

Let $\mathscr{B}$ be the region $|\boldsymbol{r}|<1$, and define the self-adjoint operator $H^{(1)}=-\Delta+V(\boldsymbol{r})$ in $L^2(\mathscr{B})$, with Dirichlet boundary conditions on $|\boldsymbol{r}|=1$ and the same boundary conditions as H at $\boldsymbol{r}=0$. Then the singular points of H are exactly the points of the essential spectrum of $H^{(1)}$.

Proof

We let $H_0^{(1)}$ be the self-adjoint operator $-\Delta$ acting in $L^2(\mathscr{B})$, with Dirichlet boundary conditions on $|\boldsymbol{r}|=1$. In a given partial-wave subspace, $H_0^{(1)}$ is unitarily equivalent to the ordinary differential operator $-\mathrm{d}^2/\mathrm{d}r^2+l(l+1)/r^2$ in $L^2(0, 1)$, with boundary conditions $f(1)=0$, and $f(0)=0$ if $l=0$. Since $H_0^{(1)}$ is known explicitly, it is straightforward to verify that $H_0^{(1)}$ has a purely discrete spectrum, and consequently that $(H_0^{(1)}-z)^{-1}$ is compact. Locally near $|\boldsymbol{r}|=1$, $H_0^{(1)}$ and $H^{(1)}$ have the same domains.

We can also define in $L^2(\mathbb{R}^3)$ a self-adjoint operator H^{D} that decouples the two regions $|\boldsymbol{r}|<1$ and $|\boldsymbol{r}|>1$. Thus $L^2(\mathbb{R}^3)=L^2(\mathscr{B})\oplus L^2(\mathbb{R}^3\setminus\mathscr{B})$, and each component of $L^2(\mathbb{R}^3)$ is made into an invariant subspace for H^{D} by applying Dirichlet boundary conditions on $|\boldsymbol{r}|=1$. Acting in $L^2(\mathscr{B})$, we can write $H^{\mathrm{D}}=H^{(1)}$. In $L^2(\mathbb{R}^3\setminus\mathscr{B})$, we define H^{D} by $H^{\mathrm{D}}=-\Delta+V(\boldsymbol{r})$ with Dirichlet boundary conditions on $|\boldsymbol{r}|=1$. We now let $\rho_0(r)$ be infinitely differentiable, with $\rho_0\equiv 1$ for $0\leq r\leq\frac{1}{3}$ and $\rho_0\equiv 0$ for $r>\frac{2}{3}$, and similarly take $\rho_\infty(r)$ such that $\rho_\infty\equiv 0$ for $0\leq r\leq 2$ and $\rho_\infty\equiv 1$ for $r\geq 3$. With $\rho^{(1)}=1-\rho_0-\rho_\infty$, we have $\rho^{(1)}\in C_0^\infty(\mathbb{R}^3\setminus\{0\})$ and $\rho^{(1)}(|\boldsymbol{r}|)\equiv 1$ near $|\boldsymbol{r}|=1$.

Since H and H^{D} have the same domains locally away from $|\boldsymbol{r}|=1$, we can use the identity

$$\rho_0(H-z)^{-1}-(H^{\mathrm{D}}-z)^{-1}\rho_0=(H^{\mathrm{D}}-z)^{-1}[-\Delta,\rho_0]\rho(H-z)^{-1},$$

where $\rho\in C_0^\infty(\mathbb{R}^3\setminus\{0\})$ and $\rho\equiv 1$ on the support of grad ρ_0, to deduce that the operator on the left-hand side of this identity is compact. Similarly, $\rho_\infty(H-z)^{-1}-(H^{\mathrm{D}}-z)^{-1}\rho_\infty$ is compact.

In order to deduce that $(H-z)^{-1}-(H^{\mathrm{D}}-z)^{-1}$ is compact, we can write $1=\rho^{(1)}+\rho_0+\rho_\infty$, and it remains only to prove that $\rho^{(1)}(H-z)^{-1}-(H^{\mathrm{D}}-z)^{-1}\rho^{(1)}$ is compact. Now $\rho^{(1)}(H-z)^{-1}$ is compact for $\rho^{(1)}\in C_0^\infty(\mathbb{R}^3\setminus\{0\})$. For the part of $(H^{\mathrm{D}}-z)^{-1}\rho^{(1)}$ in $L^2(\mathscr{B})$, we can write the adjoint of this operator as $(H_0^{(1)}-\bar{z})^{-1}[(H_0^{(1)}-\bar{z})\rho^{(1)}(H^{(1)}-\bar{z})^{-1}]$, where the operator in square brackets is bounded and the resolvent of $H_0^{(1)}$ is compact. For the part of $(H^{\mathrm{D}}-z)^{-1}\rho^{(1)}$ in $L^2(\mathbb{R}^3\setminus\mathscr{B})$, we can apply a similar argument using the compactness of the resolvent operator for $-\Delta$ acting in L^2 of the region $1<|\boldsymbol{r}|<4$, with Dirichlet boundary conditions on $|\boldsymbol{r}|=1$ and $|\boldsymbol{r}|=4$ respectively. Hence $(H^{\mathrm{D}}-z)^{-1}\rho^{(1)}$ is compact, and we have shown the compactness of $(H-z)^{-1}-(H^{\mathrm{D}}-z)^{-1}$.

For $\phi\in C_0^\infty(\mathbb{R}^3)$, the compactness of $\phi(H)-\phi(H^{\mathrm{D}})$ now follows. Defining $\phi_\varepsilon(H)$ as in (10.2.9), the operator $E_{|\boldsymbol{r}|<1}E_{|H-\lambda|<\varepsilon}$ will be compact for sufficiently small ε if and only if $E_{|\boldsymbol{r}|<1}\phi_\varepsilon(H)$ is compact for sufficiently small ε. Since $\phi_\varepsilon(H)-\phi_\varepsilon(H^{\mathrm{D}})$ is compact, the condition for λ to be a regular point becomes the condition that $E_{|\boldsymbol{r}|<1}\phi_\varepsilon(H^{\mathrm{D}})$ be compact for sufficiently small ε. However, this operator is the zero operator in $L^2(\mathbb{R}^3\setminus\mathscr{B})$, and in $L^2(\mathscr{B})$ it may be identified with $\phi_\varepsilon(H^{(1)})$. Also, $\phi_\varepsilon(H^{(1)})$ will be compact for small ε if and only if the same is true of $E_{|H^{(1)}-\lambda|<\varepsilon}$. And the condition for this is that, for small enough ε, this operator projects onto a subspace of finite dimension. Thus for λ to be a singular point of H it is necessary and sufficient that λ be a point of the essential spectrum of $H^{(1)}$. ■

Chapter 11

The Trace Method in Scattering Theory

11.1 THE TRACE THEOREM

The trace method furnishes us with a tool for attacking two of the central problems in scattering theory, namely the proof of the existence and asymptotic completeness of wave operators. The proof proceeds normally in two distinct stages. First of all, abstract criteria are obtained that guarantee the existence and completeness of the appropriate strong limits in a general Hilbert space $\mathscr{H}$. These criteria involve a pair of self-adjoint operators H_1, H_2 acting in $\mathscr{H}$, and an associated operator A, which must be of trace class. For the definition and properties of operators in the class C_1 of trace class operators, see Section 2.6. These criteria are embodied in a result known as the Trace Theorem (Theorem 11.1 of this section).

The second stage of the proof of existence and completeness of the wave operators consists in the verification, in a concrete situation involving a particular free and total Hamiltonian, that the criteria postulated by the Trace Theorem apply. In particular, the operator A must be defined and verified to be of trace class, and sometimes additional compactness or other conditions have to be verified for certain operators.

The domain of applicability of the Trace Theorem is very wide indeed. Applications exist to acoustic and electromagnetic scattering and to quantum-mechanical scattering by short-range potentials, which may be singular or non-singular. The Trace Theorem provides, within a general Hilbert-space framework, probably the simplest abstract statement of conditions guaranteeing asymptotic completeness that can be found to cover such a variety of applications. On the other hand, the trace method does have a number of disadvantages, including the following.

(i) The application of the Trace Theorem to scattering by non-spherical potentials requires a decay at infinity at least as rapid as $V \sim \text{const}/|\boldsymbol{r}|^3$ to deduce asymptotic completeness. One can improve this, by adapting the method to potentials like $V \sim \text{const}/|\boldsymbol{r}|^{2+\varepsilon}$ for $\varepsilon > 0$, but potentials that are non-spherical and that behave for large $|\boldsymbol{r}|$ like $\text{const}/|\boldsymbol{r}|^{1+\varepsilon}$ cannot be treated, as the theory stands at present. Spherically symmetric potentials with such decay at infinity present no problems, but even in the spherical case trace methods do not apply to long-range problems such as Coulomb scattering.

(ii) The many-body problem is inaccessible to trace methods, although the scattering of pairs of bound states can be dealt with below the energy threshold at which multiparticle processes begin.

(iii) The Trace Theorem does not lead to a complete spectral analysis of the Hamiltonian. For example, the absence (or otherwise) of a singular continuous spectrum cannot be proved by trace methods.

Despite these drawbacks, trace theory retains a considerable elegance and simplicity, as I shall try to demonstrate in this chapter. In Chapter 12 we shall use a quite different approach to overcome the defects listed above.

We start by defining a suitable class of vectors on which wave operators will be proved to exist. Given a self-adjoint operator H, in a Hilbert space $\mathscr{H}$, we shall say that a vector $\phi \in \mathscr{H}$ belongs to the *m-class* of vectors associated with H provided that $\phi \in \mathscr{M}_{\text{ac}}(H)$ and

$$\operatorname*{ess\,sup}_{\lambda} \frac{\mathrm{d}}{\mathrm{d}\lambda} \langle \phi, E_\lambda \phi \rangle \leq m^2, \tag{11.1.1}$$

where $H = \int \lambda \mathrm{d}E_\lambda$. On the left-hand side we write *essential* supremum rather than supremum because the derivative need not exist for all λ. We require only that (11.1.1) hold for almost all λ.

We have already met a condition like (11.1.1) when, in Section 2.6, we presented a characterization of the absolutely continuous subspace of a self-adjoint operator in terms of the corresponding evolution group and the context here is not totally dissimilar.

Given an arbitrary vector $f \in \mathscr{M}_{\text{ac}}(H)$, we let E_m denote the spectral projection of H associated with the set $\mathscr{S}_m$ of λ for which $(\mathrm{d}/\mathrm{d}\lambda)\langle f, E_\lambda f \rangle \leq m^2$. Then

$$\langle f, E_m f \rangle = \int_{\mathscr{S}_m} \mathrm{d}\langle f, E_\lambda f \rangle,$$

$$\frac{\mathrm{d}}{\mathrm{d}\lambda} \langle E_m f, E_\lambda E_m f \rangle = \chi_m(\lambda) \frac{\mathrm{d}}{\mathrm{d}\lambda} \langle f, E_\lambda f \rangle$$

where $\chi_m(\lambda)$ is the characteristic function of $\mathscr{S}_m$. Hence

$$\frac{\mathrm{d}}{\mathrm{d}\lambda} \langle E_m f, E_\lambda E_m f \rangle \leq m^2,$$

so that $E_m f$ belongs to the m-class of vectors. As m tends to intinity, an application of the Lebesgue Dominated-Convergence Theorem shows that $E_m f \to f$ strongly. Hence the union of the m-classes of vectors, as m varies, is dense in $\mathcal{M}_{\mathrm{ac}}(H)$. The inequality

$$\left|\frac{\mathrm{d}}{\mathrm{d}\lambda}\langle\phi_1, E_\lambda\phi_2\rangle\right|^2 \leq \frac{\mathrm{d}}{\mathrm{d}\lambda}\langle\phi_1, E_\lambda\phi_1\rangle \frac{\mathrm{d}}{\mathrm{d}\lambda}\langle\phi_2, E_\lambda\phi_2\rangle, \tag{11.1.2}$$

for $\phi_1, \phi_2 \in \mathcal{M}_{\mathrm{ac}}(H)$, may be used to show that the union of the m-classes is a linear set of vectors.

The following lemma gives the basic estimate that is needed for proof of the Trace Theorem.

Lemma 11.1

Let H_1 be a self-adjoint operator in a Hilbert space $\mathscr{H}$. Let $A \in C_1$ (trace-class operators) and let T be a bounded linear operator. For any bounded linear operator B, define $\gamma_a(B)$ for $a > 0$ by

$$\gamma_a(B) = \int_0^a \mathrm{d}t\, \mathrm{e}^{\mathrm{i}H_1 t} B \mathrm{e}^{-\mathrm{i}H_1 t}. \tag{11.1.3}$$

Then, for any ϕ belonging to the m-class of vectors associated with the self-adjoint operator H_1, we have

$$\int_{-\infty}^{\infty} \mathrm{d}r\, \| |A|^{1/2} \mathrm{e}^{-\mathrm{i}H_1 r}\phi\|^2 \leq 2\pi m^2 \|A\|_1, \tag{11.1.4}$$

$$|\langle\phi, \gamma_a(\mathrm{e}^{\mathrm{i}H_1 t} T A \mathrm{e}^{-\mathrm{i}H_1 s})\phi\rangle| \leq (2\pi)^{1/2} m \|T\| (\|A\|_1)^{1/2} \left(\int_s^{\infty} \mathrm{d}r \| |A|^{1/2} \mathrm{e}^{-\mathrm{i}H_1 r}\phi\|^2\right)^{1/2}. \tag{11.1.5}$$

Proof

From (2.6.17), we shall use the representation

$$A = \sum_{j=1}^{\infty} \beta_j \langle\phi_j, \cdot\rangle \psi_j, \tag{11.1.6}$$

where the sequences $\{\phi_j\}$ and $\{\psi_j\}$ are orthonormal and the β_j^2 are the positive eigenvalues of A^*A with eigenvectors ϕ_j; we take $\beta_j>0$. The norm $\|A\|_1$ of A in the space C_1 of trace class operators is then $\|A\|_1=\Sigma_{j=1}^{\infty}\beta_j$. The operator $|A|=(A^*A)^{1/2}$ may be represented as

$$|A|=\sum_{j=1}^{\infty}\beta_j\langle\phi_j,\cdot\rangle\phi_j. \tag{11.1.7}$$

The operator $\gamma_a(B)$ may be defined either as a strong limit of an approximating sum, as in the similar definition of a Riemann integral, or simply by the formula

$$\langle g,\gamma_a(B)f\rangle=\int_0^a \mathrm{d}t\,\langle \mathrm{e}^{-\mathrm{i}H_1t}g, B\mathrm{e}^{-\mathrm{i}H_1t}f\rangle.$$

These two ways of defining the integral in (11.1.3) are equivalent. We begin by obtaining the estimate (11.1.4). Substituting from (11.1.7) for $|A|$, we have

$$\begin{aligned}\||A|^{1/2}\mathrm{e}^{-\mathrm{i}H_1r}\phi\|^2&=\langle \mathrm{e}^{-\mathrm{i}H_1r}\phi,|A|\mathrm{e}^{-\mathrm{i}H_1r}\phi\rangle\\&=\sum_{j=1}^{\infty}\beta_j\langle\phi_j,\mathrm{e}^{-\mathrm{i}H_1r}\phi\rangle\langle\mathrm{e}^{-\mathrm{i}H_1r}\phi,\phi_j\rangle\\&=\sum_{j=1}^{\infty}\beta_j|\langle\phi_j,\mathrm{e}^{-\mathrm{i}H_1r}\phi\rangle|^2.\end{aligned} \tag{11.1.8}$$

We let $H_1=\int\lambda\,\mathrm{d}E_\lambda$. Now

$$\begin{aligned}\langle\phi_j,\mathrm{e}^{-\mathrm{i}H_1r}\phi\rangle&=\int\mathrm{e}^{-\mathrm{i}\lambda r}\,\mathrm{d}\langle\phi_j,E_\lambda\phi\rangle\\&=\int\mathrm{e}^{-\mathrm{i}\lambda r}\frac{\mathrm{d}}{\mathrm{d}\lambda}\langle\phi_j,E_\lambda\phi\rangle\,\mathrm{d}\lambda,\qquad\text{since }\phi\in\mathcal{M}_{\mathrm{ac}}(H_1).\end{aligned}$$

Also, from (11.1.2),

$$\begin{aligned}\left|\frac{\mathrm{d}}{\mathrm{d}\lambda}\langle\phi_j,E_\lambda\phi\rangle\right|^2&\le\frac{\mathrm{d}}{\mathrm{d}\lambda}\langle\phi_j,E_\lambda\phi_j\rangle\frac{\mathrm{d}}{\mathrm{d}\lambda}\langle\phi,E_\lambda\phi\rangle\\&\le m^2\frac{\mathrm{d}}{\mathrm{d}\lambda}\langle\phi_j,E_\lambda\phi_j\rangle,\end{aligned}$$

since ϕ is in the m-class for H_1. We are able to take $\phi_j\in\mathcal{M}_{\mathrm{ac}}(H_1)$ without loss of generality, since $\phi\in\mathcal{M}_{\mathrm{ac}}(H_1)$, and the projection P_{ac} onto $\mathcal{M}_{\mathrm{ac}}(H_1)$ commutes with E_λ; thus

$$\langle\phi_j,E_\lambda\phi\rangle=\langle\phi_j,E_\lambda P_{\mathrm{ac}}\phi\rangle=\langle P_{\mathrm{ac}}\phi_j,E_\lambda\phi\rangle.$$

Since $\|\phi_j\|=1$, we have

$$\int_{-\infty}^{\infty}\left|\frac{d}{d\lambda}\langle\phi_j,E_\lambda\phi\rangle\right|^2 d\lambda$$

$$\le m^2\int_{-\infty}^{\infty}\frac{d}{d\lambda}\langle\phi_j,E_\lambda\phi_j\rangle\,d\lambda$$

$$=m^2.$$

We have thus shown that

$$\langle\phi_j,e^{-iH_1r}\phi\rangle=\int_{-\infty}^{\infty}e^{-i\lambda r}h(\lambda)\,d\lambda,$$

where $\int_{-\infty}^{\infty}|h(\lambda)|^2\,d\lambda\le m^2$. By the unitary property of the Fourier transform, this implies that

$$\int_{-\infty}^{\infty}dr\,|\langle\phi_j,e^{-iH_1r}\phi\rangle|^2=2\pi\int_{-\infty}^{\infty}|h(\lambda)|^2\,d\lambda$$

$$\le 2\pi m^2.$$

Applying this estimate to (11.1.8), with $\|A\|_1=\Sigma_{j=1}^{\infty}\beta_j$, we arrive at (11.1.4).

We now evaluate the left-hand side of (11.1.5), substituting for A from (11.1.6), to give, on applying the Schwarz inequality to the resulting sum/integral,

$$|\langle\phi,\gamma_a(e^{iH_1t}TAe^{-iH_1s})\phi\rangle|$$

$$=\left|\sum_{j=1}^{\infty}\beta_j\int_0^a dr\langle\phi,e^{iH_1(r+t)}T\psi_j\rangle\langle\phi_j,e^{-iH_1(r+s)}\phi\rangle\right|$$

$$\le\left(\sum_{j=1}^{\infty}\beta_j\int_0^a dr\,|\langle\phi,e^{iH_1(r+t)}T\psi_j\rangle|^2\right)^{1/2}$$

$$\times\left(\sum_{j=1}^{\infty}\beta_j\int_0^a dr\,|\langle\phi_j,e^{-iH_1(r+s)}\phi\rangle|^2\right)^{1/2}. \qquad (11.1.9)$$

The second factor on the right-hand side of the inequality has been considered already, except that now we have $r+s$ replacing r, and different limits of integration. According to (11.1.8), this factor is just $(\int_0^a dr\||A|^{1/2} e^{-iH_1(r+s)}\phi\|^2)^{1/2}$. Changing integration variable from $r+s$ to r, we have an upper bound $(\int_s^\infty dr\||A|^{1/2} e^{-iH_1 r}\phi\|^2)^{1/2}$, which appears on the right-hand side of (11.1.5). The first factor on the right-hand side of the inequality (11.1.9) may be estimated in the same way as for the proof of (11.1.4), except that we now have $r+t$ instead of r and $T\psi_j$ instead of ϕ_j. Whereas $\|\phi_j\|^2=1$, we now have $\|T\psi_j\|^2 \leq \|T\|^2 \, \|\psi_j\|^2 = \|T\|^2$. Comparing with the right-hand side of (11.1.4), we obtain the estimate $(2\pi m^2\|T\|^2 \, \|A\|_1)^{1/2}$ for this factor. Combining this with our estimate for the second factor on the right-hand side of (11.1.9), the inequality (11.1.5) is now confirmed. ∎

The importance of Lemma 11.1 is that we are furnished with an estimate of integrals like (11.1.3), taken as inner products between ϕ and ϕ_j, which apply in cases where B is of trace class. The aim will be to express wave operators in terms of $\gamma_a(B)$, for suitable B, in the limit as $a\to\infty$. A useful consequence of (11.1.5) is that the bound converges to zero for large s, and does so uniformly in t. The following theorem exploits this fact to great effect.

Theorem 11.1 (Trace Theorem)

Let H_1 and H_2 be self-adjoint operators in a Hilbert space $\mathscr{H}$. Let J be a bounded linear operator such that $JD(H_1)\subseteq D(H_2)$ (i.e. such that $g\in D(H_1) \Rightarrow Jg\in D(H_2)$) and such that $A\equiv H_2J-JH_1$ is a trace-class operator. Then the strong limits

$$\Omega_\pm(H_2, H_1; J)f = \operatorname*{s-lim}_{t\to\pm\infty} e^{iH_2t} J e^{-iH_1t} f \tag{11.1.10}$$

exist for all $f\in\mathscr{M}_{ac}(H_1)$.

Proof

We let $w(t)=e^{iH_2t}Je^{-iH_1t}$, and set $Y=(w(t))^*w(s)$. Then $Y=e^{iH_1t}J^*e^{iH_2(s-t)}Je^{-iH_1s}$, and

$$\begin{aligned}[H_1,Y] &= e^{iH_1t}(H_1J^*-J^*H_2)\, e^{iH_2(s-t)}\, Je^{-iH_1s} \\ &\quad + e^{iH_1t}J^*e^{iH_2(s-t)}(H_2J-JH_1)e^{-iH_1s} \\ &= -e^{iH_1t}A^*e^{iH_2(s-t)}Je^{-iH_1s}+e^{iH_1t}J^*e^{iH_2(s-t)}Ae^{-iH_1s}.\end{aligned}$$

From (11.1.3), we have the elementary identity

$$-i\gamma_a([H_1, Y]) = Y - e^{iH_1 a} Y e^{-iH_1 a},$$

which gives, in this case,

$$w^*(t)w(s) - e^{iH_1 a} w^*(t)w(s)e^{-iH_1 a}$$
$$= i\gamma_a(e^{iH_1 t}\{A^* e^{iH_2(s-t)} J - J^* e^{iH_2(s-t)} A\} e^{-iH_1 s}).$$

Subtracting the same identity with s and t identified we have

$$w^*(t)(w(t) - w(s)) - e^{iH_1 a} w^*(t)(w(t) - w(s)) e^{-iH_1 a}$$
$$= i\gamma_a(e^{iH_1 t}\{A^* J - J^* A\} e^{-iH_1 t}$$
$$- e^{iH_1 t}\{A^* e^{iH_2(s-t)} J - J^* e^{iH_2(s-t)} A\} e^{-iH_1 s}). \tag{11.1.11}$$

We now let ϕ be any vector belonging to the m-class for the self-adjoint operator H_1. From (11.1.5) with $T = J^* e^{iH_2(s-t)}$, we see that

$$\lim_{s,t\to\infty} \langle \phi, \gamma_a(e^{iH_1 t}\{J^* e^{iH_2(s-t)} A\} e^{-iH_1 s})\phi\rangle = 0,$$

uniformly in a for $a > 0$. Similarly, from (11.1.5) with $T = J^*$, and s and t identified, we have $\lim_{t\to\infty} \langle \phi, \gamma_a(e^{iH_1 t} J^* A e^{-iH_1 t})\phi\rangle = 0$, again uniformly in a. Taking matrix elements with ϕ, the two remaining contributions to the right-hand side of (11.1.11) may be estimated in the limit as $s, t \to \infty$, using the identity $\gamma_a(e^{iH_1 t} A^* T^* e^{-iH_1 s}) = \{\gamma_a(e^{iH_1 s} TA e^{-iH_1 t})\}^*$, together with inequality (11.1.5), with s and t exchanged. We find from (11.1.11) that

$$\langle \phi, w^*(t)(w(t) - w(s))\phi\rangle - \langle \phi, e^{iH_1 a}\omega^*(t)(w(t) - w(s))e^{-iH_1 a}\phi\rangle \to 0 \tag{11.1.12}$$

as s, $t \to \infty$, uniformly in a.

Now $w(t) - w(s) = i\int_s^t dr\, e^{iH_2 r} A e^{-iH_1 r}$. Using the representation (11.1.6) for A, we may express $w(t) - w(s)$ as

$$w(t) - w(s) = i\sum_{j=1}^{\infty} \beta_j \int_s^t dr \langle e^{iH_1 r}\phi_j, \cdot\rangle e^{iH_2 r}\psi_j.$$

For an operator of rank one, we have

$$\|\langle f, \cdot\rangle g\|_1 = \|f\|\,\|g\|,$$

so that

$$\|w(t) - w(s)\|_1 \leq \sum_{j=1}^{\infty} \beta_j \int_s^t dr\, \|\phi_j\|\,\|\psi_j\| < \infty.$$

Hence $w(t) - w(s) \in C_1$, and in particular $w(t) - w(s)$ is compact. Since

$\langle h, e^{-iH_1 a}\phi\rangle = \int_{-\infty}^{\infty} e^{-i\lambda a}\,(d/d\lambda)\,\langle h, E_\lambda \phi\rangle\, d\lambda \to 0$ as $a\to\infty$ by the Riemann–Lebesgue Lemma, we know that $e^{-iH_1 a}\phi$ converges weakly to zero as $a\to\infty$. Hence $(w(t)-w(s))e^{-iH_1 a}\phi$ converges strongly to zero as $a\to\infty$. Taking first the limit $a\to\infty$ in (11.1.12), we now have $\langle \phi, w^*(t)\,(w(t)-w(s))\phi\rangle \to 0$ as $s,t\to\infty$. Adding the equivalent result with s and t interchanged gives $\langle \phi, (w^*(t)-w(s))\,(w(t)-w(s))\phi\rangle \to 0$, or

$$\lim_{s,t\to\infty} \|(w(t)-w(s))\phi\| = 0.$$

The case $s,t\to-\infty$ follows similarly. Thus $w(t)\phi$ is Cauchy in the limit as $t\to\infty$, so that, by completeness of the Hilbert space, we may deduce the convergence of $w(t)\phi$ as a strong limit. However, the m-class of vectors is dense in $\mathcal{M}_{ac}(H_1)$, so the limit may be extended to all vectors in the a.c. subspace for H_1 as stated. ■

The interested reader should tighten up the above proof of the trace theorem to his/her own satisfaction, by paying particular attention to questions of domain, which have been to some extent glossed over. The operator A is defined *a priori* only on a dense subset of $\mathcal{H}$, namely on the domain of H_1; under the hypothesis of the theorem, A may then be extended by continuity to a trace-class operator on the whole space. Alternatively, one may start by defining A as a sesquilinear form on $D(H_2)\times D(H_1)$; thus

$$\langle f_2, Af_1\rangle = \langle H_2 f_2, Jf_1\rangle - \langle f_2, JH_1 f_1\rangle.$$

The condition $JD(H_1)\subseteq D(H_2)$ is then equivalent to requiring the sesquilinear form to be the form of a bounded linear operator.

In many applications, the linear operator J in (11.1.10) has to be chosen with some care. For example, it will not be possible to take $J=I$ (identity operator) unless one has $D(H_1)\subseteq D(H_2)$. This restriction on the possible domain of H_1, as compared with the domain of H_2, may sometimes be avoided by taking J to be a spectral projection of H_2 for a finite interval. The range of J is then necessarily contained in the domain of H_2; moreover, the presence of this projection does not disturb too much our defining of a wave operator, since in (11.1.10) we then have

$$\Omega_\pm(H_2,H_1,J)f = \operatorname*{s-lim}_{t\to\pm\infty} J e^{iH_2 t} e^{-iH_1 t} f,$$

owing to the fact that J commutes with H_2. It may even be possible to remove the projection from the limit on the right-hand side, provided that we have $\text{s-lim}_{t\to\pm\infty}(1-J)e^{-iH_1 t}f=0$, and we shall see this argument put into practice in the following section.

It should be noted that the self-adjoint operators H_1 and H_2 occur almost interchangeably in the statement of the Trace Theorem. Indeed, if $A=H_2J-JH_1\in C_1$ then $-A^*=H_1J^*-J^*H_2\in C_1$, so that the same criterion that leads to the wave operators $\Omega_\pm(H_2,H_1;J)$ will provide at the same time a proof of the existence of the wave operators $\Omega_\pm(H_1,H_2,J^*)$. For this reason, trace methods are particularly well adapted to proving asymptotic completeness, which frequently presents little more difficulty than establishing the existence of wave operators by these methods, provided that the potential decays at infinity sufficiently rapidly.

11.2 WAVE OPERATORS AND COMPLETENESS BY TRACE METHODS

The following result is typical of the applications of the Trace Theorem to scattering theory.

Theorem 11.2

Suppose that $V(\boldsymbol{r})\in L^2_{\text{loc}}(\mathbb{R}^3\setminus\{0\})$, and let H be a self-adjoint extension of the operator $\hat{H}=-\Delta+V(\boldsymbol{r})$, with $D(\hat{H})=C_0^\infty(\mathbb{R}^3\setminus\{0\})$. Suppose that either

(i) $V=V(|\boldsymbol{r}|)$ is a central potential, and $V\in L^1(R,\infty)\cap L^2(R,\infty)$ for any $R>0$; or
(ii) $V(\boldsymbol{r})\in L^1(\mathscr{B}_\infty)\cap L^2(\mathscr{B}_\infty)$, where $\mathscr{B}_\infty$ is the region $|\boldsymbol{r}|>R$, for any $R>0$.

Then the wave operators $\Omega_\pm=\text{s-lim}_{t\to\pm\infty}\, e^{iHt}\, e^{-iH_0t}$ exist, and are asymptotically complete provided that H has no singular points.

Proof

We let $\rho(\boldsymbol{r})$ be a smooth non-decreasing function of $r=|\boldsymbol{r}|$, such that $\rho\equiv 0$ near $r=0$ and $\rho\equiv 1$ for sufficiently large $|\boldsymbol{r}|$. We first consider case (ii) of the theorem.

We define the linear operator J for some $M_1,M_2>0$:

$$J=E_{|H|<M_1}\rho E_{|H_0|<M_2}. \tag{11.2.1}$$

Locally, away from $\boldsymbol{r}=0$, H_0 and H have the same domains, so that $\rho D(H_0)\subseteq D(H)$ and $\rho D(H)\subseteq D(H_0)$. We can therefore write

$$\begin{aligned}HJ-JH_0&=E_{|H|<M_1}(H\rho-\rho H_0)E_{|H_0|<M_2}\\&=E_{|H|<M_1}\{-\Delta\rho-2i(\text{grad}\,\rho)\cdot\boldsymbol{P}+V\rho\}E_{|H_0|<M_2}.\end{aligned} \tag{11.2.2}$$

Note that $V\rho$ is square-integrable and $V\rho E_{|H_0|<M_2}$ is defined on the entire

Hilbert space. We define $\tilde{\rho}$ similarly to ρ, except that $\tilde{\rho}\equiv 1$ on the support of ρ. The range of $\tilde{\rho}E_{|H|<M_1}$ is contained in the domain of H_0, so that (H_0+1) $\tilde{\rho}E_{|H|<M_1}$ is bounded, by the Closed-Graph Theorem. Since $V\rho\in L^1(\mathbb{R}^3)$, we have $|V\rho|^{1/2}\in L^2(\mathbb{R}^3)$, in which case $|V\rho|^{1/2}(H_0+1)^{-1}$ is Hilbert-Schmidt. Hence $|V\rho|^{1/2}E_{|H|<M_1}=|V\rho|^{1/2}(H_0+1)^{-1}(H_0+1)\,\tilde{\rho}E_{|H|<M_1}$ is also Hilbert–Schmidt, as is $|V\rho|^{1/2}\,E_{|H_0|<M_2}$, so that

$$E_{|H|<M_1}V\rho E_{|H_0|<M_2}=[|V\rho|^{1/2}E_{|H|<M_1}]^*[\pm|V\rho|^{1/2}E_{|H_0|<M_2}]$$

is of trace class, being the product of two Hilbert–Schmidt operators. Since $\Delta\rho$, $|\text{grad }\rho|\in C_0^\infty(\mathbb{R}^3\setminus\{0\})$, both of these functions are in $L^1(\mathbb{R}^3)\cap L^2(\mathbb{R}^3)$, and it follows in the same way that the remaining terms of the right-hand side of (11.2.2) are of trace class. Thus the operator $HJ-JH_0$ is of trace class, and we can apply the Trace Theorem to deduce the existence on $\mathscr{H}$ of the strong limits

$$\operatorname*{s-lim}_{t\to\pm\infty} E_{|H|<M_1}\mathrm{e}^{\mathrm{i}Ht}\rho\mathrm{e}^{-\mathrm{i}H_0t}E_{|H_0|<M_2}.$$

Now $(H-z)\rho E_{|H_0|<M_2}$ is bounded, so that

$$\lim_{M_1\to\infty} E_{|H|\geq M_1}\rho E_{|H_0|<M_2}=\lim_{M_1\to\infty}[E_{|H|\geq M_1}(H-z)^{-1}](H-z)\rho E_{|H_0|<M_2}=0$$

(limit in operator norm). Hence we can remove the projection $E_{|H|<M_1}$ and assert the existence of the limits

$$\operatorname*{s-lim}_{t\to\pm\infty}\mathrm{e}^{\mathrm{i}Ht}\rho\mathrm{e}^{-\mathrm{i}H_0t}E_{|H_0|<M_2}.$$

Since the range of $E_{|H_0|<M_2}$ is dense in $\mathscr{H}$, as M_2 is increased, we can also remove this projection. Finally we know that freely evolving states are localized at large distances in the limits $t\to\pm\infty$, so that $\text{s-lim}_{t\to\pm\infty}$ $(1-\rho)\mathrm{e}^{-\mathrm{i}H_0t}=0$. This allows us to remove the ρ-operator, and we have deduced the existence of the wave operators as strong limits.

Exactly the same kind of arguments may be applied with the roles of H and H_0 interchanged. In this case, however, we are unable in general to remove the operator ρ, and can conclude only that

$$\Omega_\pm^*=\operatorname*{s-lim}_{t\to\pm\infty}\mathrm{e}^{\mathrm{i}H_0t}\rho\mathrm{e}^{-\mathrm{i}Ht}. \tag{11.2.3}$$

That the limit on the right-hand side does indeed give the adjoint wave operator follows from our previous analysis, since

$$\lim_{t\to\pm\infty}\langle f,\mathrm{e}^{\mathrm{i}H_0t}\rho\mathrm{e}^{-\mathrm{i}Ht}g\rangle=\lim_{t\to\pm\infty}\langle\mathrm{e}^{\mathrm{i}Ht}\rho\mathrm{e}^{-\mathrm{i}H_0t}f,g\rangle$$
$$=\langle\Omega_\pm f,g\rangle=\langle f,\Omega_\pm^* g\rangle.$$

Equation (11.2.3) holds on $\mathscr{M}_{\mathrm{ac}}(H)$. Let us suppose, then, that the range of Ω_+ is not the whole a.c. subspace of H. In this case, $\Omega_+^* g=0$ for some

$g \in \mathcal{M}_{ac}(H)$. Furthermore, by the intertwining property of wave operators, $\Omega_+^* E_{|H-\lambda|<\varepsilon} g = E_{|H_0-\lambda|<\varepsilon} \Omega_+^* g = 0$ for arbitrary λ and ε. Since the spectral support of g, with respect to H, is not a discrete set of points, points λ will exist such that $E_{|H-\lambda|<\varepsilon} g \neq 0$, however small the value of ε. Fixing some value of ε, we have, from (11.2.3),

$$\operatorname*{s-lim}_{t \to +\infty} \rho e^{-iHt} E_{|H-\lambda|<\varepsilon} g = 0,$$

so that

$$\lim_{t \to \infty} ||(1-\rho) e^{-iHt} E_{|H-\lambda|<\varepsilon} g|| = ||E_{|H-\lambda|<\varepsilon} g|| \neq 0.$$

If ρ is chosen such that $\rho \equiv 1$ in the region $|\boldsymbol{r}| \geq R$, we then have

$$||E_{|\boldsymbol{r}|<R} E_{|H-\lambda|<\varepsilon}|| \geq ||(1-\rho) e^{-iHt} E_{|H-\lambda|<\varepsilon}|| \to 1$$

as $t \to \infty$. It follows that $||E_{|\boldsymbol{r}|<R}\, E_{|H-\lambda|<\varepsilon}|| = 1$. Since both R and ε can be taken arbitrarily small, we have shown that λ is a singular point. If H has no singular points then range $(\Omega_+) = \mathcal{M}_{ac}(H)$, and similarly range $(\Omega_-) = \mathcal{M}_{ac}(H)$. Asymptotic completeness is thus proved.

In case (i) of the theorem, for which $V(\boldsymbol{r})$ is assumed to be spherically symmetric, the proof proceeds along the same lines, except that we restrict attention to a partial-wave subspace $\mathcal{H}_{lm}$. Let us consider, for example, the proof that $|V\rho|^{1/2}(H_0+1)^{-1}$ is Hilbert–Schmidt. Making use of the Fourier–Hankel transform, this operator is unitarily equivalent, in a given partial-wave subspace, to an integral operator with kernel

$$\left(\frac{2}{\pi}\right)^{1/2} |V(r)\rho(r)|^{1/2} kr j_l(kr)(k^2+1)^{-1}.$$

Since $krj_l(kr)$ is uniformly bounded, for fixed l, this operator is readily found to be Hilbert–Schmidt. Thus for central potentials a condition of square-integrability in $L^2(\mathbb{R}^3)$ may be replaced by a condition of square-integrability in $L^2(0, \infty)$. For central potentials, again, the question of existence and completeness of wave operators may be treated separately in each partial wave. Wave operators that exist and are complete in each $\mathcal{H}_{lm}$ will exist and be complete as operators in $L^2(\mathbb{R}^3)$. ∎

In the case (i) of a central potential, the result of the theorem is essentially the best possible, in the sense that we cannot expect unmodified wave operators to exist, in general, for potentials not belonging to $L^1(R, \infty)$. On the other hand, for non-spherical potentials that decay at infinity like $|\boldsymbol{r}|^{-\beta}$ for some $\beta > 1$, we should expect to prove the existence of unmodified wave operators, and completeness in the absence of singular points. Yet such potentials will not, in general, belong to $L^1(\mathcal{B}_\infty)$, and trace methods will not give us the results we seek. We shall deal with such potentials in the following chapter, by another method.

The following theorem, obtained by a further application of the Trace Theorem, applies even to scattering by long-range potentials, and shows how the possible existence of singular points for the Hamiltonian is crucial for the question of asymptotic completeness. Moreover, asymptotic completeness is related to spectral properties of the operator $H^{(1)}$ introduced in Theorem 10.4.

Theorem 11.3

Suppose that $V(\boldsymbol{r})$ is bounded in the region $|\boldsymbol{r}|\geq R$ for any $R>0$, and that $V(\boldsymbol{r})\to 0$ as $|\boldsymbol{r}|\to\infty$. For a given $a>0$, suppose that $\tilde{V}(\boldsymbol{r})$ is a potential that is bounded in the region $|\boldsymbol{r}|\leq a$, infinitely differentiable in the region $|\boldsymbol{r}|<a$, such that $\tilde{V}=V$ for $|\boldsymbol{r}|>a$. Define the self-adjoint operators $H=H_0+V$ and $\tilde{H}=H_0+\tilde{V}$. Then the following hold.

(i) Wave operators

$$\Omega_{\pm}=\operatorname*{s-lim}_{t\to\pm\infty} \mathrm{e}^{\mathrm{i}Ht}F(\boldsymbol{P},t) \tag{11.2.4}$$

exist on the entire Hilbert space if and only if

$$\tilde{\Omega}_{\pm}=\operatorname*{s-lim}_{t\to\pm\infty} \mathrm{e}^{\mathrm{i}\tilde{H}t}F(\boldsymbol{P},t) \tag{11.2.5}$$

exist on the entire Hilbert space. (We assume that the modified free evolution $F(\boldsymbol{P},t)$ converges weakly to zero.)

(ii) If H has no singular point then the wave operators $\Omega_{\pm}$ satisfy range $(\Omega_{\pm})=\mathcal{M}_{\mathrm{ac}}(H)$ if and only if range $(\tilde{\Omega}_{\pm})=\mathcal{M}_{\mathrm{ac}}(\tilde{H})$. If range $(\tilde{\Omega}_{\pm})=\mathcal{M}_{\mathrm{ac}}(\tilde{H})$, then range $(\Omega_{\pm})=\mathcal{M}_{\mathrm{ac}}(H)$ if and only if the self-adjoint operator $H^{(1)}=-\Delta+V$ in $L^2(\mathcal{B})$ $(\mathcal{B}=\{\boldsymbol{r}\colon|\boldsymbol{r}|<1\})$, with Dirichlet boundary conditions on $|\boldsymbol{r}|=1$, has no absolutely continuous spectrum.

Proof

We let $\rho(r)$ be a smooth non-decreasing function of $r=|\boldsymbol{r}|$, $\rho\equiv 1$ for $|\boldsymbol{r}|\geq\frac{2}{3}a$, $\rho\equiv 0$ for $|\boldsymbol{r}|\leq\frac{1}{3}a$. Proceeding as for Theorem 11.2, we first prove that the operator $E_{|H|<M_1}(H\rho-\rho\tilde{H})E_{|\tilde{H}|<M_2}$ is of trace class. Note that $\rho D(H)\subseteq D(\tilde{H})$ and $\rho D(\tilde{H})\subseteq D(H)$. Moreover, $(V-\tilde{V})\rho\in L^1(\mathbb{R}^3)\cap L^2(\mathbb{R}^3)$. We introduce a multiplication operator $\tilde{\rho}$, where $\tilde{\rho}\in C_0^{\infty}$ in the region $0<|\boldsymbol{r}|<a$, and $\tilde{\rho}\equiv 1$ on the support of grad ρ. Comparing with (11.2.2), the only additional difficulty is that we have to prove $E_{|H|<M_1}\tilde{\rho}(\operatorname{grad}\rho\cdot\boldsymbol{P})\tilde{\rho}E_{|\tilde{H}|<M_2}$ to be of trace

class. Now

$$\tilde{H}\tilde{\rho}E_{|\tilde{H}|<M_2}=\tilde{\rho}\tilde{H}E_{|\tilde{H}|<M_2}-(\Delta\tilde{\rho})E_{|\tilde{H}|<M_2}-2\mathrm{i}(\mathrm{grad}\ \tilde{\rho})\cdot \boldsymbol{P}\tilde{\rho}_0 E_{|\tilde{H}|<M_2},$$

where $\tilde{\rho}_0$ is C_0^∞ in $0<|\boldsymbol{r}|<a$, with $\tilde{\rho}_0\equiv 1$ on the support of $\tilde{\rho}$. Since range $(\tilde{\rho}_0 E_{|\tilde{H}|<M_2})\subseteq D(H_0)$, each term on the right-hand side is in the domain of $\boldsymbol{P}$. Moreover,

$$\tilde{H}\tilde{\rho}E_{|\tilde{H}|<M_2}=H_0\tilde{\rho}E_{|\tilde{H}|<M_2}+\tilde{V}\tilde{\rho}E_{|\tilde{H}|<M_2},$$

where $\tilde{V}\tilde{\rho}\in C_0^\infty(\mathbb{R}^3\setminus\{\boldsymbol{0}\})$. Hence range $(H_0\tilde{\rho}E_{|\tilde{H}|<M_2})\subseteq D(\boldsymbol{P})$, so that we may deduce the operator $(H_0+1)^{3/2}\ \tilde{\rho}E_{|\tilde{H}|<M_2}$ to be bounded. It also follows that $(H_0+1)\tilde{\rho}E_{|H|<M_1}$ is bounded. Since $(H_0+1)^{-1}$ $(\mathrm{grad}\,\rho)\cdot \boldsymbol{P}(H_0+1)^{-3/2}$ is factorizable into a product of Hilbert–Schmidt operators, and other terms cause no difficulty, we may conclude that $HJ-J\tilde{H}$ is of trace class, with

$$J=E_{|H|<M_1}\rho E_{|\tilde{H}|<M_2}. \tag{11.2.6}$$

Since $\tilde{V}$ is a non-singular potential, the domains of $\tilde{H}$ and H_0 are locally the same near $\boldsymbol{r}=\boldsymbol{0}$. We have, certainly, $D(H_0)\subseteq D(|\boldsymbol{r}|^{-1})$, so that $D(\tilde{H})\subseteq D(|\boldsymbol{r}|^{-1})$, from which it follows that $\tilde{H}$ can have no singular points. Thus $(1-\rho)E_{|\tilde{H}|<M_2}$ is compact. On $\mathscr{M}_{\mathrm{ac}}(H)$, we know that $\mathrm{e}^{-\mathrm{i}\tilde{H}t}$ converges weakly to zero, by the Riemann–Lebesgue Lemma; hence $(1-\rho)\mathrm{e}^{-\mathrm{i}\tilde{H}t}E_{|\tilde{H}|<M_2}$ converges strongly to zero. Applying the Trace Theorem with J given by (11.2.6), and removing projections as in the proof of Theorem 11.2, we may deduce the existence of the strong limits

$$\Omega_\pm(H,\tilde{H})=\underset{t\to\pm\infty}{\text{s-lim}}\ \mathrm{e}^{\mathrm{i}Ht}\mathrm{e}^{-\mathrm{i}\tilde{H}t} \tag{11.2.7}$$

and

$$\Omega_\pm(\tilde{H},H,\rho)=\underset{t\to\pm\infty}{\text{s-lim}}\ \mathrm{e}^{\mathrm{i}\tilde{H}t}\rho\mathrm{e}^{-\mathrm{i}Ht}, \tag{11.2.8}$$

on $\mathscr{M}_{\mathrm{ac}}(\tilde{H})$ and $\mathscr{M}_{\mathrm{ac}}(H)$ respectively. The compactness of $(1-\rho)E_{|\tilde{H}|<M_2}$ allows us to remove the operator ρ from the limit (11.2.7) but not from (11.2.8).

(i) We now suppose that the limit (11.2.4) exists. Using (11.2.8), together with transitivity, we then have the existence of the limits $\text{s-lim}_{t\to\pm\infty}$ $\mathrm{e}^{\mathrm{i}\tilde{H}t}\rho F(\boldsymbol{P},t)$. Using the compactness of $(1-\rho)E_{|H_0|<M}$, we can deduce that $(1-\rho)F(\boldsymbol{P},t)$ converges strongly to zero as $t\to\pm\infty$. Hence the existence of the limit (11.2.5) follows. Similarly, we can use (11.2.7) to show that (11.2.5) ⇒ (11.2.4).

(ii) If range $(\tilde{\Omega}_\pm)=\mathscr{M}_{\mathrm{ac}}(\tilde{H})$ then, we can deduce on $\mathscr{M}_{\mathrm{ac}}(\tilde{H})$ the existence of limits $\text{s-lim}_{t\to\pm\infty}\ F^*(\boldsymbol{P},t)\mathrm{e}^{-\mathrm{i}\tilde{H}t}$. From (11.2.8) then follows the existence on $\mathscr{M}_{\mathrm{ac}}(H)$ of $\text{s-lim}_{t\to\pm\infty}\ F^*(\boldsymbol{P},t)\rho\mathrm{e}^{-\mathrm{i}Ht}$. Since $(1-\rho)F(\boldsymbol{P},t)$ converges strongly to zero, we then have

$$\Omega_{\pm}^{*}=\operatorname*{s-lim}_{t\to\pm\infty} F^{*}(\boldsymbol{P},t)\rho \mathrm{e}^{-\mathrm{i}Ht}.$$

If range $\Omega_{+}\neq\mathcal{M}_{\mathrm{ac}}(H)$, say, there will be some vector $g\in\mathcal{M}_{\mathrm{ac}}(H)$ for which $\Omega_{+}^{*}g=0$, giving $\rho\mathrm{e}^{-\mathrm{i}Ht}g\to0$ strongly. Arguing as for Theorem 11.2, this implies the existence of a singular point for H. In the absence of singular points, we have range $(\Omega_{+})=\mathcal{M}_{\mathrm{ac}}(H)$ and similarly range $(\Omega_{-})=\mathcal{M}_{\mathrm{ac}}(H)$. The proof that range $(\Omega_{\pm})=\mathcal{M}_{\mathrm{ac}}(H)\Rightarrow$range $(\tilde{\Omega}_{\pm})=\mathcal{M}_{\mathrm{ac}}(\tilde{H})$ follows similarly, using (11.2.7).

To prove the final part of the theorem, we suppose that range $(\tilde{\Omega}_{\pm})=\mathcal{M}_{\mathrm{ac}}(\tilde{H})$, and introduce the operator H^{D} defined in the proof of Theorem 10.4. Choosing ρ_{∞} such that $\rho_{\infty}\equiv0$ for $|\boldsymbol{r}|<2$, say, we can use the Trace Theorem to prove the existence of the strong limits $\Omega_{\pm}(H,H^{\mathrm{D}},\rho_{\infty})$ and $\Omega_{\pm}(H^{\mathrm{D}},H,\rho_{\infty})$. (If $V(\boldsymbol{r})$ is not C^{∞} away from $\boldsymbol{r}=\mathbf{0}$ then the proof of the existence of these limits through the Trace Theorem may involve the use of transitivity by defining an intermediate potential $W(\boldsymbol{r})$ that is C^{∞} in some compact neighbourhood of the support of grad ρ and that is identical with $V(\boldsymbol{r})$ outside this neighbourhood.) Similarly, if ρ_{0} is C_{0}^{∞} in the region $|\boldsymbol{r}|<1$, with $\rho_{0}\equiv1$ near $\boldsymbol{r}=\mathbf{0}$, we can prove the existence of $\Omega_{\pm}(H,H^{\mathrm{D}},\rho_{0})$ and $\Omega_{\pm}(H^{\mathrm{D}},H,\rho_{0})$, in each case on $\mathcal{M}_{\mathrm{ac}}(H^{\mathrm{D}})$ and $\mathcal{M}_{\mathrm{ac}}(H)$ respectively. Using the compactness of $(1-\rho_{0}-\rho_{\infty})E_{|H|<M}$ and $(1-\rho_{0}-\rho_{\infty})E_{|H^{\mathrm{D}}|<M}$ we may deduce the existence of the wave operators $\Omega_{\pm}(H,H^{\mathrm{D}})$ and $\Omega_{\pm}(H^{\mathrm{D}},H)$. This allows a comparison between the evolutions $\mathrm{e}^{-\mathrm{i}Ht}$ and $\mathrm{e}^{-\mathrm{i}H^{\mathrm{D}}t}$ on the respective a.c. subspaces, asymptotically in the limits $t\to\infty$. For $g\in\mathcal{M}_{\mathrm{ac}}(H)$, there will be a corresponding vector $h=\Omega_{+}(H^{\mathrm{D}},H)g\in\mathcal{M}_{\mathrm{ac}}(H^{\mathrm{D}})$ such that, in the strong limit $t\to\infty$, $\mathrm{e}^{-\mathrm{i}Ht}g$ approaches $\mathrm{e}^{-\mathrm{i}H^{\mathrm{D}}t}h$. If $H^{(1)}$ has no absolutely continuous spectrum then a vector $h\in\mathcal{M}_{\mathrm{ac}}(H^{\mathrm{D}})$ must be concentrated entirely in the region $|\boldsymbol{r}|>1$, and we must also have $E_{|\boldsymbol{r}|<1}\mathrm{e}^{-\mathrm{i}H^{\mathrm{D}}t}h=0$. But this implies that $\operatorname{s-lim}_{t\to\infty}E_{|\boldsymbol{r}|<1}\mathrm{e}^{-\mathrm{i}Ht}g=0$. If $\Omega_{+}^{*}g=0$ then we have already seen that $\rho\mathrm{e}^{-\mathrm{i}Ht}g\to0$, implying that $\operatorname{s-lim}_{t\to\infty}E_{|\boldsymbol{r}|>1}\mathrm{e}^{-\mathrm{i}Ht}g=0$. In this case, $g=0$, and we have shown that range $(\Omega_{+})=\mathcal{M}_{\mathrm{ac}}(H)$. Similar arguments apply to Ω_{-}. On the other hand, let us suppose that $H^{(1)}$ has a non-trivial absolutely continuous subspace. In this case, for $h\in\mathcal{M}_{\mathrm{ac}}(H^{\mathrm{D}})$ and $E_{|\boldsymbol{r}|>1}h=0$ we have $E_{|\boldsymbol{r}|>1}\mathrm{e}^{-\mathrm{i}H^{\mathrm{D}}t}h=0$, so that with $g=\Omega_{+}(H,H^{\mathrm{D}})h$ we have $\operatorname{s-lim}_{t\to\infty}E_{|\boldsymbol{r}|>1}\mathrm{e}^{-\mathrm{i}Ht}g=0$. Such a vector g cannot be in the range of the wave operator Ω_{+}, since such states are to be found, at times $t\to\infty$, at large distance from $\boldsymbol{r}=\mathbf{0}$. We cannot then have range $\Omega_{+}=\mathcal{M}_{\mathrm{ac}}(H)$. We have thus shown that if range $(\tilde{\Omega}_{\pm})=\mathcal{M}_{\mathrm{ac}}(\tilde{H})$ then range $(\Omega_{\pm})=\mathcal{M}_{\mathrm{ac}}(H)$ if and only if $H^{(1)}$ has no absolutely continuous spectrum. ∎

Theorem 11.3 has two consequences of particular importance. The first of these is that, having seen in general the correspondence between singular points of H and points of the essential spectrum of $H^{(1)}$, we now conclude that it is points of the *absolutely continuous* spectrum of $H^{(1)}$ that are associated with any breakdown of asymptotic completeness due to the presence of local singularities. (It is still possible, however, to have singular points of the absolutely continuous spectrum of H even when range $(\Omega_{\pm})=\mathcal{M}_{ac}(H)$; see Chapter 13.) The second point to note is that, in the absence of singular points of H, the singularity of the potential becomes irrelevant to the question of asymptotic completeness. If the wave operators are complete for the *non-singular* potential $\tilde{V}(\boldsymbol{r})$ then they will be complete for $V(\boldsymbol{r})$. For this reason, in the following chapter, dealing with completeness for a wide range of potentials of short and long range, we shall be able to restrict our attention entirely to potentials that are locally non-singular.

Chapter 12

Asymptotic Completeness and the Generator of Dilations

12.1 COMPLETENESS FOR SHORT-RANGE POTENTIALS

The most successful, elegant and physically motivated proof of asymptotic completeness available to us is based on the method of V. Enss. The method, of which we present a later development motivated by ideas of E. Mourre and others, depends on a study of the space–time evolution of states under the influence of a potential, which may be either short- or long-range. Applications have extended far beyond the cases to which we shall confine ourselves here, starting with the short-range one.

A one-parameter family of unitary operators on $L^2(\mathbb{R})$ may be defined by the formula

$$(U_t g)(x) = e^{t/2} g(e^t x). \qquad (12.1.1)$$

For each value of t, the effect of the unitary operator U_t is that of a change of scale, or *dilation*. Thus x is multiplied by e^t, and the factor $e^{t/2}$ multiplying g is necessary in (12.1.1) in order to preserve the norm. It is not difficult to verify that

$$U_s U_t = U_{s+t}, \qquad (12.1.2)$$

so that $\{U_t\}$ is a one-parameter *group* of unitary operators.

The function $g(x,t) \equiv e^{t/2} g(e^t x)$ on the right-hand side of (12.1.1), provided that it is differentiable, satisfies the partial differential equation

$$-i\frac{\partial}{\partial t} g(x,t) = -ix\frac{\partial}{\partial x} g(x,t) - \tfrac{1}{2} i g(x,t),$$

together with the initial condition $g(x,0) = g(x)$. If we then define in $L^2(\mathbb{R})$

the first-order differential operator A by

$$A = -\mathrm{i}x\frac{\mathrm{d}}{\mathrm{d}x} - \tfrac{1}{2}\mathrm{i} \tag{12.1.3}$$

then we have, for a dense set of g in $L^2(\mathbb{R})$,

$$-\mathrm{i}\frac{\mathrm{d}}{\mathrm{d}t}U_t g = Ag,$$

where $U_0 g = g$. Because of this, we can write

$$U_t = \mathrm{e}^{\mathrm{i}At}, \tag{12.1.4}$$

and the linear operator A defined by (12.1.3) is called the *generator of dilations* in $L^2(\mathbb{R})$. The importance of the generator of dilations in quantum mechanics, and in particular in scattering theory, stems from the fact that

$$A = \tfrac{1}{2}(PX + XP), \tag{12.1.5}$$

where X is the position operator and $P = -\mathrm{i}\,\mathrm{d}/\mathrm{d}x$ is the momentum operator. Starting from (12.1.5) as a definition of A acting on $C_0^\infty(\mathbb{R})$, there is no difficulty in proving essential self-adjointness, for example by verifying explicitly that the differential equations $(A^* \pm \mathrm{i})h = 0$ have only the trivial solution in $L^2(\mathbb{R})$; note that (12.1.5) defines a symmetric operator on $C_0^\infty(\mathbb{R})$. From now on, A will stand for the unique self-adjoint extension of the operator given by (12.1.5) on $C_0^\infty(\mathbb{R})$. The corresponding self-adjoint operator in $L^2(\mathbb{R}^3)$ will again be denoted by A, and we have in that case

$$A = \tfrac{1}{2}(\boldsymbol{P}\cdot\boldsymbol{X} + \boldsymbol{X}\cdot\boldsymbol{P}), \tag{12.1.6}$$

where $\boldsymbol{X}$ is the three-dimensional vector position operator and $\boldsymbol{P} = -\mathrm{i}\boldsymbol{\nabla}$ is the momentum operator. Thus

$$\begin{aligned} A &= -\frac{\mathrm{i}}{2}\sum_{j=1}^{3}\left(\frac{\partial}{\partial x_j}x_j + x_j\frac{\partial}{\partial x_j}\right) \\ &= -\tfrac{1}{2}\mathrm{i}(2\boldsymbol{r}\cdot\boldsymbol{\nabla} + 3). \end{aligned} \tag{12.1.7}$$

If we use the momentum-space representation of wave functions, we may write

$$A = \tfrac{1}{2}\mathrm{i}(2\boldsymbol{k}\cdot\boldsymbol{\nabla} + 3),$$

where $\boldsymbol{k}\cdot\boldsymbol{\nabla}$ now stands for $\Sigma_{j=1}^{3} k_j\partial/\partial k_j$. Using spherical polar coordinates (k,θ,φ), this gives

$$A = \frac{\mathrm{i}}{2}\left(2k\frac{\partial}{\partial k} + 3\right). \tag{12.1.8}$$

We shall find it convenient to introduce at this point the direct integral representation of the Hilbert space; see Section 5.3 and in particular (5.3.27), which defines, for any $f \in L^2(\mathbb{R}^3)$, the one-parameter family $\{f_\lambda\}$. It is a simple exercise, using the change of variables $\lambda = k^2$, with (5.3.27), to show that the direct integral representation for A is given by

$$(Af)_\lambda = 2\mathrm{i}\lambda^{1/2}\frac{\mathrm{d}}{\mathrm{d}\lambda}(\lambda^{1/2} f_\lambda). \tag{12.1.9}$$

Classically, the observable defined by (12.1.6) corresponds to the scalar product of position with momentum. In classical scattering theory a particle far from the scattering centre, with momentum directed away from the scattering centre, will have $A>0$. Such a state, in which in fact A is approaching $+\infty$, may be described as an outgoing scattering state of the classical particle system, whereas a particle approaching the scattering centre from large distances, with $A<0$, may be said to be ingoing. By analogy with the classical situation, the positive and negative components of the self-adjoint operator A will correspond respectively to subspaces of outgoing and ingoing states. Before making these notions more precise, let us consider a unitary transformation that allows us to carry out a complete spectral analysis of the three-dimensional generator of the dilation group.

For $f \in L^2(\mathbb{R}^3)$, represented in the direct integral representation by $\{f_\lambda(\boldsymbol{\omega})\} \equiv f(\lambda, \boldsymbol{\omega})$, we define $Uf \in L^2(\mathbb{R}, L^2(\mathbb{S}^2))$ by

$$(Uf)(x, \boldsymbol{\omega}) = \sqrt{2}\,\mathrm{e}^x f(\mathrm{e}^{2x}, \boldsymbol{\omega}). \tag{12.1.10}$$

Equation (12.1.10) defines a unitary operator U that effects a change of variable $\lambda = \mathrm{e}^{2x}$, or $x = \frac{1}{2}(\log\lambda)$. The multiplicative function $\sqrt{2}\mathrm{e}^x$ ensures the preservation of norm, where in $L^2(\mathbb{R}, L^2(\mathbb{S}^2))$ the norm is defined by

$$\|h(x,\boldsymbol{\omega})\|^2 = \int_{-\infty}^{\infty} \mathrm{d}x \int \mathrm{d}\omega\, |h(x,\boldsymbol{\omega})|^2 = \int_{-\infty}^{\infty} \mathrm{d}x\,(\|h_x(\boldsymbol{\omega})\|_0)^2.$$

Thus U maps from $L^2((0,\infty), L^2(\mathbb{S}^2))$ onto $L^2(\mathbb{R}, L^2(\mathbb{S}^2))$, and the inverse transformation is given by

$$(U^{-1}h)(\lambda, \boldsymbol{\omega}) = (2\lambda)^{-1/2} h(\tfrac{1}{2}\log\lambda, \boldsymbol{\omega}). \tag{12.1.11}$$

Using (12.1.10) and (12.1.11), it is now a straightforward exercise to verify that

$$A = U^{-1}TU, \tag{12.1.12}$$

where $T = \mathrm{i}\,\mathrm{d}/\mathrm{d}x$ acting in $L^2(\mathbb{R}; L^2(\mathbb{S}^2))$. It follows that A is unitarily equivalent to the direct sum of infinitely many copies of the operator $\mathrm{i}\,\mathrm{d}/\mathrm{d}x$—in fact, of *countably* many copies, since $L^2(\mathbb{S}^2)$ has a countable basis.

The most frequently used basis is defined through the spherical harmonics $Y_{lm}(\theta, \varphi)$, and allows us to observe that each angular-momentum subspace, labelled by quantum numbers l and m, is a reducing subspace for A. We could, of course, have noticed this from the start, using the fact that A commutes with all observables of angular momentum.

Since $T = \mathrm{i}\,\mathrm{d}/\mathrm{d}x$ has absolutely continuous spectrum along the whole real line, the same will be true of A. It therefore makes sense to define projection operators P_+ and P_- associated respectively, with the positive and negative parts of the spectrum of A. These will be the projections onto the outgoing and ingoing states that we discussed previously.

Finally, we observe that $T = \mathscr{F} X \mathscr{F}^{-1}$, where $\mathscr{F}$ is the Fourier transform and X is the operator of multiplication by the independent variable in $L^2(\mathbb{R}, \mathbb{S}^2)$.

Putting the two unitary transformations together, we can now write

$$A = U^{-1} \mathscr{F} X \mathscr{F}^{-1} U. \tag{12.1.13}$$

Let us denote by $E_{x>0}$ the projection operator associated with the positive part of the spectrum of X. Then $E_{x>0}$ is just multiplication by the characteristic function of the positive real line. If A and X are unitarily equivalent then so are their projection operators, and from (12.1.13) we have

$$P_+ = U^{-1} \mathscr{F} E_{x>0} \mathscr{F}^{-1} U. \tag{12.1.14}$$

If we are to justify, in quantum mechanics, our semiclassical interpretation of the projections $P_{\pm}$, we shall have to consider, first of all, whether states governed by the free evolution $\mathrm{e}^{-\mathrm{i}H_0 t}$ lie asymptotically in the range of one of these projections. To this end, we use (12.1.14) to write down the identity

$$\mathscr{F}^{-1} U P_+ \mathrm{e}^{\mathrm{i}H_0 t} = E_{x>0} \mathscr{F}^{-1} U \mathrm{e}^{\mathrm{i}H_0 t} \tag*{(12.1.14)$'$}$$

Now the right-hand side of (12.1.14)′ may be regarded as an integral operator from $L^2(0, \infty, \mathbb{S}^2)$ to $L^2(\mathbb{R}, \mathbb{S}^2)$ with kernel

$$K(x, \lambda) = \begin{cases} (4\pi\lambda)^{-1/2} \mathrm{e}^{\mathrm{i}(x/2)\log\lambda} \mathrm{e}^{\mathrm{i}\lambda t} & (x \geq 0), \\ 0 & (x < 0) \end{cases} \tag{12.1.15}$$

That is,

$$(Kf)(x, \boldsymbol{\omega}) = \int_0^\infty K(x, \lambda) f(\lambda, \boldsymbol{\omega})\, \mathrm{d}\lambda. \tag{12.1.16}$$

We now take $f(\lambda, \boldsymbol{\omega})$ to be an infinitely differentiable function of λ and $\boldsymbol{\omega}$ whose support is a closed subset of the semi-infinite interval $0 < \lambda < \infty$.

We can then use integration by parts to evaluate the right-hand side of (12.1.16) by writing

$$e^{i(\lambda t+(x/2)\log\lambda)} = -\frac{2i\lambda}{x+2\lambda t}\frac{d}{d\lambda}(e^{i(\lambda t+(x/2)\log\lambda)}).$$

Moreover, our assumptions about the support of $f(\lambda,\boldsymbol{\omega})$ allow us to write $f(\lambda,\boldsymbol{\omega})=\phi(\lambda)u(\lambda,\boldsymbol{\omega})$, where again u is infinitely differentiable, and $\phi\in C_0^\infty(0,\infty)$. With (12.1.15), the right-hand side of (12.1.16) now becomes, for $x\geq 0$,

$$\int_0^\infty d\lambda\, e^{i(\lambda t+(x/2)\log\lambda)}\frac{\partial}{\partial\lambda}\left\{(4\pi\lambda)^{-1/2}\frac{2i\lambda}{x+2\lambda t}\phi(\lambda)u(\lambda,\boldsymbol{\omega})\right\},$$

where, to ensure the non-vanishing of the denominator $x+2\lambda t$, we must take $t>0$. There are no boundary terms coming from $\lambda=0$ and $\lambda=\infty$, because of the support properties of ϕ.

The above integral takes the form, for $x,t>0$,

$$\int_0^\infty K_0(x,\lambda)u(\lambda,\boldsymbol{\omega})\,d\lambda+\int_0^\infty K_1(x,\lambda)\frac{\partial u(\lambda,\boldsymbol{\omega})}{\partial\lambda}\,d\lambda.$$

In each term, the kernels $K_0(x,\lambda)$ and $K_1(x,\lambda)$ are square-integrable with respect to x and λ. For example,

$$K_0(x,\lambda)=e^{i(\lambda t+(x/2)\log\lambda)}\frac{\partial}{\partial\lambda}\left((4\pi\lambda)^{-1/2}\frac{2i\lambda}{x+2\lambda t}\phi(\lambda)\right)$$

so that

$$\int_0^\infty d\lambda\int_0^\infty dx|K_0(x,\lambda)|^2=\text{const}\int_0^\infty d\lambda\int_0^\infty dx\left\{\frac{\partial}{\partial\lambda}\left(\frac{\lambda^{1/2}\phi(\lambda)}{(x+2\lambda t)}\right)\right\}^2$$

$$\leq\frac{\text{const}}{t},$$

where the constant depends on ϕ. (For λ belonging to the support of ϕ, the integrand cannot exceed const$/(x+2\lambda t)^2$, and $\int_0^\infty dx/(x+2\lambda t)^2\leq$const$/t$. The integration over λ does not cause any difficulty, being effectively over a finite interval.) Similarly we have $\int_0^\infty d\lambda\int_0^\infty dx\,|K_1(x,\lambda)|^2\leq$ const$/t$, for $t>0$. We can use the same expression as before for $e^{i(\lambda t+(x/2)\log\lambda)}$ and again integrate by parts, and the right-hand side of (12.1.16) then becomes of the form

$$\int_0^\infty K_2(x,\lambda)u(\lambda,\boldsymbol{\omega})\,\mathrm{d}\lambda$$

$$+\int_0^\infty K_3(x,\lambda)\,\frac{\partial u(\lambda,\boldsymbol{\omega})}{\partial\lambda}\,\mathrm{d}\lambda$$

$$+\int_0^\infty K_4(x,\lambda)\,\frac{\partial^2 u(\lambda,\boldsymbol{\omega})}{\partial\lambda^2}\,\mathrm{d}\lambda.$$

This further integration by parts brings into the kernel another factor of $(x+2\lambda t)^{-1}$, with the result that we obtain in this case, with $j=2,3,4$.

$$\int_0^\infty \mathrm{d}\lambda \int_0^\infty \mathrm{d}x\,|K_j(x,\lambda)|^2 \leq \frac{\text{const}}{t^3}.$$

Estimates of this kind for integral operators allow us to obtain norm bounds, in the direct integral representation. For example, we have, by the Schwarz inequality,

$$\int_0^\infty \mathrm{d}x \cdot \int_{\mathbb{S}^2} \mathrm{d}\omega \left|\int_0^\infty K_2(x,\lambda)u(\lambda,\boldsymbol{\omega})\,\mathrm{d}\lambda\right|^2$$

$$\leq \int_0^\infty \mathrm{d}x \int_{\mathbb{S}^2} \mathrm{d}\omega \left(\int_0^\infty |K_2(x,\lambda)|^2\,\mathrm{d}\lambda\right)\left(\int_0^\infty |u(\lambda,\boldsymbol{\omega})|^2\,\mathrm{d}\lambda\right)$$

$$=\left(\int_0^\infty \mathrm{d}\lambda \int_0^\infty \mathrm{d}x\,|K_2(x,\lambda)|^2\right)\left(\int_0^\infty \mathrm{d}\lambda \int_{\mathbb{S}^2} \mathrm{d}\omega\,|u(\lambda,\boldsymbol{\omega})|^2\right)$$

$$\leq \frac{\text{const}}{t^3}\int_0^\infty \mathrm{d}\lambda\,(\|u_\lambda(\boldsymbol{\omega})\|_0)^2 = \frac{\text{const}}{t^3}\,\|u\|^2,$$

where $u \in L^2(\mathbb{R}^3)$ is the element of the Hilbert space that corresponds to $\{u_\lambda(\boldsymbol{\omega})\}$ in the direct integral representation.

In the same way, we obtain, for example, the estimate

$$\int_0^\infty \mathrm{d}x \int_{\mathbb{S}^2} \mathrm{d}\omega \left| \int_0^\infty K_4(x,\lambda) \frac{\partial^2 u(\lambda,\boldsymbol{\omega})}{\partial\lambda^2} \mathrm{d}\lambda \right|^2$$

$$\leq \frac{\text{const}}{t^3} \int_0^\infty \mathrm{d}\lambda \left(\left\| \frac{\partial^2 u_\lambda(\boldsymbol{\omega})}{\partial\lambda^2} \right\|_0 \right)^2.$$

Because of the support properties of $\phi(\lambda)$, every kernel $K_j(x,\lambda)$ will vanish outside some interval $\lambda_1 \leq \lambda \leq \lambda_2$ which leads to the improved estimate

$$\frac{\text{const}}{t^3} \int_{\lambda_1}^{\lambda_2} \mathrm{d}\lambda \left(\left\| \frac{\partial^2 u_\lambda(\boldsymbol{\omega})}{\partial\lambda^2} \right\|_0 \right)^2.$$

We now appeal to the bound (5.3.33), with f replaced by u, for integrals of this kind, to obtain the final estimate

$$\frac{\text{const}}{t^3} \|(|\mathbf{X}|^2+1)u\|^2.$$

Bearing in mind that the kernels $K_j(x,\lambda)$ with which we have been dealing vanish identically for $x<0$, so that all x-integrations can take place over the half-line $0<x<\infty$, we are now able to give a bound for the right-hand side of (12.1.16). After a single integration by parts, with $f(\lambda,\boldsymbol{\omega})=\phi(\lambda)u(\lambda,\boldsymbol{\omega})$, we have (norms in $L^2(\mathbb{R},\mathbb{S}^2)$)

$$\|(Kf)(x,\boldsymbol{\omega})\|^2 \leq \frac{\text{const}}{t} \|1\mathbf{X}1+1)u\|^2,$$

and after two integrations by parts we find

$$\|(Kf)(x,\boldsymbol{\omega})\|^2 \leq \frac{\text{const}}{t^3} \|(|\mathbf{X}|^2+1)u\|^2.$$

After n integrations by parts we have

$$\|(Kf)(x,\boldsymbol{\omega})\|^2 \leq \frac{\text{const}}{t^{2n-1}} \|(|\mathbf{X}|^n+1)u\|^2$$

$$\leq \frac{\text{const}}{t^{2n-1}} \|(|\mathbf{X}|+1)^n u\|^2.$$

Remembering that the direct integral representation was introduced, in the present context, as a tool for evaluating the right-hand side of (12.1.14)′, and operating on $f=\phi(H_0)u$, where $(\phi(H_0)u)_\lambda=\phi(\lambda)u_\lambda$, we have arrived at the important inequality, valid for $t>0$:

$$\begin{aligned}\|\mathscr{F}^{-1}UP_+\mathrm{e}^{\mathrm{i}H_0t}\phi(H_0)u\|^2&=\|P_+\phi(H_0)\mathrm{e}^{\mathrm{i}H_0t}u\|^2\\&=\|E_{x>0}\mathscr{F}^{-1}U\mathrm{e}^{\mathrm{i}H_0t}\phi(H_0)u\|^2\\&\le\frac{\text{const}}{t^{2n-1}}\|(|X|+1)^n u\|^2.\end{aligned}$$

Writing $u=(|X|+1)^{-n}v$, we can immediately proceed to the following lemma, which provides the basis for all that follows in this section.

Lemma 12.1

Let $\phi(\lambda)$ be infinitely differentiable, with compact support in $(0,\infty)$, and let $P_\pm$ be the projections respectively onto the positive and negative parts of the spectrum of $A=\frac{1}{2}(\boldsymbol{P}\cdot\boldsymbol{X}+\boldsymbol{X}\cdot\boldsymbol{P})$. Then, for $n=1,2,\ldots,$

$$\left.\begin{aligned}\|P_+\phi(H_0)\mathrm{e}^{\mathrm{i}H_0t}(|X|+1)^{-n}\|&\le\frac{\text{const}}{(1+t)^{n-1/2}}\quad(t>0),\\\|P_-\phi(H_0)\mathrm{e}^{\mathrm{i}H_0t}(|X|+1)^{-n}\|&\le\frac{\text{const}}{(1+|t|)^{n-1/2}}\quad(t<0).\end{aligned}\right\}\tag{12.1.17}$$

Proof

For $t>0$ the inequality is already proved. (All of the estimates that we have made apply under the assumption $v\in C_0^\infty(\mathbb{R}^3)$. Since such v are dense in the Hilbert space, the first operator-norm estimate in (12.1.17) will follow. Note that norms in (12.1.17) are uniformly bounded in t. We are then at liberty to replace t by $1+t$ on the right-hand side and thus avoid the singularity at $t=0$.) For $t<0$, with P_- instead of P_+, the proof goes through in the same way, with $E_{x<0}$ replacing $E_{x>0}$ in (12.1.14). Note that the constants on the right-hand side of (12.1.17) will depend on the value of n, and also on the function $\phi(\lambda)$ on the left-hand side. ■

The inverse powers of $1+t$ on the right-hand side of (12.1.17) are not the best possible. Moreover, for applications to scattering by short-range potentials, we need on the left-hand side an inverse power of $|X|+1$ that is close to, but slightly greater than, unity. The following lemma provides the improved estimate that we need.

Lemma 12.2

With the same assumptions and notation as for Lemma 12.1, and given any (fixed) β and β', with $0<\beta'<\beta$, we have the inequalities

$$\left.\begin{aligned} \|P_+\phi(H_0)e^{iH_0t}(|X|+1)^{-\beta}\| &\le \frac{\text{const}}{(1+t)^{\beta'}} \quad (t>0),\\ \|P_-\phi(H_0)e^{iH_0t}(|X|+1)^{-\beta}\| &\le \frac{\text{const}}{(1+|t|)^{\beta'}} \quad (t<0). \end{aligned}\right\} \tag{12.1.18}$$

Proof

We use *interpolation*, based on the equality, valid for arbitrary bounded linear operators T,

$$\|T\| = \|TT^*\|^{1/2}.$$

Applying this equality to (12.1.17), we have, for $n=1, 2, 3, \ldots,$

$$\begin{aligned} \|P_+\phi(H_0)e^{iH_0t}(|X|+1)^{-n/2}\| &= \|P_+\phi(H_0)e^{iH_0t}(|X|+1)^{-n}e^{-iH_0t}\phi(H_0)P_+\|^{1/2}\\ &\le \text{const}\,\|P_+\phi(H_0)e^{iH_0t}(|X|+1)^{-n}\|^{1/2}\\ &\le \text{const}\,(1+t)^{-(n-1/2)/2} \quad \text{for } t>0. \end{aligned}$$

Repeating this argument, we have, for $j=1, 2, 3, \ldots,$

$$\|P_+\phi(H_0)e^{iH_0t}(|X|+1)^{-n/2^j}\| \le \text{const}\,(1+t)^{-(n-1/2)/2^j}.$$

Now any positive number may be approximated arbitrarily closely by $n/2^j$, for suitable n and j sufficiently large. We choose n and j such that

$$\frac{n}{2^j} < \beta \quad \text{and} \quad \frac{(n-\frac{1}{2})}{2^j} > \beta'.$$

Since then $(|X|+1)^{-n/2^j} > (|X|+1)^{-\beta}$ and $(1+t)^{-(n-1/2)/2^j} < (1+t)^{-\beta'}$, the required result follows. The case $t<0$ with P_- is proved similarly. ■

We can now justify our interpretation of the projections $P_\pm$. For large positive times, states under the free evolution are asymptotically outgoing, i.e. in the limit they belong to the range of P_+. Ingoing states, at large negative times, are asymptotically in the range of P_-. To check the first of these statements, we have, from (12.1.18),

$$\operatorname*{s-lim}_{t\to\infty} P_- e^{-iH_0t} f = 0$$

for states f in the range of $\phi(H_0)(|X|+1)^{-\beta}$. Taking limits as $\beta\to 0$ and as $\phi(\lambda)$ "spreads out" to approach for $\lambda>0$ the constant function $\phi(\lambda)=1$, we can extend this result to the entire Hilbert space. Hence, for all $f \in \mathscr{H}$, $e^{-iH_0t} f$ asymptotically approaches the subspace corresponding to the positive spectrum of A. Similarly, $e^{-iH_0t} f$ approaches the range of P_- in the limit as $t \to -\infty$. Encouraged by this confirmation that we are physically on the right lines, we are now ready to prove compactness for certain operators involving the wave operators for short-range potentials, as a preparation for our proof of asymptotic completeness. We deal for simplicity in the first instance with non-singular potentials.

Lemma 12.3

Let $V(\boldsymbol{r})$ be a short-range non-singular potential, in the sense that

$$|V(\boldsymbol{r})| \leq \text{const}\,(1+|\boldsymbol{r}|)^{-(1+\varepsilon)} \tag{12.1.19}$$

for some $\varepsilon > 0$. Let $H = H_0 + V$. Define wave operators $\Omega_\pm$ by

$$\Omega_\pm = \operatorname*{s-lim}_{t \to \pm\infty} e^{iHt} e^{-iH_0t}. \tag{12.1.20}$$

Then the operators

$$(\Omega_\pm - 1)\phi(H_0)P_\pm \qquad \text{(both + signs, or both − signs)}$$

and

$$(\Omega_\pm - 1)\phi(H)P_\pm$$

are compact, where $P_\pm$ are the projection operators onto positive and negative parts respectively of A, and $\phi(\lambda) \in C_0^\infty(\mathbb{R})$, such that $\lambda = 0$ does not lie in the support of ϕ.

Proof

We rely on the result

$$(e^{iHt} e^{-iH_0t} - 1)\phi(H_0)P_+ = \int_0^t ds \frac{d}{ds} e^{iHs} e^{-iH_0s} \phi(H_0)P_+,$$

where d/ds denotes a derivative defined as a strong limit. Hence

$$(e^{iHt} e^{-iH_0t} - 1)\phi(H_0)P_+ = i\int_0^t ds\, e^{iHs} V e^{-iH_0s} \phi(H_0)P_+. \tag{12.1.21}$$

By (12.1.18), with $\beta = 1 + \varepsilon$ and $1 < \beta' < \beta$, we know, on taking adjoints of the operator within the norm, with $V(\boldsymbol{r})$ bounded according to (12.1.19), that

$$\int_0^\infty \mathrm{d}s\, \|\mathrm{e}^{\mathrm{i}Hs} V \mathrm{e}^{-\mathrm{i}H_0 s}\phi(H_0)P_+\| = \int_0^\infty \mathrm{d}s\, \|V\mathrm{e}^{-\mathrm{i}H_0 s}\phi(H_0)P_+\| < \infty.$$

Hence, in (12.1.21), we can take the limit $t \to \infty$ even in operator norm. Since $V\phi(H_0)$ is compact, the resulting limit is a limit, in operator norm, of compact operators, and therefore is also compact. A similar proof with P_- instead of P_+, and taking the limit as $t \to -\infty$, allows us together to assert the compactness of $(\Omega_\pm - 1)\phi(H_0)P_\pm$. The compactness of the other pair of operators follows from the fact that $\phi(H) - \phi(H_0)$ is compact. We do not insist in this result that $\phi(\lambda)$ has positive support, since any subset of $(-\infty, 0)$ contained in the support of $\phi(\lambda)$ will not contribute to the operator $\phi(H_0)$ (although there may, of course, be a contribution to $\phi(H)$). ■

Lemma 12.3 makes precise a sense in which the operator Ω_+ is "close" to the identity operator, on the outgoing subspace, provided that the range of energies is restricted to a finite interval not containing zero energy. This lemma is all that is needed in order to prove asymptotic completeness, and at the same time we obtain a result on the absence of a singular continuous spectrum.

Theorem 12.1

Let $V(\boldsymbol{r})$ be a short-range non-singular potential as in Lemma 12.3, with $H = H_0 + V$. Then H has no singular continuous spectrum, and the wave operators $\Omega_\pm$ are asymptotically complete.

Proof

Let us suppose that H has a non-trivial singular continuous subspace. We let $[a,b]$ be a closed interval not containing the origin and such that H has a singular continuous spectrum in $[a,b]$. Such an interval must exist, since a singular continuous spectrum cannot be concentrated at a single point $\lambda = 0$. We now define $\phi(\lambda)$ as in Lemma 12.3, such that $\phi(\lambda) \equiv 1$ for $\lambda \in [a,b]$. Since the range of the spectral projection of H for the interval $[a,b]$ must contain an *infinite-dimensional subspace* of $\mathcal{M}_{\mathrm{sc}}(H)$, we can construct an orthonormal sequence $\{f_n\}$ of elements of this subspace. Such a sequence $\{f_n\}$ will converge weakly to zero and satisfy $\phi(H)f_n = f_n$.

From Lemma 12.3, on taking adjoints, $P_\pm\phi(H)(\Omega_\pm^* - 1)$ is compact, so that $\text{s-lim}_{n\to\infty} P_\pm\phi(H)(\Omega_\pm^* - 1)f_n = 0$. However, $\Omega_\pm^* f = 0$ for $f \in \mathcal{M}_{\mathrm{sc}}(H)$, and

hence

$$\underset{n\to\infty}{\text{s-lim}}\, P_{\pm}\phi(H)f_n = \underset{n\to\infty}{\text{s-lim}}\, P_{\pm}f_n = 0.$$

Since $f_n = (P_+ + P_-)f_n$, this contradicts the assumption $\|f_n\| = 1$. Hence H has no singular continuous spectrum. A similar argument shows that H cannot have a limit point of eigenvalues, except possibly at zero energy.

The same argument cannot, however, be used without modification to prove asymptotic completeness, since for a vector orthogonal to the range of Ω_+, say, one has $\Omega_+^* f = 0$ but not necessarily $\Omega_-^* f = 0$. To proceed in this case, we replace the limit $n\to\infty$ by a time-dependent limit $t\to\infty$. We then suppose that f is a unit vector orthogonal to the range of Ω_+. As above, we may assume without loss of generality that f has spectral support for H concentrated in a closed interval $[a,b]$ that does not contain $\lambda = 0$. Since $e^{-iHt}f$ converges weakly to zero, we have, by Lemma 12.3 on taking adjoints, with $f\in\mathcal{M}_{\text{ac}}(\text{H})$,

$$\underset{t\to\infty}{\text{s-lim}}\, P_+\phi(H)(\Omega_+^* - 1)e^{-iHt}f = 0.$$

Since $\Omega_+^* f = 0$, and for suitable ϕ, we have $\phi(H)f = f$,

$$\underset{t\to\infty}{\text{s-lim}}\, P_+ e^{-iHt}f = 0. \tag{12.1.22}$$

On the other hand,

$$\underset{t\to\infty}{\text{s-lim}}\, P_-\phi(H)(\Omega_-^* - 1)e^{-iHt}f = 0. \tag{12.1.23}$$

Now

$$\begin{aligned}&P_-\phi(H)\Omega_-^* e^{-iHt}f\\ &\quad = P_-(\phi(H)-\phi(H_0))\Omega_-^* e^{-iHt}f + P_-\phi(H_0)\Omega_-^* e^{-iHt}f.\end{aligned}$$

On the right-hand side the first term vanishes in the limit as $t\to\infty$, since $\phi(H)-\phi(H_0)$ is compact. The second term, using the intertwining property of wave operators, is

$$P_- e^{-iH_0 t}\phi(H_0)\Omega_-^* f.$$

which again vanishes in the limit $t\to\infty$, since we have seen that, for arbitrary $g\in\mathcal{H}$, $P_- e^{-iH_0 t}g\to 0$ strongly. From (12.1.23) we may thus deduce that

$$\underset{t\to\infty}{\text{s-lim}}\, P_-\phi(H)e^{-iHt}f = \underset{t\to\infty}{\text{s-lim}}\, P_- e^{-iHt}f = 0. \tag{12.1.24}$$

Combining this result with (12.1.22), we now obtain

$$\underset{t\to\infty}{\text{s-lim}}\, e^{-iHt}f = 0,$$

which contradicts the assumption that $\|f\| = 1$. ■

12.2 RESOLVENT ESTIMATES AND COMMUTATORS

We have already seen, in expressing the wave and scattering operators as spectral integrals, the important role in scattering theory of the resolvent operator $R(z)$ as z approaches the real axis. In particular, we are often interested in considering the *boundary value*, in some sense, of the resolvent as $z \to \lambda$ for $\lambda \in \mathbb{R}$. Usually this limit is of interest for values of λ belonging to the continuous spectrum of the operator in question.

It should be noted at the outset that neither the upper boundary value $\lim_{\varepsilon \to 0+} (H-\lambda-i\varepsilon)^{-1}$ nor the lower boundary value $\lim_{\varepsilon \to 0+} (H-\lambda+i\varepsilon)^{-1}$ of the resolvent for a Hamiltonian H can be expected to exist as strong limits on the Hilbert space $\mathscr{H}$. Indeed, if either of these strong limits exist, on some vector f, then the limit will be just $(H-\lambda)^{-1} f$. And we know that $(H-\lambda)^{-1}$ is defined on the whole of $\mathscr{H}$ only if λ does *not* belong to the spectrum of H. It is often the case, however, that a suitable bounded operator B may be introduced that enables the appropriate strong limits to be taken. Thus, for example, for suitable B it may happen that the strong limits $\text{s-lim}_{\varepsilon \to 0+} B(H-\lambda \pm i\varepsilon)^{-1} B$ exist on the entire Hilbert space. Moreover, if the range of B is dense then it follows that, on a dense subset of $\mathscr{H}$, the strong limits of $B(H-\lambda \pm i\varepsilon)^{-1}$ exist. Even more may sometimes be said. The limits of $B(H-\lambda \pm i\varepsilon)^{-1} B$ may exist even in operator norm. In this case, denoting by $R_\pm(\lambda)$ the respective upper and lower boundary values of the resolvent, we can give a meaning to $BR_\pm(\lambda)B$ as bounded linear operators on $\mathscr{H}$, even though $R_\pm(\lambda)$ do not make sense as operators in $\mathscr{H}$ in their own right. We may then regard $R_\pm(\lambda)$ as bounded linear operators from one *Banach space* to another. That is, assuming B to be invertible, we make the range of B into a Banach space $\mathscr{B}_1$ with norm $\|B^{-1}y\|$ for $y \in \mathscr{B}_1$, and we define $\mathscr{B}_2$ to be the completion of $\mathscr{H}$ under the norm $\|Bz\|$ for $z \in \mathscr{B}_2$. Then we can write $z = R_+(\lambda)y$ for $y \in \mathscr{B}_1$ and $z \in \mathscr{B}_2$, where the mapping from y to z is bounded with respect to the appropriate Banach-space norms.

This approach to boundary values of the resolvent, in which one looks for a suitable pair of Banach spaces with respect to which $R_\pm(\lambda)$ are defined as bounded linear operators, is an important technique of the so-called stationary theory of scattering, and is capable of a considerable degree of extension and generalization. For example, one may consider a pair of operators B_1, B_2 and consider the boundary values of $B_1(H-\lambda \pm i\varepsilon)^{-1} B_2$, where B_1 and B_2 need not necessarily be bounded. The choice of Banach space is usually motivated by the particular problem under consideration, and one may look for a choice that is optimal in some sense.

Since we are concerned primarily with the time-dependent theory of scattering, we shall be interested in resolvent estimates mainly for the light that they throw on the asymptotic evolution of states in the limits as

$t \to \pm\infty$. How is it possible to pass from information concerning the boundary value of resolvent operators to estimates of large-time behaviour? Let us introduce a single bounded invertible operator B, which for simplicity we assume to be self-adjoint, and suppose that we have the estimate

$$\|B(H-\lambda+\mathrm{i}\varepsilon)^{-2}B\| \leq \text{const}\,\varepsilon^{-q} \tag{12.2.1}$$

for some q satisfying $\frac{1}{2} \leq q < 1$. We suppose that (12.2.1) holds in the limit as $\varepsilon \to 0+$, uniformly for all λ in some closed bounded subinterval of $\mathbb{R}$. Equation (12.2.1) is actually an estimate for $(R(z))^2$ rather than for $R(z)$. Such estimates may often be arrived at when considering the dependence of the resolvent on λ, since $(H-\lambda+\mathrm{i}\varepsilon)^{-2}$ is actually the (strong) derivative, with respect to λ, of the resolvent $(H-\lambda+\mathrm{i}\varepsilon)^{-1}$. Later in this section we shall see how to obtain estimates like (12.2.1) in the case $H=H_0+V$ for a wide class of potentials $V(\boldsymbol{r})$, which may be of short or long range.

Since, as a strong derivative, we have

$$\mathrm{i}\frac{\mathrm{d}}{\mathrm{d}\varepsilon}(H-\lambda+\mathrm{i}\varepsilon)^{-1} = (H-\lambda+\mathrm{i}\varepsilon)^{-2},$$

we can integrate with respect to ε from ε_1 to ε_2, both positive, to give

$$\mathrm{i}\{B(H-\lambda+\mathrm{i}\varepsilon_2)^{-1}B - B(H-\lambda+\mathrm{i}\varepsilon_1)^{-1}B\} = \int_{\varepsilon_1}^{\varepsilon_2} \mathrm{d}\varepsilon\, B(H-\lambda+\mathrm{i}\varepsilon)^{-2}B, \tag{12.2.2}$$

where the right-hand side may be defined, for example, as a strong limit of approximating Riemann sums. Since $q<1$, the left-hand side of (12.2.1) is integrable with respect to ε, even up to $\varepsilon=0$, so that by taking an operator-norm estimate of each side of (12.2.2), we may deduce that the norm of the left-hand side of (12.2.2) converges to zero in the joint limits $\varepsilon_1, \varepsilon_2 \to 0+$. This establishes the norm convergence to a limit, as $\varepsilon \to 0+$, of the operator $B(H-\lambda+\mathrm{i}\varepsilon)^{-1}B$, for fixed λ in the closed bounded subinterval of $\mathbb{R}$. Moreover, denoting this limit by $BR_-(\lambda)B$ (*lower* boundary value of the resolvent), we have the uniform estimate

$$\|B(H-\lambda+\mathrm{i}\varepsilon)^{-1}B - BR_-(\lambda)B\| \leq \text{const}\,\varepsilon^{1-q}. \tag{12.2.3}$$

We may also differentiate the resolvent with respect to λ and integrate the inequality (12.2.1) with respect to λ from λ_1 to λ_2, both within the closed bounded interval, to obtain

$$\|B(H-\lambda_2+\mathrm{i}\varepsilon)^{-1}B - B(H-\lambda_1+\mathrm{i}\varepsilon)^{-1}B\| \leq \text{const}\,\varepsilon^{-q}|\lambda_1-\lambda_2|. \tag{12.2.4}$$

Using (12.2.3) with $\lambda=\lambda_1, \lambda_2$ respectively, together with (12.2.4) and the triangle inequality, we now have

$$\|BR_-(\lambda_2)B - BR_-(\lambda_1)B\| \le \text{const}\,\varepsilon^{1-q} + \text{const}\,\varepsilon^{-q}|\lambda_1 - \lambda_2|.$$

Since the left-hand side of this inequality is independent of ε, we may substitute $\varepsilon = |\lambda_1 - \lambda_2|$ to give

$$\|BR_-(\lambda_2)B - BR_-(\lambda_1)B\| \le \text{const}\,|\lambda_1 - \lambda_2|^{1-q}. \tag{12.2.5}$$

Thus the operator family $BR_-(\lambda)B$ is norm-Hölder-continuous in λ, with index $1-q$. Since, by taking adjoints, we can replace ε by $-\varepsilon$, the inequality (12.2.5) also holds with $R_+(\lambda)$ in place of $R_-(\lambda)$.

According to (2.6.10), the spectral projection of H, associated with a finite interval, is obtained in the limit $\varepsilon \to 0+$ from the integral with respect to λ, over the interval, of $(2\pi i)^{-1}\{(H-\lambda-i\varepsilon)^{-1} - (H-\lambda+i\varepsilon)^{-1}\}$. Therefore if we operate to right and left by B, and take into account the above limits as $\varepsilon \to 0+$, we can write, with $H = \int \lambda\, dE_\lambda$,

$$B\frac{dE_\lambda}{d\lambda}B = \frac{1}{2\pi i}B(R_+(\lambda) - R_-(\lambda))B. \tag{12.2.6}$$

On the left-hand side of (12.2.6), $dE_\lambda/d\lambda$ is not to be interpreted as a linear operator acting in $\mathscr{H}$. Rather, the left-hand side stands for the derivative $(d/d\lambda)BE_\lambda B$, which exists even as a limit in the operator-norm topology of $\mathscr{H}$. If $dE_\lambda/d\lambda$ is to be regarded as a linear operator, it is to be defined, as for the boundary value of the resolvent, as a bounded operator from one Banach space to another. In addition, the spectral projection $E_{(a,b]}$ of H for a finite interval satisfies an equation obtained from (12.2.6) by integration, namely

$$BE_{(a,b]}B = \frac{1}{2\pi i}\int_a^b d\lambda\, B(R_+(\lambda) - R_-(\lambda))B. \tag{12.2.7}$$

(We restrict attention always to a range of values of λ for which the uniform estimate (12.2.1) holds.) For a C_0^∞ function ϕ such that (12.2.1) holds on the support of ϕ, we have, on integration of (12.2.6) with $\phi(H) = \int \phi(\lambda)\, dE_\lambda = \int \phi(\lambda)(dE_\lambda/d\lambda)d\lambda$,

$$B\phi(H)B = \frac{1}{2\pi i}\int_{-\infty}^{\infty} d\lambda\, \phi(\lambda)B(R_+(\lambda) - R_-(\lambda))B. \tag{12.2.8}$$

If we are to consider time evolution then we replace $\phi(\lambda)$ by $e^{-i\lambda t}\phi(\lambda)$ in

(12.2.8), so that

$$Be^{-iHt}\phi(H)B=\frac{1}{2\pi i}\int_{-\infty}^{\infty} d\lambda\, e^{-i\lambda t}\phi(\lambda)B(R_+(\lambda)-R_-(\lambda))B. \qquad (12.2.9)$$

The right-hand side of (12.2.9) may be thought of as the Fourier transform with respect to λ of the operator family $M(\lambda)$ given by

$$M(\lambda)=\frac{-i}{(2\pi)^{1/2}}\phi(\lambda)B(R_+(\lambda)-R_-(\lambda))B. \qquad (12.2.10)$$

For arbitrary $g\in\mathscr{H}$, we may think of $M(\lambda)g\equiv g_\lambda$ as defining an element $\{g_\lambda\}$ of a direct integral Hilbert space, in which the norm is given by $(\int\|g_\lambda\|^2\,d\lambda)^{1/2}$. By considering the operation of Fourier transform with respect to λ on elements of this space having the form $\{\psi(\lambda)h\}$ (linear combinations of finitely many such elements being dense in the space), it may be verified that the Fourier transform is a unitary operator on this direct integral space. In particular, we have, from (12.2.9),

$$\int_{-\infty}^{\infty} dt\,\|Be^{-iHt}\phi(H)Bg\|^2$$

$$=\frac{1}{2\pi}\int_{-\infty}^{\infty} d\lambda\,|\phi(\lambda)|^2\,\|B(R_+(\lambda)-R_-(\lambda))Bg\|^2$$

$$=\int_{-\infty}^{\infty} d\lambda\,\|M(\lambda)g\|^2, \qquad (12.2.11)$$

where, according to (12.2.5), and the corresponding result with R_- replaced by R_+, the operator $M(\lambda)$ is norm-Hölder-continuous with index $1-q$. The factor $\phi(\lambda)$ in $M(\lambda)$ does not affect Hölder continuity, since ϕ is even differentiable. Note that ϕ has compact support, so that the integral over $\mathbb{R}$ on the right-hand side is effectively over a finite interval.

Let us now replace $e^{-i\lambda t}$ by $e^{-i(\lambda-\delta)t}$ on the right-hand side of (12.2.9) to obtain an integral representation for $Be^{-i(H-\delta)t}\phi(H)B$. This operator is the Fourier transform of $M(\lambda+\delta)$, where $M(\lambda)$ is given by (12.2.10), and similarly $Be^{-i(H+\delta)t}\phi(H)B$ is the Fourier transform of $M(\lambda-\delta)$. By taking a linear combination of these two results, we may infer that $\sin\delta t\, Be^{-iHt}\phi(H)B$ is the Fourier transform of $[M(\lambda+\delta)-M(\lambda-\delta)]/2i$. By the corresponding result to (12.2.11) we then have

$$4\int_{-\infty}^{\infty} \mathrm{d}t\ \sin^2 \delta t\ \|B\mathrm{e}^{-\mathrm{i}Ht}\phi(H)Bg\|^2 = \int_{-\infty}^{\infty} \mathrm{d}\lambda \|(M(\lambda+\delta)-M(\lambda-\delta))g\|^2. \quad (12.2.12)$$

On account of the Hölder continuity of $M(\lambda)$, this gives, for the limit $\delta\to 0+$,

$$\int_{-\infty}^{\infty} \mathrm{d}t\ \sin^2 \delta t\ \|B\mathrm{e}^{-\mathrm{i}Ht}\phi(H)Bg\|^2 \le \text{const}\ \delta^{2(1-q)}\|g\|^2 \qquad (12.2.13)$$

Inequality (12.2.13) is not yet in a form suitable for considering the large-time evolution generated by H, because of the presence in the integrand of the factor $\sin^2 \delta t$, which may be small for large t if δt is close to a multiple of π. To get round this difficulty, we replace δ in (12.2.13) by $\beta\delta$ and integrate with respect to β from 0 to 1, to give

$$\int_{-\infty}^{\infty} \mathrm{d}t \left(1-\frac{\sin 2\delta t}{2\delta t}\right)\|B\mathrm{e}^{-\mathrm{i}Ht}\phi(H)Bg\|^2 \le \text{const}\ \delta^{2(1-q)}\|g\|^2. \quad (12.2.14)$$

Given $T>0$, we may restrict attention to the range $t\ge T$ with $\delta = T^{-1}$; we then have $1-(\sin 2\delta t)/2\delta t \ge \frac{1}{2}$, so that (12.2.14) implies that

$$\int_{T}^{\infty} \mathrm{d}t\ \|B\mathrm{e}^{-\mathrm{i}Ht}\phi(H)Bg\|^2 \le \text{const}\ T^{-2(1-q)}\|g\|^2, \qquad (12.2.15)$$

which is an estimate for the large-t norm of $B\mathrm{e}^{-\mathrm{i}Ht}\phi(H)Bg$ for arbitrary $g\in\mathscr{H}$. We have thus been able to translate Hölder continuity of $M(\lambda)$ into a decay estimate for the Fourier transform of $M(\lambda)$, in the limit as $t\to\infty$. A similar estimate to (12.2.15) holds if we replace the left-hand side by an integral from $-\infty$ to $-T$.

We can also deduce from (12.2.15) corresponding results for the integral from 0 to T, in the limit $T\to\infty$. For any p such that $\frac{1}{2}<p<1$, we define

$$J(t) = \int_{t}^{\infty} s^{-p}\|B\mathrm{e}^{-\mathrm{i}Hs}\phi(H)Bg\|\,\mathrm{d}s.$$

$J(t)$ is a bounded function of t for $0<t<\infty$, and by the Schwarz inequality, with (12.2.15), we have the estimate

$$J(t) \leq \text{const}\, t^{-(p-q+1/2)} ||g||.$$

Hence

$$\int_0^T \mathrm{d}t\, ||B\mathrm{e}^{-\mathrm{i}Ht}\phi(H)Bg|| = -\int_0^T t^p \frac{\mathrm{d}J(t)}{\mathrm{d}t}\,\mathrm{d}t$$

$$= -T^p J(T) + \int_0^T pt^{p-1} J(t)\,\mathrm{d}t,$$

where the final integral is bounded as $T \to \infty$ by const $T^{q-1/2}||g||$, and we have a similar estimate for the first term on the right-hand side. Hence we have the bound, valid in the limit $T \to \infty$,

$$\int_0^T \mathrm{d}t\, ||B\mathrm{e}^{-\mathrm{i}Ht}\phi(H)Bg|| \leq \text{const}\, T^{q-1/2} ||g||. \tag{12.2.16}$$

(The case $q = \frac{1}{2}$ requires a log T estimate on the right-hand side.) The result (12.2.16) gives, in integral form, an estimate for the time evolution of an observable $\mathrm{e}^{\mathrm{i}Ht}B\mathrm{e}^{-\mathrm{i}Ht}$ in the Heisenberg picture, with initial state $\phi(H)Bg$. The inequality will prove particularly useful in the following section, in considering the time evolution of position observables, especially for scattering by long-range potentials. The result can be generalized to allow for powers such as $|B|^\theta$ $(0 < \theta < 1)$ of the observable B. The generalization of (12.2.16) to allow for such powers leads to the result

$$\int_0^T \mathrm{d}t\, ||\,|B|^\theta \mathrm{e}^{-\mathrm{i}Ht}\phi(H)Bg|| \leq \text{const}\, T^{1+q\theta-3\theta/2} ||g||. \tag{12.2.17}$$

In order to verify (12.2.17) in the case $\theta = \frac{1}{2}$, we apply the Schwarz inequality to the left-hand side to obtain the bound

$$\text{const}\left(\int_0^T 1\,\mathrm{d}t\right)^{1/2}\left(\int_0^T \mathrm{d}t\, ||B\mathrm{e}^{-\mathrm{i}Ht}\phi(H)Bg||\, ||g||\right)^{1/2},$$

and the result follows from (12.2.16). The extension to $\theta = 2^{-n}$ follows in a similar way. To interpolate the result to $\theta=\frac{1}{2}(\theta_1+\theta_2)$, given that (12.2.17) holds for $\theta=\theta_1$, $\theta=\theta_2$ respectively, we note that

$$\begin{aligned}\| \, |B|^{(\theta_1+\theta_2)/2} h\| &= \langle |B|^{\theta_1} h, |B|^{\theta_2} h\rangle^{1/2} \\ &\leq \| \, |B|^{\theta_1} h\|^{1/2} \| \, |B|^{\theta_2} h\|^{1/2},\end{aligned}$$

and again apply the Schwarz inequality to the left-hand side of (12.2.17). Hence (12.2.17) holds at least in the case $\theta=m/2^n$, where m and n are positive integers, and this is sufficient for applications.

Having seen how a bound such as (12.2.1) may be used to derive resolvent estimates as z approaches the real axis, and thence to obtain information on bounds for the Heisenberg evolution of suitable observables, let us now consider how to prove (12.2.1) in the case of interest, namely $H=H_0+V$, where H_0 is the free Hamiltonian and V is a potential of short or long range. The method that we shall use will be to relate the boundary value of the resolvent to appropriate commutators between H_0, H and the self-adjoint operator A defined in the last section by (12.1.6). We start from the commutation relation

$$H_0 A - AH_0 = -2\mathrm{i}H_0. \qquad (12.2.18)$$

Equation (12.2.18) which expresses a fundamental relation between the generator of dilations A and the free Hamiltonian $H_0=-\Delta$ in $L^2(\mathbb{R}^3)$, may be interpreted in two ways. First of all, (12.2.18) holds if, for example, each side of the equation operates on functions belonging to $C_0^\infty(\mathbb{R}^3)$ and may be verified using the explicit expressions for H_0 and A as differential operators. Since $C_0^\infty(\mathbb{R}^3)$ is dense in $L^2(\mathbb{R}^3)$, this establishes (12.2.18) as holding on a suitable large domain of functions. Alternatively, we may interpret (12.2.18) in the sense of *forms*. That is, given any $f,g \in D(H_0) \cap D(A)$, we have

$$\langle H_0 f, Ag\rangle - \langle Af, H_0 g\rangle = -2\mathrm{i}\langle f, H_0 g\rangle. \qquad (12.2.19)$$

To verify (12.2.19), we represent f and g by $\{f_\lambda\}$ and $\{g_\lambda\}$ respectively in the direct integral decomposition of $\mathscr{H}=L^2(\mathbb{R}^3)$. H_0 then corresponds to multiplication by λ, and A is given by (12.1.9). Remembering the form of the inner product,

$$\langle f,g\rangle = \int_0^\infty \mathrm{d}\lambda \, \langle f_\lambda, g_\lambda\rangle$$

(we drop the suffix 0 for notational convenience),

we then have

$$\langle H_0 f, Ag\rangle - \langle Af, H_0 g\rangle + 2\mathrm{i}\langle f, H_0 g\rangle$$

$$= \mathrm{i}\int_0^\infty \mathrm{d}\lambda \left(\left\langle f_\lambda, \lambda g_\lambda + 2\lambda^2 \frac{\mathrm{d}g_\lambda}{\mathrm{d}\lambda}\right\rangle\right.$$

$$\left. + \left\langle \lambda f_\lambda + 2\lambda^2 \frac{\mathrm{d}f_\lambda}{\mathrm{d}\lambda}, g_\lambda\right\rangle + 2\langle f_\lambda, \lambda g_\lambda\rangle\right)$$

$$= \mathrm{i}\int_0^\infty \mathrm{d}\lambda \frac{\mathrm{d}}{\mathrm{d}\lambda}\,(2\lambda^2\langle f_\lambda, g_\lambda\rangle)$$

$$= 2\mathrm{i}[\lambda^2\langle f_\lambda, g_\lambda\rangle]_{\lambda=0}^{\lambda=\infty}. \tag{12.2.20}$$

In particular, $\lim_{\lambda\to\infty}\langle\lambda f_\lambda, \lambda g_\lambda\rangle$ exists, and since $\langle H_0 f, H_0 g\rangle = \int_0^\infty \mathrm{d}\lambda\, \langle\lambda f_\lambda, \lambda g_\lambda\rangle$, this limit can only be zero. Similarly, since $\langle f, g\rangle = \int_0^\infty \mathrm{d}\lambda\, \langle f_\lambda, g_\lambda\rangle$, $\lambda^2\langle f_\lambda, g_\lambda\rangle$ cannot converge to a non-zero limit as $\lambda\to 0$. Hence the boundary values $\lambda=\infty$ and $\lambda=0$ in (12.2.20) both vanish, and we have established (12.2.19) for $f, g\in D(H_0)\cap D(A)$.

Now if $H = H_0 + V$, we have, as a consequence of (12.2.18), the formal identity

$$(H-\lambda+\mathrm{i}\varepsilon)(A+\mathrm{i}) - (A-\mathrm{i})(H-\lambda+\mathrm{i}\varepsilon) = -2\mathrm{i}(\lambda - Q - \mathrm{i}\varepsilon), \tag{12.2.21}$$

where

$$Q = V + (2\mathrm{i})^{-1}[V, A]. \tag{12.2.22}$$

As before, (12.2.21) may be interpreted in two ways. If $V(\boldsymbol{r})$ is locally smooth then (12.2.21) will be valid if both sides act on any function belonging to $C_0^\infty(\mathbb{R}^3)$; if $V(\boldsymbol{r})$ is locally singular, say at $\boldsymbol{r}=0$, then we can allow the equation to act on $C_0^\infty(\mathbb{R}^3\setminus\{0\})$. From the expression for A as a differential operator in position space, (12.2.22) then becomes

$$Q = V + \frac{r}{2}\frac{\partial V}{\partial r}, \tag{12.2.23}$$

where $(\partial/\partial r)V(\boldsymbol{r})$ denotes partial differentiation with respect to the radial coordinate of spherical polar coordinates.

On the other hand, we may write (12.2.21) as a form identity; thus

$$\langle (H-\lambda-\mathrm{i}\varepsilon)f, (A+\mathrm{i})g\rangle - \langle (A+\mathrm{i})f, (H-\lambda+\mathrm{i}\varepsilon)g\rangle$$

$$= -2\mathrm{i}\langle f, (\lambda - Q - \mathrm{i}\varepsilon)g\rangle, \tag{12.2.24}$$

where (12.2.24) is to hold for all $f, g\in D(H)\cap D(A)$ and the right-hand side is also to be defined, through (12.2.22), in the sense of forms. Equation

(12.2.24) may be proved to hold in great generality. The following lemma gives sufficient conditions for the validity of (12.2.24), which are enough to meet the needs of most applications.

Lemma 12.4

Suppose that $V(\boldsymbol{r})$ is bounded in the region $|\boldsymbol{r}|\geq R$, for any $R>0$, so that a sesquilinear form τ may be defined for all f, $g\in D(A)$ that have compact support in $\mathbb{R}^3\setminus\{0\}$ by

$$\tau(f,g)=\langle f,Vg\rangle+(2\mathrm{i})^{-1}\{\langle Vf,Ag\rangle-\langle Af,Vg\rangle\}. \tag{12.2.25}$$

Suppose that τ is bounded above or below, and is closable; let τ' denote the form closure of τ. Assume also that $D(H)\subseteq D(H_0^{1/2})$. Then (12.2.24) holds for all f, $g\in D(H)\cap D(A)$, provided that, on the right-hand side, $\langle f,Qg\rangle$ is replaced by $\tau'(f,g)$. (That is, the operator Q is to be replaced by the form τ'.)

Proof

We let ρ denote the operator of multiplication by $\rho(|\boldsymbol{r}|)$, where $\rho(r)$ is a smooth non-decreasing function such that $\rho(r)\equiv 0$ near $r=0$ and $\rho(r)\equiv 1$ for r sufficiently large. For $f,g\in D(H)\cap D(A)$, we then have $\rho_a f,\rho_a g\in D(H)\cap D(A)$, where ρ_a denotes multiplication by $\rho(r/a)$ for any $a>0$. Moreover, $\rho_a f$, $\rho_a g\in D(V)$, so that we may use (12.2.19) and (12.2.25) to evaluate the left-hand side of (12.2.24), giving

$$\begin{aligned}&\langle (H-\lambda-\mathrm{i}\varepsilon)\rho_a f,(A+\mathrm{i})\rho_a g\rangle-\langle (A+\mathrm{i})\rho_a f,(H-\lambda+\mathrm{i}\varepsilon)\rho_a g\rangle\\ &\quad=-2\mathrm{i}\{\langle \rho_a f,(\lambda-\mathrm{i}\varepsilon)\rho_a g\rangle-\tau(\rho_a f,\rho_a g)\}.\end{aligned} \tag{12.2.26}$$

The lemma is now to be proved by taking the limit $a\to 0$, in which $\rho_a f$ and $\rho_a g$ converge strongly to f and g respectively. First we take this limit on the left-hand side of (12.2.26). Now $(H-\lambda-\mathrm{i}\varepsilon)\rho_a f=\rho_a(H-\lambda-\mathrm{i}\varepsilon)f-(\Delta\rho_a)f-2\mathrm{i}(\nabla\rho_a)\cdot\boldsymbol{P}f$, where $\boldsymbol{P}$ denotes the momentum operator; note that, by hypothesis, $f\in D(H)\Rightarrow f\in D(H_0^{1/2})$, where $D(H_0^{1/2})\subseteq D(\boldsymbol{P})$. Also, $(A+\mathrm{i})\rho_a g=\rho_a(A+\mathrm{i})g-\mathrm{i}(\boldsymbol{r}\cdot\nabla\rho_a)g$. Hence the first inner product on the left-hand side of (12.2.26) may be written as a sum of $\langle\rho_a(H-\lambda-\mathrm{i}\varepsilon)f,\rho_a(A+\mathrm{i})g\rangle$ with various contributions involving commutators, such as, for example, contributions proportional to $\langle(\Delta\rho_a)f,(\boldsymbol{r}\cdot\nabla\rho_a)g\rangle$, $\langle\nabla\rho_a\cdot\boldsymbol{P}f,\rho_a(A+\mathrm{i})g\rangle$, and so on. To verify that these additional commutator terms go to zero in the limit as $a\to 0$, we note that

(i) $|\nabla\rho_a|\leq\text{const}/a\leq\text{const}/r$ on the support of $\nabla\rho_a$; hence $|\nabla\rho_a|\leq(\text{const}/r)\chi_a$, where χ_a is the characteristic function of the support of $\nabla\rho_a$;

(ii) $|\Delta\rho_a| \leq (\text{const}/r^2)\chi_a$;

(iii) f, $g \in D(1/r)$ (i.e. f and g belong to the domain of the operator of multiplication by $1/r$ (this follows because $f, g \in D(H_0^{1/2}) \subseteq D(1/r)$));

(iv) Af, $Ag \in D(1/r)$; indeed, from the expression for A as a differential operator in position space, we have, for example,

$$\left\| \frac{1}{r} Af \right\| = \left\| \frac{1}{r}\left(\boldsymbol{r}\cdot\boldsymbol{P} - \frac{3\mathrm{i}}{2}\right)f \right\| \leq \text{const}\left(\|\boldsymbol{P}f\| + \left\| \frac{1}{r} f \right\| \right);$$

(v) $\chi_a f$ and $\chi_a g$ converge strongly to zero in the limit $a \to 0$, since, in this limit, the support of $\nabla\rho_a$ is vanishingly small; similarly for $r^{-1}\chi_a f$, and $r^{-1}\chi_a g$. We then have,

$$|\langle(\Delta\rho_a)f, (\boldsymbol{r}\cdot\nabla\rho_a)g\rangle| \leq \text{const}\left\| \frac{1}{r}\chi_a f \right\| \left\| \frac{1}{r}\chi_a g \right\| \to 0 \quad \text{as } a \to 0,$$

and

$$|\langle \nabla\rho_a \cdot \boldsymbol{P}f, \rho_a(A+\mathrm{i})g\rangle| \leq \text{const}\,\|\boldsymbol{P}f\| \left(\left\| \chi_a \frac{1}{r} Ag \right\| + \left\| \frac{1}{r}\chi_a g \right\| \right) \to 0.$$

The remaining commutator terms also approach zero, so that, in the limit $a \to 0$, the first inner product on the left-hand side of (12.2.26) becomes

$$\lim_{a\to 0} \langle \rho_a(H-\lambda-\mathrm{i}\varepsilon)f, \rho_a(A+\mathrm{i})g\rangle = \langle (H-\lambda-\mathrm{i}\varepsilon)f, (A+\mathrm{i})g\rangle.$$

The second inner product may be treated in the same way, and we obtain the left-hand side of (12.2.24). To complete the proof of the lemma, it remains only to evaluate the right-hand side of (12.2.26) in the limit $a \to 0$. This limit is clearly

$$-2\mathrm{i}\{\langle f, (\lambda-\mathrm{i}\varepsilon)g\rangle - \lim_{a\to 0} \tau(\rho_a f, \rho_a g)\}.$$

Rather similar arguments to the above show that

$$\lim_{a,b\to 0} \tau((\rho_a - \rho_b)f, (\rho_a - \rho_b)f) = 0.$$

Since $\rho_a f \to f$ as $a \to 0$, this implies that f, and similarly g, belong to the form domain of τ', and we then have

$$\lim_{a\to 0} \tau(\rho_a f, \rho_a g) = \tau'(f, g), \qquad \text{as required.}$$ ■

If $V(\boldsymbol{r})$ is continuously differentiable, except possibly at $\boldsymbol{r} = \boldsymbol{0}$, so that $Q(\boldsymbol{r})$ may be defined by (12.2.23), the conditions of the lemma mean that $Q(\boldsymbol{r})$ is bounded from above or below. In this case τ is necessarily closable, and the form domain of τ' may be identified with $D(|Q|^{1/2})$. If, for example, $Q(r)\geq 0$ then the right-hand side of (12.2.24) becomes

$$-2i\{\langle f,(\lambda-i\varepsilon)g\rangle-\langle Q^{1/2}f,Q^{1/2}g\rangle\}.$$

The condition $D(H)\subseteq D(H_0^{1/2})$ of the lemma will hold provided, for example, that the total Hamiltonian H is defined by means of the Friedrichs extension, and the lemma may be applied to a wide class of non-singular potentials, of both short and long range, as well as to potentials that are highly singular and repulsive at $r=0$. The result may be generalized still further by writing, on the right-hand side of (12.2.24),

$$\lambda-Q=\lambda-\beta^{-1}H+\beta^{-1}(H-\beta Q),$$

where β is any real non-vanishing constant. We then have

$$\lim_{a\to 0}\langle\rho_a f,(\lambda-\beta^{-1}H)\rho_a g\rangle=\langle f,(\lambda-\beta^{-1}H)g\rangle.$$

Provided that $H-\beta Q$, defined as a sesquilinear form, is bounded below and closable, we may again obtain (12.2.24) for all $f,g\in D(H)\cap D(A)$, where on the right-hand side the expression $\lambda-Q$ is to be treated as a sum of two semibounded closable forms.

Let us suppose, then, that (12.2.24) holds, in some appropriate sense, and take vectors f and g that belong respectively to the ranges of the operators $(H-\lambda-i\varepsilon)^{-1}(A-i)^{-1}$ and $(H-\lambda+i\varepsilon)^{-1}(A-i)^{-1}$. These vectors are certainly in $D(H)$. Assuming that they are also in $D(A)$, we may substitute into (12.2.24) to obtain the operator identity

$$\begin{aligned}&(H-\lambda+i\varepsilon)^{-1}(A-i)^{-1}-(A+i)^{-1}(H-\lambda+i\varepsilon)^{-1}\\&=-2i(A+i)^{-1}(H-\lambda+i\varepsilon)^{-1}(\lambda-Q-i\varepsilon)\\&\quad(H-\lambda+i\varepsilon)^{-1}(A-i)^{-1}.\end{aligned}\tag{12.2.27}$$

Equation (12.2.27) will be the basis of resolvent estimates for the total Hamiltonian H, and will hold in particular under the hypotheses of Lemma 12.4, provided that Q is interpreted suitably in the sense of forms, and under the further condition that the range of $(H-z)^{-1}$ $(A-i)^{-1}$ is contained in the domain of A. The following lemma shows that, in fact, the further domain condition is unnecessary, but may sometimes be deduced as a consequence of (12.2.27).

Lemma 12.5

Under the hypotheses of Lemma 12.4, (12.2.27) holds, provided that on the right-hand side, Q is interpreted in the sense of forms. If, in addition, for Im $(z)\neq 0$, $D(H)$ is contained in the form domain of τ', then the range of $(H-z)^{-1}$ $(A-i)^{-1}$ is contained in the domain of A.

Proof

We have already verified (12.2.24) in the case when both vectors f and g belong $D(H)\cap D(A)$. We should like to obtain a corresponding result under the more restricted assumption that $f,\ g\in D(H)$. We therefore suppose that $f,\ g\in D(H)$, and we now let ρ denote the operator of multiplication by $\rho(|\mathbf{r}|)$, where $\rho(r)$ is a smooth non-increasing function such that $\rho(r)\equiv 1$ near $r=0$ and $\rho(r)\equiv 0$ for sufficiently large r. For $R>0$, we let ρ_R denote the operator of multiplication by $\rho(r/R)$. From the domain properties of H, we know that $\rho_R f,\ \rho_R g\in D(H)$. Moreover, since $f,\ g\in D(H_0^{1/2})\subseteq D(\mathbf{P})$, and ρ_R has compact support, we know that $\rho_R f,\ \rho_R g\in D(A)$. Thus $\rho_R f,\ \rho_R g\in D(H)\cap D(A)$, so that (12.2.24) holds with f and g replaced by $\rho_R f$ and $\rho_R g$, respectively. We now follow closely the proof of Lemma 12.4, except that now the limit $a\to 0$ is replaced by a limit $R\to\infty$. Note, for example, the bounds

$$|\Delta\rho_R|\le\frac{\text{const}}{R^2},\qquad |\nabla\rho_R|\le\frac{\text{const}}{R},\qquad \|(A+\mathrm{i})\rho_R g\|\le\text{const}\,R.$$

As before, commutator contributions vanish in the limit, and we obtain, for $f,\ g\in D(H)$,

$$\lim_{R\to\infty}\{\langle\rho_R(H-\lambda-\mathrm{i}\varepsilon)f,(A+\mathrm{i})\rho_R g\rangle-\langle(A+\mathrm{i})\rho_R f,\rho_R(H-\lambda+\mathrm{i}\varepsilon)g\rangle\}$$

$$=\lim_{R\to\infty}\{-2\mathrm{i}\langle\rho_R f,(\lambda-Q-\mathrm{i}\varepsilon)\rho_R g\rangle\},$$

provided that either limit exist. We now set

$$f=(H-\lambda-\mathrm{i}\varepsilon)^{-1}(A-\mathrm{i})^{-1}f_0,$$

$$g=(H-\lambda+\mathrm{i}\varepsilon)^{-1}(A-\mathrm{i})^{-1}g_0,$$

so that the left-hand side becomes

$$\lim_{R\to\infty}\{<\rho_R(A-\mathrm{i})^{-1}f_0,(A+i)\rho_R(H-\lambda+\mathrm{i}\varepsilon)^{-1}(A-\mathrm{i})^{-1}g_0\rangle$$

$$-\langle(A+\mathrm{i})\rho_R(H-\lambda-\mathrm{i}\varepsilon)^{-1}(A-\mathrm{i})^{-1}f_0,\rho_R(A-\mathrm{i})^{-1}g_0\rangle.$$

Since $(A-\mathrm{i})\rho_R(A-\mathrm{i})^{-1}f_0$ converges strongly to f_0, and $(A-\mathrm{i})\rho_R(A-\mathrm{i})^{-1}g_0$ to g_0, the left-hand side converges to

$$\langle f_0,[(H-\lambda+\mathrm{i}\varepsilon)^{-1}(A-\mathrm{i})^{-1}-(A+\mathrm{i})^{-1}(H-\lambda+\mathrm{i}\varepsilon)^{-1}]g_0\rangle.$$

Hence, as before, we find that f and g belong to the form domain of τ', and (12.2.27) is proved. Note that

$$\langle f_0,(A+\mathrm{i})^{-1}(H-\lambda+\mathrm{i}\varepsilon)^{-1}Q(H-\lambda+\mathrm{i}\varepsilon)^{-1}(A-\mathrm{i})^{-1}g_0\rangle$$

is to be interpreted as

$$\tau'((H-\lambda-\mathrm{i}\varepsilon)^{-1}(A-\mathrm{i})^{-1}f_0,(H-\lambda+\mathrm{i}\varepsilon)^{-1}(A-i)^{-1}g_0).$$

We now suppose that $D(H)$ is contained in the form domain of τ'. Then we can define a closed form τ'' by

$$\tau''(f,g)=\tau'((H-\lambda+i\varepsilon)^{-1}f, (H-\lambda+i\varepsilon)^{-1}g),$$

for all f, $g\in\mathscr{H}$. Now τ'' is the sequilinear form related to some self-adjoint operator. By the Closed-Graph Theorem, this self-adjoint operator must be bounded. Hence, in the notation of (12.2.27), $(H-\lambda-i\varepsilon)^{-1}Q(H-\lambda+i\varepsilon)^{-1}$ is a bounded linear operator. Since $(H-\lambda+i\varepsilon)^{-1}(H-\lambda-i\varepsilon)$ is unitary, it follows that $(H-\lambda+i\varepsilon)^{-1}Q(H-\lambda+i\varepsilon)^{-1}$ is bounded. Immediately from (12.2.27) we now have that the range of $(H-\lambda+i\varepsilon)^{-1}(A-i)^{-1}$ belongs to the domain of A. ∎

Just as for Lemma 12.4, we do not necessarily need the form τ' corresponding to Q to be semibounded and closable; the result extends without difficulty to the case in which $H-\beta Q$ is bounded below and closable.

The ideas behind the proof of Lemmas 12.4 and 12.5 may be applied in a variety of situations involving commutators of differential operators. As an example, we let X_j and P_j denote the jth components of the position and momentum operators respectively, and consider, for complex numbers z_1 and z_2, the commutation relation

$$(H-z_1)(X_j-z_2)-(X_j-z_2)(H-z_1)=-2iP_j. \tag{12.2.28}$$

Equation (12.2.28) holds as an identity if both sides of the equation act in $C_0^\infty(\mathbb{R}^3\setminus\{0\})$. Or we may express (12.2.28) as a form identity

$$\langle(H-\bar{z}_1)f,(X_j-z_2)g\rangle-\langle(X_j-\bar{z}_2)f,(H-z_1)g\rangle=-2i\langle f,P_jg\rangle, \tag{12.2.29}$$

holding for all f, $g\in D(H)\cap D(X_j)$ under the assumptions that $V(\boldsymbol{r})$ is bounded for $|\boldsymbol{r}|\geq R$ and $D(H)\subseteq D(H_0^{1/2})$. The proof of (12.2.29) follows the same lines as the proof of Lemma 12.4; that is, we replace f and g by $\rho_a f$ and $\rho_a g$ and proceed to the limit $a\to 0$. Now, in (12.2.29), let us replace f and g by $\rho_R f$ and $\rho_R g$ respectively, and assume only $f,g\in D(H)$. Following the same kind of arguments as in the proof of Lemma 12.5, we can then take the limit $R\to\infty$ to obtain the identity

$$\lim_{R\to\infty}\{\langle(H-\bar{z}_1)f,(X_j-z_2)\rho_R g\rangle-\langle(X_j-\bar{z}_2)\rho_R f,(H-z_1)g\rangle\}$$
$$=-2i\langle f,P_j g\rangle,$$

which is now valid for all f, $g\in D(H)$.

Replacing f and g respectively by $(H-\bar{z}_1)^{-1}(X_j-\bar{z}_2)^{-1}f_0$ and $(H-z_1)^{-1}(X_j-z_2)^{-1}g_0$, and proceeding to the limit now gives rise to the operator identity

$$(H-z_1)^{-1}(X_j-z_2)^{-1}-(X_j-z_2)^{-1}(H-z_1)^{-1}$$
$$=-2\mathrm{i}(X_j-z_2)^{-1}(H-z_1)^{-1}P_j(H-z_1)^{-1}(X_j-z_2)^{-1}. \quad (12.2.30)$$

Since, for arbitrary $h\in\mathscr{H}=L^2(\mathbb{R}^3)$, $(X_j-z_2)^{-1}(H-z_1)^{-1}h$ belongs to the domain of the operator $A_j=\frac{1}{2}(P_jX_j+X_jP_j)$, it follows from (12.2.30) that $(H-z_1)^{-1}(X_j-z_2)^{-1}h$ also belongs to this domain. Thus $A_j(H-z_1)^{-1}(X_j-z_2)^{-1}$ is bounded, by an application of the Closed-Graph Theorem. Since $|x_j|\leq|\boldsymbol{r}|$, we may also deduce that $A_j(H-z_1)^{-1}(|\boldsymbol{r}|-z_2)^{-1}$ is bounded (where $|\boldsymbol{r}|$ now denotes the operator of multiplication by $|\boldsymbol{r}|$ in $L^2(\mathbb{R}^3)$), from which it follows, on summing over j, that $A(H-z_1)^{-1}(|\boldsymbol{r}|-z_2)^{-1}$ is bounded. It also follows in a straightforward manner from (12.2.30) that $|\boldsymbol{r}|(H-z_1)^{-1}(|\boldsymbol{r}|-z_2)^{-1}$ is bounded. Under somewhat stronger hypotheses, these operators remain bounded even in the limit $z_2\to 0$. Let us suppose, for example, that H is bounded below and that the *form* domain of H is contained in the form domain of H_0. In this case, all of the operators $X_j^{-1}(H-z_1)^{-1}$, $P_j(H-z_1)^{-1}$ and $P_j(H-z_1)^{-1}X_j^{-1}$ are bounded. We can then left-multiply both sides of (12.2.30) by the operator A_j and take the limit $z_2\to 0$ to deduce that $A_j(H-z_1)^{-1}X_j^{-1}$ is bounded. It follows easily that $A(H-z_1)^{-1}|\boldsymbol{r}|^{-1}$ is bounded under this additional hypothesis, and similarly we find that $|\boldsymbol{r}|(H-z_1)^{-1}|\boldsymbol{r}|^{-1}$ is bounded.

The following lemma establishes a simple condition under which a norm estimate for $(A+\mathrm{i})^{-1}(H-\lambda+\mathrm{i}\varepsilon)^{-1}(A-\mathrm{i})^{-1}$ may be derived.

Lemma 12.6

Suppose that (12.2.27) holds, where Q is a bounded linear operator such that $\|\lambda^{-1}Q\|<1$. Then there exists a constant γ, with $0<\gamma<1$ (in fact $\gamma=\|\lambda^{-1}Q\|$), such that

$$\|(A+\mathrm{i})^{-1}(H-\lambda+\mathrm{i}\varepsilon)^{-1}(A-\mathrm{i})^{-1}\|\leq\mathrm{const}\,\varepsilon^{-\gamma}. \quad (12.2.31)$$

If $\|\lambda_0^{-1}Q\|<1$ for some λ_0, then the estimate (12.2.31) holds uniformly in λ for λ belonging to some neighbourhood of λ_0.

Proof

Let us define the function $f(\varepsilon)$ by

$$f(\varepsilon)=\|(A+\mathrm{i})^{-1}(H-\lambda+\mathrm{i}\varepsilon)^{-1}(A-\mathrm{i})^{-1}\|. \quad (12.2.32)$$

Since $(H-\lambda+\mathrm{i}\varepsilon)^{-1}=(H-\lambda-\mathrm{i}\varepsilon)[(H-\lambda)^2+\varepsilon^2]^{-1}$, we have

$$\begin{aligned}
&\varepsilon\|(H-\lambda+\mathrm{i}\varepsilon)^{-1}(A-\mathrm{i})^{-1}h\|^2\\
&\quad=\varepsilon\langle(A-\mathrm{i})^{-1}h,[(H-\lambda)^2+\varepsilon^2]^{-1}(A-\mathrm{i})^{-1}h\rangle\\
&\quad=-\operatorname{Im}(\langle(A-\mathrm{i})^{-1}h,(H-\lambda+\mathrm{i}\varepsilon)^{-1}(A-\mathrm{i})^{-1}h\rangle)\\
&\quad\leq f(\varepsilon)\|h\|^2.
\end{aligned}$$

Hence

$$\|(H-\lambda+\mathrm{i}\varepsilon)^{-1}(A-\mathrm{i})^{-1}\|\leq\left(\frac{f(\varepsilon)}{\varepsilon}\right)^{1/2}. \tag{12.2.33}$$

We now consider the identity

$$\begin{aligned}
&(A+\mathrm{i})^{-1}(H-\lambda+\mathrm{i}\varepsilon)^{-2}(A-\mathrm{i})^{-1}\\
&\quad=\lambda^{-1}(A+\mathrm{i})^{-1}(H-\lambda+\mathrm{i}\varepsilon)^{-1}(\lambda-Q)(H-\lambda+\mathrm{i}\varepsilon)^{-1}(A-\mathrm{i})^{-1}\\
&\qquad+(A+\mathrm{i})^{-1}(H-\lambda+\mathrm{i}\varepsilon)^{-1}\lambda^{-1}Q(H-\lambda+\mathrm{i}\varepsilon)^{-1}(A-\mathrm{i})^{-1}.
\end{aligned} \tag{12.2.34}$$

Since $\|(H-\lambda+\mathrm{i}\varepsilon)^{-1}\|\leq 1/\varepsilon$, and $\|\varepsilon(H-\lambda+\mathrm{i}\varepsilon)^{-2}\|\leq 1/\varepsilon$, (12.2.27) tells us that the first term on the right-hand side of (12.2.34) is bounded in norm by const/ε. Moreover, (12.2.33) implies a similar bound for $(A+\mathrm{i})^{-1}(H-\lambda+\mathrm{i}\varepsilon)^{-1}$. Hence, with $\|\lambda^{-1}Q\|=\gamma$, the second term on the right-hand side of (12.2.34) is bounded in norm by $\gamma f(\varepsilon)/\varepsilon$. Since $(\mathrm{d}/\mathrm{d}\varepsilon)(H-\lambda+\mathrm{i}\varepsilon)^{-1}=-\mathrm{i}(H-\lambda+\mathrm{i}\varepsilon)^{-2}$, we can integrate in (12.2.34) from ε to some fixed constant value c to obtain an identity for $(A+\mathrm{i})^{-1}(H-\lambda+\mathrm{i}\varepsilon)^{-1}(A-\mathrm{i})^{-1}$. Given that $f(\varepsilon)\leq f_n(\varepsilon)$, we then find a new norm estimate of the form $f(\varepsilon)\leq f_{n+1}(\varepsilon)$, where, on taking $0<c<1$,

$$f_{n+1}(\varepsilon)=a-b\log\varepsilon+\gamma\int_\varepsilon^c\frac{f_n(\varepsilon')}{\varepsilon'}\,\mathrm{d}\varepsilon'. \tag{12.2.35}$$

The idea is to iterate this equation, starting from some initial bound, say $f_0(\varepsilon)=1/\varepsilon$. Standard estimates show that the iteration converges uniformly in any fixed closed subinterval of $(0, c]$, and we have moreover

$$f(\varepsilon)\leq\lim_{n\to\infty}f_n(\varepsilon).$$

This limiting bound for $f(\varepsilon)$ is the solution of the differential equation

$$\frac{\mathrm{d}f}{\mathrm{d}\varepsilon}=-\frac{b}{\varepsilon}-\gamma\frac{f(\varepsilon)}{\varepsilon}, \tag{12.2.36}$$

subject to the condition $f(c)=a-b\log c$. This solution is of the form $me^{-\gamma}+n$, where the constants m and n may be determined. Hence indeed

$f(\varepsilon) \leq \text{const}\, \varepsilon^{-\gamma}$ as $\varepsilon \to 0$, as required. The uniformity of the estimate in λ stated in the final part of the lemma will be left to the reader to verify. ∎

Note that the bound (12.2.31), with $0<\gamma<1$, is better than we should have anticipated. *A priori* we should expect for the left-hand side of (12.2.31) only a bound of order $1/\varepsilon$. The fact that we can do better is a consequence of the identity (12.2.27), which follows in turn from the evaluation of commutators with the generator A of dilations.

We can generalize Lemma 12.6 in various directions. The following corollary summarizes some of the main results that are available.

Corollary

Suppose that (12.2.27) holds, and that Q satisfies one of the following conditions.

(i) Q is a bounded linear operator, and either $Q \leq \lambda^-$ or $Q \geq \lambda^+$, as form inequalities, for some λ^- and λ^+ such that $\lambda^- < \lambda$ and $\lambda^+ > \lambda$.

(ii) $Q(H-z)^{-1}$ is bounded for $\text{Im}(z) \neq 0$. Moreover, $\|\lambda^{-1} E_{\mathscr{I}}(H) QE_g(H)\| < 1$ for some interval $\mathscr{I}$ containing the point λ in its interior.

(iii) $Q(H-z)^{-1}$ is bounded, and for some $\lambda^{\pm}$, either $E_{\mathscr{I}}(H)QE_{\mathscr{I}}(H) \leq \lambda^-$ or $E_{\mathscr{I}}(H)QE_{\mathscr{I}}(H) \geq \lambda^+ E_{\mathscr{I}}(H)$, as form inequalities for $\lambda^- < \lambda < \lambda^+$.

(In (ii) and (iii), $E_{\mathscr{I}}$(H) is the spectral projection of H for the interval $\mathscr{I}$.)

(iv) Q is defined as a form τ' on the form domain of H, and satisfies the form inequalities, for some g_1 and g_2,

$$H + g_1 Q \geq \text{const} \qquad (g_1 > 0),$$
$$H - g_2 Q \geq \lambda^-(1 - g_2) \quad (g_2 > 1,\ \lambda^- < \lambda \text{ if } \lambda > 0)$$

or

$$H - g_2 Q \geq \lambda^+(1 - g_2) \quad (0 < g_2 < 1,\ \lambda^+ > \lambda \text{ if } \lambda < 0).$$

Then the estimate (12.2.31) holds for some constant γ in the range $0<\gamma<1$. In each case, the estimate holds uniformly in some open interval containing λ.

Proof

(i) Let us suppose, for example, that $Q \leq \lambda^-$. For $c > -\lambda$, (12.2.34) remains valid if on the right-hand side we replace λ^{-1} by $(\lambda + c)^{-1}$ in the first term, and $\lambda^{-1}Q$ by $(\lambda + c)^{-1}(Q + c)$ in the second term. The proof of the lemma may now be extended to the present case provided that we have

$\|(\lambda+c)^{-1}(Q+c)\|<1$. The hypotheses imply that $Q+c\le\lambda^-+c$, and $Q+c\ge\text{const}>0$ for c large enough. Hence, with c large enough, $\|(\lambda+c)^{-1}(Q+c)\|\le(\lambda^-+c)/(\lambda+c)<1$. The case $Q\ge\lambda^+$ follows similarly.

(ii) We apply the method of the lemma, writing, in the second term on the right-hand side of (12.2.34),

$$\lambda^{-1}Q=\lambda^{-1}E_{\mathscr{I}}(H)QE_{\mathscr{I}}(H)+\lambda^{-1}(1-E_{\mathscr{I}}(H))QE_{\mathscr{I}}(H)+\lambda^{-1}Q(1-E_{\mathscr{I}}(H)).$$

Since

$$\begin{aligned}&Q(1-E_{\mathscr{I}}(H))(H-\lambda+\mathrm{i}\varepsilon)^{-1}\\&=Q(H-z)^{-1}\{(H-z)(1-E_{\mathscr{I}}(H))(H-\lambda+\mathrm{i}\varepsilon)^{-1}\},\end{aligned}$$

we have $\|Q(1-E_{\mathscr{I}}(H))(H-\lambda+\mathrm{i}\varepsilon)^{-1}\|\le\text{const}$ for λ belonging to the interior of $\mathscr{I}$. Hence, in replacing $\lambda^{-1}Q$ by $\lambda^{-1}E_{\mathscr{I}}(H)QE_{\mathscr{I}}(H)$ on the right-hand side of (12.2.34), any additional contributions will be bounded in norm by const/ε. Under the hypothesis that $\|\lambda^{-1}E_{\mathscr{I}}(H)QE_{\mathscr{I}}(H)\|=\gamma<1$, these additional terms will have the effect merely of changing the value of constant b in (12.2.36), and the estimate (12.2.31) may be deduced as before.

(iii) The proof of (iii) follows similar lines to the proof of (ii), if we note that in (ii) we may replace an estimate of $\lambda^{-1}E_{\mathscr{I}}(H)QE_{\mathscr{I}}(H)$ by an estimate of $E_{\mathscr{I}}(H)(\lambda+c)^{-1}(Q+c)E_{\mathscr{I}}(H)$.

(iv) There is a close connection between bounds for the forms $H+gQ$, for various values of g, and the projection-related estimates of (ii) and (iii). For example, $H-gQ\ge0$ for some $g>1$ implies that $E_{\mathscr{I}}(H)QE_{\mathscr{I}}(H)\le g^{-1}E_{\mathscr{I}}(H)HE_{\mathscr{I}}(H)\le\lambda^-$, provided that $g\lambda^-$ lies to the right of the interval $\mathscr{I}$, which will be the case provided that $\mathscr{I}$ is taken to be sufficiently small and λ^- sufficiently close to λ.

In order to prove (iv), we replace the definition (12.2.32) of $f(\varepsilon)$ by

$$f(\varepsilon)=\|(A+\mathrm{i})^{-1}(aH+b)(H-\lambda+\mathrm{i}\varepsilon)^{-1}(A-\mathrm{i})^{-1}\|.$$

Since the hypotheses of (iv) imply that H is bounded below, we can choose b in such a way that $aH+b\ge0$. Given an estimate of the form $f(\varepsilon)=O(\varepsilon^{-\gamma})$, the bound (12.2.31) follows easily, since we can write $aH+b=a(H-\lambda)+(a\lambda+b)$, and use the fact that $\|(H-\lambda)(H-\lambda+\mathrm{i}\varepsilon)^{-1}\|<1$. The inequality (12.2.33) now becomes

$$\|(aH+b)^{1/2}(H-\lambda+\mathrm{i}\varepsilon)^{-1}(A-\mathrm{i})^{-1}\|\le\left(\frac{f(\varepsilon)}{\varepsilon}\right)^{1/2}.$$

In applying the method of the lemma, with this new definition of $f(\varepsilon)$, we have to consider the integral with respect to ε of $(A+\mathrm{i})^{-1}(aH+b)(H-\lambda+\mathrm{i}\varepsilon)^{-2}(A-\mathrm{i})^{-1}$. Apart from further contributions that are of order $1/\varepsilon$, and which therefore will play no part in subsequent estimates, this operator integrand may be replaced by $(a\lambda+b)(A+\mathrm{i})^{-1}(H-\lambda+\mathrm{i}\varepsilon)^{-2}(A-\mathrm{i})^{-1}$, which, following (12.2.34) may be identified with

$$\begin{aligned}&(A+\mathrm{i})^{-1}(H-\lambda+\mathrm{i}\varepsilon)^{-1}(\lambda-Q)(H-\lambda+\mathrm{i}\varepsilon)^{-1}(A-\mathrm{i})^{-1}\\&\quad+(A+\mathrm{i})^{-1}(H-\lambda+\mathrm{i}\varepsilon)^{-1}(aH+b)^{1/2}\\&\quad\times\{(aH+b)^{-1/2}(Q+b+\lambda(a-1))(aH+b)^{-1/2}\}\\&\quad\times(aH+b)^{1/2}(H-\lambda+\mathrm{i}\varepsilon)^{-1}(A-\mathrm{i})^{-1}.\end{aligned}$$

Since the addition to Q of a term $C(H-\lambda)$ will contribute in operator norm only to order $1/\varepsilon$, the proof will be complete if we can obtain a bound

$$\|(aH+b)^{-1/2}(Q+b+\lambda(a-1)+C(H-\lambda))(aH+b)^{-1/2}\|<1.$$

We therefore look for inequalities of the form

$$-\gamma(aH+b)\leq Q+b+\lambda(a-1)+C(H-\lambda)\leq\gamma(aH+b). \tag{12.2.37}$$

where γ is a constant in the range $0<\gamma<1$. The right-hand half of (12.2.37), on setting

$$a=\frac{g_1^{-1}+g_2^{-1}}{2\gamma},\qquad C=\tfrac{1}{2}(g_1^{-1}-g_2^{-1}),$$

and dividing throughout by $g_2^{-1}=a\gamma-C$, becomes

$$H-g_2Q\geq g_2\left(b(1-\gamma)-\tfrac{1}{2}\lambda(g_1^{-1}-g_2^{-1})+\frac{\lambda(g_1^{-1}+g_2^{-1})}{2\gamma}-\lambda\right).$$

For fixed C, in the limit $\gamma\to1$, we require a lower bound for $H-g_2Q$ that approaches $\lambda(1-g_2)$. However, $\lambda^-(1-g_2)\geq\lambda(1-g_2)$ for $\lambda>0$ and $g_2>1$, whereas $\lambda^+(1-g_2)\geq\lambda(1-g_2)$ for $\lambda<0$ and $0<g_2<1$. In either case, the hypotheses of (iv) of the lemma imply that the required lower bound for $H-g_2Q$ is indeed satisfied, provided that γ is taken close enough to 1.

The remaining half of the inequality (12.2.37), on dividing through by $g_1^{-1}=C+a\gamma$, becomes

$$H+g_1Q\geq g_1\left(-b(1+\gamma)-\frac{\lambda(g_1^{-1}+g_2^{-1})}{2\gamma}+\tfrac{1}{2}\lambda(g_1^{-1}-g_2^{-1})+\lambda\right).$$

By fixing the value of b sufficiently large and then taking γ close to 1, this operator inequality is also satisfied. Hence the bound (12.2.31) has been proved in this case for some γ close to 1. The reader should have no difficulty in verifying the uniformity in λ of this bound, under each of the hypotheses (i)–(iv), in the neighbourhood of some initial value of λ. ■

Resolvent estimates such as (12.2.31), and the corresponding result

$$\|(H-\lambda+\mathrm{i}\varepsilon)^{-1}(A-\mathrm{i})^{-1}\|\leq\text{const}\,\varepsilon^{-(1+\gamma)/2}, \tag{12.2.38}$$

have a number of important applications to spectral analysis and scattering theory. For example, if λ is an eigenvalue of H then the operator norm on the left-hand side of (12.2.38) is of order ε^{-1}. Hence any of the hypotheses (i)–(iv) of the corollary imply the absence of eigenvalues of H. Again, if $E_{\mathscr{I}}(H)$ projects onto a spectral interval of length ε containing λ then we have from (12.2.38) that

$$\begin{aligned}\|E_{\mathscr{I}}(H)(A-\mathrm{i})^{-1}\| &= \|(H-\lambda+\mathrm{i}\varepsilon)E_{\mathscr{I}}(H)(H-\lambda+\mathrm{i}\varepsilon)^{-1}(A-\mathrm{i})^{-1}\| \\ &\leq \text{const}\,\varepsilon^{(1-\gamma)/2}.\end{aligned}$$

For any operator T such that $(A-\mathrm{i})(H-z)^{-1}T$ is bounded $(Im\,(z)\neq 0)$, we can deduce a similar estimate for $\|E_{\mathscr{I}}(H)T\|$, and such bounds are sometimes used in deriving localization estimates for the total energy and other observables.

Given additional conditions on the operator Q, we can strengthen the power bound (12.2.31) and deduce that the operator $(A+\mathrm{i})^{-1}(H-\lambda+\mathrm{i}\varepsilon)^{-1}(A-\mathrm{i})^{-1}$ has a limit in norm as ε approaches zero. The following lemma provides sufficient conditions for such a result to follow.

Lemma 12.7

Suppose that (12.2.27) holds, and that the bound (12.2.31) is satisfied with $0<\gamma<1$. (This will be so in particular if any of the conditions (i)–(iv) of the Corollary to Lemma 12.6 are satisfied.) Suppose further that, for some β in the range $0<\beta<1$, and $\mathrm{Im}\,(z)\neq 0$, either

(a) $(A-\mathrm{i})(H-z)^{-1}(1+|\boldsymbol{r}|)^{-1}$ and $(1+|\boldsymbol{r}|)^{\beta}Q(H-z)^{-1}$ are bounded, or
(b) $(A-\mathrm{i})(H-z)^{-1}|\boldsymbol{r}|^{-1}$ and $|\boldsymbol{r}|^{\beta}Q(H-z)^{-1}$ are bounded.

Then

$$\lim_{\varepsilon\to 0+}(A+\mathrm{i})^{-1}(H-\lambda+\mathrm{i}\varepsilon)^{-1}(A-\mathrm{i})^{-1}$$

exists as a limit in operator norm.

Proof

We first use (12.2.38), together with the corresponding result for $\|(A+\mathrm{i})^{-1}(H-\lambda+\mathrm{i}\varepsilon)^{-1}\|$, to estimate the terms of the identity (12.2.27). This gives

$$\begin{aligned}&\|(A+\mathrm{i})^{-1}(H-\lambda+\mathrm{i}\varepsilon)^{-1}(\lambda-Q)(H-\lambda+\mathrm{i}\varepsilon)^{-1}(A-\mathrm{i})^{-1}\| \\ &\leq \text{const}\,\varepsilon^{-\gamma}+\text{const}\,\varepsilon^{-(1+\gamma)/2}.\end{aligned}$$

Since γ satisfies $0<\gamma<1$, it follows that the norm on the left-hand side of this inequality is integrable, with respect to ε, from 0 to c. As in the proof of Lemma 12.6, we integrate (12.2.34) between ε and c, and we have shown that the integral of the first term on the right-hand side is bounded even in the limit $\varepsilon\to 0$. For the second term on the right-hand side of (12.2.34), we first of all note the bound

$$\begin{aligned}
&\|(A+\mathrm{i})^{-1}(H-\lambda+\mathrm{i}\varepsilon)^{-1}(1+|\boldsymbol{r}|)^{-1}\| \\
&\quad=\|(A+\mathrm{i})^{-1}(H-z)(H-\lambda+\mathrm{i}\varepsilon)^{-1} \\
&\quad\times(A-\mathrm{i})^{-1}(A-\mathrm{i})(H-z)^{-1}(1+|\boldsymbol{r}|)^{-1}\| \\
&\quad\le\|(A+\mathrm{i})^{-1}\{(H-\lambda)+(\lambda-z)\}(H-\lambda+\mathrm{i}\varepsilon)^{-1}(A-i)^{-1}\| \\
&\quad\times\|(A-\mathrm{i})(H-z)^{-1}(1+|\boldsymbol{r}|)^{-1}\| \\
&\quad\le \text{const}+\text{const}\,f(\varepsilon),
\end{aligned}$$

under condition (a) of the lemma, where $f(\varepsilon)$ is defined by (12.2.32). The same bound obtains if $(H-\lambda+\mathrm{i}\varepsilon)^{-1}$ is replaced by $(H-\lambda-\mathrm{i}\varepsilon)^{-1}$.

Given that $f(\varepsilon)\le$ const $\varepsilon^{-\gamma}$ $(0<\gamma<1)$, we can therefore deduce that

$$\|(A+\mathrm{i})^{-1}(H-\lambda+\mathrm{i}\varepsilon)^{-1}(1+|\boldsymbol{r}|)^{-1}\|\le \text{const}\,\varepsilon^{-\gamma}.$$

If we now write

$$\|(A+\mathrm{i})^{-1}(H-\lambda+\mathrm{i}\varepsilon)^{-1}(1+|\boldsymbol{r}|)^{-\theta}\|=N_\theta,$$

so that

$$N_1\le\text{const}\,\varepsilon^{-\gamma},\qquad N_0\le\text{const}\,\varepsilon^{-(1+\gamma)/2},$$

we can use an interpolation argument, along the lines indicated earlier in this section, to estimate N_θ for intermediate values in the range $0<\theta<1$. Thus, if $\theta=\frac{1}{2}(\theta_1+\theta_2)$, we have

$$\begin{aligned}
N_\theta^2&=\|(A+\mathrm{i})^{-1}(H-\lambda+\mathrm{i}\varepsilon)^{-1}(1+|\boldsymbol{r}|)^{-2\theta}(H-\lambda-\mathrm{i}\varepsilon)^{-1}(A-\mathrm{i})^{-1}\| \\
&\le\|(A+\mathrm{i})^{-1}(H-\lambda+\mathrm{i}\varepsilon)^{-1}(1+|\boldsymbol{r}|)^{-\theta_1}\| \\
&\quad\times\|\{(A+\mathrm{i})^{-1}(H-\lambda+\mathrm{i}\varepsilon)^{-1}(1+|\boldsymbol{r}|)^{-\theta_2}\}^*\| \\
&=N_{\theta_1}N_{\theta_2}.
\end{aligned}$$

For example, $N_{1/2}\le(N_0N_1)^{1/2}\le$ const $\varepsilon^{-(1+3\gamma)/4}$.

More generally, for $0\le\theta\le 1$, we find that

$$N_\theta\le\text{const}\,\varepsilon^{-((1-\theta)+\gamma(1+\theta))/2},$$

provided that θ is an integer multiple of 2^{-k} for some positive integer k. Without loss of generality, we may assume that β is such a multiple, so that

$$\begin{aligned}
N_\beta&=\|(A+\mathrm{i})^{-1}(H-\lambda+\mathrm{i}\varepsilon)^{-1}(1+|\boldsymbol{r}|)^{-\beta}\| \\
&\le\text{const}\,\varepsilon^{-((1-\beta)+\gamma(1+\beta))/2}.
\end{aligned}$$

Hence on the right-hand side of (12.2.34) we have for the operator norm of the second term the bound, on using (12.2.38),

$$\begin{aligned}&\|(A+\mathrm{i})^{-1}(H-\lambda+\mathrm{i}\varepsilon)^{-1}(1+|\boldsymbol{r}|)^{-\beta}\|\,\|(1+|\boldsymbol{r}|)^{\beta}\lambda^{-1}Q(H-z)^{-1}\|\\&\quad\times\|(H-z)(H-\lambda+\mathrm{i}\varepsilon)^{-1}(A-\mathrm{i})^{-1}\|\\&\leq\mathrm{const}\,\varepsilon^{-((1-\beta)+\gamma(1+\beta))/2}\varepsilon^{-(1+\gamma)/2}.\end{aligned}$$

If $\gamma-\frac{1}{2}\beta(1-\gamma)<0$ then this norm is integrable with respect to ε up to $\varepsilon=0$. In this case we can use the identity $(\mathrm{d}/\mathrm{d}\varepsilon)(H-\lambda+\mathrm{i}\varepsilon)^{-1}=-\mathrm{i}(H-\lambda+\mathrm{i}\varepsilon)^{-2}$ to integrate (12.2.34) up to $\varepsilon=0$, with the consequence that $(A+\mathrm{i})^{-1}(H-\lambda+\mathrm{i}\varepsilon)^{-1}(A-\mathrm{i})^{-1}$ converges in operator norm to a limit as $\varepsilon\to 0$.

On the other hand, if $\gamma-\frac{1}{2}\beta(1-\gamma)>0$ then we can integrate (12.2.34) from ε to some fixed constant c, to obtain the bound

$$\|(A+\mathrm{i})^{-1}(H-\lambda+\mathrm{i}\varepsilon)^{-1}(A-\mathrm{i})^{-1}\|\leq\mathrm{const}\,\varepsilon^{-(\gamma-\beta(1-\gamma)/2)}.$$

(The case $\gamma=\frac{1}{2}\beta(1-\gamma)$ is an intermediate one, for which a logarithmic bound is obtained.)

Comparing with the original estimate (12.3.31), we are thus able to replace the index γ in the original $\varepsilon^{-\gamma}$ bound by the strictly smaller value $\gamma-\frac{1}{2}\beta(1-\gamma)$. Starting from an arbitrary initial value of γ in the interval $0<\gamma<1$, we may successively use the integral of (12.2.34) to improve this estimate until we reduce γ to a value for which indeed $\gamma-\frac{1}{2}\beta(1-\gamma)<0$. We then obtain the required result from (12.2.34) by one further integration up to $\varepsilon=0$. This completes the proof of the lemma under hypothesis (a), and the proof for (b) is almost identical. ∎

Corollary

The hypotheses of the lemma imply respectively that

(a) $\lim_{\varepsilon\to 0+}(1+|\boldsymbol{r}|)^{-1}(H-\lambda+\mathrm{i}\varepsilon)^{-1}(1+|\boldsymbol{r}|)^{-1}$

or

(b) $\lim_{\varepsilon\to 0+}|\boldsymbol{r}|^{-1}(H-\lambda+\mathrm{i}\varepsilon)^{-1}|\boldsymbol{r}|^{-1}$

exist as limits in operator norm.

Proof

We have obtained an estimate of the form $\|(A+\mathrm{i})^{-1}(H-\lambda+\mathrm{i}\varepsilon)^{-2}(A-\mathrm{i})^{-1}\|\leq\mathrm{const}\ \varepsilon^{-q}$, where the index q may be verified to lie in the range $\frac{1}{2}<q<1$. Since *a posteriori* we can take $\gamma=0$ in (12.2.31), the bounds that we have obtained yield a value $q=1-\frac{1}{2}\beta$. It is straightforward to deduce the estimate, for Im $(z)\neq 0$,

$$\|(A+\mathrm{i})^{-1}(H-z)(H-\bar{z})(H-\lambda+\mathrm{i}\varepsilon)^{-2}(A-\mathrm{i})^{-1}\|\leq \text{const}\,\varepsilon^{-q}.$$

If we now multiply, in case (a), the operator within the norm by $(A-\mathrm{i})(H-z)^{-1}(1+|\boldsymbol{r}|)^{-1}$ on the right, and by the adjoint of this operator on the left, we obtain $\|(1+|\boldsymbol{r}|)^{-1}(H-\lambda+\mathrm{i}\varepsilon)^{-2}(1+|\boldsymbol{r}|)^{-1}\|\leq \text{const}\,\varepsilon^{-q}$, from which the conclusion of the corollary follows on integrating with respect to ε. The case (b) follows in the same way. ■

The results of this section may be applied in a variety of ways, and to scattering by a variety of potentials of both short and long range. The estimate

$$\|(H-\lambda+\mathrm{i}\varepsilon)^{-1}(A-\mathrm{i})^{-1}\|\leq \text{const}\,\varepsilon^{-1/2}$$

may be used to deduce not only the absence of eigenvalues but also the absolute continuity of the spectrum of H, for a range of energies λ. We may also follow the analysis at the beginning of this section, starting from (12.2.1) with $B=(1+|\boldsymbol{r}|)^{-1}$ or with $B=|\boldsymbol{r}|^{-1}$. The resulting bounds, such as are given by (12.2.17), for products involving the evolution operator $\mathrm{e}^{-\mathrm{i}Ht}$, are then an important ingredient in the proof of asymptotic completeness in the long-range case, to which we turn in the following section.

The case of a short-range potential may be dealt with in a relatively simple way. We let $V(\boldsymbol{r})$ be of short range, so that $V(\boldsymbol{r})=O(|\boldsymbol{r}|^{-1-\delta})$ as $|\boldsymbol{r}|\to\infty$, for some $\delta>0$, and suppose also, for simplicity, that $|V(\boldsymbol{r})|$ is bounded at the origin by const$/|\boldsymbol{r}|$. In this case $D(H)=D(H_0)$, and $V(H-z)^{-1}$ is bounded for Im $(z)\neq 0$. Since $|\boldsymbol{r}|V(\boldsymbol{r})$ is bounded and $D(H)\subseteq D(H_0^{1/2})\subseteq D(\boldsymbol{P})$, it follows that $f\in D(H)\Rightarrow f\in D(V\boldsymbol{r}\cdot\boldsymbol{P})$. From the explicit form of the operator A, we then have $D(H)\subseteq D(VA)$, so that $VA(H-z)^{-1}$ is bounded, by an application of the Closed-Graph Theorem. Indeed, we can use the fact that $|\boldsymbol{r}|V(\boldsymbol{r})\to 0$ as $|\boldsymbol{r}|\to\infty$ to show that $VA(H-z)^{-1}$ is even compact; the compactness of $(H-z)^{-1}AV$ (or rather of the closure of this operator) follows in a similar way, or by taking adjoints.

Now using the relation between Q and the commutator of V with A, we see that Q can be expressed as a sum, $Q=Q^{(1)}+Q^{(2)}$, of two operators $Q^{(1)}$ and $Q^{(2)}$, such that $Q^{(1)}(H-z)^{-1}$ and $(H-z)^{-1}Q^{(2)}$ are both bounded, even compact. Since $(H-z)E_{\mathscr{I}}(H)$ is bounded for any finite interval $\mathscr{I}$, it also follows that $E_{\mathscr{I}}(H)QE_{\mathscr{I}}(H)$ is compact. As the interval $\mathscr{I}$ shrinks to some single point $\lambda>0$ that is not an eigenvalue of H, we have $E_{\mathscr{I}}(H)\to 0$ strongly, from which it follows that $E_{\mathscr{I}}(H)QE_{\mathscr{I}}(H)$ converges to zero in operator norm. In particular, we have $\|\lambda^{-1}E_{\mathscr{I}}(H)QE_{\mathscr{I}}(H)\|<1$ for an interval $\mathscr{I}$ sufficiently small in length.

We can now apply Lemma 12.5, together with a slight modification of (ii) of the Corollary to Lemma 12.6, in order to obtain both the identity (12.2.27) and the estimate (12.2.31). In a similar way, we may slightly modify Lemma 12.7(b), using the fact that $|\boldsymbol{r}|^{\delta}Q^{(1)}(H-z)^{-1}$ and $(H-z)^{-1}Q^{(2)}|\boldsymbol{r}|^{\delta}$ are bounded, to conclude the existence of $\lim_{\varepsilon\to 0+}(A+\mathrm{i})^{-1}(H-\lambda+\mathrm{i}\varepsilon)^{-1}(A-\mathrm{i})^{-1}$ as a limit in operator norm. The operator-norm limit

$$\lim_{\varepsilon\to 0+}|\boldsymbol{r}|^{-1}(H-\lambda+\mathrm{i}\varepsilon)^{-1}|\boldsymbol{r}|^{-1}$$

then follows as before.

In obtaining comparable results in the case of long-range potentials $V(\boldsymbol{r})$, it will no longer be sufficient to treat separately the two contributions $Q^{(1)}$ and $Q^{(2)}$ coming from the commutator of V with A. Rather we shall need to evaluate this commutator using differentiability of the potential. The interested reader may wish to return to the short-range case, which we have discussed briefly here, and to consider further applications of these ideas following a reading of Section 12.3.

12.3 COMPLETENESS FOR LONG-RANGE POTENTIALS

We have seen, in Chapter 9, how to define wave operators $\Omega_{\pm}$ for scattering by long-range potentials. The results of this analysis were summarized in Theorem 9.2. The potentials with which we dealt had to satisfy a decay condition, in the limit $|\boldsymbol{r}|\to\infty$, of the form $|V(\boldsymbol{r})|=O(|\boldsymbol{r}|^{-\beta})$ for some β in the range $\frac{1}{2}<\beta<1$. In addition, we required a bound $|\partial V(\boldsymbol{r})/\partial x_j|=O(|\boldsymbol{r}|^{-(1+\beta)})$ on the derivative of the potential, together with a similar bound on the second derivatives. These conditions on the potential are listed under (9.1.25), and allow arbitrary singularities at $\boldsymbol{r}=0$.

For such potentials, wave operators of the form

$$\Omega_{\pm}=\text{s-}\lim_{t\to\pm\infty}\mathrm{e}^{\mathrm{i}Ht}F(\boldsymbol{P},t) \tag{12.3.1}$$

may be defined in terms of a modified free evolution $F(\boldsymbol{P},t)$, and exist as isometries on the entire Hilbert space $\mathscr{H}=L^2(\mathbb{R}^3)$. To prove the existence of $\Omega_{\pm}$, it was in fact more convenient to pass from (12.3.1) to the expression

$$\Omega_{\pm}=\text{s-}\lim_{t\to\pm\infty}e^{\mathrm{i}Ht}J(\boldsymbol{r},t)\mathrm{e}^{-\mathrm{i}H_0t}, \tag{12.3.2}$$

where the operator $J(\boldsymbol{r},t)$ of multiplication in position space may be given, for example, by

$$J(\boldsymbol{r}, t) = \exp\left(\frac{-\mathrm{i}tW(\boldsymbol{r})}{r}\right), \tag{12.3.3}$$

with $W(\boldsymbol{r})$ defined in terms of the potential $V(\boldsymbol{r})$ by

$$W(\boldsymbol{r}) = \int_c^r V\left(\frac{s\boldsymbol{r}}{r}\right) \mathrm{d}s. \tag{12.3.4}$$

To prove the existence of wave operators, we used the differential equation (9.1.22) satisfied by the function $J(\boldsymbol{r}, t)$ to evaluate the strong derivative

$$\frac{\mathrm{d}}{\mathrm{d}t} e^{\mathrm{i}Ht} J(\boldsymbol{r}, t) \mathrm{e}^{-\mathrm{i}H_0 t} f,$$

for suitable $f \in \mathscr{H}$, and obtained norm estimates, as a function of time t, for each term in the resulting expression. Crucial steps in the argument were (i) the replacement of $J(\boldsymbol{r}, t)$ by $\rho(\boldsymbol{r}/2t)J(\boldsymbol{r}, t)$, for a suitable smooth multiplication operator $\rho(\boldsymbol{r}/2t)$, and (ii) the estimate (9.1.29) for the norm

$$\|(x_j - 2P_j t)\mathrm{e}^{-\mathrm{i}H_0 t} f\|.$$

The proof of asymptotic completeness, although technically more complicated, will proceed upon roughly the same lines, except that we obtain $\Omega_\pm^*$ rather than $\Omega_\pm$ as a strong limit. In addition, we shall have to impose more stringent conditions on the potential at $\boldsymbol{r} = 0$, which rule out the possibility of strong local singularities. Rather than striving for the utmost generality, we shall avoid too many technical details by restricting our attention to potentials satisfying the decay condition at infinity $\beta > \frac{3}{4}$ rather than $\beta > \frac{1}{2}$. (The more general case can be arrived at by similar methods—indeed, completeness has been extended, under suitable conditions on the derivatives of V and in the absence of strong local singularities, to a larger class of potentials for which $\beta > 0$ only.) As in the short-range case, potentials *do* exist that violate asymptotic completeness, but such potentials must behave in an unusual fashion locally, and are ruled out by the present analysis.

Before proceeding to a complete proof, it may be helpful, in comparing with corresponding arguments in the short-range case, to outline the steps that are necessary in arriving at completeness for scattering by long-range potentials.

We let g be a vector in the range of the spectral projection for H associated with a finite closed subinterval $[c_1^2, c_2^2]$ of $(0, \infty)$. (We take $c_1, c_2 > 0$). Thus g has compact spectral support, for H, in the positive real line, and describes a state having strictly positive, and finite, total energy. Let us suppose that we can prove the estimate, for $j = 1,2,3$,

$$\|(x_j - 2P_j t)e^{-iHt}g\| \leq \text{const}\, t^{\beta'}, \tag{12.3.5}$$

in the limit $t \to \infty$, and for some $\beta' < \beta$. The bound (12.3.5) expresses the requirement that the time evolution of the observable $\boldsymbol{r} - 2\boldsymbol{P}t$, in the Heisenberg picture and for the state g, be such that

$$\boldsymbol{r} = 2\boldsymbol{P}t + O(t^{\beta'}).$$

This should be compared with the bound (9.1.29) that applies to the free evolution of $\boldsymbol{r} - 2\boldsymbol{P}t$ in the Heisenberg picture.

Classical analogy suggests the bound const $t^{1-\beta}$ for the norm in (12.3.5). Since, with $\beta > \frac{1}{2}$, we have $1 - \beta < \beta$, the bound (12.3.5) is certainly consistent with what we should expect from the classical theory. Note also that an eigenstate g of H would imply a linear growth of the norm with respect to time. Since we expect eigenvalues of H to occur only at negative energies, it is clear that we cannot remove the restriction that g have spectral support in the positive half-line.

Since we are taking $\beta < 1$, the bound (12.3.5) implies that $e^{iHt}(\boldsymbol{r}/t - 2\boldsymbol{P})e^{-iHt}g$ converges strongly to zero as $t \to \infty$. Since $\boldsymbol{P}^2$ and H differ only by the potential V, which, in the Heisenberg picture for an asymptotically free dynamics, we expect to converge to zero, we should expect that $\boldsymbol{r}^2/t^2 - 4H$ will also converge to zero in the Heisenberg picture. Let us therefore suppose in addition to (12.3.5), that we can obtain the result

$$\lim_{t \to \infty} \left\| \left(\frac{\boldsymbol{r}^2}{4t^2} - H \right) e^{-iHt} g \right\| = 0. \tag{12.3.6}$$

Here a precise bound for the rate at which this convergence takes place is unnecessary. We shall suppose that the same result holds for $(H - z)^{-1} g$ as for g; namely, for any z such that $\operatorname{Im}(z) \neq 0$,

$$\lim_{t \to \infty} \left\| \left(\frac{\boldsymbol{r}^2}{4t^2} - H \right) e^{-iHt} h \right\| = 0, \tag{12.3.6$'$}$$

where

$$h = (H - z)^{-1} g. \tag{12.3.7}$$

In this case, it follows from (12.3.6) and (12.3.6)′ that

$$\begin{aligned}
&\operatorname*{s-lim}_{t\to\infty} e^{iHt}\left(\frac{\boldsymbol{r}^2}{4t^2}-z\right)^{-1} e^{-iHt}g \\
&=\operatorname*{s-lim}_{t\to\infty} e^{iHt}\left(\frac{\boldsymbol{r}^2}{4t^2}-z\right)^{-1}(H-z)e^{-iHt}h \\
&=\operatorname*{s-lim}_{t\to\infty} e^{iHt}\left(\frac{\boldsymbol{r}^2}{4t^2}-z\right)^{-1}\left\{\left(\frac{\boldsymbol{r}^2}{4t^2}-z\right)e^{-iHt}h\right\} \\
&=(H-z)^{-1}g.
\end{aligned}$$

Thus, in the Heisenberg picture, the resolvent of $\boldsymbol{r}^2/4t^2$ converges to the resolvent of H, in the sense that

$$\operatorname*{s-lim}_{t\to\infty} e^{iHt}\left\{\left(\frac{\boldsymbol{r}^2}{4t^2}-z\right)^{-1}-(H-z)^{-1}\right\}e^{-iHt}g=0. \tag{12.3.8}$$

We may now use the integral formulae (cf. the derivation of (4.1.9)) for the spectral projection E_0 of a self-adjoint operator T, together with the extension to arbitrary E_λ, to deduce that if E' and E are spectral projections of $\boldsymbol{r}^2/4t^2$ and H respectively for the interval $[c_1^2, c_2^2]$ then

$$\operatorname*{s-lim}_{t\to\infty} e^{iHt}(E'-E)e^{-iHt}g=0. \tag{12.3.8$'$}$$

(Compare the argument following (4.1.7) in Chapter 4.) However, by the hypothesis concerning the H-spectral support of g, we have $\text{s-lim}_{t\to\infty}\, e^{iHt} Ee^{-iHt}g=Eg=g$. It follows immediately from (12.3.8) that

$$\operatorname*{s-lim}_{t\to\infty} e^{iHt}E'e^{-iHt}g=g.$$

In other words, asymptotically the spectral support of $e^{-iHt}g$ for the self-adjoint operator $\boldsymbol{r}^2/4t^2$ is also concentrated in the interval $[c_1^2, c_2^2]$. Equivalently, we can say that, asymptotically in the limit $t\to\infty$ in position space, the $L^2(\mathbb{R}^3)$ norm of $(e^{-iHt}g)(\boldsymbol{r})$ in the region *outside* $2c_1t\le|\boldsymbol{r}|\le 2c_2t$ must converge to zero. In accordance with our (classical) expectations, the state $e^{-iHt}g$, at large positive times, is concentrated in the region $2c_1t\le|\boldsymbol{r}|\le 2c_2t$.

The next step in the proof of asymptotic completeness for long-range potentials is to prove the existence of the adjoint wave operator as a strong limit

$$\Omega_+^*g=\operatorname*{s-lim}_{t\to\infty} e^{iH_0t}J^*(\boldsymbol{r},t)e^{-iHt}g, \tag{12.3.9}$$

where $J(\boldsymbol{r},t)$ is given by (12.3.3) and (12.3.4). Because of the asymptotic support properties of $e^{-iHt}g$ in position space, it will be sufficient to prove the existence of the limit (12.3.9) with $J(\mathbf{r}, t)$ replaced by $\rho(\mathbf{r}/2t)J(\mathbf{r}, t)$, where $\rho(\mathbf{r})$ is the operator of multiplication by a function $\rho=\rho(|\mathbf{r}|)\in C_0^\infty(\mathbb{R}^3\setminus\{0\})$ satisfying $\rho(|\mathbf{r}|)\equiv 1$ for $c_1\le|\mathbf{r}|\le c_2$. Note, as in Section 9.1, that the differential equation (9.1.22) for the modified J-function remains valid. For $J^*(\mathbf{r}, t)$, we have

$$-\mathrm{i}\frac{\partial J^*}{\partial t}=VJ^*+\mathrm{i}\frac{\mathbf{r}\cdot\operatorname{grad} J^*}{t}.$$

Corresponding to (9.1.23) and (9.1.24), we then have

$$\begin{aligned}\frac{\mathrm{d}}{\mathrm{d}t}\mathrm{e}^{\mathrm{i}H_0t}J^*(\mathbf{r}, t)\mathrm{e}^{-\mathrm{i}Ht}g&=\mathrm{i}\mathrm{e}^{\mathrm{i}H_0t}\left(H_0J^*-J^*H-\mathrm{i}\frac{\partial J^*}{\partial t}\right)\mathrm{e}^{-\mathrm{i}Ht}g\\&=\mathrm{i}[-\Delta J^*+\mathrm{i}t^{-1}(\operatorname{grad} J^*)\\&\quad\cdot(\mathbf{r}-2\mathbf{P}t)]\mathrm{e}^{-\mathrm{i}Ht}g\end{aligned}\qquad(12.3.10)$$

On the support of grad J^*, $|\mathbf{r}|$ and t may be treated as of the same order of magnitude, and, following (9.1.27) we have, as $t\to\infty$,

$$\begin{aligned}|\operatorname{grad} J^*(\mathbf{r}, t)|&\le\text{const } t^{-\beta},\\|\Delta J^*(\mathbf{r}, t)|&\le\text{const } t^{-(\beta+1/2)}+\text{const } t^{-2\beta},\end{aligned}$$

where the function J modified by multiplication by $\rho(\mathbf{r}/2t)$ is intended. With (12.3.5), we now have the norm estimate $O(t^{-(1+\beta-\beta')})$ for the right-hand side of (12.3.10) in the limit as $t\to\infty$. Since, by hypothesis, $\beta'<\beta$, the right-hand side of (12.3.10) is integrable in norm with respect to t up to $t=\infty$. Hence the strong limit on the right-hand side of (12.3.9) exists, where, by our previous argument, $J(\mathbf{r}, t)$ may be defined by (12.3.3) *without* now the factor $\rho(\mathbf{r}/2t)$. The fact that the strong limit in (12.3.9) is indeed $\Omega_+^* g$ follows easily if we consider the inner product $\langle\Omega_+ f, g\rangle=\langle f, \Omega_+^* g\rangle$ and use the formula (12.3.2) for Ω_+.

To conclude the proof of completeness, it is necessary to establish (12.3.5) and (12.3.6) for a set of vectors g that is dense in the subspace of $\mathscr{H}$ corresponding to the positive part of the spectrum of the total Hamiltonian H. (Equation (12.3.6) must also hold in the form (12.3.6)′ in which g is replaced by $(H-z)^{-1}g$; the simplest way to ensure this is to choose a dense set of vectors g that is invariant under the operation of $(H-z)^{-1}$, for Im $(z)\neq 0$.) The complex conjugates of such vectors g, in position space, then serve to define a set of vectors, again dense in the positive-energy subspace, for which the existence of Ω_-^* as a strong limit, for $t\to-\infty$, may be proved. If the set of vectors g is dense in the positive-energy subspace of $\mathscr{H}$ then we may consider strong limits of such g to deduce (12.3.9) for *all* g belonging to the positive-energy subspace. Since H_0 has a purely positive spectrum, by the intertwining property applied to the spectral projections of H_0 and H, we see that the range of Ω_+ is contained in the positive-energy subspace. If the range of Ω_+ is not the *whole* of this subspace then a vector $g \neq 0$ in this subspace will exist that is orthogonal to the range of Ω_+, and for which $\Omega_+^* g = 0$. However, from (12.3.9), with $J(\boldsymbol{r}, t)$ given by the *unitary* multiplication operator (12.3.3), we know that $\|\Omega_+^* g\| = \|g\|$, implying that $g = 0$. Hence no such vector g can exist, orthogonal to the range of Ω_+, and we have proved that the range of Ω_+ is indeed identical with the subspace of $\mathscr{H}$ corresponding to the positive part of the spectrum of H. Exactly the same argument applies to Ω_- as to Ω_+. Hence the respective ranges of Ω_+ and Ω_- may be identified, and the proof of asymptotic completeness may be concluded.

We thus have the following Lemma.

Lemma 12.8

Let $V(\boldsymbol{r})$ be a long-range potential satisfying the conditions of Theorem 9.2 (roughly, $V(\boldsymbol{r})$ must decay as $|\boldsymbol{r}| \to \infty$ more rapidly that $|\boldsymbol{r}|^{-1/2}$). Define modified wave operators as in (12.3.1) and (12.3.2). Suppose in addition that there exists a set of vectors g, invariant under the operation of $(H-z)^{-1}$ for $\operatorname{Im}(z) \neq 0$, that is dense in the $H > 0$ subspace of $\mathscr{H}$, and for which (12.3.5) and (12.3.6) are satisfied for some $\beta' < \beta$. Then the wave operators are asymptotically complete.

In order to apply Lemma 12.8 to scattering by a given long-range potential $V(\boldsymbol{r})$, it is necessary to verify (12.3.5) and (12.3.6) for some suitable set of vectors g. How is this to be done? We shall start from the identity

$$\begin{aligned} &\|(\boldsymbol{r} - 2\boldsymbol{P}t)\mathrm{e}^{-\mathrm{i}Ht}g\|^2 \\ &\quad = \langle \boldsymbol{r}\mathrm{e}^{-\mathrm{i}Ht}g, \boldsymbol{r}\mathrm{e}^{-\mathrm{i}Ht}g\rangle - 4t\langle \mathrm{e}^{-\mathrm{i}Ht}g, A\mathrm{e}^{-\mathrm{i}Ht}g\rangle \\ &\qquad + 4t^2\langle \mathrm{e}^{-\mathrm{i}Ht}g, H_0\mathrm{e}^{-\mathrm{i}Ht}\mathrm{g}\rangle. \end{aligned} \tag{12.3.11}$$

Here an expression such as $\langle \boldsymbol{r}\mathrm{e}^{-\mathrm{i}Ht}g, \boldsymbol{r}\mathrm{e}^{-\mathrm{i}Ht}g\rangle$ should be taken to mean $\Sigma_{j=1}^3 \langle x_j \mathrm{e}^{-\mathrm{i}Ht}g, x_j\mathrm{e}^{-\mathrm{i}Ht}g\rangle$. In (12.3.11) we have made use of the formulae $H_0 = \boldsymbol{P}^2$, $A = \frac{1}{2}(\boldsymbol{P}\cdot\boldsymbol{r} + \boldsymbol{r}\cdot\boldsymbol{P})$, and have assumed for the moment that $\mathrm{e}^{-\mathrm{i}Ht}g$ lies in the domain of appropriate operators in the equation. Now

$$\begin{aligned} &\frac{\mathrm{d}}{\mathrm{d}t}\langle \boldsymbol{r}\mathrm{e}^{-\mathrm{i}Ht}g, \boldsymbol{r}\mathrm{e}^{-\mathrm{i}Ht}g\rangle \\ &\quad = \mathrm{i}\{\langle \boldsymbol{r}H\mathrm{e}^{-\mathrm{i}Ht}g, \boldsymbol{r}\mathrm{e}^{-\mathrm{i}Ht}g\rangle - \langle \boldsymbol{r}\mathrm{e}^{-\mathrm{i}Ht}g, \boldsymbol{r}H\mathrm{e}^{-\mathrm{i}Ht}g\rangle\} \\ &\quad = 4\langle \mathrm{e}^{-\mathrm{i}Ht}g, A\mathrm{e}^{-\mathrm{i}Ht}g\rangle, \end{aligned}$$

where we have used the commutation relation $\boldsymbol{r}H = H\boldsymbol{r} + 2\mathrm{i}\boldsymbol{P}$ and the fact that H is symmetric. If we now differentiate (12.3.11) with respect to t, there is a partial cancellation with the derivative of the second contribution on the right-hand side. For the derivative of the final term in (12.3.11), we have

$$\begin{aligned}\frac{\mathrm{d}}{\mathrm{d}t}4t^2\langle \mathrm{e}^{-\mathrm{i}Ht}g, H_0\mathrm{e}^{-\mathrm{i}Ht}g\rangle \\ &= 8t\langle g, Hg\rangle - 8t\langle \mathrm{e}^{-\mathrm{i}Ht}g, V\mathrm{e}^{-\mathrm{i}Ht}g\rangle \\ &\quad - 4\mathrm{i}t^2\{\langle H\mathrm{e}^{-\mathrm{i}Ht}g, V\mathrm{e}^{-\mathrm{i}Ht}g\rangle - \langle V\mathrm{e}^{-\mathrm{i}Ht}g, H\mathrm{e}^{-\mathrm{i}Ht}g\rangle\}.\end{aligned}$$

Hence we have

$$\begin{aligned}\frac{\mathrm{d}}{\mathrm{d}t}\|(\boldsymbol{r}-2\boldsymbol{P}t)\mathrm{e}^{-\mathrm{i}Ht}g\|^2 \\ &= -4\mathrm{i}t\{\langle H\mathrm{e}^{-\mathrm{i}Ht}g, A\mathrm{e}^{-\mathrm{i}Ht}g\rangle - \langle A\mathrm{e}^{-\mathrm{i}Ht}g, H\mathrm{e}^{-\mathrm{i}Ht}g\rangle\} \\ &\quad + 8t\langle g, Hg\rangle - 8t\langle \mathrm{e}^{-\mathrm{i}Ht}g, V\mathrm{e}^{-\mathrm{i}Ht}g\rangle \\ &\quad - 4\mathrm{i}t^2\{\langle H\mathrm{e}^{-\mathrm{i}Ht}g, V\mathrm{e}^{-\mathrm{i}Ht}g\rangle - \langle V\mathrm{e}^{-\mathrm{i}Ht}g, H\mathrm{e}^{-\mathrm{i}Ht}g\rangle\}.\end{aligned}$$

Now the results of Section 2 of this chapter allow us to interpret the commutation relation

$$[H, A] = -2\mathrm{i}(H - Q),$$

where

$$Q = V + (2\mathrm{i})^{-1}[V, A],$$

as an equation satisfied by forms on $D(H) \cap D(A)$. Using these results and substituting $[V, A] = \mathrm{i}\boldsymbol{r}\cdot \operatorname{grad} V$, we finally obtain the identity

$$\begin{aligned}\frac{\mathrm{d}}{\mathrm{d}t}\|(\boldsymbol{r}-2\boldsymbol{P}t)\mathrm{e}^{-\mathrm{i}Ht}g\|^2 &= 4t\langle \mathrm{e}^{-\mathrm{i}Ht}g, (\boldsymbol{r}\cdot\operatorname{grad} V)\mathrm{e}^{-\mathrm{i}Ht}g\rangle \\ &\quad + 8t^2\,\mathrm{Im}\,\langle H\mathrm{e}^{-\mathrm{i}Ht}g, V\mathrm{e}^{-\mathrm{i}Ht}g\rangle. \qquad (12.3.12)\end{aligned}$$

Note that the final term in (12.3.12) may be rewritten as

$$-4t^2\frac{\mathrm{d}}{\mathrm{d}t}\langle \mathrm{e}^{-\mathrm{i}Ht}g, V\mathrm{e}^{-\mathrm{i}Ht}g\rangle.$$

If we then integrate (12.3.12) with respect to time from 0 to t and use integration by parts, we obtain the inequality, with $\|g\| = 1$,

$$\begin{aligned}&\left|\,\|(\boldsymbol{r}-2\boldsymbol{P}t)\mathrm{e}^{-\mathrm{i}Ht}g\|^2 - \|\boldsymbol{r}g\|^2\,\right| \\ &\quad \le \int_0^t \mathrm{d}s\, 4s\|(\boldsymbol{r}\cdot\operatorname{grad} V)\mathrm{e}^{-\mathrm{i}Hs}g\| \\ &\quad\quad + \int_0^t \mathrm{d}s\, 8s\|V\mathrm{e}^{-\mathrm{i}Hs}g\| + 4t^2\|V\mathrm{e}^{-\mathrm{i}Ht}g\|. \qquad (12.3.13)\end{aligned}$$

In order to provide a rigorous justification of (12.3.13), a number of conditions have to be satisfied relating to the domains of various operators involving $\boldsymbol{r}$, $\boldsymbol{P}$, V and A. We have to know, for all $t \geq 0$, that $e^{-iHt} g$ lies in the respective domains of these operators as well as the domain of H and of $\boldsymbol{r}H$, and we have to verify the equation $(d/dt)\boldsymbol{r}\, e^{-iHt} g = -i\boldsymbol{r}He^{-iHt} g$, where d/dt denotes a strong derivative.

Let us suppose, for example, that $V(\boldsymbol{r})$ satisfies the bounds

$$|V(\boldsymbol{r})| = \begin{cases} O(|\boldsymbol{r}|^{-1}), \\ O(|\boldsymbol{r}|^{-\beta}), \end{cases} \qquad |\mathrm{grad}\, V(\boldsymbol{r})| = \begin{cases} O(|\boldsymbol{r}|^{-2}) & \text{as } |\boldsymbol{r}| \to 0, \\ O(|\boldsymbol{r}|^{-(\beta+1)}) & \text{as } |\boldsymbol{r}| \to \infty. \end{cases} \tag{12.3.14}$$

We have already seen, from (12.2.30), how to deduce that the operators $\boldsymbol{r}(H-z_1)^{-1}(|\boldsymbol{r}|-z_2)^{-1}$ and $A(H-z_1)^{-1}(|\boldsymbol{r}|-z_2)^{-1}$ are bounded. More detailed estimates, following the analysis of that section, lead to a bound of order $(\mathrm{Im}\,(z_1))^{-2}$ for each of these operators in the limit as z_1 approaches the real axis. If we then take $\phi_1(z) = (z-c_1^2)^2\,(z-c_2^2)^2\psi(z)$, where $\psi(z)$ is analytic in z, we can make sense of the contour integral

$$\frac{1}{2\pi i} \oint_C A\phi_1(z_1)(H-z_1)^{-1}(|\boldsymbol{r}|-z_2)^{-1}\, dz_1$$

as a strong limit of approximating Riemann sums, where the contour of integration encircles the subinterval $[c_1^2, c_2^2]$ of the real axis once and intersects the real axis at the two points $z_1 = c_1^2$ and $z_1 = c_2^2$. Indeed, such a contour integral will define a bounded linear operator, which may be evaluated by the spectral theorem for the self-adjoint operator H (first taking matrix elements between a suitable pair of vectors) to give $A\phi_1(H)E(|\boldsymbol{r}|-z_2)^{-1}$, where E is the spectral projection of H for the interval $[c_1^2, c_2^2]$. Defining the function $\phi(\lambda)$, with support $[c_1^2, c_2^2]$, by

$$\phi(\lambda) = \begin{cases} \phi_1(\lambda) & (c_1^2 \leq \lambda \leq c_2^2), \\ 0 & \text{otherwise}, \end{cases}$$

we have shown that $A\phi(H)(|\boldsymbol{r}|-z_2)^{-1}$ is bounded. The same argument, with $\phi_1(z_1)$ replaced by $e^{-iz_1 t}\phi_1(z_1)$, shows that $Ae^{-iHt}\phi(H)(|\boldsymbol{r}|-z_2)^{-1}$ is bounded, and similarly we may deduce that $|\boldsymbol{r}|e^{-iHt}\phi(H)(|\boldsymbol{r}|-z_2)^{-1}$ is bounded. It is also not difficult to verify the formula

$$\frac{d}{dt}\boldsymbol{r}e^{-iHt}\phi(H)(|\boldsymbol{r}|-z_2)^{-1} = -i\boldsymbol{r}He^{-iHt}\phi(H)(|\boldsymbol{r}|-z_2)^{-1}.$$

In other words, if we take $g = \phi(H)f$ for any f in the domain of $|\boldsymbol{r}|$ then we have verified already that $e^{-iHt}g$ lies in the respective domains of $\boldsymbol{r}$ and A, together with the formula for the strong derivative of $\boldsymbol{r}e^{-iHt}g$. Since, for the class of potentials considered here $(V \in L^2(\mathbb{R}^3) + L^\infty(\mathbb{R}^3))$, $D(H) \subseteq D(H_0) \cap$

$D(V)$, the remaining domain properties for g are easily found to be satisfied. It can also readily be verified that the set of $g=\phi(H)f$, as the function ϕ, together with the values of c_1 and c_2 and the vector $f\in D(|\boldsymbol{r}|)$ are varied, is dense in the positive-energy subspace of H.

Let us now apply Lemma 12.7, together with (12.3.13), to obtain an estimate for $\|(\boldsymbol{r}-2\boldsymbol{P}t)\,\mathrm{e}^{-\mathrm{i}Ht}g\|$. With the assumptions that we have made for the potential $V(\boldsymbol{r})$, $Q(H-z)^{-1}$ is bounded, and even compact, for $\mathrm{Im}\,(z)\neq 0$. (If χ_R is the operator of multiplication by the characteristic function of $|\boldsymbol{r}|\leq R$ then $\chi_R Q(H_0-z)^{-1}$ is Hilbert–Schmidt. Since H and H_0 have the same domain, it follows from the Closed-Graph Theorem that $(H_0-z)(H-z)^{-1}$ is bounded. Hence $\chi_R Q(H-z)^{-1}$ is also Hilbert–Schmidt, and $Q(H-z)^{-1}$ is the norm limit, as $R\to\infty$, of this family of compact operators.) Since, with the notation of Lemma 12.6 and its corollary, $(H-z)E_{\mathscr{I}}(H)$ is bounded, it also follows that $QE_{\mathscr{I}}(H)$ is compact. If λ is not an eigenvalue then we know that $E_{\mathscr{I}}(H)$ converges strongly to zero in the limit as the length of the interval $\mathscr{I}$ containing λ approaches zero. By compactness, it follows that $E_{\mathscr{I}}(H)QE_{\mathscr{I}}(H)$ converges to zero *in operator norm.* Hence certainly, for $\lambda>0$, $\|\lambda^{-1}\,E_{\mathscr{I}}(H)QE_{\mathscr{I}}(H)\|<1$ for a small enough interval $\mathscr{I}$, and we can apply (ii), for example, of the Corollary to Lemma 12.6. Moreover, $|\boldsymbol{r}|^{\beta}Q(H-z)^{-1}$ is bounded, and, by the argument following Lemma 12.5, we also have that $(A-\mathrm{i})(H-z)^{-1}\,|\boldsymbol{r}|^{-1}$ is bounded, By Lemma 12.7, we may deduce the existence of the norm limit

$$\lim_{\varepsilon\to 0+}\,(A+\mathrm{i})^{-1}\,(H-\lambda+\mathrm{i}\varepsilon)^{-1}\,(A-\mathrm{i})^{-1}$$

The corollary to this lemma leads to the bound

$$\|\,|\boldsymbol{r}|^{-1}(H-\lambda+\mathrm{i}\varepsilon)^{-2}|\boldsymbol{r}|^{-1}\|\leq\text{const}\,\varepsilon^{-q},$$

with $q=1-\tfrac{1}{2}\beta$.

We can now apply the operator analysis of Section 12.2, starting from (12.2.1), with $B=|\boldsymbol{r}|^{-1}$. Taking (12.2.16), with g replacing $\phi(H)|\boldsymbol{r}|^{-1}g$, we have, as $t\to\infty$,

$$\int_0^t \mathrm{d}s\,\|\,|\boldsymbol{r}|^{-1}\mathrm{e}^{-\mathrm{i}Hs}g\|\leq\text{const}\,t^{(1-\beta)/2}. \tag{12.3.15}$$

(Remark 1: The bound (12.3.15) is not the best possible; we have sought to avoid the more technical arguments that would lead to a bound of order $\log t$. Remark 2: The class of allowable vectors g, from (12.2.16), is $g=\phi(H)f$, for $f\in D(|\boldsymbol{r}|)$ and $\phi\in C_0^\infty(0,\infty)$. However, reference to Section 12.2 will confirm that a larger class of functions ϕ may be admitted, including those $\phi(\lambda)$ treated above in this section.)

The extension of (12.3.15) to non-integral powers of $|\boldsymbol{r}|^{-1}$ (cf. (12.2.17)) leads to the estimate

$$\int_0^t \mathrm{d}s \,\| \, |\boldsymbol{r}|^{-\beta} \mathrm{e}^{-\mathrm{i}Hs} g \| \leq \mathrm{const}\, t^{1-\beta/2-\beta^2/2}. \tag{12.3.16}$$

In (12.3.13), by hypothesis, V and $\boldsymbol{r} \cdot \operatorname{grad} V$ are bounded at infinity by $\mathrm{const}\,|\boldsymbol{r}|^{-\beta}$, and at $\boldsymbol{r}=\mathbf{0}$ by $\mathrm{const}\,|\boldsymbol{r}|^{-1}$. Hence we can use (12.3.15) and (12.3.16) to make estimates of the integrals on the right-hand side of (12.3.13), obtaining a bound in each case of order $t^{2\beta'}$ for some $\beta'<\beta$, provided that we take $\beta \geq \frac{3}{4}$, say. (In verifying this, note for example that $1-\frac{1}{4}\beta-\frac{1}{4}\beta^2<\beta$ provided that $\beta \geq \frac{3}{4}$.)

We now apply Lemma 12.8. We have obtained a bound of order $t\beta'$ in (12.3.5), as far as the integral contributions to the right-hand side of (12.3.13) are concerned. The other term on the right-hand side of (12.3.13) causes no difficulty, since in the proof of Lemma 12.8 we need only $\int_1^\infty \mathrm{d}t\, t^{-(1+\beta)} \|(\boldsymbol{r}-2\boldsymbol{P}t)\, \mathrm{e}^{-\mathrm{i}Ht} g\| < \infty$, which, on writing the integral as

$$\int_1^\infty \mathrm{d}t\, t^{-(1+\beta)} \frac{\mathrm{d}}{\mathrm{d}t} \int_1^t \mathrm{d}s \, \|(\boldsymbol{r}-2\boldsymbol{P}t)\mathrm{e}^{-\mathrm{i}Hs} g\|$$

and integrating by parts, requires only that

$$\int_1^t \mathrm{d}s \, \|(\boldsymbol{r}-2\boldsymbol{P}t)\mathrm{e}^{-\mathrm{i}Hs} g\| \leq \mathrm{const}\, t^{1+\beta'}.$$

This inequality follows from (12.3.13) by applying the Schwarz inequality, together with the bounds (12.3.15) and (12.3.16). Hence, by this minor amendment to the proof of Lemma 12.8, in which we have replaced a pointwise estimate of $\|(\boldsymbol{r}-2\boldsymbol{P}t)\mathrm{e}^{-\mathrm{i}Ht} g\|$ by an L^1 estimate of the same norm, we have asymptotic completeness provided only that we are able to verify (12.3.6) for vectors g belonging to the appropriate class. In evaluating the right-hand side of (12.3.11), we have already obtained identities from which (12.3.6) may be derived. Thus, from a consideration of

$$\frac{\mathrm{d}}{\mathrm{d}t} \langle \boldsymbol{r}\mathrm{e}^{-\mathrm{i}Ht} g\,, \boldsymbol{r}\mathrm{e}^{-\mathrm{i}Ht} g \rangle,$$

we obtain

$$\frac{\mathrm{d}}{\mathrm{d}t} \mathrm{e}^{\mathrm{i}Ht} \boldsymbol{r}^2 \mathrm{e}^{-\mathrm{i}Ht} g = 4\mathrm{e}^{\mathrm{i}Ht} A \mathrm{e}^{-\mathrm{i}Ht} g.$$

(We leave the reader to verify that, with $g=\phi(H)f$, and $f\in D(\boldsymbol{r}^2)$, it follows that $e^{-iHt}g\in D(\boldsymbol{r}^2)$; the proof of this result may be based on (12.2.30).)

Again, we have, proceeding as before,

$$\frac{d}{dt}e^{iHt}Ae^{-iHt}g=2(Hg-e^{iHt}Qe^{-iHt}g).$$

Integrating this equation with respect to t now gives

$$e^{iHt}Ae^{-iHt}g=2Htg+o(t),$$

a norm estimate that makes use of our bound for $\int_c^t ds\,\|Qe^{-iHs}g\|$, based on (12.3.15) and (12.3.16) above. Integrating with respect to t now gives

$$e^{iHt}\boldsymbol{r}^2e^{-iHt}g=4Ht^2g+o(t^2),$$

from which (12.3.6) follows. The proof of asymptotic completeness then follows, and we can assert the following result.

Theorem 12.2

Let $V(\boldsymbol{r})$ be a long-range potential such that

$$|V(\boldsymbol{r})|,\ |\text{grad}\,V(\boldsymbol{r})|\le\text{const}\,|\boldsymbol{r}|^{-\beta}\quad\text{as}\,|\boldsymbol{r}|\to\infty$$

for some $\beta\ge\frac{3}{4}$, and

$$|V(\boldsymbol{r})|\le\text{const}\,|\boldsymbol{r}|^{-1},\quad|\text{grad}\,V(\boldsymbol{r})|\le\text{const}\,|\boldsymbol{r}|^{-2}\quad\text{as }|\boldsymbol{r}|\to0$$

and suppose that the conditions of Theorem 9.2 are satisfied. (This requires in addition only a mild condition on the rate of decay at infinity of the second derivative of the potential; a decay of order $|\boldsymbol{r}|^{-5/2}$ is certainly sufficient.) Then H has no singular continuous spectrum, and the modified wave operators $\Omega_{\pm}$ are asymptotically complete.

Proof

We have already proved completeness of the wave operators, and it remains only to consider the nature of the spectrum. At positive energies, we have already shown that states lie in the range of the wave operators, and are therefore spectrally absolutely continuous, provided that we are not at an eigenvalue of H. At negative energies, again provided that we are not at an eigenvalue, we can again make use of the bound $\|E_{\mathscr{I}}(H)QE_{\mathscr{I}}(H)\|<1$, which holds for some interval $\mathscr{I}$ containing λ. Any vector g with negative H-spectral support must satisfy $\Omega_{+}^{*}g=0$, and our previous arguments show that this implies that $g=0$, if g has spectral support within $\mathscr{I}$. Thus any

$\lambda < 0$ that is not an eigenvalue of H can have no spectrum of H within some neighbourhood. The only negative spectrum of H consists of possible isolated eigenvalues, and in particular there can be no singular continuous spectrum. ■

Chapter 13

Bound States and Scattering States

13.1 THE ASYMPTOTIC DECOMPOSITION THEOREM

Scattering theory is a study of the asymptotic evolution of states in the limits as $t \to \pm\infty$. In the theory of scattering by a potential $V(\mathbf{r})$, three possible kinds of asymptotic evolution are of interest. The simplest kind of evolution is that in which the initial state f is an eigenstate of the Hamiltonian H, or more generally in which f is a linear combination of finitely or infinitely many such eigenstates. Such a state may be described as a *bound state*, in the sense that, uniformly in time t, the probability of finding the particle at large distances from the origin is small. Mathematically, one may characterize a bound state by the property $\|E_{|\mathbf{r}|>R}\, e^{-iHt}f\| < \varepsilon$, for all t and for given ε, provided that R is taken to be sufficiently large.

There is another kind of bound state, of which we first became aware in Chapter 11. If strong asymptotic completeness fails then states g will exist, in the absolutely continuous subspace of H, such that, for any $R>0$, $\|E_{|\mathbf{r}|>R}\, e^{-iHt}g\|$ approaches zero, either in the limit $t \to \infty$ or in the limit $t \to -\infty$. Such a state will be asymptotically absorbed (perhaps "adsorbed" would be a better word), in the sense that, as $t \to \infty$ or as $t \to -\infty$, the particle approaches the origin 0 with probability 1. Given any small neighbourhood of $\mathbf{r}=0$, the probability of finding the particle *outside* that neighbourhood will asymptotically approach zero.

The third kind of asymptotic evolution, and that which is of greatest importance in scattering theory, is that of a state in the range of some wave operator $\Omega_{\pm}$. Such a state will evolve asymptotically like a free-particle state. In particular, in the limit as $t \to \pm\infty$ such a state will be localized at large distances from the origin. If the potential is of long range then we have to deal with a modified free evolution, but again such a state, in the range of the modified wave operator, will asymptotically move to large distances. Such states may be described as *scattering states*. We should note, however, that a state that is a scattering state in the limit $t \to -\infty$ will not necessarily be a scattering state as $t \to +\infty$, and similarly a state that is absorbed, say, as $t \to -\infty$, need not be absorbed as $t \to +\infty$. The condition for the scattering states as $t \to -\infty$ to be identified with those for the limit $t \to +\infty$ is that range $(\Omega_+) = $ range (Ω_-)—hence the importance of asymptotic completeness in scattering theory.

Given a potential $V(\boldsymbol{r})$, are other kinds of asymptotic evolution possible, other than the three important cases that we have mentioned? In the case of *short-range* potentials, an answer to this question is available, and is provided by the Asymptotic Decomposition Theorem. This theorem allows us to characterize all possible kinds of asymptotic evolution, and to decompose the Hilbert space with respect to them. In this section and the next, we shall prove the theorem for short-range potentials having arbitrary local singularities, and also elaborate the theory of singular points of the Hamiltonian in such cases. The following lemma presents a basic tool of the subsequent analysis.

Lemma 13.1

Let $V(\boldsymbol{r})$ be bounded in the region $|\boldsymbol{r}| > R$ for any $R > 0$, and let it satisfy

$$|V(\boldsymbol{r})| \leq \text{const}\, |\boldsymbol{r}|^{-(1+\varepsilon)} \tag{13.1.1}$$

for some $\varepsilon > 0$ and for $|\boldsymbol{r}| > R$. Let $H = H_0 + V$ and $\Omega_\pm = \text{s-lim}_{t \to \pm\infty}\, e^{iHt} e^{-iH_0 t}$. Then the operators $(\Omega_\pm - 1)\phi(H_0)P_\pm$ are compact, where $P_\pm$ are the projection operators onto positive and negative parts respectively of the generator A of dilations, and $\phi(\lambda) \in C_0^\infty(\mathbb{R})$, such that $\lambda = 0$ does not lie in the support of ϕ.

Proof

This result provides an extension of Lemma 12.3 to the more general class of potentials considered here. The additional result of Lemma 12.3 that $(\Omega_\pm - 1)\phi(H)P_\pm$ be compact follows only in the absence of singular points for H.

We follow closely the proof of the previous lemma, except that here we introduce a smooth multiplication operator $\rho(\boldsymbol{r})$ that localizes away from the origin. Note first of all, for $\beta > 0$, that the operators $(1+|\boldsymbol{r}|)^\beta (H_0+1)^{-1} (1+|\boldsymbol{r}|)^{-\beta}$ and $(1+|\boldsymbol{r}|)^\beta P_j (H_0+1)^{-1} (1+|\boldsymbol{r}|)^{-\beta}$ are bounded.

(Note that $|\boldsymbol{r}|^\beta \rho (H_0+1)^{-1}$ and $|\boldsymbol{r}|^\beta \rho P_j (H_0+1)^{-1}$ are both defined as linear operators on $C_0^\infty(\mathbb{R}^3 \setminus \{0\})$; this follows, for example, from taking $\beta = 2n$, where n is an integer (which may be arbitrarily large), and considering $|\boldsymbol{r}|^{2n}(H_0+1)^{-1}$ and $|\boldsymbol{r}|^{2n}P_j(H_0+1)^{-1}$ as operators in the momentum-space representation, with $|\boldsymbol{r}|^2=-\Delta_k$ and $(H_0+1)^{-1}$ is the operator of multiplication by $(\boldsymbol{k}^2+1)^{-1}$. We now use the commutation relation of $|\boldsymbol{r}|^\beta\rho$ with H_0 to write, for example, as operator equations on $C_0^\infty(\mathbb{R}^3\setminus\{0\})$,

$$\begin{aligned}|\boldsymbol{r}|^\beta\rho(H_0+1)^{-1}(1+|\boldsymbol{r}|)^{-\beta}=&(H_0+1)^{-1}|\boldsymbol{r}|^\beta\rho(1+|\boldsymbol{r}|)^{-\beta}\\ &+(H_0+1)^{-1}[\Delta(|\boldsymbol{r}|^\beta\rho)\\ &-2\mathrm{i}\boldsymbol{P}\cdot\operatorname{grad}(|\boldsymbol{r}|^\beta\rho)](H_0+1)^{-1}(1+|\boldsymbol{r}|)^{-\beta}.\end{aligned}$$

This equation may be verified by operating on both sides with H_0+1. In addition, the first term on the right-hand side is bounded, and in the second term, through differentiation of $|\boldsymbol{r}|^\beta\rho$, we have effectively reduced the power of $|\boldsymbol{r}|$ from β to $\beta-1$. Since the nth derivative of $|\boldsymbol{r}|^\beta\rho$ is a bounded function, for sufficiently large n, we find by repeated use of commutation relations that $|\boldsymbol{r}|^\beta\rho(H_0+1)^{-1}(1+|\boldsymbol{r}|)^{-\beta}$ is bounded on $C_0^\infty(\mathbb{R}^3\setminus\{0\})$, and hence on $L^2(\mathbb{R}^3)$ by taking the closure of this operator. A similar argument leads to the conclusion that $|\boldsymbol{r}|^\beta\rho P_j(H_0+1)^{-1}(1+|\boldsymbol{r}|)^{-\beta}$ is bounded. Since $|\boldsymbol{r}|^\beta$ is bounded in the neighbourhood of $\boldsymbol{r}=\boldsymbol{0}$, the operator ρ may then be removed.)

We now take the inequality (12.1.18), and multiply the operators within each norm, on the right, by

$$(1+|\boldsymbol{r}|)^\beta P_j(H_0+1)^{-1}(1+|\boldsymbol{r}|)^{-\beta}.$$

In addition to the inequality (12.1.18), we now have, for $0<\beta'<\beta$,

$$\left.\begin{aligned}\|P_+\phi(H_0)\mathrm{e}^{\mathrm{i}H_0t}P_j(1+|\boldsymbol{r}|)^{-\beta}\|&\le\frac{\text{const}}{(1+t)^{\beta'}}\quad(t>0),\\ \|P_-\phi(H_0)\mathrm{e}^{\mathrm{i}H_0t}P_j(1+|\boldsymbol{r}|)^{-\beta}\|&\le\frac{\text{const}}{(1+|t|)^{\beta'}}\quad(t<0).\end{aligned}\right\}\tag{13.1.2}$$

We have replaced $\phi(H_0)$ by $(H_0+1)^{-1}\phi(H_0)$ here to remove the operator H_0+1.

We now rely on the result

$$(\mathrm{e}^{\mathrm{i}Ht}\rho\mathrm{e}^{-\mathrm{i}H_0t}-\rho)\phi(H_0)P_+=\int_0^t\mathrm{d}s\frac{\mathrm{d}}{\mathrm{d}s}\mathrm{e}^{\mathrm{i}Hs}\rho\mathrm{e}^{-\mathrm{i}H_0s}\phi(H_0)P_+.$$

Hence

$$\begin{aligned}&(\mathrm{e}^{\mathrm{i}Ht}\rho\mathrm{e}^{-\mathrm{i}H_0t}-\rho)\phi(H_0)P_+\\ &=\mathrm{i}\int_0^t\mathrm{d}s\,\mathrm{e}^{\mathrm{i}Hs}[V\rho-\Delta\rho-2\mathrm{i}(\operatorname{grad}\rho\cdot\boldsymbol{P})]\mathrm{e}^{-\mathrm{i}H_0s}\phi(H_0)P_+.\end{aligned}\tag{13.1.3}$$

Now, by hypothesis, $|V\rho| \leq \text{const } (1+|\boldsymbol{r}|)^{-(1+\varepsilon)}$, and a similar inequality holds trivially for $\Delta\rho$ and the components of $\operatorname{grad}\rho$, since both are in $C_0^\infty(\mathbb{R}^3)$. By the adjoint of the first inequality in (13.1.2), and following the proof of Lemma 12.3, we may deduce the compactness of $(\Omega_+ - \rho)\phi(H_0)P_+$. (Since $(1-\rho)\mathrm{e}^{-\mathrm{i}H_0 t}$ converges to zero, note that $\Omega_+ = \text{s-lim}_{t\to\infty}\, \mathrm{e}^{\mathrm{i}Ht}\rho\mathrm{e}^{-\mathrm{i}H_0 t}$.) A similar argument leads to the conclusion that $(\Omega_- - \rho)\phi(H_0)P_-$ is compact, and the result of the lemma now follows immediately from the compactness of $(1-\rho)\phi(H_0)$. ■

We shall need to use the compactness of the operator $\rho\phi(H_0) - \phi(H)\rho$, a result that has already been obtained in Section 10.2. We shall also rely on a result from Section 12.1, that freely evolving states are asymptotically, in the limit as $t\to\infty$, to be found in the range of P_+. Thus

$$\underset{t\to\infty}{\text{s-lim}}\, P_+\mathrm{e}^{\mathrm{i}H_0 t} = \underset{t\to\infty}{\text{s-lim}}\, P_-\mathrm{e}^{-\mathrm{i}H_0 t} = 0. \tag{13.1.4}$$

The following definitions make more precise the notions of bound state, scattering state and absorbed state.

Definitions

A vector $f \in \mathscr{H}$ is said to be a *bound state* as $t\to\infty$ if, for any $\varepsilon>0$, there exists $R>0$ (depending on ε) such that

$$\|E_{|\boldsymbol{r}|>R}\mathrm{e}^{-\mathrm{i}Ht}f\| < \varepsilon \tag{13.1.5}$$

for all sufficiently large t. (Equivalently, this inequality should hold for all $t>0$. This is because, for *any* $f\in\mathscr{H}$, the inequality will hold, for fixed t, provided that R is chosen sufficiently large. Since $\mathrm{e}^{-\mathrm{i}Ht}f$ is strongly continuous in t, we can then make the inequality hold for t in some neighbourhood of any fixed t; by an application of the Heine–Borel theorem, any finite interval can be covered by a finite number of such neighbourhoods, so that, without loss of generality, the inequality may be assumed to hold on any fixed interval $0\leq t<T$.)

A vector $g\in\mathscr{H}$ is said to be a *scattering state* as $t\to\infty$ if, for any $R>0$,

$$\lim_{t\to\infty} \|E_{|\boldsymbol{r}|<R}\mathrm{e}^{-\mathrm{i}Ht}g\| = 0. \tag{13.1.6}$$

A vector $h\in\mathscr{H}$ is said to be an *absorbed state* as $t\to\infty$ if, for any $R>0$,

$$\lim_{t\to\infty} \|E_{|\boldsymbol{r}|>R}\mathrm{e}^{-\mathrm{i}Ht}h\| = 0. \tag{13.1.7}$$

The sets of bound states, scattering states and absorbed states as $t\to+\infty$ will be denoted by respectively $\mathscr{M}_\mathrm{b}^+$, $\mathscr{M}_\infty^+$ and $\mathscr{M}_0^+$, and the corresponding sets of states for the limit $t \to -\infty$ will be denoted by $\mathcal{M}_b^-$, $\mathcal{M}_\infty^-$ and $\mathcal{M}_0^-$. All of these sets define (closed) subspaces of the Hilbert space. Moreover, in each case we have $f \in \mathcal{M} \Rightarrow e^{-iHt} f \in \mathcal{M}$, so that each subspace reduces the total Hamiltonian H.

We also define a subspace $\mathcal{M}_w$, consisting of all $f \in \mathcal{H}$ such that $e^{-iHt} f$ converges weakly to zero in the limit $t \to \infty$. Again, $\mathcal{M}_w$ is a reducing subspace for H. For $f \in \mathcal{M}_w$, we have, certainly, $\lim_{t \to \infty} \langle f, e^{-iHt} f \rangle = 0$. Conversely, if $f \in \mathcal{H}$ is any vector satisfying $\lim_{t \to \infty} \langle f, e^{-iHt} f \rangle = 0$ then $\lim_{t \to \infty} \langle e^{-iHs} f, e^{-iHt} f \rangle = 0$. We let $\mathcal{M}_f$ be the subspace of $\mathcal{H}$ consisting of norm limits of linear combinations of $e^{-iHs} f$, for various s. (Thus $\mathcal{M}_f$ is the smallest subspace containing $e^{-iHs} f$ for all $s \in \mathbb{R}$.) We have $\lim_{t \to \infty} \langle h, e^{-iHt} f \rangle = 0$ for all $h \in \mathcal{M}_f$. On the other hand, any vector $h^\perp \in \mathcal{M}_f^\perp$ satisfies $\langle h, e^{-iHt} f \rangle = 0$ for all t. Since an arbitrary vector $g \in \mathcal{H}$ is of the form $g = h + h^\perp$, where $h \in \mathcal{M}_f$ and $h^\perp \in \mathcal{M}_f^\perp$, we have

$$\lim_{t \to \infty} \langle f, e^{-iHt} f \rangle = 0 \Rightarrow \lim_{t \to \infty} \langle g, e^{-iHt} f \rangle = 0$$

for all $g \in \mathcal{H}$. Hence we may alternatively define $\mathcal{M}_w$ as the set of vectors $f \in \mathcal{H}$ for which $\lim_{t \to \infty} \langle f, e^{-iHt} f \rangle = 0$. Since $\lim_{t \to -\infty} \langle f, e^{-iHt} f \rangle = \overline{\lim_{t \to \infty} \langle f, e^{-iHt} f \rangle}$, vectors will belong to $\mathcal{M}_w$ if and only if $e^{-iHt} f \to 0$ weakly as $t \to -\infty$. Hence in defining $\mathcal{M}_w$ there is no need to distinguish between the limits $t \to -\infty$ and $t \to +\infty$; for this reason, the superscripts $\pm$ may be omitted ($\mathcal{M}_w^+ = \mathcal{M}_w^-$).

The following theorem shows that the Hilbert space may be split into two mutually orthogonal components: the subspaces of bound states and scattering states respectively. Moreover, the absorbed states are just those bound states that converge weakly to zero.

Theorem 13.1 (Asymptotic Decomposition Theorem)

$$\mathcal{H} = \mathcal{M}_b^+ \oplus \mathcal{M}_\infty^+ = \mathcal{M}_b^- \oplus \mathcal{M}_\infty^-;$$

$$\mathcal{M}_\infty^+ = \text{range}\,(\Omega_+), \qquad \mathcal{M}_\infty^- = \text{range}\,(\Omega_-);$$

$$\mathcal{M}_0^+ = \mathcal{M}_b^+ \cap \mathcal{M}_w, \qquad \mathcal{M}_0^- = \mathcal{M}_b^- \cap \mathcal{M}_w.$$

Proof

(i) Given $f \in \mathcal{M}_b^+$, $g \in \mathcal{M}_\infty^+$ and $\varepsilon > 0$, we choose R and T sufficiently large that

$$\|E_{|r|>R} e^{-iHt} f\| < \varepsilon \quad \text{for } t > T.$$

Then

$$|\langle f,g\rangle|=|\langle E_{|r|>R}e^{-iHt}f,e^{-iHt}g\rangle + \langle e^{-iHt}f,E_{|r|<R}e^{-iHt}g\rangle| \le \varepsilon\|g\|+\|E_{|r|<R}e^{-iHt}g\|\,\|f\|$$

It follows from (13.1.6) that $|\langle f,g\rangle|\le\varepsilon\|g\|$. However, $\varepsilon>0$ was arbitrary, so that $\langle f,g\rangle=0$. We then have $\mathscr{M}_b^+\perp\mathscr{M}_\infty^+$, and similarly $\mathscr{M}_b^-\perp\mathscr{M}_\infty^-$. In addition, we have range $(\Omega_\pm)\subseteq\mathscr{M}_\infty^\pm$. (Let us suppose, for example, that $g=\Omega_+f\in$ range (Ω_+). Then $\text{s-lim}_{t\to\infty}(e^{-iHt}g-e^{-iH_0t}f)=0$. Since $\text{s-lim}_{t\to\infty}E_{|r|<R}e^{-iH_0t}f=0$, (13.1.6) follows, and $g\in\mathscr{M}_\infty^+$.)

To prove $\mathscr{H}=\mathscr{M}_b^\pm+\mathscr{M}_\infty^\pm$ and $\mathscr{M}_\infty^\pm=$ range $(\Omega_\pm)$, it remains only to prove that any vector orthogonal to range $(\Omega_\pm)$ must belong to $\mathscr{M}_b^\pm$. We shall prove (range $(\Omega_+))^\perp\subseteq\mathscr{M}_b^+$; the proof for range $(\Omega_\pm)$ follows similarly. We start from the identity

$$\begin{aligned}\rho\phi(H)=&\{\rho\phi(H)-\phi(H_0)\rho\}+\{P_+\phi(H_0)(\rho-\Omega_+^*)\}\\&+\{P_-\phi(H_0)(\rho-\Omega_-^*)\}+P_+\phi(H_0)\Omega_+^*\\&+P_-\phi(H_0)\Omega_-^*.\end{aligned}\tag{13.1.8}$$

Since, in the proof of Lemma 13.1, we showed that the operators $(\Omega_\pm-\rho)\phi(H_0)P_\pm$ are compact, we know on taking adjoints and noting the compactness of $\rho\phi(H)-\phi(H_0)\rho$ that each term of (13.1.8) within curly brackets is compact. We let $f\in\mathscr{H}$ be orthogonal to the range of Ω_+. Then $\Omega_+^*f=0$. From (13.1.8), we then have

$$\rho e^{-iHt}\phi(H)f=Ce^{-iHt}f+P_-e^{-iH_0t}\phi(H_0)\Omega_-^*f,\tag{13.1.9}$$

where C is a compact operator, and we have used the intertwining property of wave operators.

Let us now suppose that $\phi(H)f$ is *not* a bound state as $t\to\infty$. Then there exist sequences $\{R_n\}$ and $\{t_n\}$, with $R_n\to\infty$, $t_n\to\infty$, such that

$$\|E_{|r|>R_n}e^{iHt_n}\phi(H)f\|>\text{const}>0.\tag{13.1.10}$$

Since any bounded sequence of vectors in $\mathscr{H}$ has a weakly convergent subsequence, we may assume without loss of generality that $e^{-iHt_n}f$ converges weakly to f_1, say.

Taking $t=t_n$ with the limit $n\to\infty$ in (13.1.9), this then gives, with (13.1.4),

$$\text{s-lim}_{n\to\infty}\rho e^{-iHt_n}\phi(H)f=Cf_1.\tag{13.1.11}$$

In this case, we have

$$\underset{n\to\infty}{\text{s-lim}}\, E_{|r|>R_n} e^{-iHt_n}\phi(H)f = \underset{n\to\infty}{\text{s-lim}}\, E_{|r|>R_n}\rho e^{-iHt_n}\phi(H)f$$
$$= \underset{n\to\infty}{\text{s-lim}}\, E_{|r|>R_n} C f_1 = 0,$$

which contradicts (13.1.10). This proof by contradiction shows that $\phi(H)f \in \mathcal{M}_b^+$. Note that $\phi \in C_0^\infty(\mathbb{R}\setminus\{0\})$. By taking limits of such ϕ, approaching pointwise the characteristic function of $\mathbb{R}\setminus\{0\}$, we can deduce that $(1-P_0)f \in \mathcal{M}_b^+$, where P_0 is the projection onto (possible) eigenstates of H with eigenvalue zero. However, we certainly have $P_0 f \in \mathcal{M}_b^+$, since it is easy to verify that eigenstates of H are bound in the sense defined here. Hence $f \in \mathcal{M}_b^+$, and we have $(\text{range}\,\Omega_+)^\perp \subseteq \mathcal{M}_b^+$.

(ii) It remains to prove the result $\mathcal{M}_0^\pm = \mathcal{M}_b^\pm \cap \mathcal{M}_w$. The proof that $\mathcal{M}_0^+ \subseteq \mathcal{M}_b^+ \cap \mathcal{M}_w$ is straightforward and will be omitted. Let us suppose, conversely, that $f \in \mathcal{M}_b^+ \cap \mathcal{M}_w$, and operate by (13.1.8) on $e^{-iHt}f$. Since $f \perp \mathcal{M}_\infty^+ = \text{range}\,(\Omega_+)$ we have $\Omega_+^* f = 0$. On the other hand, by (13.1.4), $P_-\phi(H_0)\Omega_-^* e^{-iHt}f$ converges strongly to zero. Since $e^{-iHt}f$ converges weakly to zero, there will be no contribution from the terms in (13.1.8) within curly brackets, and we have $\text{s-lim}_{t\to\infty}\rho\phi(H)e^{-iHt}f = 0$. Thus (13.1.7) is satisfied with $h = \phi(H)f$. Again taking limits of such ϕ, we obtain $(1-P_0)f \in \mathcal{M}_0^+$. Since $P_0 e^{-iHt}f = P_0 f$ converges weakly to zero as $t\to\infty$, we must have $P_0 f = 0$, so that $f \in \mathcal{M}_0^+$. We have thus shown that $\mathcal{M}_0^+ = \mathcal{M}_b^+ \cap \mathcal{M}_w$, and the result with minus instead of plus follows in the same way. ∎

A number of interesting consequences follow from the Asymptotic Decomposition Theorem. One most satisfactory feature of the theorem is the identification of the range of each wave operator with the corresponding subspace of scattering states. States that are in the range of the wave operator Ω_+, for example, are just those states that move to infinity in position space at large positive times. Such states are asymptotically free, and so for short-range potentials we find that the condition for a state to be asymptotically free is simply that the particle escapes to infinity. Since the range of each wave operator is contained in $\mathcal{M}_{ac}(H)$, we see also that scattering states belong necessarily to the absolutely continuous subspace for H. This goes a long way to explaining the importance in scattering theory of subspaces of absolute continuity.

Asymptotic completeness ($\text{range}\,(\Omega_+) = \text{range}\,(\Omega_-)$) will hold whenever the subspace $\mathcal{M}_\infty^+$ of states asymptotically free at $t\to\infty$ coincides with the corresponding subspace at $t\to-\infty$. Alternatively, we can say that asymptotic completeness will hold provided that states that are bound at $t\to-\infty$ are just those states that are bound at $t\to+\infty$. This will certainly be so if the only bound states are eigenstates of H. But there may also be other kinds of bound states. States belonging to $\mathscr{M}_{sc}(H)$ are bound; however $\mathscr{M}_{sc}(H)$ does not carry a label $\pm$, so that these states are bound at $t=\pm\infty$ and cannot be associated with a failure of asymptotic completeness. Any state in $\mathscr{M}_{ac}(H)$ that is orthogonal to range (Ω_+) (or to range (Ω_-)) is also a bound state as $t\to\infty$ (respectively as $t\to-\infty$). Since, for $f\in\mathscr{M}_{ac}(H)$ we have $\lim_{t\to\pm\infty}\langle f, e^{-iHt}f\rangle=0$, such a state will lie in the subspace $\mathscr{M}_0^+$ of states that are asymptotically absorbed at the singularity $\mathbf{r}=\mathbf{0}$. The existence of states in $\mathscr{M}_0^\pm$ *may* be associated with a breakdown of asymptotic completeness. This will happen if $\mathscr{M}_0^+\neq\mathscr{M}_0^-$. In this case there will be states in the range of Ω_- that have a non-zero component in $\mathscr{M}_0^+$ (or the same may happen with plus and minus interchanged). Such a state is asymptotically free in the limit $t\to-\infty$, but there is a non-zero probability of absorption into the singularity in the limit $t\to\infty$. In other words, a particle can come in from infinity, and hit a target of zero size (the origin)! Fortunately, such a phenomenon is rare, but it cannot be ruled out without some restriction on the local behaviour of the potential. Another possibility is that of a non-trivial singular continuous subspace for H. For states f in $\mathscr{M}_{sc}(H)$ it is not known, in the present context, whether $\langle f, e^{-iHt}f\rangle$ necessarily converges to zero as $t\to\pm\infty$. If so, then f belongs necessarily to $\mathscr{M}_0^\pm$ and is asymptotically absorbed. More generally, we know that $\langle f, e^{-iHt}f\rangle$ converges to zero *in time average*, from which it follows that states in $\mathscr{M}_{sc}(H)$ are always asymptotically absorbed on time average. In particular, sequences $\{t_n\}$ always exist, with $t_n\to\infty$, such that the vector sequence $e^{-iHt_n}f$ is localized asymptotically at the origin. If, on the other hand, states $f\in\mathscr{M}_{sc}(H)$ exist that are *not* in $\mathscr{M}_w$ then $e^{-iHt_n}f$ may converge to a non-zero limit in L^2 of the region $|\mathbf{r}|>R$, for any $R>0$. Such states would certainly be an unusual kind of bound state, but so far we have no direct evidence for their existence.

13.2 CLASSIFICATION OF SINGULAR POINTS

In Chapter 10 we defined the notion of a singular point λ for a Schrödinger Hamiltonian H. We saw in Chapter 11 how the possible existence of singular points is related to the question of asymptotic completeness. If range $(\Omega_\pm)\neq\mathscr{M}_{ac}(H)$ then we shall have singular points. On the other hand, as we have seen in the last section, localization phenomena may also be associated with the presence of a singular continuous spectrum. Within the context of short-range potentials, we shall let $\lambda_0\neq0$ be a point of the singular continuous spectrum of H. We can construct a normalized sequence $\{f_n\}$ of vectors in $\mathscr{M}_{sc}(H)$ that localize asymptotically in energy at

λ; this means that $\text{s-lim}_{n\to\infty} E_{|H-\lambda_0|>\varepsilon} f_n = 0$ for all $\varepsilon > 0$. (For $f \in \mathcal{M}_{sc}(H)$, take for example $f_n = E_{|H-\lambda_0|<1/n} f / \|E_{|H-\lambda_0|<1/n} f\|$; since λ_0 is in the singular continuous spectrum of H, f can always be chosen such that the norm in the denominator does not vanish.) We choose $\phi(\lambda)$ in (13.1.8) such that $\phi(\lambda_0) = 1$, in which case $(1-\phi(H)) f_n \to 0$ strongly as $n \to \infty$. We also have $\Omega_{\pm}^* f_n = 0$, so that (13.1.8) implies

$$\text{s-lim}_{n\to\infty} \rho f_n = 0.$$

Thus $\text{s-lim}_{n\to\infty} E_{|r|>R} f_n = 0$ for any $R > 0$, and the sequence $\{f_n\}$ localizes asymptotically in position as well as energy. This implies that

$$\lim_{n\to\infty} \|E_{|r|<R} E_{|H-\lambda_0|<\varepsilon} f_n\| = 1$$

for all $R, \varepsilon > 0$, so that λ_0 is a singular point. In the same way, we may show that any limit point of eigenvalues of H is a singular point, as is any eigenvalue having infinite multiplicity.

In view of the diversity of types of singular point, it is useful to distinguish at least between singular points of the absolutely continuous spectrum of H and other singular points. The following theorem shows how this can be done.

Theorem 13.2

For $\lambda \in \mathbb{R}$, define $\gamma_{ac}(\lambda)$ by

$$\gamma_{ac}(\lambda) = \lim_{\substack{R\to 0\\ \varepsilon\to 0}} \|E_{|r|<R} E_{|H-\lambda|<\varepsilon} P_{ac}(H)\|, \tag{13.2.1}$$

where $P_{ac}(H)$ projects onto the absolutely continuous subspace of H. Then, for given λ, $\gamma_{ac}(\lambda)$ is either 0 or 1.

Proof

The proof of Theorem 10.1 may be carried through with minor amendments. We replace $\phi_\varepsilon(H)$ in (10.2.10) by $\phi_\varepsilon(H) P_{ac}(H)$. Whereas in the proof of Theorem 10.1 we made use of the compactness of $E_{|r|<1} \phi(H) E_{H_0<M}$ for suitable ϕ, we now need to know that $E_{|r|<1} \phi(H) P_{ac}(H) E_{H_0<M}$ is also compact. This will follow provided that we can show that $E_{|r|<1} \phi(H)(1-P_{ac}(H)) E_{H_0<M}$ is compact. Since $\Omega_{\pm}^*(1-P_{ac}(H)) = 0$, (13.1.8) implies that $\rho\phi(H)(1-P_{ac}(H))$ is compact. By taking adjoints, the same

applies to $\phi(H)(1-P_{ac}(H))\rho$. We can now use the identity

$$E_{|r|<1}\phi(H)(1-P_{ac}(H))E_{H_0<M}$$
$$=E_{|r|<1}\{\phi(H)(1-P_{ac}(H))\rho\}E_{H_0<M}$$
$$+E_{|r|<1}\phi(H)(1-P_{ac}(H))\{(1-\rho)E_{H_0<M}\},$$

where operators within curly brackets are compact, to obtain the result we need. The remainder of the proof follows very closely that of Theorem 10.1. ■

Under what conditions, and for what values of λ do we have $\gamma_{ac}(\lambda)=1$? We know already that if range $(\Omega_{\pm})\neq\mathcal{M}_{ac}(H)$ then H must have singular points, and it is straightforward to verify from the proof of that result that $\gamma_{ac}(\lambda)=1$ for some λ in this case. We can identify such singular points in the absolutely continuous spectrum of H by making the following definition.

Definition

Suppose, for some fixed λ, that

$$\text{range}\,(E_{|H-\lambda|<\varepsilon}\Omega_{\pm})=\text{range}\,(E_{|H-\lambda|<\varepsilon}P_{ac}(H))$$

for some $\varepsilon>0$. (Using the intertwining property, these ranges then coincide for *all* sufficiently small $\varepsilon>0$.) Then we shall say that strong asymptotic completeness holds at energy λ. Since for $\lambda<0$ we have, by intertwining, $E_{|H-\lambda|<\varepsilon}\Omega_{\pm}=0$ for all sufficiently small $\varepsilon>0$, it follows that strong asymptotic completeness will hold at a negative energy λ whenever λ is not in the absolutely continuous spectrum of H.

Our previous arguments now imply that if strong asymptotic completeness fails at energy λ then $\gamma_{ac}(\lambda)=1$. (Find f in the range of $E_{|H-\lambda|<\varepsilon}P_{ac}(H)$ such that $\Omega_{\mp}^{*}f=0$, say. Then, as $t\to\infty$, $e^{-iHt}f$ localizes asymptotically at $\boldsymbol{r}=0$, and is a state in $\mathcal{M}_{ac}(H)$ with energy localized to within ε of λ. Since $\varepsilon>0$ is arbitrarily small, we must then have $\gamma_{ac}(\lambda)=1$.) On the other hand, even if strong asymptotic completeness holds at all energies, we may still have $\gamma_{ac}(\lambda)=1$ for some λ. The following theorem gives a necessary and sufficient condition for this to happen.

Theorem 13.3

Let $\lambda\neq 0$ be fixed. Then $\gamma_{ac}(\lambda)=0$ if and only if both

(a) strong asymptotic completeness holds at energy λ; and

(b)
$$\lim_{\varepsilon \to 0} \|E_{|H_0-\lambda|<\varepsilon}[S, P_+]E_{|H_0-\lambda|<\varepsilon}\| = 0, \tag{13.2.2}$$

where $S = \Omega_+^* \Omega_-$ is the scattering operator.

Proof

Equation (13.2.2) may be interpreted as saying that S commutes with P_+ (and therefore with P_-, since $P_+ + P_- = 1$) at energy λ. Roughly, the condition means that, for states highly localized in kinetic energy at λ, S should commute with P_+.

The proof proceeds in 3 stages.

(i) $\gamma_{ac}(\lambda) = 0$ implies strong asymptotic completeness at λ. We have already shown this, since failure of strong asymptotic completeness gives $\gamma_{ac}(\lambda) = 1$.

(ii) $\gamma_{ac}(\lambda) = 0$ implies that S commutes with P_+ at energy λ. Let us suppose that $\gamma_{ac}(\lambda) = 0$. We use (13.1.8), with ϕ chosen such that $\phi = 1$ at the point λ in question. Then

$$\begin{aligned} \rho\phi(H)\Omega_- E_{|H_0-\lambda|<\varepsilon} = {} & CE_{|H_0-\lambda|<\varepsilon} \\ & + P_+\phi(H_0)SE_{|H_0-\lambda|<\varepsilon} \\ & + P_-\phi(H_0)E_{|H_0-\lambda|<\varepsilon}, \end{aligned} \tag{13.2.3}$$

where C is compact. Since $\gamma_{ac}(\lambda) = 0$, we have

$$\begin{aligned} & \lim_{\varepsilon \to 0} \|(1-\rho)\phi(H)\Omega_- E_{|H_0-\lambda|<\varepsilon}\| \\ & \quad = \lim_{\varepsilon \to 0} \|(1-\rho)E_{|H-\lambda|<\varepsilon} P_{ac}(H)\phi(H)\Omega_-\| = 0, \end{aligned}$$

so that, on taking a limit in operator norm, the multiplication operator ρ on the left-hand side of (13.2.3) may be omitted. On the right-hand side $(\phi(H_0) - 1)E_{|H_0-\lambda|<\varepsilon}$ converges in norm to zero, so that we may drop the operator $\phi(H_0)$; by intertwining, $\phi(H)$ may also be dropped from the left-hand side. We then have,

$$\lim_{\varepsilon \to 0} \|(\Omega_- - P_+S - P_-)E_{|H_0-\lambda|<\varepsilon}\| = 0. \tag{13.2.4}$$

Hence $\Omega_- E_{|H_0-\lambda|<\varepsilon}$ approaches $(P_+S + P_-)E_{|H_0-\lambda|<\varepsilon}$ in norm as ε tends to zero. A formula similar to (13.2.4) may be derived for Ω_+, showing that $\Omega_+ E_{|H_0-\lambda|<\varepsilon}$ approaches $(P_-S^* + P_+)E_{|H_0-\lambda|<\varepsilon}$ in norm. Now

$$E_{|H_0-\lambda|<\varepsilon} S E_{|H_0-\lambda|<\varepsilon} = (\Omega_+ E_{|H_0-\lambda|<\varepsilon})^* (\Omega_- E_{|H_0-\lambda|<\varepsilon}),$$

which approaches

$$E_{|H_0-\lambda|<\varepsilon}(SP_- + P_+)(P_+S + P_-)E_{|H_0-\lambda|<\varepsilon}$$
$$= E_{|H_0-\lambda|<\varepsilon}(S - [S, P_+])E_{|H_0-\lambda|<\varepsilon}$$

in norm. Equation (13.2.2) now follows.

(iii) Strong asymptotic completeness at λ, with S and P_+ commuting at λ, implies that $\gamma_{ac}(\lambda)=0$. Let us assume that both hypotheses hold, with $\gamma_{ac}(\lambda)=1$. Then we can construct a normalized sequence $\{f_n\}$ of vectors, localizing asymptotically in position at $\boldsymbol{r}=\mathbf{0}$ and in energy at λ, where $f_n \in \mathcal{M}_{ac}(H)$. Using strong asymptotic completeness at λ, we may assume that $f_n = \Omega_- y_n$ where the sequence $\{y_n\}$, with $y_n = \Omega_-^* f_n$, is localized asymptotically in *kinetic* energy at λ. We again use (13.1.8), choosing ϕ as in (ii), and considering the strong limit as $n\to\infty$ of $\rho\phi(H)f_n$. In this case, the strong limit is zero, since f_n localizes in position at $\boldsymbol{r}=\mathbf{0}$. Using also the fact that f_n converges weakly to zero, and substituting for f_n in terms of y_n, we have

$$\operatorname*{s-lim}_{n\to\infty} (P_+S + P_-)y_n = 0. \tag{13.2.5}$$

Hence both P_+Sy_n and P_-y_n converge strongly to zero. Since $\{y_n\}$ localizes in kinetic energy, (13.2.2) implies that $\varepsilon=\varepsilon(n)$ can be taken to zero so that

$$\operatorname*{s-lim}_{n\to\infty} E_{|H_0-\lambda|<\varepsilon}[S, P_+]y_n = 0.$$

Since $P_+Sy_n \to 0$ and $SP_-y_n \to 0$, this gives

$$\operatorname*{s-lim}_{n\to\infty} E_{|H_0-\lambda|<\varepsilon}Sy_n = \operatorname*{s-lim}_{n\to\infty} Sy_n = 0.$$

However, under the hypothesis of strong asymptotic completeness at energy λ, the restriction of S to the range of $E_{|H_0-\lambda|<\varepsilon}$ is unitary. It follows that $y_n \to 0$ strongly, which contradicts the assumption that $f_n = \Omega_- y_n$ is normalized. Hence $\gamma_{ac}(\lambda)=0$.

Finally, the theorem follows immediately from (i)–(iii). ■

It is interesting to consider what happens if both strong asymptotic completeness holds at energy λ and $\gamma_{ac}(\lambda)=1$. In this case, we can localize states, *in the range of* $\Omega_\pm$, to an arbitrary degree of accuracy, in both position and in total energy. Such localized states represent a wave packet, coming from large distances at $t=-\infty$, and receding to large distances at $t=+\infty$, which is nevertheless localized at time $t=0$ to an arbitrary small neighbourhood of the origin. We cannot actually hit our target, but we can get as close as we like! A further consequence of this kind of localization is the following.

Corollary

Suppose that strong asymptotic completeness holds at energy λ_0, and $\gamma_{ac}(\lambda_0)=1$. Then the scattering matrix $S(\lambda)$ is a discontinuous function of energy at λ_0.

Proof

We use the direct integral decomposition of the Hilbert space. If $S(\lambda)=1+T(\lambda)$ is norm-continuous in λ then $S(\lambda)\rightarrow S(\lambda_0)$ as $\lambda\rightarrow\lambda_0$. Defining the operator S_0 by $S_0\{f_\lambda\}=\{S_0 f_\lambda\}$, we then have $(S-S_0)E_{|H_0-\lambda|<\varepsilon}\rightarrow 0$ as $\varepsilon\rightarrow 0$. Also, S_0 commutes with A, and hence with P_+. These two facts together imply (13.2.2) with $\lambda=\lambda_0$, so that $\gamma_{ac}(\lambda_0)=0$. Hence the result is proved, by contradiction. ■

To summarize, we have identified the following phenomena that may be associated with the existence of singular points for a Schrödinger Hamiltonian $H=H_0+V$, in the case that V is of short range:

(1) H may have a singular continuous spectrum, limit points of eigenvalues, or eigenvalues of infinite multiplicity;
(2) states may exist that are asymptotically absorbed in either of the limits $t\rightarrow\pm\infty$; such states may lead to a breakdown of asymptotic completeness;
(3) scattering states may be arbitrarily localizable, simultaneously in position and energy, leading to discontinuities in the scattering matrix as a function of energy.

It is a remarkable fact that, for specific potentials belonging to the class considered in this chapter, each of the above phenomena can indeed occur.

Chapter 14

Scattering Theory in Context

The quantum theory of scattering is a branch of mathematical physics that does not exist in isolation, but that has many connections with other disciplines. The mathematical ideas and methods presented here find application in many other fields. The physical ideas that go hand in hand with the mathematics have their origins in some of the most exciting developments of twentieth-century science. Many of these important developments, which were such dramatic breaks from the past, are now to be found as part of any elementary treatment of modern physics at undergraduate level.

Scattering theory has a past and a future. The theory had its origins in the work of countless physicists, mathematicians and mathematical physicists. I shall not here attempt a definition of "mathematical physicist"; the term has too wide a variety of meanings for that. To adapt an expression borrowed from the mathematical theory itself, there is a continuous spectrum of mathematical physicists, ranging from the pure mathematician who has been drawn to some area of mathematical problems for which the original motivation came from physics, to the (almost) pure physicist who could immediately give you an order-of-magnitude estimate for the nucleon–nucleon cross-section at 1250 MeV in millibarns per steradian. Let the reader judge to which part of this spectrum the present author's work belongs. The future of scattering theory cannot be known with any certainty. There remain many unsolved problems. Much remains to be done, particularly in the many-particle theory, and a mathematically consistent relativistic field theory has yet to be developed. Even a humble problem such as to analyse in detail the spectrum of the one-dimensional Schrödinger operator $-d^2/dr^2 + \cos r^{1/2}$ has not yet been solved. Undoubtably there will continue to be surprises.

This chapter attempts to set scattering theory in the context of its past and future, and to provide a broad survey of how the theory relates to other work. There will be sufficient references to enable the reader to follow up other interesting lines of development, to acquire a broader mathematical perspective, and to see the theory as part of a whole that is continually being extended. A part of this process will be to see in a new light the more elementary aspects of the theory; part of the foundations of any subject involves studying the elementary from an advanced viewpoint.

We shall conclude this chapter, and this book, with several applications of the theory to particular examples. Traditionally, students of mathematics have always gained from applying the general theory to particular instances, and I see every reason to continue this tradition. Some of the examples may seem a little unusual, even bizarre, but I believe that this is another pointer to the future.

14.1 THE MATHEMATICAL CONTEXT

Chapter 2 is intended to provide the reader with the mathematical background that forms the basis of scattering theory. Sometimes the reader is taken further than is strictly necessary for covering the material of this book.

This is a book on *foundations*, and it is intended to equip the reader with the background and knowledge to follow up this study with other advanced texts, and, it is hoped, to extend some of the ideas beyond their original domain. Scattering theory is an area of mathematical physics that has no sharp boundaries, and any boundaries that are set up are likely to be arbitrary.

The mathematics of scattering theory is a branch of mathematics that I would describe by the term *applied functional analysis*. Of particular importance are measure theory and its applications, Hilbert space, function spaces, operator analysis, the theory of differential equations, and analytic function theory. Chapter 2 covers much of what is required, concentrating on those aspects that are particularly important in applications to scattering theory.

A number of excellent general texts are available in analysis, and can be read with profit in parallel with sections of Chapter 2. The first nine chapters of Rudin (1974), for example, treat the theory of measure and integration, L^p spaces, Hilbert and Banach spaces, and Fourier-transform theory. A good modern text is Folland (1984). The less advanced reader should consult, for example, Lang (1968). A useful analysis reference book is Hewitt and Stromberg (1965), and good introductory texts on functional analysis are Riesz and Sz.-Nagy (1955), Yosida (1974) and Rudin (1973). See also Reed and Simon (1972) for a treatment oriented towards applications in mathematical physics.

Section 2.1 introduces the basic concepts of measure theory. See also Halmos (1950); two shorter accounts that treat some of the fundamentals of the subject are Bartle (1966) and Pitt (1985). Of particular importance in scattering theory is the distinction between discrete and continuous measures, and between absolutely and singular continuous measures. I have introduced the Cantor measure as an example of a singular continuous measure. One important practical tool in scattering theory, to which attention should be drawn, is the Lebesgue Dominated-Convergence Theorem, many applications of which will be found in this book.

Section 2.2 introduces various function spaces, including the L^p spaces, and spaces of continuous or differentiable functions, together with their dual spaces. The classic treatment of the theory of distributions or generalized functions is Gelfand and Shilov (1968); for the theory of Sobolev spaces see Adams (1975), and also Folland (1984). The topology of function spaces, and in particular the identification of dense sets of functions, has an important role in characterizing the domains of operators, proving that they are closed, etc., and also in extending results by continuity to larger spaces, of which a typical example is given. The reader should consult any standard work on topology (e.g. Kelley, 1955) for some of the general-topological background. Reference to general-topological ideas such as the Heine–Borel property will usually be very special applications to $\mathbb{R}$ or $\mathbb{R}^n$ (any covering by open sets of a closed bounded subset $\mathscr{R}$ of $\mathbb{R}^n$ has a finite subcovering; i.e. a finite number of the open sets will be sufficient to cover $\mathscr{R}$).

Section 2.3 deals with the analysis of measures. We need to be able to decompose a measure into its discrete, absolutely continuous, and singular continuous parts, and for example to identify sets of points to which the restriction of the measure is absolutely continuous. Some of these results are related to complex-function theory in Section 2.4. For the characterization theorem known as the Herglotz Theorem see, for example, Weidmann (1980). For boundary values of analytic functions see Collingwood and Cartwright (1952). Note the result of Lusin and Privalov (1925), which implies that two distinct functions analytic on the open unit disc and having positive imaginary part cannot have the same boundary value at a set of positive Lebesgue measure. See also Gilbert (1984).

In Section 2.5 we look briefly at the foundations of quantum mechanics. Two books that consider the theory from the starting point of quantal proposition systems are Jauch (1968) and Piron (1975). Another book on the foundations from a different standpoint is Mackey (1969). The classic and original work is of course von Neumann (1955). Of the numerous works on quantum mechanics written more from a physicist's point of view may be cited the standard references Landau and Lifshitz (1958), Messiah (1961, 1962) and Schiff (1968). Whatever the starting point adopted, one is led to the mathematics of Hilbert space, and to the study of linear operators in such spaces (Sections 2.6, 2.7). A standard reference on linear operators is Dunford and Schwartz (1958–1971); see also Akhiezer and Glazman (1981), Kato (1976), Reed and Simon (1975, 1978) and Weidmann (1980). Each of these books is also concerned with the spectral analysis of operators, for which see also Friedrichs (1973) and Helmberg (1969). Works having a particular emphasis on differential operators are Hellwig (1967), Naimark (1968) Schechter (1971) and Hörmander (1983).

The final section of Chapter 2 deals with the Schrödinger equation, both at real and complex energies. The results of this section have application to scattering by potentials that may be singular or non-singular, of short or of long range. Most of the methods used are part of the standard mathematical equipment of those working in this field, though the results are not usually collected together in the way I have done here. A good general reference for ordinary differential equations is Coddington and Levinson (1955), and for PDE's see Schechter (1977a).

14.2 FOUNDATIONS OF SCATTERING THEORY IN HILBERT SPACE

The principle mathematical objects of scattering theory are the wave operators $\Omega_{\pm}$ and the scattering operator S. From them, physical quantities such as scattering amplitudes and cross-sections can be derived. In the simplest situation, that of single-channel unmodified wave operators, $\Omega_{\pm}$ may be defined as strong limits $\Omega_{\pm} = \text{s-lim}_{t \to \pm\infty} e^{iHt} e^{-iH_0 t}$, where the total Hamiltonian H and the free Hamiltonian H_0 are self-adjoint operators in a single Hilbert space $\mathscr{H}$. Although properties of wave operators had already been described by Møller (1945), the general definition in terms of strong limits in Hilbert space was first proposed by Jauch (1958a), who also (1958b) went on to discuss the many-channel problem. Kuroda (1959) emphasized the role of the absolutely continuous subspaces for H and H_0, and defined wave operators $\Omega_{\pm}(H_1, H_2)$ involving projections onto $\mathscr{M}_{ac}(H_2)$. Kuroda's paper also dealt with the question of asymptotic completeness and unitarity of the scattering operator, and included proofs of the intertwining and transitivity properties of wave operators.

The existence of unmodified wave operators is based on the idea that the mathematical description of a state $e^{-iHt}f$ becoming asymptotically free in the limit $t \to \pm\infty$ should entail $e^{-iHt}f$ converging in norm to a freely evolving state $e^{-iH_0 t}g$. It proved difficult to justify such norm convergence on physical grounds, and indeed Dollard (1964) was able to show that, in the important case of the Coulomb Hamiltonian, e^{iHt} $e^{-iH_D t}$ failed to converge strongly. (The weak limit does exist, but is unfortunately zero.) This led to new attempts to formulate the so-called asymptotic condition, which would lead to a correct mathematical description of asymptotic free evolution. In the particular case of the Coulomb Hamiltonian, Dollard established the existence of modified wave operators, in which the free evolution was replaced by a modified free evolution. For a formulation of the asymptotic condition in terms of convergence of density operators see Jauch, Misra and Gibson (1968). However, this condtion did not apply to the Coulomb problem, a defect that was remedied in Amrein, Martin and Misra (1970), using the so-called algebraic theory of scattering. For further applications of the algebraic theory to Coulomb-type problems see, for example, Lavine (1970), Prugovecki (1971) and Thomas (1974b). My own formulation of the asymptotic condition, in Chapter 3, is based on the physical requirement that, at large positive or negative times, no observation or sequence of observations should be able to distinguish between the state and a freely evolving state. This condition encompasses not only the Coulomb Hamiltonian, but many-channel problems as well, and has the additional feature that the free Hamiltonian may be constructed from the theory, rather than postulated *ab initio*. I have followed this analysis by a treatment of some aspects of the algebraic theory. In Chapter 4 I have presented further developments of the abstract theory of scattering, with a discussion of single-channel and many-channel completeness, and derivation of formulae for (unmodified) wave and scattering operators as spectral integrals. For the application to scattering theory of spectral integrals of operator families see Birman and Solomjak (1969), Pearson (1971), Amrein, Georgescu and Jauch (1971) and Prugovecki (1981). The formulae for wave operators define a kind of Hilbert-space version of the Lippmann–Schwinger equation.

In Section 9.2 I have outlined a mathematical model of a scattering experiment within which the relation between the scattering operator and scattering cross-sections can be established. An important ingredient in this analysis is the scattering-into-cones formula of Dollard (1969); see also Jauch, Lavine and Newton (1972). For finiteness of total cross-sections see Amrein and Pearson (1979) and Enss and Simon (1980). Explicit bounds on cross-sections can be obtained, and extended to many-particle systems (Amrein, Pearson and Sinha, 1979). For more recent results on asymptotics of scattering cross-sections see Yafaev (1986). A number of authors have proved continuity of cross-sections and scattering amplitudes as a function of energy; see, for example Davies (1980). In Section 9.4 I have proved continuity of the scattering amplitude as a function of incoming and outgoing momenta, under fairly strong conditions on the potential.

In setting up the foundations of scattering theory, precise definitions are needed of the subspaces of bound states and scattering states; see the papers of Ruelle (1969), Amrein and Georgescu (1973), Wilcox (1973) and Dollard (1977). The proof of the Asymptotic Decomposition Theorem in Chapter 13, which also incorporates the subspace of absorbed states, follows these developments; see also Amrein, Pearson and Wollenberg (1980).

For scattering by short-range potentials, the theorem also justifies the description of scattering states by means of state vectors within the absolutely continuous subspace for H; see also Sinha (1977) and Schechter (1980), both of whom refer also to the subspace of singular continuity.

14.3 FOUNDATIONS OF SPECTRAL THEORY

For scattering theory the spectral analysis of ordinary and partial differential operators is of particular importance. There is a large and growing literature on Schrödinger operators; recent references are Graffi (1984) and Cycon, Froese, Kirsch and Simon (1986); see also Jorgens (1970), Jorgens and Weidmann (1973) and Eastham and Kalf (1982). The basic results on eigenfunction expansions for the Schrödinger operator in one dimension (Chapters 6 and 7) stem from the fundamental work of Kodaira (1949, 1950). In Dunford and Schwartz (1958) some of these results are generalized. See also Coddington and Levinson (1955) for a treatment of the limit-point/limit-circle criterion. The key notion of subordinacy is due to Gilbert (1984), and has been further developed in Gilbert and Pearson (1987); see also Pearson (1982). In three dimensions the theory of localization and singular points is a development of ideas in Pearson (1984). For many-particle scattering the basic spectral result is Hunziker's Theorem, which relates, under suitable conditions on the pair potentials, the essential spectrum of H to binding energies in the scattering channels. This result follows easily from the existence of channel wave operators and asymptotic completeness, but can also be proved under much weaker assumptions. See Hunziker (1966) and Enss (1977).

A necessary preliminary to carrying out a spectral analysis of the free and total Hamiltonians is the identification of domains on which these operators are self-adjoint. There is a vast literature on the question of essential self-adjointness for Schrödinger operators; the reader should consult, for example, the volumes of Reed and Simon (1972–79) and the book of Kato (1976). The basic result on self-adjoint extensions of sums of (possibly) unbounded operators is the Rellich–Kato Theorem, for which see also Amrein, Jauch and Sinha (1977). A straightforward introduction to the theory of form extensions is to be found in Riesz and Sz-Nagy (1965), and for applications to quantum mechanics and scattering theory see Simon (1971). For references to essential self-adjointness of Schrödinger operators with singular potentials see, for example, Kato (1972), Schmincke (1972), Kalf and Walter (1972), Simon (1974) and Cycon (1981).

14.4 SCATTERING OF PARTICLES AND SYSTEMS OF PARTICLES

There are a number of excellent basic reference books on scattering theory, including Goldberger and Watson (1964), de Alfaro and Regge (1965), Newton (1966), Taylor (1972) and Joachain (1975). Books with greater emphasis on more mathematical aspects of the theory include Amrein, Jauch and Sinha (1977), Reed and Simon (1979), Amrein (1981), Baumgartel and Wollenberg (1981) and Thirring (1981)—the latter also covers the physics of atoms and molecules.

Having established a theoretical framework for scattering, there remains the task of showing how scattering of one or more particles interacting through potentials in non-relativistic quantum mechanics can be accommodated within this framework. In practice, this entails proving the existence of appropriate wave operators, establishing asymptotic completeness under certain conditions and investigating any possible breakdown of completeness, determining scattering amplitudes and phase shifts, estimating the total cross-section if it is finite, studying the effect of the potential on spectral properties, and so on.

The basic idea of proving the existence of wave operators by means of a norm estimate of $Ve^{-iH_0t}f$ is due to Cook (1957), and has been developed further by Schechter (1977b) and Simon (1977). Cook's method was originally applied to $V \in L^2(\mathbb{R}^3)$, and later extended to $V = O(|\boldsymbol{r}|^{-1-\varepsilon})$ by Hack (1958). Similar results were obtained by Kuroda (1959), who also proved asymptotic completeness in the case $V = V(|\boldsymbol{r}|)$, and in the non-central cases of $V \in L^1(\mathbb{R}^3) \cap L^2(\mathbb{R}^3)$. The simplest and most physically motivated proof of completeness in the non-central case under the assumption $V = O(|\boldsymbol{r}|^{-1-\varepsilon})$ is due to Enss, for which see the following section. Many extensions, too numerous to quote, have been made by various authors; for example the incorporation of magnetic fields, the use of pseudodifferential operators, and the extension of most of the results to n dimensions.

Following the paper of Dollard (1964), wave operators have been defined and shown to exist for a class of potentials having arbitrarily slow power decay at infinity; see, for example, Hörmander (1976) and Buslaev and Matveev (1970). Again, the method of Enss (Section 14.5) has led to simple and powerful proofs of asymptotic completeness for long-range potentials. The proof given in Chapter 12 uses a combination of commutator estimates and the Cook method, and is strongly motivated by the work of Enss.

The first general existence proof of wave operators for scattering by singular potentials is due to Kupsch and Sandhas (1966), and involves the use of smooth cutoff functions, which are so much a feature of the techniques employed throughout this book. The introduction of these cutoff functions under suitable conditions to decouple local singularities has been followed up by Deift and Simon (1976) and Combescure and Ginibre (1978), following a paper of Pearson (1975a). Local domain properties for the Hamiltonian (cf. Chapter 5) used in the latter work are due essentially to Ikebe and Kato (1962). For the completeness proof with repulsive singular potentials see Amrein and Georgescu (1974) and Pearson (1974).

In Chapter 8 I have given a fairly complete account of the theory of scattering by a central potential in a single partial wave. The analysis is based on a fundamental paper of Green and Lanford (1960); see also Kuroda (1962). I have extended the results of this paper in two directions. The first is to admit potentials that may be of long range. The second is to allow for potentials that may be highly singular at the origin; I have used the notion of subordinacy to define an appropriate eigenfunction expansion, which leads to explicit formulae for scattering phase shifts under very general conditions. For a general account of scattering theory in angular-momentum subspaces see the monograph by de Alfaro and Regge (1965). Our use of upper solutions with normalization of amplitude is related to the standard definition of Jost functions (Jost, 1947; Bargmann, 1949), which, however, has been considerably generalized in our application.

The Lippmann-Schwinger equation (Chapter 9; Lippmann and Schwinger, 1950) is really a version in three-dimensional space of the spectral integral formula for wave operators derived in Chapter 4. The equation is often used by physicists as the starting point for a stationary approach to scattering theory. Here the equation follows from resolvent estimates for the Hamiltonian, obtained by commutator methods in Chapter 12.

The existence of wave operators for many-particle systems was first proved by Hack (1959) and extended by Hunziker (1967) to allow singular potentials. The first proof of asymptotic completeness for the three-particle system is due to Faddeev (1965); see also Thomas (1975) and Ginibre and Moulin (1974). Completeness for three-particle and many-particle systems has also been proved by the Enss method. Using this in combination with other techniques, the strongest results available so far in the case of short-range n-particle systems appear to be those of Sigal and Soffer (1986); see also Sigal (1983).

14.5 TECHNIQUES OF SCATTERING THEORY

This book has drawn on a number of techniques of scattering theory that have been used in this field and in related fields over the years. Of these, the following are particularly worthy of further note.

(i) The trace method

We have used trace theory extensively in Chapter 11. The earlier result that $\Omega_{\pm}(H_1, H_2)$ exists for $H_1 - H_2$ of trace class is to be found in Rosenblum (1957), with the restriction that H_2 must have a purely absolutely continuous spectrum. This restriction was removed by Kato (1957); see also Kato (1976). A generalization that is close to the trace theorem presented here is in Birman (1968). See also Birman and Belopolskii (1968, 1969). The proof of the result in Pearson (1978) benefited from considerable simplifications following suggestions of Ginibre and Kato. In addition to applications to scattering of a single particle, the trace method has also been used (Combes, 1969; Deift and Simon, 1977) to prove asymptotic completeness for three-particle bound-state scattering at energies below the threshold for three-particle production. For a comprehensive treatment of trace theory and applications see Simon (1979). For applications to cross-sections see, for example, Martin and Misra (1973).

(ii) Smoothness and commutator methods

Although I have made no explicit reference to the theory of smooth operators, such ideas are closely connected with norm estimates of $\|Te^{-iHt}f\|$ as a function of t in $L^2(\mathbb{R})$, which were considered in Chapter 12. The basic theory was first established by Kato (1966), and led in Kato (1968) to an important relation between smoothness and commutators. These ideas were subsequently exploited in scattering theory by Lavine (1971, 1972), and applied to many-particle scattering with repulsive potentials in Lavine (1973). The derivation of resolvent estimates in Chapter 12 relies heavily on the later development of commutator estimates by Mourre (1981, 1983). The Mourre theory has been extended and applied by other authors, including Perry, Sigal and Simon (1981), Froese and Herbst (1982), Froese, Herbst and Hoffmann-Ostenhoff (1982), and Jensen, Mourre and Perry (1984).

(iii) The Enss method

I have made considerable use in Chapter 12 of the Enss method in the time-dependent theory of scattering. The method is based on a detailed phase-space or "geometrical" analysis of the asymptotic evolution of states in quantum mechanics. The original paper was Enss (1978), and contained a geometrical proof of asymptotic completeness for short-range potentials. The method was subsequently extended in Enss (1979a) to the long-range problem, and in Enss (1979b) to cluster scattering of charged particles. For the general three-particle problem see Enss (1984). The link between the Enss method and the generator of dilations, of which much use has been made in this book, was established by Mourre (1979). For an excellent treatment of scattering theory by the Enss method see Perry (1983).

(iv) Stationary methods

In regarding scattering theory as primarily a study of the asymptotic evolution of states in the limits of large positive and negative times, it is natural that I should choose in this book a time-dependent rather than stationary approach to the foundations of the subject. The two approaches are, however, complementary, and in making use, say, of the method of eigenfunction expansions, we are making use of ideas drawn from the stationary method. Indeed, in establishing in Chapter 4 expressions for the wave operators as spectral integrals, we have seen how to move from time-dependent to stationary formulae. The many contributions to scattering theory that have used stationary methods include, for example, Ikebe (1960), Alsholm and Schmidt (1971), Agmon (1971, 1975), Kato (1969), Kuroda (1973a,b,), Kato and Kuroda (1970) and Agmon and Hörmander (1976).

14.6 SPECTRAL AND SCATTERING PHENOMENA—SOME EXAMPLES

(a) A simple one-dimensional scattering problem

Most elementary quantum-mechanics textbooks deal with examples of scattering in one dimension. For the underlying theory of scattering in one dimension with potentials decaying more rapidly at infinity than $|x|^{-1}$ see Chapter 9. The long-range case can also be handled by the methods of that chapter, and the reader may find it instructive to work out the theory in that case.

The simplest example is that of a square-well potential. Let us take

$$V(x)=\begin{cases}0 & (|x|>a),\\ -V_0 & (|x|\leq a),\end{cases} \tag{14.6.1}$$

where V_0 is some positive constant. To determine transmission and reflection coefficients $T(k)$ and $R(k)$ at energy $\lambda = k^2$, with $k > 0$, it is necessary to satisfy the Schrödinger equation at this energy for $\psi_-(x, k)$, subject to the conditions

$$\psi_- = \begin{cases} e^{ikx} + Re^{-ikx} & (x < -a), \\ \alpha e^{ik_0 x} + \beta e^{-ik_0 x} & (-a \le x \le a), \\ Te^{ikx} & (x > a), \end{cases} \tag{14.6.2}$$

where $k_0 = (k^2 + V_0)^{1/2}$. The values of α, β, R and T may be determined from the continuity of ψ_- and ψ'_- at $x = \pm a$. For justification of the continuity of ψ_- and ψ'_- see Chapter 6. We then find

$$T(k) = 4k_0 k\, e^{-2ika}[(k + k_0)^2 e^{-2ik_0 a} - (k - k_0)^2 e^{2ik_0 a}]^{-1},$$

$$R(k) = (k^2 - k_0^2)\, e^{-2ika} (e^{-2ik_0 a} - e^{2ik_0 a})$$
$$[(k + k_0)^2 e^{-2ik_0 a} - (k - k_0)^2 e^{2ik_0 a}]^{-1}.$$

The reader should verify that $|R(k)|^2 + |T(k)|^2 = 1$, confirming the interpretation of $|R|^2$ and $|T|^2$ as reflection and transmission probabilities.

It is often useful to introduce the idea of *transfer matrices*. For solutions of the differential equation

$$-\psi'' + V(x)\psi = k^2\psi \quad (k > 0)$$

the transfer matrix $M(k, \mathscr{I})$ across an interval $\mathscr{I}$ is defined to be the 2×2 matrix which sends the column vector $\left[\begin{smallmatrix}\psi \\ \psi'\end{smallmatrix}\right]$ at the left-hand endpoint of the interval into the corresponding column vector at the right-hand endpoint. In the example above the transfer matrix $M_a(k)$ for the interval $[-a, a]$ must satisfy

$$M_a(k)\begin{bmatrix} \psi(-a) \\ \psi'(-a) \end{bmatrix} = \begin{bmatrix} \psi(a) \\ \psi'(a) \end{bmatrix},$$

from which it follows easily that

$$M_a(k) = \begin{bmatrix} \cos 2k_0 a & k_0^{-1} \sin 2k_0 a \\ -k_0 \sin 2k_0 a & \cos 2k_0 a \end{bmatrix}. \tag{14.6.3}$$

Note that $\det M(k) = 1$, a property that holds for arbitrary potentials $V(x)$, and which may be derived from the fact that the Wronskian of any pair of solutions is constant. Note also, as we should expect, that $\lim_{a \to 0} M_a(x) = I$, where I is the 2×2 identity matrix.

Let us, however, now define a *family* of potentials $V_a(x)$ by the formula

$$V_a(x) = \begin{cases} 0 & (|x| > a), \\ -V_0/2a & (|x| \le a), \end{cases} \tag{14.6.4}$$

and let $M_a(k, V_a)$ denote the transfer matrix for the Schrödinger equation with potential $V_a(x)$, across the interval $[-a, a]$. We then have,

$$M_a(k,V_a)=\begin{bmatrix} \cos 2a\left(k^2+\dfrac{V_0}{2a}\right)^{1/2} & \left(k^2+\dfrac{V_0}{2a}\right)^{-1/2}\sin 2a\left(k^2+\dfrac{V_0}{2a}\right)^{1/2} \\ -\left(k^2+\dfrac{V_0}{2a}\right)^{1/2}\sin 2a\left(k^2+\dfrac{V_0}{2a}\right)^{1/2} & \cos 2a\left(k^2+\dfrac{V_0}{2a}\right)^{1/2} \end{bmatrix}.$$

Taking the limit $a \to 0$ in this case, we find

$$\lim_{a\to 0} M_a(k,V_a)=\begin{bmatrix} 1 & 0 \\ -V_0 & 1 \end{bmatrix}. \tag{14.6.5}$$

The family of potentials defined by (14.6.4) represents a δ-convergent family, in the sense that $V_a(x)$ approaches $-V_0\delta(x)$, where δ is the Dirac delta function, in the sense of distributions. Note that $\int_{-\infty}^{\infty} V_a(x)\mathrm{d}x = -V_0$, whereas $V_a(x)$ converges uniformly to zero in the complement of any neighbourhood of the origin. The right-hand side of (14.6.5) then represents the transfer matrix across the singularity of a δ-function. In general, if $V(x)=g\delta(x)$ then there is a transfer matrix $\left[\begin{smallmatrix} 1 & 0 \\ g & 1 \end{smallmatrix}\right]$ at $x=0$, corresponding to the fact that ψ is continuous and there is a discontinuity $g\psi$ in ψ' at $x=0$. The Schrödinger Hamiltonian $H=-\mathrm{d}^2/\mathrm{d}x^2+g\delta(x)$ may be defined by the method of sesquilinear forms as a self-adjoint operator, and scattering theory is well defined for this potential. (For further details see Simon 1971.) The reader may confirm, by taking the limit $a+0$ in the example above with $V=V_a$, that limiting values are obtained for the transmission and reflection coefficients, namely

$$T(k)=2k(2k-\mathrm{i}V_0)^{-1}, \qquad R(k)=\mathrm{i}V_0(2k-\mathrm{i}V_0)^{-1}.$$

These are the same coefficients as could be obtained by solving the problem directly, with appropriate boundary conditions at $x=0$.

(b) Some limiting transfer matrices

Following our example of a limiting family of potentials $V_a(x)$ such that the transfer matrix approaches that corresponding to a δ-function potential, it is natural to consider the question as to whether we can define other families of potentials $V_a(x)$, converging uniformly to zero in the limit $a \to 0$, in the complement of any neighbourhood of the origin, and such that other transfer matrices are obtained in this limit? In fact a variety of limiting transfer matrices can be obtained in this way.

We start from the transfer matrix across the interval $[0, a]$ for the potential $V_1(x)=g(a)\delta(x-a)$. This potential corresponds to a δ-function at $x=a$ of strength g. We write $g=g(a)$ to indicate that we deal with a limiting family of potentials in which the value of g is allowed to vary with a. In defining a transfer matrix across $[0, a]$ we shall take that matrix which sends $\left[\begin{smallmatrix}\psi\\ \psi'\end{smallmatrix}\right]$ at $x=0$, for solutions ψ of the Schrödinger equation with potential $V_a(x)$, into the corresponding two-component vector at $x=a+$, that is just to the *right* of the singularity at $x=a$. In this case the transfer matrix is given by the product

$$\begin{bmatrix} 1 & 0 \\ g(a) & 1 \end{bmatrix}\begin{bmatrix} \cos ka & k^{-1}\sin ka \\ -k\sin ka & \cos ka \end{bmatrix}$$

$$=\begin{bmatrix} \cos ka & k^{-1}\sin ka \\ g(a)\cos ka - k\sin ka & g(a)k^{-1}\sin ka+\cos ka \end{bmatrix}$$

We shall take $g=g_1(a)=a^{-3/2}-a^{-1}$ and $g=g_2(a)=a^{-1/2}-a^{-1}$, giving rise to transfer matrices $M_a^{(1)}(k)$ and $M_a^{(2)}(k)$. Since we are interested in the limit $a\to 0$, we can expand the sine and cosine functions as series in increasing powers of a, to obtain

$$M_a^{(1)}=\begin{bmatrix} 1+O(a^2) & a+O(a^3) \\ a^{-3/2}-a^{-1}-\tfrac{1}{2}k^2a^{1/2}+O(a) & a^{-1/2}-\tfrac{1}{6}k^2a^{3/2}+O(a^2) \end{bmatrix}; \tag{14.6.6}$$

$$M_a^{(2)}=\begin{bmatrix} 1-\tfrac{1}{2}k^2a^2+O(a^{5/2}) & a+O(a^3) \\ a^{-1/2}-a^{-1}-\tfrac{1}{2}k^2a+O(a^{3/2}) & a^{1/2}-\tfrac{1}{3}k^2a^2+O(a^{5/2}) \end{bmatrix}. \tag{14.6.7}$$

Evaluation of matrix products then gives

$$M_a^{(1)}M_a^{(2)}=\begin{bmatrix} a^{1/2}+O(a^2) & a+O(a^{3/2}) \\ -\tfrac{4}{3}k^2a^{1/2}+O(a) & a^{-1/2}+O(a^{3/2}) \end{bmatrix}, \tag{14.6.8}$$

$$M_a^{(2)}M_a^{(1)}=\begin{bmatrix} a^{-1/2}+O(a^{3/2}) & a^{1/2}+O(a) \\ -\tfrac{1}{3}k^2a^{1/2}+O(a) & a^{1/2}+O(a^{3/2}) \end{bmatrix}. \tag{14.6.9}$$

Multiplying together the matrices (14.6.8) and (14.6.9) in two orders gives

$$M_a^{(1)}(M_a^{(2)})^2M_a^{(1)}=\begin{bmatrix} 1+O(a^{3/2}) & O(a) \\ -\tfrac{5}{3}k^2+O(a^{1/2}) & 1+O(a) \end{bmatrix}, \tag{14.6.10}$$

$$M_a^{(2)}(M_a^{(1)})^2M_a^{(2)}=\begin{bmatrix} 1+O(a) & 1+O(a^{1/2}) \\ O(a) & 1+O(a) \end{bmatrix}. \tag{14.6.11}$$

We can take one further matrix product, to obtain

$$M_a^{(2)}(M_a^{(1)})^2M_a^{(2)}M_a^{(1)}(M_a^{(2)})^2M_a^{(1)}=\begin{bmatrix} 1-\tfrac{5}{3}k^2+O(a^{1/2}) & 1+O(a^{1/2}) \\ -\tfrac{5}{3}k^2+O(a^{1/2}) & 1+O(a) \end{bmatrix} \tag{14.6.12}$$

The transfer matrix across $[na, (n+1)a]$ for the potential $V(x-na)$ is the same as the transfer matrix across $[0, a]$ for the potential $V(x)$, and since transfer matrices across the union of successive intervals are to be obtained by matrix multiplication, we may interpret each of the matrices in (14.6.10)–(14.6.12). Thus in (14.6.10) we have the transfer matrix across $[0, 4a]$ for the potential

$$V_a(x)=g_1\delta(x-a)+g_2\delta(x-2a)+g_2\delta(x-3a)+g_1\delta(x-4a),$$

and a similar formula applies to the potential corresponding to (14.6.11). The transfer matrix on the right-hand side of (14.6.12) corresponds to the one-parameter family of potentials

$$\begin{aligned}V_a(x)=&g_1\delta(x-a)+g_2\delta(x-2a)+g_2\delta(x-3a)+g_1\delta(x-4a)\\&+g_2\delta(x-5a)+g_1\delta(x-6a)+g_1\delta(x-7a)+g_2\delta(x-8a).\end{aligned} \tag{14.6.13}$$

In the limit $a\to 0$ the transfer matrix across $[0, 8a]$ approaches

$$M(k)=\begin{bmatrix}1-\frac{5}{3}k^2 & 1\\ -\frac{5}{3}k^2 & 1\end{bmatrix}, \tag{14.6.14}$$

whereas from (14.6.10) and (14.6.11) we derive the limiting transfer matrices

$$\begin{bmatrix}1 & 0\\ -\frac{5}{3}k^2 & 1\end{bmatrix}, \quad \begin{bmatrix}1 & 1\\ 0 & 1\end{bmatrix}$$

respectively. Since as in (a) above, we may regard δ-potentials as themselves obtained from limits of δ-convergent potentials, it is possible to construct limiting families of potentials $V_a(x)$, where V_a, for fixed a, is bounded (even continuous or differentiable), such that the three limiting transfer matrices above are obtained as $a\to 0$. In each case, the support of the potential $V_a(x)$ reduces in the limit to the single point $x=0$, and the limiting transfer matrix

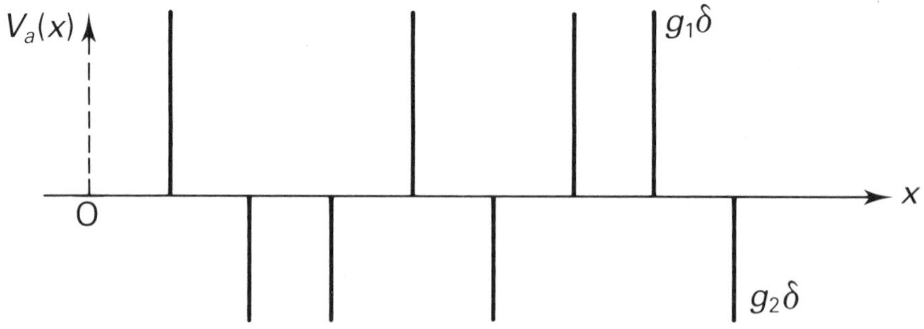

may be regarded as the transformation matrix from $x=0-$ to $x=0+$ for $\left[\begin{smallmatrix}\psi\\ \psi'\end{smallmatrix}\right]$. (Note, however, that in these cases, in contrast with the δ-potential the limiting potential does not define a generalized function or distribution.) By modifying these examples, we can arrive at a large class of limiting transfer matrices, including, for example, any transfer matrix M that is independent of k and for which $\det M=1$. The diagram represents schematically the shape of the potential V_a in the case (14.6.13)

(c) A Schrödinger operator in a finite interval, with an absolutely continuous spectrum

Under very general conditions, a Schrödinger operator defined in $L^2(0,c)$, where the interval $(0,c)$ is finite, will have a purely discrete spectrum (cf. Section 7.1). This will be so not only if $V(x)$ is in the limit-circle case at 0 and c, but also much more generally even if $V(x)$ is singular at either or both endpoints. The existence of an absolutely continuous spectrum for a Schrödinger operator in a finite interval may be regarded as an exceptional phenomenon, and we have to work quite hard to achieve it. For further details of the following example see Pearson (1975b).

Let us define the potential $V_a(x)$ by (14.6.13), with, as before,

$$g_1=g_1(a)=a^{-3/2}-a^{-1}, \qquad g_2=g_2(a)=a^{-1/2}-a^{-1}.$$

and, given a sequence $\{a_j\}, j=1,2,3,\ldots$, converging rapidly to zero as $j\to\infty$, let us define in $L^2(0,c)$ the operator $H=-\mathrm{d}^2/\mathrm{d}x^2+V(x)$, where

$$\begin{aligned} V(x)=V_{a_1}(x)+V_{a_2}(x-8a_1)+V_{a_3}(x-8(a_1+a_2)) \\ +V_{a_4}(x-8(a_1+a_2+a_3))+\ldots, \end{aligned} \tag{14.6.15}$$

and $c=\Sigma_{j=1}^{\infty}8a_j$. The potential $V(x)$ is constructed out of a sequence of δ-functions. The first block, consisting of eight δ-functions, is given by (14.6.13) with $a=a_1$. The second block of eight δ-functions is given by (14.6.13) with $a=a_2$, and shifted by $8a_1$ to the right, so that this block follows immediately after the first. The third block of δ-functions follows the second, with $a=a_3$ and so on.

The transfer matrix across the interval $[0,8a_1]$ is given in the limit $a_1\to 0$ by the right-hand side of (14.6.12), with $a=a_1$. The transfer matrix over $[0,8(a_1+a_2)]$ is the product of two such matrices, with $a=a_2$ and $a=a_1$ respectively. For the interval $[0,8(a_1+a_2+a_3)]$ we have a product of three matrices with $a=a_3$, $a=a_2$ and $a=a_1$ in that order. The asymptotic formula (14.6.14) for the single transfer matrix in the limit $a\to 0$ allows us to estimate the product of any number of these matrices, for the decreasing sequence $a=a_j$. The resulting transfer matrix across the interval $[0,c_j]$, where $c_j=\Sigma_{i=1}^{j}8a_i$ may thus be shown to have the asymptotic form, as $j\to\infty$

and $c_j \to c$.

$$M(k, c_j) \sim (M(k))^j W(k), \tag{14.6.16}$$

for some 2×2 matrix $W(k)$, which may be defined to be the limit $\lim_{j \to \infty} (M(k))^{-j} M(k, c_j)$. In proving (14.6.16) we obtain at the same time convergence of derivatives with respect to k. For example, $(d/dk)(M(k)^{-j} M(k, c_j))$ converges uniformly to $(d/dk)W(k)$ for k in closed bounded subintervals of $\mathbb{R}$. We can therefore use (14.6.16) to estimate derivatives with respect to k of the transfer matrix $M(k, c_j)$.

We now let $\psi(x, k_1)$ and $\psi(x, k_2)$ be solutions of the Schrödinger equation at energies k_1^2 and k_2^2 respectively. From the identity

$$\frac{d}{dx}(\psi(x, k_1)\psi'(x, k_2) - \psi(x, k_2)\psi'(x, k_1)) = (k_1^2 - k_2^2)\psi(x, k_1)\psi(x, k_2).$$

we have, for solutions subject to given initial values $\psi(0, k)$ and $\psi'(0, k)$,

$$\int_0^{c_j} \psi(x, k_1)\psi(x, k_2)\, dx$$

$$= (k_1^2 - k_2^2)^{-1}(\psi(c_j, k_1)\psi'(c_j, k_2) - \psi(c_j, k_2)\psi'(c_j, k_1)).$$

Taking the limit $k_1 \to k_2$ then gives

$$\int_0^{c_j} (\psi(x, k))^2\, dx = \psi'(c_j, k)\frac{\partial}{\partial k^2}\psi(c_j, k) - \psi(c_j, k)\frac{\partial}{\partial k^2}\psi'(c_j, k) \tag{14.6.17}$$

Equation (14.6.17) shows that norms in $L^2(0, c_j)$ of solutions $\psi(x, k)$, subject to initial values of ψ and ψ' at $x = 0$, may be evaluated from derivatives of ψ and ψ' with respect to energy k^2. The derivatives may be estimated, for large j, by differentiating the asymptotic formula (14.6.16) with respect to k^2. To do this, we consider the range of energies $0 \le k^2 \le \frac{12}{5}$, for which $M(k)$ has complex eigenvalues $e^{\pm i\alpha(k)}$. In this case, the jth power of the matrix $M(k)$ may be evaluated to give

$$(\sin \alpha)(M(k))^j = \begin{bmatrix} (1 - \frac{5}{3}k^2)\sin j\alpha - \sin(j-1)\alpha & \sin j\alpha \\ -\frac{5}{3}k^2 \sin j\alpha & \sin j\alpha - \sin(j-1)\alpha \end{bmatrix}. \tag{14.6.18}$$

Asymptotically, as $j \to \infty$, we have

$$\begin{bmatrix} \psi(c_j, k) \\ \psi'(c_j, k) \end{bmatrix} = M(k, c_j)\begin{bmatrix} \psi(0, k) \\ \psi'(0, k) \end{bmatrix} \sim (M(k))^j \begin{bmatrix} w_1(k) \\ w_2(k) \end{bmatrix},$$

where we have used (14.6.16) and substituted

$$\begin{bmatrix} w_1(k) \\ w_2(k) \end{bmatrix} = W(k) \begin{bmatrix} \psi(0,k) \\ \psi'(0,k) \end{bmatrix}.$$

In (14.6.18), α and k^2 are related by the formula $\cos\alpha = 1 - \frac{5}{6}k^2$. Carrying out the differentiation with respect to k^2, with $\sin\alpha\, \mathrm{d}/\mathrm{d}k^2 = \frac{5}{6}\mathrm{d}/\mathrm{d}\alpha$, introduces asymptotically a factor j in the norm estimate, and we may use (14.6.17) to show that

$$\lim_{j\to\infty} j^{-1} \int_0^{c_j} (\psi(x,k))^2\, \mathrm{d}x$$

exists for any solution $\psi(x,k)$ with $0 \le k^2 \le \frac{12}{5}$. The limit can be evaluated explicitly (Pearson, 1975b) in terms of $W(k)$ and the initial values of ψ and ψ'. Here we shall need only that the limit is non-zero except when ψ is the trivial solution.

These arguments show that the norm in $L^2(0, c_j)$ as $c_j \to c$ looks asymptotically like const $j^{1/2}$ for any non-trivial solution ψ and for some non-vanishing constant. It follows easily that, in the range of energies $(0, \frac{12}{5})$, there is no solution that is even pointwise-subordinate at $x = c$. At $x = 0$ the differential operator $H = -\mathrm{d}^2/\mathrm{d}x^2 + V(x)$, with V defined as above, is in the limit-circle case, and we take at this point the standard boundary condition $\psi(0) = 0$. Since the asymptotic analysis implies that no non-trivial solution is in $L^2(0, c)$, $x = c$ is limit-point, and the absence of pointwise subordinate solutions at this endpoint implies, from the results of Chapter 7, that the spectrum of H is purely absolutely continuous in the interval $(0; \frac{12}{5})$.

Although the potential $V(x)$ above is a combination of δ-functions, each δ-singularity may be replaced by a suitable δ-approximating function, which is bounded and may even be chosen to be continuous or differentiable. This can be done while maintaining the same asymptotic behaviour of transfer matrices (for details see Pearson). In this way, we can define a potential $V(x)$, bounded and continuous in any compact subset of $[0, c)$, but highly oscillatory and unbounded in the neighbourhood of the endpoint $x = c$, such that $H = -\mathrm{d}^2/\mathrm{d}x^2 + V(x)$, as a self-adjoint operator in $L^2(0, c)$, has non-trivial absolutely continuous subspace.

(d) A Schrödinger operator in $[0, \infty)$ with a singular continuous spectrum

We let $\{a_j\}, j = 1, 2, 3, \ldots$, be a sequence of positive numbers converging rapidly to infinity as $j \to \infty$, and define in $L^2(0, \infty)$ the operator

$H = -\mathrm{d}^2/\mathrm{d}x^2 + V(x)$, where

$$V(x) = g_1\delta(x-a_1) + g_2\delta(x-a_1-a_2) + g_3\delta(x-a_1-a_2-a_3) + \dots \quad (14.6.19)$$

The sequence $\{g_j\}$ of real numbers is chosen arbitrarily, except that we require $\Sigma_{j=1}^{\infty} g_j^2 = \infty$, and for simplicity we suppose that g_j is bounded. Thus one possibility is that $g_j = 1$ for all j. At $x=0$ we take the standard boundary condition $\psi(0)=0$. The potential $V(x)$ defined by (14.6.19) is a sequence of δ-functions, separated by gaps a_j that increase rapidly for large values of j. We define $c_j = \Sigma_{i=1}^{j} a_i$. We let $\phi \in C_0^{\infty}(0,1)$ be normalized, and define

$$\phi_j(x) = a_{j+1}^{-1/2} \mathrm{e}^{ikx} \phi((x-c_j)/a_{j+1}).$$

Then it is easy to verify that ϕ_j is normalized, and that the support of ϕ_j is contained in the interval (c_j, c_{j+1}). Hence $V(x)=0$ on the support of ϕ_j.

A straightforward evaluation of $(\mathrm{d}^2/\mathrm{d}x^2 + k^2)\phi_j$ shows that $\|(H-k^2)\phi_j\| = O(a_{j+1}^{-1}) \to 0$ as $j\to\infty$. Hence $H-k^2$ cannot have a bounded inverse, which would imply that $\phi_j \to 0$ strongly. It follows that k^2 belongs to the spectrum of H. So the spectrum of H contains the entire positive real axis. We shall see that this positive spectrum is singular continuous, for any choice of the sequence $\{a_j\}$ that increases sufficiently rapidly.

We now let $\psi(x,k)$ satisfy the Schrödinger equation at energy k^2, subject to given initial conditions at $x=0$. We shall define $\psi(c_j,k)$ and $\psi'(c_j,k)$ to be the values of ψ and ψ' just to the *right* of the singularity $x=c_j$. Then

$$\begin{bmatrix}\psi(c_j,k)\\ \psi'(c_j,k)\end{bmatrix} = \begin{bmatrix}\cos ka_j & k^{-1}\sin ka_j\\ g_j\cos ka_j - k\sin ka_j & \cos ka_j + g_j k^{-1}\sin ka_j\end{bmatrix}\begin{bmatrix}\psi(c_{j-1},k)\\ \psi'(c_{j-1},k)\end{bmatrix} \quad (14.6.20)$$

We introduce polar coordinates (R_j, θ_j) to describe the variation with j of ψ and ψ' at $x=c_j$, and we let

$$\left.\begin{aligned}\psi(c_j,k) &= k^{-1}R_j\cos\theta_j,\\ \psi'(c_j,k) &= R_j\sin\theta_j.\end{aligned}\right\} \quad (14.6.21)$$

Substituting for ψ and ψ' on the right-hand side of (14.6.20) in terms of R_{j-1} and θ_{j-1}, we may evaluate $R_j^2 = \psi'^2 + k^2\psi^2$ at $x=c_j$ to give, on simplifying.

$$\begin{aligned}R_j^2 &= R_{j-1}^2\cos^2(ka_j - \theta_{j-1})\\ &\quad + [g_j k^{-1} R_{j-1}\cos(ka_j - \theta_{j-1}) - R_{j-1}\sin(ka_j-\theta_{j-1})]^2.\end{aligned}$$

Hence

$$\frac{R_j^2}{R_{j-1}^2} = 1 + \frac{g_j^2}{2k^2} + \frac{g_j^2}{2k^2}\cos 2(ka_j - \theta_{j-1}) - \frac{g_j}{k}\sin 2(ka_j - \theta_{j-1}) \quad (14.6.22)$$

Equation (14.6.22) allows us to estimate the growth or decrease of R_j for large j. Note that $(d/dx)(\psi'^2+k^2\psi^2)=2\psi'(\psi''+k^2\psi)=0$ whenever $V=0$, so that

$$\psi'^2+k^2\psi^2=R_j^2 \quad \text{for } c_j \leq x < c_{j+1} \tag{14.6.23}$$

In this interval, we also have $\psi=k^{-1}R_j\cos k(x+\delta)$, so that the L^2 norm of ψ over the interval $c_j \leq x < c_{j+1}$ is given by

$$\|\cdot\|^2 \sim \tfrac{1}{2}k^{-2}R_j^2 a_{j+1} \tag{14.6.24}$$

as $j\to\infty$. We can now use (14.6.22) to estimate how R_j behaves for large j. The largest possible value of R_j^2/R_{j-1}^2 is

$$1+\frac{g_j^2}{2k^2}+\left[\frac{g_j^2}{k^2}+\left(\frac{g_j^2}{2k^2}\right)^2\right]^{1/2},$$

and the minimum possible value is

$$1+\frac{g_j^2}{2k^2}-\left[\frac{g_j^2}{k^2}+\left(\frac{g_j^2}{2k^2}\right)^2\right]^{1/2}=\left\{1+\frac{g_j^2}{2k^2}+\left[\frac{g_j^2}{k^2}+\left(\frac{g_j^2}{2k^2}\right)^2\right]^{1/2}\right\}^{-1}.$$

For k^2 strictly positive the right-hand side is bounded away from zero, since the g_j are bounded.

For large values of a_j the right-hand side of (14.6.22) is a rapidly oscillating function of k that alternates between values respectively greater than and less than unity. Note that θ_{j-1} is also a function of k, depending too on the values of the parameters $a_1, a_2, \ldots, a_{j-1}$. However, if a_j increases sufficiently rapidly, the oscillations of the sine and cosine functions in (14.6.22) will be controlled essentially by the term $2ka_j$ in the argument, giving a period of oscillation approximately πa_j^{-1}.

For a given value of k, we have $R_j\to\infty$ as $j\to\infty$ if and only if the infinite product $\Pi_{j=2}^{\infty} R_j^2/R_{j-1}^2$ gives $+\infty$. On taking logarithms, this requires the sum of the logarithms of the right-hand side of (14.6.22) to give $+\infty$. For the rapidly oscillating function of k in (14.6.22) in the limit of large a_j, we may define in a natural way an *average value* of the logarithm, through the formula

$$\frac{1}{2\pi}\int_0^{2\pi} dy \log(a+b\cos y+c\sin y)$$
$$=\log\left[\frac{a+(a^2-b^2-c^2)^{1/2}}{2}\right],$$

which leads to

$$\text{ave}\left(\log\left(\frac{R_j^2}{R_{j-1}^2}\right)\right)=\log\left(1+\frac{g_j^2}{4k^2}\right).$$

Since by hypothesis $\Sigma_j g_j^2=\infty$, it follows that

$$\sum_j \log\left(1+\frac{g_j^2}{4k^2}\right)=\infty$$

In Pearson (1978b), estimates of the averaged logarithm of the right-hand sides of expressions such as (14.6.22) are used to show that in fact $R_j\to\infty$ for *almost all* $k^2>0$. (Actually what is proved is slightly different, namely that $R_j\to\infty$ *in measure*; however, this implies that $R_j\to\infty$ almost everywhere if some subsequence of the c_j is taken.) On the other hand, in any interval there will be infinitely many values of k at which R_j does *not* diverge in the limit $j\to\infty$, and one may show that the spectral measure for the Hamiltonian is entirely concentrated on such values k. Even if $R_j\to 0$, we can use (14.6.24), choosing each a_{j+1} inductively, in order to show that there are no solutions $\psi(x,k)$ belonging to $L^2(0,\infty)$. It follows that the spectral measure for H is singular with respect to Lebesgue measure on the positive real line.

The above example may be further generalized. Indeed, in (14.6.19) one may replace each δ-function $\delta(x-c_j)$ by $\phi(x-c_j)$ for any function $\phi\in C_0^\infty(0,\infty)$, while still retaining singular continuity of the positive spectrum. (For details see Pearson (1978b.) This allows us to generate potentials $V(x)$ that approach zero in the limit $x\to\infty$, and for which the positive spectrum is singular continuous.

(e) A potential $V(r)$ leading to an eigenvalue embedded in the continuous spectrum

We let

$$\psi(r)=\frac{\sin r}{1+(2r-\sin 2r)^2},$$

and define $V(r)$ by the equation

$$-\frac{d^2\psi}{dr^2}+V(r)\psi=\psi.$$

Then $\psi\in L^2(0,\infty)$, and

$$V(r)=\frac{-8\sin 2r}{r}+O(r^{-2})\quad \text{as } r\to\infty.$$

The Hamiltonian $H = -\mathrm{d}^2/\mathrm{d}r^2 + V(r)$, in $L^2(0, \infty)$, has eigenvalue 1 embedded in the continuous spectrum, which extends along the positive real line. See Reed and Simon (1978). This potential is based on a celebrated example of von Neumann and Wigner (1929).

(f) Some strongly oscillatory potentials

There is a large body of literature on spectral and scattering theory for the Schrödinger operator with strongly oscillating potentials. See, for example, Combescure and Ginibre (1976). These potentials are to be distinguished from those of (c) and (d) above, in that in most cases $V(r)$ is highly oscillatory, but with smooth rather than irregular oscillations. It is an interesting phenomenon that such oscillations may actually improve the chances of the existence of (unmodified) wave operators. This happens already for potentials such as $V(r) = (\sin r)/r$. This looks like a long-range potential, but in fact with $H = H_0 + V$ and $H_0 = -\Delta$, the usual unmodified wave operators exist (and are complete); see Devinatz (1980) and Devinatz and Rejto (1983). More wildly oscillating potentials with local singularities are $V(r) = gr^{-\beta} \sin r^{\alpha}$; see Dollard and Friedman (1978). Here existence and completeness of wave operators can be proved for a range of values of α and β, and such examples have been produced in the non-spherically symmetric case. One can also treat potentials as oscillatory as $V(r) = \cos(\mathrm{e}^r)$. An even more curious example is $V(r) = \mathrm{e}^r \cos(\mathrm{e}^r)$. Here the essential spectrum of $H = -\Delta + V$ is along $[-\frac{1}{2}, \infty)$, and unmodified wave operators $\Omega_{\pm}(H, H_0 - \frac{1}{2})$ exist, with $H_0 = -\Delta$. Such examples can again be extended to non-central potentials; see Combescure (1980).

(g) A Schrödinger Hamiltonian that is (just) bounded below

The Hamiltonian $H = -\Delta - \frac{1}{4}r^{-2}$, in $L^2(\mathbb{R}^3)$, is bounded below; however, $-\Delta - g/4r^2$ is unbounded below for any $g > 1$; see Section 5.5. The semiboundedness property for Schrödinger operators defined as in (c) above is also highly unstable with respect to perturbations. For those examples we also have $-\mathrm{d}^2/\mathrm{d}x^2 + V(x)$ semibounded, but $-\mathrm{d}^2/\mathrm{d}x^2 + gV(x)$ is unbounded below for all $g > 1$.

(h) A Schrödinger Hamiltonian with gaps in the continuous spectrum

This is typically what happens in the case of periodic potentials. If $H = -\mathrm{d}^2/\mathrm{d}x^2 + g \cos x$ in $L^2(\mathbb{R})$ then the spectrum of H consists of infinitely many disjoint "bands", or non-overlapping subintervals of $\mathbb{R}$. This band structure for the spectrum is important in the theory of crystal lattices. For the spectral theory of periodic differential operators see Eastham (1973). The associated scattering theory is discussed by Thomas (1974a).

(i) A short-range potential for which the scattering operator is non-unitary

We let $-d^2/dx^2+V(x)$, acting in $L^2(0,c)$, be a Hamiltonian of the type considered in (c), which has absolutely continuous spectrum in the interval $(0,\frac{12}{5})$. We define a potential $W(r)$ by

$$W(r)=\begin{cases} V(c-r) & (0<r\le c),\\ 0 & (r>c).\end{cases}$$

Then $H_1=-d^2/dr^2-W(r)$ is limit-point at $r=0$ and at $r=\infty$, and hence uniquely defines a Hamiltonian in $L^2(0,\infty)$. The operator $-\Delta+W(|\boldsymbol{r}|)$ in L^2 of the region $|\boldsymbol{r}|<c$, with Dirichlet boundary conditions on $|\boldsymbol{r}|=c$, is unitarily equivalent, in the $l=m=0$ partial-wave subspace, to the Hamiltonian considered in (c) and therefore has a non-trivial absolutely continuous subspace. It follows that $H=-\Delta-W(|\boldsymbol{r}|)$ in $L^2(\mathbb{R}^3)$ gives rise to wave operators $\Omega_\pm(H,H_0)$ $(H_0=-\Delta)$ for which range $(\Omega_\pm)\ne\mathcal{M}_{ac}(H)$ (see Section 13.2). Thus strong asymptotic completeness fails for this Hamiltonian. In time-dependent scattering theory this breakdown may be attributed to the existence of states that are asymptotically absorbed in the limit $t\to+\infty$ or $t\to-\infty$. Further analysis (Pearson, 1975b) shows that one can go further and show that range $(\Omega_+)\ne$range (Ω_-). The scattering operator is then non-unitary.

If the restriction is removed that the potential be of short range, it is not difficult to define a single-particle Hamiltonian for which the wave operators fail to be complete. See, for example, Kato and Kuroda (1971).

Let us put $H=-\Delta+V(\boldsymbol{r})$ in $L^2(\mathbb{R}^3)$, where

$$V(\boldsymbol{r})=V(x,y,z)=V(x)+V(y)+V(z)+V(x-y)+V(y-z)+V(z-x),$$

and suppose that $-d^2/dx^2+V(x)$ has a negative eigenvalue and that $V(x)$, considered as a function of one variable, is of short range. The potential $V(\boldsymbol{r})$ is not of short range in three dimensions, because $V(\boldsymbol{r})$ does not decay to zero in all directions in the limit $|\boldsymbol{r}|\to\infty$, for example if $y,z\to\infty$ while x is held fixed.

With $H_0=-\Delta$, wave operators $\Omega_\pm(H,H_0)$ exist in this example. However, the wave operators are not asymptotically complete, in the sense that range $(\Omega_+)=$range (Ω_-). This is because there are additional asymptotic modes of evolution involving states that are bounded for a single coordinate, or relative coordinate. The Hamiltonian above is like that of a *three-particle* system, each particle moving in one dimension. The scattering theory for such a system is properly a many-channel theory. Additional scattering channels may be introduced, and the system treated within the framework of the general theory developed in Section 4.3.

(j) A long-range and a short-range potential for which the associated Hamiltonian has a singular continuous spectrum

For the long-range potential, we simply let $V(r)$ be as defined in (d) above. Then $H=-\Delta+V(|r|)$ has a singular continuous spectrum in the $l=0$ partial-wave subspace.

To construct a short-range potential giving rise to a singular continuous spectrum, it is necessary first of all to define a potential that combines some of the features of those in (c) and (d) respectively. We start with the potential $V(x)$ given by (14.6.15), with $c_j=\Sigma_{i=1}^{j} a_i$, $j=1, 2, 3, \ldots$, and let d_j, $j=1, 2, 3, \ldots$, be a subsequence of the c_j. That is, we let $d_j=c_{n_j}$, where we take $\{n_j\}$ to be a sequence of integers tending rapidly to infinity. We now define $W(x)$ by

$$W(x)=V(x)+\sum_{j=1}^{\infty} \delta(x-n_j) \quad (0<x<c)$$

$$=\sum_{i=1}^{\infty} 8a_i.$$

Then $W(x)$ is a perturbation of $V(x)$ by a series of δ-functions located at points d_j. If we measure the separation of two consecutive points d_j and d_{j+1} by means of the difference $n_{j+1}-n_j$ of the corresponding n-values, it is possible to regard the δs as becoming increasingly separated at large values of j.

The differential operator $-\mathrm{d}^2/\mathrm{d}x^2+V(x)$, in $L^2(0,c)$, has an absolutely continuous spectrum in the interval $(0,\frac{12}{5})$. Just as, in (d), the absolutely continuous spectrum of $-\mathrm{d}^2/\mathrm{d}x^2$ in $L^2(0,\infty)$ is converted into a singular continuous spectrum through the perturbation of the Hamiltonian by a series of δ-functions with rapidly increasing separation, so in the present context one may show that $H=-\mathrm{d}^2/\mathrm{d}x^2+W(x)$ has a singular continuous spectrum in $(0,\frac{12}{5})$, as an operator in $L^2(0,c)$. Thus the general effect of the δ-perturbations is to convert an absolutely continuous to a singular continuous spectrum.

Now let us define $V_{sc}(|\boldsymbol{r}|)$ by

$$V_{sc}(|\boldsymbol{r}|)=\begin{cases} W(c-|\boldsymbol{r}|)-\frac{6}{5} & (0\le|\boldsymbol{r}|\le c),\\ 0 & (|\boldsymbol{r}|>c).\end{cases} \tag{14.6.25}$$

The operator $H_{sc}=-\Delta+V_{sc}(|\boldsymbol{r}|)$ has a singular continuous spectrum in the interval $(-\frac{6}{5},0)$, in the $l=m=0$ partial-wave subspace. On the other hand, the spectrum in the interval $(0,\frac{6}{5})$ is purely absolutely continuous; it is a consequence of the spectral analysis of Schrödinger operators in Chapter 7 that, for short-range central potentials in a single partial wave, the positive spectrum is absolutely continuous. It is true, however, that the entire interval $(-\frac{6}{5},\frac{6}{5})$ consists of singular points, in the sense of Chapter 10, for the operator H_{sc}.

Detailed analysis of the above example may be carried out within the general framework of singular continuous operators established in Pearson (1978b.) By suitable modifications, the δ-functions may be replaced by locally bounded oscillations of the potential, which approximate the δ-functions as far as asymptotic limits of transfer matrices are concerned. For the use of the inverse method to obtain a singular continuous spectrum see Aronszajn (1957). For the inverse method in general see Gelfand and Levitan (1955), Chadan and Sabatier (1977).

(k) A simple example of a short-range potential giving rise to a singular point of the Hamiltonian

A number of the examples above define Hamiltonians with complete intervals of singular points, in the sense of the localization theory of Chapter 10. A much simpler example can be constructed as follows.

We let $u(r)=r^{-1}\sin^2 r^{-1}+r^{\beta}$, for some $\beta>1$. For some fixed λ_0, we define $V(r)$, $0<r<1$, by the equation

$$-\frac{d^2u(r)}{dr^2}+V(r)u(r)=\lambda_0 u(r),$$

with $V(r)=0$ for $r>1$. It is not difficult to verify that

$$\int_0^1 u^2(r)\,dr=\infty,\qquad \int_0^1\frac{dr}{u^2(r)}=\infty.$$

The function $u(r)$ is a solution, for $0<r<1$, of the Schrödinger equation at energy λ_0. A second solution of the Schrödinger equation in this interval is $u(r)\int_r^1 dt/u^2(t)$, and, because of the divergence of the second integral above, it follows that $u(r)$ is subordinate at $r=0$. Since $u\notin L^2(0,1)$ and u is subordinate, there can be no non-trivial solution in $L^2(0,1)$ at energy λ_0. This implies that λ_0 is a singular point for the Hamiltonian $H = -d^2/dr^2 + V(r)$ in $L^2(0, \infty)$. The Hamiltonian $-\Delta + V(|\mathbf{r}|)$ in $L^2(\mathbb{R}^3)$ will then have a singular point at energy λ_0. It is interesting to note, following the analysis of Chapter 13, that the scattering matrix $S(\lambda)$ for a potential like this with $\lambda_0 > 0$ is a *discontinuous* function of energy λ at $\lambda = \lambda_0$. With $\lambda_0 < 0$, similar examples of this give rise to an energy threshold at which there is a limit point of eigenvalues of the Hamiltonian. Such a concentration of eigenvalues is an unusual spectral feature for short-range Schrödinger operators, where under fairly general conditions one can prove that the essential spectrum is $[0, \infty)$. For an interesting analysis, using the inverse method, of a class of Hamiltonians for which there is an interval of singular points, in which $S(\lambda)$ is discontinuous, see Gilbert (1984).

(l) A class of Schrödinger Hamiltonians with dense point spectra

There has been considerable interest in recent years in the spectral analysis of Schrödinger operators with random potentials. This work has important applications to disordered systems; see Pastur (1980). The following example is typical of the kind of problem that can be treated.

We divide the real line $\mathbb{R}$ into subintervals $(n, n+1]$, and on each subinterval $(n, n+1]$ we let $V(x) = V_n$, where each V_n is a random variable, independently distributed for different n. For example, we may assign the constant value of V_n, in a given subinterval, to have uniform probability distribution over the interval $[0, 1]$. It can then be shown that, with probability one, the Hamiltonian $H = -d^2/dx^2 + V(x)$ in $L^2(\mathbb{R})$ has purely discrete spectrum, with eigenvalues dense in the positive real line.

Particular attention has been paid to the class of almost-periodic potentials. Here examples may be found for which the spectrum is concentrated on Cantor-like sets of positive Lebesgue measure, or for which the spectrum is singular continuous. For reviews of work in this area see Avron and Simon (1981), Bellisard and Simon (1982) and Simon (1982). It is indeed an interesting outcome that some of the phenomena that were at one time regarded as pathological examples of spectral behaviour have now found their way into the physics literature.

(m) A "simple" class of Schrödinger operators that have not yet been analysed

Although much work has been carried out on spectral and scattering theory for rapidly oscillating potentials, there is relatively little known about potentials that oscillate slowly at infinity. As an example the Hamiltonian

$H = -d^2/dx^2 + \cos x^{1/2}$ in $L^2(0, \infty)$ has an essential spectrum in the interval $[-1, \infty)$, but the nature of the spectrum in the range $[-1, 1]$ appears to be unknown. More generally, there is a need to develop sophisticated tools of spectral analysis that will cope with this kind of problem.

(n) A class of Schrödinger operators that await a theory to describe them

As yet, there is no general theory that describes the asymptotic evolution of states under the time dependence generated by strongly non-spherical Hamiltonians. We know that, for long-range potentials, asymptotic completeness demands the application of stricter conditions on the potential than does the existence of wave operators. Often these additional conditions rule out any strong dependence of the potential $V(r\boldsymbol{\omega})$ on angles $\boldsymbol{\omega}$. No general theory as yet exists relating angular dependence to asymptotic evolution, and there are even open problems in the short-range case.

This wealth of examples of spectral behaviour and their associated scattering theory is all the more striking if it is realized that we have not even touched here on systems of many particles. No doubt many interesting phenomena remain to be discovered; in the sample of potentials presented here, we are led, I hope, to a deeper understanding of the foundations of the subject. As many interesting cases have been described as there are chapters in this book!

References

Adams, R.A. (1975). *Sobolev Spaces.* Academic Press, New York.

Agmon, S. (1971). Spectral properties of Schrödinger operators and Scattering Theory. *Proc. Int. Cong. Math.* 679–683. Gauthier-Villars, Paris.

Agmon, S. (1975). *Ann. Scuola Norm. Sup. Pisa* **2**, 151–218.

Agmon, S. and Hörmander, L. (1976). *J. Anal. Math.* **30**, 1–38.

Akhiezer, N.I. and Glazman, I.M. (1981). *Theory of Linear Operators in Hilbert Space*, Vols. I and II. Pitman, London.

Alsholm, P. and Schmidt, G. (1971). *Arch. Rat. Mech. Anal.* **40**, 281–311.

Amrein, W.O. (1981). *Non-relativistic Quantum Dynamics.* Reidel, Dordrecht.

Amrein, W.O. and Georgescu, V. (1973). *Helv. Phys. Acta* **46**, 635–658.

Amrein, W.O. and Georgescu, V. (1974). *Helv. Phys. Acta* **47**, 517–533.

Amrein, W.O., Georgescu, V. and Jauch, J.M. (1971), *Helv. Phys. Acta* **44**, 407–434.

Amrein, W.O., Jauch, J.M. and Sinha, K. (1977). *Scattering Theory in Quantum Mechanics.* Benjamin, Reading, Massachusetts.

Amrein, W.O. Martin, P.A. and Misra, B. (1970). *Helv. Phys. Acta* **43**, 313–344.

Amrein, W.O. and Pearson, D.B. (1979). *J. Phys.* A: *Math. Gen.* **12**, 1469–1492.

Amrein, W.O., Pearson, D.B. and Sinha, K. (1979). *Nuovo Cim.* **52A**, 115–131.

Amrein, W.O., Pearson, D.B. and Wollenberg, M. (1980). *Helv. Phys. Acta* **53**, 335–351.

Aronszajn, N. (1957). *Am. J. Math.* **79**, 597–610.

Avron, J.A. and Simon, B. (1981). *J. Funct. Anal.* **43**, 1–31.

Bargmann, V. (1949). *Rev. Mod. Phys.* **21**, 488–493.

Bartle, R.G. (1966). *The Elements of Integration.* Wiley, New York.

Baumgartel, H. and Wollenberg, M. (1981). *Mathematical Scattering Theory.* Birkhauser, Basel.

Bellisard, J. and Simon, B. (1982). *J. Funct. Anal.* **48**, 408–419.

Birman, M.S. (1968). *Math. USSR Izv.* **2**, 879–906.

Birman, M.S. and Belopolskii (1968). *Math. USSR Izv.* **2**, 117–130.

Birman, M.S. and Belopolskii (1969). *Sov. Math. Dokl.* **10**, 393–397.

Birman, M.S. and Solomjak, M.Z. (1969). *Topics in Mathematical Physics*, **2**, 19–46. Consultants Bureau, New York.

Buslaev, V.S. and Matveev, V.B. (1970). *Theor. Math. Phys.* **2**, 266–274.

Chadan, K. and Sabatier, P.C. (1977). *Inverse Problems in Quantum Scattering Theory.* Springer, New York.

Coddington, E.A. and Levinson, N. (1955). *Theory of Ordinary Differential Equations.* McGraw-Hill, New York.

Collingwood, E. and Cartwright, M. (1952). *Acta Math.* **87**, 83–146.

Combes, J.M. (1969). *Nuovo Cim.* **64A**, 111–144.
Combescure, M. (1980). *Commun. Math. Phys.* **73**, 43–62.
Combescure, M. and Ginibre, J. (1976). *Ann. Inst. H. Poincaré* **24A**, 17–29.
Combescure, M. and Ginibre, J. (1978). *J. Funct. Anal.* **29**, 54–73.
Cook, J. (1957). *J. Math. and Phys.* **36**, 82–87.
Cycon, H.L. (1981). *J. Op. Theory* **6**, 75–76.
Cycon, H.L., Froese, R.G., Kirsch, W. and Simon, B. (1986). *Schrödinger Operators.* Springer, Berlin.
Davies, E.B. (1980). *Adv. Appl. Math.* **1**, 300–323.
de Alfaro, V. and Regge, T. (1965). *Potential Scattering.* North-Holland, Amsterdam.
Deift, P. and Simon, B. (1976). *J. Funct. Anal.* **23**, 218–238.
Deift, P. and Simon, B. (1977). *Commun. Pure Appl. Math.* **30**, 573–583.
Devinatz, A. (1980). *J. Math. Phys.* **21**, 2406–2411.
Devinatz, A. and Rejto, P. (1983). *J. Diff. Eqns.* **49**, 29–84.
Dixmier, J. (1977). *C*-Algebras.* North-Holland, Amsterdam.
Dollard, J.D. (1964). *J. Math. Phys.* **5**, 729–738.
Dollard, J.D. (1969). *Commun. Math. Phys.* **12**, 193–203.
Dollard, J.D. (1977). *J. Math. Phys.* **18**, 229–232.
Dollard, J.D. and Friedman, C. (1978). *Ann. Phys. (NY)* **111**, 251–266.
Dunford, N. and Schwartz, J.T. (1958–1971). *Linear Operators*, Vols. I and II. Interscience, New York.
Eastham, M.S.P. (1973). *The Spectral Theory of Periodic Differential Operators.* Scottish Academic Press, Edinburgh.
Eastham, M.S.P. and Kalf, H. (1982). *Schrödinger-Type Operators with Continuous Spectrum.* Pitman, London.
Enss, V. (1977). *Commun. Math. Phys.* 52, 233–238.
Enss, V. (1978). *Commun. Math. Phys.* **61**, 285–291.
Enss, V. (1979a). *Ann. Phys. (NY)* **119**, 117–132.
Enss, V. (1979b). *Commun. Math. Phys.* **65**, 151–165.
Enss, V. (1984). Scattering and spectral theory for 3-particle systems. *Differential Equations* (ed. I.W. Knowles and R.T. Lewis), 173–204. North-Holland, Amsterdam.
Enss, V. and Simon, B. (1980). *Commun. Math. Phys.* **76**, 177–209.
Faddeev, L.D. (1965). *Mathematical Aspects of the 3-Body Problem in Quantum Scattering Theory.* Israel Program of Scientific Translations, Jerusalem.
Folland, G.B. (1984). Real Analysis, Wiley, New York.
Friedrichs, K.O. (1973). *Spectral Theory of Operators in Hilbert Space.* Springer, Berlin.
Froese, R. and Herbst, I. (1982). *Duke Math. J.* **49**, 1075–1085.
Froese, R., Herbst, I. and Hoffmann-Ostenhoff, T. (1982). *J. Anal. Math.* **41** 272–284.
Gelfand, I.M. and Levitan, B.M. (1955). *Am. Math. Soc. Transl.* (2)**1**, 253–304.
Gelfand, I.M. and Shilov, G.E. (1968). *Generalised Functions.* Academic Press, New York.
Gilbert, D. (1984). Subordinacy and spectral analysis of Schrödinger operators, Ph.D. thesis, University of Hull.
Gilbert, D. and Pearson, D.B. (1987). On subordinacy and analysis of the spectrum of one-dimensional Schrödinger operators. To be published in *J. Math. Anal. Appl.*
Ginibre, J. and Moulin, M. (1974). *Ann. Inst. H. Poincaré* **21A**, 97–145.
Goldberger, M.L. and Watson, K.M. (1964). *Collision Theory*, Wiley, New York.
Graffi, S. (ed.) (1984). *Schrödinger Operators: Proc. 1984 Session of CIME.* Springer, Berlin.

Green, T.A., and Lanford, O.E. (1960). *J. Math. Phys.* **1**, 139–148.
Hack, M.N. (1958). *Nuovo Cim.* **9** 731–733.
Hack, M.N. (1959). *Nuovo Cim.* **13**, 231–236.
Halmos, P.R. (1950). *Measure Theory.* Van Nostrand, Princeton.
Hellwig, G. (1967). *Differential Operators of Mathematical Physics.* Addison-Wesley, Reading, Massachusetts.
Helmberg, G. (1969). *Introduction to Spectral Theory in Hilbert Space.* North-Holland, Amsterdam.
Hewitt, E. and Stromberg, K. (1965). *Real and Abstract Analysis.* Springer, Berlin.
Hörmander, L. (1976). *Math. Z.* **146**, 69–71.
Hörmander, L. (1983). *The Analysis of Linear Partial Differential Operators.* Springer, Berlin.
Hunziker, W. (1966). *Helv. Phys. Acta* **39**, 451–462.
Hunziker, W. (1967). *Helv. Phys. Acta* **40**, 1052–1062.
Ikebe, T. (1960). *Arch. Rat. Mech. Anal.* **5**, 1–34.
Ikebe, T. and Kato, T. (1962). *Arch. Rat. Mech. Anal.* **9**, 77–92.
Jauch, J.M. (1958a). *Helv. Phys. Acta* **31**, 127–158.
Jauch, J.M. (1958b). *Helv. Phys. Acta* **31**, 661–684.
Jauch, J.M. (1968). *Foundations of Quantum Mechanics.* Addison-Wesley, Reading, Massachusetts.
Jauch, J.M., Lavine, R. and Newton, R.G. (1972). *Helv. Phys. Acta* **45**, 325–330.
Jauch, J.M., Misra, B. and Gibson, A.G. (1968). *Helv. Phys. Acta* **41**, 513–527.
Jensen, A., Mourre, E. and Perry P.A. (1984). *Ann. Inst. H. Poincaré* **41A**, 207–225.
Joachain, C.J. (1975). *Quantum Collision Theory.* North-Holland Amsterdam.
Jorgens, K. (1970). *Spectral Theory of Schrödinger Operators.* University of Colorado lecture notes.
Jorgens, K. and Weidmann, J. (1973). *Spectral Properties of Hamiltonian Operators.* Lecture Notes in Mathematics, Vol. 313. Springer, Berlin.
Jost, R. (1947). *Helv. Phys. Acta* **20**, 250–266.
Kalf, H. and Walter, J. (1972). *J. Funct. Anal.* **10**, 114–130.
Kato, T. (1957). *Proc. Jap. Acad.* **33**, 260–264.
Kato, T. (1966). *Math. Ann.* **162**, 258–279.
Kato, T. (1968). *Stud. Math.* **31**, 535–546.
Kato, T. (1969). Some results in potential scattering. *Proc. Int. Conf. On Funct. Anal. and Related Topics*, 206–215, Tokyo.
Kato, T. (1972). *Israel J. Math.* **13**, 135–148.
Kato, T. (1976). *Perturbation Theory for Linear Operators*, 2nd Edn. Springer, New York.
Kato, T. and Kuroda, S.T. (1970). Theory of simple scattering and eigenfunction expansions. *Functional Analysis and Related Fields* 99–131. (ed. F.E. Browder), Springer, New York.
Kato, T. and Kuroda, S.T. (1971). *Rocky Mount. J. Math.* **1**, 127–171.
Kelley, J.L. (1955). *General Topology.* Van Nostrand, Princeton.
Kodaira, K. (1949). *Am J. Math.* **71**, 921–945.
Kodaira, K. (1950). *Am. J. Math.* **72**, 502–544.
Kupsch, J. and Sandhas, W. (1966). *Commun. Math. Phys.* **2**, 147–154.
Kuroda, S.T. (1959). *Nuovo Cim.* **5**, 431–454.
Kuroda, S.T. (1962). *J. Math. Phys.* **3**, 933–935.
Kuroda, S.T. (1973a). *J. Math. Soc. Jap.* **25**, 75–104.
Kuroda, S.T. (1973b). *J. Math. Soc. Jap.* **25**, 222–234.
Landau, L.D. and Lifshitz, E.M. (1958). *Quantum Mechanics: Non-Relativistic Theory.* Pergamon, Oxford.

Lang, S. (1968). *Analysis*, Vol. I. Addison-Wesley, Reading, Massachusetts.
Lavine, R. (1970). *J. Funct. Anal.* **5**, 368–382.
Lavine, R. (1971). *J. Math. Phys.* **20**, 301–323.
Lavine, R. (1972). *Indiana Univ. Math. J.* **21**, 643–656.
Lavine, R. (1973). *J. Math. Phys.* **14**, 376–379.
Lippmann, B.A. and Schwinger, J. (1950). *Phys. Rev.* **79**, 469–480.
Lusin, N. and Privalov, J. (1925). *Ann. de l'Ecole Norm. Sup.* **42**, 143–191.
Mackey, G.M. (1969). *Mathematical Foundations of Quantum Mechanics*. Benjamin, Reading, Massachusetts.
Martin, P.A. and Misra, B. (1973). *J. Math. Phys.* **14**, 997–1005.
Messiah, A. (1961, 1962). *Quantum Mechanics*, Vols. I and II. North-Holland, Amsterdam.
Møller, C. (1945). *Dan. Vid. Selsk. Mat.-Fys. Medd.* **23**, 1–48.
Mourre, E. (1979). *Commun. Math. Phys.* **68**, 91–94.
Mourre, E. (1981). *Commun. Math. Phys.* **78**, 391–408.
Mourre, E. (1983). *Commun. Math. Phys.* **91**, 279–300.
Naimark, M.A. (1968). *Linear Differential Operators*. Ungar, New York.
Newton, R.G. (1966). *Scattering Theory of Waves and Particles*. McGraw-Hill, New York.
Pastur, L.A. (1980). *Commun. Math. Phys.* **75**, 179–196.
Pearson, D.B. (1971). *Nuovo Cim.* **2A**, 853–880.
Pearson, D.B. (1974). *Helv. Phys. Acta.* **47**, 249–264.
Pearson, D.B. (1975a). *Helv. Phys. Acta.* **48**, 639–653.
Pearson, D.B. (1975b). *Commun. Math. Phys.* **40**, 125–146.
Pearson, D.B. (1978a). *J. Funct. Anal.* **28**, 182–186.
Pearson, D.B. (1978b). *Commun. Math. Phys.* **60**, 13–36.
Pearson, D.B. (1982). Spectral properties of differential equations. *Applied Mathematical Analysis: Vibration Theory*, 144–152 (ed. G.F. Roach). Shiva., Nantwich.
Pearson, D.B. (1984). *Helv. Phys. Acta* **57**, 307–320.
Perry, P.A. (1983). *Scattering Theory by the Enss method*. Harwood, Chur.
Perry, P.A., Sigal, I.M. and Simon, B. (1981). *Ann. Math.* **114**, 519–567.
Piron, C. (1975). *Foundations of Quantum Physics*. Benjamin, Reading, Massachusetts.
Pitt, H.R. (1985). *Measure and Integration for Use*. Clarendon, Oxford.
Prugovecki, E. (1971). *Nuovo Cim.* **4B**, 105–123.
Prugovecki, E. (1981). *Quantum Mechanics in Hilbert Space*. Academic Press, New York.
Reed, M. and Simon, B. (1972). *Methods of Modern Mathematical Physics*, Vol. I: *Functional Analysis*. Academic Press, New York.
Reed, M. and Simon, B. (1975). *Methods of Modern Mathematical Physics*, Vol 11: *Fourier Analysis and Self-Adjointness*. Academic Press, New York.
Reed, M. and Simon, B. (1978). *Methods of Modern Mathematical Physics*, Vol. IV: *Analysis of Operators*. Academic Press, New York.
Reed, M., and Simon, B. (1979). *Methods of Modern Mathematical Physics*, Vol. III: *Scattering Theory*. Academic Press, New York.
Riesz, F. and Sz.-Nagy, B. (1955). *Functional Analysis*. Ungar, New York.
Rosenblum, M. (1957). *Pac. J. Math.* **7**, 997–1010.
Rudin, W. (1973). *Functional Analysis*. McGraw-Hill, New York.
Rudin, W. (1974). *Real and Complex Analysis*. McGraw-Hill, New York.
Ruelle, D. (1969). *Nuovo Cim.* **61A**, 655–662.

Schechter, M. (1971). *Spectra of Partial Differential Operators.* North-Holland, Amsterdam.
Schechter, M. (1977a). *Modern Methods in Partial Differential Equations.* McGraw-Hill, New York.
Schechter, M. (1977b). *Duke Math. J.* **44**, 863–877.
Schechter, M. (1980). *Math. Proc. Camb. Phil. Soc.* **88**, 59–69.
Schiff, L.I. (1968). *Quantum Mechanics.* McGraw-Hill, New York.
Schmincke, U.W. (1972). *Math. Z.* **124**, 47–50.
Sigal, I.M. (1983). *Scattering Theory for Many Body Quantum Mechanical Systems.* Lecture Notes in Mathematics, Vol. 1011. Springer, Berlin.
Sigal, I.M. and Soffer, A. (1986). *Bull. Am. Math. Soc.* **14**, No. 1, 107–110.
Simon, B. (1971). *Quantum Mechanics for Hamiltonians Defined as Quadratic Forms.* Princeton University Press.
Simon, B. (1974). *Arch. Rat. Mech. Anal.* **52**, 44–48.
Simon, B. (1977). *Commun. Math. Phys.* **53**, 151–153.
Simon, B. (1979). *Trace Ideals and Their Applications.* Cambridge University Press.
Simon, B. (1982). *Adv. Appl. Math.* **3**, 463–490.
Sinha, K.B. (1977). *Ann. Inst. H. Poincaré.* **26A**, 263–277.
Taylor, J.R. (1972). *Scattering Theory.* Wiley, New York.
Thirring, W. (1981). *A Course of Mathematical Physics*, Vol. III: *Quantum Mechanics of Atoms and Molecules.* Springer, New York.
Thomas, L.E. (1974a). *Commun. Math. Phys.* **33**, 335–343.
Thomas, L.E. (1974b). *J. Funct. Anal.* **15**, 364–377.
Thomas, L.E. (1975). *Ann. Phys. (NY)* **90**, 127–165.
von Neumann, J. (1955). *Mathematical Foundations of Quantum Mechanics.* Princeton University Press.
von Neumann, J. and Wigner, E.P. (1929). *Z. Phys.* **30**, 465–467.
Weidmann, J. (1980). *Linear Operators in Hilbert Space.* Springer, New York.
Wilcox, C.H. (1973). *J. Funct. Anal.* **12**, 257–274.
Yafaev, D.R. (1986). *Ann. Inst. H. Poincaré* **44A**, 397–425.
Yosida, K. (1974). *Functional Analysis.* Springer, Berlin.

Index